EURYHALINE FISHES

EURYHALINE FISHES

Edited by

STEPHEN D. McCORMICK

US Geological Survey, Conte Anadromous
Fish Research Center, Turners Falls, MA
and Department of Biology, University of
Massachusetts, Amherst, Massachusetts, USA

ANTHONY P. FARRELL

Department of Zoology, and
Faculty of Land and Food Systems
The University of British Columbia
Vancouver, British Columbia
Canada

COLIN J. BRAUNER

Department of Zoology
The University of British Columbia
Vancouver, British Columbia
Canada

AMSTERDAM • BOSTON • HEIDELBERG • LONDON
NEW YORK • OXFORD • PARIS • SAN DIEGO
SAN FRANCISCO • SINGAPORE • SYDNEY • TOKYO
Academic Press is an imprint of Elsevier

Academic Press is an imprint of Elsevier
The Boulevard, Langford Lane, Kidlington, Oxford, OX5 1GB, UK
225 Wyman Street, Waltham, MA 02451, USA

First published 2013

Copyright © 2013 Elsevier Inc. All rights reserved

Portions of this book were prepared by U.S. government employees in connection with their official duties, and therefore copyright protection is not available in the United States for such portions of the book pursuant to 17 U.S.C. Section 105.

No part of this publication may be reproduced or transmitted in any form or by any means, electronic or mechanical, including photocopying, recording, or any information storage and retrieval system, without permission in writing from the publisher. Details on how to seek permission, further information about the Publisher's permissions policies and our arrangement with organizations such as the Copyright Clearance Center and the Copyright Licensing Agency, can be found at our website: www.elsevier.com/permissions

This book and the individual contributions contained in it are protected under copyright by the Publisher (other than as may be noted herein).

Notices
Knowledge and best practice in this field are constantly changing. As new research and experience broaden our understanding, changes in research methods, professional practices, or medical treatment may become necessary.

Practitioners and researchers must always rely on their own experience and knowledge in evaluating and using any information, methods, compounds, or experiments described herein. In using such information or methods they should be mindful of their own safety and the safety of others, including parties for whom they have a professional responsibility.

To the fullest extent of the law, neither the Publisher nor the authors, contributors, or editors, assume any liability for any injury and/or damage to persons or property as a matter of products liability, negligence or otherwise, or from any use or operation of any methods, products, instructions, or ideas contained in the material herein.

British Library Cataloguing-in-Publication Data
A catalogue record for this book is available from the British Library

Library of Congress Cataloging-in-Publication Data
A catalog record for this book is available from the Library of Congress

ISBN: 978-0-12-396951-4

For information on all Academic Press publications
visit our website at **store.elsevier.com**

Printed and bound by CPI Group (UK) Ltd, Croydon, CR0 4YY

Working together to grow
libraries in developing countries

www.elsevier.com | www.bookaid.org | www.sabre.org

ELSEVIER   BOOK AID
International   Sabre Foundation

CONTENTS

CONTRIBUTORS

The numbers in parentheses indicate the pages on which the authors' contributions begin.

JAMES S. BALLANTYNE *(125), Department of Integrative Biology, University of Guelph, Guelph, ON, Canada*

COLIN J. BRAUNER *(435), Department of Zoology, University of British Columbia, Vancouver, BC, Canada*

SUSAN L. EDWARDS *(1), Department of Biology, Appalachian State University, Boone, NC, USA*

DAVID I. FRASER *(125), Department of Integrative Biology, University of Guelph, Guelph, ON, Canada*

RICK J. GONZALEZ *(435), Department of Biology, University of San Diego, San Diego, CA, USA*

DIETMAR KÜLTZ *(45), Department of Animal Science, University of California, Davis, Davis, CA, USA*

WILLIAM S. MARSHALL *(1, 395), Department of Biology, St. Francis Xavier University, Antigonish, NS, Canada*

STEPHEN D. MCCORMICK *(69, 199, 477), USGS, Conte Anadromous Fish Research Center Turners Falls, MA, USA and Department of Biology, University of Massachusetts, Amherst, MA, USA*

ERIC T. SCHULTZ *(477), Ecology and Evolutionary Biology, University of Connecticut Storrs, CT, USA*

J. MARK SHRIMPTON *(327), Ecosystem Science and Management (Biology) Program, University of Northern British Columbia, Prince George, BC, Canada*

YOSHIO TAKEI *(69), University of Tokyo, Kashiwa, Chiba, Japan*

MICHAEL P. WILKIE *(253)*, Department of Biology and Institute for Water Science, Wilfred Laurier University, Waterloo, ON, Canada

JONATHAN M. WILSON *(435)*, Laboratório de Ecofisiologia, Centro Interdisciplinar de Investigação Marinha e Ambiental *(CIIMAR)*, Porto, Portugal

JOSEPH ZYDLEWSKI *(253)*, US Geological Survey, Maine Cooperative Fish and Wildlife Research Unit, Orono, ME, USA

PREFACE

The need for ion and water homeostasis is common to all life. Indeed, biology textbooks inevitably include a chapter, often their first, on the importance of water as the medium of life and associated challenges with its regulation. This volume of *Fish Physiology* considers these regulatory processes from the perspective of the enormously large range of salinities that fish can tolerate.

For fish, ion and water homeostasis is an especially important challenge for two reasons. Foremost, the high surface area of the gills, with its high exchange efficiency, is in direct contact with the water. Second, the water that the gills are in contact with can vary in salt content by over five orders of magnitude. Thus, fishes in general have adapted to cope with both dehydrating and highly dilute environments.

While most fishes are stenohaline and thus unable to move between freshwater and seawater, remarkably 3–5% of fishes are euryhaline and thus capable of surviving in a wide range of salinities. In some cases they even move regularly between freshwater and seawater. They also inhabit some of the most physiologically challenging habitats, such as hypersaline lakes and cavernous desert pools. Understandably, euryhaline fishes represent important model species for elucidating the mechanisms of salt and water balance and other physiological processes among vertebrates. Moreover, euryhaline fishes are some of the most endangered, as well as commercially and culturally important species, and include salmon, sturgeon, lamprey, tilapia, and flounder. Thus, with the advent of global climate change and rising sea levels, understanding the environmental physiology of euryhaline species is critical for resource management and any mitigative measures.

The goal of this book is to provide a single volume with the first integrative review of how fish cope with the challenge of varying and extreme environmental salinity. Although the focus is decidedly physiological, the breadth of coverage extends to aspects related to evolution, life history, behavior, endocrinology, and cellular and molecular biology. Thus, the first several chapters describe the physiological challenges of maintaining ion and water balance and the physiological systems involved in their regulation. The second series of chapters examines the challenges and responses of fish with different life histories (such as diadromy) and different habitats (intertidal zones and hypersaline lakes). In addition to salinity, these chapters consider other associated physiological challenges, including migration, temperature, and oxygen. The volume ends with a chapter on the evolution of euryhalinity, which uses for the first time a phylogenetic perspective to examine the evolutionary trends of euryhalinity in fish.

The success of this volume is primarily an outcome of the experience and scholarship of the 14 authors who worked so hard to create original and thoughtful chapters. We are grateful to all of the reviewers whose time and efforts helped to improve and refine the contents. The professional and tireless production staff at Elsevier, especially Pat Gonzalez, Kristi Gomez, and Lisa Jones, provided "spit and polish" to the final product.

The editors of this volume have had the great privilege of working with incredibly gifted and devoted scientists over their careers. We would like to dedicate this volume to the memory of Howard Bern, who died in January 2012. Howard Bern was a gifted teacher and a creative scientist. He was instrumental in founding the fields of comparative endocrinology and endocrine disruptors. He inspired three generations of scientists to understand the mechanisms and endocrine control of ion regulation in euryhaline fishes. We hope that this volume not only summarizes our current knowledge, but also helps to inspire a new generation in this field.

Stephen D. McCormick
Anthony P. Farrell
Colin J. Brauner

GLOSSARY OF TERMS

Amphihaline   Capable of surviving in freshwater and seawater.

Euryhaline   Capable of surviving in a wide range of salinity.

Stenohaline   Capable of surviving in only a narrow range of salinity.

Diadromous   Truly migratory fishes which migrate between the sea and freshwater.

Anadromous   Diadromous fishes which spend part of their lives in the sea and migrate to freshwater to breed.

Catadromous   Diadromous fishes which spend part of their lives in freshwater and migrate to the sea to breed.

Amphidromous   Diadromous fishes that reproduce in freshwater, pass to the sea as newly hatched larvae where they feed and grow, then return to freshwater as juveniles for another period of feeding and growth, followed by reproduction.

Ionocyte   Specialized cell in the gill and skin that transports ions to maintain internal (blood) homeostasis.

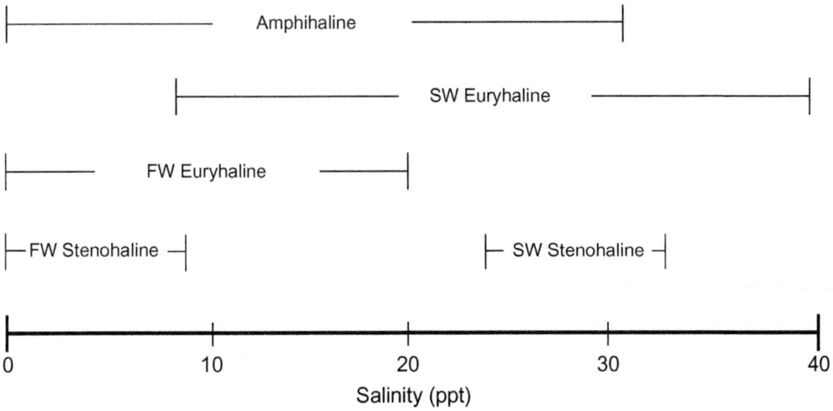

Amphihaline

SW Euryhaline

FW Euryhaline

FW Stenohaline

SW Stenohaline

Salinity (ppt)

0   10   20   30   40

McDowell (1988) gives an excellent overview of the terminology used in this field in his book *Diadromy of Fishes*. Myers (1949) was the first to explicitly define diadromy, anadromy, and catadromy; these terms were endorsed by McDowell (1988) and his definitions are used verbatim in the above glossary with only one exception: we utilize "part" rather than "most" to define the period spent in freshwater and seawater for anadromy and catadromy, respectively. The use of "most" unnecessarily excludes species that may spend more time in their initial habitat, but nonetheless have a growth phase in the second environment and return to the first for breeding. An example would be Atlantic salmon that spend on average 2.5 years in freshwater and 1–2 years in seawater. The definition of amphidromy is abbreviated from McDowell (2007).

The term euryhaline has been used by various authors to mean either the capacity to survive in a wide range of salinities (as its name implies and is used here) or the capacity to survive in freshwater and seawater. In our view there is a need to distinguish between these two meanings, and we have chosen to refer to the latter as amphihaline (able to survive in freshwater and seawater). It should be noted that the term amphihaline has been used by Fontaine (1975) to describe movement "from salt to fresh water" at "well defined stages of their life cycle". This usage does not appear to be substantially different from the types of diadromy defined above. Furthermore, since the categorization of fishes as euryhaline or stenohaline is based upon physiological capacities, the same should apply for amphihaline. Using these definitions, all amphihaline species are also euryhaline, but not all euryhaline species are amphihaline.

In the figure above, we have further categorized stenohaline and euryhaline relative to freshwater (FW) and seawater (SW) habitats. FW stenohaline fishes can survive in only a narrow range of salinity that includes FW. SW stenohaline fishes can survive in only a narrow range of salinity that includes SW (35 ppt). FW euryhaline fishes can survive in a wide range of salinity that includes FW. SW euryhaline fishes can survive in a wide range of salinity that includes SW (35 ppt). The terms defined here are used throughout the volume for consistency.

The term ionocyte is being used with increasing frequency in the physiological literature and provides a functionally relevant name for epithelial cells involved in ion transport. Therefore, ionocyte has been adopted here in favor of "chloride cell" and "mitochondrion-rich cell", terms that are still in use. "Chloride secreting cell" was the first term introduced by Keys and Willmer (1932), but its subsequent modification to the more general "chloride cell" does not correctly apply to the wide range of cells that have been found to transport other major ions such as sodium, calcium, and protons. The term "mitochondrion-rich cell" has been widely used, but does not communicate a functional attribute. Furthermore, the property of numerous mitochondria is not often used to identify ionocytes, nor are ionocytes the only cell with this property. Thus, ionocyte provides a specific and functional name for specialized cells in the gill and skin that are involved in ion homeostasis.

Stephen D. McCormick
Anthony P. Farrell
Colin J. Brauner

REFERENCES

Fontaine, M. (1975). Physiological mechanisms in the migration of marine and amphihaline fish. *Adv. Mar. Biol.* 13, 241–355.

Keys, A. B. and Willmer, E. N. (1932). Chloride secreting cells in the gills of fishes with special reference to the common eel. *J. Physiol. (Lond.)* 76, 368–378.

McDowell, R. M. (1988). *Diadromy in Fishes.* Portland, OR: Timber Press.

McDowell, R. M. (2007). On amphidromy, a distinct form of diadromy in aquatic organisms. *Fish Fish.* 8, 1–13.

Myers, G. S. (1949). Salt-tolerance of fresh-water fish groups in relation to zoogeographical problems. *Bijd. Dierk* 28, 315–322.

LIST OF ABBREVIATIONS

Abbreviations used throughout this volume are listed below. Standard abbreviations are not defined in the text. Nonstandard abbreviations are defined in each chapter and listed below for reference purposes.

14-3-3	regulatory protein 14-3-3
$\Delta\psi$	transmembrane potential (mV)
AC	accessory cell
ACE	angiotensin-converting enzyme
ACT	Atlantic Cooperative Telemetry Network
ACTH	adrenocorticotropic hormone
ADP	adenosine diphosphate
AE	anion exchanger
AM	adrenomedullin
Ang	angiotensin
ANP	atrial natriuretic peptide
AQP	aquaporin
ASIC	acid-sensing ion channel
Asn	asparagine
Asp	aspartic acid
AT1	angiotensin II receptor type 1
ATP	adenosine triphosphate
ATU	accumulated thermal unit
AVP	arginine vasopressin
BADH	betaine aldehyde dehydrogenase
BHMT	betaine homocysteine methytransferase
BMPR II	bone morphogenetic protein receptor type II
BNP	B-type natriuretic peptide
BW	brackish water

Ca^{2+}	calcium ion
CA	carbonic anhydrase
cAMP	cyclic adenosine $3',5'$-monophosphate
CaSR	calcium sensing receptor
CFTR	cystic fibrosis transmembrane conductance regulator
cGMP	cyclic guanosine monophosphate
CGRP	calcitonin gene-related peptide
ChoDH	choline dehydrogenase
Cl^-	chloride ion
ClC	chloride channel
CLR	calcitonin receptor-like receptor
CNP	C-type natriuretic peptide
CO_2	carbon dioxide
CON-A	concanavalin-A
COP	colloid osmotic pressure
COX-1,2	cyclooxygenase 1 (spontaneous), 2 (inducible)
CPS	carbamoyl phosphate synthase
CR	corticosteroid receptor
CRF	corticotropin-releasing factor
CRH	corticotropin-releasing hormone
CRHR	corticotropin-releasing hormone receptor
CYP-1a	cytochrome P450 type 1a
DGDH	dimethylglycine dehydrogenase
DOC	11-deoxycorticosterone
DOPAC	3,4-hydroxyphenylacetic acid
E2	17β-estradiol
ECF	extracellular fluid
EGF	epidermal growth factor
E_{Na}, E_{Cl}	Nernst equilibrium potential for Na, Cl (mV)
ENaC	epithelial sodium channel
EST	expressed sequence tag
F	Faraday (96.485 coulomb/mole)
FAK	focal adhesion kinase
FW	freshwater
GC-A	particulate guanylyl cyclase A
GDH	glutamate dehydrogenase
GFR	glomerular filtration rate
GH	growth hormone
Ghr	ghrelin
GHR	growth hormone receptor
GLN	guanylin
GLUTase	glutaminase

GnRH	gonadotropin releasing hormone
GR	glucocorticoid receptor
Grail	gene related to anergy in T lymphocytes
GSase	glutamine synthetase
H^+	hydrogen ion
HCO_3^-	bicarbonate ion
HPA	hypothalamic–pituitary–adrenal
hpf	hours post-fertilization
HPI	hypothalamic-pituitary-interrenal
HSF1	heat shock factor 1
HSP	heat shock protein
IBD	isolation by distance
ICF	intracellular fluid
IGF-I	insulin-like growth factor I
ILCM	interlamellar cell mass
IT	isotocin
JNK	c-Jun N-terminal kinase
KCC	K^+/Cl^- cotransporter
α-KG	α-ketoglutarate (oxoglutarate)
11KT	11-ketotestosterone
LC_{50}	median lethal concentration
MAPK	mitogen-activated protein kinase
Mg^{2+}	magnesium ion
MLCK	myosin light chain kinase
MR	mineralocorticoid receptor
MRC	mitochondrion-rich cell
mRNA	messenger ribonucleic acid
MYA	million years ago
Na^+	sodium ion
NAD^+	nicotinamide adenine dinucleotide, oxidized form
NADH	nicotinamide adenine dinucleotide, reduced form
$NADP^+$	nicotinamide adenine dinucleotide phosphate, oxidized form
NADPH	nicotinamide adenine dinucleotide phosphate, reduced form
NCC	Na^+/Cl^- cotransporter
NEFA	non-esterified fatty acid
NHE	Na^+/H^+ exchanger
NHERF1	Na^+/H^+ exchanger regulatory factor 1
NKA	Na^+/K^+-ATPase
NKCC	$Na^+/K^+/2Cl^-$ cotransporter
NP	natriuretic peptide
NPR-A	natriuretic peptide receptor A
OCLN	occludin

OLGC	*Oryzias latipes* (medaka) particulate guanylyl cyclase
OSR	osmotic stress response kinase
OSTF	osmotic stress factor protein
OTCase	ornithine transcarbamylase
OTN	Ocean Tracking Network
OUC	ornithine urea cycle
OVLT	organum vasculosum laminae terminalis
P	progesterone
PACAP	pituitary adenylate cyclase-activating peptide
P_{diff}	diffusional permeability to water
P_i	phosphate ion
PFK	phosphofructokinase
PKA, PKC	protein kinase A, C
PLA_2	phospholipase A_2
PLMPL	phospholemman-like protein
PNA	peanut agglutinin
PO_4^{2-}	phosphate ion
POMC	pro-opiomelanocortin
P_{os}	osmotic permeability to water
POST	Pacific Ocean Shelf Tracking
PP	protein phosphatase
ppt	parts per thousand
PRL	prolactin
PRLR	prolactin receptor
PrRP	prolactin-releasing peptide
PVC	pavement cell
Q_{10}	temperature coefficient (10°C change)
R	universal gas constant (8.314 Joule Kelvin^{-1} mole^{-1})
RAMP	receptor activity-modifying protein
RAS	renin–angiotensin system
RBC	red blood cell
RHP2	rhesus protein 2
RLN	relaxin
RU486	mifepristone
RVD	regulatory volume decrease
RVI	regulatory volume increase
SDH	sarcosine dehydrogenase
SGK	glucocorticoid inducible kinase
SO_4^{2-}	sulfate ion
SPAK	ste20-proline rich kinase
Sr^{2+}	strontium ion
STST	selective tidal stream transport

SW	seawater
15α-T	15α-hydroxytestosterone
T	temperature (°C)
T_3	3,5,3′-triiodothyronine
T_4	thyroxine or tetraiodothyronine
TEM	transmission electron microscopy
TEP	transepithelial potential (mV)
TGFβ	transforming growth factor-β
TL	total length TMA trimethylamine
TMAO	trimethylamine oxide
TMNAox	trimethylamine oxidase
TNFα	tumor necrosis factor-α
TRP	transient receptor potential
TRPV	transient receptor potential vallinoid type
TSC22	transforming growth factor-β-stimulated clone 22
TSH	thyroid-stimulating hormone
TU	thiourea
UFR	urinary flow rate
UI	urotensin I
UII	urotensin II
UIIR	urotensin II receptor
UT	urea transporter
V1R	vasotocin receptor type 1
VIP	vasoactive intestinal peptide
VNP	ventricular natriuretic peptide
VT	vasotocin
WNK	with-no-lysine kinase

1

PRINCIPLES AND PATTERNS OF OSMOREGULATION AND EURYHALINITY IN FISHES

SUSAN L. EDWARDS

WILLIAM S. MARSHALL

1. Introduction
2. Principles of Ion and Water Transport
   2.1. Osmosis
   2.2. Gibbs–Donnan Equilibrium
   2.3. Transmembrane Diffusion of Uncharged Solutes
   2.4. Transmembrane Ionic Diffusion
   2.5. Facilitated Transport
   2.6. Active Transport
3. Osmoregulatory Organs
   3.1. Gills
   3.2. Kidney and Urinary Bladder
   3.3. Gastrointestinal Tract
   3.4. Rectal Gland of Elasmobranchs
   3.5. Skin, Opercular Membrane, and Yolk Sac
4. Hagfishes
5. Lampreys
   5.1. Osmoregulation in Freshwater
   5.2. Osmoregulation in the Marine Environment
6. Elasmobranchs
   6.1. Ion Regulation in the Marine Environment
   6.2. Freshwater and Euryhaline Elasmobranchs
7. Teleost Fishes
   7.1. Osmoregulation in the Marine Environment
   7.2. Osmoregulation in Freshwater
8. Conclusions and Perspectives

Euryhaline fishes live in a wide salinity range from freshwater to seawater and hypersaline environments. Euryhaline fishes such as salmon, eels, and tilapia are economically important in worldwide fisheries and aquaculture. This chapter summarizes mechanisms used by hagfishes, lampreys, elasmobranchs, and teleosts to maintain ionic and osmotic homeostasis in

1

Euryhaline Fishes: Volume 32
FISH PHYSIOLOGY
Copyright © 2013 Elsevier Inc. All rights reserved
DOI: http://dx.doi.org/10.1016/B978-0-12-396951-4.00001-3

changing environmental salinity. Hyperosmoregulation in dilute environments uses gills for active ion uptake, intestine for dietary ion uptake, and kidneys for renal uptake. Extrarenal NaCl uptake occurs via high-affinity uptake mechanisms or low-affinity Na^+/Cl^- cotransport. In seawater, hypoosmoregulation involves reflexive drinking, intestinal absorption of salts and water, NaCl secretion at the gills, and renal ion excretion. Na^+ and Cl^- excretion occurs through Na^+/K^+-ATPase, the $Na^+/K^+/2Cl^-$ cotransporter, and apical cystic fibrosis transmembrane conductance regulator. These mechanisms are magnified in hypersaline conditions. Active extrarenal salt transport takes place through mitochondrion-rich ionocytes in skin and gill epithelia. The mechanisms of ion and water transport are summarized, with models presented for seawater, marine, and euryhaline conditions.

1. INTRODUCTION

Fishes account for more than half the extant vertebrate species on Earth (Nelson, 2006) and they inhabit aqueous environments that range from freshwater (FW) with very few osmolytes (0 mOsmol kg^{-1}) to approximately full-strength seawater (SW) (1000 mOsmol kg^{-1}) and greater in some cases. Over evolutionary time fishes have developed three strategies to maintain osmotic and ionic homeostasis: osmoconformity, where the osmolality of the fishes' extracellular fluid is maintained similar to that of SW; hypoosmoregulation, where fishes regulate their internal extracellular fluids below the salt concentration of the surrounding SW; and hyperosmoregulation, where fishes regulate their internal extracellular fluids at a higher salt concentration than that of the surrounding dilute environment (Table 1.1). Most fishes are restricted to a single environment, rendering these organisms either stenohaline marine or stenohaline FW organisms. A smaller number of fishes are euryhaline and have the ability to adapt to large salinity ranges, with many of these species encountering significantly large frequent alterations in environmental salinity, e.g. in the estuaries they inhabit. Lastly, there is a group of fishes, e.g. sea lamprey (*Petromyzon marinus*), eel (*Anguilla* spp.), Atlantic (*Salmo salar*) and Pacific salmon (*Oncorhynchus* spp.), shad (*Alosa sapidissima*) and others, that undertake anadromous and catadromous migrations into and out of FW associated with reproduction, and amphidromous species that migrate between FW and SW as juveniles.

Osmoconformers by definition maintain the same internal osmotic concentration as their external environment, and are therefore exempt

Table 1.1
Ionic composition of blood plasma in fishes.

	Concentration (mM kg water^{-1})									Reference
	Na$^+$	Cl$^-$	K$^+$	Mg$^+$	Ca$_2^+$	SO$_4$	Urea	TMAO	Osmolality	
Seawater	439	513	9.3	50	9.6	26	–	–	1050	Evans (1993)
Hagfish	549	563	11.1	18.9	5.1	3	2.8	0	1152.9	Currie and Edwards (2010)
Lamprey (SW)	156	159	32	7	3.5		–	–	333	Evans (1993)
Coelacanth	197	187	5.8	5.3	4.8	4.8	377	122	942	Griffith et al. (1974)
Shark	255	241	6	3	5	0.5	441	72	1118	Evans (1993)
Teleost (SW)	180	196	5.1	2.5	2.8	2.7	–	–	452	Evans (1993)
Freshwater (soft)	0.25	0.23	0.005	0.04	0.07	0.05	–	–	1	Evans (1993)
Lamprey (FW)	119.6	99.6	2.3	1.5	1.8	–	–	–	224.8	Evans (1993)
Stingray (FW)	164	151.7	4.45	–	3	–	1.1	–	282	Griffith et al. (1973)
Teleost (FW)	130	125	2.9	1.2	2.7	–	–	–	261.8	Evans (1993)
Euryhaline fishes										
Stingray (FW)	211.9	207.8	5.2	–	4.3	–	196	–	625.2	Piermarini and Evans (1998)
Stingray (SW)	310	300	6.95	–	3.1	–	394.5	–	1034	Piermarini and Evans (1998)
Teleost (FW)	124	132	2.9	–	2.7	–	–	–	274	Evans (1979)
Teleost (SW)	142	168	3.4	–	3.3	–	–	–	297	Evans (1979)

SW: seawater; FW: freshwater; TMAO: trimethylamine oxide.

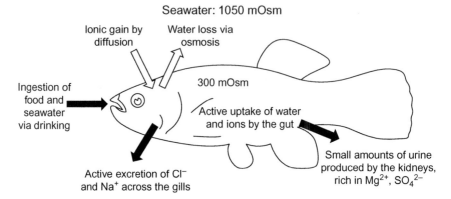

Fig. 1.1. Schematic of water and salt exchange in seawater fishes. In hypoosmotic regulation fishes counteract the osmotic loss of water across the gills and other permeable tissues, by drinking seawater. The ingested water and ions are absorbed in the esophagus and intestine. The salt load is transported via the blood to the gills and skin, where it is excreted.

from the metabolic energy requirements associated with osmoregulation. There are no osmoconforming animals in FW. Hagfishes are the only example of a true osmoconformer in the fishes, as they have the same osmolality as SW and almost the same NaCl concentration. However, as later discussed in this chapter, it appears that even hagfishes regulate plasma ions to some extent. Marine elasmobranchs, chimeras, and the coelacanths are osmoconformers that are actually slightly hyperosmotic to their environment through the retention of urea (approximately 350 mM) and trimethylamine oxide (TMAO) as osmolytes in their blood and interstitial fluids. These animals hyporegulate NaCl and other ions compared to SW (Evans, 2011; Evans et al., 2005; Marshall and Grosell, 2006).

SW teleosts and lampreys maintain their extracellular fluids (ECF) at approximately one-third the osmotic concentration of SW and thus face osmotic loss of water across their gills and other permeable external surfaces. They drink SW to maintain osmotic balance, absorbing water and salts but excreting the excess salts. The site for absorption of ingested water and salts is the esophagus and intestine. Initially there is passive uptake of salts and water, then, as the chyme becomes isotonic, there is active uptake of salts that allows osmotic uptake of water (Fig. 1.1). The salt load is transported in the blood and excess NaCl is excreted by ionocytes in the epithelium of the gills and skin. The transepithelial potential of the gill in SW is positive and favors the paracellular transport of Na^+, whereas the transport of Cl^- is carried out by an active transcellular process (Potts and Eddy, 1973; Potts, 1984; Kirschner, 1997). Any excess uptake of divalent ions is excreted by the kidney.

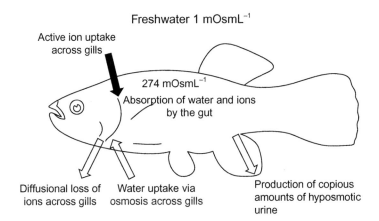

Fig. 1.2. Schematic of water and salt exchange in freshwater fishes. In hyperosmotic regulation fish face diffusional ion loss and an osmotic gain of water across the gills and other permeable tissues. To counter the water gain the animal produces large volumes of dilute urine. Ionic uptake is facilitated by the intestine through the ingestion of food and via active transport of Na^+ and Cl^- across the gill epithelium.

All FW fishes, including lampreys, elasmobranchs, and teleosts, maintain their internal salt concentration at approximately one-quarter to one-third that of SW (250–320 mOsmol kg^{-1} versus 1000 mOsmol kg^{-1}). They face diffusive ion loss and osmotic gain of water across the gill epithelium and other permeable external surfaces. To counter this, the animal produces large volumes of dilute urine and gains ions via active transport of both Na^+ and Cl^- across the gill epithelium from water and via the intestine from ingested food (Fig. 1.2).

This chapter provides a summary of the physiological strategies and mechanisms used by euryhaline fishes to maintain ionic and osmotic homeostasis. The chapter begins with a basic overview of the principles that govern the movement of water and ions. Then, the organs that are responsible for maintaining salt and water balance are described. Finally, the chapter will examine in detail the ion transport pathways used in each group of fishes (hagfishes, lampreys, elasmobranchs, and teleosts) to maintain ion homeostasis in the face of changing environmental salinity.

2. PRINCIPLES OF ION AND WATER TRANSPORT

Water has a unique physical property that enables it to act as a universal solvent for polar ions and compounds. Water is the primary body constituent, distributed between the intracellular (ICF) and the smaller ECF compartments. These two body fluids of fishes contained dissolved solutes, whose distribution is compartment specific. The primary inorganic ions in fishes are Na^+, K^+, and Cl^-, with lower concentrations of Ca^{2+}, Mg^{2+}, SO_4^{2-}, PO_3^{4-}, and HCO_3^-. In all fish that osmoregulate, these ions are all at different concentrations in the blood compared to the environment, and many differ between the ICF and ECF. The major intracellular inorganic ion is K^+ at approximately 150 mM, with low millimolar concentrations of other monovalent ions (Na^+, Cl^-, HCO_3^-). Along with the remaining intracellular organic osmolytes, total osmotic activity of ICF is approximately 300 mOsmol kg^{-1}. ICF and ECF share an identical osmotic activity even with very different ionic profiles. An osmotic similarity is important because fish cells cannot tolerate large transmembrane differences in hydrostatic and osmotic pressures.

When a fish moves from SW to a lower concentration environment, the blood osmolality decreases transiently by up to 60 mOsmol kg^{-1}, a change that induces hypoosmotic stress and increased cell volume (see review by Hoffmann et al., 2007). The cells undergo a regulatory volume decrease (RVD), and allow solutes, primarily K^+ and Cl^-, via volume-activated ion channels, and amino acids (taurine) to leave the cell along with osmotically driven water, thus restoring cell volume. Conversely, if a fish moves from FW to a higher salinity, the blood becomes temporarily more concentrated, causing cell shrinkage and regulatory volume increase (RVI) by uptake of solutes (mostly K^+ and Cl^-) by rapid activation of salt uptake by the $Na^+/K^+/2Cl^-$ cotransporter (NKCC) and slower organic osmolyte (TMAO and taurine) uptake or synthesis. These cell volume responses are limited and only help the animal cope temporarily with salinity change until the main osmoregulatory systems respond to the new environment.

In osmoregulating fishes, the internal tissues cannot cope with more than a minor disruption to ECF osmolality and therefore regulate the ECF to maintain cell volume. Cellular osmotic homeostasis can involve varying the concentration of specific intracellular organic osmolytes to maintain osmotic pressure rather than using inorganic ions that could cause detrimental cellular effects. Compatible solutes such as betaines (e.g. trimethylglycine) are osmolytes that do not interact with the surface of a protein in a detrimental way and therefore can be safely upregulated and downregulated with very little effect on cell function (recently reviewed by Singh et al., 2011). Most compatible osmolytes are neutral at physiological pH and are possibly interchangeable. Compatible osmolytes are thought to be cryoprotectants that help organisms to survive extreme osmotic stress (Yancey, 2005). In contrast, perturbing solutes are osmolytes that bond non-specifically to proteins, resulting in destabilizations through unfolding, and as a result have harmful effects on protein structure and function; urea, arginine, and inorganic ions are good examples. Finally, counteracting osmolytes are often used by organisms to reduce the detrimental effects of perturbing solutes; e.g. elasmobranchs utilize TMAO to counter the negative effects of urea (Yancey and Somero, 1979; Yancey, 1985, 2005).

Osmotic exchanges that occur between the fish and its environment can be divided into those that are under little or no physiological control and rely on gradients (e.g. osmosis and diffusion) and those that are regulated (e.g. cotransport, channel permeabilities, and active transport) and serve to maintain internal ionic and osmotic homeostasis.

2.1. Osmosis

The movement of water across cell membranes is primarily through the incorporation of water channels (aquaporins) into the plasma membrane (reviewed by Nobel laureate Agre et al. 2002) and most members of this intramembrane protein family are expressed in euryhaline fishes (reviewed by Cerda and Finn, 2010). In brief, water flows (J_v) through the aquaporin channels down its concentration gradient, an osmotic gradient, and can be regarded as either a diffusive or a hydraulic process (Zeuthen, 2010). Water flow in response to an osmotic gradient, and in the absence of hydrostatic pressure gradients that animal cell membranes cannot withstand, is described by

$$J_v = Lpa\left(\sum_i \sigma_i RT\Delta c_i\right) \tag{1}$$

where the net water flux (J_v) through membrane of area a with a hydraulic conductivity Lp moves in response to an osmotic pressure gradient that is the sum of all the concentration gradients of osmotically active particles, Δc_i, while R is the universal gas constant (0.082 L.atm.K^{-1}.mol^{-1}) and T the absolute temperature (K). The Staverman reflection coefficient, σ_i, varies from zero (for a solute as permeable as water) to 1.0 (an ideal, totally impermeant solute). A detailed treatment of the thermodynamics of osmosis is found elsewhere (Schultz, 1980; Macey, 1986).

A euryhaline teleost fish in SW has a lower blood osmotic activity (300 mOsmol kg^{-1}) than the environment (SW = 1100 mOsmol kg^{-1}); thus water is lost osmotically to the environment. Conversely, a euryhaline teleost fish in FW (1–10 mOsmol kg^{-1}) tends to gain water osmotically because its blood osmolarity is similar to that of an SW fish. To slow these diffusive water exchanges, both marine and FW fishes generally have low osmotic permeability in boundary membranes of the skin and gill epithelia (i.e. aquaporins are not expressed in the apical membranes of these cells) to minimize the transmembrane movement of water (Kirschner, 1997). In addition, further contributing to the low overall osmotic permeability of skin and gill epithelia (Wood et al., 1998), a generally low water permeability to the paracellular pathway is imparted by a preponderance of tight intercellular junctions, involving a family of structural proteins, the claudins (Bagherie-Lachidan et al., 2008, 2009). In contrast to the skin and gill, the intestine has high expression levels of four isoforms of aquaporin water channels (AQP1, 3, 8, and 10) in euryhaline fishes, reflecting the need for water absorption, especially in SW fishes (Cutler et al., 2007; Cerda and Finn, 2010; Kim et al., 2010) if a favorable osmotic gradient exists for diffusive uptake. Gut epithelia also have the capability to transport water uphill against an osmotic gradient using cotransporters and uniporters (Agre et al., 2004). Several vertebrate studies have demonstrated that active water uptake is linked to membrane proteins responsible for the flux of other substances such as solutes, e.g. NKCC1, NKCC2 (Hamann et al., 2005), and NKCC4 (*Necturus maculosus*) (recently reviewed by Zeuthen, 2010). In teleost fishes there is evidence for an involvement NKCC in intestinal water transport in the secretory direction (Marshall et al., 2002a; Whittamore, 2012).

2.2. Gibbs–Donnan Equilibrium

If diffusible solutes are separated by a membrane that is freely permeable to water and electrolytes but totally impermeable to one ionic species (e.g. a protein with a net negative charge), an unequal distribution of the diffusible solutes is generated between compartments. At Gibbs–Donnan equilibrium, the osmotically unequal distribution of solute particles forces the movement of water into the compartment with the higher osmolality (Overbeek, 1956) and it can, because of the uneven charge distribution, generate small electrical potentials across the membrane. In addition, the colloid oncotic pressure, i.e. an osmotic activity arising from the fixed charges on large impermeant colloidal particles (e.g. proteins), operates to offset the osmotic effects of Gibbs–Donnan equilibrium. A formal treatment of Gibbs–Donnan potentials and a formal development of the thermo-dynamics of water transport is found elsewhere (Schultz, 1980). Ion and water transport can be influenced by boundary conditions adjacent to the membranes; these are layers of solution wherein solute movement is by diffusion only, called unstirred layers. These layers decrease the effective transmembrane concentration difference and can reduce osmotic water flow and ion diffusion across permeable membranes. For a summary of transport types and the effects of unstirred layers on ion and water transport, see Macey (1986).

2.3. Transmembrane Diffusion of Uncharged Solutes

Passive diffusion involves the random thermal motion of suspended dissolved molecules from a region of high concentration to low concentration. The rate of diffusion for a solute can be determined by Fick's first law of diffusion:

$$J_i = J_i^{io} - J_i^{oi} = P_i(C_i^o - C_i^i) \tag{2}$$

where the net flux (J_i) is the algebraic sum of the unidirectional fluxes from inside i to outside o (J_i^{io}) and the reverse (J_i^{oi}), governed by the permeability of the membrane to the solute (P_i) and driven by the concentration gradient across the membrane. Permeability is the rate of diffusion in the membrane and the diffusion coefficient $D = RTu_i$, where R is the universal gas constant, u_i is the mobility of the solute and T the temperature (K). Because of this relation, these diffusive movements vary directly with absolute temperature. If a solute occurs on both sides of a membrane and diffusion is possible (i.e. permeability >0), it will exhibit unidirectional movement (flux) in both directions. The unidirectional flux is defined as the amount of solute passing through an area of membrane every second in a single direction. If influx and efflux are equal, then net flux is zero. If instead the flux in one direction exceeds that in the other direction, then net flux is calculated as the difference between unidirectional fluxes. An important, biologically regulated factor that helps to govern the rate of flux across a membrane is the permeability of the membrane to a particular substance. Simply, the gene expression, insertion, and activation of solute transport proteins into the apical and/or basolateral membrane of osmoregulatory organs (gill, skin, intestine, kidney, urinary bladder) govern solute transport by increasing permeability to specific ions.

2.4. Transmembrane Ionic Diffusion

Many solutes of biological importance carry an electrical charge. The movement of a charged particle is driven by its concentration gradient and the local electrical potential gradient. The permeability of the membrane to changed particles depends on both the conductance of the membrane to a specific solute and the electrical potential across the membrane. The algebraic sum of the concentration gradient and the electrical gradient determines the net electrochemical gradient for a given ion. If an ion is at equilibrium across the membrane so there is no net flux, a difference in potential will exist but it will be just enough to balance and counteract the chemical gradient acting on that specific ion. For instance, a concentration gradient of 10-fold can be opposed by an electrical gradient of approximately 60 mV; this defines the "reversal potential" for any single ion gradient in a membrane system. For a membrane permeable to one ion, i, the potential at which an ion is at electrochemical equilibrium is referred to as the equilibrium potential (as described in the Nernst equation):

$$\Delta\psi = (RT/z_iF)\ln(C_i^{\,o}/C_i^{\,i}) \tag{3}$$

where $\Delta\psi$ is the electrical potential (V), R the universal gas constant (8.3143 Coul.V.K^{-1}.mol^{-1}, temperature (T, K), ionic valence (z_i), and the ratio of the molar (mol L^{-1}) concentrations of the permeant species across the membrane from outside o to inside i ($C_i^{\,o}/C_i^{\,i}$) and typically, K$^+$ channels in cell membranes allow K$^+$ to leak out of the cell, but this leakage is opposed by the intracellular membrane potential of -50 to -80 mV, resulting in low resting efflux of K$^+$. The most important factor influencing the equilibrium potential is the ratio of ion concentrations on either side of the membrane. Therefore, it is possible to move an ion passively by diffusion against its chemical gradient if the electrical gradient across the membrane is in the opposite direction and is greater in magnitude than the chemical gradient. This situation occurs in ionocytes of SW teleost gills, where intracellular Cl$^-$ is much lower than in SW, but the negative intracellular electrical potential is large enough to passively drive Cl$^-$ efflux from cells via anion channels [cystic fibrosis transmembrane conductance regulator (CFTR)] (Kirschner, 1997; Marshall and Grosell, 2006). Importantly, diffusion-based processes vary according to the ratio of temperature in Kelvin and thus most temperature change experienced by most fish species has little influence; for example, diffusion at 30°C is only 10% faster than that at 10°C (303 K/278 K = 1.09), all other factors being equal.

2.5. Facilitated Transport

Unlike passive diffusion, passive transport uses membrane proteins specialized to facilitate the movement of solutes that cannot diffuse across the membrane, i.e. these transport proteins increase the effective permeability to specific ions. Facilitated diffusion involves reversible and loose binding of a solute with a protein that transduces the solute across the membrane. As such, facilitated transport does not directly require metabolic energy as adenosine triphosphate (ATP). However, activation of some transporters of importance to euryhalinity [notably NKCC1, NKCC2, and Na$^+$/Cl$^-$ cotransporter (NCC)] require phosphorylation of the protein by serine/threonine protein kinase (e.g. PKA) or a tyrosine kinase (e.g. SRK). The direction of the passive transport is governed by the electrochemical gradient and can stop or reverse if concentrations of the substrates change markedly. The cotransporter NKCC1 (bumetanide sensitive in a basolateral location) is important to NaCl secretion by SW ionocytes (e.g. Inokuchi et al., 2008, 2009; Flemmer et al., 2010), while NKCC2 (bumetanide sensitive in an apical location) is involved in intestinal (Marshall, 2002; Cutler et al., 2007) and renal salt absorption (e.g. Gimenez and Forbush, 2005). In addition, the NCC (thiazide sensitive in an apical location) (Hiroi et al., 2008; Furukawa et al., 2011) and a related NaCl transporter SLC12A10.2 (Wang et al., 2009) aid a low-affinity NaCl uptake by euryhaline fishes in dilute environments.

2.6. Active Transport

Euryhaline fishes, like all animals, use active transport to move many solutes across epithelial cell membranes and against electrochemical gradients. These transport enzymes require the direct input of metabolic energy, which is reliant on the cell's ability to produce ATP through catabolism. Indeed, cells involved with transport processes are large consumers of ATP; the ionocytes in the gills of fishes are no exception. Two forms of active transport exist: primary active transport, which relies on ATP-dependent membrane proteins to drive solutes against a concentration gradient; and secondary active transport, which applies the movement of a solute against an electrochemical gradient by linking it to a second substance moving down its own concentration gradient. In this case, ATP is not directly consumed by the secondary active transport mechanism but the uphill movement of the solute is driven by another primary active transport mechanism. Here, primary active transport enzymes are considered.

One of the best known and most widely studied active transport proteins is Na^+/K^+-ATPase (NKA), reviewed by Nobel laureate Skou (Skou and Esmann, 1992). The sodium pump is an electrogenic mechanism capable of creating a charge imbalance across the membrane. It is found in the plasma membrane of all cells and is concentrated on the basolateral membrane of almost all epithelial cells. In one catalytic cycle, NKA transports three Na^+ out of a cell and two K^+ into the cell, thereby creating the typical low intracellular Na^+ and high intracellular K^+ concentrations. The $3Na^+:2K^+$ stoichiometry also means that the pump is electrogenic. The largest impact of the NKA pump is the secondary active transport arising from linkage of the downhill movement of K^+ or Na^+ to the uphill transport of other solutes. In most euryhaline teleosts, NKA is upregulated by an increase in environmental salinity and in a few species by a decrease in salinity (Evans et al., 2005; Marshall and Grosell, 2006), underscoring the importance of the enzyme at both salinity extremes. Post-translational regulation also occurs, with phosphorylation by protein kinases PKA and PKC inhibiting the enzyme and protein phosphatase PP2A dephosphorylation activating the enzyme and stimulating the trafficking of the protein to the plasma membrane through interference with arrestin (Kimura et al., 2011), such that rapid changes in enzyme activity have been observed in fish during salinity transfer (Mancera and McCormick, 2000) without changes in gene expression or protein synthesis.

The second ion pump of crucial importance to euryhaline teleost fishes is the vacuolar-type $V-H^+$-ATPase, a pump that transports H^+ out of the cells, driven by catalysis of ATP, and creates pH and voltage gradients simultaneously. This pump is inhibited by bafilomycin (e.g. Brix and Grosell, 2012). In rainbow trout (*Oncorhynchus mykiss*) (Lin et al., 1994), zebrafish (*Danio rerio*) (Lin et al., 2008), and other FW fishes capable of surviving in ion-poor conditions, H^+-ATPase in expressed in the apical membrane of some gill ionocytes (reviewed by Hwang, 2009). In FW elasmobranchs (Piermarini and Evans, 2001), lampreys (Choe et al., 2004), and some euryhaline teleost fishes such as the mummichog (Katoh and Kaneko, 2003) and blennies (Uchiyama et al., 2012), the H^+-ATPase is expressed instead on the basolateral membrane of gill ionocytes. When expressed in the apical membrane, the pump theoretically enhances the apical membrane potential sufficient to provide the driving force for Na^+ uptake by a sodium channel. This pump, in parallel with an apical Na^+ channel, now thought to be one of the acid-sensing ion channels (ASICs) known in zebrafish as ASIC4 (Chen et al., 2007; Dymowska, personal communication), produces effective high-affinity Na^+ uptake from FW and simultaneous acid secretion in strongly euryhaline teleost fishes in FW. When expressed in the basolateral membrane of gill ionocytes, the H^+ pump could still aid ion transport by supporting the membrane potential or by allowing accumulation of HCO_3^- intracellularly to drive apical Cl^-/HCO_3^- exchange. In the intestine, H^+-ATPase appears in the brush border and may indirectly drive Cl^-/HCO_3^- exchange to more efficiently secrete bicarbonate into the lumen in the posterior intestine (Grosell et al., 2009a; Guffy et al., 2011).

Unlike the diffusion of ions, the effect of temperature on active ion transport is profound; enzyme turnover and transport rates approximately double with a rise of 10°C [temperature coefficient (Q_{10}) = 2]. Thus, a rise from 10°C to 30°C theoretically increases active transport about four-fold. In SW mummichog, ion transport decreases steeply in the cold, with a Q_{10} greater than 5.5 (Buhariwalla et al., 2012), while passive ion and water flows are largely unchanged. Temperature changes, therefore, place an important extra challenge on osmoregulation. A detailed treatment of temperature effects on biochemical processes is considered elsewhere (Hochachka and Somero, 2002).

3. OSMOREGULATORY ORGANS

3.1. Gills

Evans et al. (2005) extensively reviewed the multiple functions of the fish gill. The fish gill is the predominant site for osmotic and ionic regulation, acid–base regulation, and the removal of nitrogenous wastes, as well as the major site for gas exchange. The entire cardiac output perfuses the gill vasculature before re-entering the systemic circulation. The afferent arteriole feeds the posterior side of each filament and supplies the lamellae, which are the site of gas exchange. The oxygenated blood is carried by the efferent arterioles, which supply a central sinus and filamental vessels with blood that supplies the ion-transporting cells of the filament epithelium. The gills of FW lampreys, elasmobranchs, and teleosts are the major site of ion uptake from the dilute environment, whereas the gills in marine lampreys and teleost fishes are the major site of salt secretion (Evans et al., 2005).

3.2. Kidney and Urinary Bladder

The kidneys of lampreys and elasmobranchs all have glomerular nephrons, which is also true in the FW and euryhaline teleost species. In marine teleosts, the kidneys usually have reduced glomeruli numbers or, in many cases, lack glomeruli (Hickman and Trump, 1969). In FW fishes, glomerular filtration and urine flow rates are high, as copious amounts of dilute urine are used to excrete excess water gained through osmosis across the gill and minimize ion loss in the urine (Marshall and Smith, 1930). In some FW fishes, the urinary bladder enhances ion reabsorption of the urine. Its epithelium has a low osmotic permeability, ensuring a highly hypotonic urine (Marshall and Grosell, 2006). In contrast, kidneys of marine species lacking glomeruli produce very small quantities of urine, with their main osmoregulatory role thought to be the excretion of excess divalent ions (especially Mg^{2+}, SO_4^{2-}) absorbed by the intestine from imbibed SW (Beyenbach, 2004). The urinary bladder is again a site for NaCl reabsorption but its epithelium now has a high osmotic permeability and uptake of NaCl facilitates water reabsorption (Howe and Gutknecht, 1978; Loretz and Bern, 1980).

3.3. Gastrointestinal Tract

The anatomy of the gastrointestinal tract varies enormously among fish classes, largely in association with different feeding strategies (Rimmer and Wiebe, 1987; Buddington et al., 1997). Nonetheless, the gastrointestinal tract has a critical role in osmoregulation in all fishes, in part because digestion relies on secretion and absorption of electrolytes (Karasov and Hume, 1997). Marine fishes also drink water and use the gastrointestinal tract as a major osmoregulatory organ to absorb water to offset osmotic loss. This process begins with esophageal desalinization and regulation of the cardiac sphincter. (A cardiac sphincter usually regulates food passage into the stomach from the esophagus and a pyloric sphincter regulates chyme passage into the intestine.) Intestinal absorption and secretion processes follow, resulting in rectal fluid output. Specifically, the posterior intestine secretes HCO_3^-, which precipitates Mg^{2+} and Ca^{2+} carbonates in the intestinal lumen and thereby reduces the osmolality of chyme. Further water uptake is then possible (reviewed by Marshall and Grosell, 2006; Whittamore, 2012).

3.4. Rectal Gland of Elasmobranchs

The rectal gland of marine elasmobranchs secretes a highly concentrated fluid (NaCl up to 1.0 M). This secretion drains through a duct into the intestine, distal to the spiral valve. Secretory tubules are composed of numerous ionocytes that possess a highly expanded basolateral membrane and have an appearance similar to cells associated with other vertebrate salt-secreting organs in marine reptiles and birds (Doyle, 1962; Bulger, 1965).

3.5. Skin, Opercular Membrane, and Yolk Sac

Some fishes possess a highly vascularized skin containing numerous ionocytes with ion transport capabilities (Marshall, 1977, 1981; Yakota et al., 2004; LeBlanc et al., 2010). In gobies, an estimated 10–20% of total ion transport is conducted across the skin surface (Marshall, 1977, 1981; Marshall and Nishioka, 1980). Indeed, the opercular membrane of both killifish (*Fundulus heteroclitus*) and tilapia (*Oreochromis mossambicus*) is extensively used as a model preparation to study ion transport and its regulation (Marshall, 1995; Marshall et al., 1997; Buhariwalla et al., 2012).

The yolk-sac membrane prior to hatching serves as the sole ionoregulatory surface. Its numerous differentiated ionocytes have become a very informative model system for ion secretion in SW tilapia (Hiroi, 2005a, b) and seabass (*Dicentrarchus labrax*) (Sucré et al., 2011) and ion absorption in

FW zebrafish (Chang et al., 2009; Kumai and Perry, 2011) and tilapia (Fridman et al., 2011). The yolk-sac membrane has been particularly helpful in resolving regulation of FW uptake mechanisms, V-H$^+$-ATPase (Chang et al., 2009), and ammonium transport via Rhcg1 (Kumai and Perry, 2011).

4. HAGFISHES

The ionic composition of hagfishes was recently reviewed by Currie and Edwards (2010). Hagfishes are extant representative of the most ancient craniates, sharing an ECF osmotic profile similar to marine invertebrates (Schmidt-Nielsen and Schmidt-Nielsen, 1923). However, the TMAO and amino acid concentrations in hagfish liver and muscle are similar to those reported for marine elasmobranchs and are thought to play a role in isosmotic regulation of muscle (Cholette and Gagnon, 1973). Unlike elasmobranchs, TMAO is absent in hagfish plasma and urea is found at an exceptionally low concentration.

Smith (1932) first demonstrated that the blood of *Myxine glutinosa* was slightly hypertonic to SW. Robertson (1960) originally suggested that hagfishes have the ability to regulate plasma ion concentrations and subsequently established that it was only plasma Na$^+$ and K$^+$ concentrations that exceed those of SW. Sardella et al. (2009) found a clear lack of symmetry between the ionic concentrations of SW and plasma in Pacific hagfish (*Eptatretus stouti*). They had no ability to regulate plasma Na$^+$ and Cl$^-$, but regulated both Ca^{2+} and Mg^{2+} concentrations when exposed to elevated environmental concentrations of these ions through unknown mechanisms.

Despite an inability to defend plasma Na$^+$ and Cl$^-$, molecular studies have identified genes for suitable ion transport within gill tissue. In addition, immunohistochemistry has localized ion transport proteins in branchial epithelial cells, including NKA, carbonic anhydrase (CA) (Choe et al., 1999), Na$^+$/H$^+$ exchanger (NHE1, NHE2, and NHE3) (Karnaky, 1998; Choe et al., 1999, 2002; Edwards et al., 2001; Tresguerres et al., 2006), and H$^+$-ATPase (Tresguerres et al., 2006). Evans (1984) originally proposed that hagfishes may utilize an NHE localized to the branchial epithelium for acid–base homeostasis. McDonald et al. (1991) first demonstrated that *M. glutinosa* could excrete excess H$^+$ during metabolic acidosis. Edwards et al. (2001) then demonstrated increased branchial expression of NHE mRNA during metabolic acidosis. Similarly, blood pH regulation following acidosis in Pacific hagfish was probably enhanced by increased synthesis and translocation of a branchial NHE (Parks et al., 2007). Collectively, these findings suggest that hagfishes have the molecular machinery for and can regulate ions in response to acid–base and ionic disturbances, without any established role for such transporters in maintaining osmotic homeostasis, as suggested by Currie and Edwards (2010).

5. LAMPREYS

Lampreys maintain their ECF concentration in both FW and SW, and therefore can osmoregulate and ionoregulate. The fact that modern lampreys have the ability to inhabit FW (exclusively in the case of the land-locked population in the Great Lakes) or are anadromous suggests that members of this agnathan group were the first vertebrates to enter FW (Hardisty, 1979). The following is a brief review of what is currently known about the osmoregulatory mechanisms of this evolutionary keystone group of fishes.

5.1. Osmoregulation in Freshwater

The osmotic stress on FW lampreys is similar to that on FW teleost fishes: an influx of water and an efflux of ions occur across the gills and skin. Correspondingly, lamprey kidneys similarly produce copious amounts of dilute urine, to counter water loading, and actively uptake ions across the branchial epithelium to counter diffusional ion loss to the environment (Bartels and Potter, 2004).

Based on teleost literature, it is reasonable to hypothesize that FW lampreys replace lost ions by either ingestion or branchial ion uptake mechanisms. Indeed, Morris demonstrated that FW lampreys actively absorb radiolabeled Na^+ and have gill ionocytes equipped with ion pumps (Morris, 1972). NKA and V-type H^+-ATPase are in cellular positions similar to those in FW teleost fishes, suggesting similar ion uptake mechanisms (Choe et al., 2004).

Anadromous lampreys alter the branchial epithelium during the life cycle, and particularly as they migrate between FW and SW (Bartels and Potter, 2004). Larval ammocoetes in FW possess a unique lamprey mitochondrion-rich cell (MRC) that has structural and localization similarities to ionocytes in other classes of fish by being clustered on the gill lamellae and within the interlamellar regions of the filament. This cell type disappears early in metamorphosis and does not reappear in the adult stages. The absence of these cells in the downstream and upstream migrant lampreys suggests that perhaps these ammocoete MRCs are precursors to the adult MRC (Bartels et al., 1998).

A second form of branchial MRC found also in ammocoetes and adults is termed the FW mitochondrion-rich cell (fwMRC). These cells are typically single cuboidal cells and are confined to the interlamellar region at the base of the filament. The fwMRC is characterized by numerous membranous vesicles and microvilli-like structures along the apical membrane (Bartels and Potter, 2004). They degenerate on entry to SW and regenerate in FW towards the end of the lamprey life cycle (Bartels and Potter, 2004). There are two subpopulations of fwMRC: A-type, thought to be associated with Na^+ uptake; and B-type, associated with Cl^- uptake (Bartels et al., 1998; Bartels and Potter, 2004), which have been localized by immunohistochemistry in Australian pouched lamprey (*Geotria australis*). The A-type fwMRC probably has a basolateral localization with NKA immunoreactivity and possibly apical expression of the NHE gene family (Choe et al., 2004). The B-type fwMRC expressed CA immunoreactivity localized to the apical membrane with diffuse V-type H^+-ATPase immunoreactivity in the cytoplasm (Choe et al., 2004). Similar staining patterns exist in elasmobranchs, where H^+-ATPase immunoreactive cells also express an apical Cl^-/HCO_3^- exchanger (Piermarini and Evans, 2001), which led to the hypothesis that these two cell populations work in parallel to facilitate NaCl uptake and excrete acid- and base-relevant ions (Choe et al., 2004). However, to date, there are no functional data in the lamprey to support this hypothesis.

FW lampreys have urine output rates (> 10 ml kg^{-1} h^{-1}) higher than reported for some FW teleosts (Morris, 1972; Brown and Rankin, 1999), perhaps reflecting a high osmotic permeability retained from marine isotonic ancestors, but with increased glomerular filtration. The lamprey kidney has glomeruli structurally similar to those of more derived vertebrates. Over 50% of the nephron is proximal tubule, with epithelial cells possessing numerous mitochondria, microvilli, and infoldings along the basolateral membrane, features absent in hagfishes (Hentschel and Elger, 1989). The distal tubule is lined by cuboidal epithelial cells, which lack microvilli but are rich in mitochondria, similarly to other vertebrates, and is responsible for reabsorbing approximately 90% of filtered NaCl (Morris, 1972). The gill has a relatively low permeability to ions (Evans, 1979; Stinson and Mallatt, 1989). Even so, FW lampreys still lose ions to the environment via the urine, which adds to the diffusional branchial ion loss (Bartels and Potter, 2004).

5.2. Osmoregulation in the Marine Environment

Marine lampreys are hypoosmotic relative to the SW environment, which for a teleost requires decreased urine production, increased urine NaCl levels, increased branchial excretion of NaCl, and drinking of SW for osmoregulation.

Marine lamprey gills express a third type of epithelial mitochondrion-rich cell (swMRC), one that has structural similarities with branchial MRCs present in marine teleosts, and is the postulated branchial site of NaCl excretion (Bartels and Potter, 2004). Similarly to teleosts, lamprey swMRCs differ from the fwMRCs in that they lack microvilli and have reduced tight junctions, implying an increased ion permeability. The swMRCs probably increase in abundance in FW lampreys just before SW migration (Reis-Santos et al., 2008) but disappear on re-entry into FW in adults (Bartels and Potter, 2004).

Few studies have examined kidney function in SW lampreys. The most recent, Logan et al. (1980), demonstrated a 95% decrease in urine flow rate after a 2-week acclimation to SW, primarily due to a reduction of single nephron glomerular filtration rate (GFR) and increased reabsorption of water in the distal tubule. Urine Na^+ concentrations were also much lower than those in the plasma, implying kidney Na^+ reabsorption, which was hypothesized as the mechanism to increase water reabsorption (Logan et al., 1980). While evidence exists for renal NaCl excretion, the amount is insufficient to offset the gains from diffusion and ingestion.

SW-adapted lampreys ingest SW at a rate of around 1% body weight h^{-1} (Rankin, 1997), a rate much higher than that in marine teleosts (Evans, 1979). About 70–80% of the ingested water is absorbed in the intestine, while ingested ions either remain in the intestine or are absorbed into the blood and excreted (Morris, 1958; Pickering and Morris, 1970; Rankin, 1997). Only one study has localized intestinal ion transport, using radioisotopes with isolated preparations (Pickering and Morris, 1973). The majority of monovalent ion influx occurred across the anterior intestine in association with mucosal to serosal water transport. The mucosal membrane was suggested as the site of active Na^+ and Cl^- transport.

6. ELASMOBRANCHS

Elasmobranchs (sharks, skates, and rays) are widely distributed marine environments, with a number of species having euryhaline capability. Yet, only two of these species inhabit estuarine environments, while 36 species are obligate FW dwellers (Martin, 2005). These special groups are considered in detail in Chapter 4 of this volume (Ballantyne and Fraser, 2013).

6.1. Ion Regulation in the Marine Environment

Marine elasmobranchs regulate the body fluid concentrations of monovalent (Na^+, Cl^-) and divalent (Mg^{2+}, SO_4^{2-}) ions. Their plasma osmolality is similar to SW, largely due to the retention of organic nitrogenous compounds such as urea and TMAO, with urea levels comprising approximately 39% of total plasma osmolality while plasma TMAO is maintained at around 70–75 mM.

Urea and TMAO retention in the blood is critical for marine elasmobranchs. Indeed, elasmobranch gills are relatively impermeable to urea in comparison to other teleost fishes (Smith, 1936; Boylan, 1972; Pärt et al., 1998), probably owing to a high cholesterol to phospholipid ratio in the basolateral membranes of gill epithelial cells, which reduces urea permeability (Hill et al., 2004). In addition, urea is actively transported into the plasma across the basolateral membrane via an Na^+/urea antiporter to maintain a low intracellular urea concentration. Thus, the diffusional gradient for urea across the gill apical membrane is greatly reduced (Fines et al., 2001).

Elasmobranchs protect themselves from the toxic effects of high urea concentrations with TMAO, which inhibits the deleterious effects of urea on proteins (Yancey and Somero, 1979).

There is no evidence of net NaCl secretion across the branchial epithelium of marine elasmobranchs, as is found in SW teleosts (Shuttleworth, 1988; Hammerschlag, 2006). Instead, a rectal gland is thought to be the primary secretory site and regulator of NaCl homeostasis. The branchial epithelium has two types of ionocyte. One exhibits high levels of basolaterally located V-type H^+-ATPase, with high levels of NKA also located basolaterally (Piermarini and Evans, 2001; Wilson et al., 2002). These cells are apparently more involved in acid–base regulation than solute extrusion and are coupled to an apically located anion exchanger (pendrin). In the second cell type, a basolateral NKA is coupled to an apically located Na^+/H^+ exchanger, suggesting its importance in the extrusion of acid–base-relevant ions rather than ion uptake (Piermarini and Evans, 2001).

As seen in marine teleosts, marine elasmobranchs gain water osmotically across their branchial epithelium, rather than drinking SW; however, the GFR remains similar to that of FW teleosts (Hickman and Trump, 1969) but tubular reabsorption mechanisms allow for a 60–85% reabsorption of glomerular filtrate (Kempton, 1953; Schmidt-Nielsen and Rabinowitz, 1964). Yet, while kidney filtration and reabsorption is important for osmoregulation (Beyenbach, 1986, 2004), the kidney of the marine elasmobranch is not a major site for NaCl secretion. Although 90% of the filtered load of urea and TMAO is reabsorbed, only 60–70% of filtered NaCl is actively (Na^+) or passively (Cl^-) reabsorbed. The rectal gland carries the excretory load by secreting a fluid with an NaCl concentration almost twice that of the plasma.

The mechanisms of NaCl secretion in the rectal gland are similar to those seen in ionocytes of marine teleost gills. A basolateral NKA maintains intracellular Na^+ concentrations below equilibrium; this facilitates the movement of Na^+, Cl^-, and K^+ in via the NKCC. The K^+ is recycled across K^+ channels in the basolateral membrane. Cl^- exits the cell via an apical CFTR down its electrochemical gradient and Na^+ is extruded via a basolateral NKA following its electrochemical gradient and is transported to the lumen paracellularly.

6.2. Freshwater and Euryhaline Elasmobranchs

Approximately 43 species of elasmobranchs can tolerate FW for extended periods (Martin, 2005). Euryhaline and FW elasmobranchs, such as the bull shark (*Carcharhinus leucas*) and Atlantic stingray (*Dasyatis sabrina*), face different osmotic pressures compared to their marine counterparts and must reduce solute levels. Hazon et al. (2003) concluded that marine elasmobranchs have the ability to independently regulate Na^+, Cl^-, and urea concentrations in reduced salinity environments. The predominant effector of this reduction is the loss of urea. Stenohaline FW species, unlike their marine counterparts, do not retain urea in their tissues and are not ureotelic (Wood et al., 2002), but ammoniotelic. Urea and TMAO in obligate FW stingrays contribute less than 1% of total plasma osmolality while inorganic ions and osmolality are at levels similar to those of FW teleosts and much lower than those of marine elasmobranchs. The euryhaline bull shark can regulate urea, TMAO, and inorganic ion concentrations depending on its environment (Pillans and Franklin, 2004; Pillans et al., 2005, 2008; Anderson et al., 2007; Reilly et al., 2011). Transferring the euryhaline stingray from FW to SW increases its plasma NaCl, urea, and total osmolality to almost that of SW levels (Piermarini and Evans, 1998).

In dilute environments the gills play a critical role in the uptake of NaCl, functioning like those of FW teleosts. Piermarini and Evans (2001) suggested a model for ion uptake in the FW stingray that involved two types of branchial ionocytes. One exhibited a basolaterally located V-type H^+-ATPase, hypothesized to act as a base-secreting cell via an apically located Cl^-/HCO_3^- exchanger that would facilitate Cl^- uptake. The other, more prevalent type exhibited a basolaterally located NKA and was hypothesized to act as an acid-secreting cell with an apically located Na^+/H^+ exchanger facilitating Na^+ uptake (Piermarini and Evans, 2001).

FW-acclimated euryhaline elasmobranchs display smaller rectal glands than marine elasmobranchs, consistent with a reduced need for NaCl secretion in an ion-poor environment (Thorson et al., 1978, 1983; Pillans et al., 2008). Microscopically, the rectal gland has fewer glandular tubules than in marine elasmobranchs. In stenohaline FW elasmobranchs, the rectal gland is severely atrophied (Gerst, 1977).

The kidney of FW elasmobranchs is involved in retaining Na^+ and Cl^- and excreting excess water. In addition, euryhaline elasmobranchs respond to dilute environments by increasing GFR and urine flow, while reducing plasma Na^+, Cl^-, and urea concentrations (Smith, 1931; Golstein et al., 1968; Golstein and Forster, 1971; Payan and Maetz, 1973; Janech and Piermarini, 2002; Choe and Evans, 2003).

Euryhaline elasmobranchs retain the ability to store high concentrations of urea in their tissue when in FW. Consequently, the osmotic gradient between the environmental water and euryhaline elasmobranchs in FW is the highest reported for any group of fishes, at 4600 mOsmol kg^{-1} (Smith, 1931; Thorson et al., 1973; Piermarini and Evans, 1998), which increases osmotic water gain and accounts for increased urine output. The increase in urine output also results in a net loss of NaCl that is presumably compensated for by the gills.

7. TELEOST FISHES

Teleosts, the largest division of fishes, inhabit environments ranging from FW (<0.1 mOsmol kg^{-1}) through marine (~1000 mOsmol kg^{-1}) to hypersaline (2400 mOsmol kg) (Brauner et al., 2013, Chapter 9, this volume). Some are euryhaline and can adjust to a wide range of salinities. In all environments, teleosts maintain plasma osmotic concentrations between 250 and 450 mOsmol kg^{-1}. This section will explore the tissues and mechanisms responsible for this remarkable ionic and osmotic balance. Basal actinopterygians, such as sturgeon, gar, and *Amia*, appear to be similar to teleosts with regard to their strategy and mechanisms for osmoregulation in FW and SW, although these groups have not been examined in great detail.

7.1. Osmoregulation in the Marine Environment

The epithelial ionocyte of the SW teleost fish gill, first described by Keys and Wilmer in 1932, has distinct basolateral infoldings, significant NKA activity, and numerous mitochondria. Ionocytes are commonly localized in interlamellar regions of the gill filament (Evans et al., 2005), with a concave apical membrane recessed below the surface of surrounding pavement cells (PVCs) to form an apical crypt. Morphofunctional aspects of gill ionocytes have recently been reviewed by Hwang et al. (2011) and Hiroi and

McCormick (2012). The morphologically simple apical membrane has a subapical region with a complex tubular and vesicular network that forms vesicles that can be trafficked to and fused into the apical membrane (Laurent, 1984). Intercellular tight junctions between the ionocytes and PVCs are multistranded. SW ionocytes always exist as a complex with other ionocytes and accessory cells (ACs). The function of the SW AC is unclear. ACs project cytoplasmic processes into the apical membrane of the ionocyte, forming junctions that are not extensive and are considered leaky, providing a paracellular pathway for Na^+ extrusion (Sardet et al., 1979; Hootman and Philpott, 1980; Karnaky, 1992).

Ionocytes in the SW teleost gill primarily excrete NaCl through passive transport of Na^+, coupled to the secondary active transport of Cl^- (reviewed by Evans et al., 2005). Basolateral NKA maintains a low intracellular Na^+ concentration (Hirose et al., 2003), creating a gradient for Na^+ entry to Cl^- and K^+ via NKCC. K^+ entering the cell via NKA and NKCC is thought to exit via basolateral K^+ channels. Cl^- exits via an apical anion channel CFTR, which in turn generates a serosal positive transepithelial potential that powers paracellular extrusion of Na^+ through the cation-selective junctions between the AC and ionocyte (Fig. 1.3).

Marine teleosts drink SW to offset a constant osmotic water loss. Ingested SW is desalinated in the esophagus, accomplished by both passive and active NaCl transport. The esophagus is impermeable to water (Parmelee and Renfro, 1983) and so NaCl in the stomach fluid is low, reducing osmotic pressure by 50% in the anterior stomach (Grosell, 2006). The high density of capillaries in the esophagus ensures that absorbed NaCl is rapidly transported to the gills for excretion (Ando et al., 2003). NaCl-driven water absorption occurs across the intestinal lumen (Skadhauge, 1974; Mackay and Janicki, 1978; Usher et al., 1991). Extensively reviewed by Grosell (2011), the model for Na^+ uptake across the apical membrane of the intestinal cell involves either two cotransporters, NKCC2 and NCC, or an apically located NHE2 and/or NHE3. Na^+ transport across the basolateral membrane and into blood is governed by the basolaterally located NKA, which in turn drives the uptake of Na^+ across the apical membrane. Cl^- entry across the apical membrane is accomplished via the two cotransporters and a Cl^-/HCO_3^- anion exchanger, SLC2616, powered by the HCO_3^- gradient established in part by the hydration of CO_2 by CA (Wilson et al., 1996). The resulting H^+ is proposed to be removed via a basolateral H^+-ATPase (Grosell et al., 2007, 2009b) (Fig. 1.4) and Cl^- transport into the blood is accomplished by basolateral anion channels. However, immunolocalization studies have failed to demonstrate clear localization of H^+-ATPase to the basolateral membrane and an apical route for H^+ excretion has been suggested (Grosell et al., 2007), based on evidence

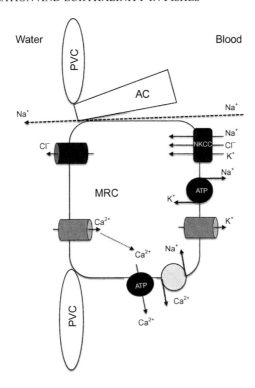

Fig. 1.3. The current gill model for NaCl secretion in marine teleost fishes. The mineralocorticoid receptor (MR) and accessory cell (AC) form a selective leaky paracellular junction that enables Na^+ to be passively secreted. In a cell membrane with a low osmotic permeability, Cl^- enters the cell using an $Na^+/K^+/2Cl^-$ (NKCC) cotransporter driven by the Na^+ gradient that is maintained by the basolateral Na^+/K^+-ATPase. Cl^- accumulates intracellularly above its electrochemical equilibrium; this is sufficient to drive its excretion across the apical membrane via cystic fibrosis transmembrane conductance regulator (CFTR)-type anion channels. This figure also shows the independent Ca^{2+} absorptive pathway that is essentially the same in seawater and in freshwater. The process involves apical Ca^{2+} entry via channels and exit via a basolaterally located Na^+/Ca^{2+} exchanger and Ca^{2+}-ATPase pump. MRC: mitochondrion-rich cell; PVC: pavement cell; ATP: adenosine triphosphate. (Modified from Marshall and Grosell, 2006.)

for an apical localization of H^+-ATPase provided by H^+-ATPase inhibitor bafilomycin resulting in a net increase in base secretion when added to luminal saline of an Ussing chamber (Grosell et al., 2009a).

Marine teleosts produce low urine flow rates and, in fish with either aglomerular or glomerular kidneys, the tonicity of the urine is similar to the extracellular fluid, with the major electrolytes being Mg^{2+}, SO_4^{2-}, and Cl^- (Beyenbach, 2004). Stenohaline marine glomerular and aglomerular kidneys have proximal tubule segments connected directly to a collecting duct (Beyenbach, 2004) (Fig. 1.5). Thus, marine teleosts mostly lack distal tubule segments and differ primarily in the presence or absence of glomeruli. The glomerular kidney has a much lower GFR than that of FW teleost fishes (Hickman and Trump, 1969; Braun and Dantzler, 1997; Baustian and Beyenbach, 1999). Even without glomeruli, urine flow rates for aglomerular species are similar to those of glomerular fishes.

Tubular secretion is important for marine fishes in secretion of electrolytes, as well as organic cations and anions (Braun and Dantzler, 1997; Dantzler, 2003). Tubular Mg^{2+} and SO_4^{2-} concentrations are elevated

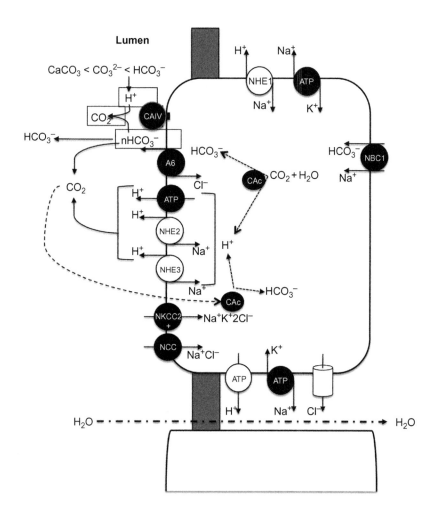

above those in plasma and therefore may also play a role in fluid secretion (see review by Marshall and Grosell, 2006).

NaCl and fluid reabsorption occurs in the late proximal tubule and in the urinary bladder, driven by NKA creating a favorable gradient for Na^+ entry from the tubular lumen across the apical membrane; this drives cotransport of glucose and amino acids too. The tubular absorption of Na^+ and Cl^- is thought to occur across the apical membrane by Cl^-/HCO_3^- exchange coupled to Na^+/H^+ exchange (Fig. 1.6) (see reviews by Beyenbach, 2004; Pelis and Renfro, 2004; Marshall and Grosell, 2006). In the European eel, NaCl uptake in the urinary bladder is via an electroneutral process mediated by apical membrane NKCC2 β-isoform (Cutler and Cramb, 2008), powered by basolateral NKA.

7.2. Osmoregulation in Freshwater

FW teleosts face diffusive osmotic gain of water across the gills (Fig. 1.2). To counter this, they excrete large volumes of very dilute urine formed by a

---

Fig. 1.4. Transport processes in the intestinal epithelium of marine teleost fishes. (Redrawn from Grosell, 2011.) Water transport (dotted lines) is driven by active NaCl absorption. Entry of Na^+ across the apical membrane is via cotransporters [$Na^+/K^+/2Cl^-$ (NKCC) and Na^+/Cl^- (NCC)] and the extrusion of Na^+ across the basolateral membrane via Na^+/K^+-ATPase. Recent evidence also suggests that Na^+/H^+ exchangers NHE2 and NHE3 are expressed in the intestinal epithelium, suggesting an alternative route for Na^+ uptake across the apical membrane. Cl^- influx across the apical membrane occurs via both cotransporters and Cl^-/HCO_3^- exchange conducted by the SLC26a6 anion exchanger, while Cl^- exits the cell via basolateral anion channels. The entry of HCO_3^- across the basolateral membrane via the Cl^-/HCO_3^- cotransporter (NBC1) provides some of the intracellular HCO_3^- for apical anion exchange. The additional HCO_3^- arises from the hydration of endogenous metabolic CO_2 facilitated by cytosolic carbonic anhydrase (CAc). The H^+ arising from CO_2 hydration are excreted predominantly across the basolateral membrane by an Na^+-dependent pathway and possibly by vacuolar H^+ pumps. The localization of NHE2 and NHE3 in the apical membrane provides a route for apical H^+ extrusion; alternatively, H^+ excretion across the apical membrane could also be via H^+ pumps. The apical secretion of HCO_3^- and its dehydration in the intestinal lumen yields molecular CO_2. Molecular CO_2 from this reaction is rehydrated in the enterocytes and resulting HCO_3^- is sensed by soluble adenylyl cyclase resulting in the stimulation of ion absorption via NKCC2 (+). The conversion of HCO_3^- to CO_2 in the intestinal lumen is via membrane-bound carbonic anhydrase, CAIV and possibly other isoforms; this results in the consumption of H^+ and, as a result, contributes to luminal alkalization and carbonate formation. The titration of luminal HCO_3^- and the formation of carbonate facilitate the creation of $CaCO_3/MgCO_3$ precipitates, which reduces luminal osmotic pressure and aids water absorption. The electrogenic SLC26a6 anion exchanger is stimulated by the hyperpolarizing effect of the H^+ pump and exports $nHCO_3^-$ in exchange for $1Cl^-$. The apical SLC26a6 and electrogenic H^+ pump constitutes a transport metabolon that possibly accounts for the active secretion of HCO_3^- and the uphill movement of Cl^- across the apical membrane.

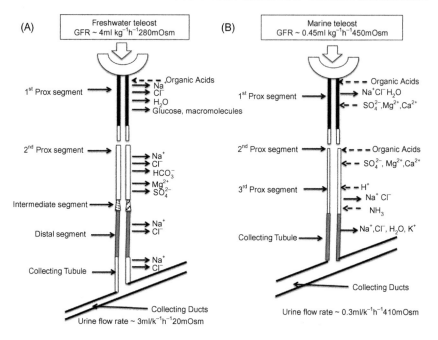

Fig. 1.5. Schematic images representative of the structure and function of the nephron in (A) freshwater and (B) marine teleost fishes. The major morphological segments of the nephron are indicated. Representative plasma filtration rates and osmotic concentrations, tubular reabsorption and secretion processes, urine flow rates, and osmolality are shown. GFR: glomerular filtration rate; Prox: proximal.

kidney specialized for electrolyte absorption. They also face diffusive ion loss, which is countered by active ion uptake across the gills and dietary absorption of salt (Kirschner, 1991, 1997, 2004).

Two distinct populations of branchial epithelial cells are associated with ion transport in FW fishes: the ionocyte and the PVCs. Pavement cells form an epithelial layer that is continuous along the lamellae and interlamellar junctions, and are connected to each other and adjacent ionocytes by tight junctions. While PVCs lack the metabolic machinery to drive ion transport (Evans et al., 2005), immunohistochemistry has localized several ion transport proteins in PVCs of FW-adapted rainbow trout (Sullivan et al., 1996; Edwards et al., 1999; Wilson et al., 2000) and killifish (Marshall et al., 2002b; Edwards et al., 2005), suggesting a role in ion transport. The participation of PVCs in overall ion and osmotic regulation is yet to be revealed. Yet, rainbow trout PVCs may be active in ammonium transport via Rhcg1 (Nawata et al., 2007) and cultured sea bass PVCs have cell

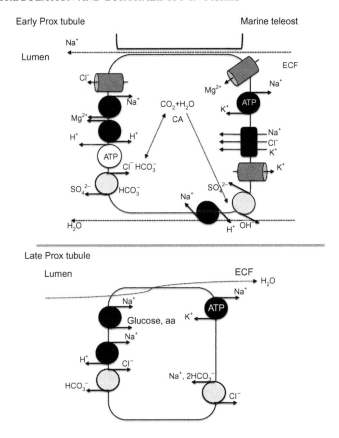

Fig. 1.6. Cellular model of both secretory and absorptive processes in the (A) early and (B) late proximal tubule of glomerular and aglomerular marine teleosts. (A) In the early proximal tubule the secretory processes include the secondary active secretion of Cl^- driven by the activity of the basolateral Na^+/K^+-ATPase which provides the electrochemical gradient for Cl^- entry via the basolateral NKCC. Cl^- is secreted from the cell across the apical membrane down its electrochemical gradient to the lumen via Cl^- channels (as yet unidentified). Na^+ is excreted via electrically coupled paracellular pathways. Mg^{2+} secretion involves the carrier-mediated entry of Mg^{2+} across the basolateral membrane down its electrochemical gradient, which provides intracellular Mg^{2+} for Na^+/Mg^{2+} exchange and H^+/Mg^{2+} exchange across the apical membrane. The H^+/Mg^{2+} exchange is driven by apical H^+-pump activity, while the Na^+/Mg^{2+} exchange process is fueled by the Na^+ gradient established by the basolateral Na^+/K^+-ATPase. The secretion of SO_4^{2-} involves anion exchange processes in both the basolateral and the apical membranes and is dependent on intracellular carbonic anhydrase (CA). (B) The driving force for solute and fluid reabsorption in the late proximal tubule is provided indirectly by the basolateral Na^+/K^+-ATPase, which creates a favorable gradient for Na^+ entry across the apical membrane and drives $Na^+/glucose$, $Na^+/amino$ acid cotransport, and Na^+/H^+ exchange. The reabsorption of Cl^- occurs via an electroneutral process driven by the electrochemical Na^+ gradient and is suggested to occur via Cl^-/HCO_3^- exchange coupled with the Na^+/H^+ exchange process. Prox: proximal; ECF: extracellular fluid. (More detail available in reviews by Beyenbach, 2004; Pelis and Renfro, 2004.)

volume regulatory transport mechanisms (Avella et al., 2009). PVCs can also cover the apical crypt of ionocytes to rapidly downregulate ion secretion when a euryhaline fish enters a dilute environment (Sakamoto et al., 2000; Daborn et al., 2001).

Rainbow trout ionocytes have two distinct but morphologically similar subtypes that differ in the presence or absence of peanut agglutinin (PNA) binding factor (Goss et al., 2001). PNA-negative cells show higher levels of H^+-ATPase activity and demonstrate a phenamil- and bafilomycin-sensitive Na^+ uptake mechanism. PNA-positive cells also have high levels of NKA but no evidence of phenamil-sensitive Na^+ transport (Reid et al., 2003), which is suggestive of an absence of an epithelial Na^+ channel (ENaC). An apical Na^+ channel homologous to ENaC is unlikely; however, ENaC orthologues or paralogues are absent in all fish genomes published to date (reviewed by Evans et al., 2005). In the stenohaline zebrafish, four distinct ionocytes are recognized, each with an apparently separate function for uptake of Na^+, Cl^-, and Ca^{2+}, and acid–base balance (Hwang et al., 2011).

Several models for Na^+ uptake are hypothesized for FW teleost gills. A generally accepted common feature is an apical H^+-ATPase electrogenically coupled to an Na^+ channel (Fig. 1.7a,b) (Reid et al., 2003). There is strong evidence for an apical H^+-ATPase in trout and zebrafish. Knockdown and pharmacological approaches (Dymowska, personal communication) have provided evidence that ASIC4 (an acid-sensing ion channel and a member of the ENaC/degenerin superfamily) may act as an Na^+ channel in rainbow trout and zebrafish. ASICs have been isolated from the teleost zebrafish (Chen et al., 2010) and are inhibited by diarylamides, particularly 4',6-diamidino-2-phenylindole (DAPI) (Chen et al., 2010). Whereas most ASICs are activated by external H^+, some are not (e.g. ASIC4.2), and some are insensitive to amiloride inhibition (Chen et al., 2007; Springauf and Greunder, 2010). Previous use of amiloride and phenamil to block Na^+/H^+ exchange as well as Na^+ channels has produced confusing results. Future investigations into DAPI inhibition of Na^+ uptake in adult FW fishes and further characterization of this channel may prove that one or more ASICs are Na^+ uptake channel(s) in FW fishes.

There is a thermodynamic argument that an Na^+/H^+ exchanger could not function in a dilute environment (Parks et al., 2008) owing to a low intracellular Na^+ content that is maintained by the basolateral NKA and Na^+/HCO_3^- cotransporter (Hirata et al., 2003). Molecular and immuno-histochemical evidence suggests a role for an apical NHE in Na^+ uptake instead (Edwards et al., 1999; Wilson et al., 2000; Hirata et al., 2003; Ivanis et al., 2008). An apical membrane ammonium transporter, Rhcg1, may function in close association with NHE to facilitate Na^+ uptake (Kumai and Perry, 2011). The discovery of the Rhesus proteins has also

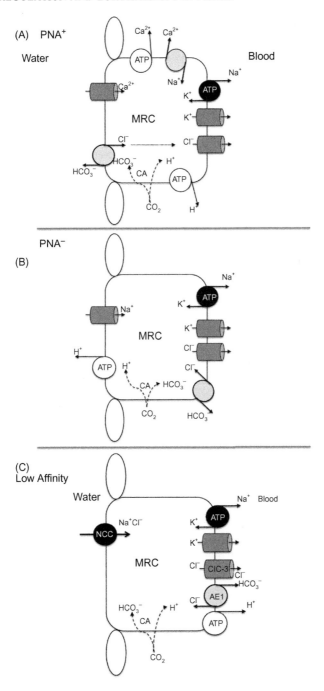

helped to solve the excretion pathway for ammonium ions, especially into acidic water (Wright and Wood, 2009; Nawata et al., 2010; Kumai and Perry, 2011). In particular, multiple mechanisms of Na^+ uptake may exist in larval zebrafish in acidic water: one in which Na^+ uptake via NHE3b is linked to ammonia excretion via Rhcg1 and another facilitated by H^+-ATPase.

Immunohistochemical and pharmacological evidence suggests that additional apical NCC cotransport may be involved in apical Na^+ uptake (Fig. 1.7c) (Hiroi et al., 1999; Kirschner, 2004; Preest et al., 2004; Hiroi and McCormick, 2012). The NCC gene, cloned from FW Mozambique tilapia (*Oreochromis mossambicus*), expresses protein exclusively in the apical membranes of yolk-sac membrane and gills of FW-acclimated fishes (Hiroi et al., 2008), seems to be restricted to convex portions of the apical membrane, and is enhanced by low Cl^- water (Inokuchi et al., 2009). Physiological studies with zebrafish provide support for this finding because metozalone, a specific NCC inhibitor, decreases Na^+ and Cl^- uptake in embryos and a morpholino against NCC reduces NaCl content (Wang et al., 2009). The NCC ionocytes are apparently responsible for NaCl uptake from FW with higher NaCl content (Inokuchi et al., 2008, 2009; Yang et al., 2011) and in tilapia are supported by the FW-adaptive hormone prolactin (Breves et al., 2010).

There are two apical membrane pathways through which apical uptake of Cl^- could occur: Cl^-/HCO_3^- exchange and a coupled Na^+–Cl^- uptake.

Fig. 1.7. Current freshwater teleost gill model representing the two morphologically similar cell types that are functionally distinct and separable with peanut agglutinin (PNA). (A) PNA^+ cells secrete base and in exchange for Cl^- uptake via an apical Cl^-/HCO_3^- exchanger driven by a proton pump situated on the basolateral membrane (V-type ATPase). Cl^- exits via a basolateral chloride channel (ClC-3) and/or cystic fibrosis transmembrane conductance regulator (CFTR) anion channel. Ca^{2+} is probably taking place in both cell types and involves apical Ca^{2+} channels and basolateral Na^+/Ca^{2+} exchange and Ca^{2+}-ATPase. (B) PNA^- cells are specialized for acid secretion and Na^+ uptake. The V-type H^+-ATPase is localized to the apical membrane and coupled to an epithelial Na^+ channel. The electrical gradient is established by the H^+-ATPase facilitated by the movement of Na^+ into the cell, and it exits via the basolateral Na^+/K^+-ATPase pump. HCO_3^- exits across the basolateral membrane in exchange for Cl^-, which is then recycled via ClC-3 and/or CFTR anion channels. (C) Low-affinity NaCl uptake from dilute brackish waters by some euryhaline species instead has NCC, the Na^+ Cl^- cotransporter, in the apical membrane to effect NaCl uptake into the cell, with Na^+/K^+-ATPase in the basolateral membrane to transport Na^+ and ClC-3 anion channel to move Cl^- into the blood. In these cases, V-type H^+-ATPase is not in the apical membrane, rather basolateral, in parallel with Cl^-/HCO_3^- exchange (AE1), probably for cellular acid–base homeostasis. MRC: mitochondrion-rich cell; CA: carbonic anhydrase.

Cl^- uptake across FW teleost gills is generally accepted to be coupled to HCO_3^- excretion. Cellular CA produces intracellular HCO_3^- that plays a role in both acid–base balance and Cl^- uptake (Gilmour and Perry, 2009). However, some FW teleost species [eel (*Anguilla anguilla*), blue gill (*Lepomis macrochirus*), and mummichog (*F. heteroclitus*)] show no evidence of gill Cl^- uptake from the environment (Tomasso and Grosell, 2005), while others do [rainbow trout (*O. mykiss*), goldfish (*Carassius auratus*), and zebrafish (*D. rerio*)] (Evans, 2011).

In rainbow trout, molecular studies have localized Cl^-/HCO_3^- exchanger (AE1, Band 3, and SLC4A1) messenger RNA (mRNA) to epithelial cells within the lamellae and filament (Sullivan et al., 1996). However, trout antibodies have failed to identify AE1 proteins in the rainbow trout (Wilson et al., 2000). Expression of SLC26A3 mRNA has been localized in adult zebrafish to epithelial cells within the filament. In addition, branchial SLC26A3 mRNA expression significantly increased following transfer to low Cl^- FW (0.02 mM) (Perry et al., 2009) and Cl^- uptake increased after transfer to media containing high HCO_3^- concentrations (10–20 mM) (Perry et al., 2009). In zebrafish, the anion exchanger AE1b is present in the gill, while the $Na^+–HCO_3^-$ cotransporter is not (Lee et al., 2011). Electrochemical gradients driving apical neutral or electrogenic Cl^-/HCO_3^- exchange have yet to be elucidated. However, the presence of a basolaterally located H^+-ATPase in two FW teleost species suggests similarities to the B-type intercalated cells of the mammalian cortical collecting duct, where basolateral H^+ excretion provides intracellular HCO_3^- to drive Cl^-/HCO_3^- exchange (Fig. 1.7c) (Evans, 2011). Apical H^+ excretion via H^+-ATPase could also provide the electrochemical gradient to facilitate Cl^-/HCO_3^- exchange (Fenwick et al., 1999; Boisen et al., 2003). Recent research suggests that the basolateral exit of Cl^- in FW ionocytes occurs via the ClC-3 anion channel (Tang et al., 2010, 2011) and basolateral ClC-3 abundance in Mozambique tilapia gills increases with acclimation to deionized water (Tang et al., 2011).

Water can be gained dietarily and must be compensated through increases renal output (Kristiansen and Rankin, 2001). Smith et al. (1989) initially suggested that dietary salt uptake may temporarily exceed the needs associated with homeostasis and therefore the gills may temporarily be required to excrete excess Na^+. However, the majority of Na^+ and K^+ in the food of FW rainbow trout was absorbed in the stomach 72 h after feeding, with the anterior intestine being the site of high NaCl secretion, although net Na^+ absorption was negligible despite the 80–90% uptake of dietary K^+ and Cl^- (Bucking and Wood, 2006).

Urine flow rate in FW teleosts is relatively high and reflects the need to excrete osmotic water gain. Urine is dilute, with Na^+ and Cl^- concentrations of <5–10 mM (Hickman and Trump, 1969). FW kidneys have a distal renal tubule, connecting the proximal segments to the collecting ducts (Hickman and Trump, 1969), unlike SW teleosts. FW teleosts mostly have glomeruli, and in contrast to terrestrial mammals, urine flow rates are directly controlled by glomerular filtration (Hickman and Trump, 1969) (Fig. 1.5). Some water is reabsorbed in the proximal tubule, but most electrolyte reabsorption takes place in the water-impermeable distal tubules (Braun and Dantzler, 1997). Monovalent ion uptake in the early distal tubules is thought to involve a basolateral NKA and an apical NKCC2. The urinary bladder is also involved in monovalent ion uptake, with Na^+ uptake facilitated by an amiloride-sensitive NHE (Fig. 1.8).

The use by euryhaline and migratory teleosts of the hypoosmotic and hyperosmotic regulatory mechanisms described above is dependent upon the environmental salinity. Euryhaline teleosts in FW are more likely to adopt a low-affinity NaCl uptake system, such as the apical NCC in FW tilapia, rather than the high-affinity uptake mechanism with apical H^+-ATPase and an Na^+ channel in the apical membrane more typical of stenohaline FW fishes such as zebrafish and anadromous rainbow trout in FW. Whereas some euryhaline teleost fishes use H^+-ATPase in FW (e.g. rainbow trout), many estuarine euryhaline teleost fishes have difficulty acclimating to ion-poor FW. Two subspecies of sheepshead minnow (*Cyprinodon variegatus*) have qualitatively similar low-affinity Na^+ uptake kinetics (K_m = 7–38 mM) when acclimated to 2 or 7 mM Na^+, but the *C. v. hubbsi* subspecies is the only one that can switch to a higher affinity system (K_m = 0.10–0.14 mM) in low-Na^+ FW (<1 mM Na^+) (Brix and Grosell, 2012). Clearly, the composition of dilute environments can drive plasticity in ion-uptake systems and different species may have different strategies that will be revealed by further investigation.

8. CONCLUSIONS AND PERSPECTIVES

Much of the research to date has focused on the molecular mechanisms enabling the successful transition between FW and SW. The ability of fishes to maintain homeostasis in the face of an osmotic stress depends on their ability to detect and quantify cellular alterations due to environmental salinity (Kültz, 2013, Chapter 2, this volume) and then coordinate an appropriate response. Osmotic stress associated with SW transfer is linked to the activation of molecular and physiological adaptations. Remodeling of the branchial epithelium takes place through cellular proliferation and differentiation, intracellular volume regulation, modulation of transport protein abundance, and more rapid alteration of their activity (Evans, 2002). The rapid and long-term physiological cellular and biochemical adjustments made by euryhaline fishes in response to salinity change are outlined in Chapter 8 of this volume (Marshall, 2013).

Recent advances in molecular biology have enabled investigators to utilize microarray techniques to identify alterations in gene expression associated with the transfer to SW. Evans and Somero (2008) identified a number of genes activated in response to SW transfer in the euryhaline goby *Gillichthys mirabilis*. The genes identified ranged from those associated with osmosensing (FKBP-51) to effectors (NKA) as well as a range of genes associated with a number of hormonally controlled signaling cascades activated through ligand receptor binding (IRS-2, IGFBP). Evans and Somero also identified the expression of several genes that encode for compensatory osmolyte production (CSAD, GluDC, and IMPase) (Evans and Somero, 2008). Recall that a key phase in hyperosmotic challenge is the maintenance of cell volume, in which case compensatory osmolytes are accumulated within the cell to equalize intracellular–extracellular tonicity and hence maintain cell volume (Yancy, 2005).

The identification of osmoregulatory associated genes is the first step in the identification of the adaptive processes. A real understanding of transcriptional alterations can only be ascertained following combined proteomic, metabolic, and physiological explorations. The increasing number of fish genomes being revealed enables a more focused experimental design to resolve the regulatory relationships among ion transport elements. Many questions remain as to how euryhaline fishes respond to frequent and large environmental changes that challenge their abilities to adapt. In particular, FW adaptive osmoregulatory mechanisms seem to be diverse. Different FW uptake mechanisms may have evolved many times in different ways. Understanding these various adaptive strategies may inform us not only about osmoregulation but also about evolutionary processes.

REFERENCES

Agre, P., King, L. S., Yasui, M., Guggino, W. B., Ottersen, O. P., Fujiyoshi, Y., Engel, A. and Nielsen, S. (2002). Aquaporin water channels – from atomic structure to clinical medicine. *J. Physiol.* 542, 3–16.

Agre, P., Nielsen, S. and Ottersen, O. P. (2004). Towards a molecular understanding of water homeostasis in the brain. *Neuroscience* 12, 849–850.

Anderson, W. G., Taylor, J. R., Good, J. P., Hazon, N. and Grosell, M. (2007). Body fluid volume regulation in elasmobrach fish. *Comp. Biochem. Physiol. A Mol. Integr. Physiol.* 148, 3–13.

Ando, M., Mukuda, T. and Kozaka, T. (2003). Water metabolism in the eel acclimated to seawater: from mouth to intestine. *Comp. Biochem. Physiol. B Biochem. Mol. Biol.* 136, 624–633.

Avella, M., Ducoudret, O., Pisani, D. F. and Poujeol, P. (2009). Swelling activated transport of taurine in cultured gill cells of seabass: physiological adaptation and pavement cell plasticity. *Am. J. Physiol. Regul. Integr. Comp. Physiol.* 296, R1149–R1160.

Bagherie-Lachidan, M., Wright, S. I. and Kelly, S. P. (2008). Claudin-3 tight junction proteins in *Tetraodon nigroviridis*: cloning, tissue-specific expression, and a role in hydromineral balance. *Am. J. Physiol. Regul. Integr. Comp. Physiol.* 294, R1638–R1647.

Bagherie-Lachidan, M., Wright, S. I. and Kelly, S. P. (2009). Claudin-8 and -27 tight junction proteins in puffer fish *Tetraodon nigroviridis* acclimated to freshwater and seawater. *Comp. Biochem. Physiol. B Biochem. Mol. Biol.* 179, 419–431.

Ballantyne, J. S. and Fraser, D. I. (2013). Euryhaline elasmobranchs. In *Fish Physiology,* Vol. 32, *Euryhaline Fishes* (eds. S. D. McCormick, A. P. Farrell and C. J. Brauner), pp. 125–198. New York: Elsevier.

Bartels, H. and Potter, I. C. (2004). Cellular composition and ultrastructure of the gill epithelium of larval and adult lampreys. Implications for osmoregulation in fresh and seawater. *J. Exp. Biol.* 207, 3447–3462.

Bartels, H., Potter, I. C., Pirlich, K. and Mallat, J. (1998). Categorization of the mitochondria-rich cells in the gill epithelium of the freshwater phases in the life cycle of lampreys. *Cell Tissue Res.* 291, 337–349.

Baustian, M. D. and Beyenbach, K. W. (1999). Natriuretic peptides and the acclimation of aglomerular toadfish to hypo-osmotic media. *J. Comp. Physiol.* 169, 507–514.

Beyenbach, K. W. (1986). Secretory NaCl and volume flow in renal tubules. *Am. J. Physiol.* 250, R753–R763.

Beyenbach, K. W. (2004). Kidneys sans glomeruli. *Am. J. Physiol.* 286, F811–F827.

Boisen, A. M., Amstrup, J., Novak, I. and Grosell, M. (2003). Sodium and chloride transport in soft water and hard water acclimated zebrafish (*Danio rerio*). *Biochim. Biophys. Acta* 1618, 207–218.

Boylan, J. W. (1972). A model for passive urea absorption in the elasmobranch kidney. *Comp. Biochem. Physiol.* 42, 27–30.

Braun, E. J. and Dantzler, W. H. (1997). *Vertebrate renal system*. In *Handbook of Physiology, Comprehensive Physiology*, Vol. 1. Bethesda, MD: American Physiological Society, pp. 481–576.

Brauner, C. J., Gonzalez, R. J. and Wilson, J. M. (2013). Extreme environments: hypersaline, alkaline, and ion-poor waters. In *Fish Physiology,* Vol. 32, *Euryhaline Fishes* (eds. S. D. McCormick, A. P. Farrell and C. J. Brauner), pp. 435–476. New York: Elsevier.

Breves, J. P., Watanabe, S., Koaneko, T., Tetsuya, H. and Gordon, G. E. (2010). Prolactin restores branchial mitochondrion-rich cells expressing Na^+/Cl^- cotransporter in hypophysectomized Mozambique tilapia. *Am. J. Physiol.* 299, R701–R710.

Brix, K. V. and Grosell, M. (2012). Comparative characterization of Na^+ transport in *Cyprinodon variegatus variegatus* and *Cyprinodon variegatus hubbsi*: a model species complex for studying teleost invasion of freshwater. *J. Exp. Biol.* 215, 1199–1209.

Brown, J. A. and Rankin, J. C. (1999). Lack of glomerular intermittency in the river lamprey *Lampetra fluviatilis* acclimated to sea water and following acute transfer to isosmotic brackish water. *J. Exp. Biol.* 202, 939–946.

Bucking, C. and Wood, C. M. (2006). Gastrointestinal processing of Na^+, Cl^-, and K^+ during digestion: implications for homeostatic balance in freshwater rainbow trout. *Am. J. Physiol.* 291, R1764–R1772.

Buddington, R. K., Kroghdal, A. and Bakke-McKellep, A. M. (1997). The intestine of carnivorous fishes: structure and functions and relations with diet. *Acta Physiol. Scand.* 161, 67–80.

Buhariwalla, H. E. C., Osmond, E. M., Barnes, K. R., Cozzi, R. R. F., Roberston, G. N. and Marshall, W. S. (2012). Control of ion transport by mitochondrion-rich chloride cells of eurythermic teleost fish: cold shock vs. cold acclimation. *Comp. Biochem. Physiol. A Mol. Integr. Physiol.* 162, 234–244.

Bulger, R. E. (1965). Roles of the rectal gland and the kidneys in salt and water excretion in the spiny dogfish. *Physiol. Zool.* 38, 191–196.

Cerda, J. and Finn, R. N. (2010). Piscine aquaporins: an overview of recent aquaporins. *J. Exp. Zool. A Ecol. Genet. Physiol.* 313, 623–650.

Chang, W., Horng, J., Yan, J., Hsiao, C. and Hwang, P. (2009). The transcription factor, glial cell missing 2, is involved in differentiation and functional regulation of H^+-ATPase-rich cells in zebrafish (*Danio rerio*). *Am. J. Physiol.* 296, R1192–R1201.

Chen, X., Polleichtner, G., Kadurin, I. and Gruender, S. (2007). Zebrafish acid-sensing ion channel (ASIC) 4, characterization of homo- and heteromeric channels, and identification of regions important for activation by H. *J. Biol. Chem.* 282, 30406–30413.

Chen, X., Qui, L., Li, M., Duerrnagel, S., Orser, B. A., Xiong, Z. and MacDonald, J. F. (2010). Diarylamidines: high potency inhibitors of acid-sensing ion channels. *Neuropharmacology* 58, 1045–1053.

Choe, K. P. and Evans, D. H. (2003). Compensation for hypercapnia by a euryhaline elasmobranch: effect of salinity and roles of gills and kidneys in fresh water. *J. Exp. Zool. A* 297, 52–63.

Choe, K. P., Edwards, S. L., Morrison-Shetlar, A. I., Toop, T. and Claiborne, J. B. (1999). Immunolocalisation of Na^+/K^+-ATPase in mitochondrion-rich cells of the Atlantic hagfish (*Myxine glutinosa*) gill. *Comp. Biochem. Physiol. A Mol. Integr. Physiol.* 124, 161–168.

Choe, K. P., Morrison-Shetlar, A. I., Wall, B. P. and Claiborne, J. B. (2002). Immunological detection of Na^+/H^+ exchangers in the gills of a hagfish, *Myxine glutinosa*, an elasmobranch, *Raja erinacea* and a teleost, *Fundulus heteroclitus*. *Comp. Biochem. Physiol. A Mol. Integr. Physiol.* 131, 375–385.

Choe, K. P., O'Brien, S., Evans, D. H., Toop, T. and Edwards, S. L. (2004). Immunolocalization of Na^+/K^+-ATPase, carbonic anhydrase II and vacuolar H^+ ATPase in the gills of freshwater adult lampreys *Geotria australis*. *J. Exp. Zool.* 301, 654–665.

Cholette, C. and Gagnon, A. (1973). Isoosmotic adaptation in *Myxine glutinosa* L. II. Variations of the free amino acids, trimethylamine oxide and potassium of the blood and muscle cells. *Comp. Biochem. Physiol. A* 45, 1009–1021.

Currie, S. and Edwards, S. L. (2010). The curious case of the chemical composition of hagfish tissues – 50 years on. *Comp. Biochem. Physiol. A Mol. Integr. Physiol.* 157, 111–115.

Cutler, C. P. and Cramb, G. (2008). Differential expression of absorptive intestinal and renal tissues of the cation–chloride-cotransporters in the European eel *Anguilla anguilla*. *Comp. Biochem. Physiol. B Biochem. Mol. Biol.* 149, 63–73.

Cutler, C. P., Martinez, A. S. and Cramb, G. (2007). The role of aquaporin 3 in teleost fish. *Comp. Biochem. Physiol. A Mol. Integr. Physiol.* 148, 82–91.

Daborn, K., Cozzi, R. R. F. and Marshall, W. S. (2001). Dynamics of pavement cell–chloride cell interactions during abrupt salinity change in *Fundulus heteroclitus*. *J. Exp. Biol.* 204, 1889–1899.

Dantzler, W. H. (2003). Regulation of renal proximal and distal tubule transport: sodium, chloride and organic ions. *Comp. Biochem. Physiol. A Mol. Integr. Physiol.* 136, 453–478.

Doyle, W. L. (1962). Tubular cells of the rectal salt gland of *Urlophus*. *Am. J. Anat.* 111, 223–237.

Edwards, S. L., Tse, C.-M. and Toop, T. (1999). Immunolocalisation of NHE3-like immunoreactivity in the gills of the rainbow trout (*Oncorhynchus mykiss*) and the blue throated wrasse (*Pseudolabrus tetrious*). *J. Anat.* 195, 465–469.

Edwards, S. L., Claiborne, J. B., Morrison-Shetlar, A. I. and Toop, T. (2001). NHE mRNA expression in the gill of the Atlantic hagfish (*Myxine glutinosa*) in response to metabolic acidosis. *Comp. Biochem. Physiol. A Mol. Integr. Physiol.* 130, 81–91.

Edwards, S. L., Wall, B. P., Morrison-Shetlar, A., Sligh, S., Weakley, J. C. and Claiborne, J. B. (2005). The effect of environmental hypercapnia and salinity on the expression of NHE-like isoforms in the gills of a euryhaline fish (*Fundulus heteroclitus*). *J. Exp. Zool.* 303, 464–475.

Evans, D. H. (1979). Fish. In *Comparative Physiology of Osmoregulation in Animals* (ed. G. M. O. Maloiy), pp. 305–390. Orlando, FL: Academic Press.

Evans, D. H. (1984). Gill Na^+/H^+ and Cl^-/HCO_3^- exchange systems evolved before the vertebrates entered fresh water. *J. Exp. Biol.* 113, 464–469.

Evans, D. H. (1993). Osmotic and ionic regulation. In *Physiology of Fishes* (ed. D. H. Evans), pp. 331–341. Boca Raton, FL: CRC Press.

Evans, D. H. (2002). Cell signaling and ion transport across the fish gill epithelium. *J. Exp. Zool.* 293, 336–347.

Evans, D. H. (2011). Freshwater fish gill ion transport: August Krough to morpholinos and microprobes. *Acta Physiol.* 202, 349–359.

Evans, D. H., Piermarini, P. M. and Choe, K. P. (2005). The multifunctional fish gill: dominant site of gas exchange, osmoregulation, acid–base regulation, and excretion of nitrogenous waste. *Physiol. Rev.* 85, 97–177.

Evans, T. G. and Somero, G. N. (2008). A microarray based transcriptomic time-course of hyper- and hyosomotic stress signaling events in the euryhaline fish *Gillichthys mirabilis*: osmosensors to effectors. *J. Exp. Biol.* 211, 3636–3649.

Fenwick, J. C., Wendelaar Bonga, S. E. and Flik, G. (1999). *In vivo* bafilomycin-sensitive Na^+ uptake in young freshwater fish. *J. Exp. Biol.* 202, 3659–3666.

Fines, G. A., Ballantyne, J. S. and Wright, P. A. (2001). Active urea transport and an unusual basolateral membrane composition in the gills of a marine elasmobranch. *Am. J. Physiol.* 280, R16–R24.

Flemmer, A. W., Monette, M. Y., Djurisic, M., Dowd, B., Darman, R., Gimenez, I. and Forbush, B. (2010). Phosphorylation state of the $Na^+-K^+-Cl^-$ cotransporter (NKCC1) in the gills of Atlantic killifish (*Fundulus heteroclitus*) during acclimation to water of varying salinity. *J. Exp. Biol.* 213, 1558–1566.

Fridman, S., Bron, J. E. and Rana, K. J. (2011). Ontogenetic changes in location and morphology of chloride cells during early life stages of the Nile tilapia *Oreochromis niloticus* adapted to fresh and brackish water. *J. Fish. Biol.* 79, 597–614.

Furukawa, F., Watanabe, S., Inokuchi, M. and Toyoji, K. (2011). Responses of gill mitochondria-rich cells in Mozambique tilapia exposed to acidic environments (pH 4.0) in combination with different salinities. *Comp. Biochem. Physiol. A Mol. Integr. Physiol.* 158, 468–476.

Gerst, J. (1977). Effects of saline acclimation on plasma electrolytes, urea excretion, and hepatic urea biosynthesis in a freshwater stingray, *Potamotrygon sp.* Garman, 1877. *Comp. Biochem. Physiol. A* 56, 87–93.

Gilmour, K. M. and Perry, S. F. (2009). Carbonic anhydrase and acid–base regulation in fish. *J. Exp. Biol.* 212, 1647–1661.

Gimenez, I. and Forbush, B. (2005). Regulatory phosphorylation sites in the NH_2 terminus of the renal NaKCl cotransporter (NKCC2). *Am. J. Physiol.* 289, F1341–F1345.

Golstein, L. and Forster, R. P. (1971). Osmoregulation and urea metabolism in the little skate *Raja erinacea*. *Am. J. Physiol.* 220, 743–746.

Golstein, L., Oppelt, W. W. and Maren, T. H. (1968). Osmotic regulation and urea metabolism in the lemon shark *Negaprion brevirostris*. *Am. J. Physiol.* 215, 1493–1497.

Goss, G., Adamia, S. and Galvez, F. (2001). Peanut lectin binds to a subpopulation of mitochondria-rich cells in the rainbow trout gill epithelium. *Am. J. Physiol.* 281, R1718–R1725.

Griffith, R. W., Pang, P. L. T., Srivastava, A. K. and Pickford, G. E. (1973). Serum composition of freshwater stingrays *(Potamotrygonidae)* adapted to freshwater and dilute seawater. *Bio. Bull.* 144, 304–320.

Griffith, R. W., Umminger, B. L., Grant, B. F., Pang, P. L. T. and Pickford, G. E. (1974). Serum composition of the coelacanth *Latimeria chalumnae*. *J. Exp. Zool.* 187, 87–102.

Grosell, M. (2006). Intestinal anion exchange in marine fish osmoregulation. *J. Exp. Biol.* 209, 2813–2827.

Grosell, M. (2011). Intestinal anion exchange in marine fish. *Acta Physiol.* 202, 421–434.

Grosell, M., Gilmour, K. and Perry, S. F. (2007). Intestinal carbonic anhydrase, bicarbonate and proton carriers play a role in the acclimation of rainbow trout to seawater. *Am. J. Physiol.* 293, R2099–R2111.

Grosell, M., Genz, J., Taylor, J. R., Perry, S. F. and Gilmour, K. M. (2009a). The involvement of H^+-ATPase and carbonic anhydrase in intestinal HCO_3^- secretion in seawater acclimated-rainbow trout. *J. Exp. Biol.* 212, 1940–1948.

Grosell, M., Mager, E. M., Williams, C. and Taylor, J. R. (2009b). High rates of HCO_3^- secretion and Cl^- absorption against adverse gradients in the marine teleost intestine: the involvement of an electrogenic anion exchanger and H^+ pump metabolon? *J. Exp. Biol.* 212, 1684–1696.

Guffy, S., Esbaugh, A. and Grosell, M. (2011). Regulation of apical H^+-ATPase activity and intestinal HCO_3^- secretion in marine fish osmoregulation. *Am. J. Physiol.* 301, R1682–R1691.

Hamann, S., Herrera-Perez, J., Bundgaard, M., Alvarez-Leefmans, F. J. and Zeuthen, T. (2005). Water permeability of Na^+–K^+–$2Cl^-$ cotransporters in mammalian epithelial cells. *J. Physiol.* 568, 123–135.

Hammerschlag, N. (2006). Osmoregulation in elasmobranchs: a review for fish biologists, behaviourists and ecologists. *Mar. Freshw. Behav. Physiol.* 39, 209–228.

Hardisty, M. W. (1979). *Biology of the Cyclostomes*. London: Chapman and Hall.

Hazon, N., Wells, A., Pillans, R. D., Good, J. P., Gary Anderson, W. and Franklin, C. E. (2003). Urea based osmoregulation and endocrine control in elasmobranch fish with special reference to euryhalinity. *Comp. Biochem. Physiol. B Biochem. Mol. Biol.* 136, 685–700.

Hentschel, H. and Elger, M. (1989). Morphology of glomerular and aglomerular kidneys. In *Structure and Function of the Kidney* (ed. R. K.H. Kinne), pp. 1–72. Basel: Karger.

Hickman, C. P. and Trump, B. F. (1969). The kidney. In *Fish Physiology*, Vol. 1 (eds. W. S. Hoar and D. J. Randall), pp. 91–227. London: Academic Press.

Hill, W. G., Mathai, J. C., Gensure, R. H., Zeidel, J. D., Apodaca, G., Saenz, J. P., Kinne-Saffran, E., Kinne, R. and Zeidel, M. L. (2004). Permeabilities of teleost and elasmobranch gill apical membranes: evidence that lipid bilayers alone do not account for barrier function. *Am. J. Physiol.* 287, C235–C242.

Hirata, T., Kaneko, T., Ono, T., Nakazato, T., Furukawa, N., Hasegawa, S., Wakabayashi, S., Shigekawa, M., Chang, M.-H., Romero, M. F. and Hirose, S. (2003). Mechanism of acid adaptation of a fish living in a pH 3.5 lake. *Am. J. Physiol.* 284, R1199–R1212.

Hiroi, J. and McCormick, S. D. (2012). New insights into gill ionocyte and ion transporter function in euryhaline and diadromous fish. *Respir. Physiol. Neurobiol.* 184(3), 257–268

Hiroi, J., Kaneko, T., Seikai, T. and Tanaka, M. (1999). *In vivo* sequential changes in chloride cell morphology in the yolk-sac membrane of the Mozambique tilapia (*Oreochromis mossambicus*) embryos and larvae during seawater adaptation. *J. Exp. Biol.* 202, 3485-3495.

Hiroi, J., McCormick, S. D., Ohtani-Kaneko, R. and Kaneko, T. (2005a). Functional classification of mitochondrion rich cells in euryhaline Mozambique tilapia (*Oreochromis mossambicus*) embryos, by means of triple immunofluorescence staining for $Na^+/K^+ATPase$, $Na^+/K^+/2Cl^-$ cotransporter and CFTR anion channel. *J. Exp. Biol.* 208, 2023-2036.

Hiroi, J., Miyazaki, H., Katoh, F., Ohtani-Kaneko, R. and Kaneko., T. (2005b). Chloride turnover and ion transporting activities of yolk-sac preparations (yolk balls) separated from Mozambique tilapia embryos and incubated in freshwater and seawater. *J. Exp. Biol.* 208, 3851-3858.

Hiroi, J., Yasumasu, S., McCormick, S. D., Hwang, P. P. and Kaneko, T. (2008). Evidence for an apical NaCl cotransporter involved in ion uptake in a teleost fish. *J. Exp. Biol.* 211, 2684-2599.

Hirose, S., Kaneko, T., Naito, N. and Takei, Y. (2003). Molecular biology of major components of chloride cells. *Comp. Biochem. Physiol. B Biochem. Mol. Biol.* 136, 593-620.

Hochachka, P. W. and Somero, G. N. (2002). *Biochemical Adaptation: Mechanism and Process in Physiological Evolution.* New York: Oxford University Press.

Hoffmann, E. K., Schettino, T. and Marshall, W. S. (2007). The role of volume-sensitive ion transport systems in regulation of epithelial transport. *Comp. Biochem. Physiol. A Mol. Integr. Physiol.* 148, 29-43.

Hootman, S. R. and Philpott, C. W. (1980). Accessory cells in teleost branchial epithelium. *Am. J. Physiol.* 238, R199-R206.

Howe, D. and Gutknecht, J. (1978). Role of urinary bladder in osmoregulation in marine teleost *Opsanus tau. Am. J. Physiol.* 235, R48-R54.

Hwang, P. P. (2009). Ion uptake and acid secretion in zebrafish (*Danio rerio*). *J. Exp. Biol.* 212, 1745-1752.

Hwang, P. P., Lee, T. H. and Lin, L. Y. (2011). Ion regulation in fish gills: recent progress in the cellular and molecular mechanisms. *Am. J. Physiol.* 301, R28-R47.

Inokuchi, M., Hiroi, J., Watanabe, S., Lee, K. M. and Kaneko, T. (2008). Gene expression and morphological localization of NHE3, NCC and NKCC1a in branchial mitochondria-rich cells of Mozambique tilapia (*Oreochromis mossambicus*) acclimated to a wide range of salinities. *Comp. Biochem. Physiol. A Mol. Integr. Physiol.* 151, 151-158.

Inokuchi, M., Hiroi, J., Watanabe, S., Hwang, P. P. and Kaneko, T. (2009). Morphological and functional classification of ion-absorbing mitochondria-rich cells in the gills of Mozambique tilapia. *J. Exp. Biol.* 212, 1003-1010.

Ivanis, G., Esbaugh, A. J. and Perry, S. F. (2008). Branchial expression and localization of SLC9A2 and SLC9A3 sodium/hydrogen exchangers and their possible role in acid-base regulation in the freshwater rainbow trout (*Oncorhynchus mykiss*). *J. Exp. Biol.* 211, 2467-2477.

Janech, M. G. and Piermarini, P. M. (2002). Renal and solute excretion in the Atlantic stingray in freshwater. *J. Fish Biol.* 61, 1053-1057.

Karasov, W. H. and Hume, I. D. (1997). Vertebrate gastrointestinal system. In *The Handbook of Physiology: Comparative Physiology*, Vol. 1 (ed. W. H. Dantzler), pp. 409-480. Oxford: Oxford University Press.

Karnaky, K. J., Jr. (1992). Teleost osmoregulation: changes in the tight junction in response to the salinity of the environment. In *Tight Junctions* (ed. M. Cereijido), pp. 175-185. Boca Raton, FL: CRC Press.

Karnaky, K. J., Jr. (1998). Osmotic and ionic regulation. In *Physiology of Fishes* (ed. D. H. Evans), pp. 157–176. Boca Raton, FL: CRC Press.

Katoh, F. and Kaneko, T. (2003). Short-term transformation and long-term replacement of branchial chloride cells in killifish transferred from seawater to freshwater, revealed by morphofunctional observations and a newly established time-differential double fluorescent staining' technique. *J. Exp. Biol* 206, 4113–4123.

Kempton, R. T. (1953). Studies of the elasmobranch kidney II: Reabsorption of urea by the smooth dogfish. *Mustelus canis. Biol. Bull.* 104, 45–56.

Keys, A. B. and Wilmer, E. N. (1932). Chloride secreting cells in the gills of fishes with special reference to the common eel. *J. Physiol* 76, 368–377.

Kim, Y. K., Watanabe, S., Kaneko, T., Do, H. M. and Park, S. I. (2010). Expression of aquaporins 3, 8 and 10 in the intestines of freshwater- and seawater-acclimated Japanese eels *Anguilla japonica. Fish. Sci.* 76, 695–702.

Kimura, T., Han, W. S., Pagel, P., Nairn, A. C. and Caplan, M. J. (2011). Protein phosphatase 2A interacts with the Na^+,K^+-ATPase and modulates its trafficking by inhibition of its association with arrestin. *PLoS ONE* 6.

Kirschner, L. B. (1991). Water and ions. In *Comparative Animal Physiology, Environmental and Metabolic Animal Physiology* (ed. C. L. Prosser), pp. 13–107. New York: Wiley-Liss.

Kirschner, L. B. (1997). *Extrarenal Mechanisms in Hydromineral and Acid–Base Regulation in Aquatic Vertebrates.* New York: John Wiley and Sons.

Kirschner, L. B. (2004). The mechanism of sodium chloride absorption in hyperregulating aquatic animals. *J. Exp. Biol.* 207, 1439–1452.

Kristiansen, H. R. and Rankin, J. C. (2001). Discrimination between endogenous and exogenous water sources in juvenile rainbow trout fed extruded dry feed. *Aquat. Living Resour.* 14, 359–366.

Kültz, D. (2013). Osmosensing. In *Fish Physiology, Vol. 32, Euryhaline Fishes* (eds. S. D. McCormick, A. P. Farrell and C. J. Brauner), pp. 45–68. New York: Elsevier.

Kumai, Y. and Perry, S. F. (2011). Ammonia excretion via Rhcg1 facilitates Na^+ uptake in larval zebrafish, *Danio rerio*, in acidic water. *Am. J. Physiol.* 301, R1517–R1528.

Kumai, Y., Ward, M. and Perry, S. F. (2012). β-Adrenergic regulation of Na^+ uptake by larval zebrafish, *Danio rerio*, in acidic and ion poor environments. *Am. J. Physiol. Regul. Integr. Comp. Physiol.* 303, R1031–R1041.

Laurent, P. (1984). Gill internal morphology. In *Fish Physiology, XA*, pp. 73–183.

LeBlanc, D. M., Wood, C. M., Fudge, D. S. and Wright, P. A. (2010). A fish out of water: gill and skin remodeling promotes osmo- and ionoregulation in the Mangrove killifish *Kryptolebias marmoratus. Physiol. Biochem. Zool.* 83, 932–949.

Lee, Y., Yan, J. J. and Cruz, S. A. (2011). Anion exchanger 1b, but not sodium-bicarbonate cotransporter 1b, plays a role in transport functions of zebrafish H^+-ATPase-rich cells. *Am. J. Physiol.* 300, C295–C307.

Lin, H., Pfeiffer, D. C., Vogl, A. W., Pan, J. and Randall, D. J. (1994). Immunolocalisation of H^+-ATPase in the gill epithelia of rainbow trout. *J. Exp. Biol.* 195, 169–183.

Lin, T. Y., Liao, B. K., Horng, J. L., Yan, J. J., Hsiao, C. D. and Hwang, P. P. (2008). Carbonic anhydrase 2-like a and 15a are involved in acid–base regulation and Na^+ uptake in zebrafish H^+-ATPase-rich cells. *Am. J. Physiol.* 294, C1250–C1260.

Logan, A. G., Morris, R. and Rankin, J. C. (1980). A micropuncture study of kidney function in the river lamprey, *Lampetra fluviatilis*, adapted to sea water. *J. Exp. Biol.* 88, 239–247.

Loretz, C. A. and Bern, H. A. (1980). Ion transport by the urinary bladder of the gobiid teleost *Gillichthys mirabilis. Am. J. Physiol.* 239, R415–R423.

Macey, R. I. (1986). Mathematical models of membrane transport processes. In *Physiology of Membrane Disorders* (eds. T. E. Andreoli, J. F. Hoffmann, D. D. Fanestil and S. G. Schultz), pp. 111–131. New York: Plenum Press.

Mackay, W. C. and Janicki, R. (1978). Changes in the eel intestine during seawater adaptation. *Comp. Biochem. Physiol. A* 62, 757–761.

Mancera, J. M. and McCormick, S. D. (2000). Rapid activation of gill Na^+,K^+-ATPase in the euryhaline teleost *Fundulus heteroclitus*. *J. Exp. Zool.* 287, 263–274.

Marshall, E. K. and Smith, H. W. (1930). Glomerular development of the vertebrate kidney in relation to habitat. *Biol. Bull.* 59, 135–153.

Marshall, W. S. (1977). Transepithelial potential and shortcircuit current across the isolated skin of *Gillichthys mirabilis* (Teleostei: Gobiidae) acclimated to 5% and 100% seawater. *J. Comp. Physiol* 114, 157–165.

Marshall, W. S. (1981). Sodium dependency of active chloride transport across isolated fish skin (*Gillichthys mirabilis*). *J. Physiol. Lond.* 319, 165–178.

Marshall, W. S. (1995). Transport processes in teleost epithelia. In *Cellular and Molecular Approaches to Fish Ionic Regulation*, Vol. 14 (eds. C. M. Wood and T. J. Shuttleworth), pp. 1–23. San Diego: Academic Press.

Marshall, W. S. (2002). Na^+, Cl^-, Ca^{2+} and Zn^{2+} transport by fish gills: retrospective review and prospective synthesis. *J. Exp. Zool.* 293, 264–283.

Marshall, W. S. (2013). Osmoregulation in estuarine and intertidal fishes. In *Fish Physiology*, Vol. 32, *Euryhaline Fishes* (eds. S. D. McCormick, A. P. Farrell and C. J. Brauner), pp. 395–434. New York: Elsevier.

Marshall, W. S. and Grosell, M. (2006). Ion transport, osmoregulation and acid–base balance. In *The Physiology of Fishes* (eds. D. H. Evans and J. B. Claiborne), pp. 177–230. Boca Raton, FL: CRC Press.

Marshall, W. S. and Nishioka, R. S. (1980). Relation of mitochondria rich chloride cells to anion transport by marine teleost skin. *J. Exp. Zool.* 214, 147–156.

Marshall, W. S., Bryson, S. E., Darling, P., Whitten, C., Patrick, M., Wilkie, M., Wood, C. M. and Buckland-Nicks, J. (1997). NaCl transport and ultrastructure of opercular epithelium from a freshwater-adapted euryhaline teleost, *Fundulus heteroclitus*. *J. Exp. Zool.* 277, 23–37.

Marshall, W. S., Howard, J. A., Cozzi, R. R. F. and Lynch, E. M. (2002a). NaCl and fluid secretion by the intestine of the teleost *Fundulus heteroclitus*: involvement of CFTR. *J. Exp. Biol.* 205, 745–758.

Marshall, W. S., Lynch, E. M. and Cozzi, R. R. F. (2002b). Redistribution of immunofluorescence of CFTR anion channel and NKCC cotransporter in chloride cells during adaptation of the killifish *Fundulus heteroclitus* to seawater. *J. Exp. Biol.* 205, 1265–1273.

Martin, R. A. (2005). Conservation of freshwater and euryhaline elasmobranchs: a review. *J. Mar. Biol. Assoc. U.K* 85, 1049–1073.

McDonald, D. G., Cavdek, V., Calvert, L. and Milligan, C. L. (1991). Acid–base regulation in the Atlantic hagfish *Myxine glutinosa*. *J. Exp. Biol.* 161, 201–215.

Morris, R. (1958). The mechanism of marine osmoregulation in the lampern, *Lampetra fluviatilis* L. and the cause of its breakdown during the spawning migration. *J. Exp. Biol.* 35, 649–665.

Morris, R. (1972). Osmoregulation. In *The Biology of Lampreys*, Vol. 2 (eds. M. W. Hardisty and I. C. Potter), pp. 193–239. London: Academic Press.

Nawata, C. M., Hung, C. C. Y., Tsui, T. K. N., Wilson, J. M., Wright, P. A. and Wood, C. M. (2007). Ammonia excretion in rainbow trout (*Oncorhynchus mykiss*): evidence for Rh glycoprotein and H^+ATPase involvement. *Physiol. Genomics* 31, 463–474.

Nawata, C. M., Wood, C. M. and O'Donnell, M. J. (2010). Functional characterization of Rhesus glycoproteins from an ammoniotelic teleost, the rainbow trout, using oocyte expression and SIET analysis. *J. Exp. Biol.* 213, 1049–1059.

Nelson, J. S. (2006). *Fishes of the World.* Hoboken, NJ: John Wiley & Sons.

Overbeek, J. T. G. (1956). The Donnan equilibrium. *Progr. Biophys.* 6, 57–84.

Parks, S. K., Tresguerres, M. and Goss, G. G. (2007). Blood and gill responses to HCl infusions in the Pacific hagfish (*Eptatretus stoutii*). *Can. J. Zool.* 85, 855–862.

Parks, S. K., Tresguerres, M. and Goss, G. (2008). Theoretical considerations underlying Na^+ uptake mechanisms in freshwater fishes. *Comp. Biochem. Physiol. C Toxicol. Pharmacol.* 148, 411–418.

Parmelee, J. T. and Renfro, J. L. (1983). Esophageal desalination of seawater in flounder: role of active sodium transport. *Am. J. Physiol.* 245, R888–R893.

Pärt, P., Wright, P. A. and Wood, C. M. (1998). Urea and water permeability in dogfish (*Squalus acanthias*). *Comp. Biochem. Physiol. A Mol. Integr. Physiol.* 119, 117–123.

Payan, P. and Maetz, J. (1973). Branchial sodium transport mechanisms in *Scyliorhinus canicula*: evidence for a Na^+/NH_4^+ and Na^+/H^+ exchanges and a role for carbonic anhydrase. *J. Exp. Biol.* 58, 487–502.

Pelis, R. M. and Renfro, J. L. (2004). Role of tubular secretion and carbonic anhydrase in vertebrate renal sulfate secretion. *Am. J. Physiol.* 287, R479–R501.

Perry, S. F., Vulesevic, B., Grosell, M. and Bayaa, M. (2009). Evidence that SLC26 anion transporters mediate branchial chloride uptake in adult zebrafish (*Danio rerio*). *Am. J. Physiol.* 297, R988–R997.

Pickering, A. D. and Morris, R. (1970). Osmoregulation of *Lampetra fluviatilis* and *Petromyzon marinus* in hypertonic solutions. *J. Exp. Biol.* 53, 231–243.

Pickering, A. D. and Morris, R. (1973). Localization of ion-transport in the intestine of the migrating river lamprey, *Lampetra fluviatilis* L. *J. Exp. Biol.* 58, 165–176.

Piermarini, P. M. and Evans, D. (1998). Osmoregulation of the Atlantic stingray (*Dasyatis sabina*) from the freshwater Lake Jesup of the St. Johns River, Florida. *Physiol. Zool.* 71, 553–560.

Piermarini, P. M. and Evans, D. H. (2001). Immunochemical analysis of the vacuolar proton-ATPase B-subunit in the gills of a euryhaline stingray (*Dasyatis sabina*): effects of salinity and relation to Na^+/K^+-ATPase. *J. Exp. Biol.* 204, 3251–3259.

Pillans, R. D. and Franklin, C. E. (2004). Plasma osmolyte concentrations and rectal gland mass of a bullsharks *Carchahinus leucas* captured along a salinity gradient. *Comp. Biochem. Physiol. A Mol. Integr. Physiol.* 138, 363–371.

Pillans, R. D., Good, J. P., Anderson, W. G., Hazon, N. and Franklin, C. E. (2005). Freshwater to seawater acclimation of juvenile bull sharks (*Carcharhinus leucas*): plasma osmolytes and Na^+/K^+-ATPase activity in gill, rectal gland, kidney and intestine. *J. Comp. Physiol. B* 175, 37–44.

Pillans, R. D., Good, J. P., Anderson, W. G., Hazon, N. and Franklin, C. E. (2008). Rectal gland morphology of freshwater and seawater acclimated bull sharks *Carcharhinus leucas*. *J. Fish Biol.* 72, 1559–1571.

Potts, W. T. W. (1984). Transepithelial potentials in fish gills. In *Fish Physiology,* Vol. XB, *Gills: Ion and Water Transfer* (eds. W. S. Hoar and D. J. Randall), pp. 105–176. Orlando, FL: Academic Press.

Potts, W. T. W. and Eddy, F. B. (1973). Gill potentials and sodium fluxes in the flounder *Platichthys flesus*. *J. Comp. Physiol.* 87, 29–48.

Preest, A. M., Gonzalez, R. J. and Wilson, R. W. (2004). A pharmacological examination of Na^+ and Cl^- transport in two species of freshwater fish. *Physiol. Biochem. Zool.* 72, 259–272.

Rankin, C. (1997). Osmotic and ionic regulation in cyclostomes. In *Ionic Regulation in Animals: A Tribute to Professor W.T.W. Potts* (eds. N. Hazon, F. B. Eddy and G. Flik), pp. 50–69. Berlin: Springer.

Reid, S. D., Hawkings, G. S., Galvez, F. and Goss, G. (2003). Localization and characterization of phenamil-sensitive Na^+ influx in isolated rainbow trout epithelial cells. *J. Exp. Biol.* 206, 551–559.

Reilly, B. D., Cramp, R. L., Wilson, J. M., Campbell, H. A. and Franklin, C. E. (2011). Branchial osmoregulation in the euryhaline bull shark, *Carcharhinus leucas*: a molecular analysis of ion transporters. *J. Exp. Biol.* 214, 2883–2895.

Reis-Santos, P., McCormick, S. D. and Wilson, J. M. (2008). Ionoregulatory changes during metamorphosis and salinity exposure of juvenile sea lamprey (*Petromyzon marinus* L.). *J. Exp. Biol.* 211, 978–988.

Rimmer, D. W. and Wiebe, W. J. (1987). Fermentative microbial digestion in herbivorous fishes. *J. Fish Biol.* 31, 229–236.

Robertson, J. D. (1960). Studies on the chemical composition of muscle tissue I. The muscles of the hagfish *Myxine glutinosa* L and the Roman eel *Muraena helena* L. *J. Exp. Biol* 37, 879–888.

Sakamoto, T., Yokota, S. and Ando, M. (2000). Rapid morphological oscillation of mitochondrion rich cell in estuarine mudskipper following salinity changes. *J. Exp. Zool.* 286, 666–669.

Sardella, B. A., Baker, D. W. and Brauner, C. J. (2009). The effects of variable water salinity and ionic composition on the plasma status of the Pacific hagfish (*Eptatretus stouti*). *J. Comp. Physiol. B* 179, 721–728.

Sardet, C., Pisan, M. and Maetz, J. (1979). The surface epithelial of teleostean fish gills. Cellular and tight junctional adaptations of the chloride cell in relation to salt adaptations. *J. Cell. Biol.* 80, 96–117.

Schmidt-Nielsen, B. and Rabinowitz, L. (1964). Methylurea and acetamide: active reabsorption by elasmobranch renal tubules. *Science* 146, 1587–1588.

Schmidt-Nielsen, S. and Schmidt-Nielsen, S. (1923). *Beiträge zur Kenntnis des osmotischen Druckes der Fische. Det Kgl. Norske Vidensk. Selsk. Skrifter.* Aktietr.

Schultz, S. G. (1980). *Basic Principles of Membrane Transport.* Cambridge: Cambridge University Press.

Shuttleworth, T. J. (1988). Salt and water balance – extrarenal mechanisms. In *Physiology of Elasmobranch Fishes* (ed. T. J. Shuttleworth), pp. 171–199. Berlin: Springer.

Singh, L. R., Poddar, N. K., Dar, T. A., Kumar, R. and Ahmad, F. (2011). Protein and DNA destabilization by osmolytes: the other side of the coin. *Life Sci.* 88, 117–125.

Skadhauge, E. (1974). Coupling of transmural flows of NaCl and water in the intestine of the eel (*Anguilla anguilla*). *J. Exp. Biol.* 60, 535–546.

Skou, J. C. and Esmann, M. (1992). The Na,K-ATPase. *J. Bioenerg. Biomembr.* 24, 249–261.

Smith, H. (1931). The absorption and excretion of water and salts by elasmobranch fishes: II. Marine elasmobranchs. *Am. J. Phsyiol.* 98, 296–310.

Smith, H. W. (1932). Water regulation and its evolution in the fishes. *Q. Rev. Biol.* VII, 1–26.

Smith, H. W. (1936). The retention and physiological role of urea in elasmobranchii. *Biol. Rev.* 11, 48–82.

Smith, N. F., Talbot, C. and Eddy, F. B. (1989). Dietary salt intake and its relevance to ionic regulation in freshwater salmonids. *J. Fish Biol.* 35, 749–753.

Springauf, A. and Greunder, S. (2010). An acid-sensing ion channel from shark (*Squalus acanthias*) mediates transient and sustained responses to protons. *J. Physiol. Lond.* 588, 809–820.

Stinson, C. M. and Mallatt, J. (1989). Branchial ion fluxes and toxicant extraction efficiency in lamprey (*Petromyzon marinus*) exposed to methylmercury. *Aquat. Toxicol.* 15, 237–252.

Sucré, E., Charmantier-Daures, M., Grousset, E. and Cucchi-Mouillot, P. (2011). Embryonic ionocytes in the European sea bass (*Dicentrarchus labrax*): structure and functionality. *Dev. Growth Differ.* 53, 26–36.

Sullivan, G. V., Fryer, J. N. and Perry, S. F. (1996). Localization of mRNA for the proton pump (H$^+$-ATPase) and Cl$^-$/HCO$_3^-$ exchanger in the rainbow trout gill. *Can. J. Zool.* 74, 2095–2103.

Tang, C. H., Hwang, L. Y. and Lee, T. H. (2010). Chloride channel ClC-3 in gills of the euryhaline teleost, *Tetraodon nigroviridis*: expression, localization and the possible role of chloride absorption. *J. Exp. Biol.* 213, 683–693.

Tang, C. H., Hwang, L. Y., Shen, I. D., Chiu, Y. H. and Lee, T. H. (2011). Immunolocalization of chloride transporters to gill epithelia of euryhaline teleosts with opposite salinity-induced Na$^+$/K$^+$-ATPase responses. *Fish Physiol. Biochem.* 37, 709–724.

Thorson, T. B., Cowan, C. M. and Watson, D. E. (1973). Body fluid solutes of juveniles and adults of the euryhaline bull shark *Carcharhinus leucas* from freshwater and saline environments. *Physiol. Zool.* 46, 29–42.

Thorson, T. B., Wotton, R. M. and Georgi, T. A. (1978). Rectal gland of freshwater stingrays, *Potamotrygon* spp. (Chondrichthyes: Potamotrygonidae). *Biol. Bull.* 154, 508–516.

Thorson, T. B., Langhammer, J. K. and Oetinger, M. I. (1983). Reproduction and development of the South American freshwater stingrays, *Potamotrygon circularis* and *P. Motoro*. *Environ. Biol. Fish* 9, 3–24.

Tomasso, J. R., Jr. and Grosell, M. (2005). Physiological basis for large differences in resistance to nitrite among freshwater and freshwater acclimated euryhaline fishes. *Environ. Sci. Technol* 39, 98–102.

Tresguerres, M., Parks, S. K. and Goss, G. G. (2006). V-H$^+$ATPase, Na$^+$/K$^+$-ATPase and NHE2 immunoreactivity in the gill epithelium of the Pacific hagfish (*Eptatretus stoutii*). *Comp. Biochem. Physiol. A Mol. Integr. Physiol.* 145, 312–321.

Uchiyama, M., Komiyama, M., Yoshizawa, H., Shimizu, N., Konno, N. and Matsuda, K. (2012). Structures and immunolocalization of Na$^+$, K$^+$-ATPase, Na$^+$/H$^+$ exchanger 3 and vacuolar-type H$^+$-ATPase in the gills of blennies (Teleostei: Blenniidae) inhabiting rocky intertidal areas. *J. Fish Biol.* 80, 2236–2252.

Usher, M. L., Talbot, C. and Eddy, F. B. (1991). Intestinal water transport in juvenile Atlantic salmon (*Salmo salar* L.) during smolting and following transfer to seawater. *Comp. Biochem. Physiol. A* 100, 813–818.

Wang, Y. F., Tseng, Y. C., Yan, J. J., Hiroi, J. and Hwang, P. P. (2009). Role of SLC12A10.2, a NaCl cotransporter-like protein, in a Cl uptake mechanism in zebrafish (*Danio rerio*). *Am. J. Physiol.* 296, R1650–R1660.

Whittamore, J. M. (2012). Osmoregulation and epithelial water transport: lessons from the intestine of marine teleost fish. *J. Comp. Physiol. B* 182, 1–39.

Wilson, J. M., Laurent, P., Tufts, B., Benos, D. J., Donowitz, M, Vogl, A. W. and Randall, D. J. (2000). NaCl uptake by the freshwater teleost fish branchial epithelium. An immunological approach to ion transport protein localization. *J. Exp. Biol.* 203, 2279–2296.

Wilson, J. M., Morgan, J. D., Vogl, A. W. and Randall, D. J. (2002). Branchial mitochondria-rich cells in the dogfish *Squalus acanthias*. *Comp. Biochem. Physiol. A Mol. Integr. Physiol.* 132, 365–374.

Wilson, R. W., Gilmour, K., Henry, R. and Wood, C. (1996). Intestinal base excretion in the seawater adapted rainbow trout: a role in acid/base balance? *J. Exp. Biol.* 199, 2331–2343.

Wood, C. M., Gilmour, K. M. and Pärt, P. (1998). Passive and active transport properties of a gill model, the cultured branchial epithelium of the freshwater rainbow trout (*Oncorhynchus mykiss*). *Comp. Biochem. Physiol. A Mol. Integr. Physiol.* 119, 87–96.

Wood, C. M., Matsuo, A. Y. O., Gonzalez, R. Y., Wilson, R. W., Patrick, M. L. and Val, A. L. (2002). Mechanisms of ion transport in *Potamotrygon*, a stenohaline freshwater elasmobranch native to the ion-poor black waters of the Rio Negro. *J. Exp. Biol.* 205, 3039–3054.

Wright, P. A. and Wood, C. M. (2009). A new paradigm for ammonia excretion in aquatic animals: role of Rhesus (Rh) glycoproteins. *J. Exp. Biol.* 212, 2303–2312.

Yakota, S., Iwata, K., Fujii, Y. and Ando, M. (2004). Ion transport across the skin of the mudskipper *Periophthalmus modestus*. *Comp. Biochem. Physiol. A Mol. Integr. Physiol.* 118, 903–910.

Yancey, P. H. (1985). Organic osmotic effectors in cartilaginous fishes. In *Transport Processes, Iono- and Osmoregulation* (eds. R. Gilles and M. Gilles-Ballien), pp. 424–436. Berlin: Springer.

Yancey, P. H. (2005). Organic osmolytes as compatible, metabolic and counteracting cryprotectants in high osmolality and other stresses. *J. Exp. Biol.* 208, 2819–2830.

Yancey, P. H. and Somero, G. N. (1979). Counteraction of urea destabilization of protein structure by methylamine osmoregulatory compounds of elasmobranch fishes. *Biochem. J.* 182, 317–323.

Yang, W. K., Kang, C. K., Chen, T. Y., Chang, W. B. and Lee, T. H. (2011). Salinity-dependent expression of the branchial $Na^+/K^+/2Cl^-$ cotransporter and Na^+/K^+-ATPase in the sailfin molly correlates with hypoosmoregulatory endurance. *J. Comp. Physiol. B Biochem. Mol. Biol.* 181, 953–964.

Zeuthen, W. (2010). Water transporting proteins. *J. Membr. Biol.* 234, 57–73.

2

OSMOSENSING

DIETMAR KÜLTZ

Euryhaline fishes can perceive and compensate for changes in environmental salinity. Osmosensing is the physiological process of perceiving a change in environmental salinity. Osmosensors are molecules that mediate such perception by having two critical properties: their conformation depends directly (ionic, osmotic, and water activity effects) or indirectly (macromolecular crowding, cytoskeletal strain, macromolecular damage) on salinity; upon salinity-induced conformational change they alter signal transduction pathways that control osmoregulatory effector mechanisms. At the organismal level, specialized osmoreceptor cells monitor body fluid osmolality, and perhaps external salinity, to provide feedback to maintain osmotic homeostasis. At the cellular level, molecular osmosensors control intracellular signaling pathways needed for cell-type specific (compensatory) responses during osmotic stress. The interaction of multiple osmosensory elements at both levels converges into a specific set of signals that encodes information on environmental salinity, ionic composition, and salinity

Euryhaline Fishes: Volume 32
FISH PHYSIOLOGY
Copyright © 2013 Elsevier Inc. All rights reserved
DOI: http://dx.doi.org/10.1016/B978-0-12-396951-4.00002-5

change. This chapter reviews the main elements and principles of osmosensing in euryhaline fishes.

1. INTRODUCTION

Euryhaline fishes are able to perceive and compensate for large changes in environmental salinity. This physiological trait depends on a multitude of regulatory mechanisms that can be mapped on to three levels: (1) sensory mechanisms that allow perception of salinity changes; (2) signal transduction mechanisms that carry information from sensors to effectors; and (3) effector mechanisms that mediate adjustment of osmotic homeostasis to meet the new requirements brought about by environmental salinity change.

Osmosensing is the physiological process of perceiving a change in environmental salinity. Osmosensors are molecules that mediate such perception by having two critical properties. First, their conformation is directly or indirectly dependent on salinity. Direct dependence follows from ionic effects, osmotic effects, and water activity effects. Indirect dependence follows from macromolecular crowding, cytoskeletal strain, and macromolecular damage. Second, salinity-induced conformational change alters signal transduction pathways that control osmoregulatory effector mechanisms. Osmoregulatory effector mechanisms in euryhaline fishes include ion transport across gill, gastrointestinal tract, renal, and other epithelia, water transport in the gastrointestinal tract, regulation of organic osmolyte levels, regulation of epithelial tight junctions, and regulation of urine composition and volume. Much is known about these effector mechanisms, which are discussed in detail in Chapter 1 of this volume (Edwards and Marshall, 2013). The present chapter is concerned with the events preceding the regulation of effector mechanisms during salinity stress. These events concern osmosensing and osmotic stress signaling. Current knowledge of osmosensing and osmotic stress signaling is quite limited, compared to our vast knowledge of osmoregulatory effector mechanisms.

Recent work on osmosensory signal transduction in fish and other animals paints a picture of relatively low specificity but high interactivity of animal osmosensors. Many molecular osmosensors of animals, some of which are discussed below, seem to be logically integrated into a combinatorial network that allows highly specific signaling outcomes despite the lack of exclusively specific osmosensors (Kültz, 2011). A combinatorial architecture of osmosensory networks not merely allows for on/off responses to occur but provides an avenue for fine-grained quantitation of the degree and direction of salinity change. Such quantitative assessment of environmental salinity is critical for euryhaline fishes to be able to match the degree of compensatory responses mediated by osmoregulatory effector mechanisms with the degree of environmental salinity change. For instance, euryhaline fishes encountering a change in environmental salinity from hyperosmotic to hypoosmotic relative to blood need to switch direction from active NaCl secretion to active NaCl absorption across the gill epithelium. But in addition to such qualitative change they need to finely tune the amount of active ion absorption (or secretion) to the exact extent of the osmotic gap between their blood and the environment (Fiol and Kültz, 2007).

During evolution towards highly complex vertebrates such as fishes the combinatorial architecture of sensory and signaling networks, including those for osmosensing, has become increasingly complex and sophisticated. The driving forces for increasing complexity were manifold and include limited amount of genomic space leading to increased multifunctionality of gene products to support further cellular and tissue specialization and differentiation (Koonin, 2011), limited cell volume and capacity for accommodating a finite number of macromolecules (Garner and Burg, 1994), and selection for increased redundancy and robustness of essential signaling networks (Gerhart and Kirschner, 1997). These evolutionary driving forces have yielded a mosaic of processes and molecules that combine to provide the capacity for osmosensing in euryhaline fishes. The following is a summary of the current knowledge about this mosaic and its recognized main elements.

2. WHOLE-ORGANISM (SYSTEMIC) OSMOSENSING

All teleost fishes are osmoregulators, meaning that they maintain a constant osmolality of their body fluids (in most cases approximately $300 \text{ mOsmol kg}^{-1}$, 10 ppt). This is true for marine teleosts that live in a hyperosmotic environment relative to blood of approximately $1050 \text{ mOsmol kg}^{-1}$ (35 ppt) as well as freshwater teleosts, for which the environment is hypoosmotic relative to blood ($<50 \text{ mOsmol kg}^{-1}$, <1 ppt). During exposure of euryhaline fishes to severe osmotic stress, e.g. transfer from freshwater (FW) to seawater (SW) or vice versa, the internal osmolality setpoint of about $300 \text{ mOsmol kg}^{-1}$ can deviate temporarily by as much as $\pm 100 \text{ mOsmol kg}^{-1}$ (Kültz and Somero, 1995; DiMaggio et al., 2010; Kammerer et al., 2010; Velan et al., 2011). Smaller environmental salinity changes lead to proportionally smaller changes in the osmolality of blood plasma and other body fluids and to shorter periods of deviation of body fluid osmolality from the optimal setpoint. Thus, in addition to sensing environmental salinity changes directly through external tissues such as gills, euryhaline fishes can indirectly recognize environmental salinity change through its impact on internal body fluid osmolality (Fiol and Kültz, 2007).

Because the constancy of the internal milieu is pivotal and tightly regulated in all vertebrates, special osmoreceptor cells have evolved to monitor body fluid and electrolyte homeostasis with exquisite sensitivity and provide fine-grained regulatory feedback during even the slightest deviation from the optimal setpoint (Noble, 2008). These central osmoreceptor cells add an additional layer of complexity regarding osmosensing networks in fishes and other vertebrates, which is lacking in unicellular organisms and osmoconforming animals that do not regulate their internal (body fluid) osmolality.

2.1. Brain Osmoreceptor Cells

Like other vertebrates, fishes have specialized cells in the brain and in the peripheral system that are capable of sensing plasma osmolality and volume, both of which are a function of body fluid homeostasis and can be altered by changes in environmental salinity. Because environmental salinity change alters the concentration of permeable solutes (mostly NaCl) in the body fluids that surround all tissues of euryhaline fishes, all cells have the potential to perceive salinity stress via alterations in cell volume, macromolecular crowding, intracellular ionic (mostly K^+) strength and/or membrane/cytoskeletal strain (Strange, 1994; Kültz and Burg, 1998b; Kültz, 2007). However, certain cell types are better at transducing such signals into specific physiological outcomes than are others. In the brain, such cells are neurons referred to as osmoreceptor cells (or osmoreceptors) (Bourque et al., 1994). In mammals, they are localized to distinct circumventricular brain areas including the organum vasculosum laminae terminalis (OVLT), subfornical organ, area postrema, and the posterior pituitary (Bourque, 2008). In euryhaline fishes, the OVLT is present and well developed but has not been studied with regard to its role in osmosensing (Gomez-Segade et al., 1991). The area postrema and subfornical organ are known to participate in the control of osmoregulatory actions of atrial natriuretic peptide and renin–angiotensin hormone systems in euryhaline eels (Takei, 2000; Tsukada et al., 2007; Nobata et al., 2010). Thus, it appears that osmoreceptor cells in the brain are conserved in all vertebrate classes, consistent with the conserved physiological capacity for maintaining a constant osmolality of the internal milieu. However, there is much yet to be learned about central osmoreceptors in fish and it still needs to be established whether these osmoreceptors are indeed universally conserved in diverse species of euryhaline fish.

The posterior pituitary (neurohypophysis) thought to harbor mammalian osmoreceptor cells is derived from neural ectoderm and represents an extension of the brain containing a collection of hypothalamic neuronal projections. In fishes, however, the most prominent pituitary osmoreceptor cells have been identified in the adenohypophysis, specifically in the proadenohypophysis (rostral pars distalis), which is derived from oral ectoderm (Olivereau and Ball, 1964; Emmart et al., 1966; Bern, 1967; Grau et al., 1981). Compared to our limited knowledge of fish brain osmosensors, the role of the pituitary rostral pars distalis for osmosensing in euryhaline fishes is well understood. This discrete tissue consists to a large degree of prolactin-producing cells, which can be isolated and cultured *in vitro* (Emmart and Mossakow, 1967). Experiments on such isolated cells provided proof that they are functioning as osmoreceptors. Prolactin cells are capable of sensing subtle extracellular osmolality changes in the range of plasma osmolality variation experienced in euryhaline fishes during salinity stress (Grau et al., 1981). These cells swell in response to hypoosmotic stress that decreases extracellular fluid osmolality. Such cell swelling, in turn, triggers release of the hormone prolactin, which is an important hormone for coordinating acclimation responses to hypoosmotic environments relative to blood (Weber et al., 2004). The swelling-induced prolactin release is mediated by stretch-activated ion channels, transient receptor potential vanilloid 4 (TRPV4) channels, aquaporin 3, and the second messengers calcium and cyclic adenosine monophosphate (cAMP) (Seale et al., 2003a,b, 2011, 2012; Watanabe et al., 2009) (Fig. 2.1). Of note, the osmotic responsiveness of prolactin cells depends on the acclimation status of fish (i.e. whether they are acclimated to SW or FW) (Seale et al., 2012) and in some euryhaline fishes (e.g. underyearling and second year Coho salmon) they may not be osmosensitive within a physiological range of plasma osmolality but, rather, primarily respond to circulating levels of cortisol (Kelley et al., 1990).

2.2. Systemic Osmoreceptor Cells

Mammalian plasma volume/pressure receptor cells are located in the carotid and aortic arches, which are evolutionarily derived from fish gill arches (Milsom and Burleson, 2007). In mammals, these baroreceptors are thought to contribute (at least in the short term) to body fluid osmotic and volume homeostasis by signaling to osmoregulatory effector tissues that control blood pressure and osmolality via adjustment of water intake and urinary excretion (de Castro, 2009; Kougias et al., 2010; Grassi et al., 2012). These systemic baroreceptors appear to employ similar cell volume-

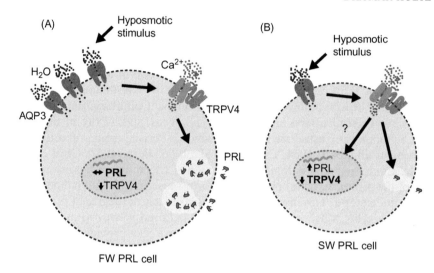

Fig. 2.1. Responses of prolactin cells to hyposmotic stress (decrease in plasma osmolality) differ in tilapia acclimated to freshwater (FW) (A) and seawater (SW) (B). This difference is due to differences in expression and availability of the major elements involved in the prolactin secretion pathway. The main elements shown are prolactin (PRL), transient receptor potential vanilloid 4 (TRPV4), and aquaporin 3 (AQP3). Reproduced with permission from Seale et al. (2012). **SEE COLOR PLATE SECTION.**

dependent mechanosensory cation channels for monitoring blood pressure, as do brain osmoreceptor cells for sensing osmolality (Chapleau et al., 2007). Baroreceptor neurons of euryhaline fishes have been described in the vasculature of branchial arches in several teleost and elasmobranch species (Burleson, 2009).

In addition to these endothelial baroreceptor neurons of gill arches, euryhaline fishes may have hepatic osmoreceptors. Although not much is known about the role of the liver for fish osmosensing, experiments with hepatocytes from several species of euryhaline fish have clearly demonstrated that they potently act as cellular osmometers and it is possible that neurons innervating the liver can do so as well (Ballatori and Boyer, 1997; Espelt et al., 2000; Ollivier et al., 2006b). The liver, in particular the portal vein system, represents a strategic location for rapidly sensing the osmotic strength of ingested matter and is capable of inducing very rapid osmoregulatory responses in mammals (reviewed in Bourque, 2008).

Because drinking rates of euryhaline fishes and the corresponding osmolality of the ingested water are directly proportional to environmental salinity (Maetz and Skadhaug, 1968), it seems plausible that hepatic/portal vein osmoreceptors are also present in euryhaline fishes. Although direct evidence for the existence of hepatic osmoreceptors is lacking, it has been demonstrated that the enzyme iodothyronine deiodinase type 2 (D2) is rapidly (within 2 h) transcriptionally induced during hypoosmotic stress in *Fundulus heteroclitus* liver via a pathway that terminates at a specific tandem osmotic response element (Lopez-Bojorquez et al., 2007). This enzyme controls conversion of T4 to the more active T3, which modulates osmoregulatory effectors.

In summary, current knowledge of systemic osmoreceptor neurons in euryhaline fishes and their potential role for coordinating and synchronizing the whole organism physiological response to salinity stress is rather limited. The following section will briefly discuss an additional avenue for systemic osmosensing, which complements neuronal networks.

2.3. Role of Epithelial Cells and Cytokines for Osmosensing

The prevailing paradigm for osmosensing in vertebrates is that it is mostly mediated by brain and systemic osmoreceptor neurons, as discussed in the preceding paragraphs. These neurons convert an osmotic stimulus into an electrical signal that can be transduced along axons and then converted into a neurotransmitter response at efferent nerve endings. However, experiments on isolated gill cells of euryhaline fishes have shown that they are capable of responding directly to osmotic stress to immediately change cell function and activate genes involved in cytokine production, including c-fos and osmotic stress transcription factor 1 (OSTF1) (Kültz, 1996; Fiol et al., 2006b). In addition, proteome analysis of euryhaline sharks (Squalus acanthias, *Triakis semifasciata*) suggested that the cytokine tumor necrosis factor-α (TNF-α) is an important element in the osmotic stress response network of gill and rectal gland epithelium (Kültz, 1996; Dowd et al., 2010). Thus, it has been proposed that cytokines, which act as paracrine, autocrine, and/or endocrine messengers, represent an additional avenue for transducing osmotic signals within and between cells and tissues (Kültz, 2011). It appears that epithelial cells directly facing the external milieu, such as gill epithelial cells and cells lining the gastrointestinal tract including the esophagus, are prime candidates for sensing changes in environmental salinity. These cells may evoke rapid anticipatory responses even before disruption of physiological homeostasis.

Euryhaline fishes harbor a special cell type, neuroepithelial cells, in their gills that merits mention in the context of osmosensing (Dunel-Erb et al., 1982; Bailly et al., 1984). These cells are embedded in the branchial epithelium and some of them can be in direct apical contact with the external environment (Zaccone et al., 1992). A prominent feature of these progenitor cells in tetrapods is an apically located primary cilium, which is a major signaling structure of epithelial cells and has been implicated in osmosensing of tetrapod osmoregulatory tissues (Abou Alaiwi et al., 2009; Goetz and Anderson, 2010; LaRusso and Masyuk, 2011; Nauli et al., 2011; Willardsen and Link, 2011). Moreover, neuroepithelial cells are known to produce and respond to cytokines (Nakashima et al., 2001; Sun et al., 2002) and they were suggested to have paracrine functions in euryhaline fishes (Gonia-kowska-Witalinska et al., 1995; Monteiro et al., 2010). Currently, it is not known whether neuroepithelial cells of euryhaline fishes have a primary cilium, although it has been observed that "Apical processes from the neuroepithelial cells occasionally contact the water on the surface of the filament epithelium" (Bailly et al., 1984). Because of their potential for linking neuronal and cytokine-based signaling mechanisms neuroepithelial cells represent an exciting and promising avenue for future research on mechanisms of osmosensing in euryhaline fishes.

3. MOLECULAR MECHANISMS OF CELLULAR OSMOSENSING

It is important to re-emphasize here that all cells have the potential to sense osmolality changes via associated alterations in cell volume/macro-molecular density, intracellular ionic strength, and cytoskeletal/membrane strain. However, different cell types display different sensitivity towards osmolality changes (e.g. the osmoreceptor cells discussed above are highly sensitive while other cell types are less sensitive). In addition, the competency for activation of specific intracellular signaling pathways/networks in response to osmotic stimuli differs depending on cell type. The cell-type-specific osmosensitivity and capacity for translating osmotic signals into specific intracellular signaling events illustrate the combinatorial nature of osmosensing at the cellular level. This means that osmolality or salinity is not just sensed by a single cell or particular cell type but by multiple cell types, which have different sensitivity ranges towards perceiving the osmotic signal and activate distinct sets of signaling pathways (e.g. prolactin secretion by the osmoreceptive prolactin cells in the pituitary gland). The information about the exact degree of change in body fluid osmolality/environmental salinity is encoded in the specific combination of those signals and their relative strengths. The remainder of this chapter will illustrate that the same principle also applies to osmosensing at the molecular level. That is, within individual cells multiple molecular osmosensors perceive changes in extracellular osmolality within different ranges of sensitivity leading to activation of a multitude of common intracellular signaling pathways that converge into a unique combination to provide stressor-specific and quantitatively appropriate output (Fig. 2.2).

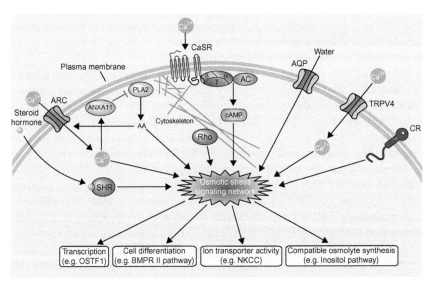

Fig. 2.2. Osmosensing relies on multiple sensors, some of which are depicted in this schematic drawing. The combination of stimuli received and transduced by those sensors triggers appropriate acclimatory responses via an osmotic stress signaling network and activation of osmoregulatory effector mechanisms. Such effector mechanisms include changes in osmoprotective gene and protein expression, cell differentiation, ion transport, and organic osmolyte metabolism. ARC: Arachidonic acid-regulated Ca^{2+} selective channel; ANXA11: annexin A11; PLA2: phospholipase A_2; CaSR: calcium sensing receptor; AC: Adenylate cyclase; AQP: aquaporin; TRPV4: transient receptor potential vanilloid 4; CR: corticosteroid receptor; SHR: Steroid hormone receptor; Rho: Rho family of small GTPases; cAMP: cyclic adenosine monophosphate; OSTF1: osmotic stress transcription factor 1; BMPR II: bone morphogenetic protein receptor type II; NKCC: $Na^+/K^+/2Cl^-$ cotransporter. Reproduced with permission from Kültz (2012). **SEE COLOR PLATE SECTION.**

3.1. Ion Sensing

Multiple mechanisms for sensing concentrations of inorganic ions and intracellular ionic strength have evolved, most of which depend on sensing cations (Na^+, Ca^{2+}, K^+) (Kültz and Burg, 1998b; Kültz, 2011). These three cation species (along with Mg^{2+}) constitute the most prevalent solutes in ocean water (Millero et al., 2008). In addition to cations, sensing of external Cl^- evokes the drinking reflex in Japanese eels (Hirano, 1974; Mayer-Gostan and Hirano, 1976). Because life originated in the primordial ocean, evolution has favored mechanisms for monitoring the concentrations of these cations by all cells. Euryhaline fishes inhabiting estuarine and other environments characterized by large salinity fluctuations have increased the utility of such mechanisms. Sensory perception of Na^+, Ca^{2+}, and intracellular ionic strength (or K^+) represents an important branch of input into the intracellular combinatorial osmosensing network. Although much remains to be learned on how ion sensing occurs in cells of euryhaline fishes, transient receptor potential (TRP) cation channels are involved in sensing Na^+ and Ca^{2+} and in osmosensing in euryhaline fishes and other animals (Choe and Strange, 2007; Liedtke, 2007, 2008; Chen et al., 2008; Gomis et al., 2008; Martinac, 2008; Zhang et al., 2008; Bossus et al., 2011; Pan et al., 2011; Xiao and Xu, 2011). Several isoforms of TRP channels are present in stenohaline zebrafish and osmosensory TRPV4 channels were localized to the developing kidney in this species (Mangos et al., 2007; Moeller et al., 2008). Moreover, the western clawed frog (*Xenopus tropicalis*) genome harbors six copies of TRPV4, which raises the possibility that diversification of osmosensory TRPV4 has occurred in parallel to adaptation to semi-aquatic and highly osmotically fluctuating environments (Saito and Shingai, 2006). Despite these indirect implications for a potential role of TRP channels in osmosensing of euryhaline fishes, direct evidence is currently still lacking.

The most comprehensively characterized ion sensor protein expressed by cells of euryhaline fishes is the calcium sensing receptor (CaSR). First cloned in the Hebert Lab in 1994, this protein has become the focus of studies on calcium homeostasis as well as osmoregulation in euryhaline fishes (Hebert and Brown, 1995; Brown et al., 1998; Abbink et al., 2006; Loretz, 2008). The CaSR is a membrane glycoprotein that is regulated directly by extracellular calcium (and to some extent also other polyvalent cations, notably Mg^{2+}) (Loretz, 2008). In a hyperosmotic environment relative to blood, the effective calcium concentration for CaSR regulation corresponds to the range found in the external milieu, leading to the proposal that this protein is a salinity sensor in fish (Nearing et al., 2002). CaSR is expressed in diverse osmoregulatory tissues of many euryhaline fishes, including elasmobranchs and teleosts (Ingleton et al., 2000, 2002; Fellner and Parker, 2002, 2004; Flanagan et al., 2002; Loretz et al., 2009). It has been localized to gill mitochondrion-rich ionocytes in winter flounder (*Pleuronectes americanus*), Atlantic salmon (*Salmo salar*), Mozambique tilapia (*Oreochromis mossambicus*), and Japanese eel (*Anguilla japonica*) (Nearing et al., 2002; Loretz, 2008). CaSR is also expressed in the rostral pars distalis and the pars intermedia of the pituitary gland (Loretz et al., 2009), where its messenger RNA (mRNA) decreased by about 50% within hours after transfer of both Nile and Mozambique tilapia from FW to 15 ppt (Velan et al., 2011). In Mozambique tilapia and flounder gill and in corpuscles of Stannius, CaSR abundance was unaffected by acclimation to either FW or SW (Greenwood et al., 2009; Breves et al., 2010a). Nevertheless, in the corpuscles of Stannius the CaSR protein displayed a difference in molecular mass in FW- and SW-acclimated flounder, suggesting that it is post-translationally modified during salinity acclimation of this euryhaline species (Greenwood et al., 2009). Stable expression and possible post-translational modification of CaSR are consistent with a proximal sensory role rather than downstream effector role because a sensor needs to be present and ready to respond immediately when a signal arrives. CaSR has been shown to activate downstream signaling pathways in euryhaline tilapia in response to salinity change, including phospholipase C-dependent and mitogen-activated protein kinase (MAPK)-dependent signaling pathways (Loretz et al., 2004).

Calcium signaling pathways are an obvious downstream target of CaSR, and intracellular calcium concentration is important for controlling osmotic stress signaling pathways in euryhaline fishes. For instance, hypoosmotic stimulation of prolactin secretion by rostral pars distalis cells isolated from the pituitary gland of Mozambique tilapia requires stretch-activated ion channels and changes in intracellular calcium levels (Seale et al., 2003b). In addition, the osmoregulatory hormone cortisol inhibits prolactin secretion via reduction of free intracellular calcium (Hyde et al., 2004), whereas angiotensin II, another osmoregulatory hormone, increases free intracellular calcium in fish tissues (Russell et al., 2001). A central role of intracellular calcium in fish osmotic stress signaling also emerged by modeling a signaling network based on immediate-early genes that rapidly respond to salinity stress in gills of euryhaline fishes (Fiol et al., 2006a; Evans and Somero, 2008; Evans, 2010).

3.2. Cytoskeletal Strain

Cytoskeleton strain has been proposed to represent one of the main inputs into combinatorial intracellular osmosensing (Kültz and Burg, 1998b; Kültz, 2011). Consistent with this notion, several studies suggest an important role of the cytoskeleton for osmosensing. Calcium ion binding annexins and two other immediate-early salinity response genes identified in Mozambique tilapia gills (gelsolin, galectin 4) (Fiol et al., 2006a) are known regulators of actin-based cytoskeleton remodeling. An mRNA screen in Nile tilapia identified β-actin as downregulated and Ca^{2+}-ATPase as upregulated after step-wise acclimation from FW to SW over a period of 10 days (Rengmark et al., 2007). Unfortunately, in this study brain and gill samples were pooled, which makes it difficult to interpret the results. Five out of six proteins found overexpressed in gill cells of European sea bass (*Dicentrarchus labrax*) after 3-month acclimation from FW to SW are isoforms of actin and tubulin (Ky et al., 2007). Moreover, the actin cytoskeleton has been shown to participate in the salinity-dependent regulation of the $Na^+/$

$K^+/2Cl^-$ (NKCC) cotransporter (Lionetto and Schettino, 2006). Furthermore, changes in ion transport across the intestinal epithelium of euryhaline eels require intact F-actin and microtubules (Lionetto et al., 2002) and the open/closed state of apical crypts of ionocytes also depends on the actin cytoskeleton (Daborn et al., 2001). Actin remodeling is involved in changing the morphology of tight junctions in response to salinity stress. In primary cultures of trout gill cells osmolality-induced changes of tight junctions required claudin 28b and the supporting F-actin ring (Sandbichler et al., 2011). Moreover, in primary cultures of Japanese eel gill cells hypertonic induction of the Na^+/Cl^-/taurine cotransporter, which represents an important effector of osmotic stress signaling pathways, depended on myosin light-chain kinase (MLCK) (Chow et al., 2009). MLCK-dependent phosphorylation of a major cytoskeletal building block that associates with tight junctions (MLC) is accompanied by redistribution of F-actin to the cell periphery (Chow et al., 2009), illustrating the interdependence of cytoskeletal organization and tight junction ultrastructure.

Another possible link between cytoskeletal strain and osmoregulatory effectors is focal adhesion kinase (FAK), which is a non-receptor tyrosine kinase that responds to cytoskeletal strain by promoting stabilization of the cytoskeleton (Lunn and Rozengurt, 2003, 2006; von Wichert et al., 2008; Alisi et al., 2011; Fabry et al., 2011). Bill Marshall's group has shown that in gill and opercular epithelia of euryhaline killifish, FAK is dephosphorylated in an integrin-dependent manner upon exposure to hypotonic stress (Marshall et al., 2005). They also demonstrated that FAK modulates the activity of the NKCC cotransporter and the cystic fibrosis transmembrane conductance regulator (CFTR) Cl^- channel in branchial and opercular ionocytes in response to salinity change (Marshall et al., 2008, 2009). The stenohaline FW zebrafish show a somewhat muted response to osmotic stress, which involves a micro-RNA that controls the activity of Na^+/H^+ exchanger regulatory factor 1 (NHERF1), an adaptor protein interacting directly with the actin cytoskeleton (Flynt and Patton, 2010). The findings summarized above do not provide causative proof, but nonetheless support a major role of cytoskeletal strain for osmosensing in euryhaline fishes.

3.3. Mitogen-Activated Protein Kinase Cascades

Further evidence for a role of the cytoskeleton in osmosensing comes from the osmotic activation of MAPK cascades that are regulated by cytoskeleton dynamics. MAPKs belong to the cell-cycle dependent protein kinase superfamily and they are important for osmosensory signal transduction in yeast, plant, and animal cells (Kültz, 1998; Kültz and Burg,

1998a). Despite the very strong evolutionary conservation of MAPKs, many of their activators and substrates are not conserved between yeast, plants, and animals (Fiol and Kültz, 2007). MAPK cascades seem to have evolved as a major component of a cell signaling node that is capable of integrating diverse inputs to produce diverse types of output depending on the particular combination of inputs received. In gill epithelium of euryhaline fishes, the activity (dual phosphorylation) of all three major MAPKs [including p38 and c-Jun N-terminal kinase kinase (JNK), whose upstream regulators include small GTPases that are linked to the cytoskeleton] is rapidly altered in response to salinity stress *in vivo* (Kültz and Avila, 2001). Osmotic regulation of p38 and JNK phosphorylation was also observed in killifish isolated opercular epithelium. In such *in vitro* preparations chloride secretion is inhibited by a pharmacological p38 blocker (Marshall et al., 2005). In addition, regulatory volume decrease of isolated hepatocytes from turbot (*Scophthalmus maximus*) depends on p38 MAPK (Ollivier et al., 2006a). Furthermore, an upstream regulator of MAPK cascades, mitogen-activated protein kinase kinase kinase 7 (MAPKKK7) interacting protein 2 (TAK 1 binding protein 2 = TAB 2) is induced as an immediate-early gene during hyperosmotic stress in tilapia gill epithelium (Fiol et al., 2006a). TAB2 induction is very rapid (within 2 h at the mRNA level) and transient, which is consistent with a role in osmotic stress signaling. In euryhaline medaka, activation of the JNK pathway following osmotic stress is required for induction of OSTF1b, thereby providing a link between osmosensing and downstream transcriptional regulation of osmoregulatory effector genes (Tse et al., 2011).

3.4. Osmotic Stress Transcription Factor 1

OSTF1 was first cloned from Mozambique tilapia (Fiol and Kültz, 2005) and later shown to be a homologue of mammalian transforming growth factor-β (TGF-β) stimulated clone 22 3b (TSC22-3b) (Fiol et al., 2006b, 2007). The initial discovery resulted from a suppression subtractive hybridization screen designed to detect immediate-early genes that are induced by hyperosmotic salinity stress in tilapia gill epithelium. Since then OSTF1 has also been cloned from and shown to be osmotically regulated in other euryhaline fishes, including black porgy (*Acanthopagrus schlegeli*) (Choi and An, 2008), Japanese eel (*Anguilla rostrata*) (Tse et al., 2008; Chow and Wong, 2011), Nile tilapia (*Oreochromis niloticus*) (Breves et al., 2010b), and Japanese medaka (*Oryzias latipes*) (Tse et al., 2011). In Mozambique tilapia exposed to hyperosmotic stress, OSTF1 protein levels increase in gill epithelial cells with a slight delay following the induction of mRNA (Fiol and Kültz, 2005). In gills of this euryhaline species the mechanism of OSTF1 upregulation includes both mRNA stabilization and transcriptional induction (Fiol et al., 2006b). Based on its leucine zipper DNA binding domain and known functions of mammalian homologues, OSTF1 appears to be a transcription factor but transcriptional target genes of OSTF1 have yet to be identified in any species of euryhaline fish.

3.5. Macromolecular Crowding and Damage

Two additional types of sensory input contributing to the combinatorial signaling logic in cells encountering osmotic stress are macromolecular crowding versus dilution and macromolecular damage, both of which are direct physical consequences of salinity stress. These effects are mostly encountered towards the severe end of osmotic stress, although it cannot be excluded that highly sensitive proteins or other macromolecules exist in cells that respond to minor changes in intracellular ionic strength resulting from osmotic stress. Changes in macromolecular density are intimately linked to cell volume changes. Hypertonicity (increase in extracellular salinity) leads to passive loss of water from cells along the osmotic gradient, which causes cell shrinkage and macromolecular crowding. Conversely, hypotonicity (decreases in extracellular salinity) causes cell swelling and macromolecular dilution. Experiments with primary cultures of teleost fish cells isolated from a wide variety of tissues have shown an abundance of molecular responses to osmotically induced changes in cell volume, and the corresponding literature is too extensive to be reviewed in this chapter (e.g. King and Goldstein, 1983; Leguen and Prunet, 1998; Duranton et al., 2000; Leguen et al., 2001; Hallgren et al., 2003; Tse et al., 2007; Avella et al., 2009; Chara et al., 2011; Wormser et al., 2011). However, some general conclusions can be drawn from *in vitro* studies with primary fish cell cultures. First, for epithelial cells it is unlikely that changes in cell volume lead to overall changes in membrane tension such as stretch because these cells are characterized by a high degree of basolateral and apical plasma membrane infolding (at least in a confluent state). Instead, it is more likely that cell volume changes are sensed via changes in intracellular macromolecular density in general or in certain localized areas (such as adjacent to specialized membrane lipid rafts) (Garner and Burg, 1994). Second, fish cells are capable of sensing osmotic stress and accompanying cell volume changes autonomously and independent of changes in systemic factors. Third, isolated fish cells respond to anisotonic osmotic stress by activating a multitude of diverse signaling pathways and adaptive effector mechanisms for counteracting the osmotic stress. Resolving the numerous interactions and coordination between systemic factors (osmoregulatory hormones, neurotransmitters, and cytokines) and autonomous responses of fish cells to salinity stress remains a challenging task.

DNA and protein damage (and macromolecular damage/functional disruption of cell metabolism in general) results from perturbed macromolecular density and suboptimal intracellular ionic strength and represents an important signal for cellular recognition of severe environmental stress, including osmotic stress (Kültz, 2005). Thus, mechanisms for protein and nucleic acid stabilization are part of osmotic stress acclimation in euryhaline fishes. Not much is known about osmotic effects on DNA integrity in fish, but in mammalian kidney cells (mIMCD3, HEK293) it has been shown that hypertonic salinity stress causes DNA double-strand breaks (Kültz and Chakravarty, 2001). Destabilizing and damaging effects of osmotic stress on proteins are well documented for many organisms, including fishes (Somero, 1986). To counteract destabilization and damage to macromolecules during osmotic stress, cells regulate the intracellular concentration of compatible organic osmolytes such as myo-inositol, sorbitol, glycine betaine, taurine, trimethylamine oxide (TMAO), or glycerophosphocholine (Yancey et al., 1982). When damaged proteins accumulate they have to be removed to avoid formation of non-specific aggregates, and the proteins discussed below are involved in this protective process.

In gills of euryhaline Mozambique tilapia, a ubiquitin E3 ligase (a homologue of Grail/Goliath E3 ligases) is rapidly upregulated at the mRNA level after transfer of fish from FW to SW (Fiol et al., 2006a, 2011). E3 ubiquitin ligases are enzymes that tag specific substrate proteins for degradation by attaching an ubiquitin tail or changing substrate function via monoubiquitination. Another ubiquitin E3 ligase, Rbx1 homologue Shop21, is induced by hyperosmotic salinity stress in salmon (Pan et al., 2002). In addition, several classes of heat shock proteins (HSPs) and components of the proteasome system are induced by hyperosmotic stress in various tissues of euryhaline fishes (e.g. Kültz, 1996; Pan et al., 2000; Deane et al., 2002; Niu et al., 2008; Dowd et al., 2010; Tine et al., 2010). Because these proteins stabilize damaged proteins (chaperone function) and/or tag them for degradation by the proteasome their osmotic induction indicates an increase in protein damage during salinity stress in euryhaline fishes. In addition to their "clean-up" function, ubiquitin E3 ligases and molecular chaperones may sense protein damage by quantifying the amount of substrates that they encounter and then passing this information on to osmosensory signaling networks, but this idea remains to be tested. One way of quantifying the amount of unfolded proteins in cells is via competition of such proteins with heat shock factor 1 (HSF1) for binding to HSPs and other genes containing a heat shock enhancer element (Anckar and Sistonen, 2011).

4. CONCLUSIONS AND PERSPECTIVES

Osmosensing is a complex physiological process that involves many players and is evident at multiple levels of organization in euryhaline fishes and other osmoregulating animals. At the whole organism level this process is coordinated via neuronal, endocrine, paracrine, and autocrine signals emanating from specialized cell types (osmoreceptor cells) that are highly sensitive towards osmolality change. The combination of the specific osmoreceptor cell types that are activated and to what degree, their dynamic range for perceiving changes in osmolality, and the temporal scale of activation encodes a specific overall stimulus permitting a finely tuned organismal response that is proportional to the degree of environmental salinity change. At the cellular level, the same combinatorial principle applies except that the combination of players consists of specific molecular osmosensors and intracellular signaling pathways (in lieu of the systemic osmoreceptor cells and neuronal/humoral messengers). Unraveling the major principles governing the osmotic activation of these numerous elements contributing to osmosensing as well as their modes of interaction represents a great challenge. Euryhaline fishes, because of their unique physiological capacity for tolerating a wide range of salinity and their aquatic habitat, will be invaluable assets for solving this complex puzzle.

ACKNOWLEDGMENT

Supported by NSF grant IOB-0542755.

REFERENCES

Abbink, W., Bevelander, G. S., Hang, X. M., Lu, W. Q., Guerreiro, P. M., Spanings, T., Canario, A. V. M. and Flik, G. (2006). PTHrP regulation and calcium balance in sea bream (*Sparus auratus* L.) under calcium constraint. *J. Exp. Biol.* 209, 3550–3557.

Abou Alaiwi, W. A., Lo, S. T. and Nauli, S. M. (2009). Primary cilia: highly sophisticated biological sensors. *Sensors* 9, 7003–7020.

Alisi, A., Arciello, M., Petrini, S., Conti, B. and Balsano, C. (2011). HCV infection promotes cytoskeletal reorganization in hepatocytes and a pro-fibrogenic phenotype in hepatic stellate cells by the activation of focal adhesion kinase. *J. Hepatol.* 54, S409.

Anckar, J. and Sistonen, L. (2011). Regulation of HSF1 function in the heat stress response: implications in aging and disease. *Annu. Rev. Biochem.* 80, 1089–1115.

Avella, M., Ducoudret, O., Pisani, D. F. and Poujeol, P. (2009). Swelling-activated transport of taurine in cultured gill cells of sea bass: physiological adaptation and pavement cell plasticity. *Am. J. Physiol. Reg. Int. Comp. Physiol.* 296, R1149–R1160.

Bailly, Y., Dunelerb, S. and Laurent, P. (1984). Neuroepithelial branchial cells in the fish. *J. Physiologie* 79, A39.

Ballatori, N. and Boyer, J. L. (1997). ATP regulation of a swelling-activated osmolyte channel in skate hepatocytes. *J. Exp. Zool.* 279, 471–475.

Bern, H. A. (1967). Hormones and endocrine glands in fishes. *Science* 158, 455–462.

Bossus, M., Charmantier, G. and Lorin-Nebel, C. (2011). Transient receptor potential vanilloid 4 in the European sea bass *Dicentrarchus labrax*: a candidate protein for osmosensing. *Comp. Biochem. Physiol. A Mol. Integr. Physiol.* 160, 43–51.

Bourque, C. W. (2008). Central mechanisms of osmosensation and systemic osmoregulation. *Nat. Rev. Neurosci.* 9, 519–531.

Bourque, C. W., Oliet, S. H. R. and Richard, D. (1994). Osmoreceptors, osmoreception, and osmoregulation. *Front. Neuroendocrinol.* 15, 231–274.

Breves, J. P., Fox, B. K., Pierce, A. L., Hirano, T. and Grau, E. G. (2010a). Gene expression of growth hormone family and glucocorticoid receptors, osmosensors, and ion transporters in the gill during seawater acclimation of Mozambique tilapia. *Oreochromis mossambicus. J. Exp. Zool. A Ecol. Genet. Physiol.* 313, 432–441.

Breves, J. P., Hasegawa, S., Yoshioka, M., Fox, B. K., Davis, L. K., Lerner, D. T., Takei, Y., Hirano, T. and Grau, E. G. (2010b). Acute salinity challenges in Mozambique and Nile tilapia: differential responses of plasma prolactin, growth hormone and branchial expression of ion transporters. *Gen. Comp. Endocrinol.* 167, 135–142.

Brown, E. M., Pollak, M. and Hebert, S. C. (1998). The extracellular calcium-sensing receptor: its role in health and disease. *Annu. Rev. Med.* 49, 15–29.

Burleson, M. L. (2009). Sensory innervation of the gills: O_2-sensitive chemoreceptors and mechanoreceptors. *Acta Histochem.* 111, 196–206.

de Castro, F. (2009). Towards the sensory nature of the carotid body: Hering, De Castro and Heymans. *Front. Neuroanat.* 3, 23.

Chapleau, M. W., Lu, Y. and Abboud, F. M. (2007). Mechanosensitive ion channels in blood pressure-sensing baroreceptor neurons. In *Mechanosensitive Ion Channels, Part B* (ed. O. P. Hamill), Vol. 59, p. 541.

Chara, O., Espelt, M. V., Krumschnabel, G. and Schwarzbaum, P. J. (2011). Regulatory volume decrease and P receptor signaling in fish cells: mechanisms, physiology, and modeling approaches. *J. Exp. Zool. A Ecol. Genet. Physiol.* 315, 175–202.

Chen, L., Liu, C. and Liu, L. (2008). Changes in osmolality modulate voltage-gated calcium channels in trigeminal ganglion neurons. *Brain Res.* 1208, 56–66.

Choe, K. P. and Strange, K. (2007). Molecular and genetic characterization of osmosensing and signal transduction in the nematode *Caenorhabditis elegans*. *FEBS J.* 274, 5782–5789.

Choi, C. Y. and An, K. W. (2008). Cloning and expression of Na^+/K^+-ATPase and osmotic stress transcription factor 1 mRNA in black porgy, *Acanthopagrus schlegeli* during osmotic stress. *Comp. Biochem. Physiol. B Biochem. Mol. Biol.* 149, 91–100.

Chow, S. C. and Wong, C. K. C. (2011). Regulatory function of hyperosmotic stress-induced signaling cascades in the expression of transcription factors and osmolyte transporters in freshwater Japanese eel primary gill cell culture. *J. Exp. Biol.* 214, 1264–1270.

Chow, S. C., Ching, L. Y., Wong, A. M. F. and Wong, C. K. C. (2009). Cloning and regulation of expression of the Na^+–Cl^-–taurine transporter in gill cells of freshwater Japanese eels. *J. Exp. Biol.* 212, 3205–3210.

Daborn, K., Cozzi, R. R. F. and Marshall, W. S. (2001). Dynamics of pavement cell–chloride cell interactions during abrupt salinity change in *Fundulus heteroclitus. J. Exp. Biol.* 204, 1889–1899.

Deane, E. E., Kelly, S. P., Luk, J. C. and Woo, N. Y. (2002). Chronic salinity adaptation modulates hepatic heat shock protein and insulin-like growth factor I expression in black sea bream. *Mar. Biotechnol.* 4, 193–205.

DiMaggio, M. A., Ohs, C. L., Grabe, S. W., Petty, B. D. and Rhyne, A. L. (2010). Osmoregulatory evaluation of the Seminole killifish after gradual seawater acclimation. *North Am. J. Aquacult.* 72, 124–131.

Dowd, W. W., Harris, B. N., Cech, J. J., Jr. and Kültz, D. (2010). Proteomic and physiological responses of leopard sharks (*Triakis semifasciata*) to salinity change. *J. Exp. Biol.* 213, 210–224.

Dunel-Erb, S., Bailly, Y. and Laurent, P. (1982). Neuroepithelial cells in fish gill primary lamellae. *J. Appl. Physiol.* 53, 1342–1353.

Duranton, C., Mikulovic, E., Tauc, M., Avella, M. and Poujeol, P. (2000). Potassium channels in primary cultures of seawater fish gill cells. II. Channel activation by hypotonic shock. *Am. J. Physiol. Reg. Int. Comp. Physiol.* 279, R1659–R1670.

Edwards, S. L. and Marshall, W. S. (2013). Principles and patterns of osmoregulation and euryhalinity in fishes. In *Fish Physiology*, Vol. 32, *Euryhaline Fishes* (eds. S. D. McCormick, A. P. Farrell and C. J. Brauner), pp. 1–44. New York: Elsevier.

Emmart, E. W. and Mossakow, M. J. (1967). Localization of prolactin in cultured cells of rostral pars distalis of pituitary of *Fundulus heteroclitus* (Linnaeus). *Gen. Comp. Endocrinol.* 9, 391–400.

Emmart, E. W., Pickford, G. E. and Wilhelmi, A. E. (1966). Localization of prolactin within pituitary of a cyprinodont fish *Fundulus heteroclitus* (Linnaeus) by specific fluorescent antiovone prolactin globulin. *Gen. Comp. Endocrinol.* 7, 571–583.

Espelt, M. V., Amodeo, G., Krumschnabel, G. and Schwarzbaum, P. J. (2000). Volume regulation in hepatocytes from goldfish. *J. Physiol.* 523P, 86P.

Evans, T. G. (2010). Co-ordination of osmotic stress responses through osmosensing and signal transduction events in fishes. *J. Fish Biol.* 76, 1903–1925.

Evans, T. G. and Somero, G. N. (2008). A microarray-based transcriptomic time-course of hyper- and hypo-osmotic stress signaling events in the euryhaline fish *Gillichthys mirabilis*: osmosensors to effectors. *J. Exp. Biol.* 211, 3636–3649.

Fabry, B., Klemm, A. H., Kienle, S., Schaffer, T. E. and Goldmann, W. H. (2011). Focal adhesion kinase stabilizes the cytoskeleton. *Biophys. J.* 101, 2131–2138.

Fellner, S. K. and Parker, L. (2002). A Ca^{2+}-sensing receptor modulates shark rectal gland function. *J. Exp. Biol.* 205, 1889–1897.

Fellner, S. K. and Parker, L. (2004). Ionic strength and the polyvalent cation receptor of shark rectal gland and artery. *J. Exp. Zool. A Comp. Exp. Biol.* 301, 235–239.

Fiol, D. F. and Kültz, D. (2005). Rapid hyperosmotic coinduction of two tilapia (*Oreochromis mossambicus*) transcription factors in gill cells. *Proc. Natl. Acad. Sci. U. S. A.* 102, 927–932.

Fiol, D. F. and Kültz, D. (2007). Osmotic stress sensing and signaling in fishes. *FEBS J.* 274, 5790–5798.

Fiol, D. F., Chan, S. Y. and Kültz, D. (2006a). Identification and pathway analysis of immediate hyperosmotic stress responsive molecular mechanisms in tilapia (*Oreochromis mossambicus*) gill. *Comp. Biochem. Physiol. D Genomics Proteomics* 1, 344–356.

Fiol, D. F., Chan, S. Y. and Kültz, D. (2006b). Regulation of osmotic stress transcription factor 1 (OSTF1) in tilapia (*Oreochromis mossambicus*) gill epithelium during salinity stress. *J. Exp. Biol.* 209, 3257–3265.

Fiol, D. F., Mak, S. K. and Kültz, D. (2007). Specific TSC22 domain transcripts are hypertonically induced and alternatively spliced to protect mouse kidney cells during osmotic stress. *FEBS J.* 274, 109–124.

Fiol, D. F., Sanmarti, E., Lim, A. H. and Kültz, D. (2011). A novel GRAIL E3 ubiquitin ligase promotes environmental salinity tolerance in euryhaline tilapia. *Biochim. Biophys. Acta* 1810, 439–445.

Flanagan, J. A., Bendell, L. A., Guerreiro, P. M., Clark, M. S., Power, D. M., Canario, A. V. M., Brown, B. L. and Ingleton, P. M. (2002). Cloning of the cDNA for the putative calcium-sensing receptor and its tissue distribution in sea bream (*Sparus aurata*). *Gen. Comp. Endocrinol.* 127, 117–127.

Flynt, A. S. and Patton, J. G. (2010). Crosstalk between planar cell polarity signaling and miR-8 control of NHERF1-mediated actin reorganization. *Cell Cycle* 9, 235–237.

Garner, M. M. and Burg, M. B. (1994). Macromolecular crowding and confinement in cells exposed to hypertonicity. *Am. J. Physiol.* 266, C877–C892.

Gerhart, J. and Kirschner, M. (1997). *Cells, Embryos and Evolution.* Oxford: Blackwell.

Goetz, S. C. and Anderson, K. V. (2010). The primary cilium: a signalling centre during vertebrate development. *Nat. Rev. Genet.* 11, 331–344.

Gomez-Segade, P., Segade, L. A. G. and Anadon, R. (1991). Ultrastructure of the organum vasculosum laminae terminalis in the advanced teleost *Chelon labrosus* (RISSO, 1826). *J. Hirnforsch.* 32, 69–77.

Gomis, A., Soriano, S., Belmonte, C. and Viana, F. (2008). Hypoosmotic- and pressure-induced membrane stretch activate TRPC5 channels. *J. Physiol.* 586, 5633–5649.

Goniakowska-Witalinska, L., Zaccone, G., Fasulo, S., Mauceri, A., Licata, A. and Youson, J. (1995). Neuroendocrine cells in the gills of the bowfin *Amia calva*: an ultrastructural and immunocytochemical study. *Folia Histochem. Cytobiol.* 33, 171–177.

Grassi, G., Bertoli, S. and Seravalle, G. (2012). Sympathetic nervous system: role in hypertension and in chronic kidney disease. *Curr. Opin. Nephrol. Hypertens.* 21, 46–51.

Grau, E. G., Nishioka, R. S. and Bern, H. A. (1981). Effects of osmotic pressure and calcium ion on prolactin release *in vitro* from the rostral pars distalis of the tilapia *Sarotherodon mossambicus. Gen. Comp. Endocrinol.* 45, 406–408.

Greenwood, M. P., Flik, G., Wagner, G. F. and Balment, R. J. (2009). The corpuscles of Stannius, calcium-sensing receptor, and stanniocalcin: responses to calcimimetics and physiological challenges. *Endocrinology* 150, 3002–3010.

Hallgren, N. K., Busby, E. R. and Mommsen, T. P. (2003). Cell volume affects glycogen phosphorylase activity in fish hepatocytes. *J. Comp. Physiol. B Biochem. Syst. Environ. Physiol.* 173, 591–599.

Hebert, S. C. and Brown, E. M. (1995). The extracellular calcium receptor. *Curr. Opin. Cell Biol.* 7, 484–492.

Hirano, T. (1974). Some factors regulating water intake by eel. *Anguilla japonica. J. Exp. Biol.* 61, 737–747.

Hyde, G. N., Seale, A. P., Grau, E. G. and Borski, R. J. (2004). Cortisol rapidly suppresses intracellular calcium and voltage-gated calcium channel activity in prolactin cells of the tilapia (*Oreochromis mossambicus*). *Am. J. Physiol. Endocrinol. Metab.* 286, E626–E633.

Ingleton, P. M., Bendell, L. A., Flanagan, J. A., Clark, M. S., Elgar, G., Teitsma, C. and Balment, R. J. (2000). Calcium-sensing receptors and parathyroid hormone-related protein (PTHRP) in the caudal neurosecretory system of the flounder. *Platichthys flesus. J. Endocrinol.* 167, P54.

Ingleton, P. M., Bendell, L. A., Flanagan, J. A., Teitsma, C. and Balment, R. J. (2002). Calcium-sensing receptors and parathyroid hormone-related protein in the caudal neurosecretory system of the flounder (*Platichthys flesus*). *J. Anat.* 200, 487–497.

Kammerer, B. D., Cech, J. J., Jr. and Kültz, D. (2010). Rapid changes in plasma cortisol, osmolality, and respiration in response to salinity stress in tilapia (*Oreochromis mossambicus*). *Comp. Biochem. Physiol. A Mol. Integr. Physiol.* 157, 260–265.

Kelley, K. M., Nishioka, R. S. and Bern, H. A. (1990). *In vitro* effect of osmotic pressure and cortisol on prolactin cell physiology in the Coho salmon (*Oncorhynchus kisutch*) during the parr–smolt transformation. *J. Exp. Zool.* 254, 72–82.

King, P. A. and Goldstein, L. (1983). Organic osmolytes and cell volume regulation in fish. *Mol. Physiol.* 4, 53–66.

Koonin, E. V. (2011). Are there laws of genome evolution? *PLoS Comput. Biol.* 7, 8.

Kougias, P., Weakley, S. M., Yao, Q. Z., Lin, P. H. and Chen, C. Y. (2010). Arterial baroreceptors in the management of systemic hypertension. *Med. Sci. Mon.* 16, RA1–RA8.

Kültz, D. (1996). Plasticity and stressor specificity of osmotic and heat shock responses of *Gillichthys mirabilis* gill cells. *Am. J. Physiol.* 271, C1181–C1193.

Kültz, D. (1998). Phylogenetic and functional classification of mitogen- and stress-activated protein kinases. *J. Mol. Evol.* 46, 571–588.

Kültz, D. (2005). Molecular and evolutionary basis of the cellular stress response. *Annu. Rev. Physiol.* 67, 225–257.

Kültz, D. (2007). Osmotic stress sensing and signaling in animals. *FEBS J.* 274, 5781.

Kültz, D. (2011). Osmosensing. In *Encyclopedia of Fish Physiology: From Genome to Environment,*, Vol. 2 (ed. A. P. Farrell), pp. 1373–1380. San Diego: Academic Press.

Kültz, D. (2012). The combinatorial nature of osmosensing in fishes. *Physiology (Bethesda)* 27, 259–275.

Kültz, D. and Avila, K. (2001). Mitogen-activated protein kinases are *in vivo* transducers of osmosensory signals in fish gill cells. *Comp. Biochem. Physiol. B Biochem. Mol. Biol.* 129, 821–829.

Kültz, D. and Burg, M. (1998a). Evolution of osmotic stress signaling via MAP kinase cascades. *J. Exp. Biol.* 201, 3015–3021.

Kültz, D. and Burg, M. B. (1998b). Intracellular signaling in response to osmotic stress. *Cell Vol. Regul.* 123, 94–109.

Kültz, D. and Chakravarty, D. (2001). Hyperosmolality in the form of elevated NaCl but not urea causes DNA damage in murine kidney cells. *Proc. Natl. Acad. Sci. U. S. A.* 98, 1999–2004.

Kültz, D. and Somero, G. N. (1995). Ion transport in gills of the euryhaline fish *Gillichthys mirabilis* is facilitated by a phosphocreatine circuit. *Am. J. Physiol. Reg. Int. Comp. Physiol.* 268, R1003–R1012.

Ky, C. L., de Lorgeril, J., Hirtz, C., Sommerer, N., Rossignol, M. and Bonhomme, F. (2007). The effect of environmental salinity on the proteome of the sea bass (*Dicentrarchus labrax* L.). *Anim. Genet.* 38, 601–608.

LaRusso, N. F. and Masyuk, T. V. (2011). The role of cilia in the regulation of bile flow. *Dig. Dis.* 29, 6–12.

Leguen, I. and Prunet, P. (1998). Effect of hypotonic shock on cell volume and intracellular calcium of trout gill cells. *Bull. Franc. Peche Piscicult.* 350–351, 521–528.

Leguen, I., Cravedi, J. P., Pisam, M. and Prunet, P. (2001). Biological functions of trout pavement-like gill cells in primary culture on solid support: pH$_i$ regulation, cell volume regulation and xenobiotic biotransformation. *Comp. Biochem. Physiol. A Mol. Integr. Physiol.* 128, 207–222.

Liedtke, W. B. (2007). RPV channels' function in osmo- and mechanotransduction. In *TRP Ion Channel Function in Sensory Transduction and Cellular Signaling Cascades* (eds. W. B. Liedtke and S. Heller), Chapter 22. Available from: http://www.ncbi.nlm.nih.gov/books/NBK5262/. Boca Raton, FL: CRC Press.

Liedtke, W. (2008). TRPV ion channels and sensory transduction of osmotic and mechanical stimuli in mammals. In *Sensing with Ion Channels,* (ed. B. Martinac), Vol. 11, p. 85.

Lionetto, M. G. and Schettino, T. (2006). The Na^+–K^+–$2Cl^-$ cotransporter and the osmotic stress response in a model salt transport epithelium. *Acta Physiol.* 187, 115–124.

Lionetto, M. G., Pedersen, S. F., Hoffmann, E. K., Giordano, M. E. and Schettino, T. (2002). Roles of the cytoskeleton and of protein phosphorylation events in the osmotic stress response in eel intestinal epithelium. *Cell. Physiol. Biochem.* 12, 163–178.

Lopez-Bojorquez, L., Villalobos, P., Garcia-G, C., Orozco, A. and Valverde-R, C. (2007). Functional identification of an osmotic response element (ORE) in the promoter region of the killifish deiodinase 2 gene (FhDio2). *J. Exp. Biol.* 210, 3126–3132.

Loretz, C. A. (2008). Extracellular calcium-sensing receptors in fishes. *Comp. Biochem. Physiol. A Mol. Integr. Physiol.* 149, 225–245.

Loretz, C. A., Pollina, C., Hyodo, S., Takei, Y., Chang, W. H. and Shoback, D. (2004). cDNA cloning and functional expression of a Ca^{2+}-sensing receptor with truncated C-terminal tail from the Mozambique tilapia (*Oreochromis mossambicus*). *J. Biol. Chem.* 279, 53288–53297.

Loretz, C. A., Pollina, C., Hyodo, S. and Takei, Y. (2009). Extracellular calcium-sensing receptor distribution in osmoregulatory and endocrine tissues of the tilapia. *Gen. Comp. Endocrinol.* 161, 216–228.

Lunn, J. A. and Rozengurt, J. E. (2003). Hyperosmotic stress stimulates phosphorylation of tyrosines 397 and 577 on focal adhesion kinase: role of src, RHO family GTPases and actin cytoskeletal organization. *Gastroenterology* 124, A607.

Lunn, J. A. and Rozengurt, E. (2006). Focal adhesion kinase (FAK) and C-SRC protect mammalian cells from hyperosmotic stress-stimulated apoptosis: evidence that FAK signaling lies upstream of the actin cytoskeleton in IEC-18 intestinal epithelial cells responses to this apoptotic stimulus. *Gastroenterology* 130, A532–A533.

Maetz, J. and Skadhaug, E. (1968). Drinking rates and gill ionic turnover in relation to external salinities in eel. *Nature* 217, 371–373.

Mangos, S., Liu, Y. and Drummond, I. A. (2007). Dynamic expression of the osmosensory channel trpv4 in multiple developing organs in zebrafish. *Gene Expr. Patterns* 7, 480–484.

Marshall, W. S., Ossum, C. G. and Hoffmann, E. K. (2005). Hypotonic shock mediation by p38 MAPK, JNK, PKC, FAK, OSR1 and SPAK in osmosensing chloride secreting cells of killifish opercular epithelium. *J. Exp. Biol.* 208, 1063–1077.

Marshall, W. S., Katoh, F., Main, H. P., Sers, N. and Cozzi, R. R. F. (2008). Focal adhesion kinase and beta 1 integrin regulation of $Na^+,K^+, 2Cl^-$ cotransporter in osmosensing ion transporting cells of killifish, *Fundulus heteroclitus*. *Comp. Biochem. Physiol. A Mol. Integr. Physiol.* 150, 288–300.

Marshall, W. S., Watters, K. D., Hovdestad, L. R., Cozzi, R. R. F. and Katoh, F. (2009). CFTR Cl^- channel functional regulation by phosphorylation of focal adhesion kinase at tyrosine 407 in osmosensitive ion transporting mitochondria rich cells of euryhaline killifish. *J. Exp. Biol.* 212, 2365–2377.

Martinac, B. (2008). *Sensing with Ion Channels. Springer Series in Biophysics*, Vol. 11. Berlin: Springer.

Mayer-Gostan, N. and Hirano, T. (1976). Effects of transecting 9th and 10th cranial nerves on hydromineral balance in eel *Anguilla anguilla*. *J. Exp. Biol.* 64, 461–475.

Millero, F. J., Feistel, R., Wright, D. G. and McDougall, T. J. (2008). The composition of standard seawater and the definition of the reference composition salinity scale. *Deep Sea Res. I Oceanograph. Res. Pap.* 55, 50–72.

Milsom, W. K. and Burleson, M. L. (2007). Peripheral arterial chemoreceptors and the evolution of the carotid body. *Respir. Physiol. Neurobiol.* 157, 4–11.

Moeller, C. C., Mangos, S., Drummond, I. A. and Reiser, J. (2008). Expression of trpC1 and trpC6 orthologs in zebrafish. *Gene Expr. Patterns* 8, 291–296.

Monteiro, S. M., Oliveira, E., Fontainhas-Fernandes, A. and Sousa, M. (2010). Fine structure of the branchial epithelium in the teleost *Oreochromis niloticus*. *J. Morphol.* 271, 621–633.

Nakashima, K., Takizawa, T., Ochiai, W., Yanagisawa, M. and Taga, T. (2001). Cytokine-mediated cell fate modulation in the developing brain. *J. Neurochem.* 78, 114.

Nauli, S. M., Haymour, H. S., Aboualaiwi, W. A., Lo, S. T. and Nauli, A. M. (2011). Primary cilia are mechanosensory organelles in vestibular tissues. In *Mechanosensitivity and Mechanotransduction*, Vol. 4, *Mechanosensitivity in Cells and Tissues* (eds. A. Kamkin and I. Kiseleva), pp. 317–350. Berlin: Springer.

Nearing, J., Betka, M., Quinn, S., Hentschel, H., Elger, M., Baum, M., Bai, M., Chattopadyhay, N., Brown, E. M., Hebert, S. C. and Harris, H. W. (2002). Polyvalent cation receptor proteins (CaRs) are salinity sensors in fish. *Proc. Natl. Acad. Sci. U. S. A.* 99, 9231–9236.

Niu, C. J., Rummer, J. L., Brauner, C. J. and Schulte, P. M. (2008). Heat shock protein (Hsp70) induced by a mild heat shock slightly moderates plasma osmolarity increases upon salinity transfer in rainbow trout (*Oncorhynchus mykiss*). *Comp. Biochem. Physiol. C Toxicol. Pharmacol.* 148, 437–444.

Nobata, S., Tsukada, T. and Takei, Y. (2010). Area postrema in hindbrain is a target of hormones that regulate drinking in eels. *Endocr. J.* 57, S254.

Noble, D. (2008). Claude Bernard, the first systems biologist, and the future of physiology. *Exp. Physiol.* 93, 16–26.

Olivereau, M. and Ball, J. N. (1964). Contribution a l'histophysiologie de l'hopophyse des teleosteens, en particulier de celle de *Poecilia* species. *Gen. Comp. Endocrinol.* 4, 523–532.

Ollivier, H., Pichavant-Rafini, K., Puill-Stephan, E., Calves, P., Nonnotte, L. and Nonnotte, G. (2006a). Effects of hyposmotic stress on exocytosis in isolated turbot, *Scophthalmus maximus*, hepatocytes. *J. Comp. Physiol. B Biochem. Syst. Environ. Physiol.* 176, 643–652.

Ollivier, H., Pichavant, K., Puill-Stephan, E., Roy, S., Calves, P., Nonnotte, L. and Nonnotte, G. (2006b). Volume regulation following hyposmotic shock in isolated turbot (*Scophthalmus maximus*) hepatocytes. *J. Comp. Physiol. B Biochem. Syst. Environ. Physiol.* 176, 393–403.

Pan, F., Zarate, J. M., Tremblay, G. C. and Bradley, T. M. (2000). Cloning and characterization of salmon hsp90 cDNA: upregulation by thermal and hyperosmotic stress. *J. Exp. Zool.* 287, 199–212.

Pan, F., Zarate, J. and Bradley, T. M. (2002). A homolog of the E3 ubiquitin ligase Rbx1 is induced during hyperosmotic stress of salmon. *Am. J. Physiol. Regul. Integr. Comp. Physiol.* 282, R1643–R1653.

Pan, Z., Yang, H. and Reinach, P. S. (2011). Transient receptor potential (TRP) gene superfamily encoding cation channels. *Hum. Genom.* 5, 108–116.

Rengmark, A. H., Slettan, A., Lee, W. J., Lie, O. and Lingaas, F. (2007). Identification and mapping of genes associated with salt tolerance in tilapia. *J. Fish Biol.* 71, 409–422.

Russell, M. J., Klemmer, A. M. and Olson, K. R. (2001). Angiotensin signaling and receptor types in teleost fish. *Comp. Biochem. Physiol. A Mol. Integr. Physiol.* 128, 41–51.

Saito, S. and Shingai, R. (2006). Evolution of temperature receptor homologs in vertebrates: diversification of mammalian temperature- and osmoreceptor homologs in western clawed frog (*Xenopus tropicalis*). *Genes Genet. Syst.* 81, 417.

Sandbichler, A. M., Egg, M., Schwerte, T. and Pelster, B. (2011). Claudin 28b and F-actin are involved in rainbow trout gill pavement cell tight junction remodeling under osmotic stress. *J. Exp. Biol.* 214, 1473–1487.

Seale, A. P., Richman, N. H., Hirano, T., Cooke, I. and Grau, E. G. (2003a). Cell volume increase and extracellular Ca^{2+} are needed for hyposmotically induced prolactin release in tilapia. *Am. J. Physiol. Cell Physiol.* 284, C1280–C1289.

Seale, A. P., Richman, N. H., Hirano, T., Cooke, I. and Grau, E. G. (2003b). Evidence that signal transduction for osmoreception is mediated by stretch-activated ion channels in tilapia. *Am. J. Physiol. Cell Physiol.* 284, C1290–C1296.

Seale, A. P., Mita, M., Hirano, T. and Grau, E. G. (2011). Involvement of the cAMP messenger system and extracellular Ca^{2+} during hyposmotically-induced prolactin release in the Mozambique tilapia. *Gen. Comp. Endocrinol.* 170, 401–407.

Seale, A. P., Watanabe, S. and Grau, E. G. (2012). Osmoreception: perspectives on signal transduction and environmental modulation. *Gen. Comp. Endocrinol.* 176, 354–360.

Somero, G. N. (1986). Protons, osmolytes, and fitness of internal milieu for protein function. *Am. J. Physiol.* 251, R197–R213.

Strange, K. (1994). Are all cell volume changes the same? *News Physiol. Sci.* 9, 223–228.

Sun, W., Funakoshi, H. and Nakamura, T. (2002). Localization and functional role of hepatocyte growth factor (HGF) and its receptor c-met in the rat developing cerebral cortex. *Mol. Brain Res.* 103, 36–48.

Takei, Y. (2000). Comparative physiology of body fluid regulation in vertebrates with special reference to thirst regulation. *Jpn. J. Physiol.* 50, 171–186.

Tine, M., Bonhomme, F., McKenzie, D. J. and Durand, J.-D. (2010). Differential expression of the heat shock protein Hsp70 in natural populations of the tilapia, *Sarotherodon melanotheron*, acclimatised to a range of environmental salinities. *BMC Ecol.* 10, 11.

Tse, W. K. F., Au, D. W. T. and Wong, C. K. C. (2007). Effect of osmotic shrinkage and hormones on the expression of Na^+/H^+ exchanger-1, $Na^+/K^+/2Cl^-$ cotransporter and Na^+/K^+-ATPase in gill pavement cells of freshwater adapted Japanese eel, *Anguilla japonica*. *J. Exp. Biol.* 210, 2113–2120.

Tse, W. K. F., Chow, S. C. and Wong, C. K. C. (2008). The cloning of eel osmotic stress transcription factor and the regulation of its expression in primary gill cell culture. *J. Exp. Biol.* 211, 1964–1968.

Tse, W. K. F., Lai, K. P. and Takei, Y. (2011). Medaka osmotic stress transcription factor 1b (Ostf1b/TSC22D3-2) triggers hyperosmotic responses of different ion transporters in medaka gill and human embryonic kidney cells via the JNK signalling pathway. *Int. J. Biochem. Cell B* 43, 1764–1775.

Tsukada, T., Nobata, S., Hyodo, S. and Takei, Y. (2007). Area postrema, a brain circumventricular organ, is the site of antidipsogenic action of circulating atrial natriuretic peptide in eels. *J. Exp. Biol.* 210, 3970–3978.

Velan, A., Hulata, G., Ron, M. and Cnaani, A. (2011). Comparative time-course study on pituitary and branchial response to salinity challenge in Mozambique tilapia (*Oreochromis mossambicus*) and Nile tilapia (*O. niloticus*). *Fish Physiol. Biochem.* 37, 863–873.

Watanabe, S., Hirano, T., Grau, E. G. and Kaneko, T. (2009). Osmosensitivity of prolactin cells is enhanced by the water channel aquaporin-3 in a euryhaline Mozambique tilapia (*Oreochromis mossambicus*). *Am. J. Physiol. Regul. Integr. Comp. Physiol.* 296, R446–R453.

Weber, G. M., Seale, A. P., Richman, N. H., Stetson, M. H., Hirano, T. and Grau, E. G. (2004). Hormone release is tied to changes in cell size in the osmoreceptive prolactin cell of a euryhaline teleost fish, the tilapia, *Oreochromis mossambicus*. *Gen. Comp. Endocrinol.* 138, 8–13.

von Wichert, G., Krndija, D., Schmid, H., Wichert, G., Haerter, G., Adler, G., Seufferlein, T. and Sheetz, M. P. (2008). Focal adhesion kinase mediates defects in the force-dependent reinforcement of initial integrin–cytoskeleton linkages in metastatic colon cancer cell lines. *Eur. J. Cell Biol.* 87, 1–16.

Willardsen, M. I. and Link, B. A. (2011). Cell biological regulation of division fate in vertebrate neuroepithelial cells. *Dev. Dyn.* 240, 1865–1879.

Wormser, C., Mason, L. Z., Helm, E. M. and Light, D. B. (2011). Regulatory volume response following hypotonic stress in Atlantic salmon erythrocytes. *Fish Physiol. Biochem.* 37, 745–759.

Xiao, R. and Xu, X. Z. S. (2011). *C. elegans* TRP channels. In *Transient Receptor Potential Channels,*, Vol. 704 (ed. M. S. Islam), pp. 323–339. Berlin: Springer.

Yancey, P. H., Clark, M. E., Hand, S. C., Bowlus, R. D. and Somero, G. N. (1982). Living with water stress: evolution of osmolyte systems. *Science* 217, 1214–1222.

Zaccone, G., Lauweryns, J. M., Fasulo, S., Tagliafierro, G., Ainis, L. and Licata, A. (1992). Immunocytochemical localization of serotonin and neuropeptides in the neuroendocrine paraneurons of teleost and lungfish gills. *Acta Zool.* 73, 177–183.

Zhang, X.-F., Chen, J., Faltynek, C. R., Moreland, R. B. and Neelands, T. R. (2008). Transient receptor potential A1 mediates an osmotically activated ion channel. *Eur. J. Neurosci.* 27, 605–611.

3

HORMONAL CONTROL OF FISH EURYHALINITY

YOSHIO TAKEI
STEPHEN D. McCORMICK

Hormones play a critical role in maintaining body fluid balance in euryhaline fishes during changes in environmental salinity. The neuroendocrine axis senses osmotic and ionic changes, then signals and coordinates tissue-specific responses to regulate water and ion fluxes. Rapid-acting hormones, e.g. angiotensins, cope with immediate challenges by controlling drinking rate and the activity of ion transporters in the gill, gut, and kidney. Slow-acting hormones, e.g. prolactin and growth hormone/insulin-like

Euryhaline Fishes: Volume 32
FISH PHYSIOLOGY
Copyright © 2013 Elsevier Inc. All rights reserved
DOI: http://dx.doi.org/10.1016/B978-0-12-396951-4.00003-7

growth factor-1, reorganize the body for long-term acclimation by altering the abundance of ion transporters and through cell proliferation and differentiation of ionocytes and other osmoregulatory cells. Euryhaline species exist in all groups of fish, including cyclostomes, and cartilaginous and teleost fishes. The diverse strategies for responding to changes in salinity have led to differential regulation and tissue-specific effects of hormones. Combining traditional physiological approaches with genomic, transcriptomic, and proteomic analyses will elucidate the patterns and diversity of the endocrine control of euryhalinity.

1. INTRODUCTION

Euryhalinity is originally an ecological term meaning an ability to live in broad (*eurys*) salinity (*halinos*) environments. Therefore, euryhaline fishes are those inhabiting estuaries, where salinity changes regularly, and those migrating between rivers [freshwater (FW)] and seas [seawater (SW)] during their lifespan. Osmoregulator is a physiological term for an organism with the ability to maintain body fluid osmolality at a certain level (usually around one-third of SW) irrespective of environmental salinities. Therefore, euryhaline fishes are always osmoregulators or ionoregulators. The endocrine system plays a pivotal role in manifesting euryhalinity, as it mediates homeostatic regulation to maintain ionic and water balance of the internal milieu. Various hormones have been implicated in acclimation to diverse environmental salinities in fishes, and while these are often conveniently grouped as FW-acclimating or SW-acclimating hormones (McCormick, 2001; Takei and Loretz, 2006), it should be noted that their function may differ among species or that they may even have dual functions depending on their interaction with other systems. As fish live in water, osmoregulation in the hypoosmotic FW environment is achieved by limiting osmotic water influx across body surfaces and by excreting via the kidney excess water that unavoidably enters the body (Marshall and Grosell, 2006). Obligatory loss of ions from the gills and kidney is compensated for by accelerating ion uptake by the gills and intestine. By contrast, acclimation to a hyperosmotic SW environment is achieved by increasing water gain by drinking environmental SW and subsequent water absorption by the intestine to compensate for osmotic water loss from the body surfaces (Marshall and Grosell, 2006). Excess monovalent ions (Na^+ and Cl^-) that enter the body surfaces (mainly the gills) and the intestine are actively secreted by ionocytes (also called mitochondrion-rich cells or chloride cells) in the gills and opercular epithelia. Therefore, euryhaline fishes must change drinking rate and reverse water and ion fluxes at these major osmoregulatory organs (gills, intestine, and kidney) when they encounter hypoosmotic and hyperosmotic media, and this ability is the key to euryhalinity. Details of these mechanisms can be found in Edwards and Marshall (2013), Chapter 1, this volume.

The degree of euryhalinity (salinity tolerance) often changes during the lifespan of fishes. Highly migratory diadromous fishes spawn either in FW (anadromous) or in SW (catadromous) and the tolerance may differ in early life stages compared with fish preparing for migration. Fish may also lose amphihalinity after the end of migration. Many migratory fishes experience drastic changes in body functions, for example smoltification (salmonids) or silvering (eels), before migration into completely the opposite osmotic environment (see McCormick, 2013, Chapter 5, this volume). Thus, the ontogenic change in euryhalinity during early life stages and during maturation is an important theme in studying euryhalinity.

Euryhaline fishes are often capable of surviving direct transfer from FW to SW or vice versa known as amphihalinity. In the acute phase of acclimation, the sympathetic nervous system responds immediately and usually changes drinking rate and blood supply to the osmoregulatory organs such as gills, intestine, and kidney to regulate water and ion fluxes (Marshall, 2003). The nervous system also activates the hormonal system, which consists of rapid- and slow-acting hormones, to cope with the changes in a coordinated fashion. Rapid-acting hormones are amine or oligopeptide hormones, which are secreted immediately (seconds to minutes) upon the environmental changes for the rapid acclimation (minutes to hours), and removed quickly from the circulation (McCormick and Bradshaw, 2006; Takei, 2008). The rapid-acting hormones also act on the brain to regulate drinking and on the peripheral osmoregulatory organs to change the activity of various transport molecules (transporters, channels, pumps, and cell adhesion molecules) that are already present in the transport epithelium. Importantly, the rapid-acting hormones usually stimulate the secretion of slow-acting hormones that are involved in chronic acclimation to a new osmotic environment. The slow-acting hormones induce reorganization of osmoregulatory organs by *de novo* synthesis of transport molecules on the epithelial cell membrane and intercellular junctions, and by proliferation and differentiation of stem cells or morphogenesis of cell types to reverse the direction of ion and water fluxes. They are hormones that are secreted slowly (hours to days) and stay in the circulation for longer acclimation to a new osmotic medium (days to weeks). The interaction among the sympathetic nervous system, and the rapid and slow hormonal systems is crucial for permitting euryhalinity in fish. It should be noted that the categorization of hormones into rapid- and slow-acting hormones cannot be absolute and some hormones may fall into both categories, as discussed below.

This chapter will review the hormonal regulation and induced mechanisms that permit reversal of ion and water regulation associated with euryhalinity. The chapter will be divided into sections that examine the role of rapid- and slow-acting hormones in controlling osmoregulation (Sections 2 and 3), the target organs for these hormones (from brain to kidney, Section 4), ontogeny (from eggs to adults, Section 5), and phylogeny (from cyclostomes to teleosts, Section 6), and the information will be integrated in relation to euryhalinity. The authors will attempt a complete summary and overview of the current state of knowledge and point out areas in need of more research. It should be noted that given the large number of fish species that exist and the relatively few that have been studied to date, these generalizations may not necessarily apply to all species.

2. RAPID-ACTING HORMONES

To cope with the sudden changes in environmental salinity, euryhaline fishes immediately shut off active transport machinery and activate existing transporters that often reverse the direction of ion and water transport via the

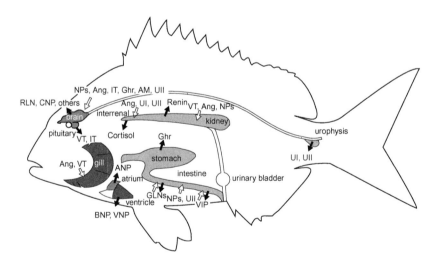

Fig. 3.1. Schema showing secretion (closed arrow) and action (open arrow) of rapid-acting hormones in teleost fishes. Ang: angiotensin II; ANP: atrial natriuretic peptide; AM: adrenomedullin; BNP: B-type natriuretic peptide; CNP: C-type natriuretic peptide; Ghr: ghrelin; GLN: guanylin; IT: isotocin; NP: natriuretic peptide; RLN: relaxin; VIP: vasoactive intestinal peptide; UI: urotensin I; UII: urotensin II.

nervous and endocrine systems (Marshall, 2003; Wood, 2011). In this acute phase, rapid- or short-acting hormones such as angiotensins, natriuretic peptides, neurohypophysial hormones, urotensins, and guanylins play a major role (McCormick, 2001; Takei, 2008; see Takei and Loretz, 2006, for details on the structure of these hormones). These oligopeptide hormones are secreted from the endocrine organs or produced in plasma by enzyme actions (renin–angiotensin system and kallikrein–kinin system) soon after encountering changes in environmental salinity. However, these hormones are metabolized by various peptidases in plasma or those associated with endothelial cells and disappear quickly from the circulation. Thus, the rapid-acting hormones take the role of combining the actions of the sympathetic nervous system and the slow-acting hormones. The tissues of synthesis/secretion and of action of rapid-acting hormones are summarized in Fig. 3.1.

Since plasma levels of some rapid-acting hormones increase before changes in plasma osmolality and blood volume occur, there must be an external sensor that detects the change in environmental osmolality or ion concentrations (see Kültz, 2013, Chapter 2, this volume), which transmits this information to the sympathetic nervous system and to the endocrine organs. Several proteins have been suggested to have osmosensing and/or Na^+-sensing functions, including adenylyl cyclase G (Saran and Schaap, 2004), transient receptor potential vallinoid type (TRPV) such as TRPV1, 2, and 4 (Liedtke and Kim, 2005), Nax channels (Shimizu et al., 2007), Ca-sensing receptor (Quinn et al., 1998) and aquaporin 4 (Venero et al., 1999). Osmosensing proteins that have been identified in fish (Fiol and Kültz, 2007) include Ca-sensing receptor in the dogfish (*Squalus acanthias*) (Nearing et al., 2002), gilthead seabream (*Sparus aurata*) (Flanagan et al., 2002), and Mozambique tilapia (*Oreochromis mossambicus*) (Loretz et al., 2004), and TRPV4 in sea bass (*Dicentrarchus labrax*) (Bossus et al., 2011) and Mozambique tilapia (Seale et al., 2011).

It is of interest to note that vasodepressor and natriferic (Na^+-extruding) hormones are much more diversified in fishes than in tetrapods, possibly to facilitate the complex osmoregulatory demands of aquatic life, as discussed in detail by Takei et al. (2007). For instance, vasodilatory (vasodepressor) hormones change blood flow to the epithelial cells of osmoregulatory organs (gills, intestine, and kidney) to regulate water and ion fluxes. Owing to their existence in water where they are much less influenced by the effects of gravity, fish have evolved low arterial pressure which may require dominant control by vasodepressor hormones. The reason for the diversified natriferic hormones is not clear as fishes in FW must retain Na^+ for body fluid homeostasis, but it may be related to the radiation of teleost fishes after they re-entered the marine environment around 160 million years ago (Nelson, 2006).

2.1. Angiotensins

The renin–angiotensin system (RAS) is an important hormonal system for the maintenance of water and ion balance in vertebrates (Kobayashi and Takei, 1996). Renin is a highly specific aspartyl proteinase that is secreted from the juxtaglomerular cells of the kidney and acts on angiotensinogen in plasma to cleave off an N-terminal decapeptide angiotensin I (Ang I). Subsequently, angiotensin-converting enzyme (ACE), a dipeptidyl carboxy-peptidase, removes a C-terminal dipeptide from Ang I during the passage in the gill circulation to form the biologically active Ang II. Tetrapods have [Asp[1]] Ang II (except for a highly aquatic clawed toad, *Xenopus laevis*), but all fishes have [Asn[1]] Ang II, including lungfishes that are ancestral to tetrapods (Takei et al., 2004b). In recent years, the molecular and functional characterization of the RAS has rapidly expanded (Fyhrquist and Saijonmaa, 2008); not only N-terminal truncated Ang III [Ang II (2–8)] and Ang IV [Ang II (3–8)], but also C-terminal truncated Ang-(1–7) have been shown to have a range of biological actions. The second converting enzyme (ACE2) has been identified, which is a critical enzyme for Ang-(1–7) production (Xu et al., 2011). In addition to the new Ang peptides, new RAS receptors have been discovered (Fyhrquist and Saijonmaa, 2008). Previously, most biological actions of Ang II and Ang III were thought to be mediated by the AT1 and AT2 receptors, but it is now known that there is a Mas receptor for Ang-(1–7) (Xu et al., 2011), an insulin-regulated aminopeptidase receptor for Ang IV (Albiston et al., 2001), and a prorenin/renin receptor (Nguyen et al., 2002). Receptors for Ang IV and Ang-(1–7) mediate the specific actions of each ligand, and the prorenin/renin receptor appears to activate prorenin to form Ang II on the cell surface.

The pattern of changes in plasma osmolality after FW to SW transfer differs greatly among euryhaline species; plasma osmolality increases gradually with a peak after 1–2 days in eels (*Anguilla japonica*) (Okawara et al., 1987; Takei et al., 1998), but a peak in a few hours in Mozambique tilapia (Breves et al., 2010b) and striped bass (*Morone saxatilis*) (Tipsmark et al., 2007). As eels can readily survive direct transfer from FW to SW, there may be differences in the ability to tolerate acute salinity changes among euryhaline species in which rapid-acting hormones play critical roles. Plasma Ang II concentration increases transiently when eels are transferred from FW to SW, and the pattern of the increase exactly parallels that of plasma NaCl concentration (Okawara et al., 1987; Tierney et al., 1995; Wong and Takei, 2012). As NaCl loading inhibits renin secretion in mammals and birds (Takei et al., 1988a), this inhibitory mechanism may be absent in eels. Consistently, injection of hypertonic NaCl solution into the circulation profoundly increased plasma Ang II concentration in eels (Takei et al., 1988b). It is now known that increased Cl^- excretion by the kidney is sensed by the macula densa of distal tubule to inhibit renin release in tetrapods (Kobayashi and Takei, 1996), but the macula densa is absent in teleosts. The data confirm the role of macula densa in renin secretion from the comparative viewpoint.

After the initial increase following SW transfer, plasma Ang II levels usually return to those in FW after SW acclimation in euryhaline fishes (Kobayashi and Takei, 1996), but are higher in fish acclimated to double-strength SW (Wong et al., 2006; Wong and Takei, 2012), indicating the role of Ang II in hyperosmotic acclimation. Detailed analysis of immunoreactive Ang II revealed that eel plasma contains comparable amounts of [Asn^1] Ang II, [Asp^1] Ang II, Ang III and Ang IV (Wong and Takei, 2012). Asparaginase appears to convert native [Asn^1] Ang II to [Asp^1] Ang II in the liver and kidney. Thus, previous studies using radioimmunoassay for Ang II measured all these Ang peptides as Ang II. It is necessary to compare the biological activity of these Ang peptides in fish. In eels, [Asn^1] Ang II was two-fold more potent than [Asp^1] Ang II and Ang III was much less potent for the vasopressor activity (Takei et al., 2004b).

When injected into the circulation, Ang II has a biphasic action on drinking in eels, causing an initial burst of drinking followed by prolonged inhibition (Takei et al., 1979), while constant acceleration of drinking was induced after an intracranial injection (Nobata and Takei, 2011). After peripheral injection, Ang II lowered net sodium reabsorption in the eel kidney (Nishimura and Sawyer, 1976), indicating its important role in SW acclimation. Ang II infusion reduced the glomerular filtration rate (GFR) of individual nephrons in rainbow trout (*Oncorhynchus mykiss*) (Brown et al., 1980). Ang II stimulated the activity of Na^+/K^+-ATPase (NKA) in isolated gill cells and renal tissues of eels (Marsigliante et al., 2000b), but inhibited it in isolated enterocytes (Marsigliante et al., 2001). The inhibition was mediated by a transient increase in intracellular calcium and subsequent protein kinase C activation. Furthermore, Ang II induced cortisol secretion *in vivo* and *in vitro* in several teleost species (Perrott and Balment, 1990). As cortisol is known to increase NKA abundance in most euryhaline fishes, Ang II appears to be involved indirectly in the chronic acclimation to SW (see Section 3.3, this chapter).

2.2. Natriuretic Peptides

The natriuretic peptide (NP) family consists of atrial, B-type, and ventricular natriuretic peptide (ANP, BNP, and VNP) secreted from the heart, and four C-type natriuretic peptides (CNP1–4) synthesized principally in the brain (Toop and Donald, 2004). Comparative genomic analyses have indicated that CNP4 is an ancestral molecule of the NP family, from which CNP3 was first duplicated (Inoue et al., 2003). In the cyclostomes, only CNP4 is present and it is synthesized in both heart and brain (Kawakoshi et al., 2006). In elasmobranchs, however, CNP4 does not exist and only CNP3 is synthesized and secreted from the heart and brain (Kawakoshi et al., 2001). The cardiac ANP, BNP, and VNP were generated by tandem duplication from CNP3 on the same chromosome, while CNP1 and CNP2 were produced by block duplication from CNP3 (Inoue et al., 2003). It is apparent that all seven NPs existed when ray-finned and lobe-finned bony fishes diverged, as all types of NPs are present in the extant tetrapods with some deletions in different classes. For instance, ANP, BNP, and CNP4 are present in mammals, BNP, VNP, CNP1, and CNP3 in birds, ANP, BNP, CNP3, and CNP4 in amphibians, and ANP, BNP, VNP, and CNP1–4 in teleosts, with some exceptions. Although ray-finned fishes have the most diversified NPs among vertebrates, VNP is present only in some migratory teleosts such as eels and salmonids and in more basal ray-finned fishes such as bichir (*Polypterus endlicheri*) and sturgeon (*Acipenser transmontanus*). However, most ray-finned fishes have four CNPs, which indicates their important functions in this fish group.

Four types of NP receptors have been identified in eels. Whereas NPR-A (GC-A) and NPR-B (GC-B) have guanylyl cyclase in the cytoplasmic domain and use cyclic guanosine monophosphate cGMP as a second messenger, NPR-C and NPR-D have only a short cytoplasmic domain, and their function is thought to be ligand clearance, adjusting the concentration of NPs at various target tissues. However, new functions of NPR-C have been suggested via inhibition of cyclic adenosine 3′,5′-monophosphate (cAMP) production or stimulation of phospholipase C (Anand-Srivastava, 2005). Cardiac ANP and VNP bind NPR-A with high affinity, CNP1 binds NPR-B specifically, and all NPs bind NPR-C and NPR-D with high affinities in eels (Hirose et al., 2001). In medaka (*Oryzias latipes*), two types of NPR-A (OLGC2 and 7) and a single NPR-B (OLGC1) have been identified (Yamagami and Suzuki, 2005). Medaka has only BNP as a cardiac NP and four CNPs (Inoue et al., 2003). CNP1, 2, and 4 bind to OLGC1, CNP3 binds to OLGC2, and BNP binds to OLGC7 with high affinities (Inoue et al., 2003, 2005). Of note is the high affinity of CNP3, a direct ancestor of cardiac NPs, to an NPR-A type receptor.

Using a specific radioimmunoassay, it was shown that plasma ANP concentration increased transiently after transfer of eels from FW to SW and then returned to an FW level (Kaiya and Takei, 1996a). In mammals, hypervolemia is a primary stimulus for ANP secretion (Toop and Donald, 2004), but hypovolemia occurs after SW transfer in teleost fishes. It was shown that an increase in plasma osmolality is the primary stimulus for ANP secretion in eels (Kaiya and Takei, 1996b). In rainbow trout, however, hypervolemia appears to be a potent stimulus for ANP secretion as in mammals (Cousins and Farrell, 1996). Research on more species is needed to determine whether such a difference is due to the particular adaptive strategy, which may differ among species.

Accumulating evidence indicates that ANP limits NaCl entry from the environment and facilitates SW acclimation. It was shown that ANP strongly inhibits drinking even at a non-depressor, physiological dose in eels (Tsuchida and Takei, 1998). Following SW transfer, there is an initial burst of drinking, but a transient suppression follows before the normal high drinking rate in SW fishes is re-established. The transient inhibition of drinking mirrors the transient increase in plasma ANP concentration after SW transfer (Kaiya and Takei, 1996a). Thus, ANP may delay an immediate increase in plasma osmolality that would otherwise occur, thereby promoting the initial phase of SW acclimation. Furthermore, ANP is also involved in the chronic inhibition of excess drinking because removal of circulating ANP by immunoneutralization enhanced the drinking rate and plasma Na^+ concentration in SW acclimated eels (Tsukada and Takei, 2006). As mentioned above, injection of hypertonic solutions, which is the most potent dipsogenic stimulus in tetrapods (Fitzsimons, 1998), clearly suppressed drinking in SW eels (Takei et al., 1988b). This may be due to the profound ANP secretion after hyperosmotic stimulus (Kaiya and Takei, 1996b). These results show that ANP is a major regulator of drinking in eels and probably in other teleost species. Recently, the potency order was shown to be $ANP = VNP > BNP = CNP3 > > CNP1 = CNP4$ in eels (Miyanishi et al., 2011). CNP3 exhibited a stronger antidipsogenic effect than other CNPs.

ANP inhibits intestinal NaCl absorption to further limit NaCl entry into the body in a few teleost species (O'Grady et al., 1985; Ando et al., 1992; Loretz et al., 1997). The inhibitory effect is highly potent at physiological concentrations (two to three orders more potent than other hormones) and highly efficacious (nullified short-circuit current) at high concentrations. Thus, it is most likely that ANP is a physiological regulator of intestinal NaCl absorption. The effect is mediated by the inhibition of $Na^+/K^+/2Cl^-$ cotransporter type 2 (NKCC2), a major transporter that facilitates water absorption (O'Grady et al., 1985). Thus, water absorption may be inhibited by ANP, which in the long term is disadvantageous for SW acclimation. This may reflect the fact that eels primarily regulate plasma osmolality over blood volume for body fluid regulation (Takei and Balment, 2009).

Concerning NaCl excretion, ANP increases urine Na^+ concentration but reduces urine volume in SW eels, resulting in a constant rate of total NaCl excretion by the kidney (Takei and Kaiya, 1998). This effect may be unique in eels because ANP induces profound natriuresis and diuresis in the trout (Duff and Olson, 1986). It was shown that 99.5% of NaCl is excreted via the gills in SW eels, but ANP failed to increase ^{22}Na excretion into the medium, indicating no direct action of ANP on the branchial Na^+ excretion (Tsukada et al., 2005). However, ANP may indirectly stimulate the branchial NaCl excretion via secretion of slow-acting hormones. It has been shown that ANP stimulates cortisol secretion *in vivo* in the flounder and *in vitro* from the interrenal tissue of trout (Arnold-Reed and Balment, 1991). In the eel, ANP also stimulates cortisol secretion *in vivo* (Li and Takei, 2003), but it failed to stimulate cortisol secretion *in vitro* from the interrenal tissue (Ventura et al., 2011). It is now found that ANP was effective *in vivo* in eels as it enhances the steroidogenic action of adrenocorticotropic hormone (ACTH). The steroidogenic effect was observed only in SW-acclimated eels. ANP also stimulates growth hormone (GH) secretion from the dispersed pituitary cells of Mozambique tilapia (Fox et al., 2007). Since cortisol and GH work in concert to promote differentiation of SW-type ionocytes in salmonid fishes (Madsen, 1990; McCormick, 2001), ANP can increase branchial NaCl secretion through slow-acting hormones in the late stage of SW acclimation.

In contrast to ANP, CNP appears to be an FW-acclimating hormone (Takei and Hirose, 2002). CNP was first isolated from the brain of killifish and eel, and as the CNP sequence was almost identical to mammalian CNP, this was initially thought to be a teleostean homologue of mammalian CNP. However, comparative genomic analyses later showed that the first teleostean CNP is CNP1 and mammalian CNP is an orthologue of teleostean CNP4 (Inoue et al., 2003). Plasma CNP1 concentration measured by radio-immunoassay for eel CNP1 was higher in FW eels than in SW eels (Takei et al., 2001). However, as CNP3 and CNP4 were later identified in the eel (Nobata et al., 2010), it is possible that the radioimmunoassay measured these CNPs also as CNP1. On the other hand, CNP1 infusion increased ^{22}Na uptake from the environment and increased plasma Na^+ concentration in FW eels (Takei and Hirose, 2002). In dispersed pituitary cells of Mozambique tilapia, CNP1 has no effect on prolactin secretion (Fox et al., 2007). As CNP3 is expressed abundantly in the pituitary (Nobata et al., 2010), the effect of CNP3 on prolactin secretion for FW acclimation needs to be examined.

2.3. Guanylins

Guanylin has a dual identity as an endocrine hormone and exocrine ectohormone because it is secreted into the intestinal lumen that is outside the internal milieu. The guanylin family is also diversified in teleost fishes. Three guanylins have been identified in the eel; guanylin is synthesized only in the intestine, while uroguanylin and renoguanylin are produced also in other segments of the digestive tracts and the kidney (Yuge et al., 2003; Kaljnaia et al., 2009). Guanylin is synthesized by the goblet cells of the eel intestine and secreted into the lumen with mucus (Yuge et al., 2003), and uroguanylin may be synthesized by the enterochromaffin cells and secreted in both directions (lumen and circulation) as in mammals (Nakazato, 2001). Uroguanylin is resistant against degrading enzymes on the brush-border membrane of renal proximal tubules, but renoguanylin may be metabolized quickly if it is secreted into the lumen of renal tubules. Therefore, uroguanylin may act on renal tubules from the luminal side and the final urine contains significant amounts of uroguanylin in teleost fishes as in mammals (Forte et al., 2000). Two types of guanylin receptors have been identified in eels, GC-C1 and GC-C2 (so named as they have a GC domain intracellularly), and uroguanylin has higher affinity to GC-C1 while guanylin and renoguanylin have higher affinities to GC-C2 (Yuge et al., 2006). This coincides with the higher expression of the GC-C1 gene in the kidney. The affinity of guanylin to GC-C is lower than that of NPs to GC-A and GC-B (NPR-A and NPR-B), probably because NPs are secreted into the circulation and guanylins into the intestinal lumen.

None of the hormones discussed so far as being involved in SW acclimation shows changes in messenger RNA (mRNA) expression in response to SW transfer. For instance, the ANP mRNA levels were not significantly increased after transfer of eels from FW to SW (H. Kaiya, unpublished data). However, intestinal guanylin mRNA expression increased five-fold 24 h following SW transfer relative to levels in FW-acclimated eels (Yuge et al., 2003). Furthermore, the expression of GC-C1 and GC-C2 genes was also upregulated in the intestine of SW eels compared to FW eels (Yuge et al., 2006). The higher expression of both hormone and receptor genes in acute and chronic phases of SW acclimation strongly suggests their important roles in SW acclimation.

Like cardiac ANP and VNP, guanylins inhibited short-circuit current in a dose-dependent manner when applied to the luminal (mucosal) side of intestinal epithelia and reversed the current at high doses (Yuge and Takei, 2007). It was shown that the reversal is due to Cl^- secretion via the cystic fibrosis transmembrane conductance regulator (CFTR)-type Cl^- channel as in mammals. Interpretation of these observations has been outlined by Takei and Yuge (2007) as follows. After imbibed SW is desalted in the esophagus, which is highly permeable to NaCl, but not water (Hirano and Mayer-Gostan, 1976), and further diluted to isotonicity in the stomach, water is absorbed together with monovalent ions in the anterior intestine from the isotonic luminal fluid. The major transporter for ion absorption is NKCC2 on the apical membrane of absorptive epithelial cells. However, as SW contains similar concentrations of Na^+ and Cl^-, Cl^- becomes deficient in the luminal fluid after absorption of one Na^+ and two Cl^- by NKCC2, which depresses NKCC operation and, thus, decreases water absorption. In fact, there seems to be Cl^- secretion into the lumen as judged by the maintained luminal Cl^- concentration along the intestine compared with Na^+ (Tsukada and Takei, 2006). It is possible that guanylin is involved in the active Cl^- secretion into the lumen to supplement luminal Cl^- to ensure water absorption in the posterior intestine. It has also been shown that guanylin stimulates HCO_3^- secretion through CFTR to precipitate concentrated Ca^{2+} and Mg^{2+} ions after water absorption (M. Ando and Y. Takei, unpublished data). There is accumulating evidence showing that active secretion of HCO_3^- into the lumen precipitates $Mg/CaCO_3$ and decreases luminal fluid osmolality, thereby facilitating water absorption (Groscll et al., 2009; Wilson et al., 2009). It remains to be determined whether guanylins are secreted in response to high luminal NaCl concentration or osmolality of imbibed SW as reported in mammals (Nakazato, 2001), and whether guanylin actually promotes SW acclimation by secretion of Cl^- and HCO_3^- into the lumen.

2.4. Neurohypophysial Hormones

Neurohypophysial hormones, vasotocin (VT) and isotocin (IT), are likely to be important osmoregulatory hormones in teleost fishes, as vasopressin, an orthologue of VT, is critical for body fluid regulation in terrestrial animals through its effect on tubular water reabsorption in the kidney (Babey et al., 2011). A recent *in silico* study has shown that teleosts possess at least five distinct receptors [two V1-type receptors (V1R), two V2-type receptors (V2R) and an oxytocin-type receptor] for VT and IT (Daza et al., 2012). A V2R of medaka and bichir (*P. senegalus*) transiently expressed in culture cells responded to VT with cAMP accumulation, consistent with what has been observed with the mammalian V2R (Konno et al., 2010). As mentioned in detail below (Section 4.3), urine volume is primarily regulated by GFR in teleost fishes and the nephron is unable to concentrate urine above plasma because of the lack of a countercurrent system formed by the loop of Henle (Nishimura and Fan, 2003). Therefore, the tubular effect of VT for water reabsorption may be minor in teleost fishes.

VT may be involved in acclimation to high-salinity environments, but the data are still somewhat controversial. An initial study suggested the involvement of VT in FW acclimation as the VT mRNA levels decreased for 2 weeks after transfer of rainbow trout from FW to 80% SW and increased again following transfer back to FW (Hyodo and Urano, 1991). The IT gene did not exhibit obvious changes after either transfer. However, hypothalamic VT mRNA levels increased 4 h after transfer of euryhaline flounder (*Platichthys flesus*) from FW to SW, with concomitant increases in plasma VT concentration (Warne et al., 2005). Plasma VT concentration decreased after transfer of the flounder from SW to FW (Bond et al., 2002). In primary culture of sea bass gill pavement cells grown on a permeable support, VT and IT decreased short-circuit current (Cl^- secretion) in a dose-dependent manner with IT more potent and efficacious than VT (Guibbolini and Avella, 2003). V1R antagonist blocked the VT effect, but V2R agonist and antagonist had no effect, suggesting mediation by V1R. Chronic VT treatment increased gill NKA activity and plasma cortisol levels in the gilthead sea bream (Sangiago-Alvarellos et al., 2006). VT at physiological concentrations caused a dose-dependent reduction in urine flow and reduced filtering population of glomeruli to one-third in the *in situ* perfused trout kidney (Amer and Brown, 1995). In eels, IT stimulated drinking through its relaxing effect on the esophageal sphincter muscle while VT inhibited drinking, antagonizing the IT effect (Ando et al., 2000; Watanabe et al., 2007).

2.5. Urotensins

Urotensins (UI and UII) were first expected to have osmoregulatory functions in fishes because they are secreted from the urophysis, which is upstream and in direct circulatory contact with the kidney and intestine (McCrohan et al., 2007). UI, the orthologue of which in mammals was named urocortin, is a member of the corticotropin-releasing hormone (CRH) family and thus stimulates ACTH secretion, resulting in cortisol secretion (Lovejoy and Balment, 1999). UI (urocortin) and CRH bind to CRH type 1 (CRHR1) and type 2 (CRHR2) receptors. The CRHR1 has similar affinities to both ligands, whereas the CRHR2 exhibits higher affinity to UI. The goby (*Gillichthys mirabilis*), transferred from SW to FW, increased urophyseal UI content after 24 h, showing osmotic sensitivity of the urophysis (Larson and Madani, 1991). Initial physiological studies reported that UI inhibited water and NaCl absorption in isolated anterior intestinal segments of FW-acclimated, but not SW-acclimated, Mozambique tilapia (Mainoya and Bern, 1982). In the flounder, UI appears to stimulate cortisol secretion directly and interact synergistically with ACTH (Kelsall and Balment, 1998). Given cortisol's role in both FW and SW acclimation (see Section 3.3, this chapter), these results implicate a possible role of UI in FW acclimation.

UII is a member of the somatostatin superfamily and thus likely to be involved in the inhibition of GH secretion (Tostivint et al., 2008). Expression of the UII gene and its receptor (UT) gene was reduced after acute transfer of euryhaline flounder from SW to FW, and the UT gene expression in the gills and kidney is downregulated in FW-acclimated fish, although plasma UII levels do not differ between SW and FW fishes (Lu et al., 2006). However, plasma UII concentration was reduced for some time after transfer of the flounder from SW to FW, showing the rapid-acting nature of UII (Bond et al., 2002). Initial physiological studies revealed that UII has a direct action on the intestine to increase water and NaCl absorption in SW-acclimated tilapia (Mainoya and Bern, 1982) and 5% SW-acclimated goby (Loretz et al., 1983) and on the urinary bladder of SW-acclimated goby (Loretz and Bern, 1981). UII inhibited prolactin secretion from the rostral part of Mozambique tilapia pituitary (Grau et al., 1982). These results suggest a role of UII in SW acclimation. However, UII inhibited short-circuit current in the goby skin, suggesting an inhibition of active Cl^- secretion, probably via ionocytes (Marshall and Bern, 1981). As UI reversed the effect of UII in the goby skin, UI and UII may be involved in the acclimation to opposite osmotic environments. More recently, UII was found to potently inhibit drinking in eels through its action on the brain (Nobata et al., 2011). UII stimulated cortisol secretion from the interrenal cells of rainbow trout (Arnold-Reed and Balment, 1994).

2.6. Adrenomedullins

Adrenomedullin (AM) is a member of the calcitonin gene-related peptide (CGRP) family that consists of CGRP, AM, and amylin (Lópetz and Martínez, 2002). However, five AMs (AM1–5) form a subfamily in teleost fishes, of which AM in mammals is an orthologue of teleost AM1 (Ogoshi et al., 2003). This finding in teleosts led to the discovery of AM2 and AM5 in mammals (Takei et al., 2004a, 2008; Ogoshi et al., 2006). Accordingly, it is now generally accepted that the CGRP family is comprised of CGRP, AM, AM2, AM5, and amylin in fishes and tetrapods, and AM4 and AM3 are generated from AM1 and AM2, respectively, by the whole-genome duplication that occurred only in the teleost lineage (Ogoshi et al., 2006). AM1 binds calcitonin receptor-like receptor (CLR) associated with receptor activity-modifying protein (RAMP) 2 or 3, and uses cAMP as a second messenger (Hay et al., 2005). AM2 and AM5 bind to CLR and RAMP3 complex with low affinity, and thus the specific receptor to these peptides may be present. Judging from the retained multiple paralogues after genome duplication and potent vasodepressor and natriferic actions in eels (Nobata et al., 2008; Ogoshi et al., 2008), the AM peptides may have important functions for SW acclimation as observed with the NP peptides.

The tissue distribution of the AM mRNA differs among teleost species. In tiger pufferfish (*Takifugu rubripes*), AM1/4 is ubiquitously expressed in various tissues, AM2/3 mostly in the brain, and AM5 in the spleen and gills (Ogoshi et al., 2003). In the eel, AM1 is expressed in the heart, kidney, and red body of air bladder, AM2/3 in a large number of tissues, and AM5 in the spleen and red body (Nobata et al., 2008). The expression of the AM genes in the major osmoregulatory organs did not change after transfer of pufferfish from SW to FW (M. Ogoshi and Y. Takei, unpublished results) or between FW- and SW-acclimated eels (Nobata et al., 2008). AMs are highly efficacious vasodepressor hormones in eels with a potency order of AM2>AM5>>AM1 (Nobata et al., 2008), which is different from the results in mammals where the effects of AM1 and AM2 are comparable (Takei et al., 2004a). Concerning their renal action, AM2 and AM5 infused into the circulation of FW eels caused antidiuresis without changes in blood pressure but AM1 infusion caused antinatriuresis (Ogoshi et al., 2008). AM2 and AM5 induced drinking as potently as Ang II when infused into the circulation of FW eels, but intracranial injection failed to affect drinking, although Ang II was effective when injected into the same site. Further investigations are necessary to define the role of AMs in fish osmoregulation.

2.7. Other Peptide Hormones

Relaxins (RLN1/2 and 3) belong to the insulin superfamily that contains insulin, insulin-like peptides, and insulin-like growth factors (Wilkinson et al., 2005). Three RLNs were identified in humans, but RLN1 and 2 were generated by tandem duplication only in primates. Thus, two RLNs generally exist in mammals, of which RLN1/2 is principally a peripheral hormone involved in cardiovascular regulation and reproduction, while RLN3 is a neuropeptide whose function has not yet been fully elucidated. In teleost fishes, three relaxins (RLN1/2, RLN3a, and RLN3b) are present in all species thus far examined (Good-Avila et al., 2009). All three teleost RLNs are produced in the brain of eels and the sequences are highly conserved even between RLN1/2 and RLN3 (Hu et al., 2011). RLNs may be involved in body fluid regulation since RLN1/2 was shown to have a potent dipsogenic effect in mammals (Sunn et al., 2002). However, expression of the three RLN genes does not differ between FW and SW eels (Hu et al., 2011). More research is needed to evaluate the role of RLNs in osmoregulation in teleost fishes.

Vasoactive intestinal peptide (VIP) is a member of the secretin superfamily and a close paralogue of pituitary adenylate cyclase-activating peptide (PACAP) (Sherwood et al., 2000). VIP and PACAP share the PAC1 and VPAC receptors and act through stimulation of intracellular cAMP (Cardoso et al., 2007). VIP is duplicated in some teleost species, while two PACAP peptides exist in all species thus far examined (Takei, 2008). VIP has long been known to affect osmoregulation, where it has been shown to stimulate Cl^- secretion from the opercular epithelia of SW-acclimated Mozambique tilapia (Foskett et al., 1982). Furthermore, VIP was shown to inhibit the short-circuit current (NaCl absorption) in the tilapia intestine (Mainoya and Bern, 1984) and in the SW eel intestine (Ando et al., 2003). In the intestine of winter flounder, VIP stimulated Cl^- secretion through cAMP acting as a second messenger (O'Grady and Wolters, 1990). More recently, VIP was found to have no effect on drinking when injected centrally and peripherally in the eel, although it is dipsogenic in the rat after central injection (Ando et al., 2003). VIP and PACAP regulate the secretion of pituitary hormones, renin, and adrenal steroids in mammals (Sherwood et al., 2000; Vaudry et al., 2009), but in teleost fishes only PACAP has been implicated in GH release (Canosa et al., 2007).

3. SLOW-ACTING HORMONES

Hormones such as prolactin, GH/insulin-like growth factor-1 (IGF-I), and cortisol act to alter the overall capacity for osmoregulation in teleost fishes. These protein hormones and steroid hormones have relatively long half-lives in plasma and reorganize osmoregulatory organs primarily by *de novo* synthesis of transport proteins, transporter/channel/pumps, and intercellular matrix, and through regulation of cell proliferation and differentiation.

3.1. Prolactin

Fifty years ago, Grace Pickford found that removal of the pituitary resulted in mortality in FW, but not in SW, and that survival in FW could be restored by treatment with prolactin in the killifish (*Fundulus heteroclitus*) (Pickford and Phillips, 1959). Since this classic study, a great deal of evidence has been generated supporting the involvement of prolactin in the process of FW acclimation in teleost fishes. This evidence includes changes in prolactin gene expression and circulating levels in response to salinity change, localization and regulation of prolactin receptors, and further studies on the osmoregulatory mechanisms induced with prolactin treatment (see reviews by McCormick, 2001; Manzon, 2002; Sakamoto and McCormick, 2006).

For most euryhaline teleosts, gene transcription, synthesis, secretion, and plasma levels of prolactin all increase following exposure to FW (Manzon, 2002; Lee et al., 2006). Stenohaline FW fishes also appear to adjust prolactin production in response to low ion concentrations (Liu et al., 2006; Hoshijima and Hirose, 2007), even at very early developmental stages. In some euryhaline fishes lactotrophs are directly responsive to osmolality (Seale et al., 2002), although this is not universal (Kelley et al., 1990). Cortisol has a negative effect on prolactin secretion that can be both rapid and sustained (Kelley et al., 1990; Borski et al., 2002). Metabolic clearance rates of prolactin in salmonids are also increased following FW acclimation (Sakamoto et al., 1991), suggesting increased utilization, metabolism, and/or excretion. As in mammals, prolactin-releasing peptide (PrRP) has been identified in teleost hypothalamus; this peptide increases prolactin secretion and is expressed at higher levels in FW than in SW (Moriyama et al., 2002; Sakamoto et al., 2005).

Prolactin receptor transcription and abundance are high in osmoregulatory organs such as the gill, intestine, and kidney, and are normally in greater abundance in FW than SW (Fryer, 1979; Dauder et al., 1990; Auperin et al., 1995). High levels of prolactin receptor transcription have been found in gill ionocytes and enterocytes, cells specifically involved in osmoregulation (Sandra et al., 2000). In zebrafish (*Danio rerio*), gene knockout experiments have shown that prolactin receptors are necessary for early development of pituitary function and the capacity to respond to low ion environments with prolactin transcription (Liu et al., 2006). Multiple prolactin receptor isoforms have been described in several species and are likely to differ in their physiological functions, although the nature of these differences has yet to be determined. Expression of tilapia prolactin receptor (PRLR) 1 and 2 in a heterologous expression system indicates that the receptors activate different downstream signaling pathways with different actions on cell ion regulatory capacity (Fiol et al., 2009). Furthermore, PRLR isoforms in zebrafish and tilapia are tissue specific in their expression and respond differentially to changes in environmental ion concentrations (Breves et al., 2011; J. P. Breves, personal communication).

The response of fish in FW to removal of the pituitary is not universal. Some FW species such as rainbow trout and goldfish are able to survive in FW for sustained periods following hypophysectomy. Amphihaline eels (*Anguilla* sp.) spend the majority of their life cycle in FW and can survive in FW after hypophysectomy (Hirano, 1969). In the case of goldfish (*Carassius auratus*), hypophysectomy results in lower levels of plasma ions that can be restored by prolactin. Amphihaline Mozambique tilapia die in FW after hypophysectomy, but can survive in SW (Breves et al., 2010b). When these patterns across species are considered together, it seems plausible that FW species have a level of constitutive (prolactin-independent) ion uptake capacity that may be absent in brackish water and marine species, and an additional prolactin-dependent regulation that allows for regulation in the face of increased demand for ion uptake such as ion poor and acidic conditions. The sea bream does not change circulating prolactin levels after exposure to FW, nor does the isolated pituitary alter prolactin secretion in response to changes in osmotic pressure (Fuentes et al., 2010), suggesting that prolactin may have a limited role in controlling osmoregulation in stenohaline marine species.

A large number of prolactin's actions are associated directly or indirectly with cell proliferation and/or apoptosis (Sakamoto and McCormick, 2006). Prolactin has been shown to affect ionocytes, both by inhibiting the development of secretory ionocytes (Herndon et al., 1991; Kelly et al., 1999) and by promoting the morphology and functional attributes of ion uptake cells (Pisam et al., 1993). Prolactin also regulates permeability characteristics of epithelial tissue (Manzon, 2002). For example, prolactin has been shown to reduce transcellular permeability, characteristic of exposure to FW, in an *in vitro* gill pavement cell culture system (Kelly and Wood, 2002b). More recent studies indicate that proteins involved in active ion transport by the gill are regulated by prolactin, including downregulation of the SW isoform of NKA (NKAα1b) (Tipsmark and Madsen, 2009), upregulation of the FW isoform NKAα1a (T. O. Nilsen, S. Stefansson, and S. D. McCormick, unpublished results), and gill NKA activity (Shrimpton and McCormick, 1998) in Atlantic salmon. In tilapia, gill mRNA levels of the apical Na$^+$/Cl$^-$ cotransporter (NCC) and NKAα1a, both of which are involved in ion uptake, are reduced after hypophysectomy and restored by prolactin treatment (Breves et al., 2010c; Tipsmark et al., 2011). Intestinal claudins 15 and 25b that are upregulated by SW exposure are downregulated by prolactin treatment (Tipsmark et al., 2010b). Prolactin treatment also reduces ion and water permeability of the esophagus and intestine, a response that normally occurs during acclimation to FW, perhaps acting through regulation of cell apoptosis and proliferation (Takahashi et al., 2006).

There is some evidence for the interaction of prolactin and cortisol in promoting ion uptake, and it is possible that prolactin is acting as a "switch" for promoting ion uptake in the same way that the GH–IGF-I axis interacts with cortisol to promote salt secretion (McCormick, 2001). In hypophysecto-mized and/or interrenalectomized fish, prolactin and/or ACTH or cortisol are necessary to completely restore ion and water balance in FW (McCormick, 2001). In hypophysectomized channel catfish (*Ictalurus punctatus*), prolactin and cortisol in combination cause a greater restoration of plasma ions than either acting alone (Eckert et al., 2001). Also, cortisol and prolactin together have a greater effect than either hormone alone on promoting the transepithelial resistance and potential and ion influx of an *in vitro* gill cell preparation from rainbow trout (Zhou et al., 2003). In hypophysectomized tilapia in FW, ovine prolactin and cortisol stimulated fxyd-11 expression, a subunit gene of NKA involved in modulation of its activity, in a synergistic manner (Tipsmark et al., 2011). These studies support an interaction between prolactin and cortisol in controlling ion uptake in fish, although the universality of this model is unclear, and there is little understanding of the cellular pathways that may be involved. Prolactin decreases transcription of liver IGF-I (Tipsmark and Madsen, 2009), providing a possible pathway for how prolactin may interfere with signals that promote salt secretion.

3.2. Growth Hormone/Insulin-Like Growth Factor-1

D. C. W. Smith (1956) observed that multiple injections of GH could increase the capacity of brown trout to tolerate exposure to SW. This was at first attributed to the growth effect of GH because size confers greater salinity tolerance in salmonids. More recently it was found that a single injection of GH in unfed fish was sufficient to increase salinity tolerance, indicating a relatively rapid effect that was independent of body size (Bolton et al., 1987). This effect of GH on salinity tolerance has been found in two other phylogenetically disparate euryhaline species, tilapia and killifish (Sakamoto et al., 1997; Mancera and McCormick, 1998).

A major route of the osmoregulatory action of GH is through its capacity to increase circulating levels and local tissue production of IGF-I. Exogenous treatment of IGF-I has been found to increase the salinity tolerance of rainbow trout, Atlantic salmon, and killifish (McCormick, 2001). GH cannot directly increase NKA activity in cultured gill tissues (McCormick and Bern, 1989), whereas IGF-I can (Madsen and Bern, 1993). The ability of prior GH treatment to increase *in vitro* responsiveness of gill tissue to IGF-I further suggests an indirect action of GH on gill tissue, and a direct action of IGF-I (Madsen and Bern, 1993).

Increased gene expression, secretion, circulating levels, and metabolic clearance rate of GH and IGF-I after exposure to SW provide strong evidence for their hypoosmoregulatory actions in salmonids (Sakamoto et al., 1990, 1993; Sakamoto and Hirano, 1993). Binding proteins may also play a role, as circulating levels of the 21, 42, and 50 kDa IGF-I binding proteins change after SW exposure of rainbow trout (Shepherd et al., 2005). *In vitro*, the tilapia pituitary responds to physiologically relevant elevations of extracellular osmolality with increased GH secretion (Seale et al., 2002), and plasma GH levels have been found to increase after SW exposure (Yada et al., 1994; Breves et al., 2010a). Plasma GH levels also increase in stenohaline FW catfish following exposure to brackish water (Drennon et al., 2003). GH transcription has also been detected in osmoregulatory organs (Yang et al., 1999), opening up the possibility that GH is acting in an autocrine or a paracrine manner in these tissues. IGF-I mRNA levels in liver, gill, and kidney increase following GH injection and exposure to SW in rainbow trout (Sakamoto and Hirano, 1993). Similar responses were seen in the gill of tilapia with increased transcription of IGF-I, IGF-2, and GH receptor, whereas SW caused decreased transcription of these genes in the kidney (Link et al., 2010). IGF-I has been found at higher levels in gill ionocytes than in other cell types in the gill (Reinecke et al., 1997).

High levels of GH receptors as measured by GH binding have been found in the liver, gill, gut, and kidney of euryhaline fishes. The proportion of hepatic GH receptors bound by GH increases following exposure to SW (Sakamoto and Hirano, 1991). GH receptor transcription also has been detected at high levels in osmoregulatory organs, and in the gill is upregulated by environmental salinity in salmonids, Nile tilapia, and flounder (Kiilerich et al., 2007b; Nilsen et al., 2008; Meier et al., 2009; Breves et al., 2010b). Specific high-affinity, high-capacity IGF-I receptors have been found in gill tissue of salmon and tilapia (S. D. McCormick and A. Regish, unpublished results), and have been immunocytochemically localized to gill ionocytes in striped bass, where their transcription is increased by SW exposure (Tipsmark et al., 2007).

The GH–IGF-I and cortisol axes interact to regulate salt secretion in teleosts. Simultaneous treatment with GH and cortisol increases salinity tolerance and gill NKA activity in salmonids and killifish to a greater extent than either hormone alone (Madsen, 1990; Mancera and McCormick, 1998; McCormick, 2001). Cortisol treatment of Atlantic salmon in FW causes an increase in both the FW and SW isoforms of NKA, but treatment with GH and cortisol causes the SW isoform to increase to an even greater extent, and the FW isoform to decrease (S.D. McCormick, unpublished results). These findings suggest that GH is acting as a switch for the effects of cortisol, shifting its actions away from ion uptake and towards salt secretion (Fig. 3.2). At least some of the interaction of GH and cortisol is through GH's capacity to upregulate the number of gill cortisol receptors (Shrimpton and McCormick, 1998), which makes the tissue more responsive to cortisol. Cortisol also increases gill transcription of GH and IGF-I receptors in Atlantic salmon, providing another potential pathway for interaction (Tipsmark and Madsen, 2009).

GH–IGF-I has been shown to regulate several specific ion transporters and other proteins involved in osmoregulation in SW. It has been known for some time that GH and IGF-I can increase the activity of NKA in the

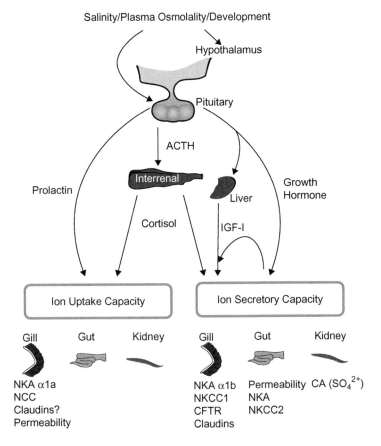

Fig. 3.2. Effect of slow-acting hormones on the ion uptake and salt secretory capacity of euryhaline fishes, emphasizing the interaction of prolactin and cortisol in promoting ion uptake and the interaction of the GH–IGF-I axes and cortisol in promoting salt secretion. The lower part of the figure shows proteins involved in osmoregulation that are known to be under the control of these endocrine systems for gill, gut, and kidney. PrP: prolactin-releasing peptide; GH: growth hormone; ACTH: adrenocorticotropic hormone; IGF-I: insulin-like growth factor-1; NKA: Na$^+$/K$^+$-ATPase; NCC: Na$^+$/Cl$^-$ cotransporter; NKCC: Na$^+$/K$^+$/2Cl$^-$ cotransporter; CFTR: cystic fibrosis transmembrane regulator; CA = carbonic anhydrase. **SEE COLOR PLATE SECTION**

gill, which also increases after SW acclimation in most teleosts (McCormick, 2001). More recently, the abundance and/or transcription of the three transport proteins most directly involved in salt secretion, NKAa1b, NKCC, and CFTR (see Edwards and Marshall, 2013, Chapter 1, this volume) have been found to be increased by the GH–IGF-I axis (Pelis and McCormick, 2001; Tipsmark and Madsen, 2009). These genes are also regulated by cortisol, which may reflect the important interaction between these two endocrine axes. Since cortisol and IGF-I can upregulate many of these targets *in vitro*, it appears that cortisol-dependent and -independent pathways are present. GH also increases the gill transcription of FXYD-11 (Tipsmark et al., 2010a). In the gut, claudins 15 and 25b are upregulated by exposure to SW and GH treatment in FW, but these are inhibited by cortisol (Tipsmark et al., 2010b). These findings suggest that the positive interaction of the GH–IGF-I and cortisol axes that have been found for the gill may not be present in the gut and perhaps the kidney.

GH increases mitotic activity in several cell types in the gill of rainbow trout. Cortisol has no effect on mitotic activity but increases the number of ionocytes, suggesting that cortisol acts primarily to promote their differentiation. Therefore, another pathway for GH–IGF-I and cortisol interaction is stimulation of stem/progenitor cell proliferation by GH and/or IGF-I, creating more stem cells that can then be acted on by cortisol. The GH–IGF-I axis and cortisol may also interact at "higher" regulatory pathways, such as the hypothalamus and pituitary. *In vivo* and *in vitro* exposure to GH increases the sensitivity of interrenal tissue to ACTH, causing increased release of cortisol (Young, 1988). CRH is a potent stimulator of *in vitro* GH release in eels (Rousseau et al., 1999).

To date, a relatively small number of teleosts have been examined for the physiological impact of the GH–IGF-I axis on osmoregulation. Exogenous treatments have been found to affect most salmonids, Mozambique tilapia, and killifish. Convincing evidence for endocrine and paracrine actions of the GH–IGF-I axis comes from circulating hormones, local production, and from salmonids and tilapia, but there is relatively little information in this area from other teleosts, especially stenohaline species, which would offer a contrast to euryhaline models. However, there is no apparent effect of exogenous GH on several osmoregulatory parameters in the euryhaline gilthead sea bream (*Sparus aurata*) (Mancera et al., 2002), and osmoregulatory effects on euryhaline silver sea bream (*Sparus sarba*) are not consistent with an SW acclimating impact (Kelly et al., 1999). Pituitary GH and liver IGF-I mRNA levels in sea bream were lower after exposure to both hypersaline and hyposaline conditions (Deane and Woo, 2004). Similarly, GH may not play an osmoregulatory role in the eel (Sakamoto et al., 1993). Mozambique tilapia can effectively activate branchial ionoregulatory machinery to tolerate SW transfer following hypophysectomy (Breves et al., 2010c), whereas coho salmon (*Oncorhynchus kisutch*) cannot (Björnsson et al., 1987). Even closely related species show differences, as the Nile tilapia that we maintain in our aquarium cannot survive in higher salinity than half strength SW, so we call them stenohaline fish. Nile tilapia increases plasma GH and gill GH receptor transcription after acute SW exposure, which is not observed in the more euryhaline Mozambique tilapia

(Breves et al., 2010b). Species variation linked to phylogeny or life history differences in ion regulatory capacity may have determined whether and to what extent the GH–IGF-I axis is involved in osmoregulation.

3.3. Mineralocorticoids

Cortisol is the major corticosteroid produced in teleost fishes and has a well-established role in salt secretion in euryhaline fishes. The close interaction between cortisol and GH has been detailed above (Section 3.2). In addition to its osmoregulatory function, cortisol plays a role in intermediary metabolism, growth, stress, and immune function (Mommsen et al., 1999). Treatment of a number of euryhaline fishes with cortisol in FW improves their subsequent survival and capacity to maintain low levels of plasma ions after exposure to SW. This effect is due to increases in the size and abundance of gill ionocytes, which have been demonstrated *in vivo* and *in vitro* (McCormick, 2001). Cortisol has also been shown to increase the transcription and abundance of the major transport proteins involved in salt secretion by the gill, especially the SW-type NKAα1b isoform in those species known to contain it, NKCC1 and CFTR (Singer et al., 2003; McCormick et al., 2008; Tipsmark and Madsen, 2009), although it should be noted that these responses are not universal among euryhaline species (Madsen et al., 2007; Tipsmark et al., 2011). Cortisol increases transepithelial resistance and decreases paracellular permeability in the gill, probably working through regulation of specific occludins and claudins (Kelly and Wood, 2002a; Tipsmark et al., 2009; Chasiotis and Kelly, 2011). The effect of cortisol on permeability and tight junction proteins is greater for euryhaline trout than for stenohaline goldfish (Chasiotis and Kelly, 2011). The effect of exogenous cortisol generally requires several days to reach its peak, suggesting that changes in gene expression, cell proliferation, and differentiation are required for its complete action. However, recent studies suggest that some osmoregulatory effects of cortisol may be relatively rapid, less than 1 h (Babitha and Peter, 2010), suggesting a non-genomic action. Rapid, non-genomic actions of corticosteroids in vertebrates are now well established, although the mechanism(s) for these effects, including the presence of a corticosteroid membrane receptor, remain(s) controversial (Losel and Wehling, 2008).

In the intestine, exogenous cortisol stimulates ion and water absorption, thus improving acclimation to high environmental salinity (Hirano and Utida, 1968; Cornell et al., 1994; Veillette et al., 1995). Specific ion transporters that are upregulated by cortisol in the intestine include NKA, NKCC2, and aquaporin 1 and 3 (Seidelin et al., 1999; Martinez et al., 2005; Veillette and Young, 2005; Cutler et al., 2007). Intestinal expression of claudins 15 and 25b in Atlantic salmon is upregulated by SW exposure but, surprisingly, inhibited by cortisol treatment in FW (Tipsmark et al., 2010b). These authors suggest that contact with imbibed SW may be necessary for full induction of intestinal transport capacity. An increased drinking response after SW transfer has been observed in salmonids treated with cortisol in FW (Fuentes et al., 1996), although whether this is a direct effect on the brain or indirect through other endocrine pathways has not been determined.

Surprisingly little work has been done on the impact of salinity on the kidney. FW exposure resulted in increased NKA activity in the euryhaline mullet (*Chelon labrasus*), but no effect of cortisol was found (Gallis et al., 1979). Cortisol treatment in North African catfish (*Clarias gariepinus*) caused an increased NKA activity in the short term (20 min) but decreased it in the long term (5 days). In an *in vitro* preparation of renal proximal tubules of winter flounder, cortisol significantly increased carbonic anhydrase activity and sulfate secretion (Pelis et al., 2003).

Changes in circulating cortisol in response to increased environmental salinity are reported for many teleost species (Mommsen et al., 1999). The clearance rate of cortisol also increases in SW, suggesting increased utilization by osmoregulatory target tissues. The release of cortisol from the interrenal is primarily controlled by ACTH, although other endocrine factors may also be involved (see Section 2, this chapter). Although there is evidence for salinity activation of pituitary ACTH cells *in vitro*, salinity effects on circulating levels of ACTH have not been detected. ACTH production by isolated pituitary does not appear to be directly responsive to changes in osmolality (Seale et al., 2002). The increase in cortisol during osmotic stress occurs in both stenohaline and euryhaline fishes and may be part of a general stress response. Thus, the regulation of cortisol receptors may represent a critical component of osmoregulation in euryhaline fishes.

The classical signaling action of steroids begins with transport/diffusion into the cell, followed by binding to a cytosolic receptor, which is then translocated into the nucleus. There, the steroid/receptor complex binds to specific genes to increase or decrease their expression. Several studies on cortisol binding in fish tissues have found evidence for only a single class of corticosteroid receptors (CRs) present in high concentrations in gill, gut, and kidney (Mommsen et al., 1999). More recently, two isoforms with differing isoelectric points have been found in gill tissue of the eel, and these are differentially regulated by salinity (Marsigliante et al., 2000a). During exposure to increased salinity, intracellular cortisol and CR levels in the gill shift from the cytosol to the nucleus, indicative of CR binding and translocation (Weisbart et al., 1987). Consistent with direct osmoregulatory action, high concentrations of CR have been found in gill ionocytes (Uchida et al., 1998).

In the past several years, molecular techniques have demonstrated the presence of two homologues of the mammalian glucocorticoid receptor (GR) and one homologue of the mineralocorticoid receptor (MR) in several teleost species (Bury and Sturm, 2007). The two isoforms of fish "GR-like" genes have different activation affinities for cortisol (Greenwood et al., 2003; Stolte et al., 2006). In addition, at least one cichlid species (*Haplochromis burtoni*) has splice variants of GR2 that have different tissue distributions and cortisol transactivation characteristics (Greenwood et al., 2003). Expression of fish MRs in mammalian cell lines indicated high binding and transactivation efficiency for both aldosterone and 11-deoxycorticosterone (DOC), similar to the binding characteristics of the mammalian MR (Sturm et al., 2005; Stolte et al., 2008). The divergent binding and expression patterns of the GRs and MR in fish suggest different physiological functions, although these have yet to be established.

It has been suggested that DOC, present in the plasma of some teleosts at levels that could activate the fish MR, might be a second mineralocorticoid in fish (Prunet et al., 2006). Injection studies indicate that DOC cannot carry out the SW-adapting functions of cortisol and that cortisol (but not DOC) stimulated both the FW- and SW-dependent NKA isoforms (McCormick et al., 2008). *In vitro* studies indicate that DOC and cortisol have distinct effects on gill transport proteins that vary with salinity, species, and developmental stage (Kiilerich et al., 2011b, c). In these studies cortisol and DOC had similar effective concentrations. Since DOC is present at much lower concentrations than cortisol and does not respond to changes in environmental salinity (Kiilerich et al., 2011a), it seems unlikely that DOC is involved in osmoregulation, at least in rainbow trout. This is supported by studies on an *in vitro* gill preparation of rainbow trout in which cortisol but not DOC increased transepithelial resistance, and both GR and MR antagonists were required to completely block the actions of cortisol (Kelly and Chasiotis, 2011). It should also be noted that there are high mRNA levels of the 11-β hydroxysteroid (including corticoids) metabolizing enzyme genes present in gill tissue, which may have a role in regulating intracellular corticosteroid actions (Nilsen et al., 2008). To date, the weight of evidence indicates that cortisol carries out all or most of the osmoregulatory effects of corticosteroids and acts primarily through a GR, but with some effects occurring through the MR.

Cortisol has been regarded as a SW-acclimating hormone in a large number of teleost species, but there is increasing evidence that cortisol is also involved in ion uptake, indicating that it has dual osmoregulatory functions. Plasma cortisol levels decrease following transfer of salmonids from SW to

FW or after exposure of FW fishes to ion-poor FW (McCormick, 2001). Cortisol treatment of a number of teleost species held in FW increases the surface area of gill ionocytes and the influx of Na^+ and Cl^- (Perry et al., 1992). Survival and plasma ion levels of FW fish that have had their pituitary removed are increased by treatment with ACTH, which can be presumed to be acting through its stimulation of cortisol release from the interrenal (McCormick, 2001). Cortisol is also required to maintain water movement across the gut of FW eels. Cortisol treatment significantly increases the ion regulatory capacity of marine fishes during exposure to low salinity (Mancera et al., 1994) and the ability of acid-resistant fishes to maintain plasma Na^+ levels after exposure to acidic water (Yada and Ito, 1999). Cortisol also upregulates transcription and protein abundance of the FW-dependent NKAα1a isoform in Atlantic salmon gills (Kiilerich et al., 2007a; McCormick et al., 2008). Since cortisol also upregulates the SW-dependent NKAα1b isoform in Atlantic salmon, this is further evidence of a dual osmoregulatory role of cortisol. At least in zebrafish, cortisol also plays an important role in Ca^{2+} balance, and regulates the expression of an epithelial calcium channel (TRPV6) to support active Ca^{2+} uptake by ionocytes (Lin et al., 2011). These studies provide evidence that in at least some teleosts cortisol has a physiological role in acclimation to FW and ion-poor environments. This function of cortisol has not been fully appreciated owing to an emphasis on the role of cortisol in salt secretion.

3.4. Thyroid and Sex Steroid Hormones

There is conflicting evidence regarding the role of thyroid hormones (T_3 and T_4) in osmoregulation, but most studies suggest that they have an indirect role in regulating ion uptake or secretory capacity (McCormick, 2001). Prolonged treatment with T_4 or T_3 accelerates smolt-related increases in gill ionocytes in Atlantic salmon with variable effects on gill NKA activity. Physiological levels of exogenous T_4 and T_3 in Mozambique tilapia result in increased ionocyte size, gill NKA activity, and plasma Na^+ and Cl^- levels, suggesting that thyroid hormones may have a role in ion uptake in this species (Peter et al., 2000). T_3 treatment altered the distribution of gill ionocytes and increased gill NKA activity in FW- and SW-acclimated air-breathing fish *Anabas testudineus* (Peter et al., 2011). Thyroid hormones play at least a supportive role in SW acclimation, and may interact with both the GH–IGF-I and cortisol axes. Inhibition of the thyroid axis with thiourea in killifish caused increased plasma ions in SW but had no effect in FW (Knoeppel et al., 1982). T_4 treatment alone has no effect, but potentiates the action of cortisol on gill NKA activity in Mozambique tilapia (Dange, 1986), and the action of GH on gill NKA activity in Atlantic salmon

(McCormick, 2001). Inhibiting the conversion of T_4 to T_3 interferes with normal and GH-induced SW acclimation in rainbow trout. T_3 treatment increases the number of gill cortisol receptors in trout and salmon (Leloup and Lebel, 1993). Thyroid hormones thus appear to exert their influence on salt secretory mechanisms primarily through an interaction with cortisol and the GH–IGF-I axis.

Sex steroids have been found to have a negative impact on salinity tolerance in salmonids and tilapia (Madsen et al., 1997; Vijayan et al., 2001). Estrogenic compounds have been found to decrease circulating IGF-I levels and gill NKA activity (McCormick et al., 2005). Early developmental exposure can have effects on salinity tolerance a year after exposure, suggesting a possible epigenetic effect (Lerner et al., 2007). The effects of estrogens and androgens in salmonids may be related to their anadromous life history in which departure from the ocean is associated with maturation and elevated androgens and estrogens. In sockeye salmon increases in gill NKAα1a mRNA levels are observed as fish move from the open ocean to coastal waters but before they enter FW, suggesting a preparation for FW entry (Shrimpton et al., 2005). A more complete discussion associated with migration in adult anadromous fishes can be found in Chapter 7 of this volume (Shrimpton, 2013).

4. TARGET TISSUES

Euryhaline fishes maintain water and ion balance in either FW or SW by modulating water and ion exchange at each osmoregulatory organ. As mentioned above, various rapid-acting and slow-acting hormones work in concert to regulate water and ion balance by modulating drinking, intestinal absorption, branchial fluxes, and renal excretion during the whole process of acclimation to changing environmental salinity. This section provides a summary of the hormonal regulation of water and ion trafficking at each osmoregulatory organ.

4.1. Brain Control of Drinking

Drinking is usually suppressed in FW fishes because water enters the body by osmosis and fish are exposed to a constant threat of over-hydration (Takei, 2002; Takei and Balment, 2009). Reflecting the dominance of inhibitory mechanisms in fish, most hormones that regulate drinking are anti-dipsogenic hormones (Ando et al., 2003; Kozaka et al., 2003), except for Ang II, AM2, and IT, as mentioned above. Importantly, the dipsogenic potency of Ang II in teleosts is 1/100th that of mammals but anti-dipsogenic potency of ANP is 1000-fold more potent in teleosts than in mammals (Takei, 2002). However, drinking is as important in SW fishes as in terrestrial animals because it is the major route for obtaining water that has been passively lost to the environment. In fact, eels with esophageal fistulae die of cellular and extracellular dehydration within 5 days after SW transfer if ingested water is not reintroduced into the stomach (Takei et al., 1998b). Prior to death these eels drank at a rate twice that of intact fish, probably owing to severe hypovolemia and increased plasma Ang II (Takei and Balment, 2009). It is possible that Ang II is involved in the immediate drinking after SW exposure as an ACE inhibitor, SQ14225 (captopril), abolished the immediate drinking induced by SW transfer (Tierney et al., 1995), although the effect of captopril could be due to increased plasma concentration of antidipsogenic bradykinin (Takei and Tsuchida, 2000). It remains to be determined whether the immediate drinking is caused simply by a neural reflex of swallowing triggered by activation of a Cl^- sensor (Hirano, 1974) or by some fast-acting hormones.

The hormone that suppresses excess drinking in SW may be ANP, as its removal from plasma increased drinking in SW eels (Tsukada and Takei, 2006). However, it is possible that ghrelin, another potent antidipsogenic hormone in eels (Kozaka et al., 2003), is secreted from the stomach in response to the increased luminal osmolality or distension by SW drinking and acts on the area postrema to inhibit drinking (Nobata and Takei, 2011). These hormones, together with the nervous signal from the distended stomach, seem to suppress excess drinking and promote SW acclimation (Takei and Hirose, 2002). However, the interaction with other inhibitory hormones in salinity acclimation in euryhaline fishes remains to be determined.

4.2. Intestinal Ion and Water Absorption

As the lumen of the digestive tract is a kind of external environment within the body, water and ions in the lumen become body fluid for the first time after absorption by the intestine. When fish are in ion-poor FW, the intestine actively absorbs ions from food but not water. When they are in SW, ions must also be absorbed in order to obtain water. To achieve this, ingested SW is processed during passage through the digestive tract and water is absorbed in association with monovalent ions when luminal fluid becomes isotonic to the plasma. The initial step of this process is removal of NaCl (desalting) by the esophagus (Hirano and Mayer-Gostan, 1976; Parmelee and Renfro, 1983), but nothing is known about its hormonal regulation. As is the case for regulation of drinking, several inhibitory hormones have been identified for intestinal absorption in SW fishes.

However, no hormone has been elucidated thus far that enhances intestinal absorption of water and ions when administered in isolation *in vitro*, although somatostatin, neuropeptide Y, and catecholamines slightly restore the absorption if it has been suppressed by pretreatment with inhibitory hormones (Ando et al., 2003). Hormones that act alone to facilitate water and ion uptake have not been identified in mammals either, as the absorption is maximally activated to maintain water and ion balance in the desiccative terrestrial environment. The lack of information about the facilitative hormones in fish may be partly because experiments on the intestinal absorption have been conducted in SW fishes in relation to water absorption for SW acclimation. To identify hormones that facilitate ion absorption, FW fish intestine may be a good model as ion absorption is usually suppressed to avoid excess water absorption.

It has been shown that NKCC, NCC, CFTR, and anion exchanger (AE) on the apical side of epithelial cells play major roles in ion absorption by the intestine of SW fishes (Marshall and Grosell, 2006; Edwards and Marshall, 2013, Chapter 1, this volume). There is accumulating evidence showing that active secretion of HCO_3^- into the lumen by the intestinal epithelial cells precipitates $Mg/CaCO_3$ and decreases these ions, thereby decreasing osmolality and further facilitating water absorption (Grosell et al., 2009; Wilson et al., 2009). AE secretes HCO_3^- into the lumen in exchange of Cl^-, which facilitates water absorption. However, if NKCC, NCC, and AE transport Cl^- into the epithelial cells, Cl^- ions in the luminal fluid become deficient, which halts the operation of NKCC and NCC. Therefore, CFTR may supplement Cl^- in the luminal fluid as mentioned above.

It has been shown that ANP inhibits NKCC to decrease NaCl absorption in the intestine of winter flounder (O'Grady et al., 1985). Guanylin stimulates CFTR to secrete Cl^- and HCO_3^- into the lumen of SW-acclimated eels (Yuge and Takei, 2007). Both ANP and guanylin utilize cGMP as an intracellular messenger, although ANP acts from the basolateral side and guanylin from the luminal side. Thus, it is likely that both hormones regulate the same transporters, but only guanylin induces Cl^- secretion as judged by the reversal of short-circuit current. Guanylin may also inhibit NKCC, but ANP does not seem to activate CFTR for Cl^- secretion. It has been suggested that in mammalian intestine guanylin-induced cGMP production not only activates protein kinase G type II but also inhibits phosphodiesterase type III to increase cAMP, and finally activates protein kinase A (Sindić and Schlatter, 2006). This result coincides with the observations that hormones that increase intracellular cAMP such as VIP are inhibitory to NaCl absorption and those that inhibit cAMP production such as somatostatin reverse the inhibitory effect in the teleost intestine.

4.3. Renal Regulation

In terrestrial animals, including mammals, the kidney is the primary organ for body fluid regulation, particularly renal tubules where various hormones are involved in the regulation of ion and water reabsorption. In teleosts, however, urine volume is primarily determined by GFR, and glomerular intermittency (shutting off of some glomeruli) occurs to decrease urine volume by vascular actions. There are some hormones that alter water and NaCl excretion by the kidney, but the changes are mostly ascribed to the glomerular effects. Furthermore, the kidney plays less important roles in volume regulation in SW fishes than in FW fishes. Probably reflecting this difference, ANP exerts only a small antidiuretic effect in SW eels (Takei and Kaiya, 1998) but induces profound diuresis and natriuresis in FW rainbow trout (Duff and Olson, 1986). In addition, Ang II and VT are only mildly antidiuretic via glomerular effects in teleost fishes.

The SW teleost kidney is the site of excretion of divalent ions (Mg^{2+}, Ca^{2+}, and SO_4^{2-}), and these ions are secreted into the proximal tubular lumen via various transporters (Beyenbach, 1995; Renfro, 1999; Edwards and Marshall, 2013, Chapter 1, this volume). Mg^{2+} and SO_4^{2-} concentrations in SW are 50- and 30-fold higher, respectively, than those in plasma. Passive influx is generally negligible but unavoidable influx is balanced by the active secretion in the kidney. Recently, the transporters involved in SO_4^{2-} uptake in FW fishes and SO_4^{2-} excretion in SW fishes have been identified and localized in the kidney: FW eel (Nakada et al., 2005), SW euryhaline pufferfish (*Takifugu obscures*) (Kato et al., 2009), and SW eel (Watanabe and Takei, 2011b). Cortisol has been shown to increase SO_4^{2-} excretion in an *in vitro* preparation of the renal proximal tubule of winter flounder (Pelis et al., 2003). It is likely that renal regulation of divalent ions is of primary importance for acclimation in both FW and SW (euryhalinity), but little else is known about the hormonal regulation of divalent ions in the kidney of fish.

4.4. Branchial Regulation

It is obvious that the gills are a critical osmoregulatory organ in fishes as this organ is directly exposed to environmental water with monolayer respiratory epithelia and thus serves as a window for communication with the environment. Several slow-acting hormones (GH/IGF-I and cortisol) have been implicated in the transformation of ionocytes between an absorptive FW type and an excretory SW type, but the information about the role of rapid-acting hormones in acute regulation of ion fluxes is still limited. It has been reported that VT and IT inhibited cAMP production in the gill cell membranes from rainbow trout (Guibbolini and Lahlou, 1987), and ANP and UII increase Cl^- secretion in the killifish opercular membrane, probably via ionocytes (Marshall and Bern, 1981; Scheide and Zadunaisky, 1988). However, the effects are controversial in other species and remain to be established. As several rapid-acting hormones stimulate GH and/or cortisol secretion, they may have indirect actions on the gills to promote differentiation of ionocytes between FW and SW type, as mentioned above.

As noted in detail above (Sections 3.2 and 3.3), cortisol and the GH/IGF-I axis interact to promote the differentiation of salt-secreting ionocytes in the gill, including the upregulation of the three major transport proteins involved: NKA, NKCC1, and CFTR. This interaction is determined at least in part by the effect of GH on cortisol receptors and responsiveness of the interrenal to ACTH. Prolactin and cortisol are each involved in ion uptake, and act to promote the differentiation of FW ionocytes, and at least some of the transporters involved, such as the apical NCC. There is increasing evidence of an interaction between cortisol and prolactin in regulating gill ion uptake, although the universality and mechanism of this interaction remain to be established.

5. DEVELOPMENTAL (ONTOGENIC) ASPECTS

Diadromous fishes are amphihaline, but the degree of euryhalinity or adaptability to different salinity environments changes during development. During the parr–smolt transformation of salmonids or the "silvering" stage of eels, there are large increases in SW tolerance in association with downstream migration (McCormick, 2013; Zydlewski and Wilkie, 2013, Chapters 5 and 6, this volume). Salmonids and eels have been widely used for the study of osmoregulation and euryhalinity, and it is often found that the endocrine regulation of osmoregulation differs between the species. This may derive from the fact that salmonids probably have their evolutionary origins in FW and are anadromous, whereas eels have an SW origin and are catadromous. Salmonids have been widely used as a model for developmental changes in hypoosmoregulatory ability, and differences in the timing of smolt development provide useful comparative approaches. Smolting appears to be a "pan-hyperendocrine" event, and details of the hormones involved can be found in Chapter 5 of this volume (McCormick, 2013). Eels can be used for the developmental aspect of hyperosmoregulatory ability, but research on the complete aquaculture of eels from eggs has yet to be completed (Tanaka et al., 2003).

Medaka (genus *Oryzias*) are mostly non-migratory fishes of FW origin, but there are a variety of species that exhibit distinct adaptability to SW (Inoue and Takei, 2002). *Oryzias mormoratus* is a species endemic to a highland lake of Sulawesi Island, Indonesia, and fertilized eggs and adult fish cannot survive in more than half-strength SW. *Oryzias javanicus* is distributed around the coastal area of the Indonesian Islands and Malay Peninsula, and even fertilized eggs develop normally in SW. *Oryzias latipes* lives in the rice fields of Japan and is intermediate in terms of salinity tolerance. It cannot survive direct transfer from FW to SW but can live in double-strength SW if acclimated gradually. The eggs can be fertilized in SW and develop normally until hatching (Inoue and Takei, 2002). It has been shown that the yolk-sac membrane of euryhaline Mozambique tilapia embryos has many ionocytes, which are transformed from FW-type to complex SW-type after transfer of embryos to SW (Hiroi et al., 1999). As embryos cannot gain water by drinking and intestinal absorption, they seem to limit water efflux from the body surfaces and produce sufficient oxidative water by active yolk metabolism to balance the obligatory osmotic water loss.

It is interesting to examine how medaka embryos cope with dehydration during early developmental stages and which hormones are involved in the development of adaptability. For this purpose, a knockdown technology using antisense oligonucleotide (gripNA) was applied to BNP, as medaka have only BNP as cardiac NPs and therefore ANP and VNP cannot compensate for the BNP function in this species (see Section 2.2, this chapter). The BNP gene starts to express in the primordial ventricular tissue at 48 h postfertilization (hpf) and its receptor (OLGC7) gene transcripts appear even earlier, at 10 hpf (Miyanishi et al., 2013a). CNP3, which is an ancestral molecule of cardiac NPs but usually synthesized in the extracardiac tissue, is expressed in the developing atrium at 35 hpf and its receptor (OLGC2) at 10 hpf. After knockdown of the BNP gene, no change was observed in the developing embryos, but a double knockdown of the BNP and OLGC7 genes severely impaired the normal ventricular development (Miyanishi et al., 2013a). By contrast, knockdown of the CNP3 gene alone caused abnormal atrial development. These results clearly show that BNP and CNP3 are important not only for osmoregulation but also for normal development of cardiac tissues, and that CNP3 compensates for the function of ANP for atrial development in medaka embryos. Furthermore, knockdown of BNP or CNP3 gene increased body fluid osmolality of embryos kept in SW compared to controls (Miyanishi et al., 2013b). The increase was found to be due to impaired blood flow to the yolk-sac membrane, which leads to malfunction of NaCl secretion by ionocytes and reduced yolk metabolism for metabolic water production. However, expression of the major transporter genes involved in NaCl secretion by ionocytes (NKA, NKCC1a, and CFTR) and expression of the key metabolic enzymes for energy metabolism were not suppressed by the knockdown. The increase in body fluid osmolality was further exaggerated in the later stage of development after CNP3 knockdown, which was due to the failure to suppress the expression of aquaporin (AQP3, 4, and 9) genes by the loss of CNP3 (Miyanishi et al., 2013b) as observed in the ANP/NPR-A knockout mouse (Kishimoto et al., 2011).

6. EVOLUTIONARY (PHYLOGENETIC) ASPECTS

6.1. Cyclostomes (Lampreys)

Early vertebrates are thought to have evolved in the SW environment near the coast, but basal fishes might have once entered the FW environment and then re-entered the sea (Carroll, 1988; Schultz and McCormick, 2013, Chapter 10, this volume). However, one of the two most basal extant vertebrates, hagfishes, does not appear to have experienced FW during their evolutionary history, as judged by their simple kidney structure. Therefore, they live in the deep sea and are strictly stenohaline with a plasma ion composition almost identical to SW except for divalent ions. By contrast, lampreys are FW or anadromous species that have an osmoregulatory strategy similar to teleosts. Such large differences in basic physiology support the idea that the two cyclostome species diverged long ago in vertebrate phylogeny. Nonetheless, the mechanisms of euryhalinity and their hormonal regulation in lampreys may provide us with a prototype of the earliest aspects of osmoregulatory control in vertebrates.

Embryonic anadromous lamprey hatch in rivers and are known as ammocoetes in their FW stage; they cannot survive in water with salinity higher than their body fluids (Reis-Santos et al., 2008). However, metamorphosing juveniles (transformers) obtain excellent euryhalinity and can maintain low plasma osmolality even in SW. Transformers have much more abundant (approximately 10-fold) NKA activity in the gills than do ammocoetes, which may in part explain their SW adaptability, while H^+-ATPase and carbonic anhydrase may be responsible for ion uptake in FW (Reis-Santos et al., 2008). Several rapid-acting, peptide hormones such as VT (Suzuki et al., 1995), Ang II (Wong and Takei, 2011), NPs (Kawakoshi et al., 2006), and AMs (Wong and Takei, 2009) have been identified in lampreys, but their osmoregulatory functions have not been examined yet.

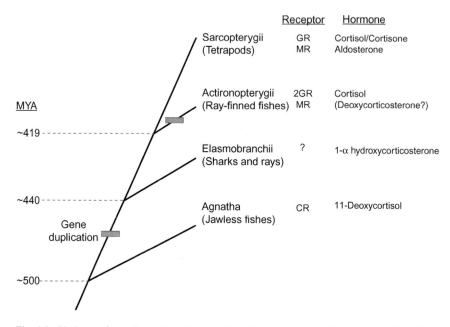

Fig. 3.3. Phylogenetic analysis of corticosteroids and their receptors in vertebrates (modified from Close et al., 2010). Blue boxes indicate hypothesized whole genome duplication events. MYA, million years ago; CR, corticosteroid receptor; GR, glucocorticoid receptor; MR, mineralocorticoid receptor.

It has recently been determined that 11-deoxycortisol is the major corticosteroid of lamprey and has both osmoregulatory and metabolic functions (Close et al., 2010). Lampreys appear to have a single CR, consistent with their origination prior to the gene duplication event of other vertebrates (Fig. 3.3). Heterologous expression studies suggested that this receptor can respond to a variety of corticosteroids (Bridgham et al., 2006), but binding and physiological studies indicate that the receptor is highly specific for 11-deoxycortisol (Close et al., 2010). Unlike other metamorphic events, thyroid hormones are inhibitory to the metamorphosis of ammocoetes up to the parasitic transformer stage (Youson, 2003). It will therefore be of interest to examine the relationship between 11-deoxycortisol and thyroid hormone during events when salinity tolerance develops.

Lampreys apparently have only one protein hormone of the GH-PRL-SL family (Kawauchi et al., 2010). This has been suggested to be GH based on its sequence similarity with other vertebrate GHs and its ability to induce IGF-I production. To date, there is no published information on the possible role of this hormone or its receptor in osmoregulation.

6.2. Elasmobranchs

Although most elasmobranchs are stenohaline marine species, a number of euryhaline species migrate between FW and SW (Ortega et al., 2009; Evans et al., 2010) or even live wholly in FW (Ballantyne and Fraser, 2013, Chapter 4, this volume). Elasmobranchs and holocephalans display an osmoregulatory strategy in which they accumulate urea and trimethylamine oxide in their body fluids to counter osmotic water loss in SW. Euryhaline elasmobranchs retain NaCl and urea in plasma at high concentrations after acclimation to FW (Piermarini and Evans, 1998; Pillans and Franklin, 2004), which differs from stenohaline SW species such as *Scyliorhynus canicula*, in which transfer to diluted SW decreases plasma NaCl and urea concentrations linearly (Wells et al., 2002). Stenohaline FW rays in the Amazon River (*Potamotrygon* spp.) have lost plasma urea and maintain plasma NaCl at much lower concentration than that of euryhaline species in FW (Ballantyne and Robinson, 2010). Therefore, it appears that the ability to maintain high plasma NaCl and urea concentrations in low-salinity media is the key to euryhalinity in elasmobranchs.

It has been shown that maintenance of plasma NaCl at a concentration lower than SW is achieved by rectal gland secretion (Piermarini and Evans, 2000) and that high plasma urea concentration is maintained by active synthesis of urea in the liver and skeletal muscle (Kajimura et al., 2006) and facilitated reabsorption of urea by the collecting tubule of the kidney (Hyodo et al., 2004a; Yamaguchi et al., 2009). Elasmobranch gills also play important roles in body fluid regulation as a site of ion uptake in FW and as a barrier against the loss of ammonia and urea (Hazon et al., 2003). Various transporters have been identified in these osmoregulatory organs (Silva et al., 1997; Piermarini et al., 2002; Hyodo et al., 2004a). Detailed accounts of transporters involved in elasmobranch osmoregulation can be found in Chapter 4 of this volume (Ballantyne and Fraser, 2013), and thus only hormonal regulation will be described in detail here.

There are relatively few studies on the hormonal control of elasmobranch osmoregulation, and thus only few firm conclusions are possible (Good and Hazon, 2009). Plasma Ang II concentration increased transiently 12 h after transfer of bull shark from FW to 75% SW (Anderson et al., 2006). Ang II has been shown to participate in various aspects of osmoregulation in elasmobranchs (Hazon et al., 1999). Ang II is dipsogenic in *Triakis scyllium*, and pharmacological activation or inhibition of the RAS increased or decreased drinking rate, respectively, in *S. canicula* (Anderson et al., 2001). Dense Ang II binding sites have been identified in the *Triakis* gill (Tierney et al., 1997). More recently, an Ang II receptor (AT1-like) has been identified in the euryhaline Atlantic stingray (*Dasyatis sabina*), the gene of which is most abundant in the interrenal tissue, followed by the kidney, gills, and rectal gland (Evans et al., 2010). Ang II infused into the *in situ* perfused trunk preparation of *S. canicula* caused glomerular antidiuresis with a parallel decrease in perfusate flow, indicating constriction of preglomerular afferent arteries (Wells et al., 2006). Ang II had no effect on rectal gland secretion in *S. canicula* (Anderson et al., 2002).

CNP3 is the sole member of the NP family in elasmobranchs and its gene is expressed in both the brain and heart (Kawakoshi et al., 2001). Plasma CNP3 concentration does not change for 4 days after transfer of bull shark from FW to 75% SW but it is significantly higher in SW-acclimated fishes (Anderson et al., 2005, 2006). The CNP3 receptor (NPR-B) gene is expressed most abundantly in the rectal gland, followed by the kidney and interrenal gland of Atlantic stingray (Evans et al., 2010). CNP3 is a powerful stimulator for rectal gland secretion in *S. acanthias* (Solomon et al., 1992) and *S. canicula* (Anderson et al., 2002). In contrast to Ang II, CNP3 induced glomerular diuresis in the perfused trunk preparation of *S. canicula* (Wells et al., 2006).

Hypothalamic VT gene expression is enhanced in *T. scyllium* after transfer of fish from SW to 130% SW, with a parallel increase in plasma VT concentration (Hyodo et al., 2004b). Plasma VT concentration also increased 3 days after transfer of bull shark from FW to 75% SW (Anderson et al., 2006). VT infused into the *in situ* perfused trunk preparation caused glomerular antidiuresis with a parallel decrease in perfusate flow, indicating constriction of preglomerular afferent arteriole as observed with Ang II (Wells et al., 2002). In addition, VIP has been reported to act as a potent secretagogue in the rectal gland of *S. acanthias* (Stoff et al., 1979). A stimulatory gut peptide for rectal gland secretion was first reported in *S. canicula* and named rectin (Shuttleworth and Thorndyke, 1984). Anderson et al. (1995) purified and sequenced the possible rectin from the intestine of *S. canicula* and showed this factor to be the previously identified intestinal tachylinin, scyliorhinin II. The possible role of these gut peptides in elasmobranch osmoregulation has been reviewed recently (Good and Hazon, 2009; Takei and Loretz, 2011).

A single, unique corticosteroid, 1α-hydroxycorticosterone, appears to be the major corticosteroid of elasmobranchs (Turscott and Idler, 1968). Receptors to this hormone are present in the gills and rectal gland (Hazon et al., 2003). Surgical removal of the interrenal in the ray (*Raja ocellata*) resulted in reduced ion secretion by the rectal gland which was restored by 1α-hydroxycorticosterone (Holt and Idler, 1975), suggesting an osmoregulatory role for this corticosteroid. To date, there is no published information on how circulating levels of this hormone change in response to external salinity, or its mechanism(s) of action in osmoregulatory tissues. Similarly, there is little information on how pituitary hormones may be involved in osmoregulation in elasmobranchs. Hypophysectomy results in reduced branchial water permeability and urine flow; branchial permeability is restored by both ACTH and prolactin treatment, whereas only ACTH restores urine flow (Payan and Maetz, 1971). Much remains to be investigated for hormonal control of the unique osmoregulation in cartilaginous fishes.

7. CONCLUSIONS AND PERSPECTIVES

Fish have the greatest species diversity among vertebrates and have acclimated to almost all aquatic habitats. Teleosts are a favored food item and often cultured in a wide variety of salinities, so it is not surprising that most of the research on the hormonal control of osmoregulation has been performed in euryhaline teleost species. However, our understanding of euryhalinity will be broadened by an increased understanding of the physiological mechanisms of ionoregulation in stenohaline FW and marine species for comparison with those of euryhaline species. In addition, migratory lampreys have a similar life history and osmoregulatory strategy to anadromous salmon, and comparison of the major hormones that allow them to adapt to both FW and SW should give us important clues to the evolutionary changes in hormone function. Similarly, increased understanding of the hormonal control of osmoregulation in other basal fishes such as holocephalans and chondrosteans will help to fill the large knowledge gaps in our understanding of the evolution of euryhalinity.

In the study of euryhalinity, the regulation of monovalent ions (Na^+ and Cl^-) has been the major focus as these ions are the dominant extracellular ions that change quickly after transfer between FW and SW. However, the concentration gradient of divalent ions, Mg^{2+} and SO_4^{2-}, between plasma and the external environment is even greater than that of Na^+ and Cl^-. Therefore, amphihaline fishes must alter the way in which divalent ions are regulated even more drastically when they migrate between FW and SW (Watanabe and Takei, 2011a). In the teleost kidney, cortisol has been shown to stimulate SO_4^{2-} secretion by stimulation of carbonic anhydrase (Pelis et al., 2003), and SO_4^{2-} transporters involved in reabsorption in FW and excretion in SW have been identified (Nakada et al., 2005; Kato et al., 2009; Watanabe and Takei, 2011b). Determining the endocrine pathways that regulate renal transporters is a critical next step of investigation.

There are closely related species within the same genus whose genome and phenotype are highly similar but that have very different degrees of euryhalinity. These include the genus *Oreochromis* (the stenohaline Nile tilapia and the euryhaline Mozambique tilapia), *Oryzias* (the stenohaline *marmoratus* medaka and the euryhaline *javanicus* medaka), anadromous salmonids that migrate at different times after hatching, and *Fundulus* (stenohaline marine and FW killifish and euryhaline mummichog; see Griffith, 1974). Migratory and landlocked salmonids of the same species also exist. Genome projects in some of these species are ongoing, and the recent development of next generation sequencing will accelerate the completion of the projects. Comparison of hormone genes and their promoter regions between the euryhaline and stenohaline species (such as occurrence of mutation in the osmoregulatory hormone genes) will enable us to gain new insights into the key hormones for euryhalinity. Furthermore, transcriptome analysis using next generation sequencers to compare the gene expression profiles after salinity challenges between euryhaline and stenohaline species will elucidate the major genes responsible for rapid and slow responses including hormone genes (Whitehead et al., 2011). Proteomics and other approaches that allow the accurate measurement of small quantities of hormone transcripts should also provide a needed link between genomics and the more physiologically relevant hormone and protein production. There have been many recent advances in imaging techniques allowing analysis of ion movements, hormone secretion, and endocrine-driven differentiation of cell types within osmoregulatory tissues. We expect that the application of new technologies combined with classic endocrine approaches to fish studies will open a new field of research for understanding the hormonal control of euryhalinity.

ACKNOWLEDGMENTS

We thank Keigo Kakumura, Tara Duffy, and Andrew Weinstock for their help with drawing Figs 3.1, 3.2, and 3.3 respectively. We also thank all of the past and present members of our labs for their contributions to the research cited here and our many discussions over the years that have contributed to the ideas presented in this chapter.

REFERENCES

Albiston, A. L., McDowall, S. G., Matsacos, D., Sim, P., Clune, E., Mustafa, T., Lee, J., Mendelsohn, F. A. O., Simpson, R. J., Connolly, L. M. and Chai, S. Y. (2001). Evidence that the angiotensin IV (AT_4) receptor is the enzyme insulin-regulated aminopeptidase. *J. Biol. Chem.* 276, 48623–48626.

Amer, S. and Brown, J. A. (1995). Glomerular actions of arginine vasotocin in the *in situ* perfused trout kidney. *Am. J. Physiol.* 269, R775–R780.

Anand-Srivastava, M. B. (2005). Natriuretic peptide receptor-C signaling and regulation. *Peptides* 26, 1044–1059.

Anderson, W. G., Conlon, J. M. and Hazon, N. (1995). Characterization of the endogenous intestinal peptide that stimulates the rectal gland of *Scyliorhinus canicula. Am. J. Physiol.* 268, R1359–R1364.

Anderson, W. G., Takei, Y. and Hazon, N. (2001). The dipsogenic effect of the renin–angiotensin system in elasmobranch fish. *Gen. Comp. Endocrinol.* 125, 300–307.

Anderson, W. G., Good, J. P. and Hazon, N. (2002). Changes in secretion rate and vascular perfusion in the rectal gland of the European lesser spotted dogfish (*Scyliorhinus canicula* L.) in response to environmental and hormonal stimuli. *J. Fish Biol.* 60, 1580–1590.

Anderson, W. G., Hyodo, S., Tsukada, T., Meischke, L., Pillans, R. D., Good, J. P., Takei, Y., Cramb, G., Franklin, C. G. and Hazon, N. (2005). Sequence, circulating levels, and expression of C-type natriuretic peptide in a euryhaline elasmobranch, *Carcharhinus leucas. Gen. Comp. Endocrinol.* 144, 90–98.

Anderson, W. G., Pillans, R. D., Hyodo, S., Tsukada, T., Good, J. P., Takei, Y., Franklin, C. E. and Hazon, N. (2006). The effects of freshwater to seawater transfer on circulating levels of angiotensin II, C-type natriuretic peptide and arginine vasotocin in the euryhaline elasmobranch *Carcharhinus leucas. Gen. Comp. Endocrinol.* 147, 39–46.

Ando, M., Kondo, K. and Takei, Y. (1992). Effects of eel atrial natriuretic peptide on NaCl and water transport across the intestine of the seawater eel. *J. Comp. Physiol.* B 162, 436–439.

Ando, M., Fujii, Y., Kadota, T., Kozaka, T., Mukuda, T., Takase, I. and Kawahara, A. (2000). Some factors affecting drinking behavior and their interactions in seawater-acclimated eels. *Anguilla japonica. Zool. Sci.* 17, 171–178.

Ando, M., Mukuda, T. and Kozaka, T. (2003). Water metabolism in the eel acclimated to sea water: from mouth to intestine. *Comp. Biochem. Physiol. B Biochem. Mol. Biol.* 136, 621–633.

Arnold-Reed, D. E. and Balment, R. J. (1991). Atrial natriuretic factor stimulates *in-vivo* and *in-vitro* secretion of cortisol in teleosts. *J. Endocrinol.* 128, R17–R20.

Arnold-Reed, D. E. and Balment, R. J. (1994). Peptide hormones influence interrenal secretion of cortisol in the trout *Oncorhynchus mykiss. Gen. Comp. Endocrinol.* 96, 85–91.

Auperin, B., Rentierdelrue, F., Martial, J. A. and Prunet, P. (1995). Regulation of gill prolactin receptors in tilapia (*Oreochromis niloticus*) after a change in salinity or hypophysectomy. *J. Endocrinol.* 145, 213–220.

Babey, M., Kopp, P. and Robertson, G. L. (2011). Familial forms of diabetes insipidus: clinical and molecular characteristics. *Nat. Rev. Endocrinol.* 7, 701–714.

Babitha, G. S. and Peter, M. C. S. (2010). Cortisol promotes and integrates the osmotic competence of the organs in North African catfish (*Clarias gariepinus* Burchell): evidence from *in vivo* and *in situ* approaches. *Gen. Comp. Endocrinol.* 168, 14–21.

Ballantyne, J. S. and Fraser, D. I. (2013). Euryhaline elasmobranchs. In *Fish Physiology, Vol. 32, Euryhaline Fishes* (eds. S. D. McCormick, A. P. Farrell and C. J. Brauner), pp. 125–198. New York: Elsevier.

Ballantyne, J. S. and Robinson, J. W. (2010). Freshwater elasmobranchs: a review of their physiology and biochemistry. *J. Comp. Physiol.* B 180, 475–493.

Beyenbach, K. W. (1995). Secretory electrolyte transport in renal proximal tubules of fish. In *Fish Physiology,* Vol. 14, *Cellular and Molecular Approaches to Fish Ionic Regulation* (eds. C. M. Wood and T. J. Shuttleworth), pp. 85–105. San Diego: Academic Press.

Björnsson, B. Th., Yamauchi, K., Nishioka, R. S., Deftos, L. J. and Bern, H. A. (1987). Effects of hypophysectomy and subsequent hormonal replacement therapy on hormonal and osmoregulatory status of coho salmon *Oncorhynchus kisutch. Gen. Comp. Endocrinol.* 68, 421–430.

Bolton, J. P., Collie, N. L., Kawauchi, H. and Hirano, T. (1987). Osmoregulatory actions of growth hormone in rainbow trout (*Salmo gairdneri*). *J. Endocrinol.* 112, 63–68.

Bond, H., Winter, M. J., Warne, J. M., McCrohan, C. R. and Balment, R. J. (2002). Plasma concentrations of arginine vasotocin and urotensin II are reduced following transfer of the euryhaline flounder (*Platichthys flesus*) from seawater to fresh water. *Gen. Comp. Endocrinol.* 125, 113–120.

Borski, R. J., Hyde, G. N. and Fruchtman, S. (2002). Signal transduction mechanisms mediating rapid, nongenomic effects of cortisol on prolactin release. *Steroids* 67, 539–548.

Bossus, M., Charmantier, G. and Lorin-Nebel, C. (2011). Transient receptor potential vanilloid 4 in the European sea bass *Dicentrarchus labrax*: a candidate protein for osmosensing. *Comp. Biochem. Physiol. A Mol. Integr. Physiol.* 160, 43–51.

Breves, J. P., Fox, B. K., Pierce, A. L., Hirano, T. and Grau, E. G. (2010a). Gene expression of growth hormone family and glucocorticoid receptors, osmosensors, and ion transporters in the gill during seawater acclimation of Mozambique tilapia *Oreochromis mossambicus*. *J. Exp. Zool. A* 313, 432–441.

Breves, J. P., Hasegawa, S., Yoshioka, M., Fox, B. K., Davis, L. K., Lerner, D. T., Takei, Y., Hirano, T. and Grau, E. G. (2010b). Acute salinity challenges in Mozambique and Nile tilapia: differential responses of plasma prolactin, growth hormone and branchial expression of ion transporters. *Gen. Comp. Endocrinol.* 167, 135–142.

Breves, J. P., Watanabe, S., Kaneko, T., Hirano, T. and Grau, E. G. (2010c). Prolactin restores branchial mitochondrion-rich cells expressing Na^+/Cl^- cotransporter in hypophysectomized Mozambique tilapia. *Am. J. Physiol.* 299, R702–R710.

Breves, J. P., Seale, A. P., Helms, R. E., Tipsmark, C. K., Hirano, T. and Grau, E. G. (2011). Dynamic gene expression of GH/PRL-family hormone receptors in gill and kidney during freshwater-acclimation of Mozambique tilapia. *Comp. Biochem. Physiol. A Mol. Integr. Physiol.* 158, 194–200.

Bridgham, J. T., Carroll, S. M. and Thornton, J. W. (2006). Evolution of hormone-receptor complexity by molecular exploitation. *Science* 312, 97–101.

Brown, J. A., Oliver, J. A. and Henderson, I. W. (1980). Angiotensin and single nephron glomerular function in the trout *Salmo gairdneri*. *Am. J. Physiol.* 239, R509–R514.

Bury, N. R. and Sturm, A. (2007). Evolution of the corticosteroid receptor signalling pathway in fish. *Gen. Comp. Endocrinol.* 153, 47–56.

Canosa, L. F., Chang, J. P. and Peter, R. E. (2007). Neuroendocrine control of growth hormone in fish. *Gen. Comp. Endocrinol.* 151, 1–26.

Cardoso, J. C. R., Vieira, F. A., Gomes, A. S. and Power, D. M. (2007). PACAP, VIP and their receptors in the metazoan: insights about the origin and evolution of the ligand–receptor pair. *Peptides* 28, 1902–1919.

Carroll, R. L. (1988). *Vertebrate Paleontology and Evolution*. New York: W. H. Freeman and Company, pp. 16–25.

Chasiotis, H. and Kelly, S. P. (2011). Effect of cortisol on permeability and tight junction protein transcript abundance in primary cultured gill epithelia from stenohaline goldfish and euryhaline trout. *Gen. Comp. Endocrinol.* 172, 494–504.

Close, D. A., Yun, S. S., McCormick, S. D., Wildbill, A. J. and Li, W. M. (2010). 11-Deoxycortisol is a corticosteroid hormone in the lamprey. *Proc. Natl. Acad. Sci. U. S. A.* 107, 13942–13947.

Cornell, S. C., Portesi, D. M., Veillette, P. A., Sundell, K. and Specker, J. L. (1994). Cortisol stimulates intestinal fluid uptake in Atlantic salmon (*Salmo salar*) in the post-smolt stage. *Fish Physiol. Biochem.* 13, 183–190.

Cousins, K. L. and Farrell, A. P. (1996). Stretch-induced release of atrial natriuretic factor from the heart of rainbow trout (*Oncorhynchus mykiss*). *Can. J. Zool.* 74, 380–387.

Cutler, C. P., Phillips, C., Hazon, N. and Cramb, G. (2007). Cortisol regulates eel (*Anguilla anguilla*) aquaporin 3 (AQP3) mRNA expression levels in gill. *Gen. Comp. Endocrinol.* 152, 310–313.

Dange, A. D. (1986). Branchial Na$^+$,K$^+$-ATPase activity in freshwater or saltwater acclimated tilapia *Oreochromis* (*Sarotherodon*) *mossambicus*: effects of cortisol and thyroxine. *Gen. Comp. Endocrinol.* 62, 341–343.

Dauder, S., Young, G., Hass, L. and Bern, H. A. (1990). Prolactin receptors in liver, kidney, and gill of the tilapia (*Oreochromis mossambicus*): characterization and effect of salinity on specific binding of iodinated ovine prolactin. *Gen. Comp. Endocrinol.* 77, 368–377.

Daza, D. O., Lewicka, M. and Larhammar, D. (2012). The oxytocin/vasopressin receptor family has at least five members in the gnathostome lineage, including two distinct V2 subtypes. *Gen. Comp. Endocrinol.* 175, 135–143.

Deane, E. E. and Woo, N. Y. S. (2004). Differential gene expression associated with euryhalinity in sea bream (*Sparus sarba*). *Am. J. Physiol.* 287, R1054–R1063.

Drennon, K., Moriyama, S., Kawauchi, H., Small, B., Silverstein, J., Parhar, I. and Shepherd, B. (2003). Development of an enzyme-linked immunosorbent assay for the measurement of plasma growth hormone (GH) levels in channel catfish (*Ictalurus punctatus*): assessment of environmental salinity and GH secretogogues on plasma GH levels. *Gen. Comp. Endocrinol.* 133, 314–322.

Duff, D. W. and Olson, K. R. (1986). Trout vascular and renal responses to atrial natriuretic factor and heart extract. *Am. J. Physiol.* 251, R639–R642.

Eckert, S. M., Yada, T., Shepherd, B. S., Stetson, M. H., Hirano, T. and Grau, E. G. (2001). Hormonal control of osmoregulation in the channel catfish *Ictalurus punctatus*. *Gen. Comp. Endocrinol.* 122, 270–286.

Edwards, S. L. and Marshall, W. S. (2013). Principles and patterns of osmoregulation and euryhalinity in fishes. In *Fish Physiology*, Vol. 32, *Euryhaline Fishes* (eds. S. D. McCormick, A. P. Farrell and C. J. Brauner), pp. 1–44. New York: Elsevier.

Evans, A. N., Henning, T., Gelsleicher, J. and Nunetz, B. S. (2010). Molecular classification of an elasmobranch angiotensin receptor: quantification of angiotensin receptor and natriuretic peptide receptor mRNAs in saltwater and freshwater populations of the Atlantic stingray. *Comp. Biochem. Physiol. B Biochem. Mol. Biol.* 157, 423–431.

Fiol, D. F. and Kültz, D. (2007). Osmotic stress sensing and signaling in fishes. *FEBS J.* 274, 5790–5798.

Fiol, D. F., Sanmarti, E., Sacchi, R. and Kültz, D. (2009). A novel tilapia prolactin receptor is functionally distinct from its paralog. *J. Exp. Biol.* 212, 2007–2015.

Fitzsimons, J. T. (1998). Angiotensin, thirst, and sodium appetite. *Physiol. Rev.* 78, 583–686.

Flanagan, J. A., Bendell, L. A., Guerreiro, P. M., Clark, M. S., Power, D. M., Canario, A. V. M., Brown, B. L. and Ingleton, P. M. (2002). Cloning of the cDNA for the putative calcium-sensing receptor and its tissue distribution in sea bream (*Sparus aurata*). *Gen. Comp. Endocrinol.* 127, 117–127.

Forte, L. R., London, R. M., Freeman, R. H. and Krause, W. J. (2000). Guanylin peptides: renal actions mediated by cyclic GMP. *Am. J. Physiol.* 278, F180–F191.

Foskett, J. K., Hubbard, G. M., Machen, T. E. and Bern, H. A. (1982). Effects of epinephrine, glucagon and vasoactive intestinal polypeptide on chloride secretion by teleost opercular membrane. *J. Comp. Physiol. B* 146, 27–34.

Fox, B. K., Naka, T., Inoue, K., Takei, Y., Hirano, T. and Grau, G. E. (2007). In vitro effects of homologous natriuretic peptides on growth hormone and prolactin release in the tilapia, *Oreochromis mossambicus*. *Gen. Comp. Endocrinol.* 150, 270–277.

Fryer, J. N. (1979). Prolactin-binding sites in tilapia (*Sarotharodon mossambicus*) kidney. *Gen. Comp. Endocrinol.* 39, 397–403.

Fuentes, J., Bury, N. R., Carroll, S. and Eddy, F. B. (1996). Drinking in Atlantic salmon presmolts (*Salmo salar* L.) and juvenile rainbow trout (*Oncorhynchus mykiss* walbaum) in response to cortisol and sea water challenge. *Aquaculture* 141, 129–137.

Fuentes, J., Brinca, L., Guerreiro, P. M. and Power, D. M. (2010). PRL and GH synthesis and release from the sea bream (*Sparus auratus* L.) pituitary gland *in vitro* in response to osmotic challenge. *Gen. Comp. Endocrinol.* 168, 95–102.

Fyhrquist, F. and Saijonmaa, O. (2008). Renin–angiotensin system revisited. *J. Intern. Med.* 264, 224–236.

Gallis, J.-L., Lasserre, P. and Belloc, F. (1979). Freshwater adaptation in the euryhaline teleost, *Chelon labrosus*. I. Effects of adaptation, prolactin, cortisol and actinomycin D on plasma osmotic balance and (Na^+-K^+)ATPase in gill and kidney. *Gen. Comp. Endocrinol.* 38, 1–10.

Good, J. P. and Hazon, N (2009). Osmoregulation in elasmobranchs. In *Osmoregulation and Ion Transport: Integrating Physiological, Molecular and Environmental Aspects* (eds. R. D. Handy, N. Bury and G. Flik), pp. 19–61. London: Society for Experimental Biology Press.

Good-Avila, S. V., Yegorov, S., Harron, S., Bogerd, I., Glen, P., Ozon, J. and Wilson, B. C. (2009). Relaxin gene family in teleosts: phylogeny, syntenic mapping, selective constraint, and expression analysis. *BMC Evol. Biol.* 9, 293.

Grau, E. G., Nishioka, R. S. and Bern, H. A. (1982). Effects of somatostatin and urotensin II on tilapia pituitary prolactin release and interactions between somatostatin, osmotic pressure, Ca^{++}, and adenosine $3',5'$-monophosphate in prolactin release *in vitro*. *Endocrinology* 110, 910–915.

Greenwood, A. K., Butler, P. C., White, R. B., DeMarco, U., Pearce, D. and Fernald, R. D. (2003). Multiple corticosteroid receptors in a teleost fish: distinct sequences, expression patterns, and transcriptional activities. *Endocrinology* 144, 4226–4236.

Griffith, R. W. (1974). Environment and salinity tolerance in the genus *Fundulus*. *Copeia* 1974 (2), 319–331.

Grosell, M., Mager, E. M., Williams, C. and Taylor, J. R. (2009). High rate of HCO_3^- secretion and Cl^- absorption against adverse gradient in the marine teleost intestine: the involvement of an electrogenic anion exchanger and H^+-pump metabolon? *J. Exp. Biol.* 212, 1684–1696.

Guibbolini, M. E. and Avella, M. (2003). Neurohypophysial hormone regulation of Cl^- secretion: physiological evidence for V1-type receptors in sea bass gill respiratory cells in culture. *J. Endocrinol.* 176, 111–119.

Guibbolini, M. E. and Lahlou, B. (1987). Neurohypophysial hormone inhibition of adenylate cyclase activity in fish gills. *FEBS Lett.* 220, 98–102.

Hay, D. L., Christopoulos, G., Christopoulos, A., Poyner, D. R. and Sexton, P. M. (2005). Pharmacological discrimination of calcitonin receptor: receptor activity-modifying protein complexes. *Mol. Pharmacol.* 67, 1655–1665.

Hazon, N., Tierney, M. L. and Takei, Y. (1999). The renin–angiotensin system in elasmobranch fish: a review. *J. Exp. Zool.* 284, 526–534.

Hazon, N., Wells, A., Pillans, R. D., Good, J. P., Anderson, W. G. and Franklin, C. E. (2003). Urea based osmoregulation and endocrine control in elasmobranch fish with special reference to euryhalinity. *Comp. Biochem. Physiol. A Mol. Integr. Physiol.* 136, 685–700.

Herndon, T. M., McCormick, S. D. and Bern, H. A. (1991). Effects of prolactin on chloride cells in opercular membrane of seawater-adapted tilapia. *Gen. Comp. Endocrinol.* 83, 283–289.

Hirano, T. (1969). Effects of hypophysectomy and salinity change on plasma cortisol concentration in Japanese eel *Anguilla japonica*. *Endocrinol. Japon.* 16, 557–560.

Hirano, T. (1974). Some factors regulating drinking by the eel *Anguilla japonica*. *J. Exp. Biol.* 61, 737–747.

Hirano, T. and Mayer-Gostan, N. (1976). Eel esophagus as an osmoregulatory organ. *Proc. Natl. Acad. Sci. U. S. A.* 73, 1348–1350.

Hirano, T. and Utida, S. (1968). Effects of ACTH and cortisol on water movement in isolated intestine of the eel *Anguilla japonica*. *Gen. Comp. Endocrinol.* 11, 373–380.

Hiroi, J., Kaneko, T. and Tanaka, M. (1999). In vitro sequential changes in chloride cell morphology in the yolk-sac membrane of Mozambique tilapia (*Oreochromis mossambicus*) embryos and larvae during seawater adaptation. *J. Exp. Biol.* 202, 3485–3495.

Hirose, S., Hagiwara, H. and Takei, Y. (2001). Comparative molecular biology of natriuretic peptide receptors. *Can. J. Physiol. Pharmacol.* 79, 665–672.

Holt, W. F. and Idler, D. R. (1975). Influence of the interrenal gland on the rectal gland of a skate. *Comp. Biochem. Physiol. C Toxicol. Pharmacol.* 50, 111–119.

Hoshijima, K. and Hirose, S. (2007). Expression of endocrine genes in zebrafish larvae in response to environmental salinity. *J. Endocrinol.* 193, 481–491.

Hu, G.-B., Kusakabe, M. and Takei, Y. (2011). Localization of diversified relaxin gene transcripts in the brain of eels. *Gen. Comp. Endocrinol.* 172, 430–439.

Hyodo, S. and Urano, A. (1991). Changes in expression of probasotocin and proisotocin genes during adaptation to hyper- and hypo-osmotic environments in rainbow trout. *J. Comp. Physiol.* B 161, 549–556.

Hyodo, S., Katoh, F., Kaneko, T. and Takei, Y. (2004a). A facilitative urea transporter is localized in the renal collecting tubule of dogfish *Triakis scyllia*. *J. Exp. Biol.* 207, 347–356.

Hyodo, S., Tsukada, T. and Takei, Y. (2004b). Neurohypophysial hormones of dogfish, *Triakis scyllium*: structures and salinity-dependent secretion. *Gen. Comp. Endocrinol.* 138, 97–104.

Inoue, K. and Takei, Y. (2002). Diverse adaptability in *Oryzias* species to high environmental salinity. *Zool. Sci.* 19, 727–734.

Inoue, K., Naruse, K., Yamagami, S., Mitani, H., Suzuki, N. and Takei, Y. (2003). Four functionally distinct C-type natriuretic peptides found in fish reveal new evolutionary history of the natriuretic system. *Proc. Natl. Acad. Sci. U. S. A.* 100, 10079–10084.

Inoue, K., Sakamoto, T., Yuge, S., Iwatani, H., Yamagami, S., Tsutsumi, M., Hori, H., Cerra, M. C., Tota, B., Suzuki, N., Okamoto, N. and Takei, Y. (2005). Structural and functional evolution of three cardiac natriuretic peptides. *Mol. Biol. Evol.* 22, 2428–2434.

Kaiya, H. and Takei, Y. (1996a). Changes in plasma atrial and ventricular natriuretic peptide concentrations after transfer of eels from fresh water and seawater or vice versa. *Gen. Comp. Endocrinol.* 104, 337–345.

Kaiya, H. and Takei, Y. (1996b). Osmotic and volaemic regulation of atrial and ventricular natriuretic peptide secretion in conscious eels. *J. Endocrinol.* 149, 441–447.

Kajimura, M., Walsh, P. J., Mommsen, T. P. and Wood, C. M. (2006). The dogfish shark (*Squalus acanthias*) increases both hepatic and extrahepatic ornithine urea cycle enzyme activities for nitrogen conservation after feeding. *Physiol. Biochem. Zool.* 79, 602–613.

Kaljnaia, S., Wilson, G. D., Feilen, A. L. and Cramb, G. (2009). Guanylin-like peptides, guanylate cyclase and osmoregulation in the European eel (*Anguilla anguilla*). *Gen. Comp. Endocrinol.* 161, 103–114.

Kato, A., Chang, M. H., Kurita, Y., Nakada, T., Ogoshi, M., Nakazato, T., Doi, H., Hirose, S. and Romero, M. F. (2009). Identification of renal transporters involved in sulfate excretion in marine teleost fish. *Am. J. Physiol.* 297, R1647–R1659.

Kawakoshi, A., Hyodo, S. and Takei, Y. (2001). CNP is the only natriuretic peptide in an elasmobranch fish, *Triakis scyllia*. *Zool. Sci.* 18, 861–868.

Kawakoshi, A., Hyodo, S., Nozaki, M. and Takei, Y. (2006). Identification of a single natriuretic peptide (NP) in cyclostomes, lamprey and hagfish: CNP-4 is an ancestral gene of the NP family. *Gen. Comp. Endocrinol.* 148, 41–47.

Kawauchi, H., Suzuki, K., Yamazaki, T., Moriyama, S., Nozaki, M., Yamaguchi, K., Takahashi, A., Youson, J. and Sower, S. A. (2010). Identification of growth hormone in the sea lamprey, an extant representative of a group of the most ancient vertebrates. *Endocrinology* 143, 4916–4921.

Kelley, K. M., Nishioka, R. S. and Bern, H. A. (1990). In vitro effect of osmotic pressure and cortisol on prolactin cell physiology in the coho salmon (*Oncorhynchus kisutch*) during the parr–smolt transformation. *J. Exp. Zool.* 254, 72–82.

Kelly, S. P. and Chasiotis, H. (2011). Glucocorticoid and mineralocorticoid receptors regulate paracellular permeability in a primary cultured gill epithelium. *J. Exp. Biol.* 214, 2308–2318.

Kelly, S. P. and Wood, C. M. (2002a). Cultured gill epithelia from freshwater tilapia (*Oreochromis niloticus*): effect of cortisol and homologous serum supplements from stressed and unstressed fish. *J. Membr. Biol.* 190, 29–42.

Kelly, S. P. and Wood, C. M. (2002b). Prolactin effects on cultured pavement cell epithelia and pavement cell plus mitochondria-rich cell epithelia from freshwater rainbow trout gills. *Gen. Comp. Endocrinol.* 128, 44–56.

Kelly, S. P., Chow, I. K. and Woo, N. S. (1999). Effects of prolactin and growth hormone on strategies of hypoosmotic adaptation in a marine teleost. *Sparus sarba. Gen. Comp. Endocrinol.* 113, 9–22.

Kelsall, C. J. and Balment, R. J. (1998). Native urotensins influence cortisol secretion and plasma cortisol concentration in the euryhaline flounder. *Platichthys flesus. Gen. Comp. Endocrinol.* 112, 210–219.

Kiilerich, P., Kristiansen, K. and Madsen, S. S. (2007a). Cortisol regulation of ion transporter mRNA in Atlantic salmon gill and the effect of salinity on the signaling pathway. *J. Endocrinol.* 194, 417–427.

Kiilerich, P., Kristiansen, K. and Madsen, S. S. (2007b). Hormone receptors in gills of smolting Atlantic salmon, *Salmo salar*: expression of growth hormone, prolactin, mineralocorticoid and glucocorticoid receptors and 11 beta-hydroxysteroid dehydrogenase type 2. *Gen. Comp. Endocrinol.* 152, 295–303.

Kiilerich, P., Milla, S., Sturm, A., Valotaire, C., Chevolleau, S., Giton, F., Terrien, X., Fiet, J., Fostier, A., Debrauwer, L. and Prunet, P. (2011a). Implication of the mineralocorticoid axis in rainbow trout osmoregulation during salinity acclimation. *J. Endocrinol.* 209, 221–235.

Kiilerich, P., Pedersen, S. H., Kristiansen, K. and Madsen, S. S. (2011b). Corticosteroid regulation of Na^+,K^+-ATPase alpha 1-isoform expression in Atlantic salmon gill during smolt development. *Gen. Comp. Endocrinol.* 170, 283–289.

Kiilerich, P., Tipsmark, C. K., Borski, R. J. and Madsen, S. S. (2011c). Differential effects of cortisol and 11-deoxycorticosterone on ion transport protein mRNA levels in gills of two euryhaline teleosts, Mozambique tilapia (*Oreochromis mossambicus*) and striped bass (*Morone saxatilis*). *J. Endocrinol.* 209, 115–126.

Kishimoto, I., Tokudome, T., Nakao, K. and Kangawa, K. (2011). Natriuretic peptide system: an overview of studies using genetically engineered animal models. *FEBS J.* 278, 1830–1841.

Knoeppel, S. J., Atkins, D. L. and Packer, R. K. (1982). The role of the thyroid gland in osmotic and ionic regulation in *Fundulus heteroclitus* acclimated to freshwater and seawater. *Comp. Biochem. Physiol. A* 73, 25–29.

Kobayashi, H. and Takei, Y. (1996). *Zoophysiology*, Vol. 35, *The Renin–Angiotensin System – Comparative Aspects*. Berlin: Springer, pp. 1–245.

Konno, N., Kurosawa, M., Kaiya, H., Miyazato, M., Matsuda, K. and Uchiyama, M. (2010). Molecular cloning and characterization of V2-type receptor in two ray-finned fish, gray bichir, *Polypterus senegalus* and medaka, *Oryzias latipes*. *Peptides* 31, 1273–1279.

Kozaka, T., Fujii, Y. and Ando, M. (2003). Central effects of various ligands on drinking behavior in eels acclimated to seawater. *J. Exp. Biol.* 206, 687–692.

Kültz, D. (2013). Osmosensing. In *Fish Physiology*, Vol. 32, *Euryhaline Fishes* (eds. S. D. McCormick, A. P. Farrell and C. J. Brauner), pp. 45–68. New York: Elsevier.

Larson, B. A. and Madani, Z. (1991). Increased urotensin I and urotensin II immunoreactivity in the urophysis of *Gillichthys milabilis* transferred to low salinity water. *Gen. Comp. Endocrinol.* 83, 379–387.

Lee, K. M., Kaneko, T., Katoh, F. and Aida, K. (2006). Prolactin gene expression and gill chloride cell activity in fugu *Takifugu rubripes* exposed to a hypoosmotic environment. *Gen. Comp. Endocrinol.* 149, 285–293.

Leloup, J. and Lebel, J. M. (1993). Triiodothyronine is necessary for the action of growth hormone in acclimation to seawater of brown (*Salmo trutta*) and rainbow trout (*Oncorhynchus mykiss*). *Fish Physiol. Biochem.* 11, 165–173.

Lerner, D. T., Bjornsson, B. T. and McCormick, S. D. (2007). Larval exposure to 4-nonylphenol and 17 beta-estradiol affects physiological and behavioral development of seawater adaptation in Atlantic salmon smolts. *Environ. Sci. Technol.* 41, 4479–4485.

Li, Y.-Y. and Takei, Y. (2003). Ambient salinity-dependent effects of homologous natriuretic peptides (ANP, VNP and CNP) on plasma cortisol levels in the eel. *Gen. Comp. Endocrinol.* 130, 317–323.

Liedtke, W. and Kim, C. (2005). Functionality of the TRPV subfamily of TRP ion channels: add mechano-TRP and osmo-TRP to the lexicon! *Cell Mol. Life Sci.* 62, 2985–3001.

Lin, C. H., Tsai, I. L., Su, C. H., Tseng, D. Y. and Hwang, P. P. (2011). Reverse effect of mammalian hypocalcemic cortisol in fish: cortisol stimulates Ca^{2+} uptake via glucocorticoid receptor-mediated vitamin D3 metabolism. *PLoS ONE* 6 (8), e23689.

Link, K., Berishvili, G., Shved, N., D'Cotta, H., Baroiller, J. F., Reinecke, M. and Eppler, E. (2010). Seawater and freshwater challenges affect the insulin-like growth factors IGF-I and IGF-II in liver and osmoregulatory organs of the tilapia. *Mol. Cell. Endocrinol.* 327, 40–46.

Liu, N. A., Liu, Q., Wawrowsky, K., Yang, Z. A., Lin, S. and Melmed, S. (2006). Prolactin receptor signaling mediates the osmotic response of embryonic zebrafish lactotrophs. *Mol. Endocrinol.* 20, 871–880.

López, J. and Martínez, A. (2002). Cell and molecular biology of the multifunctional peptide, adrenomedullin. *Int. Rev. Cytol.* 221, 1–92.

Loretz, C. A. and Bern, H. A. (1981). Stimulation of sodium transport across the teleost urinary bladder by urotensin II. *Gen. Comp. Endocrinol.* 43, 325–330.

Loretz, C. A., Freel, R. W. and Bern, H. A. (1983). Specificity of response of intestinal ion transport systems to a pair of natural peptide hormone analogs: somatostatin and urotensin II. *Gen. Comp. Endocrinol.* 52, 198–206.

Loretz, C. A., Pollina, C., Kaiya, H., Sakaguchi, H. and Takei, Y. (1997). Local synthesis of natriuretic peptides in the eel intestine. *Biochem. Biophys. Res. Commun.* 238, 817–822.

Loretz, C. A., Pollina, C., Hyodo, S., Chang, W., Pratt, S., Shoback, D. and Takei, Y. (2004). cDNA cloning and functional expression of a Ca^{2+}-sensing receptor with truncated carboxyterminal tail from the Mozambique tilapia (*Oreochromis mossambicus*). *J. Biol. Chem.* 279, 53288–53297.

Losel, R. M. and Wehling, M. (2008). Classic versus non-classic receptors for nongenomic mineralocorticoid responses: emerging evidence. *Front. Neuroendocrinol.* 29, 258–267.

Lovejoy, D. A. and Balment, R. J. (1999). Evolution and physiology of the corticotropin-releasing factor (CRF) family of neuropeptides in vertebrates. *Gen. Comp. Endocrinol.* 115, 1–22.

Lu, W., Greenwood, M., Dow, L., Yuill, J., Worthington, J., Brierley, M. J., McCrohan, C. R., Riccardi, D. and Balment, R. J. (2006). Molecular characterization and expression of urotensin II and its receptor in the flounder (*Platichthys flesus*): a hormone system supporting body fluid homeostasis in euryhaline fish. *Endocrinology* 147, 3692–3708.

Madsen, S. S. (1990). The role of cortisol and growth hormone in seawater adaptation and development of hypoosmoregulatory mechanisms in sea trout parr (*Salmo trutta trutta*). *Gen. Comp. Endocrinol.* 79, 1–11.

Madsen, S. S. and Bern, H. A. (1993). In vitro effects of insulin-like growth factor-I on gill Na^+, K^+-ATPase in coho salmon, *Oncorhynchus kisutch*. *J. Endocrinol.* 138, 23–30.

Madsen, S. S., Mathiesen, A. B. and Korsgaard, B. (1997). Effects of 17-beta-estradiol and 4-nonylphenol on smoltification and vitellogenesis in Atlantic salmon (*Salmo salar*). *Fish Physiol. Biochem.* 17, 303–312.

Madsen, S. S., Jensen, L. N., Tipsmark, C. K., Kiilerich, P. and Borski, R. J. (2007). Differential regulation of cystic fibrosis transmembrane conductance regulator and Na^+, K^+-ATPase in gills of striped bass, *Morone saxatilis*: effect of salinity and hormones. *J. Endocrinol.* 192, 249–260.

Mainoya, J. R. and Bern, H. A. (1982). Effects of teleost urotensins on intestinal absorption of water and NaCl in tilapia, *Saratherodon mossambicus*, adapted to fresh water or sea water. *Gen. Comp. Endocrinol.* 47, 54–58.

Mainoya, J. R. and Bern, H. A. (1984). Influence of vasoactive intestinal peptide and urotensin II on the absorption of water and NaCl by the anterior intestine of tilapia, *Sarotherodon mossambicus*. *Zool. Sci.* 1, 100–105.

Mancera, J. M. and McCormick, S. D. (1998). Evidence for growth hormone/insulin-like growth factor I axis regulation of seawater acclimation in the euryhaline teleost *Fundulus heteroclitus*. *Gen. Comp. Endocrinol.* 111, 103–112.

Mancera, J. M., Perez-Figares, J. M. and Fernandez-Lebrez, P. (1994). Effect of cortisol on brackish water adaptation in the euryhaline gilthead sea bream (*Sparus aurata* L.). *Comp. Biochem. Physiol. A* 107, 397–402.

Mancera, J. M., Carrion, R. L. and del Rio, M. D. M. (2002). Osmoregulatory action of PRL, GH, and cortisol in the gilthead seabream (*Sparus aurata* L.). *Gen. Comp. Endocrinol.* 129, 95–103.

Manzon, L. A. (2002). The role of prolactin in fish osmoregulation: a review. *Gen. Comp. Endocrinol.* 125, 291–310.

Marshall, W. S. (2003). Rapid regulation of NaCl secretion by estuarine teleost fish: coping with strategies for short-duration freshwater exposure. *Biochim. Biophys. Acta* 1618, 95–105.

Marshall, W. S. and Bern, H. A. (1981). Active chloride transport by the skin of a marine teleost is stimulated by urotensin I and inhibited by urotensin II. *Gen. Comp. Endocrinol.* 43, 484–491.

Marshall, W. S. and Grosell, M. (2006). Ion transport, osmoregulation, and acid–base balance in homeostasis and reproduction. In *The Physiology of Fishes* (eds. D. H. Evans and J. B. Claiborne), 3rd edn, pp. 177–230. Boca Raton, FL: CRC Press.

Marsigliante, S., Barker, S., Jimenez, E. and Storelli, C. (2000a). Glucocorticoid receptors in the euryhaline teleost *Anguilla anguilla*. *Mol. Cell. Endocrinol.* 162, 193–201.

Marsigliante, S., Muscella, A., Barker, S. and Storelli, C. (2000b). Angiotensin II modulates the activity of the Na$^+$/K$^+$ATPase in eel kidney. *J. Endocrinol.* 165, 147–156.

Marsigliante, S., Muscella, A., Greco, S., Elia, M. G., Vilella, S. and Storelli, C. (2001). Na$^+$/K$^+$ATPase activity inhibition and isoform-specific translocation of protein kinase C following angiotensin II administration in isolated eel enterocytes. *J. Endocrinol.* 168, 339–346.

Martinez, A. S., Cutler, C. P., Wilson, G. D., Phillips, C., Hazon, N. and Cramb, G. (2005). Regulation of expression of two aquaporin homologues in the intestine of the European eel: effects of seawater acclimation and cortisol treatment. *Am. J. Physiol.* 288, R1733–R1743.

McCormick, S. D. (2001). Endocrine control of osmoregulation in teleost fish. *Am. Zool.* 41, 781–794.

McCormick, S. D. (2013). Smolt physiology and endocrinology. In *Fish Physiology, Vol. 32, Euryhaline Fishes* (eds. S. D. McCormick, A. P. Farrell and C. J. Brauner), pp. 199–251. New York: Elsevier.

McCormick, S. D. and Bern, H. A. (1989). In vitro stimulation of Na$^+$,K$^+$-ATPase activity and ouabain binding by cortisol in coho salmon gill. *Am. J. Physiol.* 256, R707–R715.

McCormick, S. D. and Bradshaw, D. (2006). Hormonal control of salt and water balance in vertebrates. *Gen. Comp. Endocrinol.* 147 (3–8), 2006.

McCormick, S. D., O'Dea, M. F., Moeckel, A. M., Lerner, D. T. and Bjornsson, B. T. (2005). Endocrine disruption of parr–smolt transformation and seawater tolerance of Atlantic salmon by 4-nonylphenol and 17 beta-estradiol. *Gen. Comp. Endocrinol.* 142, 280–288.

McCormick, S. D., Regish, A., O'Dea, M. F. and Shrimpton, J. M. (2008). Are we missing a mineralocorticoid in teleost fish? Effects of cortisol, deoxycorticosterone and aldosterone on osmoregulation, gill Na$^+$,K$^+$-ATPase activity and isoform mRNA levels in Atlantic salmon. *Gen. Comp. Endocrinol.* 157, 35–40.

McCrohan, C. R., Lu, W., Brierley, M. J., Dow, L. and Balment, R. J. (2007). Fish caudal neurosecretory system: a model for the study of neuroendocrine secretion. *Gen. Comp. Endocrinol.* 153, 243–250.

Meier, K. M., Figueiredo, M. A., Kamimura, M. T., Laurino, J., Maggioni, R., Pinto, L. S., Dellagostin, O. A., Tesser, M. B., Sampaio, L. A. and Marins, L. F. (2009). Increased growth hormone (GH), growth hormone receptor (GHR), and insulin-like growth factor I (IGF-I) gene transcription after hyperosmotic stress in the Brazilian flounder *Paralichthys orbignyanus*. *Fish Physiol. Biochem.* 35, 501–509.

Miyanishi, H., Nobata, S. and Takei, Y. (2011). Relative dipsogenic potencies of six natriuretic peptides in eels. *Zool. Sci.* 28, 719–726.

Miyanishi, H., Okubo, K. and Takei, Y. (2013a). Natriuretic peptides in developing medaka embryos: Implication in cardiac development by loss-of-function studies. *Endocrinology* 154, (in press) doi: 10.1210/en.2012-1730.

Miyanishi, H., Okubo, K. and Takei, Y. (2013b). Cardiac natriuretic peptide in seawater adaptation in medaka embryos as revealed by loss-of-function analysis. *Am. J. Physiol.* (in press).

Mommsen, T. P., Vijayan, M. M. and Moon, T. W. (1999). Cortisol in teleosts: dynamics, mechanisms of action, and metabolic regulation. *Rev. Fish Biol. Fish.* 9, 211–268.

Moriyama, S., Ito, T., Takahashi, A., Amano, M., Sower, S. A., Hirano, T., Yamamori, K. and Kawauchi, H. (2002). A homolog of mammalian PRL-releasing peptide (fish arginyl–phenylalanyl–amide peptide) is a major hypothalamic peptide of PRL release in teleost fish. *Endocrinology* 143, 2071–2079.

Nakada, T., Zandi-Nejad, K., Kurita, Y., Kudo, H., Broumand, V., Kwon, C. Y., Mercado, A., Mount, D. B. and Hirose, S. (2005). Roles of Slc13a1 and Slc26a1 sulfate transporters of eel kidney in sulfate homeostasis and osmoregulation in freshwater. *Am. J. Physiol.* 289, 575–585.

Nakazato, M. (2001). Guanylin family: new intestinal peptides regulating electrolyte and water homeostasis. *J. Gastroenterol.* 36, 219–225.

Nearing, J., Betka, M., Quinn, S., Hentschel, H., Elger, M., Baum, M., Bai, M., Chattopadyhay, N., Brown, E. M., Hebert, S. H. and Harris, H. W. (2002). Polyvalent cation receptor proteins (CaRs) are salinity sensors in fish. *Proc. Natl. Acad. Sci. U. S. A.* 99, 9231–9236.

Nelson, J. S. (2006). *Fishes of the World* (4th edn.). Hoboken, NJ: John Wiley & Sons, pp. 4–5

Nguyen, G., Delarue, F., Burckle, C., Bouzhir, L., Giller, T. and Sraer, J. D. (2002). Pivotal role of the renin/prorenin receptor in angiotensin II production and cellular responses to renin. *J. Clin. Invest.* 109, 1417–1427.

Nilsen, T. O., Ebbesson, L. O. E., Kiilerich, P., Bjornsson, B. T., Madsen, S. S., McCormick, S. D. and Stefansson, S. O. (2008). Endocrine systems in juvenile anadromous and landlocked Atlantic salmon (*Salmo salar*): seasonal development and seawater acclimation. *Gen. Comp. Endocrinol.* 155, 762–772.

Nishimura, H. and Fan, Z. (2003). Regulation of water movement across vertebrate renal tubules. *Comp. Biochem. Physiol. A Mol. Integr. Physiol.* 136, 479–498.

Nishimura, H. and Sawyer, W. H. (1976). Vasopressor, diuretic, and natriuretic responses to angiotensins by the American eel. *Anguilla rostrata. Gen. Comp. Endocrinol.* 29, 337–348.

Nobata, S. and Takei, Y. (2011). The area postrema in hindbrain is a central player for regulation of drinking behavior in eels, *Anguilla japonica. Am. J. Physiol.* 300, R1569–R1577.

Nobata, S., Ogoshi, M. and Takei, Y. (2008). Potent cardiovascular actions of homologous adrenomedullins in eel. *Am. J. Physiol.* 294, R1544–R1553.

Nobata, S., Ventura, A., Kaiya, H. and Takei, Y. (2010). Diversified cardiovascular actions of six homologous natriuretic peptides (ANP, BNP, VNP, CNP1, CNP3 and CNP4) in conscious eels. *Am. J. Physiol.* 298, R1549–R1559.

Nobata, S., Donald, J. A., Balment, R. J. and Takei, Y. (2011). Potent cardiovascular effects of homologous urotensin II (UII) and UII-related peptide in conscious eels after peripheral and central injections. *Am. J. Physiol.* 300, R437–R446.

O'Grady, S. M. and Wolters, P. J. (1990). Evidence for chloride secretion in the intestine of the winter flounder. *Am. J. Physiol.* 258, C243–C247.

O'Grady, S. M., Field, M., Nash, N. T. and Rao, M. C. (1985). Atrial natriuretic factor inhibits NaKCl cotransport in teleost intestine. *Am. J. Physiol.* 249, C531–C534.

Ogoshi, M., Inoue, K. and Takei, Y. (2003). Identification of a novel adrenomedullin gene family in teleost fish. *Biochem. Biophys. Res. Commun.* 311, 1072–1077.

Ogoshi, M., Inoue, K., Naruse, K. and Takei, Y. (2006). Evolutionary history of the calcitonin gene-related peptide family in vertebrates revealed by comparative genomic analyses. *Peptides* 27, 3154–3164.

Ogoshi, M., Nobata, S. and Takei, Y. (2008). Potent osmoregulatory actions of peripherally and centrally administered homologous adrenomedullins in eels. *Am. J. Physiol.* 295, R2075–R2083.

Okawara, Y., Karakida, T., Aihara, M., Yamaguchi, K. and Kobayashi, H. (1987). Involvement of angiotensin in water intake in the Japanese eel *Anguilla japonica. Zool. Sci.* 4, 523–528.

Ortega, L. A., Heupel, M. R., Beynen, P. V. and Motta, P. J. (2009). Movement patterns and water quality preferences of juvenile bull sharks (*Carcharhinus leucas*) in a Florida estuary. *Environ. Biol. Fish.* 84, 361–373.

Parmelee, J. T. and Renfro, J. L. (1983). Esophageal desalination of seawater in flounder: role of active sodium transport. *Am. J. Physiol.* 245, R888–R893.

Payan, P. and Maetz, J. (1971). Water balance in elasmobranchs: arguments in favour of an endocrine control. *Gen. Comp. Endocrinol.* 16, 535–554.

Pelis, R. M. and McCormick, S. D. (2001). Effects of growth hormone and cortisol on Na⁺K⁺2Cl⁻ cotransporter localization and abundance in the gills of Atlantic salmon. *Gen. Comp. Endocrinol.* 124, 134–143.

Pelis, R. M., Goldmeyer, J. E., Crivello, J. and Renfro, J. L. (2003). Cortisol alters carbonic anhydrase-mediated renal sulfate secretion. *Am. J. Physiol.* 285, R1430–R1438.

Perrott, M. N. and Balment, R. J. (1990). The renin–angiotensin system and the regulation of plasma cortisol in the flounder *Platichthys flesus. Gen. Comp. Endocrinol.* 78, 414–420.

Perry, S. F., Goss, G. G. and Laurent, P. (1992). The interrelationships between gill chloride cell morphology and ionic uptake in four freshwater teleosts. *Can. J. Zool.* 70, 1775–1786.

Peter, M. C. S., Leji, J. and Peter, V. S. (2011). Ambient salinity modifies the action of triiodothyronine in the air-breathing fish *Anabas testudineus* Bloch: effects on mitochondria-rich cell distribution, osmotic and metabolic regulations. *Gen. Comp. Endocrinol.* 171, 225–231.

Peter, M. S., Lock, R. C. and Bonga, S. W. (2000). Evidence for an osmoregulatory role of thyroid hormones in the freshwater Mozambique tilapia *Oreochromis mossambicus. Gen. Comp. Endocrinol.* 120, 157–167.

Pickford, G. E. and Phillips, J. R. (1959). Prolactin, a factor in promoting survival of hypophysectomised killifish in fresh water. *Science* 130, 454–455.

Piermarini, P. M. and Evans, D. H. (1998). Osmoregulation of the Atlantic stingray (*Dasyatis sabina*) from the freshwater lake Jesup of the St. Johns River, Florida. *Physiol. Zool.* 71, 553–560.

Piermarini, P. M. and Evans, D. H. (2000). Effects of environmental salinity of Na⁺/K⁺-ATPase in the gills and rectal gland of a euryhaline elasmobranch (*Dasyatis sabina*). *J. Exp. Biol.* 203, 2957–2966.

Piermarini, P. M., Verlander, J. W., Royaux, I. E. and Evans, D. H. (2002). Pendrin immunoreactivity in the gill epithelium of a euryhaline elasmobranch. *Am. J. Physiol.* 283, R983–R992.

Pillans, R. R. and Franklin, C. E. (2004). Plasma osmolyte concentrations and rectal gland mass of bull sharks *Carcharhinus leucas*, captured along a salinity gradient. *Comp. Biochem. Physiol. A Mol. Integr. Physiol.* 138, 363–371.

Pisam, M., Auperin, B., Prunet, P., Rentierdelrue, F., Martial, J. and Rambourg, A. (1993). Effects of prolactin on alpha and beta chloride cells in the gill epithelium of the saltwater adapted tilapia *Oreochromis niloticus. Anat. Rec.* 235, 275–284.

Prunet, P., Sturm, A. and Milla, S. (2006). Multiple corticosteroid receptors in fish: from old ideas to new concepts. *Gen. Comp. Endocrinol.* 147, 17–23.

Quinn, S. J., Kifor, O., Trivedi, S., Diaz, R., Vassilev, P. and Brown, E. (1998). Sodium and ionic strength sensing by the calcium receptor. *J. Biol. Chem.* 273, 19579–19586.

Reinecke, M., Schmid, A., Ermatinger, R. and Loffing-Cueni, D. (1997). Insulin-like growth factor I in the teleost *Oreochromis mossambicus*, the tilapia: gene sequence, tissue expression, and cellular localization. *Endocrinology* 138, 3613–3619.

Reis-Santos, P., McCormick, S. D. and Wilson, J. M. (2008). Ionoregulatory changes during metamorphosis and salinity exposure of juvenile sea lamprey (*Petromyzon marinus* L.). *J. Exp. Biol.* 211, 978–988.

Renfro, J. L. (1999). Recent developments in teleosts renal transport. *J. Exp. Zool.* 283, 653–661.

Rousseau, K., Le, B. N., Marchelidon, J. and Dufour, S. (1999). Evidence that corticotropin-releasing hormone acts as a growth hormone-releasing factor in a primitive teleost, the European eel (*Anguilla anguilla*). *J. Neuroendocrinol.* 11, 385–392.

Sakamoto, T. and Hirano, T. (1991). Growth hormone receptors in the liver and osmoregulatory organs of rainbow trout: characterization and dynamics during adaptation to seawater. *J. Endocrinol.* 130, 425–433.

Sakamoto, T. and Hirano, T. (1993). Expression of insulin-like growth factor-I gene in osmoregulatory organs during seawater adaptation of the salmonid fish: possible mode of osmoregulatory action of growth hormone. *Proc. Natl. Acad. Sci. U. S. A.* 90, 1912–1916.

Sakamoto, T. and McCormick, S. D. (2006). Prolactin and growth hormone in fish osmoregulation. *Gen. Comp. Endocrinol.* 147, 24–30.

Sakamoto, T., Ogasawara, T. and Hirano, T. (1990). Growth hormone kinetics during adaptation to a hyperosmotic environment in rainbow trout. *J. Comp. Physiol. B* 160, 1–6.

Sakamoto, T., Iwata, M. and Hirano, T. (1991). Kinetic studies of growth hormone and prolactin during adaptation of coho salmon, *Oncorhynchus kisutch*, to different salinities. *Gen. Comp. Endocrinol.* 82, 184–191.

Sakamoto, T., McCormick, S. D. and Hirano, T. (1993). Osmoregulatory actions of growth hormone and its mode of action in salmonids: a review. *Fish Physiol. Biochem.* 11, 155–164.

Sakamoto, T., Shepherd, B. S., Madsen, S. S., Nishioka, R. S., Siharath, K., Richman, N. H, Bern, H. A. and Grau, E. G. (1997). Osmoregulatory actions of growth hormone and prolactin in an advanced teleost. *Gen. Comp. Endocrinol.* 106, 95–101.

Sakamoto, T., Amano, M., Hyodo, S., Moriyama, S., Takahashi, A., Kawauchi, H. and Ando, M. (2005). Expression of prolactin-releasing peptide and prolactin in the euryhaline mudskippers (*Periophthalmus modestus*): prolactin-releasing peptide as a primary regulator of prolactin. *J. Mol. Endocrinol.* 34, 825–834.

Sandra, O., Le Rouzic, P., Cauty, C., Edery, M. and Prunet, P. (2000). Expression of the prolactin receptor (tiPRL-R) gene in tilapia *Oreochromis niloticus*: tissue distribution and cellular localization in osmoregulatory organs. *J. Mol. Endocrinol.* 24, 215–224.

Sangiago-Alvarellos, S., Polakof, S., Arjona, F. J., Kleszczynska, A., Marin del Rio, M. P., Miguez, J. M., Soengas, J. L. and Mancera, J. M. (2006). Osmoregulatory and metabolic changes in the gilthead sea bream *Sparus auratus* after arginine vasotocin (AVT) treatment. *Gen. Comp. Endocrinol.* 148, 348–358.

Saran, S. and Schaap, P. (2004). Adenylyl cyclase G is activated by an intramolecular osmosensor. *Mol. Biol. Cell* 15, 1479–1486.

Scheide, J. I. and Zadunaisky, J. A. (1988). Effect of atriopeptin II on isolated opercular epithelium of *Fundulus heteroclitus*. *Am. J. Physiol.* 254, R27–R32.

Schultz, E. T. and McCormick, S. D. (2013). Euryhalinity in an evolutionary context. In *Fish Physiology*, Vol. 32, *Euryhaline Fishes* (eds. S. D. McCormick, A. P. Farrell and C. J. Brauner), pp. 477–533. New York: Elsevier.

Seale, A. P., Riley, L. G., Leedom, T. A., Kajimura, S., Dores, R. M., Hirano, T. and Grau, E. G. (2002). Effects of environmental osmolality on release of prolactin, growth hormone and ACTH from the tilapia pituitary. *Gen. Comp. Endocrinol.* 128, 91–101.

Seale, A. P., Watanabe, S. and Grau, E. G. (2011). Osmoreception: perspectives on signal transduction and environmental modulation. *Gen. Comp. Endocrinol.* 176, 354–360.

Seidelin, M., Madsen, S. S., Byrialsen, A. and Kristiansen, K. (1999). Effects of insulin-like growth factor-I and cortisol on Na^+,K^+-ATPase expression in osmoregulatory tissues of brown trout (*Salmo trutta*). *Gen. Comp. Endocrinol.* 113, 331–342.

Shepherd, B. S., Drennon, K., Johnson, J., Nichols, J. W., Playle, R. C., Singer, T. D. and Vijayan, M. M. (2005). Salinity acclimation affects the somatotropic axis in rainbow trout. *Am. J. Physiol.* 288, R1385–R1395.

Sherwood, N. M., Krueckl, S. L. and McRory, J. E. (2000). The origin and function of the pituitary adenylate cyclase-activating polypeptide (PACAP)/glucagon superfamily. *Endocr. Rev.* 21, 619–670.

Shimizu, H., Watanabe, E., Hiyama, T. Y., Nagakura, A., Fujikawa, A., Okado, H., Yanagawa, Y., Obata, K. and Noda, M. (2007). Glial Nax channels control lactate signaling to neurons for brain [Na^+] sensing. *Neuron* 54, 59–72.

Shrimpton, J. M. (2013). Seawater to freshwater transitions in diadromous fishes. In *Fish Physiology*, Vol. 32, *Euryhaline Fishes* (eds. S. D. McCormick, A. P. Farrell and C. J. Brauner), pp. 327–393. New York: Elsevier.

Shrimpton, J. M. and McCormick, S. D. (1998). Regulation of gill cytosolic corticosteroid receptors in juvenile Atlantic salmon: interaction effects of growth hormone with prolactin and triiodothyronine. *Gen. Comp. Endocrinol.* 112, 262–274.

Shrimpton, J. M., Patterson, D. A., Richards, J. G., Cooke, S. J., Schulte, P. M., Hinch, S. G. and Farrell, A. P. (2005). Ionoregulatory changes in different populations of maturing sockeye salmon *Oncorhynchus nerka* during ocean and river migration. *J. Exp. Biol.* 208, 4069–4078.

Shuttleworth, T. J. and Thorndyke, J. L. (1984). An endogenous peptide stimulates secretory activity in the elasmobranch rectal gland. *Science* 225, 319–321.

Silva, P., Solomon, R. J. and Epstein, F. H. (1997). Transport mechanisms that mediate the secretion of chloride by the rectal gland of *Squalus acanthias*. *J. Exp. Zool.* 279, 504–508.

Sindić, A. and Schlatter, E. (2006). Cellular effects of guanylin and uroguanylin. *J. Am. Soc. Nephrol.* 17, 607–616.

Singer, T. D., Finstad, B., McCormick, S. D., Wiseman, S. B., Schulte, P. M. and McKinley, R. S. (2003). Interactive effects of cortisol treatment and ambient seawater challenge on gill Na^+,K^+-ATPase and CFTR expression in two strains of Atlantic salmon smolts. *Aquaculture* 222, 15–28.

Smith, D. C. W. (1956). The role of the endocrine organs in the salinity tolerance of trout. *Mem. Soc. Endocrinol.* 5, 83–101.

Solomon, R., Protter, A., McEnroe, G., Potter, J. G. and Silva, P. (1992). C-type natriuretic peptides stimulate chloride secretion in the rectal gland of *Squalus acanthias*. *Am. J. Physiol.* 262, R707–R711.

Stoff, J. S., Rosa, R., Hallac, R., Silva, P. and Epstein, F. H. (1979). Hormonal regulation of active chloride transport in the dogfish rectal gland. *Am. J. Physiol.* 237, F138–F144.

Stolte, E. H., van Kemenade, B. M. L. V., Savelkoul, H. F. J. and Flik, G. (2006). Evolution of glucocorticoid receptors with different glucocorticoid sensitivity. *J. Endocrinol.* 190, 17–28.

Stolte, E. H., de Mazon, A. F., Leon-Koosterziel, K. M., Jesiak, M., Bury, N. R., Sturm, A., Savelkoul, H. F. J., van Kemenade, B. M. L. V. and Flik, G. (2008). Corticosteroid receptors involved in stress regulation in common carp *Cyprinus carpio*. *J. Endocrinol.* 198, 403–417.

Sturm, A., Bury, N., Dengreville, L., Fagart, J., Flouriot, G., Rafestion-Oblin, M. E. and Prunet, P. (2005). 11-Deoxycorticosterone is a potent agonist of the rainbow trout (*Oncorhynchus mykiss*) mineralocorticoid receptor. *Endocrinology* 146, 47–55.

Sunn, N., Egli, M., Burazin, T. C. D., Burns, P., Colvill, L., Davern, P., Denton, D. A., Oldfield, B. J., Weisinger, R. S., Rauch, M., Schmid, H. A. and McKinley, M. J. (2002). Circulating relaxin acts on subfornical organ neurons to stimulate water drinking in the rat. *Proc. Natl. Acad. Sci. U. S. A.* 99, 1701–1706.

Suzuki, M., Kubokawa, K., Nagasawa, H. and Urano, A. (1995). Sequence analysis of vasotocin cDNAs of the lamprey, *Lampetra japonica*, and the hagfish, *Eptatretus burgeri*: evolution of cyclostome vasotocin precursors. *J. Mol. Endocrinol.* 14, 67–77.

Takahashi, H., Sakamoto, T. and Narita, K. (2006). Cell proliferation and apoptosis in the anterior intestine of an amphibious, euryhaline mudskipper (*Periophthalmus modestus*). *J. Comp. Physiol. B* 176, 463–468.

Takei, Y. (2002). Hormonal control of drinking in the eel: an evolutionary approach. In *Osmoregulation and Drinking in Vertebrates* (eds. N. Hazon and G. Flik), pp. 61–82. Oxford: BIOS Scientific Publishers.

Takei, Y. (2008). Exploring novel hormones essential for seawater adaptation in teleost fish. *Gen. Comp. Endocrinol.* 157, 3–13.

Takei, Y. and Balment, R. J. (2009). The neuroendocrine regulation of fluid intake and fluid balance. In *Fish Neuroendocrinology* (eds. N. J. Bernier, G. Van Der Kraak, A. P. Farrell, and C. J. Brauner), pp. 366–419. San Diego: Academic Press.

Takei, Y. and Hirose, S. (2002). The natriuretic peptide system in eel: a key endocrine system for euryhalinity? *Am. J. Physiol.* 282, R940–R951.

Takei, Y. and Kaiya, H. (1998). Antidiuretic effect of eel ANP infused at physiological doses in seawater-adapted eels, *Anguilla japonica. Zool. Sci.* 15, 399–404.

Takei, Y. and Loretz, C. A. (2006). Endocrinology. In *The Physiology of Fishes* (eds. D. H. Evans and J. B. Claiborne), 3rd edn, pp. 271–318. Boca Raton, FL: CRC Press.

Takei, Y. and Loretz, C. A. (2011). The gastrointestinal tract as an endocrine/neuroendocrine/ paracrine organ: organization, chemical messengers and physiological targets. In *The Multifunctional Gut of Fish* (eds. M. Grosell, A. P. Farrell and C. J. Brauner), pp. 261–317. San Diego: Academic Press.

Takei, Y. and Tsuchida, T. (2000). Role of the renin–angiotensin system in drinking of seawater-adapted eels, *Anguilla japonica*: a reevaluation. *Am. J. Physiol.* 279, R1105–R1111.

Takei, Y. and Yuge, S. (2007). The intestinal guanylin system and seawater adaptation in eels. *Gen. Comp. Endocrinol.* 152, 339–351.

Takei, Y., Hirano, T. and Kobayashi, H. (1979). Angiotensin and water intake in the Japanese eel *Anguilla japonica. Gen. Comp. Endocrinol.* 38, 446–475.

Takei, Y., Okawara, Y. and Kobayashi, H. (1988a). Drinking induced by cellular dehydration in the quail *Coturnix coturnix japonica. Comp. Biochem. Physiol. A* 90, 291–296.

Takei, Y., Okubo, J. and Yamaguchi, K. (1988b). Effect of cellular dehydration on drinking and plasma angiotensin II level in the eel *Anguilla japonica. Zool. Sci.* 5, 43–51.

Takei, Y., Tsuchida, T. and Tanakadate, A. (1998). Evaluation of water intake in seawater adaptation in eels using a synchronized drop counter and pulse injector system. *Zool. Sci.* 15, 677–682.

Takei, Y., Inoue, K., Ando, K., Ihara, T., Katafuchi, T., Kashiwagi, M. and Hirose, S. (2001). Enhanced expression and release of C-type natriuretic peptide by the heart of freshwater eels *Anguilla japonica. Am. J. Physiol.* 280, R1727–R1735.

Takei, Y., Inoue, K., Ogoshi, M., Kawahara, T., Bannai, H. and Miyano, S. (2004a). Mammalian homolog of fish adrenomedullin 2: Identification of a novel cardiovascular and renal regulator. *FEBS Lett.* 556, 53–58.

Takei, Y., Joss, J. M. P., Kloas, W. and Rankin, J. C. (2004b). Identification of angiotensin I from several vertebrate species: its structural and functional evolution. *Gen. Comp. Endocrinol.* 135, 286–292.

Takei, Y., Ogoshi, M. and Inoue, K. (2007). A "reverse" phylogenetic approach for identification of novel osmoregulatory and cardiovascular hormones in vertebrates. *Front. Neuroendocrinol.* 28, 143–160.

Takei, Y., Hashimoto, H., Inoue, K., Osaki, T., Yoshizawa-Kumagaye, K., Watanabe, T. X., Minamino, N. and Ueta, Y. (2008). Central and peripheral cardiovascular actions of adrenomedullin 5, a novel member of the calcitonin gene-related peptide family, in mammals. *J. Endocrinol.* 197, 391–400.

Tanaka, H., Kagawa, H., Ohta, H., Unuma, T. and Nomura, K. (2003). The first production of glass eel in captivity: fish reproductive physiology facilitates great progress in aquaculture. *Fish Physiol. Biochem.* 28, 493–497.

Tierney, M. L., Luke, G., Cramb, G. and Hazon, N. (1995). The role of the renin–angiotensin system in the control of blood pressure and drinking in the European eel, *Anguilla anguilla. Gen. Comp. Endocrinol.* 100, 39–48.

Tierney, M. L., Takei, Y. and Hazon, N. (1997). The presence of angiotensin receptors in elasmobranchs. *Gen. Comp. Endocrinol.* 105, 9–17.

Tipsmark, C. K. and Madsen, S. S. (2009).). Distinct hormonal regulation of Na^+,K^+-ATPase genes in the gill of Atlantic salmon (*Salmo salar* L.). *J. Endocrinol.* 203, 301–310.

Tipsmark, C. K., Luckenbach, J. A., Madsen, S. S. and Borski, R. J. (2007). IGF-I and branchial IGF receptor expression and localization during salinity acclimation in striped bass. *Am. J. Physiol.* 292, R535–R543.

Tipsmark, C. K., Jorgensen, C., Brande-Lavridsen, N., Engelund, M., Olesen, J. H. and Madsen, S. S. (2009). Effects of cortisol, growth hormone and prolactin on gill claudin expression in Atlantic salmon. *Gen. Comp. Endocrinol.* 163, 270–277.

Tipsmark, C. K., Mahmmoud, Y. A., Borski, R. J. and Madsen, S. S. (2010a). FXYD-11 associates with Na^+K^+ATPase in the gill of Atlantic salmon: regulation and localization in relation to changed ion-regulatory status. *Am. J. Physiol.* 299, R1212–R1223.

Tipsmark, C. K., Sorensen, K. J., Hulgard, K. and Madsen, S. S. (2010b). Claudin-15 and -25b expression in the intestinal tract of Atlantic salmon in response to seawater acclimation, smoltification and hormone treatment. *Comp. Biochem. Physiol. A Mol. Integr. Physiol.* 155, 361–370.

Tipsmark, C. T., Breves, J. P., Seale, A. P., Lerner, D. T., Hirano, T. and Grau, E. G. (2011). Switching of $Na(+), K(+)$-ATPase isoforms by salinity and prolactin in the gill of a cichlid fish. *J. Endocrinol.* 209, 237–244.

Toop, T. and Donald, J. A. (2004). Comparative aspects of natriuretic peptide physiology in non-mammalian vertebrates: a review. *J. Comp. Physiol. B* 174, 189–204.

Tostivint, H., Lihrmann, I. and Vaudry, H. (2008). New insight into the molecular evolution of the somatostatin family. *Mol. Cell. Endocrinol.* 286, 5–17.

Tsuchida, T. and Takei, Y. (1998). Effects of homologous atrial natriuretic peptide on drinking and plasma angiotensin II level in eels. *Am. J. Physiol.* 275, R1605–R1610.

Tsukada, T. and Takei, Y. (2006). Integrative approach to osmoregulatory action of atrial natriuretic peptide in seawater eels. *Gen. Comp. Endocrinol.* 147, 31–38.

Tsukada, T., Rankin, J. C. and Takei, Y. (2005). Mechanisms underlying hyponatremic effect of atrial natriuretic peptide in seawater eels: physiological significance of drinking and intestinal absorption. *Zool. Sci.* 22, 77–85.

Turscott, B. and Idler, D. R. (1968). The widespread occurrence of a corticosteroid 1α-hydroxycorticosterone. *J. Endocrinol.* 40, 515–526.

Uchida, K., Kaneko, T., Tagawa, M. and Hirano, T. (1998). Localization of cortisol receptor in branchial chloride cells in chum salmon fry. *Gen. Comp. Endocrinol.* 109, 175–185.

Vaudry, D., Falluel-Morel, A., Bourgault, S., Basille, M., Burel, D., Wurtz, O., Fournier, A., Chow, B. K. C., Hashimoto, H., Galas, L. and Vaudry, H. (2009). Pituitary adenylate cyclase-activating polypeptide and its receptors: 20 years after the discovery. *Pharmacol. Rev.* 61, 283–357.

Veillette, P. A. and Young, G. (2005). Tissue culture of sockeye salmon intestine: functional response of Na^+-K^+-ATPase to cortisol. *Am. J. Physiol.* 288, R1598–R1605.

Veillette, P. A., Sundell, K. and Specker, J. L. (1995). Cortisol mediates the increase in intestinal fluid absorption in Atlantic salmon during parr smolt transformation. *Gen. Comp. Endocrinol.* 97, 250–258.

Venero, J. L., Vizuete, M. L., Ilundain, A. A., Machado, A., Echevarria, M. and Cano, J. (1999). Detailed localization of aquaporins-4 messenger RNA in the CNS: preferential expression in periventricular organs. *Neuroscience* 94, 239–250.

Ventura, A., Kusakabe, M. and Takei, Y. (2011). Distinct natriuretic peptides interact with ACTH for cortisol secretion from interrenal tissue of eels in different salinities. *Gen. Comp. Endocrinol.* 173, 129–138.

Vijayan, M. M., Takemura, A. and Mommsen, T. P. (2001). Estradiol impairs hyposmoregulatory capacity in the euryhaline tilapia *Oreochromis mossambicus. Am. J. Physiol.* 281, R1161–R1168.

Warne, J. M., Bond, H., Weybourne, E., Sahajpal, V., Lu, W. and Balment, R. J. (2005). Altered plasma and pituitary arginine vasotocin and hypothalamic provasotocin expression in flounder (*Platichthys flesus*) following hypertonic challenge and distribution of vasotocin receptors within the kidney. *Gen. Comp. Endocrinol.* 144, 240–247.

Watanabe, T. and Takei, Y. (2011a). Environmental factors responsible for switching of the SO_4^{2-} excretory system in the kidney of seawater eels. *Am. J. Physiol.* 301, R402–R411.

Watanabe, T. and Takei, Y. (2011b). Molecular physiology and functional morphology of sulfate excretion by the kidney of seawater-adapted eels. *J. Exp. Biol.* 214, 1783–1790.

Watanabe, Y., Sakihara, T., Mukuda, T. and Ando, M. (2007). Antagonistic effects of vasotocin and isotocin on the upper esophageal sphincter muscle of the eel acclimated to seawater. *J. Comp. Physiol. B* 177, 867–873.

Weisbart, M., Chakraborti, P. K., Gallivan, G. and Eales, J. G. (1987). Dynamics of cortisol receptor activity in the gills of the brook trout, *Salvelinus fontinalis*, during seawater adaptation. *Gen. Comp. Endocrinol.* 68, 440–448.

Wells, A., Anderson, W. G. and Nazon, N. (2002). Development of an *in situ* perfused kidney preparation for elasmobranch fish: action of arginine vasotocin. *Am. J. Physiol.* 282, R1636–R1642.

Wells, A., Anderson, W. G., Cains, J. E., Cooper, M. W. and Hazon, N. (2006). Effects of angiotensin II and C-type natriuretic peptide on the *in situ* perfused trunk preparation of the dogfish, *Scyliorhinus canicula. Gen. Comp. Endocrinol.* 145, 109–115.

Whitehead, A., Roach, J. L., Zhang, S. J. and Galvez, F. (2011). Genomic mechanisms of evolved physiological plasticity in killifish distributed along an environmental salinity gradient. *Proc. Natl. Acad. Sci. U. S. A.* 108, 6193–6198.

Wilkinson, T. N., Speed, T. P., Tregear, G. W. and Bathgate, R. A. D. (2005). Evolution of the relaxin-like peptide family. *BMC Evol. Biol.* 5, 14–31.

Wilson, R. W., Millero, F. J., Taylor, J. R., Walsh, P. J., Christensen, V., Jennings, S. and Grosell, M. (2009). Contribution of fish to the marine inorganic carbon cycle. *Science* 323, 359–362.

Wong, M. K. S. and Takei, Y. (2009). Cyclostome and chondrichthyan adrenomedullins reveal ancestral features of the adrenomedullin family. *Comp. Biochem. Physiol. B Biochem. Mol. Biol.* 154, 317–325.

Wong, M. K. S. and Takei, Y. (2011). Characterization of a native angiotensin from an anciently diverged serine-protease inhibitor in lamprey. *J. Endocrinol.* 209, 127–137.

Wong, M. K. S. and Takei, Y. (2012). Changes in plasma angiotensin subtypes in Japanese eel acclimated to various salinities from deionized water to double-strength seawater. *Gen. Comp. Endocrinol.* 178, 250–258.

Wong, M. K. S., Takei, Y. and Woo, N. Y. S. (2006). Differential status of the renin angiotensin system in silver seabream (*Sparus sarba*) in different salinities. *Gen. Comp. Endocrinol.* 149, 81–89.

Wood, C. M. (2011). Rapid regulation of Na^+ and Cl^- flux rates in killifish after acute salinity challenge. *J. Exp. Mar. Biol. Ecol.* 409, 62–69.

Xu, P., Sriramula, S. and Lazartigues, E. (2011). ACE2/ANG-(1–7)/Mas pathway in the brain: the axis of good. *Am. J. Physiol.* 300, R804–R817.

Yada, T. and Ito, F. (1999). Sodium-retaining effects of cortisol, prolactin, and estradiol-17 beta in medaka *Oryzias latipes* exposed to acid water. *Fish. Sci.* 65, 405–409.

Yada, T., Hirano, T. and Grau, E. G. (1994). Changes in plasma levels of the two prolactins and growth hormone during adaptation to different salinities in the euryhaline tilapia, *Oreochromis mossambicus. Gen. Comp. Endocrinol.* 93, 214–223.

Yamagami, S. and Suzuki, N. (2005). Diverse forms of guanylyl cyclases in medaka fish – their genomic structure and phylogenetic relationships to those in vertebrates and invertebrates. *Zool. Sci.* 22, 819–835.

Yamaguchi, Y., Takaki, S. and Hyodo, S. (2009). Subcellular distribution of urea transporter in the collecting tubule of shark kidney is dependent on environmental salinity. *J. Exp. Zool. A* 9, 705–718.

Yang, B. Y., Green, M. and Chen, T. T. (1999). Early embryonic expression of the growth hormone family protein genes in the developing rainbow trout. *Oncorhynchus mykiss. Mol. Reprod. Dev.* 53, 127–134.

Young, G. (1988). Enhanced response of the interrenal of coho salmon (*Oncorhynchus kisutch*) to ACTH after growth hormone treatment *in vivo* and *in vitro. Gen. Comp. Endocrinol.* 71, 85–92.

Youson, J. H. (2003). The biology of metamorphosis in sea lampreys: endocrine, environmental, and physiological cues and events, and their potential application to lamprey control. *J. Great Lakes Res.* 29, 26–49.

Yuge, S. and Takei, Y. (2007). Regulation of ion transport in eel intestine by the homologous guanylin family of peptides. *Zool. Sci.* 24, 1222–1230.

Yuge, S., Inoue, K., Hyodo, S. and Takei, Y. (2003). A novel guanylin family (guanylin, uroguanylin and renoguanylin) in eels: possible osmoregulatory hormones in intestine and kidney. *J. Biol. Chem.* 278, 22726–22733.

Yuge, S., Yamagami, S., Inoue, K., Suzuki, N. and Takei, Y. (2006). Identification of two functional guanylin receptors in eel: multiple hormone–receptor system for osmoregulation in fish intestine and kidney. *Gen. Comp. Endocrinol.* 149, 10–20.

Zhou, B. S., Kelly, S. P., Ianowski, J. P. and Wood, C. M. (2003). Effects of cortisol and prolactin on Na^+ and Cl^- transport in cultured branchial epithelia from FW rainbow trout. *Am. J. Physiol.* 285, R1305–R1316.

Zydlewski, J. and Wilkie, M. P. (2013). Freshwater to seawater transitions in migratory fish. In *Fish Physiology,* Vol. 32, *Euryhaline Fishes* (eds. S. D. McCormick, A. P. Farrell and C. J. Brauner), pp. 253–326. New York: Elsevier.

4

EURYHALINE ELASMOBRANCHS

J.S. BALLANTYNE

D.I. FRASER

This chapter explores the physiology of freshwater, marine, and euryhaline elasmobranchs. In seawater, the blood of elasmobranchs is slightly hyperosmotic and the need for water balance is limited. Salt balance is handled partly by a rectal gland that removes salt from the plasma and secretes it into the rectum for elimination. In freshwater this gland is nonfunctional, and salt balance is handled by gill uptake mechanisms and kidney retention. A significant portion of plasma and cellular osmolarity in seawater is comprised of organic solutes, e.g. urea. The small number of

Euryhaline Fishes: Volume 32
FISH PHYSIOLOGY
Copyright © 2013 Elsevier Inc. All rights reserved
DOI: http://dx.doi.org/10.1016/B978-0-12-396951-4.00004-9

freshwater and euryhaline elasmobranch species may be explained by several factors, including: the osmoregulatory system (rectal gland cannot reverse its function in freshwater); temperature (freshwater distribution limited to warmer latitudes), reproductive and sensory limitations (reduced electro-reception in freshwater); and the use of urea as the main organic osmolyte in seawater (may be limiting in freshwater). Further studies of euryhaline elasmobranch physiology and biochemistry are needed to fully understand the scarcity of these species.

1. INTRODUCTION

The elasmobranchs (sharks, skates, and rays) comprise a monophyletic, primarily marine, group of Chondrichthyan fishes characterized, in the marine forms, by the accumulation of osmotically large amounts of urea in all body fluid compartments. Although elasmobranchs are found in marine, freshwater (FW), and variable salinity environments, their capacity for living in lower salinities is limited and the reason for this forms the basis for this chapter. A related group of Chondrichthyans, the Holocephalans or chimaeras, are found only in marine environments and will not be discussed further.

Although the focus of this chapter is euryhalinity in elasmobranchs, it is also necessary to discuss the physiology of fully marine and fully FW species to understand the issues related to elasmobranch adaptation to salinity changes. Stenohaline marine species are the largest elasmobranch group including many pelagic species. A few tropical species can tolerate hypersaline conditions. At the other extreme are stenohaline FW species that include about 30 species of South American stingrays. Another group includes species living and reproducing in FW but with some capacity for higher salinities. This group includes a handful of Southeast Asian and African species. Based on the available data (Table 4.1), only 5% of known elasmobranchs can enter or live in FW (Helfman et al., 1997) and only 3–4% of all elasmobranch species live permanently in FW (Martin, 2005) compared to 40% of teleost species (Cohen, 1970) (Table 4.1, Fig. 4.1). Truly euryhaline species, i.e. species that can live in seawater (SW) or FW, are the rarest of all, with only a handful of species known: bull shark (*Carcharhinus leucas*), Atlantic stingray (*Dasyatis sabina*), and at least two sawfish, smalltooth sawfish (*Pristis microdon*) and largetooth sawfish (*Pristis*

Table 4.1

Documented locations of elasmobranchs in freshwater.

Scientific name	Common name	Known freshwater incursions	Map locality[a]	Reference
Carcharhinus leucas	Bull shark	Zambezi River, Southern Africa	1	Skelton (2001)
		Limpopo River, Southern Africa	2	Skelton (2001)
		Lower Shire River, Malawi	3	Skelton (2001)
		Phongolo River, Kwazulu-Natal	4	Skelton (2001)
		Adelaide River, Northern Territory, Australia	5	Taniuchi and Shimizu (1991)
		Daly River, Northern Territory, Australia	6	Taniuchi and Shimizu (1991)
		Wenlock River, Queensland, Australia	34	Reilly et al. (2011)
		Brisbane River, Australia	67	Pillans et al. (2006)
		Sepik River, Papua New Guinea	7	Taniuchi and Shimizu (1991)
		Usumacinta River, Tabasco, Mexico	8	Sosa-Nishizaki (1998)
		Betsiboka River West Madagascar	9	Taniuchi et al. (2003)
		Amazon River, Brazil	10	Thorson (1972); Taniuchi et al. (2003)
		Lake Bayano Panama	11	Montoya and Thorson (1982)
		Lake Nicaragua, Nicaragua	59	Thorson (1982)
		Mississippi River, Illinois, USA	12	Thomerson et al. (1977)
		Hoogly River, India	13	Venkataraman et al. (2003)
		Tigris River, Iraq	65	Coad and Papahn (1988)
		Karun River, Iran	66	Coad and Papahn (1988)
		Batang Hari Basin, Sumatra, Indonesia	38	Tan and Lim (1998)
Carcharhinus melanopterus	Blacktip shark	Perak River, Indonesia	14	Smith (1931a)
Glyphis fowlerae	Borneo river shark	Kinabatangan River, Borneo, Malaysia	15	Compagno et al. (2010)
Glyphis garricki	Northern river shark	East and South Alligator Rivers, Northern Territory, Australia	16	Last et al. (2008)
		Adelaide River, Northern Territory, Australia	5	Last et al. (2008)
		Doctor's Creek Western Australia, Australia	17	Last et al. (2008)
		Baimuru, New Guinea	40	Last et al. (2008)
Glyphis gangeticus	Ganges shark	Ganges-Hoogly River system, India	13	Venkataraman et al. (2003)
Dasyatis sabina	Atlantic stingray	St. Johns River system, Florida, USA	18	Johnson and Snelson (1996)

(Continued)

Table 4.1 (Continued)

Scientific name	Common name	Known freshwater incursions	Map locality[a]	Reference
Dasyatis garouaensis	Smooth freshwater stingray	Lake Ezanga, Niger River, Nigeria	19	Thorson and Watson (1975); Compagno and Roberts (1984)
		Cross River, Nigeria	20	
		Sanaga River Cameroun	21	Taniuchi (1991)
Dasyatis ukpam	Thorny freshwater stingray	Zaire River, Zaire	22	Compagno and Roberts (1984)
		Benue River, Nigeria	23	Compagno and Roberts (1984)
		Ogooue River, Gabon,	24	Compagno and Roberts (1984)
		Old Calabar, Nigeria	25	Compagno and Roberts (1984)
		Sanaga River, Cameroun	26	Last and Stevens (1994); Taniuchi (1991)
Dasyatis bennetti	Bennett's stingray	Indragiri River. Indonesia	27	Taniuchi (1979)
Dasyatis (Himantura) (Amphotistius) laosensis	Mekong stingray	Mekong River, Laos	28	Rainboth (1996); Kottelat (2001); Vidthayanon (2010)
		Mekong River, Cambodia	29	Rainboth (1996); Kottelat (2001); Vidthayanon (2010)
		Mekong River, Thailand	28	Rainboth (1996); Kottelat (2001); Vidthayanon (2010)
Himantura chaophraya	Giant freshwater whipray	Kinabatangan River, Borneo, Malaysia	15	Yano et al. (2005)
		Mahakam Basin, Borneo, Indonesia	30	Kottelat et al. (1993)
		Mekong River, Laos, Cambodia, Thailand	28	Rainboth (1996); Kottelat (2001); Vidthayanon (2010)
		Chao Phraya River, Thailand	31	Vidthayanon (2010)
Himantura dalyensis	NA	Daly and Roper Rivers, Northern Territory, Australia	32	Taniuchi and Shimizu (1991); Last et al. (2008)
		Normanby River, Queensland, Australia	33	Taniuchi and Shimizu (1991); Last et al. (2008)
		Wenlock River, Queensland, Australia	34	Taniuchi and Shimizu (1991); Last et al. (2008)
		Gilbert Rivers, Queensland, Australia	35	Taniuchi and Shimizu (1991); Last et al. (2008)
		Fitzroy River, Western Australia, Australia	36	Taniuchi and Shimizu (1991); Last et al. (2008)
		Ord River, Western Australia, Australia	32	Taniuchi and Shimizu (1991); Last et al. (2008)

Species	Common name	Location		Reference
		Fitzroy River, Western Australia, Australia	36	Taniuchi and Shimizu (1991); Last et al. (2008)
		Pentecost River, Western Australia, Australia	32	Taniuchi and Shimizu (1991); Last et al. (2008)
Himantura bleekeri	Bleeker's whipray	Tapi River, Thailand	33	Vidthayanon (2010)
		Tongle Sap, Thailand	33	Vidthayanon (2010)
Himantura krempfi	Marbled freshwater whipray	Mekong River, Cambodia	28	Rainboth (1996)
Himantura signifer	White-rimmed stingray	Jelai River, Pahang River, Malaysia (Peninsular)	33	Yano et al. (2005)
		Kapuas River, Borneo, Indonesia	37	Roberts (1989)
		Batang Hari Basin, Sumatra, Indonesia	38	Kottelat et al. (1993); Tan and Lim (1998)
		Chao Phraya River Thailand	31	Chatchavalvanich et al. (2005)
Himantura kittipongi	Kittipong's stingray	Maekhlong River, Thailand	39	Vidthayanon and Roberts (2005)
Himantura (Dasyatis) uarnak	Honeycomb stingray	Perak River, Sumatra, Indonesia	14	Smith (1931a)
		Indragiri River, Sumatra, Indonesia	27	Otake (1991)
		Oriomo River, Papua New Guinea	40	Otake (1991)
Pastinachus (Hypolophus) (Dasyatis) sephen	Cowtail stingray	Indragiri River, Sumatra, Indonesia	27	Taniuchi (1979)
Potamotrygon boesmani	NA	Corantijn River, Surinam	41	Rosa et al. (2008)
Potamotrygon brachyura	Short-tailed river stingray	Rio Parana, Brazil	42	Rosa (1985)
		Rio Colastine Sur, Argentina	42	Rosa (1985)
		Rio Cuiaba, Argentina	43	Rosa (1985)
		Rio Uruguay, Uruguay	44	Rosa (1985)
Potamotrygon castexi	Largespot river stingray	Rio Manu, Peru	45	Lovejoy et al. (1998)
		Rio San Alejandro, Peru	63	Lovejoy et al. (1998)
		Rio Pachitea, Peru	63	Lovejoy et al. (1998)
		Rio Guapore, Bolivia	61	Lovejoy et al. (1998)
		Rio Yutupis, Peru	69	Lovejoy et al. (1998)
		Rio Santiago, Peru	56	Lovejoy et al. (1998)
		Rio Colastine Sur, Argentina	42	Lovejoy et al. (1998)
		Rio Solimoes, Argentina	46	Lovejoy et al. (1998)
		Rio Paragui, Brazil	71	Lovejoy et al. (1998)
		Rio Itenez, Brazil	61	Lovejoy et al. (1998)

(Continued)

Table 4.1 (Continued)

Scientific name	Common name	Known freshwater incursions	Map locality[a]	Reference
Potamotrygon constellata	Thorny river stingray	Rio Solimoes, Brazil	46	Rosa (1985)
Potamotrygon dumerilii	Smoothback river stingray	Rio Parana, Rio Araguaia, Argentina	47	Rosa (1985)
Potamotrygon falkneri	Largespot river stingray	Parana River, Brazil	42	Toffoli et al. (2008)
Potamotrygon henlei	Bigtooth river stingray	Rio Tocatins, Rio Itacaiunas, Brazil	48	Rosa (1985)
Potamotrygomhistrix	Porcupine river stingray	Rio Parana, Rio Colastine Sur, Argentina	42	Rosa (1985)
Potamotrygon humerosa	NA	Rio Tapajos, Brazil	49	Rosa (1985)
		Rio Amazonas, Brazil	50	Rosa (1985)
		Rio Para, Brazil	62	Rosa (1985)
		Rio Canuma, Brazil	50	Rosa (1985)
Potamotrygon motoro	South American freshwater stingray	Rio Napo, Ecuador	72	Rosa (1985); Lovejoy et al. (1998); Toffoli et al. (2008)
		Rio Paraguay, Rio Aquidabanat, Paraguay	51	Rosa (1985); Lovejoy et al. (1998); Toffoli et al. (2008)
		Rio Cuyuni, Guyana	41	Rosa (1985); Lovejoy et al. (1998); Toffoli et al. (2008)
		Rio Paragui, Rio Tocatins, Rio Solimoes, Rio Amazonas, Brazil	58	Rosa (1985); Lovejoy et al. (1998); Toffoli et al. (2008)
		Rio Negro, Brazil	46	Rosa (1985); Lovejoy et al. (1998); Toffoli et al. (2008)
		Rio Purus, Brazil	49[a]	Rosa (1985); Lovejoy et al. (1998); Toffoli et al. (2008)
		Rio Parana, Brazil	42	Rosa (1985); Lovejoy et al. (1998); Toffoli et al. (2008)
		Rio Tapajos, Brazil	47	Rosa (1985); Lovejoy et al. (1998); Toffoli et al. (2008)
		Rio Trombetas, Brazil	68	Rosa (1985); Lovejoy et al. (1998); Toffoli et al. (2008)
		Rio Parana, Rio Colastine, Argentina	42	Rosa (1985); Lovejoy et al. (1998); Toffoli et al. (2008)
		Rio Uruguay, Uruguay	44	Rosa (1985); Lovejoy et al. (1998); Toffoli et al. (2008)

Species	Common name	Locality		Reference
		Rio Ucayali, Peru	63	Rosa (1985); Lovejoy et al. (1998); Toffoli et al. (2008)
		Rio Orinoco, Venezuela	52	Rosa (1985); Lovejoy et al. (1998); Toffoli et al. (2008)
Potamotrygon magdalenae	Magdalena river stingray	Rio Magdalena, Rio Sucio, Rio Truando, Columbia	53	Rosa (1985)
Potamotrygon leopoldi	White-blotched river stingray	Xingu River, Brazil	55	Toffoli et al. (2008)
Potamotrygon marinae	NA	Maroni River, French Guiana	54	Deynat (2006)
Potamotrygon ocellata	Red-blotched river stingray	Rio Pedreira, Brazil	56	Rosa (1985)
Potamotrygon orbignyi	Smooth back river stingray	Rio Orinoco, Rio Cinarucu, Rio Orituco, Venezuela	52	Lovejoy et al. (1998); Toffoli et al. (2008)
		Rio Tocatins, Rio Amazonas, Rio Souve, Brazil	58	Lovejoy et al. (1998); Toffoli et al. (2008)
		Rio Meta, Columbia	70	Lovejoy et al. (1998); Toffoli et al. (2008)
		Approuage River, French Guiana	58	Lovejoy et al. (1998); Toffoli et al. (2008)
		Essequibo River, Guyana	54	Lovejoy et al. (1998); Toffoli et al. (2008)
Potamotrygon schroederi	Rosette river stingray	Rio Cauro River, Venezuela	57	Toffoli et al. (2008)
Potamotrygon schuemacheri	NA	Rio Colastine Sur, Argentina	52	Rosa (1985)
		Rio Paraguay, Paraguay	51	Rosa (1985)
Potamotrygon scobina	Raspy river stingray	Rio Para, Rio Trombetas, Brazil	62	Toffoli et al. (2008)
Potamotrygon signata	Parnaiba River stingray	Rio Sambito, Rio Poti, Brazil	67	Rosa (1985)
Potamotrygon yepezi	Maracaibo River stingray	Lake Maracaibo, Venezuela	52	Rosa (1985); Lovejoy et al. (1998)
Paratrygon aiereba	Discus ray	Rio San Alejandro, Rio Ucayali, Peru	63	Rosa (1985); Lovejoy et al. (1998)
		Rio Tocatins, Brazil	55	Rosa (1985); Lovejoy et al. (1998)
		Rio Negro, Brazil	46	Rosa (1985); Lovejoy et al. (1998)
		Rio Branco, Brazil	68	Rosa (1985); Lovejoy et al. (1998)
		Rio Para, Brazil	62	Rosa (1985); Lovejoy et al. (1998)
		Rio Jurua, Brazil	48	Rosa (1985); Lovejoy et al. (1998)
		Rio Tapajos, Brazil	49[a]	Rosa (1985); Lovejoy et al. (1998)
		Rio Solimoes, Brazil	46	Rosa (1985); Lovejoy et al. (1998)
		Rio Xingu, Brazil	49	Rosa (1985); Lovejoy et al. (1998)
		Rio Baurer, Bolivia	61	Rosa (1985); Lovejoy et al. (1998)
		Rio Itenez, Bolivia	61	Rosa (1985); Lovejoy et al. (1998)

(Continued)

Table 4.1 (Continued)

Scientific name	Common name	Known freshwater incursions	Map locality[a]	Reference
Plesiotrygon iwamae	Long-tailed river stingray	Solimoes River, Brazil	46	Rosa (1985); Lovejoy et al. (1998); Toffoli et al. (2008)
Heliotrygon gomesi	Gomes round ray	Rio Napo, Ecuador	72	De Carvalho and Lovejoy (2011)
Heliotrygon rosai	Rosa's round ray	Rio Jamari, Amazonas, Brazil	61	De Carvalho and Lovejoy (2011)
Pristis microdon	Smalltooth sawfish	Rio Nanay, Amazonas, Peru	63	De Carvalho and Lovejoy (2011)
		Perak River, Sumatra, Indonesia	14	Smith (1931a)
		Mekong River, Laos, Cambodia	28	Rainboth (1996); Kottelat (2001)
		Lower Shire River, Malawi	3	Skelton (2001)
		Phongolo River, Kwazulu-Natal	4	Skelton (2001)
		Kapuas River, Indonesia	37	Kottelat et al. (1993)
		Gilbert River, Queensland, Australia	35	Taniuchi and Shimizu (1991)
		Daly River, Northern Territory, Australia	32	Taniuchi and Shimizu (1991)
		Sepik River, Papua New Guinea	7	Taniuchi and Shimizu (1991)
		Lake Murray, Papua New Guinea	64	Taniuchi and Shimizu (1991)
		Betsiboka River, West Madagascar	9	Taniuchi et al. (2003)
		Indragiri River, Sumatra, Indonesia	27	Otake (1991)
		Batang Hari Basin, Sumatra, Indonesia	38	Tan and Lim (1998)
		Lake Santani, Papua New Guinea	60	Boeseman (1956)
Pristis perotteti	Largetooth sawfish	Amazon, Brazil	50	Thorson (1974)
		Lake Bayano, Panama	11	Montoya and Thorson (1982)
		Lake Nicaragua, Nicaragua	59	Thorson (1982)

NA: not applicable.
[a]Nearest location (Fig. 4.1). Some locations are omitted for clarity.

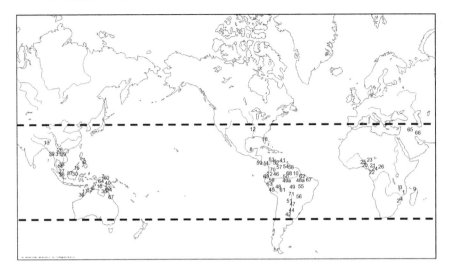

Fig. 4.1. Map of documented incursions into fresh or brackish water. Refer to Table 4.1 for the exact locations and references. The dashed lines indicate approximate positions of 35°N and 35°S latitude.

perotteti). This represents less than 0.5% of elasmobranch species. A list of all species known to enter or live in FW as well as their locations is provided here (Table 4.1; Fig. 4.1) to update the original listing of Smith (1931a) in his physiological studies of FW elasmobranchs.

The lack of success in penetrating FW has been attributed to multiple factors (Ballantyne and Robinson, 2010). These include the requirement of some proteins for high urea levels for optimal function, which may constrain FW adaptation. Since the sensory modalities of elasmobranchs include electroreception and since the conductivity of the environment changes with salinity, the capacity of this sense to adapt to low salinity may also be problematic. Another constraint may relate to reproductive physiology. All elasmobranchs reproduce by internal fertilization, either laying eggs or giving birth to live young. Changing salinity imposes limits on reproduction, perhaps due to issues of sperm viability or poor viability of eggs released in FW. These constraints will be discussed in more detail in subsequent sections and further limitations introduced.

The earliest physiological studies of the effects of changing salinity on elasmobranchs were those of Garrey (1905) and Dakin (1908), who showed a decrease in plasma osmotic pressure in elasmobranchs placed in FW. The specimens used were stenohaline marine species and died within a few hours of transfer. Changes in plasma osmolarity were observed but the basis for the changes was not determined. The first substantive investigations of euryhaline elasmobranchs were those of Homer Smith (1931a). He traveled widely to study elasmobranchs in FWs of Southeast Asia, and Central and South America. His studies of *C. leucas* in Lake Nicaragua were followed by those of Thorson, who further expanded our understanding of euryhaline elasmobranchs (Thorson, 1962, 1967; Thorson et al., 1973). Thorson was the first to demonstrate that the South American FW potamotrygonid stingrays could not produce significant amounts of urea and had a very limited capacity to tolerate higher salinities (Thorson et al., 1967; Thorson, 1970) and thus differed from all other elasmobranchs in FW. These early studies are important not only as the earliest investigations of elasmobranchs in FW, but as they may well become the only investigations of certain species since some of the species examined are rare (Martin, 2005) and difficult if not impossible to obtain for physiological studies.

This chapter will focus on the biochemical and physiological processes that are affected by salinity challenge. it will also examine some of the broader aspects of the biology of elasmobranchs and osmoregulatory challenges as they relate to understanding euryhalinity and its relative rarity in elasmobranchs.

2. DISTRIBUTION

Elasmobranchs found in FW are largely confined to tropical and subtropical latitudes between about 35°N and 35°S, with the greatest diversity close to the equator (Fig. 4.1). The current distribution of elasmobranchs in FW has been substantially impacted by changing sea levels due to glaciation/deglaciation cycles. There are two main foci of diversity in FW incursions: tropical South America and Southeast Asia. The South American potamotrygonid stingrays have speciated in rivers that have been isolated by repeated marine incursions over the past 25 million years. This has resulted in the evolution of more than 20 species. More recently, in Southeast Asia, the Sunda shelf was inundated by a 100 m rise in sea level ending 10,000 years ago (Hanebuth et al., 2000). This event isolated species such as the white-edged whipray (*Himatura signifer*) and the giant river ray (*H. chaophraya*) in rivers of Sumatra, Borneo, and Thailand. Such a short and recent history of isolation may explain why so few FW species of elasmobranchs are found in Southeast Asia. Speciation on the scale of the South American potamotrygonids would take much longer. Similarly, one dasyatid in North America (*D. sabina*) may have been isolated in FW in Florida by the same sea level rise in the late Pleistocene (Johnson and Snelson, 1996). The diversity of African FW dasyatid species may not have been fully documented and currently appears to be low.

3. PHYLOGENY OF EURYHALINE ELASMOBRANCHS

The ability to tolerate reduced salinity and FW environments appears to have evolved several times among extant elasmobranchs (Fig. 4.2). According to Martin (2005), there are 13 euryhaline and 36 obligate FW elasmobranchs described. Of these, the majority (around 76%) are confined within the order Myliobatiformes. The current authors estimate that approximately 84% of elasmobranchs documented in FW habitats belong to the Myliobatiformes. Furthermore, they found that euryhaline/FW habitat use occurs in 10 extant genera, of which seven are from the order Myliobatiformes. This is substantial when compared to estimates of only 5% of all elasmobranchs being able to commonly enter FW environments (Helfman et al., 1997). Within the Myliobatiformes, FW and euryhaline rays are confined to just two families: Dasyatidae (accounting for around 30% of FW or euryhaline species reported here) and Potamotrygonidae (around 52%). It is worth noting that of only seven species within the family Pristidae, Martin (2005) describes six as being euryhaline. The authors of this chapter have adopted a more restricted definition of euryhalinity. Regardless, tolerance of reduced salinity environments appears to be a major defining trait of this family. The stenohaline FW family Potamo-trygonidae is found only in FW habitats (Thorson et al., 1967).

Establishing phylogenetic relationships within Myliobatiformes is difficult. The exact relationship between the FW-tolerant genera *Himantura*, *Dasyatis*, and *Potamotrygon* has yet to be determined. Of particular note, some datasets suggest that *Himantura* and *Potamotrygon* are sister clades, while others do not support this relationship (Dunn et al., 2003). This is of interest in determining whether euryhalinity (as in some dasyatid rays) precedes evolution of the complete life cycle in FW (potamotrygonid rays) as suggested by Lovejoy (1996). Alternatively, as suggested by Brooks et al. (1981), invasion and adaptation to FW can occur directly from a marine stock. For a more detailed review of phylogenetic studies pertaining to the relationship of these three genera, see Rosa et al. (2010). The complexities of the relationships in this group remain unresolved. As noted by Wueringer et al. (2009), the phylogenetic relationship of sawfishes to other extant batoids remains unclear, owing in part to the absence of pristids from

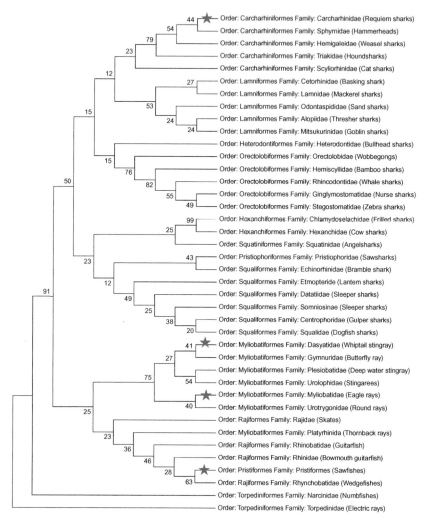

Fig. 4.2. The evolutionary history of the Euselachii was inferred using mitochondrial *col* sequence data and the maximum parsimony (MP) method. The bootstrap consensus tree inferred from 100 replicates is taken to represent the evolutionary history of the taxa analyzed (Felsenstein, 1985). Branches corresponding to partitions reproduced in less than 50% bootstrap replicates are collapsed. The percentage of replicate trees in which the associated taxa clustered together in the bootstrap test (100 replicates) is shown next to the branches (Felsenstein, 1985). The MP tree was obtained using the close-neighbor-interchange algorithm (Nei and Kumar, 2000, p. 128) with search level 1 in which the initial trees were obtained with the random addition of sequences (10 replicates). The analysis involved 40 nucleotide sequences. Codon positions included were 1st+2nd+3rd+non-coding. All positions containing gaps and missing data were eliminated. There was a total of 618 positions in the final dataset. Evolutionary analyses were conducted in MEGA5 (Tamura et al., 2011).

molecular studies of the evolutionary systematics of elasmobranchs. The present authors' phylogenetic analysis using DNA barcodes (Mt-*co1*) placed the Pristiformes with the Rajiformes (Fig. 4.2).

In contrast to the stingrays, only 10% of species reported here as documented in FW belong to the family Carcharhinidae. Martin (2005) reports 23% of euryhaline species as being carcharhinid sharks. Besides the bull shark (*C. leucas*) and possibly blacktip reef sharks (*C. melanopterus*), all other examples of FW sharks are restricted to the poorly known genus *Glyphis* (Compagno, 1984). Although *Glyphis* spp. have a history of being misidentified as *C. leucas*, probably due to their presence in FW habitats, analysis of the mitochondrial gene *co1* showed *Glyphis* and *Carcharhinus* to be two distinct clades (Wynen et al., 2009). Vélez-Zuazo and Agnarsson (2011) did not find *C. leucas* and *Glyphis* to be especially close relatives. Instead, river sharks (*Glyphis* spp.) appear to be more closely related to the broadfin shark (*Lamiopsis temminckii*) and lemon sharks (*Negaprion* spp.). This is consistent with the authors' findings using *co1* sequence data and maximum parsimony analysis (Fig. 4.3).

4. OSMOREGULATION

Osmoregulation in marine elasmobranchs is very different from that of teleost fishes. In SW they retain high levels of organic solutes in their tissues to raise their osmolarity close to that of SW. The main organic solute is urea but high levels of methylamines are also accumulated. In FW, two situations occur: one group of elasmobranchs retains lower but osmotically significant amounts of urea and methylamines, while another group (the potamo-trygonid stingrays of South America) has eliminated urea and methylamines altogether. In FW, the euryhaline elasmobranchs belong to the urea-retaining group. The basis for this strategy is explained in more detail below. In FW, all elasmobranchs face the need to take up ions from the environment, while in SW excess ions must be excreted.

Marine elasmobranchs are almost always hyperosmotic and this is central to understanding their water relations. The only instances of hypoosmotic conditions occur in situations where elasmobranchs suddenly enter higher salinities (Table 4.2). To properly document their osmotic status with respect to the environment, environmental salinity must be measured at the point of capture. Unfortunately, many studies have not been rigorous in doing this. The difference in osmolarity between the marine environment and the plasma is usually very small (Table 4.2) and the hyperosmotic state can be easily overlooked, especially if water samples

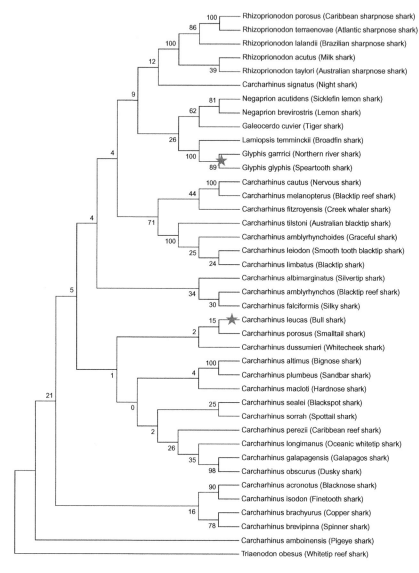

Fig. 4.3. The evolutionary history of the Carcharhiniformes was inferred using mitochondrial *col* sequence data and the maximum parsimony (MP) method. The bootstrap consensus tree inferred from 100 replicates is taken to represent the evolutionary history of the taxa analyzed (Felsenstein, 1985). Branches corresponding to partitions reproduced in less than 50% bootstrap replicates are collapsed. The percentage of replicate trees in which the associated taxa clustered together in the bootstrap test (100 replicates) is shown next to the branches (Felsenstein, 1985). The MP tree was obtained using the close-neighbor-interchange algorithm (Nei and Kumar, 2000, p. 128) with search level 1 in which the initial trees were obtained with the random addition of sequences (10 replicates). The analysis involved 40 nucleotide sequences. Codon positions included were 1st+2nd+3rd+non-coding. All positions containing gaps and missing data were eliminated. There was a total of 639 positions in the final dataset. Evolutionary analyses were conducted in MEGA5 (Tamura et al., 2011).

Table 4.2
Plasma osmolality versus environmental osmolality for valid paired samples.

Species	Plasma osmolarity (mOsm L^{-1})	Environmental osmolarity (mOsm L^{-1})	Osmolarity Above (+) or below (−) ambient	Reference
Carcharhinus leucas	1067	1047	+20	Pillans and Franklin (2004)
	642	3	+639	Pillans and Franklin (2004)
	798 ± 16	706 ± 114	+92	Reilly et al. (2011)
	640 ± 14	3.2 ± 0.6	+637	Reilly et al. (2011)
Negaprion brevirostris	1117	1130	+13	Goldstein et al. (1968)
Sphyrna tiburo	733	615	+118	Mandrup-Poulsen (1981)
	933	896	+37	Mandrup-Poulsen (1981)
	1095	1154	−59	Mandrup-Poulsen (1981)
Triakis scyllium	1000	998	+2	Hyodo et al. (2004b)
	1267	1269	−2	Hyodo et al. (2004b)
	734	567	+167	Hyodo et al. (2004b)
Scyliorhinus canicula	1173	1136	+37	Robertson (1989)
Poroderma africanum	960	940	+20	Haywood (1975)
Hemiscyllium plagiosum	1032	1143	−111	Chan and Wong (1977)
	882	877	+5	Chan and Wong (1977)
	545	412	+133	Chan and Wong (1977)
	448	288	+160	Chan and Wong (1977)
	435	169	+266	Chan and Wong (1977)
Squalus acanthias	973 osmolal	967 osmolal	+6	Wood et al. (2007)
	993 osmolal	965 osmolal	+28	Robertson (1975)
Dasyatis sabina	1021	1018	+3	de Vlaming and Sage (1973)
	1048	986	+62	de Vlaming and Sage (1973)
	892	813	+79	de Vlaming and Sage (1973)
	898	623	+275	de Vlaming and Sage (1973)
	790	556	+234	de Vlaming and Sage (1973)
	787	354	+433	de Vlaming and Sage (1973)
	629	262	+367	de Vlaming and Sage (1973)
Leucoraja erinacea	1033	995	+38	Smith (1931b)

(Continued)

Table 4.2 (Continued)

Species	Plasma osmolarity (mOsm L^{-1})	Environmental osmolarity (mOsm L^{-1})	Osmolarity Above (+) or below (−) ambient	Reference
Trygonorhina	1132	1080	+52	Browning (1978)
fasciata	920	930	−10	Browning (1978)
	1090	960	+130	Browning (1978)
	1405	1420	−15	Browning (1978)

are not taken from the site of capture. A slightly hyperosmotic condition is important as it implies that elasmobranchs do not need to drink significant amounts of SW and thus have less salt (NaCl) to get rid of. This can be accomplished by an internalized salt gland (the rectal gland). On the other hand, hypoosmotic teleost fishes in SW must drink constantly and have a much greater need to excrete NaCl, which is accomplished by specialized ionocytes in the gill (Edwards and Marshall, 2013, Chapter 1, this volume).

Before the evolutionary appearance of the elasmobranchs, most osmoconforming marine invertebrates and early craniates such as hagfish used organic compounds such as amino acids and methylamines in their cells to replace the main inorganic ions Na^+ and Cl^-, while the plasma resembled SW in the levels of all major ions. The ancestral elasmobranch replaced the high levels of amino acids with urea in the cells while retaining the methylamines. The blood was similarly modified by replacing large concentrations of Na^+ and Cl^- with urea and methylamines.

The urea-retaining strategy of elasmobranchs requires adaptation of many aspects of their anatomy, physiology, and metabolism. The first measurements of urea in elasmobranchs were conducted by Staedeler and Fredrichs (1859). At that time urea was largely regarded as a waste product because of its excretion by mammals. In the 1930s Homer Smith recognized urea as a "special metabolite" rather than a waste product (Smith, 1929, 1931a). Urea is indeed a special metabolite and its "choice" as the major solute is intriguing. Thus, while urea plays a role in buoyancy and acid–base regulation in marine elasmobranchs (see Withers, 1998, for a review), it is worth examining its other properties to gain some insight into why this particular molecule has such an important osmotic role.

The ideal solute to replace inorganic ions in the blood and organic compounds inside the cells should be uncharged, small, and rapidly diffusing, with a high solubility. Urea has all of these properties. It is a small, uncharged molecule with a diffusion coefficient (1.35×10^5 cm^2 s^{-1}) (Gosting and Akeley, 1952) about 60% that of water (2.2×10^5 cm^2 s^{-1} in physiological concentrations of urea) (Easteal, 1990). Osmotic adjustment would be facilitated with such a molecule partly through solvent drag. This would be beneficial in situations of changing external salinity such as inshore environments including estuaries. It would thus appear that the useful physical properties of urea in solution outweigh the high cost of synthesis of urea in elasmobranchs [5 adenosine triphosphate (ATP) per urea]. The high cost of synthesis of urea means that losses to the environment must be minimized. Thus, the tissues interfacing with the environment need to be impermeable to urea while all other membranes within the animal must be highly permeable.

The disruptive effects of urea on protein structure are well documented. Both direct and indirect effects are involved (Das and Mukhopadhyay, 2009). The direct effect involves urea binding to the protein, displacing water and denaturing the protein. The indirect effect involves the effect of urea on the structure of water, weakening the hydrophobic interactions necessary for protein structure. In marine elasmobranchs the urea levels are high enough to affect protein structure and function. A role of methylamines such as trimethylamine *N*-oxide (TMAO) in counteracting the disruptive effects of urea has been documented (Yancey and Somero, 1978, 1979; Yancey et al., 1982). TMAO acts to counteract the effects of urea indirectly by altering the water structure in the opposite way to urea (Zou et al., 2002). In general, methylamines like TMAO compact proteins and activate them, while urea unfolds them and inhibits them. The effects of urea and methylamines do not always act in a counteracting mode, especially when the compact forms of enzymes are not the most active (Mashino and Fridovich, 1987). A ratio of two urea molecules to one methylamine has been suggested as being optimal (Yancey and Somero, 1979). This ratio has been supported by thermodynamic measurements of ribonuclease that indicate equal but opposite contributions to protein denaturation at 2:1 (Lin and Timasheff, 1994). *In vivo*, the ratio of urea to counteracting solutes in elasmobranch tissues is close to 2:1, but both methylamines and β-amino acids must be tallied to achieve this ratio (Treberg et al., 2006). Some species deviate from this ideal ratio (Treberg et al., 2006).

As environmental salinity decreases, solutes must be eliminated, and urea and methylamines are the main osmolytes that are reduced. Ideally, in FW urea and methylamines should be completely eliminated, but this is not the case for euryhaline elasmobranchs (Table 4.3). Indeed, some species that apparently reside exclusively in FW (e.g. *H. signifer*) also retain significant

Table 4.3

Plasma and urine major ion and osmolyte levels in elasmobranchs in freshwater (FW), seawater (SW), and dilute SW.

Species	Na⁺	K⁺	Cl⁻	Mg²⁺	Ca²⁺	PO₄²⁻	SO₄²⁻	Urea	TMAO	Osmolarity
Leucoraja erinacea										
Plasma SW (Stolte et al., 1977)	2666±6	5.4±0.5	301±13	1.2±0.2	1.7±0.2	1.9±0.3	5.4±0.5			968±16
Plasma SW (Forster and Goldstein, 1976)	299±5	4.96±0.4						361±18	39±3	965±18
Urine SW (Stolte et al., 1977)	180±16	32±9	209±16	154±29	10.4±1.6	100±23	61±11			967±19
Plasma 50% SW (Forster and Goldstein, 1976)	217±5	4.25±0.2						264±12	30±3	719±14
Plasma 75% SW (Stolte et al., 1977)	182±11	3.9±1.1								811±40
Urine 75% SW (Stolte et al., 1977)	79±22	44±6.5								
Dasyatis americana										
Plasma SW (Cain et al., 2004)	315	5	342		4.12	1.5		444		1065
Dasyatis sabina										
Plasma SW (1018 mM) (de Vlaming and Sage, 1973)	301±3		310±7					383±8		
Plasma SW (853 mM) (Janech et al., 2006b)	278±13		280±13					349±19		
Plasma FW (Piermarini and Evans, 1998)	211±3		207±3					196±8		
Plasma FW (Janech and Piermarini, 2002)	187±7	2±0	167±8					181±9		557±21
Urine FW (Janech and Piermarini, 2002)	8±1							20±2		53±2
Hemiscyllium plagiosum										
Plasma SW (Chan and Wong, 1977)	272±13	5.02±0.4	238±10	2.24±0.23	4.5±0.2			468±25		
Plasma dilute SW 169 mOsm (Chan and Wong, 1977)	135±12	2.74±0.4	128±0.4	0.53±0.11	1.5±0.08			99±11		

Urine SW (Wong and Chan, 1977)	249±18	3.6±0.7	225±16	51±15	3.3±0.65			248±21	90±9	797±34
Urine 0.5 SW (Chan and Wong, 1977)	147±8	3.7±1.2	147±11	6.16±1.3	1.43±0.11			138±10	72.4	450±8
Squalus acanthias										
Plasma SW (Bedford, 1983)	268±4	5.0±0.1	248±3	5.2±0.4	8.6±0.5	10.9±0.9		376±24	90±9	987±6
Plasma SW (Robertson, 1975)								308		
Plasma SW (Wood et al., 1995)								318±6		
Plasma SW (Thurau et al., 1969)	266							348	71	
Plasma SW (Goldstein and Palatt, 1974)								274		
Plasma ~50% SW (Thurau et al., 1969)	234									
Plasma SW (Burger, 1967)	250	4	240	1.2	3.5	0.97	0.5	350	70	1000
Urine SW (Burger, 1967)	240	2	240	40	3	33	70	100	10	800
Urine SW (Wood et al., 1995)								63±15		
Urine SW (Thurau et al., 1969)	281							94		
Urine ~50% SW (Thurau et al., 1969)	124							142		
Scyliorhinus canicula										
Plasma SW (Robertson, 1989)	307±4	6.2±0.24	289±4	3.1±0.5	3.4±0.13	4.4±0.5		407±6	53±6	1115
Ginglymostoma cirratum										
Plasma SW (Goldstein and Palatt, 1974)									75	
Negaprion brevirostris										
Plasma SW (Goldstein and Palatt, 1974)									69	
Plasma SW (Goldstein et al., 1968)			310±5					421±2	76±4	
Mustelus antarcticus										
Plasma SW (Trivett et al., 2001)	296		294		4.28			489		

(Continued)

Table 4.3 (Continued)

Species	Na^+	K^+	Cl^-	Mg^{2+}	Ca^{2+}	PO_4^{2-}	SO_4^{2-}	Urea	TMAO	Osmolarity
Plasma 80% SW (Trivett et al., 2001)	296		296		4.75			473		
Carcharhinus leucas										
Plasma SW (Pillans et al., 2005)	304±3	5.8±0.3	315±3	1.8±0.1	4.4±0.3			293±9	47±4.5	
Plasma FW (Thorson, 1967)	245.8	6.4	219.3	3.2	8.9			180		
Plasma FW (Reilly et al., 2011)	234±2	4.1±0.1	231±2	1.3±0.04	3.1±0.07			168±7		
Plasma FW (Anderson et al., 2006)	213±3		215±5					165±7		
Plasma FW (Pillans et al., 2005)	221±4	4.2±0.2	220±4	1.3±0.1	3.0±0.1			151±5	19±1.3	
Trygonoptera testacea										
Plasma SW (Cooper and Morris, 1998)		4.7		0.58	6.25				54	
Plasma 75% SW (Cooper and Morris, 1998)		3.8		0.58	5.09					
Plasma 50% SW (Cooper and Morris, 1998)		4.6		0.35	4.20				38	
Heterodontus portusjacksoni										
Plasma SW (Withers et al., 1994)	317±7	7.0±0.3	306±5	2.1±0.4	4.0±0.03			353±5	81±.05	
Plasma SW (Cooper and Morris, 1998)		4.5		1.14	4.21				74	
Plasma 75% SW (Cooper and Morris, 1998)		4.1		0.96	4.23				44	
Plasma 50% SW (Cooper and Morris, 1998)		3.06		0.6	3.24				48	

Orectolobus ornatus									
Plasma SW (Otway et al., 2011)	287	5.2	277	1.9	4.6				396
Sphyrna tiburo									
Plasma SW (Harms et al., 2002)	282	7.3	290		4.2				358
Plasma SW 20 ppt (Mandrup-Poulsen, 1981)	242±4	4.8±0.2	264±6	5.8±0.4	3.6±0.5			179±33	45±14
Plasma SW 30 ppt (Mandrup-Poulsen, 1981)	258±3	6.0±1.8	279±8	6.6±1.8	4.5±0.9			290±30	67±17
Plasma SW 40 ppt (Mandrup-Poulsen, 1981)	319±14	6.1±1.3	354±24	10±1.3	5.3±0.6			354±24	95±14
Pristis microdon									
Plasma FW (Ishihara et al., 1991; Otake, 1991)	156	10.6	166	1.2	2.8			87	553
Plasma FW (Smith, 1931a)			170					130	
Urine FW (Smith, 1931a)	2.2	6.3		1.3	1.7	6.8	0.33	14	55
Pristis perotteti									
Plasma FW (Gerzeli et al., 1969)	205±14	3.8±0.4	236±28	4.4±0.6	4.8±0.9	1.01±0.15		116±7	
Plasma FW (Thorson, 1967)	216.6	6.5	193.1	1.7	8.3				
Dasyatis bennetti									
Plasma FW (Ballantyne and Robinson, 2010)								140	
Himantura uarnak									
Plasma FW (Otake, 1991)								75	

(Continued)

Table 4.3 (Continued)

Species	Na^+	K^+	Cl^-	Mg^{2+}	Ca^{2+}	PO_4^{2-}	SO_4^{2-}	Urea	TMAO	Osmolarity
Himantura signifier										
Plasma FW (Chew et al., 2006)								78		
Dasyatis garouaensis										
Plasma FW (Thorson and Watson, 1975)								212		
Potamotrygon hystrix										
Plasma FW (Bittner and Lang, 1980)								0.68		
Potamotrygon motoro										
Plasma FW (Tam et al., 2003)								0.65 ± 0.17		
Potamotrygon magdalenae										
Plasma FW (Ogawa and Hirano, 1982)	141 ± 2	8.5 ± 0.29	147 ± 5	1.3 ± 0.06	3.0 ± 0.1			0.7 ± 0.1		358 ± 7
Paratrygon aiereba										
Plasma FW (Duncan et al., 2009)	156 ± 0.7	5.7 ± 0.4	178 ± 2		2.9 ± 0.4			4.2 ± 0.3		305 ± 16

All values are in mM except for osmolarity, which is in mOsmol.
TMAO: trimethylamine oxide.

amounts of urea (Table 4.3). The basis for this is not known but may lie in the apparent requirement of some proteins for urea. For example, the M4 lactate dehydrogenase of marine elasmobranchs requires urea to maintain an optimal K_m for pyruvate (Yancey and Somero, 1978). Similarly, an eye lens protein of marine elasmobranchs precipitates below $10°C$ in the absence of urea (Zigman et al., 1965). If all proteins do not have the same susceptibility or requirement for urea (Ballantyne, 1997) this could present an important constraint on the colonization of FW environments. Only a few key proteins would need to be rendered non-functional in the absence of urea to make the total elimination of urea in FW a lethal strategy. This may be one of the main constraints on colonizing FW and may explain why all but the potamotrygonids retain significant levels of urea in FW (Table 4.3). Thus, euryhaline elasmobranchs in FW may need to retain enough urea to maintain key protein conformation in spite of the osmotic disadvantages of this strategy.

4.1. Ion Regulation

Euryhaline elasmobranchs in general do not regulate ion levels in plasma or intracellular compartments as well as euryhaline teleost fishes. Marine elasmobranchs maintain plasma Na^+ and Cl^- levels higher than those of most teleost fishes and these are decreased by about one-third in FW in euryhaline species such as *D. sabina* and *C. leucas* (Table 4.3). The levels of inorganic ions in tissues are rarely reported, but in *Hemiscyllium plagiosum* intracellular K^+ and Na^+ decrease by about two-thirds with environmental dilution from 1147 mOsm L^{-1} to 169 mOsm L^{-1} (Chan and Wong, 1977). These large changes in plasma and intracellular ion levels with entry into FW in euryhaline species may compromise membrane potentials and protein properties. The capacity of non-osmoregulatory tissues to deal with osmotic imbalances varies. Some parts of the brain, such as the medulla oblongata, have no capacity to deal with osmotic disturbances, while others, such as the telencephalon, can effectively regulate ions (Cserr et al., 1983). Such limitations may prove disruptive and contribute to poorer performance in FW.

Various tissues are exposed to the environment and thus have the potential to act as osmoregulatory organs. Of these, the gills have the largest interface. Unlike teleosts, marine elasmobranchs do not use the gills as the main site of ion. The rectal gland is a specialized salt-secreting tissue that is not exposed directly to the environment. The intestine also plays a role in ion regulation but owing to low rates of drinking its role in osmoregulation is more limited. The physiology of these tissues and their roles in salinity acclimation is discussed in more detail below.

4.2. Gills

The three primary functions of the gills of elasmobranchs are gas exchange, acid–base regulation, and urea retention. Without a substantial role in salt secretion in the gill, the activity of Na^+/K^+-ATPase (NKA) in marine elasmobranch gill can be an order of magnitude lower that than of marine teleost fishes (Jampol and Epstein, 1970). Gill respiration rate accounts for 9.8% of basal metabolic rate in *Squalus acanthias* but NaCl transport (ouabain-sensitive respiration) accounts for only 0.14% (Morgan et al., 1997). This compares to 2.4% of whole body respiration for the gills of a marine teleost, with the ouabain-sensitive component accounting for 0.9% of whole body respiration (Morgan and Iwama, 1999). Based on measurements of two marine elasmobranchs and three marine teleosts, the gills of elasmobranchs are half as permeable to water as those of marine teleosts (Table 4.4). The permeability of the gills does not change with transfer from SW to lower salinities (Goldstein and Forster, 1971a). The turnover of water in both marine and FW elasmobranchs is higher than that of marine or FW teleosts (Ballantyne and Robinson, 2010) (Table 4.5), implying other sources of water uptake.

4.2.1. UREA RETENTION IN THE GILLS

The large surface area of the gill in contact with the environment makes it the major site for urea loss. More than 90% of the loss of urea occurs at the gills, while only 2% is lost at the kidney (Wood et al., 1995). In spite of this, the elasmobranch gill is 50–100-fold less leaky than that of a teleost fish gill (Part et al., 1998) (Table 4.4). Thus, an important structural characteristic of the gill is a low permeability to urea. Since the gradient of urea concentrations between the plasma and the environment is infinite, maintaining low urea concentrations in the gill cells would facilitate urea retention. Although urea levels in gill cells have not been measured they have been inferred in several studies (Part et al., 1998; Fines et al., 2001). Low intracellular urea concentrations in gill cells are thought to occur via a sodium-dependent urea transporter in the basolateral membrane that pumps urea out of gill cells to the plasma side in exchange for extracellular sodium (Fines et al., 2001). NKA in the basolateral membrane thus provides the energetic basis for urea retention. In addition to the involvement of transporters in retaining urea at the gills a low permeability is achieved by high membrane cholesterol levels in the basolateral membrane (Fines et al., 2001). High cholesterol content of membranes correlates with low permeability to urea (Pugh et al., 1989). Gill apical and basolateral membrane vesicles have similar low permeabilities to water and urea (Hill et al., 2004) (Table 4.4).

Table 4.4

Permeabilities of elasmobranch and teleost epithelia and membranes to urea and water.

Species	Tissue	Permeability coefficient water ($cm\ s^{-1}$)	Permeability coefficient urea ($cm\ s^{-1}$)	Reference
Squalus acanthias	Rectal gland basolateral membrane	$4.3 \pm 1.3 \times 10^{-3}$	$4.2 \pm 0.8 \times 10^{-7}$	Zeidel et al. (2005)
	Rectal gland apical membrane	$7.5 \pm 1.6 \times 10^{-4}$	$2.2 \pm 0.4 \times 10^{-7}$	Zeidel et al. (2005)
	Gill (whole)	6.55×10^{-6}	3.2×10^{-8}	Part et al. (1998)
	Gill whole	$6.6 - 7.6 \times 10^{-6}$	$3.2 - 7.5 \times 10^{-8}$	Hill et al. (2004)
	Gill whole	7.6×10^{-6}	7.5×10^{-8}	Boylan (1967)
	Gill apical	$7.4 \pm 0.7 \times 10^{-4}$	$4.3 \pm 1.7 \times 10^{-7}$	Hill et al. (2004)
	Gill basolateral	$14 \pm 2 \times 10^{-4}$	$6.5 \pm 1.1 \times 10^{-7}$	Hill et al. (2004)
Scyliorhinus canicula	Gill whole	5.6×10^{-6}	0.004×10^{-6}	Payan and Maetz (1971)
Anguilla anguilla SW	Gill whole	14.6×10^{-6}	2.7×10^{-6}	Steen and Stray-Pedersen (1975)
Oncorhynchus mykiss SW	Gill whole	$7.7 \pm 0.9 \times 10^{-6}$	$2.6 \pm 0.8 \times 10^{-6}$	Isaia (1982)
Oncorhynchus mykiss FW	Gill whole	$8.8 \pm 1.4 \times 10^{-6}$	$1.7 \pm 0.8 \times 10^{-6}$	Isaia (1982)
Pleuronectes americanus SW	Gill whole	$1.5 - 1.6 \times 10^{-5}$	$2.2 - 2.6 \times 10^{-6}$	Hill et al. (2004)
	Gill apical	$6.6 \pm 0.8 \times 10^{-4}$	$5.9 \pm 0.5 \times 10^{-7}$	Hill et al. (2004)

SW: seawater; FW: freshwater.

Table 4.5
Turnover of Na$^+$ Cl$^-$ and water in marine and freshwater elasmobranchs and teleosts.

Species	Na$^+$ (% h^{-1})	Cl$^-$ (% h^{-1})	Water (% h^{-1})	Reference
Seawater elasmobranchs				
Raja montagu			167 ± 11	Payan and Maetz (1971)
Raja erinacea			64 ± 1.9	Payan et al. (1973)
Ginglymostoma cirrhatum	0.46	1.52	81	Carrier and Evans (1972)
Scyliorhinus canicula			157 ± 13	Payan and Maetz (1971)
Torpedo marmorata			97 ± 8.5	Payan and Maetz (1971)
Poroderma africanum			97	Haywood (1974)
Freshwater elasmobranchs				
Potamotrygon sp.	0.28	0.22	96	Carrier and Evans (1973)
Seawater teleosts				
Platichthys flesus			36	Evans (1969)
	47 ± 2			Motais et al. (1966)
Serranus scriba			24 ± 1	Motais et al. (1969)
	42 ± 7			Motais et al. (1966)
Anguilla Anguilla			34	Evans (1969)
	0.26	0.26		Isaia and Masoni (1976)
Freshwater teleosts				
Platichthys flesus	0.4			Motais et al. (1966)
Carassius auratus	51 ± 3			Motais et al. (1969)

At lower salinities urea loss declines as a result of the reduced urea concentration and concomitant lower urea gradient across the gill (Wong and Chan, 1977; Part et al., 1998). Part of the maintenance of low permeability to urea at different tissues levels may relate to the changing gill membrane structure. Studies with *H. signifer* indicate that the cholesterol content of the gill membranes increases with increasing salinity, consistent with the need to reduce permeability in the face of higher urea gradients at high salinities (J. S. Ballantyne, unpublished data).

The gills of marine elasmobranchs are more impermeable to ammonia than those of teleost fishes. It has been suggested that a system for retaining ammonia at the gills may exist to salvage nitrogen for urea synthesis (Wood et al., 1995).

Table 4.6
Drinking rates of marine elasmobranchs and marine and freshwater teleosts.

Species	Drinking rate (ml 100 g^{-1} h^{-1})	Reference
Seawater elasmobranchs		
Triakis scyllia (80% SW)	100–150	Anderson et al. (2002)
Scyliorhinus canicula (80% SW)	10–15	Anderson et al. (2002)
Triakis scyllia (100% SW 24 h)	60	Anderson et al. (2002)
Scyliorhinus canicula (100% SW 6 h)	10	Anderson et al. (2002)
Scyliorhinus canicula	29	Hazon et al. (1989)
Freshwater teleosts		
Salmo salar	15	Fuentes and Eddy (1997)
Anguilla japonica	20–40	Hirano (1974)
Seawater teleosts		
Anguilla japonica	10	Hirano (1974)
Salmo salar	38.9	Fuentes and Eddy (1997)
Scophthalmus maximus	0.08	Carroll et al. (1994)
Pleuronectes flesus	0.09	Carroll et al. (1994)
Tilapia mossambica	0.975	Evans (1968)
Pholis gunnelus	0.051	Evans (1968)

4.2.2. ION TRANSPORT

In SW, ion transport at the gills is of limited importance in elasmobranchs. Marine elasmobranchs are slightly hyperosmotic and do not drink significant amounts of SW (Table 4.6). Their need to excrete NaCl is thus substantially reduced and largely handled by the rectal gland. The main ion transporters found in the gill in SW include an Na$^+$/H$^+$ exchanger and NKA, which colocalize in one mitochondrion-rich cell (MRC) type, and pendrin (HCO$_3^-$/Cl$^-$ exchanger) and V-H$^+$ATPase, which colocalize in another MRC type (Reilly et al., 2011). Na$^+$/H$^+$ and pendrin are apical and NKA and V-H$^+$-ATPase are basolateral (Reilly et al., 2011). These transporters are found in gills of both FW and SW-adapted euryhaline species as well as in stenohaline marine species such as *S. acanthias* (Choe et al., 2007). Thus, in SW these transporters are likely to have a greater role in acid–base regulation (see below).

In FW, ions, especially Na$^+$ and Cl$^-$, need to be taken up from the environment. Based on data for *D. sabina* (Evans et al., 2005), the mechanism involves two cell types (Fig. 4.4). In one cell type NKA in

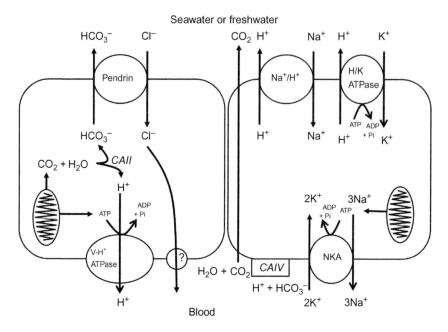

Fig. 4.4. Diagram of elasmobranch branchial mitochondrion-rich cells. Carbonic anhydrase (CA) is present in two forms: a soluble form, CAII, found in the cytosol; and a membrane-bound form, CAIV. See text for explanation of function.

the basolateral membrane moves Na^+ out of the cell, driving the apical uptake of Na^+ in exchange for cytosolic H^+ via the Na^+/H^+ exchanger. Chloride uptake takes place in the other cell type, driven by a basolateral V-H^+-ATPase that pumps protons out of the cell, raising the bicarbonate levels. This allows apical bicarbonate exchange for external Cl^- via pendrin. Cytosolic carbonic anhydrase facilitates the production of HCO_3^-. Several issues remain to be resolved for this mechanism, including identification of the mechanism of Cl^- efflux across the basolateral membrane.

The expression of Na^+/H^+ and NKA messenger RNA (mRNA) increases in both *C. leucas* (Reilly et al., 2011) and *D. sabina* in FW (Choe et al., 2005). Activity and immunoreactivity of NKA increase, as does the number of MRCs in FW *D. sabina* (Piermarini and Evans, 2000), but studies of wild-caught *C. leucas* show no change in gill NKA activity associated with moving between FW and SW (Pillans et al., 2005); the basis for this is unknown. In FW V-H^+-ATPase (Piermarini and Evans, 2001) and pendrin (Piermarini et al., 2002), immunoreactivity increases in *D. sabina*. Thus, in FW, higher expression of Na^+/H^+ and NKA is needed for Na^+ uptake, and higher V-H^+-ATPase and pendrin for Cl^- uptake (Fig. 4.4). An H^+/K^+-

ATPase may function in K^+ uptake in FW since expression increases with FW adaptation (Choe et al., 2004).

Gill NKA characteristics of elasmobranchs appear to be similar to those of teleost fishes. The molecular weight of the active protein is 360,000–381,000 (Hastings and Reynolds, 1979). It binds ouabain in a 1:1 ratio, the same as the teleost enzyme, but is isolated with higher levels of phospholipids due to phosphatidylcholine and phosphatidylethanolamine (Perrone et al., 1973). Associated sphingomyelin is 10-fold higher in the elasmobranch enzyme (Perrone et al., 1973). Membrane cholesterol activates NKA up to 10 mol% in the *S. acanthias* enzyme (Cornelius, 1995). The specific activity (activity per mole of enzyme) of the elasmobranch enzyme is similar to that of teleost electric organ or mammalian kidney (Ottolenghi, 1975) and the stoichiometry is 3:2 for Na:K (Schwarz and Gu, 1988). The enzyme is regulated by a phospholemman-like (PLMLP) protein via interactions of the phosphorylated form of PLMLP with the α-subunit (Mahmmoud et al., 2003). There is evidence for multiple isoforms of the α-subunits of NKA in elasmobranchs, but mRNA and protein expression in gills are unaffected by transfer between FW and SW in *C. leucas* (Meischke and Cramb, 2005).

The design of the ion transport systems of the elasmobranch gill differs substantially from that of teleost fishes. The need for Na^+ extrusion in SW is taken over by the rectal gland in elasmobranchs, so there is no need for the Cl^- channel [cystic fibrosis transmembrane regulatory channel (CFTR)] in the gill. Similarly, the $Na^+/K^+/2Cl^-$ (NKCC) transport protein is not found in the elasmobranch gill. The CFTR and NKCC are found in the rectal gland, where they function much as they do in the SW teleost gill (see below). NKA is not upregulated in SW as in teleost fishes, but is upregulated in FW. V-H^+-ATPase is not in pavement cells as in teleosts (Piermarini and Evans, 2001). In euryhaline teleost fishes, expression of the α1a and α1b subunits of NKA changes reciprocally with acclimation from FW to SW (Bystriansky and Ballantyne, 2007), but this does not occur in elasmobranchs. It is not immediately obvious how these differences may account for the low incidence of euryhalinity in elasmobranchs.

4.2.3. ACID–BASE REGULATION BY THE GILL

In elasmobranchs, acid–base regulation occurs primarily at the gill (Heisler, 1988) and involves Na^+/H^+ and HCO_3^-/Cl^- (pendrin) exchangers, H^+-ATPase, and potentially H^+/K^+ ATPase (Fig. 4.4). As indicated above, the Na^+/H^+ exchanger is apical and resides in a separate cell type from apical pendrin. Acidosis of environmental or metabolic origin requires secretion of acid. Elasmobranchs lack a gill apical H^+-ATPase and thus cannot secrete acid directly. Acid secretion is Na^+ dependent and amiloride sensitive (Evans et al., 1979; Evans, 1982). Two forms of Na^+/H^+ (NAH2 and NAH3) have been found in elasmobranch gills (Edwards et al., 2002). In *S. acanthias* NAH2 is involved in acid secretion (Tresquerres et al., 2005). Since H^+ must be transported out of the gill cells in exchange for environmental Na^+, this is energetically more favorable in SW owing to the high levels of Na^+. The Na^+ taken up is removed from the cells by the basolateral NKA. Thus, the elevated NKA observed in FW-adapted euryhaline elasmobranchs may be needed to facilitate acid–base regulation as well as for Na^+ uptake. Two forms of carbonic anhydrase are found in elasmobranch gills: a cytosolic form and a membrane-bound form. The role of a membrane-bound carbonic anhydrase (CA IV) in *S. acanthias* gill basolateral membrane provides another route for removing acid. CA IV is accessible to the blood and functions to facilitate the excretion of carbon dioxide by catalyzing the dehydration of plasma HCO_3^- (Gilmour et al., 2007) (Fig. 4.4). The enzyme colocalizes with NKA but is also found in pillar cells (Gilmour et al., 2007).

Alkaline stress can occur after feeding as part of the "alkaline tide" associated with digestion (Tresquerres et al., 2007). A cytosolic CA hydrates CO_2 into H^+ and HCO_3^-. The HCO_3^- is excreted via pendrin and the H^+ is pumped into the blood via a $V-H^+$-ATPase, reducing the alkalosis (Fig. 4.4). Alkaline stress results in a relocalization of H^+-ATPase from cytosolic sequestration sites to the basolateral membrane in *S. acanthias* (Tresquerres et al., 2006). The abundance of pendrin carriers increases in FW *D. sabina* (Piermarini et al., 2002) and although this is due to a greater need for Cl^- uptake in FW it would contribute to the capacity for acid–base regulation.

Recovery from acidosis is slower in FW than in SW in *D. sabina* (Choe and Evans, 2003) in spite of these changes to transporter abundance, and this could place further limitations on the successful adaptation to FW.

4.3. Rectal Gland

The rectal gland is a tubular outpocketing of the rectum that is responsible for most of the NaCl excretion in marine elasmobranchs (Fig. 4.5A). The activity of NKA in the rectal gland is seven times higher than that of the gill in *D. sabina* in SW (Piermarini and Evans, 2000). Respiration of the rectal gland represents 1% of the whole animal basal metabolic rate in marine species such as *S. acanthias*, with NaCl secretion (ouabain-sensitive respiration) accounting for a smaller proportion (0.5%) (Morgan et al., 1997). The rectal gland has similar permeabilities to water and urea as the gill in marine elasmobranchs (Table 4.4). This may be due to high levels of cholesterol in the membranes of the rectal glands of elasmobranchs in SW. Gerzeli et al. (1969) found the cholesterol content

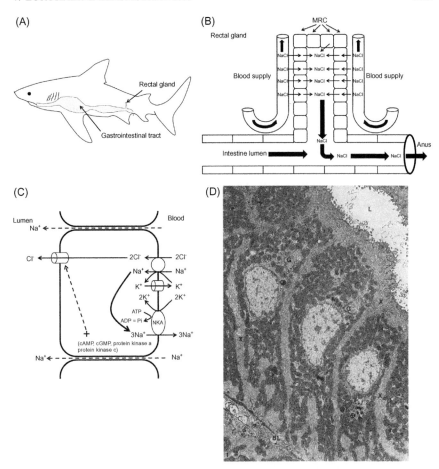

Fig. 4.5. Marine elasmobranchs. (A) Location of the rectal gland in the gastrointestinal tract. (B) Diagram of the cellular structure of the rectal gland. (C) Simplified diagram of a mitochondrion-rich cell with transporters involved in NaCl transport. CFTR: cystic fibrosis transmembrane regulatory channel; (D) Electron micrograph of the rectal gland from *Squalus acanthias* (Ernst et al., 1981). BL: basal lamina; X: infoldings of basolateral membrane; L: lumen; G: Golgi membranes.

of the rectal glands of *P. perotteti* and *C. leucas* to decrease in FW, consistent with a reduced need for urea retention in FW. The same study noted much lower levels of sulfolipids in the rectal glands of both species in FW. Sulfolipids have been implicated in regulating NKA in salt-transporting epithelia and teleost fish gills (Lingwood et al., 2004, 2006). Further research into the remodeling of membrane lipids in the rectal gland of euryhaline elasmobranchs is needed.

The rectal gland has long been a model for studying NaCl secretion owing to its convenient anatomy and simple organization (Silva et al., 1996, 1997). A counter-current arrangement of the blood flow with the lumen fluid flow occurs in at least part of the gland (Fig. 4.5B). There is a considerable body of information on the role and regulation of the rectal gland during salinity challenge. In euryhaline species such as *C. leucas*, the size (Pillans and Franklin, 2004) and morphology (Pillans et al., 2008) change little at lower salinities, probably owing to the need to maintain functionality for re-entry into SW. The activity of NKA is lower in the rectal gland of *D. sabina* (Piermarini and Evans, 2000) and *C. leucas* (Pillans et al., 2005) in FW compared to the activity in SW. This mirrors the changes in teleost fish gill NKA with adaptation to FW. In FW potamotrygonids the rectal gland is vestigial and non-functional (Thorson et al., 1978). Its role in salt secretion cannot be reversed for salt absorption in FW because of its internal location. This indicates how the design of the ion regulatory system of marine elasmobranchs is ill suited to the ion regulatory needs in FW.

The rectal gland secretes a fluid that is isosmotic and isoionic (Na^+ and Cl^-) with SW but with negligible urea (Burger and Hess, 1960). In this way the rectal gland can reduce the NaCl burden without the need to form concentrated urine. In 50% SW the rectal gland of *H. plagiosum* continues to excrete a fluid higher in NaCl than is found in plasma (Wong and Chan, 1977).

The salt transport function is carried out by chloride (i.e. mitochondrion-rich) cells. The mechanism of NaCl excretion is similar to that of teleost chloride cells and many other salt-secreting epithelia. The basolateral NKA creates an Na^+ electrochemical gradient. Na^+ re-enters the cell from the blood along with K^+ and 2 Cl^- via the NKCC cotransporter. Chloride ion concentrations in the cells are thus elevated, resulting in an apical efflux of Cl^- along its electrochemical gradient into the lumen via CFTR. (Fig. 4.5C). This Cl^- efflux is electrogenic with the apical side electronegative with respect to the blood side. This favors Na^+ diffusion via a paracellular pathway into the lumen (Greger et al., 1986). The net result is NaCl extrusion into the lumen, where it exits the body via the anus (Fig. 4.5B).

An Na^+/H^+ exchanger has been found in the rectal gland of *S. acanthias* but its function has not been established (Claiborne et al., 2008). An Na^+/Ca^{2+} exchanger has also been identified in the rectal gland that may be involved in ion regulation (Bleich et al., 1999).

The rate-limiting step in NaCl secretion is the CFTR protein (Riordan et al., 1994). CFTR is upregulated by protein kinase A in elasmobranchs as in humans and downregulated by membrane phosphatases (Hanrahan et al., 1996). Several hormones are involved in regulating salt secretion in the rectal gland. Both vasoactive intestinal peptide (VIP) and cardiac natriuretic peptide (CNP) activate NaCl secretion by the rectal gland by different but synergistic mechanisms. An increase in blood volume results in secretion of CNP from the heart, sensitizing the Cl^- channel via guanylate cyclase (Silva et al., 1996). Other factors are needed to effect increased Cl^- secretion. CNP stimulation of VIP release by nerves of the rectal gland itself activates adenosine monophosphate (AMP) kinase, resulting in cyclic AMP (cAMP) release, which stimulates Cl^- secretion (Lytle and Forbush, 1992a; Epstein and Silva, 2005). cAMP may act by stimulating protein kinase A to phosphorylate CFTR, opening its channel. Other factors such as protein kinases C may be important in maximizing stimulation (Silva et al., 1999; Epstein and Silva, 2005). Neuropeptide Y inactivates Cl^- transport by a mechanism that does not involve cAMP (Silva et al., 1993). Cell shrinkage activates the NKCC cotransporter, but not via cAMP (Lytle and Forbush, 1992a). NKCC is regulated by phosphorylation of serine and threonine residues (Lytle and Forbush, 1992b). Dilution or concentration of the medium stimulates this phosphorylation to activate the transporter (Lytle and Forbush, 1992b).

4.4. Kidney

As in most vertebrate kidneys, urine formation is the result of filtration, secretion, and reabsorption. The urine of marine elasmobranchs is slightly hypoosmotic to the plasma and is lower in Na^+ and Cl^- than plasma, with high concentrations of Mg^{2+}, phosphate, and sulfate (Table 4.3). The kidneys of marine elasmobranchs have a complex structure, the basis for which is not well understood. Figure 4.6 summarizes the basic structure of the elasmobranch kidney tubules and the known localization of various transporters. There is a glomerulus and four loops of the tubules before the urine enters the bladder. In addition, there are two regions, the sinus zone and the bundle zone. Loops 1 and 3 are in the bundle zone and loops 2 and 4 are in the sinus zone (Fig. 4.6). In the FW potamotrygonids the kidney is much simpler, with only two loops and no sinus and bundle zones (Lacy and Reale, 1995). The structure of the kidney of other FW elasmobranchs has not been examined. The kidneys of elasmobranchs play roles in the regulation of plasma Na^+ and Cl^-, excretion of Mg^{2+}, PO_4^{2-}, and SO_4^{2-}, as well as urea, ammonia, and methylamine retention. The kidneys play a very limited role in acid–base regulation and the pH of urine is kept low (5–6) in the face of acid or base loads (Rodler et al., 1955; Murdaugh and Robin, 1967).

A variety of renal functions increase in FW. These include urine flow rate (Wong and Chan, 1977) and glomerular filtration rate (GFR) (Smith, 1931a; Janech and Piermarini, 2002) (Table 4.7). The urine flow rate of *D. sabina* in FW is the highest reported for any fish (Table 4.7). The turnover of Na^+ and

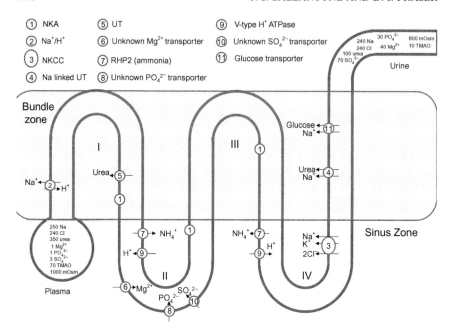

Fig. 4.6. Diagram of the elasmobranch kidney. Urine solute levels are approximated from Burger (1967) for *Squalus acanthias*. Filtrate solute levels are plasma values from Burger (1967), except for plasma sulfate, which is from Robertson (1975). Solutes do not add up to measured osmolarities. The locations of various transporters are based on references cited in the text. NKA: Na$^+$/K$^+$-ATPase; NKCC: Na$^+$/K$^+$/2Cl$^-$ cotransporter; UT: urea transporter; RHP2: rhesus protein 2; TMAO: trimethylamine oxide.

Cl$^-$ is lower in FW species than in marine species owing to reabsorption by the kidney (Table 4.5). The kidney excretes a greater proportion of urea in 50% SW (38%) than in 100% SW (5% of total) (Wong and Chan, 1977).

4.4.1. FILTRATION

In order for plasma to be filtered in the glomerulus, the balance of the hydrostatic and osmotic forces must be favorable. The force promoting filtration is the blood pressure at the glomerulus and this must overcome two opposing forces, the colloid osmotic pressure (COP) and hydrostatic pressure of the capsular fluid. Since none of these forces has been measured in elasmobranch kidneys it is only possible to speculate on their magnitude. The postbranchial blood pressure of many elasmobranchs is low compared to that of teleost fishes (Butler and Metcalfe, 1988) and one could assume that it would be correspondingly low at the glomerulus. This would reduce the rate of filtration were it not for the fact that the COP (equal to oncotic pressure) of marine elasmobranchs is also low (2.27 mmHg, mean of 11

Table 4.7
Kidney characteristics in elasmobranchs and teleosts.

Species	Glomerular filtration rate (ml kg^{-1} h^{-1})	Urine flow rate (ml kg^{-1} h^{-1})
Seawater elasmobranchs		
Squalus acanthias	1.89±0.16 (Hayslett et al., 1973)	0.21±0.04 (Wood et al., 1995)
	1.29±0.21 (Benyajati and Yokota, 1988)	0.5 (Burger, 1967)
	1–4 (Swenson and Maren, 1986)	0.5–1.0 (Swenson and Maren, 1986)
	3.3 (Forster, 1967)	1.27 (Forster, 1967)
		0.57 (Cross et al., 1969)
90% SW	2.00±0.38 (Benyajati and Yokota, 1988)	
	1.96 (Thurau et al., 1969)	0.38 (Thurau et al., 1969)
~50% SW	2.56 (Thurau et al., 1969)	1.35 (Thurau et al., 1969)
Scyliorhinus canicula	1.13 (Henderson et al., 1988)	0.19 (Henderson et al., 1988)
Scyliorhinus stellaris		0.1–1.8 (Holeton and Heisler, 1983)
Mustelus canis	2.38 (Kempton, 1953)	0.66 (Kempton, 1953)
Hemiscyllium plagiosum		0.36 (Wong and Chan, 1977)
Dasyatis sabina	3.8±0.08 (Janech et al., 2006b)	0.9±0.2 (Janech et al., 2006b)
Dasyatis sabina 50% SW	12.4±0.4 (Janech et al., 2006b)	8.1±0.3 (Janech et al., 2006b)
Raja erinacea		0.19±0.02 (Goldstein and Forster, 1971a)
Raja erinacea	0.62±0.18 (Goldstein and Forster, 1971a)	0.29±0.02 (Payan et al., 1973)
Raja erinacea (50% SW)	2.6±1.1 (Goldstein and Forster, 1971a)	1.3±0.3 (Goldstein and Forster, 1971a)
Freshwater elasmobranchs		
Dasyatis sabina		10 (Janech and Piermarini, 2002)
		14.7 (Choe and Evans, 2003)
Pristis microdon		10 (Smith, 1931a)
Potamotrygon sp.	8.3 (Goldstein and Forster, 1971b)	
Seawater teleosts		
Anguilla rostrata	2.1±0.2 (Schmidt-Nielsen and Renfro, 1975)	
Salmo salar SW	1.49 (Talbot et al., 1989)	0.72 (Talbot et al., 1989)
Pleuronectes americanus	1.4±0.3 (Renfro, 1980)	
	0.61±0.08 (Elger et al., 1987)	0.11±0.01 (Elger et al., 1987)

(Continued)

Table 4.7 (Continued)

Species	Glomerular filtration rate (ml kg^{-1} h^{-1})	Urine flow rate (ml kg^{-1} h^{-1})
Freshwater teleosts		
Anguilla rostrata	2.2±0.4 (Schmidt-Nielsen and Renfro, 1975)	
Salmo salar	2.61 (Talbot et al., 1989)	
Ictalurus punctatus	8.2±1.1 (Nishimura, 1977)	3.75 (Cameron and Kormanik, 1982)
Carassius auratus		7.7 (King and Goldstein, 1983b)

References are in parentheses after values.
SW: seawater.

species of elasmobranchs) (Olson, 1992). This is about six-fold lower than the COP of teleosts (12.8 mmHg, mean of 28 species of SW teleosts) (Olson, 1992). The lower COP is in part due to the absence in marine elasmobranchs of albumin, a protein that makes up the largest portion of protein in the plasma of most vertebrates. The hydrostatic pressure of the capsular fluid is unknown but is likely to be similar to that of other animals. Taken together, the low COP would compensate for the low blood pressure and indeed the GFR of marine elasmobranchs is not substantially different from that of marine teleost fishes (Nishimura and Fan, 2003) (Table 4.7). Blood pressure remains constant at lower salinities in *D. sabina* but GFR increases three-fold in 50% SW (Janech et al., 2006b). The large increase in GFR may be due to an increased recruitment of non-filtering glomeruli (Janech et al., 2006b). Plasma protein does not change with salinity acclimation in *C. leucas* (Cowan, 1971) but it does in *T. semifasciatus* (Dowd et al., 2010) (Table 4.8). It has been suggested that Mg^{2+} in the tubular fluid acts the same as albumin in the blood, as an oncotic agent facilitating fluid entry into the tubule (Beyenbach, 1995). Thus, the low oncotic pressure due to lack of plasma albumin in elasmobranchs may be offset by the high Mg^{2+} content in the tubular fluid. Plasma protein of FW potamotrygonids is lower than that of most marine species (Table 4.8), which may aid in filtration in FW species.

4.4.2. REABSORPTION

In addition to reabsorption of ions and useful organic molecules, the elasmobranch kidney has an important role in reabsorbing urea. Most of the filtered urea is reabsorbed (70–99.5%) (Kempton, 1953) with a stoichiometry of 1.6 urea per Na^+ (Schmidt-Nielsen et al., 1972). Sodium-linked urea

Table 4.8
Plasma protein levels in elasmobranchs at different salinities.

Species	Total protein % g 100 ml^{-1}	Reference
Seawater		
Dasyatis americana	2.6	Cain et al. (2004)
Raja radiata	3.4	Larsson et al. (1976)
Raja batis	3.2	Larsson et al. (1976)
Raja lintea	3.6	Larsson et al. (1976)
Etmopterus spinax	1.7	Larsson et al. (1976)
Squalus acanthias	3.1	Larsson et al. (1976)
Somniosus microcephalus	2.5	Larsson et al. (1976)
Caracharhinus leucas	2.2	Urist and Van de Putte (1967)
	3.3	Cowan (1971)
	1.66	Gerzeli et al. (1969)
Heterodontus portusjacksoni	1.3+0.04	Withers et al. (1994)
Cephaloscyllium ventriosum	3.4	Heckly and Herald (1970)
Rhinobatos productus	5.9	Heckly and Herald (1970)
Notorynchus maculates	1.9	Heckly and Herald (1970)
Heterodontus francisci	5.3	Heckly and Herald (1970)
Heterodontus portusjacksoni	1.32	Withers et al. (1994)
Platyrhinoidis triseriata	0.64	Heckly and Herald (1970)
Sphyrna tiburo	2.9	Harms et al. (2002)
Orectolobus ornatus	4.6	Otway et al. (2011)
Triakis semifasciata	5 (estimated from figure)	Dowd et al. (2010)
	2.75	Heckly and Herald (1970)
60% SW	3 (estimated from figure)	Dowd et al. (2010)
Freshwater		
Caracharhinus leucas	2.07±0.25	Gerzeli et al. (1969)
Paratrygon aiereba	1.2±0.2	Duncan et al. (2009)
Potamotrygon sp.	0.83±0.22	Griffith et al. (1973)
	3.4	Urist and Van de Putte (1967)
	3.1	Cowan (1971)
	1.69±0.09	Wood et al. (2002)
Pristis perotteti	3.7	Thorson (1967)
	2.06±0.21	Gerzeli et al. (1969)

SW: seawater.

reabsorption may be a major energy cost for the marine elasmobranch kidney owing to the need for high NKA activities in that tissue. A facilitated urea transporter has been identified (ShUT) with high expression in kidney (Smith and Wright, 1999; Janech et al., 2003). Three transcripts of a urea transporter decreased in little skates *Leucoraja erinacea* exposed to 50% SW (Morgan et al., 2003a). In *D. sabina*, only one of four transcripts decreased in FW (Janech et al., 2006a).

Functionally, two forms of urea transport have been found in skate kidney (Morgan et al., 2003b). Reabsorption is facilitated by a non-saturable form in apical membranes of cells in the dorsal bundle region and an Na^+-linked form in the apical membranes of cells in the ventral blood sinus region (Morgan et al., 2003b) (Fig. 4.6). Basolateral membrane urea transporters have not been identified. Based on localization of a facilitated urea transporter in *Triakis* kidney, it has been suggested that the collecting duct is the main site of urea reabsorption (Hyodo et al., 2004a; Yamaguchi et al., 2009). The K_m for urea of the Na^+-linked UT is very low (0.7 mM) in *L. erinacea*, implying a high capacity to recover urea at the low levels that may occur in the collecting duct (Morgan et al., 2003b). Euryhaline and stenohaline elasmobranchs seem to express the same highly conserved UT (Janech et al., 2008).

Dasyatis sabina has been demonstrated to alter urea clearance, with FW individuals having a 2.5–10 times greater urine flow rate, with urea being the primary osmolyte compared to FW teleosts (Janech and Piermarini, 2002). Reduced reabsorption may be responsible. The levels of UT protein change in proportion to salinity in *Triakis*, consistent with a greater need to recover urea at high salinities and reduce urea levels at low salinities (Yamaguchi et al., 2009). Messenger RNA levels for the UT do not change with salinity, suggesting that intracellular sequestration may explain the changing apical levels of protein (Yamaguchi et al., 2009). This contrasts with a decrease in mRNA in skate kidney with environmental dilution (Morgan et al., 2003a).

Ammonia may need to be conserved especially during periods of starvation, as a nitrogen source for urea synthesis, and the kidney may play a role in this process. Ammonia excretion rates are negligible during prolonged fasting in *S. acanthias* (Wood et al., 2010). Thus, ammonia transport may need to be regulated along with the need for changing rates of urea synthesis. An ammonia transporter, the Rhesus glycoprotein P2 (RHP2), has been found in elasmobranch kidney and localized to the basolateral membranes of the second and fourth loops in the sinus zone (Nakada et al., 2010) (Fig. 4.6). Expression increased at higher and decreased at lower salinities, consistent with a role in recovering ammonia for urea synthesis (Nakada et al., 2010). RHP2 colocalized with V-type H^+-

ATPase, and it has been suggested that this ATPase passes H^+ from NH_4^+ to NH_3+H^+ from tubular cells into the urine (Nakada et al., 2010) (Fig. 4.6).

Aside from the reabsorption of urea, little is known of the reabsorption of other organic molecules. An Na^+-linked glucose transporter has been identified in shark kidney and localized to the collecting duct (Althoff et al., 2006). This unusual localization may be due to the alternative role of this carrier in urea transport (Althoff et al., 2006). TMAO is filtered at the glomerulus and reabsorbed with high efficiency (95–98%) (Cohen et al., 1958; Forster, 1967) but it is not known in what part of the kidney this occurs or whether it changes with salinity acclimation. *Potamotrygon* sp. cannot reabsorb urea in the kidney (Goldstein and Forster, 1971b) and lack the loop (4) (Fig. 4.6) associated with this function in other elasmobranchs.

NaCl is reabsorbed in the proximal tubule via an Na^+/H^+ exchanger (Fig. 4.6). NKCC is involved in NaCl transport in elasmobranch kidney, being particularly abundant in the apical membranes of distal tubules, and exists in multiple forms (Biemesderfer et al., 1996) (Fig. 4.6).

NKA is not found in the collecting duct (Hyodo et al., 2004a). In FW elasmobranchs urine Na^+ and Cl^- are very low, indicating substantial reabsorption (Smith, 1931a). The higher activity of NKA in kidneys of FW-acclimated *C. leucas* suggests that they play a greater role in salt retention in FW than in salt excretion in SW (Pillans et al., 2005).

4.4.3. SECRETION

In marine elasmobranchs, the reduction of Na^+ and Cl^- compared to plasma levels and the recovery of urea leaves an osmotic gap that is filled by the secretion of divalent cations and anions. Mg^{2+}, SO_4^{2-}, and PO_4^{2-} are secreted into the urine to high levels (Table 4.3). These have been localized to the second loop in the sinus zone (Hentschel et al., 2003). Secretion in FW elasmobranchs has not been examined.

4.5. Gastrointestinal Tract

The digestive system of fish may be a site of significant water and salt uptake depending on the diet and the rate of drinking. In SW, it should not be necessary for elasmobranchs to drink because they are slightly hyperosmotic, unlike the situation with marine teleost fishes. Smith (1931b) contended that marine elasmobranchs do not drink; however, recent work (Anderson et al., 2002) indicates that small amounts are taken in, but these are 100–1000-fold lower than rates in marine teleosts (Table 4.6). Consistent with a very low rate of drinking in SW, the activity of NKA in the intestine does not change with salinity acclimation (Pillans et al., 2005), unlike the situation in teleost fishes where the activity increases in SW (Jampol and Epstein, 1970).

The intestine is also a site of urea loss. While urea does pass into the gut lumen, by the time the gut fluid reaches the colon little urea remains (Anderson et al., 2010). Urea uptake from the gut fluid occurs with the aid of a urea transporter, although this may be a minor route (Anderson et al., 2010). The urea that passes into the intestine is degraded to ammonia by the gut bacteria (Lloyd and Goldstein, 1969; Walsh et al., 1994). This ammonia may be recovered by the Rhesus-like protein (Knight et al., 1988; Anderson et al., 2010). It has been suggested that the colon may be important for water retention under hyperosmotic conditions (Anderson et al., 2010).

4.6. Cell Volume Regulation

In addition to the specialized tissues for ion regulation and osmoregulation, most animal cells have their own capacity for dealing with osmotic issues. Amino acids are known to contribute greatly to intracellular osmolarity (approximately 19%), with β-alanine, sarcosine, and taurine being the most important in the muscle of little skates (*L. erinacea*) (King and Goldstein, 1983a). Reductions in environmental salinity result in corresponding reductions in cellular amino acid concentrations in muscle (King and Goldstein, 1983a) and brain (Cserr et al., 1983).

The mechanisms used by cells to regulate volume in elasmobranchs have been investigated in red blood cells (RBCs) (McConnell and Goldstein, 1988; Goldstein et al., 1990), rectal gland cells (Ziyadeh et al., 1988; Goldstein et al., 1990), hepatocytes (Ballatori and Boyer, 1992a, 1992b), and heart cells (Goldstein et al., 1993). Erythropoietin stimulates taurine transport under isosmotic conditions (Musch et al., 1996). In erythrocytes, dilution of the external environment is signaled via protein kinase C and inositol phosphate systems, with ion and organic solutes transported to effect volume change (McConnell and Goldstein, 1988). ATP released from hypotonically stressed cells may stimulate taurine efflux (Goldstein et al., 2003). One of the main solutes for cell volume regulation in rectal gland (Ziyadeh et al., 1988; Goldstein et al., 1990) and RBCs (Goldstein et al., 1990) is taurine. Band 3 anion transporter is involved in taurine efflux in RBCs (Goldstein et al., 1990; Goldstein and Brill, 1991) and rectal gland cells (Goldstein et al., 1990). Changes in the association of dimers of Band 3 into tetramers may stimulate taurine transport in hypoosmotic stress (Musch et al., 1994). Cell swelling causes Band 3 to move from internalized cholesterol-rich rafts to the plasma membrane to facilitate taurine efflux (Musch et al., 2004). Tyrosine phosphatase is involved in regulating Band 3

(Musch et al., 1998). Two tyrosine kinases, $p72^{syk}$ and $p56^{lyn}$, are also involved in activation of Band 3 (Musch et al., 1999).

TMAO is involved in cell volume regulation in RBCs, with transport occurring by two mechanisms: an Na^+-dependent and an Na^+-independent mechanism (Wilson et al., 1999). TMAO efflux following cell swelling seems to be mediated by the same processes as for taurine (Koomoa et al., 2001). Hypoosmotic media stimulate TMAO uptake by RBCs (Wilson et al., 1999).

Following exposure to elevated salinity conditions, glycine, glutamine, and glutamate increase significantly in the muscle and liver of the ocellate river stingray, *Potamotrygon motoro* (Ip et al., 2009). FW elasmobranchs, except for *D. sabina*, also accumulate β-amino acids in place of methylamines (Treberg et al., 2006). This could play a role in cell volume regulation, with β-alanine content increasing in the muscle of *H. signifer* following acclimation to 20% SW (Tam et al., 2003). Liver mitochondria from *L. erinacea* oxidize sarcosine at greater rates at lower osmolality (Ballantyne et al., 1986); likewise, glutamate oxidation is affected by osmolality and the presence of TMAO (Ballantyne and Moon, 1986; Ballantyne et al., 1986). The shift from needing amino acids for this purpose could alter their use in energy production or protein synthesis.

4.7. Temperature Effects

Elasmobranchs in FW are confined to relatively warm water between latitudes 35°N and 35°S (Fig. 4.1). It has been speculated that the lack of FW incursions in temperate and polar latitudes is due to a temperature limitation on the ability to supply energy for the transport of solutes to counteract the relatively high rates of diffusion at low temperatures (Ballantyne and Robinson, 2010). The Q_{10} values for diffusion and leaks are low (1–2), while the Q_{10} values for many metabolic processes are closer to 2–3 (Sidell and Hazel, 1987; Raynard and Cossins, 1991; Cossins et al., 1995). As temperature falls, the reduction in pump activity exceeds the reduction in leak activity, resulting in osmotic stress. Thus, to achieve effective osmoregulation in the cold, enhanced pump activity is needed. Elasmobranchs may lack the capacity to upregulate key transport systems in the cold. Further research on the effects of temperature on elasmobranch osmoregulation is needed.

5. METABOLISM

The ability of elasmobranchs to live in environments that differ widely in salinity requires metabolic changes to coincide with changes in organic solute levels, primarily urea and methylamines. While control of losses is one way to maintain levels of these molecules, regulation of synthetic rates is also important.

5.1. Urea and Amino Acid Metabolism

It has been estimated that urea synthesis and retention accounts for 10.3–14.6% of whole animal respiration rates (Kirschner, 1993). Carrier and Evans (1972) reported urea turnover rates of 0.14% h^{-1}, which amounts to 3.36% day^{-1}. Marine species need to synthesize enough urea to replace losses in urine and across their gills. Indeed, turnover rates of urea match closely with published rates of urea production (Goldstein, 1967). Urea synthesis is energetically expensive, costing 5 ATP for each mole of urea formed *de novo* (Kirschner, 1993). Consequently, ureogenic fishes could be under pressure to reduce unnecessary urea synthesis. The pathway for urea synthesis involves the ornithine urea cycle (OUC) (Fig. 4.7). Most urea

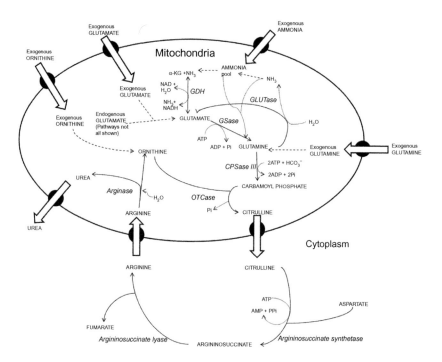

Fig. 4.7. Diagram of the ornithine urea cycle in the liver and skeletal muscle of elasmobranch fishes. OTCase: ornithine transcarbamylase; CPSase III: carbamoyl phosphate synthetase III; GSase: glutamine synthetase; GLUTase: glutaminase; GDH: glutamate dehydrogenase. Modified from Ballantyne (1997).

synthesis occurs in liver but some may occur extrahepatically in muscle (Steele et al., 2005). In elasmobranchs the compartmentalization of the pathway differs from that of other vertebrates, with arginase being mitochondrial rather than cytosolic, and thus urea is generated in the mitochondria. The rationale for this is not known and mitochondria consequently require a mechanism for efflux of urea. Efflux is via a phloretin-sensitive high-affinity urea transporter that has been characterized kinetically in the liver mitochondria of *L. erinacea* (Rodela et al., 2008). From its site of synthesis urea must be distributed to all tissues across all body fluid compartments (Ballantyne and Moon, 1986). It is likely that most urea transport is not by simple diffusion across the lipid bilayer. Urea transport in elasmobranch hepatocytes (Walsh et al., 1994) and RBCs (Rabinowitz and Gunther, 1973; Walsh et al., 1994) occurs by a non-saturable mechanism, probably facilitated diffusion. The nature of the channel has not been established. Five aquaporin isoforms have been found in elasmobranchs (Cutler, 2007) and localized to all osmoregulatory tissues (Cutler and Cramb, 2009) but it is not known which of these can transport urea as well as water. Ideally these carriers should transport both. The UT3 transports both urea and water and has been suggested to facilitate urea solvent drag in mammal kidney (Yang and Verkman, 1998). Solvent drag minimizes cell volume changes and would be of obvious benefit in elasmobranchs.

Upon environmental dilution changes in urea synthesis may or may not occur. Upon exposure to 50% SW *Negaprion brevirostris* alters urea levels by increasing urea clearance three-fold while urea biosynthesis remains unchanged (Goldstein et al., 1968). In contrast, the euryhaline *C. leucas* can reduce or increase urea levels through altered biosynthesis following change in environmental salinity (Anderson et al., 2005). *Himatura signifer*, a largely FW species, possesses a functional OUC in the liver with the activity of CPSase III, glutamine synthetase (GSase), ornithine transcarbamoylase, and arginase increasing significantly when exposed to 20 ppt SW (Tam et al., 2003). This increase in OUC enzyme activity was accompanied by an increase in *de novo* urea production (Tam et al. 2003). In contrast, obligate FW potamotrygonid rays have OUC enzyme activities that are between 5% and 50% of that of marine species (Goldstein and Forster, 1971b), an ability to synthesize urea at a rate of only 1% of that in marine species (Schooler et al., 1966; Goldstein and Forster, 1971b) and accumulate negligible amounts of urea (Thorson et al., 1967), and possess a greatly reduced ability to retain injected urea compared with marine species (Goldstein and Forster, 1971b). Little is known of the compartmentalization of the urea cycle in this group. Potamotrygonids are ammonotelic and thus resemble FW teleosts more than they do other elasmobranch groups in FW.

Unlike other ureotelic animals, marine elasmobranch urea synthesis requires glutamine as a nitrogen donor instead of using ammonia directly (Anderson and Casey, 1984). This is a consequence of the differing substrate preferences of carbamoyl phosphate synthase I (CPS I: ammonia) and CPS III (glutamine). The result is a substantial need for glutamine to be fed into the OUC (Fig. 4.7). In marine environments this is met by high levels in the liver of GSase, an enzyme catalyzing glutamine formation from glutamate and ammonia (Ballantyne, 1997). In contrast, nitrogen is not channeled into urea production in potamotrygonid rays. Consequently, the nitrogen and amino acid metabolism of FW and euryhaline elasmobranchs living in FW is likely to be different from that of marine species. In fact, nitrogen metabolism may have more in common with that of FW teleosts. In support of this, activity levels of GSase are comparable in potamotrygonid rays and FW teleosts, being much lower than in marine elasmobranchs (Ballantyne and Robinson, 2010). Furthermore, the activity of CPS III is nearly absent in the livers of adult rainbow trout (Wright et al., 1995), a representative teleost species, and potamotrygonid rays (Anderson, 1980; Wright et al., 1995).

Plasma glutamine levels are generally lower in marine elasmobranchs than in other vertebrates (Ballantyne, 1997). Ip et al. (2009) report plasma levels of glutamine in *P. motoro* at 0.09 ± 0.01 mM in FW and 0.12 ± 0.02 mM in 15% SW. These are comparable to values reported for *H. signifer* (0.010 ± 0.002 mM in FW and 0.020 ± 0.003 mM in 20% SW) and marine species (Ballantyne, 1997; Tam et al., 2003) even though the FW species have no need for glutamine for urea synthesis. Glutamine is known to be an important oxidative substrate in elasmobranch muscle mitochondria, with oxidative rates higher than those of other vertebrates (Ballantyne, 1997). The role of glutamine may have shifted from urea production to oxidative substrate in FW-dwelling species. Measurements of glutaminase activity in FW as well as mitochondria respiration rates for amino acid substrates would be of great interest.

Glutamate dehydrogenase (GDH) can work either in the aminating direction, producing glutamate from α-ketoglutarate, ammonia, and the reduced form of nicotinamide adenine dinucleotide (NADH), or in a deaminating direction (Fig. 4.7). Liver GDH activity correlates inversely with tissue urea levels, with highest activity in the obligate FW and the marginally euryhaline elasmobranch (*H. signifer*) (Speers-Roesch et al., 2006). This indicates that GDH is not important for glutamate synthesis for urea synthesis via glutamine formation. Glutamate for glutamine and ultimately urea synthesis may thus come from transaminases involving numerous amino acids. Higher activity of GDH in FW species is likely to allow the oxidative deamination of amino acids as oxidative substrates (Speers-Roesch et al., 2006).

Studies of the response to pH disturbance in SW stenohaline elasmobranchs indicate that urea synthesis is not linked to acid–base regulation (Wood et al., 1995). The pH of the blood of *D. sabina* in FW is higher (7.81) than in FW (7.66) and it has been suggested that this is due to reduced urea synthesis in FW resulting in less HCO_3^- being removed (Choe and Evans, 2003).

5.2. Methylamine and β-Amino Acid Metabolism

There are two general pathways for TMAO biosynthesis. The first involves trimethylalkylammonium compounds (e.g. choline) that are degraded to trimethylamine (TMA), which is oxidized to TMAO, generally requiring gut microbes. While this may occur in elasmobranchs, another pathway involving conversion of choline to betaine within mitochondria can produce TMA without microbial involvement (Fig. 4.8) (Seibel and Walsh, 2002). TMAO can be synthesized from TMA by a nicotinamide adenine dinucleotide phosphate (NADPH)-dependent oxidase (TMAox) that is distinct from P-450 (Goldstein and Dewitt-Harley, 1973). The enzyme is microsomal (Fig. 4.8) and may be related to mixed function oxidase (Goldstein and Dewitt-Harley, 1973). Mitochondria could also be involved in TMAO synthesis if betaine is produced as an intermediate. There is considerable variation in the capacity for TMAO synthesis in elasmobranchs. Some species obtain sufficient amounts in the diet and cannot synthesize it *de novo*. Treberg et al. (2006) found measurable TMAox activity in the brownbanded bamboo shark (*Centroscyllium punctatum*) but comparably low activity of betaine aldehyde. Thus, TMA may not be derived from betaine, at least not in *C. punctatum*. Baker et al. (1963) found that TMA oxidation in liver homogenate from *S. acanthias* generally does not occur. Owing to oxidation of TMA in a single animal, Baker et al. (1963) suggested that biosynthesis may only occur intermittently, being active when TMAO stocks need to be replenished. For example, juvenile *C. leucas* transferred from FW to SW increased TMAO content along with urea over a 10 day period (Pillans et al., 2006). This was despite being deprived of food and thus an exogenous source of TMAO. Relying on exogenous TMAO and retention could be important for adjusting TMAO content during salinity change. TMAO concentrations were found to be maintained for 45 days in unfed winter skates (*Leucoraja ocellata*) (Treberg and Driedzic, 2006). This was attributed to a reduced loss of TMAO and not TMAO biosynthesis, with efflux of TMAO only being measured in fed skates. Similarly, prolonged fasting (2 months) did not alter plasma methylamine levels in *S. acanthias* maintained in SW (Wood et al., 2010).

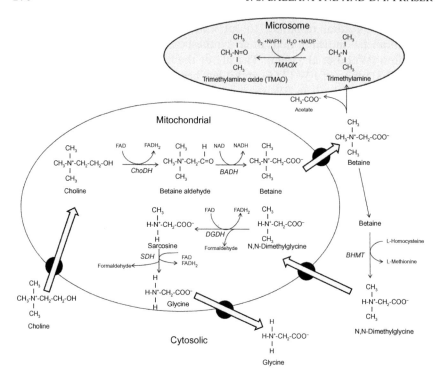

Fig. 4.8. Possible pathways of methylamine synthesis in the liver inferred from studies working with elasmobranch fishes (Goldstein and Dewitt-Harley, 1973; Moyes et al., 1986; Treberg and Driedzic, 2007: DGDH, SDH, and BHMT; Goldstein and Dewitt-Harley, 1973: TMAOx; and Treberg and Driedzic, 2007: ChoDH and BADH). ChoDH: choline dehydrogenase; BADH: betaine aldehyde dehydrogenase; TMAOx: trimethylamine oxidase; BHMT: betaine homocysteine methyltransferase; DGDH: dimethylglycine dehydrogenase; SDH: sarcosine dehydrogenase.

Further research is required to properly map out the mechanism by which TMAO is produced in both marine and euryhaline elasmobranchs.

The inability to synthesize TMAO from TMA does not appear critical to euryhaline and FW existence, as indicated by the lack of synthetic capacity in euryhaline rays (*D. sabina, H. signifer*) and the obligate FW ray (*P. motoro*) (Treberg et al., 2006). The alternative use of betaine and/or sarcosine to counteract the effects of changing urea levels appears sufficient to permit euryhaline existence. Betaine could take on an increased role in euryhaline species spending extended periods in FW, much like the St. Johns River system population of *D. sabina*. The lack of dietary TMAO in FW may play a role in the preferential use of other solutes.

5.3. Lipid and Ketone Body Metabolism in Euryhaline Elasmobranchs

Marine elasmobranchs are known to possess a reduced ability to utilize lipids as extrahepatic oxidative substrates (Zammit and Newsholme, 1979; Ballantyne, 1997). Low extrahepatic lipid oxidation is accompanied by a greater use of ketone bodies and amino acids as oxidative substrates (Ballantyne, 1997). The oxidation rate of β-hydroxybutyrate in muscle mitochondria of elasmobranchs is exceptionally high when compared with rates in other vertebrates (Ballantyne, 1997). This has been suggested as a consequence of accumulating urea and methylamines in high concentrations. Albumin, absent in elasmobranchs, is responsible for binding and transporting non-esterified fatty acids (NEFAs) in the plasma of other vertebrates (Ballantyne, 1997; Metcalf and Gemmell, 2005). Urea has been suggested to inhibit NEFA binding to albumin, according to the "urea hypothesis" coined by Speers-Roesch et al. (2006). Thus urea may account for low levels of NEFA in the plasma of elasmobranchs (Ballantyne, 1997). Whereas a carrier protein is required for transport of NEFAs in the plasma, ketone bodies are soluble and do not require the involvement of protein carriers. For this reason it could be expected that a reduction in urea content associated with FW existence could have consequences on the lipid and ketone body metabolism of elasmobranchs. However, reduction of urea in euryhaline and obligate FW rays does not alter the heavy reliance on ketone bodies as extrahepatic oxidative substrates (Speers-Roesch et al., 2006). Griffith et al. (1973) and Duncan et al. (2009) demonstrated the presence of albumin in the plasma of potamotrygonid stingrays, with stingray albumin being electrophoretically similar to that of mammalian albumin. This albumin, however, may not bind fatty acids since NEFAs remain low in the plasma of FW species including *P. motoro* (Speers-Roesch et al., 2008). Lipoproteins have been suggested as alternative transport vehicles for fatty acids in elasmobranchs (Metcalf and Gemmell, 2005). Further study of lipoproteins, particularly in FW species, is of interest.

The changing diet encountered in FW may affect the availability of certain essential fatty acids as FW fishes have lower n-3/n-6 ratios than marine fishes (Henderson and Tocher, 1987), the only study to examine the fatty acids of FW elasmobranchs found low n-3/n-6 ratios in two FW species; however, levels were similar to those of tropical marine elasmobranchs (Speers-Roesch et al., 2008). In contrast, there was a noticeable difference with n-3/n-6 ratios being lower in tropical marine species than in temperate marine species (Speers-Roesch et al., 2008). This indicates, at least in elasmobranchs, that temperature may play a larger role than salinity in the need for certain fatty acids. Diet can influence the plasma NEFAs of elasmobranchs (Semeniuk et al., 2007) but whether this has an impact on osmoregulatory capacity is not known.

6. SENSORY BIOLOGY

6.1. Electroreception

Electroreception is important in elasmobranchs for prey detection, predator avoidance, mate detection, geonavigation, and even communication (Tricas and Sisneros, 2004). The difference in the conductivity of SW and FW greatly affects electroreception and may be a limiting factor in FW invasion. Structurally, the ampullae of Lorenzini, the sensory organs involved in electroreception, differ among obligate FW rays, euryhaline elasmobranchs, and stenohaline marine species (Table 4.9).

Andres and von Düring (1988) divided ampullae of Lorenzini into three broad categories: (1) macroampullae – found in marine elasmobranchs, with canal lengths up to 20 cm; (2) miniampullae – found in holocephalans and hexanchid sharks, with canal lengths ranging from 1.5 to 10 mm [these were originally termed microampullae by Andres and von Düring (1988), but are reported as miniampullae here for consistency with the literature]; and (3) microampullae – found in obligate FW rays, with canal lengths ranging between 300 and 500 μm. FW specialist potamotrygonid rays possess the smallest "microampullae" (Szabo et al., 1972). The smooth FW stingray

Table 4.9

Comparison between the average length of ampullary organs from obligate freshwater, euryhaline, and marginally euryhaline batoids.

Species	Habitat	Length of ampullary organ (mm)	Disc width (cm)	Length as proportion of disc width (%)	Reference
Potamotrygon circularis	Freshwater	0.3–0.5	15–23	0.2–0.22	Szabo et al. (1972)
Dasyatis garouaensis	Freshwater	0.7–2.1	23–32	0.3–0.65	Raschi and Mackanos (1989)
Himantura signifer	Freshwater, brackish	1.0–2.3	21.5	0.46–1.07	Raschi et al. (1997)
Aptychotrema rostrata	Marine, brackish	4.7–55.5	24	1.96–23.15	Wueringer and Tibbits (2008)
Glaucostegus typus	Marine, freshwater, brackish	6.7–53.4	15.1	4.44–35.36	Wueringer and Tibbits (2008)

(*Dasyatis garouaensis*), an African obligate FW ray, also possesses comparatively small miniampullae (Raschi and Mackanos, 1989). Likewise, *H. signifer,* a FW ray from Asia, possesses miniampullae (Raschi et al., 1997). Features of the ampullary system in *D. garouaensis* and *H. signifer* are intermediate to those of marine elasmobranchs and potamotrygonid rays (Table 4.9). Smaller ampullary organs, specifically canals leading to sensory epithelium, are suited to allow impedance matching the increased resistance of FW compared to SW (Szamier and Bennett, 1980). Even among fully euryhaline species such as *D. sabina* and *C. leucas,* the structure of electroreceptive organs differs (Whitehead, 2002).

Receptor cells of the ampullary epithelium are less sensitive in FW rays than in marine elasmobranchs (Szamier and Bennett, 1980). McGowan and Kajiura (2009) demonstrated that the electroreception of *D. sabina* is greatly reduced in FW. Szamier and Bennett (1980) suggest that the background noise is greater from geoelectric and bioelectric sources in FW, preventing greater sensitivity from being advantageous.

As a case study, low conductivity has been linked to reproductive failure and increased mortality in Lake Monroe populations of *D. sabina* during the 1991/1992 season (Johnson and Snelson, 1996). Conductivity readings were especially low during this period. Electroreception is critical for locating prey by rays and elasmobranchs in general. The inability to locate food could account for the observation that many of the live rays captured by Johnson and Snelson (1996) were emaciated with reduced hepatic lipid reserves. Electroreception in *D. sabina* has also been linked to mate detection and thus reproductive fitness. Specifically, androgen steroid levels have been linked to changes in the response properties of the electrosensory system of *D. sabina* (Tricas and Sisneros, 2004). Low sensitivity of electroreceptors and a poor capacity to change the anatomical structure of these receptors could be constraining factors for living in FW.

In stenohaline marine skates even small changes in environmental salinity (1–3%) result in a noticeable change in the response of ampullae of Lorenzini (Akoev et al., 1980). Akoev et al. (1980) suggest that these electroreception organs may be involved in salinity detection. Katsuki et al. (1969) suggested a role of pit organs in the detection of salinity. However, others have pointed out that this may not be their primary role owing to the presence of pit organs in most elasmobranchs (Peach and Marshall, 2000). The sensitivity of pit organs to changes in salinity could still be relevant to euryhaline species. Likewise, calcium polyvalent cation-sensing receptors, expressed in the kidney, brain, gills, gastrointestinal tract, olfactory lamellae, and rectal gland of elasmobranchs, appear to act as salinity sensors (Nearing et al., 2002).

6.2. Vision

A shift in visual sensitivity to shorter wavelengths (blue-shifted) is characteristic of pelagic marine and deep-water existence. In contrast, fish living in shallow, coastal environments such as estuaries and rivers are red-shifted or more sensitive to longer wavelengths. Two primary visual pigments are found in fish: rhodopsin and porphyrodopsin (Wald, 1939). Rhodopsin is generally found in marine fishes and porphyropsin in FW fishes (Wald, 1939). In addition, euryhaline teleosts are known to have a mixture of these two visual pigments, the proportions of which can change during their life cycle (Beatty, 1966).

In elasmobranchs porphyrodopsin does not appear to be as common as in teleosts. Intact retina from the FW stingray *P. motoro* absorb maximally at 510 nm (Muntz et al., 1973). However, this was not found to be due to the presence of either a red-shifted rhodopsin or a porphyropsin. Instead, a gold-colored tapetum behind the retina was implicated in shifting the effective sensitivity to longer wavelengths (Muntz et al., 1973). In this respect FW elasmobranchs do not appear to have ideal visual adaptations to FW. However, a porphyropsin-type pigment has been identified in juvenile *N. brevirostris*, while adults of the same species have a rhodopsin-type pigment (Cohen et al., 1990; Cohen, 1991). This was the first time porphyropsin had been identified in an elasmobranch. This shift in visual pigments is believed to be an adaptation to coastal, shallower waters by juveniles occupying an inshore nursery, whereas blue-shifted vision is an adult adaptive in open water (Cohen, 1991).

Juvenile *C. leucas*, a fully euryhaline species, possess red-shifted rods and cones (λ_{max} 518 and 554 nm, respectively), an adaptation to coastal nurseries and shallow FW habitats (Hart et al., 2011). This is the result of having a mixture of both rhodopsin and porphyropsin chromophores (Hart et al., 2011). It is not known whether this is a trait exclusive to juveniles or true of adult sharks as well. Likewise, *Negaprion acutidens*, a tropical inshore shark able to tolerate brackish conditions, also has red-shifted visual pigments (λ_{max} 513 nm rod, cone unknown) (Compagno, 1984; Hart et al., 2011).

7. BEHAVIOR

The ability to tolerate a wide range of salinities influences the behavior of elasmobranchs. Where stable populations exist (Johnson and Snelson, 1996), euryhaline elasmobranchs are also known to move freely back and forth between SW and FW environments (Thorson, 1971); however, the reasons are not always known. In some cases movement is related to the parturition of pups in coastal nurseries and the subsequent habitat partitioning, with young, vulnerable sharks remaining in shallow coastal nurseries while subadult and adult sharks move into more open habitats (Snelson et al., 1984; Simpfendorfer et al., 2005). Furthermore, movement patterns within nursery environments may be explained by a number of factors including food availability (Castro, 1993), climate conditions (Heupel et al., 2003), predation (Guttridge et al., 2012), physical characteristics of the nursery (Morrissey and Gruber, 1993), and abiotic conditions such as temperature and salinity (Heupel and Simpfendorfer, 2008; Ubeda et al., 2009). (Nurseries are known to have great food abundance but selection of microhabitats within a nursery based on food availability has never been demonstrated.)

It has been known for some time that elasmobranchs utilize coastal nursery areas (Meek, 1916). Coastal nurseries are typically characterized by an abundance of food and protection from predation (Castro, 1993). The use of coastal nurseries has even been demonstrated in extinct species based on the fossil record (Fischer et al., 2011). Some nurseries are characterized by areas of low or reduced salinity as well as areas of variable salinity such as rivers and estuaries (Simpfendorfer et al., 2005; Heupel and Simpfendorfer, 2008; Ubeda et al., 2009). *Dasyatis sabina* and *C. leucas* are two euryhaline species known to use reduced salinity nurseries (Snelson et al., 1984; Johnson and Snelson, 1996; Simpfendorfer et al., 2005; Heupel and Simpfendorfer, 2008).

Despite spending much of their early life in reduced salinity environments, juvenile *C. leucas* are able to tolerate SW conditions (Pillans et al., 2005, 2006). In fact, fetal and maternal urea levels in the serum are typically close regardless of whether a pregnant *C. leucas* was acclimatized to SW or to FW (Thorson and Gerst, 1972). In this respect, even before birth *C. leucas* appear to be fully euryhaline. Despite this, juvenile *C. leucas* demonstrate salinity preference, influencing their behavior within nursery habitats. Tracking studies of juvenile *C. leucas* from the Caloosahatchee River (Florida) found that they avoided areas with very low (<7 ppt) salinity, instead altering their movement patterns to coincide with areas of moderate or elevated salinity (Simpfendorfer et al., 2005; Heupel and Simpfendorfer, 2008). The strength of the relationship between salinity and the location was reduced with age (Heupel and Simpfendorfer, 2008). This may relate to habitat partitioning to avoid intraspecific predation. Likewise, acoustic monitoring of *Sphyrna tiburo*, a species with marginal salinity tolerance, from Pine Island Sound estuary (Florida), demonstrated that sharks alter movement patterns in response to changes in salinity (Ubeda et al., 2009).

The apparent increased osmoregulatory challenge faced by *C. leucas* during acclimation from 75% to 100% SW compared with acclimation from FW to 75% SW (Pillans et al., 2006) is noteworthy. Simfpendorfer et al. (2005) and Ortega et al. (2009) noted changes in the distribution of young *C. leucas* following periods of great salinity decrease. This supports the idea that *C. leucas* have a preferred salinity range despite being able to survive in a wider range of salinity values.

As suggested by previous authors, microhabitat selection within coastal nurseries may be beneficial to reduce energy costs associated with osmoregulation, particularly for very young sharks. More energy dedicated to growth than to osmoregulation would help to reduce predation in the long term, while seeking out food in an area of unfavorable salinity may cost more energy than it's worth. Meloni et al. (2002) observed a noticeable increase in standard metabolic rate in bat rays following exposure to reduced salinity (25 ppt). However, no similar change in oxygen consumption rate was observed in Port Jackson sharks (*Heterodontus portusjacksoni*) exposed to reduced salinity (17 ppt) (Cooper and Morris, 2004). Further studies on the energetic costs of osmoregulation in elasmobranchs are needed. Regardless, changes in energy expenditure as part of the costs of osmoregulation may be a factor influencing neonate shark behavior, although the specific changes in cost are likely to be species specific (Piermarini and Evans, 2000; Pillans et al., 2005).

Further considerations relate to the time-frame of salinity change, and how the preference for a certain salinity relates to other needs of young sharks, such as food and avoidance of predation. As pointed out by Ubeda et al. (2009), rapid changes in salinity appear to have a greater effect on movement patterns than absolute values. *Carcharhinus leucas* are capable of surviving sudden changes in salinity (Pillans et al., 2006) but would be faced with a greater osmoregulatory challenge during more rapid changes in salinity. Consequently, behavioral changes are to be expected. Work with blacktip sharks (*Carcharhinus limbatus*) in a coastal nursery found that the sharks spent the majority of their time in regions of the habitat where food was less abundant (Heupel and Hueter, 2002). The authors suggest that other factors may therefore be more influential on shark movement. Likewise, Simfpendorfer et al. (2005) suggested that *C. leucas* may expose themselves to greater predation risk to avoid very low salinities. In this respect, there could be a hierarchy of factors influencing movement patterns, with salinity/osmoregulatory concerns being the most important to euryhaline sharks, followed by predation avoidance and food availability. Migration between environments may also be sex specific in some species, such as *Himantura dalyensis* (Campbell et al., 2012).

8. REPRODUCTION

8.1. Behavioral Considerations: Freshwater Copulation and Parturition of
Young

Approximately 43% of extant chondrichthyans display oviparous
reproduction (Compagno, 1990). Oviparity within elasmobranchs is
generally believed to be restricted to marine environments. The major
challenge faced by oviparous FW elasmobranchs relates to the permeability
of mermaid purses (egg cases) to urea (Foulley and Mellinger, 1980). Only
small amounts of urea (<20 mM) are maintained within egg capsules
containing developing embryos, while ion concentrations are high, albeit
lower than those of SW (Evans, 1981). Evans (1981) reported the total
concentration of Na^+, K^+, Cl^-, and urea levels at approximately 780 mOsm
kg^{-1} in the egg case of stenohaline little skates (*L. erinacea*) compared with
approximately 950 mOsm kg^{-1} in SW. This, however, is only true when egg
cases are in SW. In FW, an influx of FW would more drastically alter the
ionic and osmotic properties of the fluid within egg cases and may prove too
much of an osmotic challenge for developing embryos. Clearnose skate
(*Raja eglanteria*) eggs removed from capsules prior to day 20 of
development die when exposed to SW, indicating their inability to
osmoregulate during early development (Libby, 1959). In this respect,
exposure to FW early in development could be detrimental to embryos and
a major factor limiting reproductive success in reduced salinity environments. Price and Daiber (1967) note that a major selection pressure for the
evolution of viviparity among elasmobranchs relates to the difficultly in
regulating urea and osmotic pressure during early development. As noted by
Ballantyne and Robinson (2010), all Rajiformes are viviparous except for
skates. Ballantyne and Robinson (2010) go on to suggest that the reliance on
oviparity may explain why skates, unlike stingrays and sawfish, have never
colonized FW habitats.

The mechanism of sperm transfer may also be problematic in FW. In
sharks, copulation is assisted by muscular contraction of siphon sacs
associated with the claspers of males. Siphon sacs are filled with SW during
copulation and help to flush sperm into the uterus of the females (Gilbert
and Heath, 1972). Sperm from the banded houndshark (*Triakis scyllium*)
was demonstrated to acquire motility upon dilution in electrolyte solutions.
Motility was especially high when it was diluted in solutions with ionic
and osmotic properties comparable to those of SW (Minamikawa and
Morisawa, 1996). In this respect SW or at least brackish conditions may be
required for successful copulation in sharks. In contrast, SW is not used to

flush sperm into the female reproductive tract in batoid elasmobranchs (Babel, 1967). This is important as euryhaline batoids such as *D. sabina* and *P. perotteti* can successfully copulate in FW, unlike *C. leucas* (Jensen, 1976; Thorson, 1976; Johnson and Snelson, 1996). Thorson et al. (1983) note that macroscopically the clasper gland and sac of potamotrygonid rays do not appear to differ appreciably from those of their marine relatives. It would be interesting to determine the otolith Sr:Ca ratios of *Glyphis* river sharks to determine how dependent they are on brackish and marine environments. This would help to resolve the question of whether they copulate in FW, since direct observation of copulation is rare.

Himatura signifer from Southeast Asian rivers have been demonstrated to be tolerant of brackish conditions (Tam et al., 2003), opening the possibility that they may migrate between FW and habitats with elevated salinity. Fluctuations in the otolith Sr:Ca ratios of *H. signifer* from Chao Phraya river support the idea that these rays enter brackish or even marine environments at some point (Otake et al., 2005). However, the lowest otolith Sr:Ca ratio was found in a neonatal ray, supporting a FW birth.

FW populations of *D. sabina* likewise live in a river system that is connected to the ocean (Johnson and Snelson, 1996). Seasonal migration has been documented among coastal, predominantly marine rays of this species (Snelson et al., 1988). Migration has been linked to changes in temperature and is related to reproductive timing (Snelson et al., 1988). However, the likelihood of large-scale migration from permanent FW populations in the southern parts of the river system is small (Johnson and Snelson, 1996). Large migrations (300–500 km) would be required to reach estuarine and marine habitats. It thus seems unlikely that migratory, coastal rays mate with residents of permanent FW populations. Regardless, *D. sabina* is capable of copulating and parturition of young in FW, indicating its adaptiveness to low-salinity habitats. It is worth noting that Charvet-Almeida et al. (2005) suggest that the reproductive cycle of obligate FW potamotrygonid rays has been shaped by the hydrological cycle of the Amazon river basins. However, a similar effect has not occurred in permanent FW populations of *D. sabina* living in the St. Johns River system, as evidenced by the lack of divergence in reproductive life history traits with coastal marine populations (Johnson and Snelson, 1996). This may be due to a shorter divergence time in FW for *D. sabina* compared with potamotrygonid rays.

In *C. leucas*, copulation and pup parturition generally does not normally occur in FW habitats (Jensen, 1976). This is despite adults being able to live in FW long term and fetal and neonatal sharks being fully euryhaline (Thorson, 1971; Thorson and Gerst, 1972; Pillans et al., 2006). As noted earlier, low conductivity makes reproductive success more difficult. Furthermore, as previously noted, it is probably more energetically costly for the pups to be born in FW than in SW, owing to the osmotic differences.

9. CONCLUSIONS AND PERSPECTIVES

Elasmobranchs can be divided into three groups based on their osmotic capacities. These are stenohaline marine species, stenohaline FW species, and euryhaline species. The osmotic relations of these are summarized in

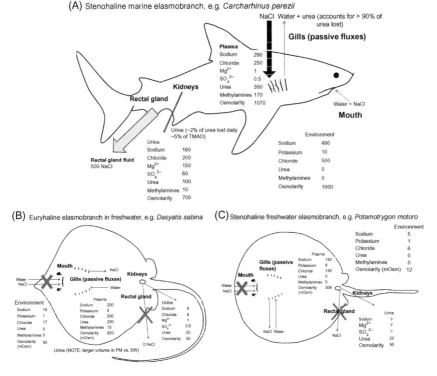

Fig. 4.9. Water and solute fluxes and levels in (A) stenohaline marine elasmobranch, (B) euryhaline elasmobranch, and (C) stenohaline freshwater elasmobranch. Values are approximations based on values in Table 4.3 and sources mentioned in the text. Thickness of lines indicates relative size of fluxes. TMAO: trimethylamine oxide.

Fig. 4.9. The vast majority of elasmobranchs live in SW and cannot survive in FW. The fact that against all energetic and osmotic concerns, high levels of urea are retained by all known euryhaline species in FW and even some species mostly found in FW, indicates the likely importance of urea in maintaining the function of critical systems such as proteins. Thus, the basic urea-retaining osmotic strategy of elasmobranchs may be the major limitation. Future research of a functional nature could reveal which proteins have a requirement for urea, but considering that only one or a few proteins may be limiting the process would be hit or miss and require a very large number of proteins to be examined.

The most successful euryhaline species must be considered the species that can not only survive in FW but also reproduce there. *Dasyatis sabina* and *P. perotteti* are the only euryhaline species known to reproduce in FW. The ability of the batoids to copulate successfully in FW seems to be due to the anatomical design of the male copulatory organs. Unlike other elasmobranchs they do not use ambient water to pump sperm during internal fertilization. Elasmobranch sperm may not be able to survive in FW, unlike teleost sperm. Since not all batoids can tolerate FW it should be stressed that this adaptation is not sufficient to permit euryhalinity. All elasmobranchs that reproduce in FW give birth to live young. Egg-laying species may be constrained by the inability of the developing embryos to osmoregulate in FW.

Another limiting factor is the design and function of the ion regulatory system. An internal salt-secreting gland (rectal gland) cannot be reversed in FW for ion uptake. The mechanism for salt uptake in the gills in FW differs from that of teleost fishes (Evans et al., 2005) and may be limiting. Certainly, the capacity to reduce ion and osmotic gradients in FW is much poorer in elasmobranchs than in euryhaline teleost fishes. The restriction of euryhaline species to relatively warm water indicates a further limitation in the capacity of the transport systems to handle the osmotic and ionic fluxes. Studies of the thermal sensitivity of the transport systems of elasmobranchs versus teleosts are needed.

Finally, the lower conductivity of FW limits the capacity for electro-sensory function. The lack of capacity to alter the anatomical structure of the ampullae of Lorenzini in FW to compensate for this may limit the ability to detect prey, especially in murky waters.

A summary of the factors that may be limiting to elasmobranchs in FW is given in Table 4.10. Understanding how the extant euryhaline elasmobranchs

Table 4.10
Summary of limiting aspects of elasmobranch physiology that may constrain existence in
freshwater and euryhalinity in general.

Limitation	Comments
1. Urea-retaining osmotic strategy	Obligate urea requirement of some proteins results in improper protein conformation and function if urea reduced. Osmotic and energetic costs higher in freshwater due to need for urea retention
2. Ion regulation	Internal anatomical location of the salt-secreting rectal gland. Cannot be used for salt uptake in freshwater. Poor ion regulation results in large changes in ion levels with changing salinity that disrupt protein function and nerve conduction. Low temperature limits the ion transport system's capacity to deal with ion leaks
4. Electroreception in freshwater	Poor conductivity in freshwater: reduced sensitivity and behavioral constraints
5. Reproduction	Poor sperm viability if ambient water used for sperm pumping during internal fertilization. Poor capacity of embryos

survive in FW and SW will require more detailed molecular and physiological studies. Whole-genome DNA sequence comparisons of euryhaline and stenohaline species may provide clues but may be limited by the small number of euryhaline species. Although beyond the scope of this review, other non-physiological factors may also play a role in limiting FW adaptation.

REFERENCES

Akoev, G. N., Volpe, N. O. and Zhadan, G. G. (1980). Analysis of effects of chemical and thermal stimuli on the ampullae of Lorenzini of the skates. *Comp. Biochem. Physiol. A* 65, 193–201.

Althoff, T., Hentschel, H., Luig, J., Schutz, H., Kasch, M. and Kinne, R. K. H. (2006). Na^+-D-Glucose cotransporter in the kidney of *Squalus acanthias*: molecular identification and intrarenal distribution. *Am. J. Physiol.* 290, R1094–R1104.

Anderson, P. M. (1980). Glutamine- and N-acetylglutamate-dependent carbamoyl phosphate synthetase in elasmobranchs. *Science* 208, 291–293.

Anderson, P. M. and Casey, C. A. (1984). Glutamine-dependent synthesis of citrulline by isolated hepatic mitochondria from *Squalus acanthias*. *J. Biol. Chem.* 259, 456–462.

Anderson, W. G., Takei, Y. and Hazon, N. (2002). Osmotic and volaemic effects on drinking rate in elasmobranch fishes. *J. Exp. Biol.* 205, 1115–1122.

Anderson, W. G., Good, J. P., Pillans, R. D., Hazon, N. and Franklin, C. E. (2005). Hepatic urea biosynthesis in the euryhaline elasmobranch *Carcharhinus leucas*. *J. Exp. Zool. A* 303, 917–921.

Anderson, W. G., Pillans, R. D., Hyodo, S., Tsukada, T., Good, J. P., Takei, Y., Franklin, C. E. and Hazon, N. (2006). The effects of freshwater to seawater transfer on circulating levels of angiotensin II, C-type natriuretic peptide and arginine vasotocin in the euryhaline elasmobranch *Carcharhinus leucas*. *Gen. Comp. Endocrinol.* 147, 39–46.

Anderson, W. G., Dasiewicz, P. J., Liban, S., Ryan, C., Taylor, J. R., Grosell, M. and Weihrauch, D. (2010). Gastro-intestinal handling of water and solutes in three species of elasmobranch fish, the white-spotted bamboo shark, *Chiloscyllium plagiosum*, little skate, *Leucoraja erinacea* and the clearnose skate *Raja eglanteria*. *Comp. Biochem. Physiol. A Mol. Integr. Physiol.* 155, 493–502.

Andres, K. H. and von Düring, M. (1988). Comparative anatomy of vertebrate electro-receptors. *Prog. Brain Res.* 74, 113–131.

Babel, J. S. (1967). Reproduction, life history and ecology of the round stingray *Urolophus halleri* Cooper. *Calif. Dept. Fish Game Fish. Bull.* 137, 1–104.

Baker, J. R., Struempler, A. and Chaykin, S. (1963). A comparative study of trimethylamine-*N*-oxide biosynthesis. *Biochim. Biophys. Acta.* 71, 58–64.

Ballantyne, J. S. (1997). Jaws, the inside story. The metabolism of elasmobranch fishes. *Comp. Biochem. Physiol. B Biochem. Mol. Biol.* 118, 703–742.

Ballantyne, J. S. and Moon, T. W. (1986). Solute effects on mitochondria from an elasmobranch (*Raja erinacea*) and teleost (*Pseudopleuronectes americanus*). *J. Exp. Zool.* 239, 319–328.

Ballantyne, J. S. and Robinson, J. W. (2010). Freshwater elasmobranchs: a review of their physiology and biochemistry. *J. Comp. Physiol. B* 180, 475–493.

Ballantyne, J. S., Moyes, C. D. and Moon, T. W. (1986). Osmolarity affects oxidation of sarcosine by isolated hepatocytes and mitochondria from a euryhaline elasmobranch. *J. Exp. Zool.* 238, 267–271.

Ballatori, N. and Boyer, J. L. (1992a). Taurine transport in skate hepatocytes. I. Uptake and efflux. *Am. J. Physiol.* 262, G445–G450.

Ballatori, N. and Boyer, J. L. (1992b). Taurine transport in skate hepatocytes. II. Volume activation, energy and sulfhydryl dependence. *Am. J. Physiol.* 262, G451–G460.

Beatty, D. D. (1966). A study of the succession of visual pigments in the Pacific salmon (*Oncorhynchus*). *Can. J. Zool.* 44, 429–455.

Bedford, J. J (1983). The effect of reduced salinity on tissue and plasma composition of the dogfish, *Squalus acanthias*. *Comp. Biochem. Physiol. A* 76, 81–84.

Benyajati, S. and Yokota, S. D. (1988). Hormonal control of glomerular filtration rate in a marine elasmobranch (*Squalus acanthias*). *Bull. Mount Desert Island Biol. Lab.* 27, 55–58.

Beyenbach, K. W. (1995). Secretory electrolyte transport in renal proximal tubules of fish. In *Cellular and Molecular Approaches to Fish Ionic Regulation* (eds. C. M. Wood and T. J. Shuttleworth), pp. 85–105. San Diego: Academic Press.

Biemesderfer, D., Payne, J. A., Lytle, C. Y. and Forbush, B. (1996). Immunocytochemical studies of the NaKCl cotransporter of shark kidney. *Am. J. Physiol.* 270, F927–F936.

Bittner, A. and Lang, S. (1980). Some aspects of the osmoregulation of Amazonian freshwater stingrays (*Potamotrygon hystrix*) – I. Serum osmolality, sodium and chloride content, water content, hematocrit and urea level. *Comp. Biochem. Physiol. A* 67, 9–13.

Bleich, M., Hug, M. J., Heitzmann, D., Warth, R. and Greger, R. (1999). Evidence for Na^+/Ca^{2+} exchange in the rectal gland of *Squalus acanthias*. *Pflugers Arch.* 439, 49–51.

Boeseman, M. (1956). Fresh-water sawfishes and sharks in Netherlands New Guinea. *Science* 123, 222–223.

Boylan, J. W. (1967). Gill permeability in *Squalus acanthias*. In *Sharks, Skates and Rays* (eds. P. W. Gilbert, R. F. Mathewson and D. P. Rall), pp. 197–206. Baltimore, MD: Johns Hopkins Press.

Brooks, D. R., Thorson, T. B. and Mayes, M. A. (1981). Fresh-water stingrays (Potamotrygonidae) and their helminth parasites: testing hypotheses of evolution and coevolution. In *Advances in Cladistics* (eds. V. A. Funk and D. R. Brooks), pp. 147–175. New York: New York Botanical Gardens.

Browning, J. (1978). Urea levels in plasma and erythrocytes of the southern fiddler skate, *Tryogogorhina fasciata guanerius*. *J. Exp. Zool.* 203, 325–330.

Burger, J. W. (1967). Problems in the electrolyte economy of the spiny dogfish, *Squalus acanthias*. In *Sharks, Skates and Rays* (eds. P. W. Gilbert, R. F. Mathewson and D. P. Rall), pp. 177–185. Baltimore, MD: Johns Hopkins Press.

Burger, J. W. and Hess, W. N. (1960). Function of the rectal gland in the spiny dogfish. *Science* 131, 670–671.

Butler, P. J. and Metcalfe, J. D. (1988). Cardiovascular and respiratory systems. In *Physiology of Elasmobranch Fishes* (ed. T. J. Shuttleworth), pp. 1–47. New York: Springer.

Bystriansky, J. S. and Ballantyne, J. S. (2007). Gill Na^+K^+-ATPase activity correlates with basolateral membrane lipid composition in seawater but not freshwater acclimated Arctic char (*Salvelinus alpinus*). *Am. J. Physiol.* 292, R1043–R1051.

Cain, D. K., Harms, C. A. and Segars, A. (2004). Plasma biochemistry reference values of wild-caught southern stingrays (*Dasyatis americana*). *J. Zoo Wildl. Med.* 35, 471–476.

Cameron, J. N. and Kormanik, G. A. (1982). The acid–base responses of gills and kidneys to infused acid and base loads in the channel catfish, *Ictalurus punctatus*. *J. Exp. Biol.* 99, 143–160.

Campbell, H. A., Hewitt, M., Watts, M. E., Peverell, S. and Franklin, C. E. (2012). Short- and long-term movement patterns in the freshwater whipray (*Himantura dalyensis*) determined by the signal processing of passive acoustic telemetry data. *Mar. Freshw. Res.* 63, 341–350.

Carrier, J. C. and Evans, D. H. (1972). Ion, water and urea turnover rates in the nurse shark, *Ginglymostoma cirratum*. *Comp. Biochem. Physiol. A* 41, 761–764.

Carrier, J. C. and Evans, D. H. (1973). Ion and water turnover in the fresh-water elasmobranch *Potamotrygon* sp. *Comp. Biochem. Physiol. A* 45, 667–670.

Carroll, S., Kelsall, C., Hazon, N. and Eddy, F. B. (1994). Effect of temperature on the drinking rates of two species of flatfish, flounder and turbot. *J. Fish Biol.* 44, 1097–1099.

Castro, J. I. (1993). The shark nursery of Bulls Bay South Carolina, with review of the shark nurseries of the southeastern coast of the United States. *Environ. Biol. Fish.* 38, 37–48.

Chan, D. K. O. and Wong, T. M. (1977). Physiological adjustments to dilution of the external medium in the lip-shark *Hemiscyllium plagiosum* (Bennett) I. Size of body compartments and osmolyte composition. *J. Exp. Zool.* 200, 71–84.

Charvet-Almeida, P., Goes de Araujo, M. L. and de Almeida, M. P. (2005). Reproductive aspects of freshwater stingrays (Chondrichthyes: Potamotrygonidae) in the Brazilian Amazon Basin. *J. Northw. Atl. Fish. Sci.* 35, 165–171.

Chatchavalvanich, K., Thongpan, A. and Nakai, M. (2005). Structure of the testis and genital duct of freshwater stingray *Himantura signifer* (Elasmobranchii: Myliobatiformes: Dasyatidae). *Ichthyol. Res.* 52, 123–131.

Chew, S. F., Poothodiyil, N. K., Wong, W. P. and Ip, Y. K. (2006). Exposure to brackish water, upon feeding, leads to enhanced conservation of nitrogen and increased urea synthesis and retention in Asian freshwater stingray *Himantura signifer*. *J. Exp. Biol.* 209, 484–492.

Choe, K. P. and Evans, D. H. (2003). Compensation for hypercapnia by a euryhaline elasmobranch: effect of salinity and roles of gills and kidneys in freshwater. *J. Exp. Zool.* 297A, 52–63.

Choe, K. P., Verlander, J. W., Wingo, C. S. and Evans, D. H. (2004). A putative H^+–K^+-ATPase in the Atlantic stingray, *Dasyatis sabina*: primary sequence and expression in the gills. *Am. J. Physiol.* 287, R981–R991.

Choe, K. P., Kato, A., Hirose, S., Plata, C., Sindic, A., Romero, M. F., Claiborne, J. B. and Evans, D. H. (2005). NHE3 in an ancestral vertebrate: primary sequence, distribution, localization and function in gills. *Am. J. Physiol.* 289, R1520–R1534.

Choe, K. P., Edwards, S. L., Claiborne, J. B. and Evans, D. H. (2007). The putative mechanism of Na^+ absorption in euryhaline elasmobranchs exists in the gills of a stenohaline marine elasmobranch, *Squalus acanthias. Comp. Biochem. Physiol. A Mol. Integr. Physiol.* 146, 155–162.

Claiborne, J. B., Choe, K. P., Morrison-Shetlar, A. I., Weakley, J. C., Havird, J., Freiji, A., Evans, D. H. and Edwards, S. L. (2008). Molecular detection and immunological localization of gill Na^+/H^+ exchanger in the dogfish (*Squalus acanthias*). *Am. J. Physiol.* 294, R1092–R1102.

Coad, B. W. and Papahn, F. (1988). Shark attacks in the rivers of southern Iran. *Environ. Biol. Fish.* 23, 131–134.

Cohen, D. M. (1970). How many recent fishes are there? *Proc. Calif. Acad. Sci.* 38, 341–346.

Cohen, J. J., Krupp, M. A. and Chidsey, C. A. (1958). Renal conservation of trimethylamine oxide by the spiny dogfish, *Squalus acanthias. Am. J. Physiol.* 194, 229–235.

Cohen, J. L. (1991). Adaptations for scotopic vision in the lemon shark (*Negaprion brevirostris*). *J. Exp. Zool.* 256 (Suppl. 5), 76–84.

Cohen, J. L., Hueter, R. E. and Organisciak, D. T. (1990). The presence of a porphyropsin-based visual pigment in the juvenile lemon shark (*Negaprion brevirostris*). *Vision Res.* 30, 1949–1953.

Compagno, L. J. V. (1984). FAO species catalogue: Vol. 4. Sharks of the world. An annotated and illustrated catalogue of shark species known to date. Part 2 – Carcharhiniformes. *FAO Fish. Synop,* 125, 251–655.

Compagno, L. J. V. (1990). Alternative life-history styles of cartilaginous fishes in time and space. *Environ. Biol. Fish.* 28, 33–75.

Compagno, L. J. V. and Roberts, T. R. (1984). Marine and freshwater stingrays (Dasyatidae) of West Africa, with description of a new species. *Proc. Calif. Acad. Sci.* 43, 283–300.

Compagno, L. J. V., White, W. T. and Cavanagh, R. D. (2010). *Glyphis fowlerae* sp. nov., a new species of river shark (Carcharhiniformes; Carcharhinidae) from northeastern Borneo. In *Descriptions of New Sharks and Rays from Borneo* (eds. P. R. Last, W. T. White and J. J. Pogonoski), pp. 29–44. Hobart: CSIRO Marine and Atmospheric Research. Paper 32.

Cooper, A. R. and Morris, S. (1998). Osmotic, ionic and haematological response of the Port Jackson shark *Heterodontus portusjacksoni* and the common stingaree *Trygonoptera testacea* upon exposure to diluted seawater. *Mar. Biol.* 132, 29–42.

Cooper, A. R. and Morris, S. (2004). Haemoglobin function and respiratory status of the Port Jackson shark, *Heterodontus portusjacksoni*, in response to lowered salinity. *J. Comp. Physiol. B* 174, 223–236.

Cornelius, F. (1995). Cholesterol modulation of molecular activity of reconstituted shark Na^+, K^+-ATPase. *Biochim. Biophys. Acta* 1235, 205–212.

Cossins, A. R., Schwarzbaum, P. J. and Wieser, W. (1995). Effects of temperature on cellular ion regulation and membrane transport systems. In *Biochemistry and Molecular Biology of Fishes,* Vol. 5, *Environmental and Ecological Biochemistry* (eds. P. W. Hochachka and T. P. Mommsen), pp. 101–126. New York: Elsevier.

Cowan, C. M. (1971). Serum protein variation in the bull shark, *Carcharhinus leucas* Muller and Henle, 1841. *Int. J. Biochem.* 2, 691–696.

Cross, C. E., Packer, B. S., Linta, J. M., Murdaugh, H. V. and Robin, E. D. (1969). H^+ buffering and excretion in response to acute hypercapnia in the dogfish *Squalus acanthias. Am. J. Physiol.* 216, 440–452.

Cserr, H. F., Bradbury, M. W. B., Mackie, K. and Moody, E. J. (1983). Control of extracellular ions in skate brain during osmotic disturbances. *Am. J. Physiol.* 245, R853–R859.

Cutler, C. (2007). Aquaporin channels in teleost and elasmobranch fish. *Comp. Biochem. Physiol. A Mol. Integr. Physiol.* 146, S89.

Cutler, C. P. and Cramb, G. (2009). Immunohistochemical localization and expression of aquaporin water channel membrane transport protein homologues in the osmoregulatory tissues of the dogfish (*Squalus acanthias*). *Comp. Biochem. Physiol. A Mol. Integr. Physiol.* 153, S65.

Dakin, W. J. (1908). Variations in the osmotic concentration of the blood and coelomic fluids of aquatic animals caused by changes in the external medium. *Biochem. J.* 3, 473–490.

Das, A. and Mukhopadhyay, C. (2009). Urea-mediated protein denaturation: a consensus view. *J. Phys. Chem. B* 113, 12816–12824.

De Carvalho, M. R. and Lovejoy, N. R. (2011). Morphology and phylogenetic relationships of a remarkable new genus and two new species of neotropical freshwater stingrays from the Amazon basin (Chondrichthyes: Potamotrygonidae). *Zootaxa* 2776, 13–48.

Deynat, P. (2006). *Potamotrygon marinae* n. sp. une nouvelle espèce de raies d'eau douce de Guyane (Myliobatiformes, Potamotrygonidae). *Compt. Rend. Biol.* 329, 483–493.

Dowd, W. W., Harris, B. N., Cech, J. J. and Kuma, K. (2010). Proteomic and physiological responses of leopard sharks (*Triakis semifasciata*) to salinity change. *J. Exp. Biol.* 213, 210–224.

Duncan, W. P., Costa, O. T. F., Araujo, M. L. G. and Fernandes, M. N. (2009). Ionic regulation and Na$^+$–K$^+$-ATPase activity in gills and kidney of the freshwater stingray *Paratrygon aiereba* living in white and blackwaters in the Amazon basin. *J. Fish Biol.* 74, 956–960.

Dunn, K. A., McEachran, J. D. and Honeycutt, R. L. (2003). Molecular phylogenetics of myliobatiform fishes (Chondrichthyes: Myliobatiformes), with comments on the effects of missing data on parsimony and likelihood. *Mol. Phylogen. Evol.* 27, 259–270.

Easteal, A. J. (1990). Tracer diffusion in aqueous sucrose and urea solutions. *Can. J. Chem.* 68, 1611–1615.

Edwards, S. L. and Marshall, W. S. (2013). Principles and patterns of osmoregulation and euryhalinity in fishes. In *Fish Physiology*, Vol. 32, *Euryhaline Fishes* (eds. S. D. McCormick, A. P. Farrell and C. J. Brauner), pp. 1–44. New York: Elsevier.

Edwards, S. L., Donald, J. A., Toop, T., Donowitz, M. and Tse, C. M. (2002). Immunolocalization of sodium/proton exchanger-like proteins in the gills of elasmobranchs. *Comp. Biochem. Physiol. A Mol. Integr. Physiol.* 131, 257–265.

Elger, E., Elger, B., Hentschel, H. and Stolte, H. (1987). Adaptation of renal function to hypotonic medium in the winter flounder (*Pseudopleuronectes americanus*). *J. Comp. Physiol. B* 157, 21–30.

Epstein, F. H. and Silva, P. (2005). Mechanisms of rectal gland secretion. *Bull. Mount Desert Island Biol. Lab.* 44, 1–5.

Ernst, S. A., Hootman, S. R., Schreiber, J. H. and Riddle, C. V. (1981). Freeze-fracture and morphometric analysis of occluding junctions in rectal glands of elasmobranch fishes. *J. Membr. Biol.* 58, 101–114.

Evans, D. H. (1968). Measurement of drinking rates in fish. *Comp. Biochem. Physiol.* 25, 751–753.

Evans, D. H. (1969). Studies on the permeability to water of selected marine, freshwater and euryhaline teleosts. *J. Exp. Biol.* 50, 689–703.

Evans, D. H. (1981). The egg case of the oviparous elasmobranch *Raja erinacea*, does osmoregulate. *J. Exp. Biol.* 92, 337–340.

Evans, D. H. (1982). Mechanisms of acid extrusion by two marine fishes: the teleost, *Opsanus beta*, and the elasmobranch, *Squalus acanthias*. *J. Exp. Biol.* 97, 289–299.

Evans, D. H., Kormanik, G. A. and Krasny, E. J. (1979). Mechanism of ammonia and acid extrusion by the little skate, *Raja erinacea*. *J. Exp. Zool.* 208, 431–437.

Evans, D. H., Piermarini, P. M. and Choe, K. P. (2005). The multifunctional fish gill: dominant site of gas exchange, osmoregulation, acid–base regulation, and excretion of nitrogenous waste. *Physiol. Rev.* 85, 97–177.

Felsenstein, J. (1985). Confidence limits on phylogenies: an approach using the bootstrap. *Evolution* 39, 783–791.

Fines, G. A., Ballantyne, J. S. and Wright, P. A. (2001). Active urea transport and an unusual basolateral membrane composition in the gills of a marine elasmobranch. *Am. J. Physiol.* 280, R16–R24.

Fischer, J., Voigt, S., Schneider, J. W., Buchwitz, M. and Voigt, S. (2011). A selachian freshwater fauna from the Triassic of Kyrgyzstan and its implication for Mesozoic shark nurseries. *J. Vert. Paleontol.* 31, 937–953.

Forster, R. P. (1967). Osmoregulatory role of the kidney in cartilaginous fishes (Chondrichthyes). In *Sharks, Skates and Rays* (eds. P. W. Gilbert, R. F. Mathewson and D. P. Rall), pp. 187–195. Baltimore, MD: Johns Hopkins Press.

Forster, R. P. and Goldstein, L. (1976). Intracellular osmoregulatory role of amino acids and urea in marine elasmobranchs. *Am. J. Physiol.* 230, 925–931.

Foulley, M. and Mellinger, J. (1980). La diffusion de l'eau tritiees, de l'uree ^{14}C et d'autres substances a travers la coque de l'ouef de Rousette, *Scyliorhinus canicula*. *C. R. Acad. Sci. Paris* 290, 427–430.

Fuentes, J. and Eddy, F. B. (1997). Effect of manipulation of the renin–angiotensin system in control of drinking in juvenile Atlantic salmon (*Salmo salar* L.) in freshwater and after transfer to sea water. *J. Comp. Physiol. B* 167, 438–443.

Garrey, W. E. (1905). The osmotic pressure of sea water and the blood or marine animals. *Biol. Bull.* 8, 257–269.

Gerzeli, G., De Stefano, G. F., Bolognani, L., Koenig, K. W., Gervaso, M. V. and Omodeo-Sale, M. F. (1969). The rectal gland in relation to the osmoregulatory mechanisms of marine and freshwater elasmobranchs. *Boll. Zool.* 36, 399–400.

Gilbert, P. W. and Heath, G. W. (1972). The clasper-siphon sac mechanism in *Squalus acanthias* and *Mustelus canis*. *Comp. Biochem. Physiol. A* 42, 97–119.

Gilmour, K. M., Bayaa, M., Kenney, B., McNeill, B. and Perry, S. F. (2007). Type IV carbonic anhydrase is present in the gills of spiny dogfish (*Squalus acanthias*). *Am. J. Physiol.* 292, R556–R567.

Goldstein, L. (1967). Urea biosynthesis in elasmobranchs. In *Sharks, Skates and Rays* (eds. P. W. Gilbert, R. F. Mathewson and D. P. Rall), pp. 207–214. Baltimore, MD: Johns Hopkins Press.

Goldstein, L. and Brill, S. R. (1991). Volume-activated taurine efflux from skate erythrocytes: possible band 3 involvement. *Am. J. Physiol.* 260, R1014–R1020.

Goldstein, L. and Dewitt-Harley, S. (1973). Trimethylamine oxidase of nurse shark liver and its relation to mammalian mixed function amine oxidase. *Comp. Biochem. Physiol. B* 45, 895–903.

Goldstein, L. and Forster, R. P. (1971a). Osmoregulation and urea metabolism in the little skate *Raja erinacea*. *Am. J. Physiol.* 220, 742–746.

Goldstein, L. and Forster, R. P. (1971b). Urea biosynthesis and excretion in freshwater and marine elasmobranchs. *Comp. Biochem. Physiol. B* 39, 415–421.

Goldstein, L. and Palatt, P. J. (1974). Trimethylamine oxide excretion rates in elasmobranchs. *Am. J. Physiol.* 227, 1268–1272.

Goldstein, L., Oppelt, W. W. and Maren, T. H. (1968). Osmotic regulation and urea metabolism in the lemon shark *Negaprion brevirostris*. *Am. J. Physiol.* 215, 1493–1497.

Goldstein, L., Brill, S. R. and Freund, E. V. (1990). Activation of taurine efflux in hypotonically stressed elasmobranch cells: inhibition by stilbene disulfonates. *J. Exp. Zool.* 254, 114–118.

Goldstein, L., Luer, C. A. and Blum, P. C. (1993). Taurine transport characteristics of the embryonic skate (*Raja eglanteria*) heart. *J. Exp. Biol.* 182, 291–295.

Goldstein, L., Koomoa, D. L. and Musch, M. W. (2003). ATP release from hypotonically stressed skate RBC: potential role in osmolyte channel regulation. *J. Exp. Zool. A* 296, 160–163.

Gosting, L. J. and Akeley, D. F. (1952). A study of the diffusion of urea in water at 25° with the Guoy interference method. *J. Am. Chem. Soc.* 74, 2058–2060.

Greger, R., Schlatter, E. and Gogelein, H. (1986). Sodium chloride secretion in rectal gland of dogfish, *Squalus acanthias*. *NIPS* 1, 134–136.

Griffith, R. W., Pang, P. K. T., Srivastava, K. and Pickford, G. E. (1973). Serum composition of freshwater stringrays (Potamotrygonidae) adapted to freshwater and dilute seawater. *Biol. Bull.* 144, 304–320.

Guttridge, T. L., Gruber, S. H., Franks, B. R., Kessel, S. T., Gledhill, K. S., Uphill, J., Krause, J. and Sims, D. W. (2012). Deep danger: intra-specific predation risk influences habitat use and aggregation formation of juvenile lemon sharks, *Negaprion brevirostris*. *Mar. Ecol. Prog. Ser.* 445, 279–291.

Hanebuth, T., Stattegger, K. and Grootes, P. M. (2000). Rapid flooding of the Sunda shelf: a late-glacial sea level record. *Science* 288, 1033–1035.

Hanrahan, J. W., Mathews, C. J., Grygorczyk, R., Tabcharani, J. A., Grzelczak, Z., Chang, X. B. and Riordan, J. R. (1996). Regulation of the CFTR chloride channel from humans and elasmobranchs. *J. Exp. Zool.* 275, 283–291.

Harms, C. A., Ross, T. and Segars, A. (2002). Plasma biochemistry reference values of wild bonnethead sharks *Sphyrna tiburo*. *Vet. Clin. Pathol.* 31, 111–115.

Hart, N. S., Theiss, S. M., Harahush, B. K. and Collin, S. P. (2011). Microspectrophotometric evidence for cone monochromacy in sharks. *Naturwissenschaften* 98, 193–201.

Hastings, D. F. and Reynolds, J. A. (1979). Molecular weight of (Na^+,K^+) ATPase from shark rectal gland. *Biochemistry* 18, 817–821.

Hayslett, J. P., Jampol, L. M., Forrest, J. N., Epstein, M., Murdaugh, H. V. and Myers, J. D. (1973). Role of NaK-ATPase in the renal absorption of sodium in the elasmobranch *Squalus acanthias*. *Comp. Biochem. Physiol. A* 44, 417–422.

Haywood, G. P. (1974). The exchangeable ionic space, and salinity effects upon ion, water, and urea turnover rates in the dogfish *Poroderma africanum*. *Mar. Biol.* 26, 69–75.

Haywood, G. P. (1975). A preliminary investigation into the roles played by the rectal gland and kidneys in the osmoregulation of the striped dogfish *Poroderma africanum*. *J. Exp. Zool.* 193, 167–176.

Hazon, N., Balment, R. J., Perrott, M. and O'Toole, L. B. (1989). The renin–angiotensin system and vascular and dipsogenic regulation in elasmobranchs. *Gen. Comp. Endocrinol.* 74, 230–236.

Heckly, R. J. and Herald, E. S. (1970). Size and distribution of proteins in elasmobranch plasma. *Proc. Calif. Acad. Sci.* 38, 415–420.

Heisler, N. (1988). Acid–base regulation. In *Physiology of Elasmobranch Fishes* (ed. T. J. Shuttleworth), pp. 213–252. New York: Springer.

Helfman, G. S., Collette, B. B. and Facey, D. E. (1997). *The Diversity of Fishes* (3rd edn.). Malden, MA: Blackwell Science.

Henderson, I. W., O'Toole, L. B. and Hazon, N. (1988). Kidney function. In *Physiology of Elasmobranch Fishes* (ed. T. J. Shuttleworth), pp. 201–214. New York: Springer.

Henderson, R. J. and Tocher, D. R. (1987). The lipid composition and biochemistry of freshwater fish. *Prog. Lipid Res.* 26, 281–347.

Hentschel, H., Nearing, J., Harris, H. W., Betka, M., Baum, M., Hebert, S. C. and Elger, M. (2003). Localization of Mg^{2+}-sensing shark kidney calcium receptor SKCaR in kidney of spiny dogfish, *Squalus acanthias. Am. J. Physiol.* 285, F430–F439.

Heupel, M. R. and Hueter, R. E. (2002). Importance of prey density in relation to the movement patterns of juvenile blacktip sharks (*Carcharhinus limbatus*) within a coastal nursery area. *Mar. Freshw. Res.* 53, 543–550.

Heupel, M. R. and Simpfendorfer, C. A. (2008). Movement and distribution of young bull sharks *Carcharhinus leucas* in a variable estuarine environment. *Aquat. Biol.* 1, 277–289.

Heupel, M. R., Simpfendorfer, C. A. and Hueter, R. E. (2003). Running before the storm: blacktip sharks respond to falling barometric pressure associated with tropical storm Gabrielle. *J. Fish Biol.* 63, 1357–1363.

Hill, W. G., Mathai, J. C., Gensure, R. H., Zeidel, J. D., Apodaca, G., Saenz, J. P., Kinne-Saffran, E., Kinne, R. and Zeidel, M. L. (2004). Permeabilities of teleost and elasmobranch gill apical membranes: evidence that lipid bilayers alone do not account for barrier function. *Am. J. Physiol.* 287, C235–C242.

Hirano, T. (1974). Some factors regulating water intake by the eel, *Anguilla japonica. J. Exp. Biol.* 61, 737–747.

Holeton, G. F. and Heisler, N. (1983). Contribution of net ion transfer mechanisms to acid–base regulation after exhausting activity in the larger spotted dogfish (*Scyliorhinus stellaris*). *J. Exp. Biol.* 103, 31–46.

Hyodo, S., Katoh, F., Kaneko, T. and Takei, Y. (2004a). A facilitative urea transporter is localized in the renal collecting tubule of the dogfish *Triakis scyllia. J. Exp. Biol.* 207, 347–356.

Hyodo, S., Tsukada, T. and Takei, Y. (2004b). Neurohypophysial hormones of dogfish, *Triakis scyllium*: structures and salinity-dependent secretion. *Gen. Comp. Endocrinol.* 138, 97–104.

Ip, Y. K., Loong, A. M., Ching, B., Tham, G. H. Y., Wong, W. P. and Chew, S. F. (2009). The freshwater Amazonian stingray, *Potamotrygon motoro*, up-regulates glutamine synthetase activity and protein abundance, and accumulates glutamine when exposed to brackish (15o/oo) water. *J. Exp. Biol.* 212, 3828–3836.

Isaia, J. (1982). Effects of environmental salinity on branchial permeability of rainbow trout, *Salmo gairdneri. J. Physiol.* 326, 297–307.

Isaia, J. and Masoni, A. (1976). The effects of calcium and magnesium on water and ionic permeabilities in the sea water adapted eel, *Anguilla anguilla* L. *J. Comp. Physiol.* 109, 221–233.

Ishihara, H., Taniuchi, T., Sano, M. and Last, P. R. (1991). Record of *Pristis clavata* Garman from the Pentecost River, Western Australia, with brief notes on its osmoregulation, and comments on the systematics of Pristidae. *Univ. Mus. Univ. Tokyo Nat. Cult.* 3, 43–53.

Jampol, L. M. and Epstein, F. M. (1970). Sodium–potassium-activated adenosinetriphosphatase and osmotic regulation by fishes. *Am. J. Physiol.* 218, 607–611.

Janech, M. G. and Piermarini, P. M. (2002). Renal water and solute excretion in the Atlantic stingray in freshwater. *J. Fish Biol.* 61, 1053–1057.

Janech, M. G., Fitzgibbon, W. R., Chen, R., Nowak, M. W., Miller, D. H., Paul, R. V. and Ploth, D. W. (2003). Molecular and functional characterization of a urea transporter from the kidney of the Atlantic stingray. *Am. J. Physiol.* 284, F996–F1005.

Janech, M. G., Fitzgibbon, W. R., Nowak, M. W., Miller, D. H., Paul, R. V. and Ploth, D. W. (2006a). Cloning and functional characterization of a second urea transporter from the kidney of the Atlantic stingray, *Dasyatis sabina. Am. J. Physiol.* 291, R844–R853.

Janech, M. G., Fitzgibbon, W. R., Ploth, D. W., Lacy, E. R. and Miller, D. H. (2006b). Effect of low environmental salinity on plasma composition and renal function of the Atlantic stingray, a euryhaline elasmobranch. *Am. J. Physiol.* 291, F770–F780.

Janech, M. G., Gefroh, H. A., Cwengros, E. E., Sulikowski, J. A., Ploth, D. W. and Fitzgibbon, W. R. (2008). Cloning of urea transporters from the kidneys of two batoid elasmobranchs: evidence for a common elasmobranch urea transporter isoform. *Mar. Biol.* 153, 1173–1179.

Jensen, N. H. (1976). Reproduction of the bull shark, *Carcharhinus leucas*, in the Lake Nicaragua–Rio San Juan system. In *Investigations of the Ichthyofauna of Nicaraguan Lakes* (ed. T. B. Thorson), pp. 539–559. Lincoln, NE: School of Life Sciences, University of Nebraska.

Johnson, M. R. and Snelson, F. F. (1996). Reproductive life history of the Atlantic stingray *Dasyatis sabina* (Pisces, Dasyatidae), in the freshwater St. Johns River, Florida. *Bull. Mar. Sci.* 59, 74–88.

Katsuki, Y., Yanagisawa, K., Tester, A. L. and Kendall, J. I. (1969). Shark pit organs: response to chemicals. *Science* 163, 405–407.

Kempton, R. T. (1953). Studies on the elasmobranch kidney. II. Reabsorption of urea by the smooth dogfish, *Mustelus canis*. *Biol. Bull.* 104, 45–56.

King, P. A. and Goldstein, L. (1983a). Organic osmolytes and cell volume regulation in fish. *Mol. Physiol.* 4, 53–66.

King, P. A. and Goldstein, L. (1983b). Renal ammonia excretion and production in goldfish, *Carassius auratus*, at low environmental pH. *Am. J. Physiol.* 245, R590–R599.

Kirschner, L. B. (1993). The energetics of osmotic regulation in ureotelic and hypoosmotic fishes. *J. Exp. Zool.* 267, 19–26.

Knight, C. A., Hallett, J. and DeVries, A. L. (1988). Solute effects on ice recrystallization: an assessment technique. *Cryobiology* 25, 55–60.

Koomoa, D. L. T., Musch, M. W., Maclean, A. V. and Goldstein, L. (2001). Volume-activated trimethylamine oxide efflux in red blood cells of spiny dogfish (*Squalus acanthias*). *Am. J. Physiol.* 281, R803–R810.

Kottelat, M. (2001). *Fishes of Laos*. Columbo, Sri Lanka: WHT Publications (PTE).

Kottelat, M., Whitten, A. J., Kartikasari, S. N. and Wirjoatmodjo, S. (1993). *Freshwater fishes of western Indonesia and Sulawesi*. Jakarta: Periplus.

Lacy, E. R. and Reale, E. (1995). *Functional morphology of the elasmobranch nephron and retention of urea. Cellular and Molecular Approaches to Fish Ionic Regulation.* New York: Academic Press, 107–146.

Larsson, A., Johansson-Sjobeck, M. and Fange, R. (1976). Comparative study of some haematological and biochemical blood parameters in fishes from the Skagerrak. *J. Fish Biol.* 9, 425–440.

Last, P. R. and Stevens, J. D. (1994). *Sharks and rays of Australia*. Melbourne: CSIRO.

Last, P. R., White, W. T. and Pogonoski, J. J. (2008). *Descriptions of new Australian Chondrichthyans*. Hobart: CSIRO Marine and Atmospheric Research.

Libby, E. L. (1959). Miracle of the mermaid's purse. *Natl. Geog. Mag.* 116, 413–420.

Lin, T. and Timasheff, S. N. (1994). Why do some organisms use a urea–methylamine mixture as osmolyte? Thermodynamic compensation of urea and trimethylamine N-oxide interactions with protein. *Biochemistry* 33, 12695–12701.

Lingwood, D. D., Fisher, L. J., Callanhan, J. W. and Ballantyne, J. S. (2004). Sulfatide and Na^+-K^+ ATPase: a salinity-sensitive relationship in the gill basolateral membrane of rainbow trout. *J. Membr. Biol.* 200, 1–8.

Lingwood, D., Harauz, G. and Ballantyne, J. S. (2006). Decoupling the Na^+–K^+-ATPase *in vivo*: a possible new role in the gills of freshwater fishes. *Comp. Biochem. Physiol. A Mol. Integr. Physiol.* 144, 451–457.

Lloyd, K. W. and Goldstein, L. (1969). Permeability and metabolism of urea in the intestine of the elasmobranch, *Squalus acanthias*. *Bull. Mount Desert Island Biol. Lab.* 9, 22–23.

Lovejoy, N. R. (1996). Systematics of myliobatoid elasmobranchs: with emphasis on the phylogeny and historical biogeography of neotropical freshwater stingrays (Potamotrygonidae: Rajiformes). *Zool. J. Linn. Soc. Lond.* 117, 207–257.

Lovejoy, N. R., Bermingham, E. and Martin, A. P. (1998). Marine incursion into South America. *Nature* 396, 421–422.

Lytle, C. and Forbush, B. (1992a). NaKCl cotransport in the shark rectal gland. II. Regulation in isolated tubules. *Am. J. Physiol.* 262, C1009–C1017.

Lytle, C. and Forbush, B. (1992b). The NaKCl cotransport protein of shark rectal gland. II. Regulation by direct phosphorylation. *J. Biol. Chem.* 267, 25438–25443.

Mahmmoud, Y. A., Cramb, G., Maunsbach, A. B., Cutler, C. P., Meischke, L. and Cornelius, F. (2003). Regulation of Na,K-ATPase by PLMS, phospholemman-like protein from a shark. Molecular cloning sequence expression, cellular distribution and functional effects of PLM. *J. Biol. Chem.* 278, 37427–37438.

Mandrup-Poulsen, J. (1981). Changes in selected blood serum constituents, as a function of salinity variations, in the marine elasmobranch, *Sphyrna tiburo*. *Comp. Biochem. Physiol. A* 70, 127–131.

Martin, R. A. (2005). Conservation of freshwater and euryhaline elasmobranchs. *J. Mar. Biol. Assoc. U. K.* 85, 1049–1073.

Mashino, T. and Fridovich, I. (1987). Effects of urea and trimethylamine-*N*-oxide on enzyme activity and stability. *Arch. Biochem. Biophys.* 258, 356–360.

McConnell, F. M. and Goldstein, L. (1988). Intracellular signals and volume regulatory response in skate erythrocytes. *Am. J. Physiol.* 255, R982–R987.

McGowan, D. W. and Kajiura, S. M. (2009). Electroreception in the euryhaline stingray, *Dasyatis sabina*. *J. Exp. Biol.* 212, 1544–1552.

Meek, A. (1916). *The Migrations of Fish*. London: Edward Arnold.

Meischke, L. and Cramb, G. (2005). Cloning and expression of Na,K-ATPase isoforms from the euryhaline bullshark *Carcharhinus leucas*. *J. Gen. Physiol.* 69A–70A.

Meloni, C. J., Cech, J. J. and Katzman, S. M. (2002). Effect of brackish salinities on oxygen consumption of bat rays (*Myliobatis californica*). *Copeia* 2, 462–465.

Metcalf, V. J. and Gemmell, N. J. (2005). Fatty acid transport in cartilaginous fish: absence of albumin and possible utilization of lipoproteins. *Fish Physiol. Biochem.* 31, 55–64.

Minamikawa, S. and Morisawa, M. (1996). Acquisition, initiation and maintenance of sperm motility in the shark, *Triakis scyllia*. *Comp. Biochem. Physiol. A* 113, 387–392.

Montoya, R. V. and Thorson, T. B. (1982). The bull shark (*Carcharhinus leucas*) and largetooth sawfish (*Pristis perotteti*) in Lake Bayano, a tropical man-made impoundment in Panama. *Environ. Biol. Fish.* 7, 341–347.

Morgan, J. D. and Iwama, G. K. (1999). Energy cost of NaCl transport in isolated gills of cutthroat trout. *Am. J. Physiol.* 277, 631–639.

Morgan, J. D., Wilson, J. M. and Iwama, G. K. (1997). Oxygen consumption and Na$^+$,K$^+$-ATPase activity of rectal gland and gill tissue in the spiny dogfish, *Squalus acanthias*. *Can. J. Zool.* 75, 820–825.

Morgan, R. L., Ballantyne, J. S. and Wright, P. A. (2003a). Regulation of a renal urea transport with salinity in a marine elasmobranch *Raja erinacea*. *J. Exp. Biol.* 206, 3285–3292.

Morgan, R. L., Ballantyne, J. S. and Wright, P. A. (2003b). Urea transporter in kidney brush-border membrane vesicles from a marine elasmobranch, *Raja erinacea*. *J. Exp. Biol.* 206, 3202–3293.

Morrissey, J. F. and Gruber, S. H. (1993). Habitat selection by juvenile lemon sharks, *Negaprion brevirostris*. *Environ. Biol. Fish.* 38, 311–319.

Motais, R., Romeu, F. G. and Maetz, J. (1966). Exchange diffusion effect and euryhalinity in teleosts. *J. Gen. Physiol.* 49, 391.

Motais, R., Isaia, J., Rankin, J. C. and Maetz, J. (1969). Adaptive changes of the water permeability of the teleostean gill epithelium in relation to external salinity. *J. Exp. Biol.* 51, 529–546.

Moyes, C. D., Moon, T. W. and Ballantyne, J. S. (1986). Osmotic effects on amino acid oxidation in skate liver mitochondria. *J. Exp. Biol.* 125, 181–195.

Muntz, W. R. A., Church, E. and Dartnall, H. J. A. (1973). Visual pigment of the freshwater stingray, *Paratrygon motoro*. *Nature* 246, 517.

Murdaugh, H. V. and Robin, E. D. (1967). Acid–base metabolism in the dogfish shark. In *Sharks, Skates and Rays* (eds. P. W. Gilbert, R. F. Mathewson and D. P. Rall), pp. 249–264. Baltimore, MD: Johns Hopkins Press.

Musch, M. W., Davis, E. M. and Goldstein, L. (1994). Oligomeric forms of skate erythrocyte band 3. Effect of volume expansion. *J. Biol. Chem.* 269, 19683–19686.

Musch, M. W., Davis-Amaral, E. M. and Goldstein, L. (1996). Eythropoietin stimulates tyrosine phosphorylation and taurine transport in skate erythrocytes. *J. Exp. Zool.* 274, 81–92.

Musch, M. W., Davis-Amaral, E. M., Leibowitz, K. L. and Goldstein, L. (1998). Hypotonic-stimulated taurine efflux in skate erythrocytes: regulation by tyrosine phosphatase activity. *Am. J. Physiol.* 274, R1677–R1686.

Musch, M. W., Hubert, E. M. and Goldstein, L. (1999). Volume expansion stimulates $p72^{syk}$ and $p56^{lyn}$ in skate erythrocytes. *J. Biol. Chem.* 274, 7923–7928.

Musch, M. W., Koomoa, D. L. and Goldstein, L. (2004). Hypotonicity-induced exocytosis of the skate anion exchanger skAE1. Role of lipid raft regions. *J. Biol. Chem.* 279, 39447–39453.

Nakada, T., Westhoff, C. M., Yamaguchi, Y., Hyodo, S., Muro, T., Kato, A., Nakamura, N. and Hirose, S. (2010). Rhesus glycoprotein P2 (Rhp2) is a novel member of the Rh family of ammonia transporters highly expressed in shark kidney. *J. Biol. Chem.* 285, 2653–2664.

Nearing, J., Betka, M., Quinn, S., Hentschel, H., Elger, M., Baum, M., Bai, M., Chattopadhyay, N., Brown, E. M., Hebert, S. C. and Harris, H. W. (2002). Polyvalent cation receptor proteins (CaRs) are salinity sensors in fish. *Proc. Natl. Acad. Sci. U. S. A.* 99, 9231–9236.

Nei, M. and Kumar, S. (2000). *Molecular Evolution and Phylogenetics*. New York: Oxford University Press.

Nishimura, H. (1977). Renal responses to diuretic drugs in freshwater catfish *Ictalurus punctatus*. *Am. J. Physiol.* 232, F278–F285.

Nishimura, H. and Fan, Z. (2003). Regulation of water movement across vertebrate renal tubules. *Comp. Biochem. Physiol. A Mol. Integr. Physiol.* 136, 479–498.

Ogawa, M. and Hirano, T. (1982). Studies of the nephron of a freshwater stingray, *Potamotrygon magdalenae*. *Zool. Mag.* 91, 101–105.

Olson, K. R. (1992). Blood and extracellular fluid volume regulation: role of the renin–angiotensin system, kallikrein–kinin system, and atrial natriuretic peptides. In *Fish Physiology*, Vol. XIIB, *The Cardiovascular System* (eds. W. S. Hoar, D. J. Randall and A. P. Farrell), pp. 135–254. San Diego: Academic Press.

Ortega, L. A., Heupel, M. R., Van Beynen, P. and Motta, P. J. (2009). Movement and water quality preferences of juvenile bull sharks (*Carcharhinus leucas*) in a Florida estuary. *Environ. Biol. Fish.* 84 (4), 361–373.

Otake, T. (1991). Serum composition and nephron structure of freshwater elasmobranchs collected from Australia and Papua New Guinea. *Univ. Mus. Univ. Tokyo Nat. Cult.* 3, 55–62.

Otake, T., Ishii, T. and Tanaka, S. (2005). Otolith strontium:calcium ratios in a freshwater stingray, *Himantura signifer* Compagno and Roberts, 1982, from the Chao Phraya River, Thailand. *Coast. Mar. Sci.* 29, 147–153.

Ottolenghi, P. (1975). The reversible depilidation of a solubilized sodium-plus-potassium ion-dependent adenosine triphosphatase from the salt gland of the spiny dogfish. *Biochem. J.* 151, 61–66.

Otway, N. M., Ellis, M. T. and Starr, R. (2011). Serum biochemical reference intervals for wild dwarf ornate wobbegong sharks (*Orectolobus ornatus*). *Vet. Clin. Pathol* 40, 361–367.

Part, P., Wright, P. A. and Wood, C. M. (1998). Urea and water permeability in dogfish (*Squalus acanthias*) gills. *Comp. Biochem. Physiol. A Mol. Integr. Physiol.* 119, 117–123.

Payan, P. and Maetz, J. (1971). Balance hydrique chez les elasmobranches: arguments en faveur d'un controle endocrinien. *Gen. Comp. Endocrinol.* 16, 535–554.

Payan, P., Goldstein, L. and Forster, R. P. (1973). Gills and kidneys in ureosmotic regulation in euryhaline skates. *Am. J. Physiol.* 224, 367–372.

Peach, M. B. and Marshall, N. J. (2000). The pit organs of elasmobranchs: a review. *Philos. Trans. R. Soc. Lond.* 355B, 1131–1134.

Perrone, J. R., Hackney, J. F., Dixon, J. F. and Hokin, L. E. (1973). Molecular properties of purified (sodium+potassium)-activated adenosine triphosphatases and their subunits from the rectal gland of *Squalus acanthias* and the electric organ of *Electrophorus electricus*. *J. Biol. Chem.* 250, 4178–4184.

Piermarini, P. M. and Evans, D. H. (1998). Osmoregulation of the Atlantic stingray (*Dasyatis sabina*) from the freshwater Lake Jesup of the St. Johns River, Florida. *Physiol. Zool.* 71, 553–560.

Piermarini, P. M. and Evans, D. H. (2000). Effects of environmental salinity on Na^+/K^+-ATPase in the gills and rectal gland of a euryhaline elasmobranch (*Dasyatis sabina*). *J. Exp. Biol.* 203, 2957–2966.

Piermarini, P. M. and Evans, D. H. (2001). Immunochemical analysis of the vacuolar proton-ATPase B-subunit in the gills of a euryhaline stingray (*Dasyatis sabina*): effects of salinity and relation to Na^+/K^+-ATPase. *J. Exp. Biol.* 204, 3251–3259.

Piermarini, P. M., Verlander, J. W., Royaux, I. E. and Evans, D. H. (2002). Pendrin immunoreactivity in the gill epithelium of a euryhaline elasmobranch. *Am. J. Physiol.* 283, R983–R992.

Pillans, R. D. and Franklin, C. E. (2004). Plasma osmolyte concentrations and rectal gland mass of bull sharks *Carcharhinus leucas*, captured along a salinity gradient. *Comp. Biochem. Physiol. A Mol. Integr. Physiol.* 138, 363–371.

Pillans, R. D., Good, J. P., Anderson, W. G., Hazon, N. and Franklin, C. E. (2005). Freshwater to seawater acclimation of juvenile bull sharks (*Carcharhinus leucas*): plasma osmolytes and Na^+/K^+-ATPase activity in gill, rectal gland, kidney and intestine. *J. Comp. Physiol. B* 175, 37–44.

Pillans, R. D., Anderson, W. G., Good, J. P., Hyodo, S., Takei, Y., Hazon, N. and Franklin, C. E. (2006). Plasma and erythrocyte solute properties of juvenile bull sharks, *Carcharhinus leucas*, acutely exposed to increasing environmental salinity. *J. Exp. Mar. Biol. Ecol.* 331, 145–157.

Pillans, R. D., Good, J. P., Anderson, W. G., Hazon, N. and Franklin, C. E. (2008). Rectal gland morphology of freshwater and seawater acclimated bull sharks *Carcharhinus leucas*. *J. Fish Biol.* 72, 1559–1571.

Price, K. S. and Daiber, F. C. (1967). Osmotic environments during fetal development of dogfish, *Mustelus canis* (Mitchill) and *Squalus acanthias* Linnaeus, and some comparisons with skates and rays. *Physiol. Zool.* 40, 248–260.

Pugh, E. L., Bittman, R., Fugler, L. and Kates, M. (1989). Comparison of steady-state fluorescence polarization and urea permeability of phosphatidyl choline and phosphatidylsulfocholine liposomes as a function of sterol structure. *Chem. Phys. Lipids* 50, 43–50.

Rabinowitz, L. and Gunther, R. A. (1973). Urea transport in elasmobranch erythrocytes. *Am. J. Physiol.* 224, 1109–1115.

Rainboth, W. J. (1996). *Fishes of the Cambodian Mekong*. Rome: Food and Agriculture Organization of the United Nations.

Raschi, W. and Mackanos, L. A. (1989). The structure of the ampullae of Lorenzini in *Dasyatis garouaensis* and its implications on the evolution of freshwater electroreceptive systems. *J. Exp. Zool.* 252 (Suppl. 2), 101–111.

Raschi, W., Keithan, E. D. and Rhee, W. C. H. (1997). Anatomy of the ampullary electroreceptor in the freshwater stingray, *Himantura signifer*. *Copeia* 1997, 101–107.

Raynard, R. S. and Cossins, A. R. (1991). Homeoviscous adaptation and thermal compensation of sodium pump of trout erythrocytes. *Am. J. Physiol.* 260, R916–R924.

Reilly, B. D., Cramp, R. L., Wilson, J. M., Campbell, H. A. and Franklin, C. E. (2011). Branchial osmoregulation in the euryhaline shark, *Carcharhinus leucas*: a molecular analysis of ion transporters. *J. Exp. Biol.* 214, 2883–2895.

Renfro, J. L. (1980). Relationship between renal fluid and Mg secretion in a glomerular marine teleost. *Am. J. Physiol.* 238, F92–F98.

Riordan, J. R., Forbush, B. and Hanrahan, J. W. (1994). The molecular basis of chloride transport in shark rectal gland. *J. Exp. Biol.* 196, 405–418.

Roberts, T. R. (1989). The freshwater fishes of Western Borneo (Kalimantan Barat, Indonesia). *Mem. Calif. Acad. Sci.* 14, 1–210.

Robertson, J. D. (1975). Osmotic constituents of the blood plasma and parietal muscle of *Squalus acanthias* L. *Biol. Bull.* 148, 303–319.

Robertson, J. D. (1989). Osmotic constituents of the blood plasma and parietal muscle of *Scyliorhinus canicula* (L.). *Comp. Biochem. Physiol. A* 93, 799–805.

Rodela, T. M., Ballantyne, J. S. and Wright, P. A. (2008). Carrier-mediated urea transport across the mitochondrial membrane of an elasmobranch (*Raja erinacea*) and a teleost (*Oncorhynchus mykiss*) fish. *Am. J. Physiol.* 294, R1947–R1957.

Rodler, J., Heinemann, H. O., Fishman, A. P. and Smith, H. W. (1955). The pH and carbonic anhydrase activity in the marine dogfish. *Am. J. Physiol.* 183, 155–162.

Rosa, R. S. (1985). *A systematic revision of the South American freshwater stingrays (Chondrichthyes: Potamotrygonidae)*. Williamsberg, VA: Dissertation, College of William and Mary.

Rosa, R. S., De Carvalho, M. R. and Wanderley, C. D. A. (2008). *Potamotrygon boesemani* (Chondrichthyes: Myliobatiformes: Potamotrygonidae), a new species of neotropical freshwater stingray from Surinam. *Neotrop. Ichthyol.* 6, 1–8.

Rosa, R. S., Charvet-Almeida, P. and Quijada, C. C. D. (2010). Biology of the South American potamotrygonid stingrays. In *Sharks and their Relatives II: Biodiversity, Adaptive Physiology and Conservation* (eds. J. C. Carrier, J. A. Musick and M. R. Heithaus), pp. 241–281. Boca Raton, FL: CRC Press.

Schmidt-Nielsen, B. and Renfro, J. L. (1975). Kidney function of the American eel *Anguilla rostrata*. *Am. J. Physiol.* 228, 420–431.

Schmidt-Nielsen, B., Truniger, B. and Rabinowitz, L. (1972). Sodium-linked urea transport by the renal tubule of the spiny dogfish *Squalus acanthias*. *Comp. Biochem. Physiol. A* 42, 13–25.

Schooler, J. M., Goldstein, L., Hartman, S. C. and Forster, R. P. (1966). Pathways of urea synthesis in the elasmobranch, *Squalus acanthias*. *Comp. Biochem. Physiol.* 18, 271–281.

Schwarz, W. and Gu, Q. (1988). Characteristics of the Na^+/K^+-ATPase from *Torpedo californica* expressed in *Xenopus* oocytes: a combination of tracer flux measurements with electrophysiological measurements. *Biochim. Biophys. Acta* 945, 167–174.

Seibel, B. A. and Walsh, P. J. (2002). Trimethylamine oxide accumulation in marine animals: relationship to acylglycerol storage. *J. Exp. Biol.* 205, 297–306.

Semeniuk, C. A. D., Speers-Roesch, B. and Rothley, K. D. (2007). Using fatty-acid profile analysis as an ecologic indicator in the management of tourist impacts on marine wildlife: a case of stingray-feeding in the Caribbean. *Environ. Manage.* 40, 665–677.

Sidell, B. D. and Hazel, J. R. (1987). Temperature affects the diffusion of small molecules through cytosol of fish muscle. *J. Exp. Biol.* 129, 191–203.

Silva, P., Epstein, F. H., Karnaky, K. J., Reichlin, S. and Forrest, J. N. (1993). Neuropeptide Y inhibits chloride secretion in the shark rectal gland. *Am. J. Physiol.* 265, R439–R446.

Silva, P., Solomon, R. J. and Epstein, F. H. (1996). The rectal gland of *Squalus acanthias*: a model for the transport of chloride. *Kidney Int.* 49, 1552–1556.

Silva, P., Solomon, R. J. and Epstein, F. H. (1997). Transport mechanisms that mediate the secretion of chloride by the rectal gland of *Squalus acanthias*. *J. Exp. Zool.* 279, 504–508.

Silva, P., Solomon, R. J. and Epstein, F. H. (1999). Mode of activation of salt secretion by C-type natriuretic peptide in the shark rectal gland. *Am. J. Physiol.* 277, R1725 R1732.

Simpfendorfer, C. A., Freitas, G. G., Wiley, T. R. and Heupel, M. R. (2005). Distribution and habitat partitioning of immature bullsharks (*Carcharhinus leucas*) in a southwest Florida estuary. *Estuaries* 28, 78–85.

Skelton, P. (2001). *Freshwater Fishes of Southern Africa*. Cape Town: Struik.

Smith, C. P. and Wright, P. A. (1999). Molecular characterization of an elasmobranch urea transporter. *Am. J. Physiol.* 276, R622–R626.

Smith, H. W. (1929). The composition of body fluids of elasmobranchs. *J. Biol. Chem.* 81, 407–419.

Smith, H. W. (1931a). The absorption and excretion of water and salts by the elasmobranch fishes. I. Freshwater elasmobranchs. *Am. J. Physiol.* 98, 279–295.

Smith, H. W. (1931b). The absorption and excretion of water and salts by the elasmobranch fishes. II. Marine elasmobranchs. *Am. J. Physiol.* 98, 296–310.

Snelson, F. F., Mulligan, T. J. and Williams, S. E. (1984). Food habits, occurrence, and population structure of the bull shark, *Carcharhinus leucas*, in Florida coastal lagoons. *Bull. Mar. Sci.* 34, 71–80.

Snelson, F. F., Williams-Hooper, S. E. and Schmid, T. H. (1988). Reproduction and ecology of the Atlantic stingray, *Dasyatis sabina*, in Florida coastal lagoons. *Copeia* 1988, 729–739.

Sosa-Nishizaki, O., Taniuchi, T., Ishihara, H. and Shimizu, M. (1998). The bull shark, *Carcharhinus leucas* (Valenciennes, 1841), from the Usumacinta River, Tabasco, Mexico, with notes on its serum composition and osmolarity. *Cienc. Mar.* 24, 183–192.

Speers-Roesch, B., Ip, Y. K. and Ballantyne, J. S. (2006). Metabolic organization of freshwater, euryhaline, and marine elasmobranchs: implications for the evolution of energy metabolism in sharks and rays. *J. Exp. Biol.* 209, 2495–2508.

Speers-Roesch, B., Ip, Y. K. and Ballantyne, J. S. (2008). Plasma non-esterified fatty acids of elasmobranchs: comparisons of temperate and tropical species and effects of environmental salinity. *Comp. Biochem. Physiol. A Mol. Integr. Physiol.* 149, 209–216.

Staedeler, G. and Frerichs, F. T. (1859). Ueber das vorkommen von harnstoff, taurin und scyllit in den organen der plagiostomen. *J. Prakt. Chem.* 73, 48–55.

Steele, S. L., Yancey, P. H. and Wright, P. A. (2005). The little skate *Raja erinacea* exhibits an extrahepatic ornithine urea cycle in the muscle and modulates nitrogen metabolism during low salinity challenge. *Physiol. Biochem. Zool.* 78, 216–226.

Steen, J. B. and Stray-Pedersen, S. (1975). The permeability of fish gills with comments on the osmotic behaviour of cellular membranes. *Acta Physiol. Scand.* 95, 6–20.

Stolte, H., Galaske, R. G., Eisenbach, G. M., Lechene, C., Schmidt-Nielsen, B. and Boylan, J. W. (1977). Renal tubule ion transport and collecting duct function in the elasmobranch little skate, *Raja erinacea*. *J. Exp. Zool.* 199, 403–410.

Swenson, E. R. and Maren, T. H. (1986). Dissociation of CO_2 hydration and renal acid secretion in the dogfish, *Squalus acanthias*. *Am. J. Physiol.* 250, F288–F293.

Szabo, T., Kalmijn, A. J., Enger, P. S. and Bullock, T. H. (1972). Microampullary organs and a submandibular sense organ in the freshwater ray, *Potamotrygon*. *J. Comp. Physiol.* 79, 15–27.

Szamier, R. B. and Bennett, M. V. L (1980). Ampullary electroreceptors in the freshwater ray, *Potamotrygon*. *J. Comp. Physiol. A* (138), 225–230.

Talbot, C., Eddy, F. B., Potts, W. T. W. and Primmett, D. R. N. (1989). Renal function in migrating adult Atlantic salmon, *Salmo salar* L. *Comp. Biochem. Physiol. A* 92, 241–245.

Tam, W. L., Wong, W. P., Loong, A. M., Hiong, K. C., Chew, S. F., Ballantyne, J. S. and Ip, Y. K. (2003). The osmotic response of the Asian freshwater stingray (*Himantura signifer*) to increased salinity: a comparison with marine (*Taeniura lymma*) and Amazonian freshwater (*Potamotrygon motoro*) stingrays. *J. Exp. Biol.* 206, 2931–2940.

Tamura, K., Peterson, D., Peterson, N., Stecher, G., Nei, M. and Kumar, S. (2011). MEGA5: molecular evolutionary genetics analysis using maximum likelihood, evolutionary distance, and maximum parsimony methods. *Mol. Biol. Evol.* 28, 2731–2739.

Tan, H. H. and Lim, K. K. P. (1998). Freshwater elasmobranchs from Batang Hari Basin of central Sumatra, Indonesia. *Raffles J. Zool* 46, 425–429.

Taniuchi, T. (1979). Freshwater elasmobranchs from Lake Naujan, Perak River, and Indragiri River, Southeast Asia. *Jpn. J. Ichthyol.* 25, 273–277.

Taniuchi, T. (1991). Occurrence of two species of stingrays of the genus *Dasyatis* (Chondrichthyes) in the Sanaga Basin, Cameroun. *Environ. Biol. Fish* 31, 95–100.

Taniuchi, T. and Shimizu, M. (1991). Elasmobranchs collected from seven river systems in Northern Australia and Papua New Guinea. *Univ. Mus. Univ. Tokyo Nat. Cult.* 3, 3–10.

Taniuchi, T., Ishihara, H., Tanaka, S., Hyodo, S., Murakami, M. and Seret, B. (2003). Occurrence of two species of elasmobranchs, *Carcharhinus leucas* and *Pristis microdon* in Betsiboka River, West Madagascar. *Cybium* 27, 237–241.

Thomerson, J. E., Thorson, T. B. and Hempel, R. L. (1977). The bull shark, *Carcharhinus leucas*, from the upper Mississippi River near Alton, Illinois. *Copeia* 1977, 166–167.

Thorson, T. B. (1962). Partitioning of body fluids in the Lake Nicaragua shark and three marine sharks. *Science* 138, 688–690.

Thorson, T. B. (1967). Osmoregulation in fresh-water elasmobranchs. In *Sharks, Skates and Rays* (eds. P. W. Gilbert, R. F. Mathewson and D. P. Rall), pp. 265–270. Baltimore, MD: Johns Hopkins Press.

Thorson, T. B. (1970). Freshwater stingrays, *Potamotrygon* spp.: failure to concentrate urea when exposed to saline medium. *Life Sci.* 9, 893–900.

Thorson, T. B. (1971). Movement of bull sharks, *Carcharhinus leucas*, between Caribbean Sea and Lake Nicaragua demonstrated by tagging. *Copeia* 1971, 336–338.

Thorson, T. B. (1972). The status of the bull shark *Carcharhinus leucas*, in the Amazon River. *Copeia* 1972, 601–605.

Thorson, T. B. (1974). Occurrence of the sawfish *Pristis perotteti*, in the Amazon River, with notes on *P. pectinatus*. *Copeia* 1974, 560–564.

Thorson, T. B. (1976). Observations on the reproduction of the sawfish, *Pristis perotteti*, in Lake Nicaragua, with recommendations for its conservation. In *Investigations of the Ichthyofauna of Nicaraguan Lakes* (ed. T. B. Thorson), pp. 641–650. Lincoln, NE: University of Nebraska-Lincoln.

Thorson, T. B. (1982). Life history implications of a tagging study of the largetooth sawfish, *Pristis perotteti*, in the Lake Nicaragua–Rio San Juan system. *Environ. Biol. Fish* 7, 207–228.

Thorson, T. B. and Gerst, J. W. (1972). Comparison of some parameters of serum and uterine fluid of pregnant, viviparous sharks (*Carcharhinus leucas*) and serum of their near-term young. *Comp. Biochem. Physiol. A* 42, 33–40.

Thorson, T. B. and Watson, D. E. (1975). Reassignment of the African freshwater stingray, *Potamotrygon garouaensis*, to the genus *Dasyatis*, on physiologic and morphologic grounds. *Copeia* 1975, 701–712.

Thorson, T. B., Cowan, C. M. and Watson, D. E. (1967). *Potamotrygon* spp.: elasmobranchs with low urea content. *Science* 158, 375–377.

Thorson, T. B., Cowan, C. M. and Watson, D. E. (1973). Body fluid solutes of juveniles and adults of the euryhaline bull shark *Carcharinus leucas* from freshwater and saline environments. *Physiol. Zool.* 46, 29–42.

Thorson, T. B., Wotton, R. M. and Georgi, T. A. (1978). Rectal gland of freshwater stingrays, *Potamotrygon* spp. (Chondrichthyes: Potamotrygonidae). *Biol. Bull.* 154, 508–516.

Thorson, T. B., Langhammer, J. K. and Oetinger, M. I. (1983). Reproduction and development of the South American freshwater stingrays, *Potamotrygon circularis* and *P. motoro*. *Environ. Biol. Fish.* 9, 3–24.

Thurau, K., Antkowiak, D. and Boylan, J. W. (1969). Demonstration of a renal osmoregulatory mechanism in the spiny dogfish, *Squalus acanthias*. *Bull. Mount Desert Island Biol. Lab* 9, 63–64.

Toffoli, D., Hrbek, T., de Araujo, M. L. G., de Almeida, M. P., Charvet-Almeida, P. and Farias, I. P. (2008). A test of the utility of barcoding in the radiation of the freshwater stingray genus *Potamotrygon* (Potamotrygonidae, Myliobatiformes). *Genet. Mol. Biol.* 31, 324–336.

Treberg, J. R. and Driedzic, W. R. (2006). Maintenance and accumulation of trimethylamine oxide by winter skate (*Leucoraja ocellata*): reliance on low whole animal losses rather than synthesis. *Am. J. Physiol.* 291, R1790–R1798.

Treberg, J. R. and Driedzic, W. R. (2007). The accumulation and synthesis of betaine in winter skate (*Leucoraja ocellata*). *Comp. Biochem. Physiol. A Mol. Integr. Physiol.* 147, 475–483.

Treberg, J. R., Speers-Roesch, B., Piermarini, P. M., Ip, Y. K., Ballantyne, J. S. and Driedzic, W. R. (2006). The accumulation of methylamine counteracting solutes in elasmobranchs with differing levels of urea: a comparison of marine and freshwater species. *J. Exp. Biol.* 209, 860–870.

Tresquerres, M., Katoh, F., Fenton, H., Jasinska, E. and Goss, G. G. (2005). Regulation of branchial V-H$^+$-ATPase, Na$^+$/K$^+$-ATPase and NHE2 in response to acid and base infusions in the Pacific spiny dogfish (*Squalus acanthias*). *J. Exp. Biol.* 208, 345–354.

Tresquerres, M., Parks, S. K., Katoh, F. and Goss, G. G. (2006). Microtubule-dependent relocation of branchial V-H$^+$-ATPase to the basolateral membrane in the Pacific spiny dogfish (*Squalus acanthias*): a role in base secretion. *J. Exp. Biol.* 209, 599–609.

Tresquerres, M., Parks, S. K., Wood, C. M. and Goss, G. G. (2007). V-H$^+$-ATPase translocation during blood alkalosis in dogfish gills: interaction with carbonic anhydrase and involvement in the postfeeding alkaline tide. *Am. J. Physiol.* 292, R2012–R2019.

Tricas, T. C. and Sisneros, J. A. (2004). Ecological functions and adaptations of the elasmobranch electrosense. In *The Senses of Fish: Adaptations for the Reception of Natural Stimuli* (eds. G. Emde, J. Mogdans and B. J. Kapoor), pp. 308–329. New Delhi: Narosa.

Trivett, M. K., Walker, T. I., Clement, J. G., Ho, P. M. W., Martin, T. J. and Danks, J. A. (2001). Effects of water temperature and salinity on parathyroid hormone-related protein in the circulation and tissues of elasmobranchs. *Comp. Biochem. Physiol. A Mol. Integr. Physiol.* 129, 327–336.

Ubeda, A. J., Simpfendorfer, C. A. and Heupel, M. R. (2009). Movements of bonnetheads, *Sphyrna tiburo,* as a response to salinity challenge in a Florida estuary. *Environ. Biol. Fish.* 84, 293–303.

Urist, M. R. and Van de Putte, K. A. (1967). Comparative biochemistry of the blood of fishes: identification of fishes by the chemical composition of serum. In *Sharks, Skates and Rays* (eds. P. W. Gilbert, R. F. Mathewson and D. P. Rall), pp. 271–285. Baltimore, MD: Johns Hopkins Press.

Vélez-Zuazo, X. and Agnarsson, I. (2011). Shark tales: a molecular species-level phylogeny of sharks (Selachimorpha, Chondrichthyes). *Mol. Phylogen. Evol.* 58, 207–217.

Venkataraman, K., Milton, M. C. J. and Raghuram, K. P. (2003). *Handbook on Sharks of Indian Waters. Diversity, Fishery Status, Trade and Conservation.* New Delhi: Zoological Survey of India.

Vidthayanon, C. (2010). *Freshwater Fish (of Thailand).* Bangkok: Sarakadee Press. In Thai.

Vidthayanon, C. and Roberts, T. R. (2005). *Himantura kittipongi,* a new species of freshwater whiptailed stingray from the Maekhlong River of Thailand (Elasmobranchii, Dasyatidae). *Nat. Hist. Bull. Siam Soc.* 53, 123–132.

de Vlaming, V. L. and Sage, M. (1973). Osmoregulation in the euryhaline elasmobranch, *Dasyatis Sabina. Comp. Biochem. Physiol. A* 45, 31–44.

Wald, G. (1939). The porphyropsin visual system. *J. Gen. Physiol.* 22, 775–794.

Walsh, P. J., Wood, C. M., Perry, S. F. and Thomas, S. (1994). Urea transport by hepatocytes and red blood cells of selected elasmobranch and teleost fishes. *J. Exp. Biol.* 193, 321–335.

Whitehead, D. L. (2002). Ampullary organs and electroreception in freshwater *Carcharhinus leucas. J. Physiol. (Paris)* 96, 391–395.

Wilson, E. D., McGunn, M. R. and Goldstein, L. (1999). Trimethylamine oxide transport across plasma membranes of elasmobranch erythrocytes. *J. Exp. Zool.* 284, 605–609.

Withers, P. C. (1998). Urea: diverse functions of a "waste" product. *Clin. Exp. Pharmacol. Physiol.* 25, 722–727.

Withers, P. C., Morrison, G. and Guppy, M. (1994). Buoyancy role of urea and TMAO in an elasmobranch fish, the Port Jackson shark, *Heterodontus portusjacksoni. Physiol. Zool.* 67, 693–705.

Wong, T. M. and Chan, D. K. O. (1977). Physiological adjustments to dilution of the external medium in the lip-shark *Hemiscyllium plagiosum* (Bennett) II. Branchial, renal and rectal gland function. *J. Exp. Zool.* 200, 85–96.

Wood, C. M., Part, P. and Wright, P. A. (1995). Ammonia and urea metabolism in relation to gill function and acid–base balance in a marine elasmobranch, the spiny dogfish (*Squalus acanthias*). *J. Exp. Biol.* 198, 1545–1558.

Wood, C. M., Matsuo, A. Y. O., Gonzalez, R. J., Wilson, R. W., Patrick, M. L. and Val, A. L. (2002). Mechanisms of ion transport in *Potamotrygon,* a stenohaline freshwater elasmobranch native to the ion-poor blackwaters of the Rio Negro. *J. Exp. Biol.* 205, 3039–3054.

Wood, C. M., Kajimura, M., Bucking, C. and Walsh, P. J. (2007). Osmoregulation, ionoregulation and acid–base regulation by the gastrointestinal tract after feeding in the elasmobranch *Squalus acanthias. J. Exp. Biol.* 210, 1335–1349.

Wood, C. M., Walsh, P. J., Kajimura, M., McClelland, G. B. and Chew, S. F. (2010). The influence of feeding and fasting on plasma metabolites in the dogfish shark (*Squalus acanthias*). *Comp. Biochem. Physiol. A Mol. Integr. Physiol.* 155, 435–444.

Wright, P. A., Felskie, A. and Anderson, P. M. (1995). Induction of ornithine–urea cycle enzymes and nitrogen metabolism and excretion in rainbow trout (*Oncorhynchus mykiss*) during early life stages. *J. Exp. Biol.* 198, 127–135.

Wueringer, B. E. and Tibbits, I. R. (2008). Comparison of the lateral line and ampullary systems of two species of shovelnose ray. *Rev. Fish Biol. Fish.* 18, 47–64.

Wueringer, B. E., Squire, L. and Collin, S. P. (2009). The biology of extinct and extant sawfish (Batoidea: Sclerorhynchidae and Pristidae). *Rev. Fish Biol. Fish.* 19, 445–464.

Wynen, L., Larson, H., Thorburn, D., Peverell, S., Morgan, D., Field, I. and Gibb, K. (2009). Mitochondrial DNA supports the identification of two endangered river sharks (*Glyphis glyphis* and *Glyphis garricki*) across northern Australia. *Mar. Freshw. Res.* 60, 554–562.

Yamaguchi, Y., Takaki, S. and Hyodo, S. (2009). Subcellular distribution of urea transporter in the collecting tubule of shark kidney is dependent on environmental salinity. *J. Exp. Zool. A* 311, 705–718.

Yancey, P. H. and Somero, G. N. (1978). Urea-requiring lactate dehydrogenases of marine elasmobranch fishes. *J. Comp. Physiol. B* 125, 135–141.

Yancey, P. H. and Somero, G. N. (1979). Counteraction of urea destabilization of protein structure by methylamine osmoregulatory compounds of elasmobranch fishes. *Biochem. J.* 183, 317–323.

Yancey, P. H., Clark, M. E., Hand, S. C., Bowlus, R. D. and Somero, G. N. (1982). Living with water stress: evolution of osmolyte systems. *Science* 217, 1214–1222.

Yang, B. and Verkman, A. S. (1998). Urea transporter UT3 functions as an efficient water channel. Direct evidence for a common water/urea pathway. *J. Biol. Chem.* 273, 9369–9372.

Yano, K., Ali, A., Gambang, A. C., Hamid, I. A., Razak, S. A. and Zainal, A. (2005). *Sharks and Rays of Malaysia and Brunei Darussalam*. Kuala Terengganu, Malaysia: Marine Fishery Resources Development and Management Department, Southeast Asian Fisheries Development Center.

Zammit, V. A. and Newsholme, E. A. (1979). Activities of enzymes of fat and ketone body metabolism and effects of starvation on blood concentrations of glucose and fat fuels in teleost and elasmobranch fish. *Biochem. J.* 184, 313–322.

Zeidel, J. D., Mathai, J. C., Campbell, J. D., Ruiz, W. G., Apodaca, G. L., Riordan, J. and Zeidel, M. L. (2005). Selective permeability barrier to urea in shark rectal gland. *Am. J. Physiol.* 289, F83–F89.

Zigman, S., Munro, J. and Lerman, S. (1965). Effects of urea on the cold precipitation of protein in the lens of the dogfish. *Nature* 207, 414–415.

Ziyadeh, F. N., Feldman, G. M., Booz, G. W. and Kleinzeller, A. (1988). Taurine and cell volume maintenance in the shark rectal gland: cellular fluxes and kinetics. *Biochim. Biophys. Acta* 943, 43–52.

Zou, Q., Bennion, B. J., Daggett, V. and Murphy, K. P. (2002). The molecular mechanism of stabilization of proteins by TMAO and its ability to counteract the effects of urea. *J. Am. Chem. Soc.* 124, 1192–1202.

5

SMOLT PHYSIOLOGY AND ENDOCRINOLOGY

STEPHEN D. McCORMICK

The parr–smolt transformation of anadromous salmonids is a suite of behavioral, morphological, and physiological changes that are preparatory for downstream migration and seawater entry. The timing of smolt development varies among species, occurring soon after hatching in pink and chum salmon and after one to several years in Atlantic salmon. In many species the transformation is size dependent and occurs in spring, mediated through photoperiod and temperature cues. Smolt development is stimulated by several hormones including growth hormone, insulin-like growth

Euryhaline Fishes: Volume 32
FISH PHYSIOLOGY

Copyright © 2013 Elsevier Inc. All rights reserved
DOI: http://dx.doi.org/10.1016/B978-0-12-396951-4.00005-0

factor-1, cortisol, and thyroid hormones, whereas prolactin is generally inhibitory. Increased salinity tolerance is one of the most important and tractable changes, and is caused by alteration in the function of the major osmoregulatory organs, the gill, gut, and kidney. Increased abundance of specific ion transporters (Na^+/K^+-ATPase, $Na^+/K^+/Cl^-$ cotransporter and apical Cl^- channel) in gill ionocytes results in increased salt secretory capacity, increased growth and swimming performance in seawater, and higher marine survival.

1. INTRODUCTION

After spending up to several years in freshwater (FW), often in a small territory within a single stream, juvenile salmon abandon their FW life in favor of a long downstream migration followed by an even longer ocean journey. This journey and its accompanying changes have fascinated biologists for centuries. The developmental shift from the stream-dwelling parr to the downstream migrating smolt is known as the parr–smolt transformation, smolting, or smoltification. This transformation includes changes in physiology, behavior, and morphology that are adaptive for downstream migration and an ocean existence. Although not a true metamorphosis, smolting shares the feature of being a radical alteration in development that accompanies a dramatic niche shift.

Smolt development has received substantial research attention in the past 30 years. This research initially was directed at understanding smolt development as it related to the developing aquaculture industry, and in particular the need to control the timing and quality of smolt development for transfer of juveniles into ocean net pens. More recently, the increasing threats to many salmon populations coupled with a realization of the importance of smolt recruitment to natural salmon populations has led to interest in understanding the "natural" biology of smolts and their sensitivity to environmental disturbance (McCormick et al., 2009a). In addition to the influential reviews by William Hoar (1976, 1988), several excellent reviews of smolt biology have appeared (Boeuf, 1993; McCormick et al., 1998; Stefansson et al., 2008). Rather than go over well-tilled ground, wherever possible this review will emphasize new research findings and synthesize the existing literature.

There is large variation in the timing of ocean entry both among and within salmonid species, with some salmon such as pink and chum (*Oncorhynchus gorbuscha*, *O. keta*) entering seawater (SW) soon after hatching, whereas others such as coho (*O. kisutch*), steelhead trout (*O. mykiss*), and Atlantic salmon (*Salmo salar*) may spend one to several years in FW. In Atlantic salmon, fishes that do not reach the minimum size for smolt development in their first year will wait one year (or more) to smolt, allowing for direct comparison of these two life stages. Although this review will emphasize common elements of smolt development among salmonids, it should be noted that important aspects of timing and environmental effectors differ among salmon species or even within a species; these will be discussed in detail in Section 7.

2. MORPHOLOGY

The morphological differences between parr and smolt can be striking (Fig. 5.1). Parr usually have strong vertical bands, known as parr marks, and spotting that varies among species. During smolting parr marks are reduced and there is development of extensive silvering and darkened fin margins, especially on the caudal, dorsal, and pectoral fins. Silvering is the result of increased deposition of the crystalline purines guanine and hypoxanthine in skin and scales (Johnston and Eales, 1967). Darkened fin margins develop gradually over a period of several weeks owing to the expansion of melanophores (Mizuno et al., 2004).

Smolts are slimmer than parr, and this appears to be due to an increased rate of linear relative to mass growth. This relative slimness is manifested in

Fig. 5.1. Morphological differences between Atlantic salmon parr (top) and smolt reared in the wild. Note the vertical bands and spots on the sides of parr, and the presence of intense silvering and darkened caudal, pectoral, and dorsal fin margins in smolt. Photo credit: S.D. McCormick. **SEE COLOR PLATE SECTION**

decreased condition factor (weight to length ratio) of smolts during spring, and contrasts with parr which generally increase condition factor in late spring. The reduced condition factor of smolts may in part be due to the energetic demands of smolt development, which include increased basal metabolic rate, increased activity, and reduced lipid content (McCormick and Saunders, 1987). It is also possible that the rapid growth in length allows for greater overall swimming capacity and increased predator avoidance during migration and ocean entry. Shape change analysis has revealed clear differences during smolt development (Beeman et al., 1994) that may be a consequence of the altered pattern of linear and mass growth or perhaps more complex processes.

In addition to these obvious morphological changes, several more subtle changes have been observed. Coho salmon (*O. kisutch*) smolts develop teeth on the maxilla, mandible, and tongue, exhibit increased size of the integumentary folds adjacent to the cloacal opening, and develop a larger auxiliary appendage of the pelvic fin (Gorbman et al., 1982). The relative length of pectoral, pelvic, and anal fins (but not dorsal or caudal) is smaller in Atlantic salmon smolts than in parr (Pelis and McCormick, 2003). Surprisingly, there have been no other morphological analyses to determine whether these or other changes are common to all smolting salmonids.

3. MIGRATION

Downstream migration and movement into the ocean is a critical part of the anadromous life history and thus elemental to smolt biology. A number of behavioral changes occur together as part of downstream migration of smolts, including increased negative rheotaxis to flow (i.e. downstream orientation), decreased agonistic and territorial behavior, and increased schooling and salinity preference (Hoar, 1988; Iwata, 1995). There appears to be a link between altered behavior and the physiological changes that occur during smolting. Gibson (1983) found that Atlantic salmon parr and smolts in artificial streams had similar agonistic behaviors that did not change in spring, and concluded that smolt-related increases in buoyancy and seasonal increases in water velocity were necessary to cause decreased aggression and downstream migration. This finding, along with others, indicates that physiological changes precede and are a requirement for subsequent changes in downstream migratory behavior. Baggerman (1960) suggested that "migration occurs only when the animals are in the proper physiological condition (migration disposition), and at the same time under the influence of appropriate external stimuli which act as 'releasers'". Thus, there are preparatory changes cued by photoperiod and temperature that result in "migratory readiness" of smolts that are then acted on by environmental factors such as temperature, turbidity, and water flow that act as releasing factors to initiate downstream migration of smolts (McCormick et al., 1998). In addition to these environmental factors, other factors such as social cues (i.e. the presence of other migrants) may play a role in initiating downstream migration (Hansen and Jonsson, 1985). The synchronous nature of smolt migration (often occurring over a 3–6 week period) is due to each individual smolt's reaction to these releasing factors and may have the adaptive value of overwhelming predators by sheer numbers.

It may be useful to divide smolt migration into four major components: initiation, downstream, estuarine, and ocean migrations. Initiation is the initial departure from the FW rearing area. The factors involved in the initiation of migration are probably the most critical in determining the overall timing of smolt migration. Environmental factors that trigger initiation of migration (such as temperature) may not be the same as those that affect patterns or rates of the other components of smolt migration. For instance, initiation of migration may not be strongly regulated by flow (discharge), whereas rates of downstream migration are likely to be strongly influenced by flow (Sykes and Shrimpton, 2010). Indeed, most studies of smolt migration in the wild are unable to distinguish environmental impacts on initiation and downstream movement, as fish counting stations are often far downstream of rearing locations (which are often unknown), and thus will incorporate both initiation and downstream movement. However, laboratory studies (Zydlewski et al., 2005; Sykes and Shrimpton, 2010) and those in the wild in which counting stations are immediately below rearing habitats should be able to determine factors that affect the initiation of migration. For many smolts the initiation of migration will result in SW entry within days or weeks. However, this pattern is not universal, as some species in long river systems may take many weeks before they enter the marine environment. There is some evidence that in some populations smolt migration (or at least a downstream dispersal) may begin a whole year before actual entry into SW (Rimmer et al., 1983). This diversity may have a genetic origin and thus may be adaptive for the river system in which it occurs.

In addition to these four major components of smolt migration, one could add "termination" (thus initiation, downstream migration, and termination), which is usually an artificial construct that can be observed in fish under laboratory conditions. Smolts appear to maintain downstream migratory behavior for only a limited period, 10–20 days (dependent on temperature) in the case of Atlantic salmon (Zydlewski et al., 2005). Under normal (wild) conditions this timing is likely to be sufficient to allow fish to enter the estuary and begin ocean migration. The initiation and termination of smolt migration is part of a "smolt window", a limited period of high SW preparedness, which will be discussed in detail below.

There is now a voluminous literature on smolt migration and an exhaustive review of smolt migratory behavior is beyond the scope of this chapter, although such an undertaking would be a valuable contribution. The influence of specific environmental factors on smolt migration, especially the initiation of migration, will be addressed in Section 7 (Developmental and Environmental Regulation) of this chapter. The downstream migration and early ocean entry is a period of high mortality that can be critical to overall survival and strongly affected by FW, estuarine and coastal environmental conditions (Thorstad et al., 2012). Environmental conditions in streams and rivers such as temperature, migratory delays (Rechisky et al., 2009), and contaminants can greatly influence the marine survival of smolts (McCormick et al., 2009a). The comparative life histories including smolt migration strategies of Atlantic salmon, brown trout, and Arctic char have been reviewed by Klemetsen et al. (2012).

Levings (1994) and Thorpe (1994b) document the highly variable use of estuaries among Atlantic and Pacific salmon. The estuarine residence time of Atlantic salmon is relatively brief, generally lasting only one or two tidal cycles (Hansen and Quinn, 1998). In contrast, some species of Pacific salmon may spend many weeks in estuaries before moving out into the open ocean. Schools of postsmolts in estuaries reside in and are displaced with the surface current, and movement is influenced by the tide and the direction of the water flow. Smolts in the upper estuary move by selective ebb tide transport but often display active swimming in the lower estuary or bays.

Recent advances in tagging technology and extensive array systems such as Pacific Ocean Shelf Tracking (POST), the Ocean Tracking Network (OTN), and the Atlantic Cooperative Telemetry Network (ACT) have led to a greater understanding of the early coastal and ocean migratory behavior and survival of smolts (Welch et al., 2011; Tucker et al., 2012; Thorstad et al., 2012), although much remains to be determined. Physiologists have a substantial role to play in this effort by determining the factors that contribute to growth and survival of smolts and providing physiological and endocrine indicators of health.

Several early studies indicated that smolts have reduced swimming ability relative to parr, which led to the hypothesis that downstream migration may be a passive phenomenon in which smolts are no longer capable of maintaining their position in the face of increasing flow rates that normally occur in spring (Thorpe and Morgan, 1978). More recent studies indicate that swimming speeds of Atlantic salmon smolts are the same as or slightly greater than those of parr when standardized for body length, but substantially greater when expressed in absolute speeds (cm s^{-1}) (Peake and McKinley, 1998). Purely passive processes are therefore unlikely to be the basis for downstream movement of smolts, and "activation" of migration seems likely. This is further supported by estimates of smolt migration speeds that are greater than average flow speeds.

Several physiological and biochemical changes occur in muscle during smolt development that may relate to migration and swimming capacity. Wild Atlantic salmon smolts exhibit a 70% increase in relative heart mass in spring, a phenomenon that is absent in parr under the same conditions (Leonard and McCormick, 2001). A three-fold increase in white muscle phosphofructokinase (PFK) was also observed in smolts, indicative of an upregulation of glycolytic pathways that could supply energy for sustained or repeated burst swimming. White muscle buffering capacity is greater in masu salmon smolts relative to parr, which may minimize pH disturbances associated with anaerobic metabolism (Ogata et al., 1998). Mizuno et al. (2007) observed lower levels of hematocrit, hemoglobin, and burst swimming speed in hatchery masu salmon compared to wild fish, all of which could be remedied by dietary iron supplementation. While these data suggest that burst swimming ability may be high in smolts, an analysis of changes in burst swimming ability during smolt development or a comparison between parr and smolt has yet to be undertaken.

Salinity preference increases during smolt development and is probably associated with the development of increased salinity tolerance (Iwata, 1995). Swimming and predator avoidance ability have been shown to decrease in the first several days after exposure to SW (Jarvi, 1990; Brauner et al., 1992), and are probably due to osmotic perturbations that occur in this period. These observations underline the importance of the heightened salinity tolerance that normally develops during smolting.

4. IMPRINTING

Adult salmon return to their natal stream with very high fidelity, seldom straying to other river systems, a phenomenon that is largely dependent upon olfactory cues (Hasler and Scholz, 1983). The now classic work of Arthur Hasler used artificial odorants to determine that fish could imprint upon a chemical signature during the final stage of smolt development, that fish exposed to odorants had a long-term memory for these compounds (detected through electroencephalography), and that this memory was used for upstream migration. Shifting the location of the imprinted artificial odorants into different streams would shift the final migration destination of adult salmon. Similarly, fish reared in one water source such as a hatchery and released as smolts into a different river system will return as adults to the river where they were released as smolts (Ueda, 2011). Dittman et al., (1996) further demonstrated that relatively little imprinting occurred before the parr–smolt transformation, and that the migration process itself appeared to be critical for full imprinting to occur. It has been suggested that imprinting may be even more complex than just memorizing the odor of a single location, and that smolts may memorize a series of odorant "waypoints" during their downstream migration and ocean entry (Quinn et al., 1989). Given the diversity of migratory strategies in salmon, it seems likely that some imprinting occurs outside smolting, especially in those populations that begin downstream migration long before SW entry. Extensive reviews of the timing, cues, and mechanisms involved in imprinting can be found in Dittman and Quinn (1996) and Ueda (2011).

4.1. Chemical Nature of Imprinting Odorants

The olfactory system of salmon consists of the sensory olfactory rosettes located in the nares, the olfactory nerve, and the olfactory bulb located in the forebrain. Early studies on electrical responses of the olfactory bulb to exposure of the olfactory rosettes indicated that salmon respond strongly to natal water sources (Ueda, 2011). The compounds inducing these responses were heat stable, dialyzable, and non-volatile. Hara (1994) concluded that, compared to terrestrial animals, salmon can detect only a limited number of chemicals, which largely consist of amino acids, steroids, bile acids, and prostaglandins.

Studies by Hiroshi Ueda and colleagues have determined that amino acids are the most likely sources of the olfactory cues used by salmon in their homing migration. They demonstrated that streams have large natural variation in their composition of specific amino acids, bile acids, and salts. Using olfactory nerve responses as an endpoint, masu salmon were shown to be responsive to the stream water in which they were reared, but not to water from neighboring streams (Shoji et al., 2000). Furthermore, this response could be mimicked by an artificial mixture of amino acids characteristic of their home stream, but not by an artificial mixture of bile acids or salts. Further studies have demonstrated that during their return migration adult male chum salmon are strongly attracted to artificial amino acid mixtures that mimicked those of the stream in which they were reared (Shoji et al., 2003).

Nordeng (1977) proposed that Atlantic salmon and Arctic char (*Salvelinus alpinus*) were imprinting on juveniles of the same species, and potentially those that were most closely related. This process could not apply to all salmon, as juveniles of pink, chum, and ocean-type Chinook salmon are not present in streams when adults are returning. Experiments with artificial odorants noted above indicate that the presence of juveniles is not necessary for precise homing. While there is evidence for adult sea lamprey (*Petromyzon marinus*) using unique bile acids secreted by conspecific juveniles as attractants (Li et al., 1995), experimental evidence in Pacific and Atlantic salmon does not provide support for conspecific attractants (Brannon and Quinn, 1990). Although it cannot be ruled out that juvenile salmon contribute to the odorants used for imprinting, the weight of evidence indicates that salmon do not use pheromones for homing and that natural amino acids are the primary olfactory cue for the final stage of adult homing in salmon (Brannon and Quinn, 1990; Ueda, 2011).

4.2. Mechanism of Imprinting

Some of the mechanisms involved in olfactory imprinting have been determined by examining changes during smolt development, often in the presence of exposure to natural and artificial odorants. Atlantic salmon smolts have an increased number of filaments and more developed lamellae in the olfactory rosettes compared to parr (Bertmar, 1983). In Chinook salmon (*O. tshawytscha*) there is a rapid increase in "organizational and structural maturity" of the olfactory bulb early in development, and while there is no obvious change associated with smolting, the volume of the olfactory bulb relative to other parts of the brain undergoes a continuous increase from smolting through the adult stage (Jarrard, 1997). Two peaks of olfactory sensitivity and odor learning were detected in Atlantic salmon smolts, one with low response threshold early in smolt development and a second with greater threshold at the peak of smolting (Morin and Doving, 1992). Sockeye salmon (*O. nerka*) can be imprinted on a single amino acid during smolt development (March to June) but not immediately following (July), and a period of 8–14 days was required for imprint memory to be retained (Yamamoto et al., 2010).

There is evidence that some aspects of olfactory imprinting memory are due to retained changes in the olfactory rosettes. Olfactory epithelium of coho salmon imprinted with an artificial odorant during smolting had increased sensitivity to the odorant a year later compared to odorant-naïve fish (Nevitt et al., 1994). Using a similar approach, Dittman et al. (1997) demonstrated that imprinting to an artificial odorant during smolting resulted in greater activity of the second messenger guanylyl cyclase in olfactory cilia following exposure to an artificial odorant, an effect that is only detectable 2 years after imprinting when upstream migration occurs. Altered sensitivity and peripheral memory may also be driven by changes in specific olfactory receptors; in coho salmon the basic amino acid receptor increases during smolting, decreases in postsmolts and then increases again as adults (A. Dittman, personal communication). The xenobiotic-metabolizing enzyme glutathione-S-transferase is located in olfactory receptor cells and is present at higher levels in FW juveniles and returning adult kokanee salmon (*O. nerka*) than it is in subadults in SW (Shimizu et al., 1993). The period of highest responsiveness of sockeye salmon to single amino acids described in the previous paragraph is correlated with the messenger RNA (mRNA) levels of the salmon olfactory-imprinting gene, whose function is currently unknown but which is in a class of compounds involved in olfactory neuron signaling.

It also seems likely that there are "olfactory memories" established in the olfactory bulb during smolting, although the associated mechanisms have not been established. Blood oxygen level-dependent magnetic resonance imaging has revealed that strong responses of adult sockeye salmon to natal stream water occurred in the lateral area of the dorsal telencephalon (Bandoh et al., 2011). The observation that animals become more responsive to previously imprinted olfactory cues during sexual maturation as noted above suggests that the recognition and/or responsiveness to olfactory memory are increased in association with upstream migration and possibly signaled by reproductive hormones.

Recent advances in molecular methods have led to an increasing list of candidate genes that may be involved in imprinting. Transcription of an olfactory receptor gene (SORB) and two vomeronasal receptors (SVRA and SVRC) was found to transiently increase during smolt development of Atlantic salmon (Dukes et al., 2004). Using a microarray approach, Johnstone et al. (2011) found seven members of the OlfC gene family that are putative olfactory receptors have differential transcription levels between juveniles (parr and smolts) and returning adult anadromous Atlantic salmon. No differences in parr and smolt were detected, however, and there were no life history stage differences within a non-anadromous population of the same species. Comparison of Atlantic salmon parr and smolts sampled in spring found 88 genes that were differentially expressed in olfactory rosettes by at least 1.2-fold (Robertson and McCormick, 2012). Upregulated genes that may be specifically involved in olfactory detection and imprinting include olfactomedin, rhodopsin, crytallins, and ubiquitin and ubiquitin-like protein. Olfactomedin is an extracellular matrix glycoprotein that is specifically expressed in the mucus of olfactory neuroepithelium. Rhodopsin and crystallins are involved in eye development, but crystallin genes are expressed in a variety of non-lens tissues and may be involved in neuron growth, and rhodopsin belongs to a family of G-protein coupled receptors that includes the olfactory receptors. Ubiquitin is involved in protein recycling and its increased transcription may be related to cell turnover and/or cell proliferation in the olfactory rosette during smolting. It should be noted that annotated olfactory receptor genes are not well represented on the GRASP 16K microarray chip used in these studies, and more targeted approaches or larger microarrays will be necessary for a more complete analysis of transcription differences between parr and smolt.

In summary, our current understanding of the mechanisms of imprinting indicates that specific olfactory receptors increase during smolt development, resulting in greater sensitivity of the olfactory epithelium. Exposure to river-specific amino acids during smolting results in formation of a peripheral (rosette-specific) memory, and possibly additional memory formation in the olfactory bulb. These memories are stimulated at the time of upstream migration (possibly by reproductive hormones), leading to high fidelity homing to the natal stream or site of imprinting. Many of the mechanisms and control of these processes have yet to be elucidated.

5. OSMOREGULATION

All teleosts maintain a nearly constant internal osmotic concentration irrespective of their external environment (Edwards and Marshall, 2013, Chapter 1, this volume). The ability to absorb water and secrete salts is therefore critical to the survival of ocean migrating smolts. The capacity for SW osmoregulation increases dramatically during smolt development. Parr generally have only a limited ability to secrete salts, and direct transfer to SW of salinity greater than 30 ppt is usually lethal. Smolts develop increased salinity tolerance over a period of weeks before their entry into SW, the result of coordinated development of the major osmoregulatory organs, the gill, gut, and kidney, as will be described in detail below. These preparatory adaptations for salt secretion result in a high level of salt secretory ability that minimizes osmotic perturbations in SW and allows for rapid movements through estuaries and into full-strength SW.

Increased salinity tolerance is often measured by either increased survival or lower plasma ions and osmolality after direct transfer from FW to high salinity (an SW challenge) (Clarke et al., 1996). Previous work has established that the peak of plasma ion and osmolality changes occurs within 24–48 h after transfer to SW. An example of changes in salinity tolerance can be seen in Fig. 5.2(A); plasma osmolality is high after transfer to 35 ppt early in smolt development (January), but the capacity to maintain low plasma osmolality increases throughout smolt development until it reaches a peak (in mid-May in this case), normally coincident with downstream migration. The levels of plasma osmolality are similar to those seen in FW salmon, indicating that smolts experience almost no osmotic

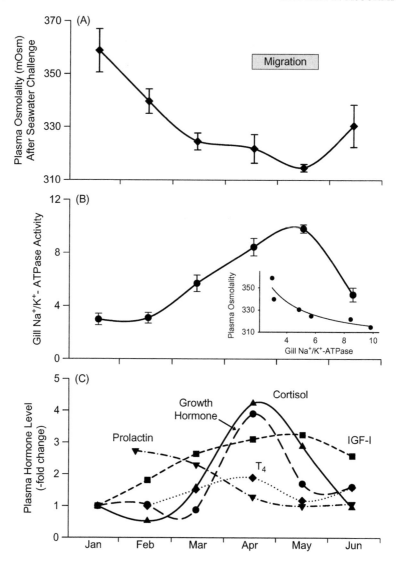

Fig. 5.2. Changes in (A) salinity tolerance, (B) gill NKA activity, and (C) plasma hormone levels during smolt development of Atlantic salmon. Data for plasma hormones are normalized to the changes relative to the first sampling in January and error bars removed for clarity. Salinity tolerance is measured as changes in plasma osmolality 24 h after a direct transfer to 35 ppt seawater. Data are from a single study (McCormick et al., 2009a), with the exception of prolactin (Prunet et al., 1989). NKA: Na$^+$/K$^+$-ATPase; IGF-I: insulin-like growth factor-1; T$_4$: thyroxine.

perturbations when transferred at the peak of smolt development. The increase in salinity tolerance that occurs during smolting is also accompanied by an increase in several other metrics of SW performance, including swimming ability, predator avoidance, and growth (McCormick et al., 2009a).

5.1. Gill

The gill is the site of ion uptake in FW and salt secretion in SW that allows euryhaline teleosts to maintain control of their internal salt and water balance (see Evans et al., 2005, for review). Ion transport is primarily carried out by specialized cells in the gill that have been termed ionocytes, chloride cells, or mitochondrion-rich cells. Three major transport proteins are involved in salt secretion and localized to ionocytes. Na^+/K^+-ATPase (NKA) is located in the basolateral membrane and provides low Na^+ levels and a negative charge within the ionocyte that is used for net Cl^- and Na^+ secretion. The $Na^+/K^+/2Cl^-$ cotransporter (NKCC) is also located in the basolateral membrane and utilizes low Na^+ to transport Cl^- ions into the ionocyte. Chloride can then leave on a "downhill" electrical gradient through an apically located Cl^- channel, the cystic fibrosis transmembrane regulator (CFTR). Sodium leaves the ionocyte through the action of NKA and exits the gill by a paracellular pathway.

It has been known for some time that NKA activity increases in smolts coincident with the development of salinity tolerance (Zaugg and McLain, 1970). The strong positive relationship between gill NKA activity and salinity tolerance (Fig. 5.2) has led to the use of the former as a metric for assessing smolt development and salinity tolerance. Molecular biology studies in rainbow trout (*O. mykiss*) found that there are several isoforms of the catalytic α-subunit of NKA present in the gill, and two of these are differentially regulated by salinity (Richards et al., 2003). Atlantic salmon have two major NKAα isoforms present in distinct ionocytes in the gill: NKAα1a is most abundant in FW, whereas NKAα1b predominates in SW (McCormick et al., 2009b). Gill mRNA levels of NKAα1b increase during smolting of Atlantic salmon, whereas gill NKAα1a mRNA decreases (Nilsen et al., 2007). Antibodies that are highly specific for each of these salinity-dependent isoforms in Atlantic salmon have recently been developed and have provided new insight into the reorganization of the gill during smolt development (Fig. 5.3; McCormick et al., 2012b). NKAα1a predominates in FW, and is present in both filamental and lamellar ionocytes. NKAα1b is present at low levels in FW and is localized to small filamental ionocytes that appear to be below pavement cells and thus not in contact with the external environment. The increase in salinity

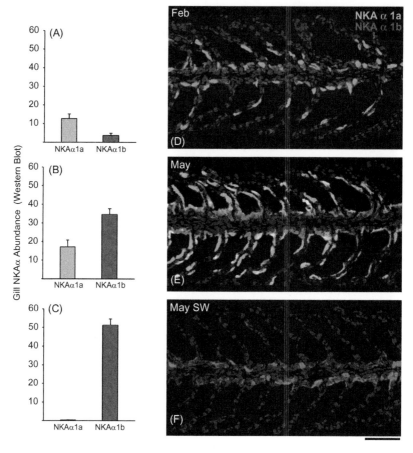

Fig. 5.3. Changes in branchial protein abundance (A–C) and immunolocalization (D–F) of NKAα1a (green) and NKAα1b (red) during smolt development and seawater (SW) exposure of Atlantic salmon (McCormick et al., 2013a). Fish were sampled in freshwater (FW) in February (A,D), FW at the peak of smolting in May (B,E), and after 2 weeks in SW in mid-May (C,F). NKA: Na^+/K^+-ATPase. **SEE COLOR PLATE SECTION**

tolerance in smolts is accompanied by large increases in NKAα1b abundance (10-fold) and cell numbers. A large number of ionocytes with both NKAα1a and NKAα1b are present in smolts in FW, suggesting that NKAα1a ionocytes are gradually transforming to NKAα1b ionocytes. After exposure to SW NKAα1b abundance increases further, there are almost no ionocytes expressing both isoforms, and only a small number of NKAα1a ionocytes remain. These results indicate that increased NKAα1b develops coincident with and is likely to be causal to increased salinity tolerance of smolts. Gill NKAα1a protein abundance does not change during smolting but is slightly lower in smolts than in parr. FXYD-11, a potential regulator of NKA activity, is present in ionocytes and increases during smolt development and after SW acclimation (Tipsmark et al., 2010a).

There are two major isoforms of NKCC in vertebrates: the secretory NKCC1 isoform that is present on the basolateral membrane, and the absorptive NKCC2 isoform present on the apical surface. As might be expected from the secretory function of the gill, the prevailing evidence is that an NKCC1 is present in salt-secreting ionocytes of the gill. Using an antibody that recognizes both isoforms, Pelis et al. (2001) observed that the abundance of NKCC and NKCC-positive ionocytes increases during smolt development. Ultrastructural localization of the NKCC to the basolateral membrane suggested that this was the NKCC1 isoform. Partial cloning of a NKCC in Atlantic salmon gill found high sequence similarity with other vertebrate NKCC1 isoforms (>60%) and lower similarity to NKCC2 (<40%), and the mRNA levels of this putative NKCC1 isoform increased five-fold during smolt development (Nilsen et al., 2007). Using an isoform-specific antibody, Christensen and McCormick (unpublished results) found that the NKCC1 is expressed specifically in the gill (and not the intestine), is present primarily in ionocytes, and increases during smolt development. Gill NKCC transcription and abundance are stimulated under normal daylength but not under continuous light; the latter treatment also inhibits development of salinity tolerance, providing further evidence of the importance of NKCC to SW tolerance (Stefansson et al., 2007).

Information on the apical CFTR Cl⁻ channel in smolts comes primarily from transcription studies. There are at least two isoforms of CFTR expressed in the gill. CFTR I mRNA increases substantially during Atlantic salmon smolt development, whereas CFTR II does not change (Nilsen et al., 2007). After SW exposure CFTR I mRNA increased steadily over 2 weeks, whereas CFTR II increased only transiently (Singer et al., 2002). These results suggest that CFTR I is the major gill CFTR isoform associated with increased salt secretory capacity of smolts, although further work on protein abundance and immunolocalization is warranted. Although the full sequence of CFTR is known, production of homologous antibodies has not yielded antibodies with expected distribution patterns on the apical membrane of ionocytes (Regish and McCormick, unpublished results). However, recent screening of several antibodies derived from mammalian CFTR has found one which shows clear apical staining in SW ionocytes but not FW ionocytes of Atlantic salmon (Christensen, Regish, and McCormick, unpublished results). During smolt development there is a large increase in the number of CFTR-positive ionocytes, indicating that all three major transporters are present in gill ionocytes of smolts before exposure to SW.

The secretion of Na^+ through a paracellular pathway between ionocytes and adjacent cells suggests that intercellular junctions are important for the development of salt secretory capacity. Claudins are a family of membrane proteins that form tight junctions and thus determine transepithelial resistance and ion permeability. Tipsmark et al. (2008) have shown that there are five major claudins expressed in the gills of Atlantic salmon. Claudin 10e mRNA levels increase during smolt development and after exposure to SW, suggesting their involvement in ion secretion. Although the other isoforms did not change during smolt development, claudins 27a and 30 decrease after exposure to SW, suggesting that they are involved in ion uptake.

Aquaporins are a family of proteins involved in cellular and transcellular water movement in vertebrates. Their role in osmoregulation in the gill is not entirely clear as water permeability does not appear to differ substantially in FW and SW in salmonids. However, aquaporins may have a role in regulatory volume decrease or other signaling processes in ionocytes and other cell types in the gill. Three aquaporins have been found to be expressed in the gill of Atlantic salmon (Tipsmark et al., 2010c). An observed increase in aquaporin (AQP)1b and decrease in AQP3 transcription during smolting and after SW acclimation suggests a physiological role in salt secretion and ion uptake, respectively.

Increased numbers of gill ionocytes during smolt development have been observed in coho and Atlantic salmon (Richman et al., 1987; Lubin et al., 1989). Ionocytes are the site of both ion uptake in FW and secretion in SW, and most previous histological studies have been unable to distinguish between these functional forms of ionocytes. This is an important knowledge gap as ion uptake demands may also be changing in relation to or independent of the increased salinity tolerance of smolts. Pisam et al. (1988) found that while the total number of ionocytes did not change during smolting of Atlantic salmon, the number of large ionocytes with a well-developed tubular system and association with accessory cells increased in FW during smolt development while the number of small ionocytes with paler mitochondria decreased. Furthermore, there was an increase in the number of NKCC-positive ionocytes during smolt development of Atlantic salmon, which was interpreted as an indicator of increased salt secretory capacity (Pelis et al., 2001).

The capacity to visualize the abundance and location of the FW and SW isoforms of the NKA α-subunit has provided a useful tool for examining functionally distinct ionocytes during smolt development (McCormick et al., 2013a). A hypothetical model of the development of SW ionocytes during smolting is presented in Fig. 5.4. During smolt development there is an increase in the number of NKAα1b ionocytes that also contain NKCC, CFTR, and other functional attributes of salt secretory ionocytes. Based on the colocalization of NKAα1a and NKAα1b in many ionocytes during smolt development and only a small increase in the total number of ionocytes, it may be hypothesized that NKAα1a ionocytes gradually transform to NKAα1b ionocytes. NKAα1b ionocytes of smolts in FW probably lie quiescent below the surface of the gill filament until they are rapidly activated by exposure to SW, thus providing the high salt secretory capacity and salinity tolerance that is characteristic of smolts. The rapid exposure of ionocytes is supported by the observation that the number of ionocytes increases during smolt development of Atlantic salmon but the number of surface-exposed ionocytes does not change until SW exposure, increasing within 1 day and continuing through 5 days (Lubin et al., 1989). Further increases in NKAα1b cell size and abundance after exposure to SW presumably confer on smolts an even greater level of salt secretory capacity, a built-in safety factor to allow for ion transport demands under a wide variety of conditions.

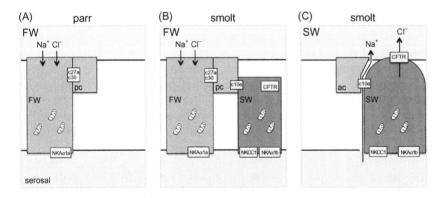

Fig. 5.4. Schematic of gill NKAα1a and NKAα1b ionocyte changes during smolt development in freshwater (FW) and after seawater (SW) exposure. In FW parr (A), NKAα1a ionocytes predominate. During smolt development the numbers of NKAα1b and colabeled cells increase (B). These cells appear to be beneath pavement cells and thus likely to be inactive. After exposure to SW (C), NKAα1b increase in size and now have clear contact with the external environment, while NKAα1a cells disappear. NKCC, CFTR, and claudin 10e are also upregulated in FW during smolt development and increased further after exposure to SW. Claudins 27a and 30 are downregulated during smolt development and exposure to SW. NKA: Na$^+$/K$^+$-ATPase; NKCC: Na$^+$/K$^+$/Cl$^-$ cotransporter; CFTR: cystic fibrosis transmembrane regulator; ac: accessory cell. **SEE COLOR PLATE SECTION**

5.2. Gut

In SW teleosts drink more than in FW in order to absorb water lost to the hyperosmotic medium. In order to take up water the gut first absorbs Na^+ and Cl^- to reduce the osmotic pressure of the intestinal fluid. The latter is accomplished by the coordinated action of an apical, absorptive $Na^+/K^+/2Cl^-$ cotransporter (NKCC2) and a basolateral NKA. Active alkalinization further reduces the osmotic pressure of the gut by causing the precipitation of Mg^{2+}, Ca^{2+}, and SO_4^{2-}. Paracellular and transcellular water absorption then takes place, for most species in the anterior intestine (Grossell, 2011; Edwards and Marshall, 2013).

In coho and Atlantic salmon fluid transport (Jv) of the isolated posterior intestine increases during smolting and after SW exposure (Collie and Bern, 1982; Veillette et al., 1993). Both water transport capacity and NKA activity increased in anterior (pyloric ceca) and posterior segments of the intestine of yearling Chinook salmon during early summer just before the development of salinity tolerance (Veillette and Young, 2004). In a smolting population of brown trout (*Salmo trutta*) intestinal water transport capacity increased five-fold but was not accompanied by increased intestinal NKA activity (Nielsen et al., 1999). Sundell et al. (2003) found increased intestinal NKA activity and paracellular permeability (as measured by transepithelial resistance) in association with increased Jv in the anterior intestine of Atlantic salmon smolts. These authors observed a decrease in transepithelial resistance after SW exposure, suggesting a shift from paracellular to transcellular routes of water uptake.

Recent work on gene expression has suggested some of the underlying transporters that may be involved in altered intestinal water transport capacity of smolts. Tipsmark et al. (2010b) observed increased AQP8b mRNA levels during smolting, whereas AQP1a, 1b, and 8b transcription all increased after SW acclimation. Although no changes were detected in intestinal claudins during smolt development, claudins 15 and 25b were upregulated after exposure to SW (Tipsmark et al., 2010b), suggesting their involvement in paracellular permeability that is altered after SW exposure.

5.3. Kidney and Urinary Bladder

The kidney of FW teleosts creates a highly dilute urine to excrete excess water, whereas in SW the urine is isosmotic with plasma but contains excess divalent cations (Edwards and Marshall, 2013, Chapter 1, this volume). Surprisingly little work has been done to quantify renal function or changes in ion transporter abundance or other proteins involved in salt and water transport in the kidney during smolt development. The function of the kidney is complex and highly regionalized, and there are no well-established biochemical markers for function in FW and SW. Urine production of Atlantic salmon smolts increased in spring coincident with increased gill NKA activity (Eddy and Talbot, 1985). However, kidney NKA activity does not appear to change in Atlantic salmon smolts or juveniles exposed to SW (McCormick et al., 1989).

The urinary bladder also has an important role in ion reabsorption in FW in many teleost species. Sodium and Cl^- reabsorption by the urinary bladder of coho salmon smolts held in FW did not change between March and June (Loretz et al., 1982). When transferred to SW in May, poor survival was accompanied by high levels of urinary bladder Na^+ and Cl^- reabsorption, characteristic of FW fishes and maladaptive for SW. High survival after SW transfer in June was accompanied by low levels of Na^+ and Cl^- reabsorption. Thus, while no obvious functional change had occurred in FW, there was a developmental increase in the ability of the urinary bladder to respond appropriately to SW exposure.

5.4. Is Freshwater Osmoregulation Compromised during Smolting?

The previous sections have shown that the underlying mechanisms of water absorption by the gut and salt secretion by the gill change during smolt development and are associated with a high level of salinity tolerance not present in earlier developmental stages. However, these changes may impair the physiological capacity of smolts to regulate water and salts while they still reside in FW, before and during their seaward migration. While most studies report minimal changes in the levels of the major plasma ions during smolt development (McCormick and Saunders, 1987), smolts appear to be more sensitive to external factors that affect osmoregulation. Identical handling stresses of parr and smolts result in greater loss of plasma ions in the latter (Carey and McCormick, 1998). There is evidence that Atlantic salmon experience altered Na^+ flux levels during smolting (Primmett et al., 1988), although some or all of these changes may be due to the stress of confinement inherent in these measurements. Atlantic salmon smolts lose ions and have lower survival than other life stages after exposure to acid and aluminum, which negatively affects gill ionocytes and NKA (Rosseland et al., 2001; Monette and McCormick, 2008). Many heavy metals also exert their main toxic effect through their impacts on osmoregulation. Coho salmon smolts in May were found to be more sensitive to zinc and copper than they were as parr in November (Lorz and McPherson, 1976). While the FW NKAα1a isoform does not change in abundance over the course of smolt development, it is slightly lower in smolts than in parr, and the number of NKAα1a ionocytes decreases, suggesting a possible decrease in the capacity for ion uptake (McCormick et al., 2013a). Thus, it would appear that while smolting does not necessarily impair ionoregulatory homeostasis in FW, smolts may be more sensitive to environmental factors that affect ion regulation.

Thorpe (1994a) has argued that smolt development should be viewed as a maladaptation to or abandonment of FW. The recent findings that the major changes in NKA in the gill during smolting are due to increases in the "SW isoform" (NKAα1b) support a more adaptationist interpretation of smolting; that is, that it is primarily an adaptation for increased osmoregulatory capacity in SW and other adaptive changes for SW performance (McCormick et al., 2013a). Increased sensitivity of smolts to osmoregulatory disturbance probably represents constraints on the overall capacity of osmoregulatory systems.

6. ENDOCRINE CONTROL

Smolting has been described as a pan-hyperendocrine state (Bern, 1978), where many hormones that have different physiological actions are increasing during development, although not necessarily at the same time or rate (Fig. 5.5). This makes smolt development different from metamorphic events where most of the developmental events that occur are often controlled by one stimulatory hormone and possibly one inhibitory hormone. It can be argued that this allows flexibility among the various aspects of development (physiology, morphology, behavior) that occur during smolting. This flexibility may be particularly important to the fitness of populations and individuals, in that behavioral changes such as the initiation of migration may be occurring at different times in relation to other developmental events such as SW entry. However, the extent of population variability in various aspects of smolt development has yet to be established.

6.1. Growth Hormone/Insulin-Like Growth Factor-1

Growth hormone (GH) and insulin-like growth factor (IGF-I) both appear to be central to the smolting process. Increases in circulating levels of GH during smolt development have been shown for Atlantic and coho salmon (Björnsson et al., 2011) (Fig. 5.2C). Increased plasma GH is due to an increase in pituitary secretion early in smolt development and later by increases in GH synthesis and secretion (Agustsson et al., 2001). No change in the number of growth hormone-releasing hormone (GHRH) neurons was observed in the brain of chum salmon during smolting (Parhar and Iwata,

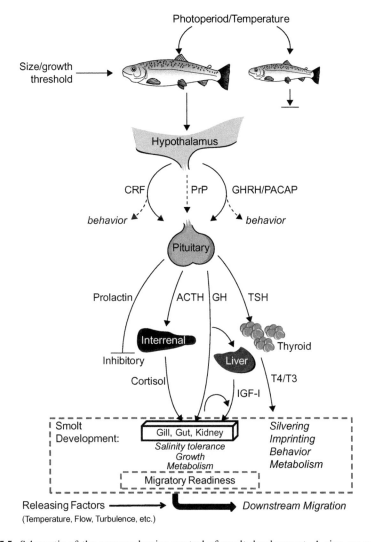

Fig. 5.5. Schematic of the neuroendocrine control of smolt development. A size- or growth-related threshold results in increased response of the light–brain–pituitary axis to stimulate circulating levels of growth hormone (GH), cortisol, and thyroid hormones. GH and cortisol interact to control hyperosmoregulatory mechanisms in the gill, gut, and kidney, resulting in increased salinity tolerance, as well as changes in growth and metabolism. Thyroid hormones have a role in morphological changes such as silvering, imprinting, metabolism, and possibly behavior. Prolactin is generally thought to be inhibitory to most aspects of smolt development. There is also substantial interaction among these endocrine components, including the regulation of hormone secretion and peripheral receptors. It should be noted that early development of salinity tolerance and other aspects of smolting in pink and chum salmon appear to be independent of photoperiod cues and to be driven primarily by developmental processes. CRF: corticotropin-releasing factor; PrP: prolactin-releasing peptide; GHRH: growth hormone-releasing hormone; PACAP: pituitary adenylate cyclase-activating peptide; ACTH: adrenocorticotropic hormone; TSH: thyroid-stimulating hormone. **SEE COLOR PLATE SECTION**

1996). In Atlantic salmon photoperiod appears to be the major factor regulating increased GH, with temperature affecting the rate of response to photoperiod (McCormick et al., 2000). There is a strong association between increased plasma GH levels and osmoregulatory indicators of smolt development. During spring Atlantic salmon parr have constant, low levels of plasma GH and cortisol; however, both increase dramatically in smolts (McCormick et al., 2007). In landlocked Atlantic salmon in the spring there is no increase in plasma GH at a time when it is elevated in anadromous conspecifics, and this is associated with poor salinity tolerance (Nilsen et al., 2008). GH appears to be the first hormone to increase following photoperiod manipulation and has an important controlling or interaction effect with other endocrine axes during smolting (see below); GH therefore appears to be critical to the initiation and overall timing of smolt development.

Plasma IGF-I also increases during smolting (Fig. 5.2C), although the pattern is usually not identical to that of circulating GH. Although GH is generally recognized as the major secretagogue for plasma IGF-I that is primarily released from the liver, it is clear that other factors are also involved. In Atlantic salmon plasma IGF-I has been shown to increase in both parr and smolts in spring, with smolts exhibiting higher levels throughout (McCormick et al., 2007), but not all studies have found increasing plasma IGF-I during smolting (Nilsen et al., 2008). Increased IGF-I occurs in association with smolting in both spring and autumn in the highly plastic Chinook salmon (Beckman and Dickhoff, 1998). Plasma IGF-I levels and gill NKA activities in spring Chinook salmon smolts reared at different hatcheries were significant predictors of the adult return rates to the hatcheries (Beckman et al., 1999), indicating that environmental conditions acting through smolt developmental pathways are important for marine survival. In addition to changes in circulating levels, local production of IGF-I by target tissues is also likely to be involved in smolt development. Liver and gill IGF-I mRNA increase during smolting of coho salmon (Sakamoto et al., 1995). Increased transcription of both IGF-I and IGF-I receptor in the gill has been found in anadromous Atlantic salmon smolts but is absent in a landlocked strain (Nilsen et al., 2008). Transcription of branchial GH receptor also increases during smolt development of Atlantic salmon (Kiilerich et al., 2007b), which, along with increased circulating GH, may explain the observed increases in branchial production of IGF-I. There is relatively little information on regulation of the GH–IGF-I axis in pink and chum salmon that smolt early in development. This is likely to be due to measurement limitations associated with their small size, although the advent of molecular approaches should make it possible to examine transcriptional changes.

Exogenous treatment with GH and IGF-I can increase the salinity tolerance of juvenile trout and salmon (Fig. 5.5) (see also Takei and McCormick, 2013, Chapter 3, this volume). Increased transcription and/or protein abundance have been observed for the major gill transport proteins involved in ion secretion (NKAα1b, NKCC1, and CFTR) (Tipsmark and Madsen, 2009; McCormick et al., 2013a), suggesting that GH controls a differentiation program for salt-secreting ionocytes. There is an important interaction between the GH–IGF-I axis and cortisol that will be discussed in detail below.

GH and IGF-I are important regulators of growth and metabolism in teleost fishes (Wood et al., 2005). GH and cortisol are lipolytic in salmon (Sheridan, 1986) and their combined increase is probably responsible for the increased linear growth relative to mass growth (lower condition factor) that occurs during smolt development. Surprisingly, the role of the GH–IGF-I axis in the restoration of condition factor and the increased scope for growth that occurs after smolting has not been examined. Reduced liver GH receptor and IGF-I production accompany the "stunting" phenomenon, where impaired growth is observed in some individuals after SW exposure (Duan et al., 1995). It has been suggested that this occurs when fish are exposed to SW before the development of SW tolerance, although pathways involved in the genesis of this phenomenon have not been established.

There is intriguing evidence that the GH–IGF-I axis is involved in behavioral changes during smolting. GH treatment increases the salinity preference of coho salmon, an effect that is augmented with coincident thyroxine (T_4) treatment (Iwata et al., 1990). Intracerebral treatment with GHRH stimulated both downstream movement and schooling behavior in chum salmon fry, while corticotropin-releasing hormone (CRH), melatonin, and serotonin stimulated downstream movement only (Ojima and Iwata, 2009). Behavioral experience may also feed back onto the GH–IGF-I axis.

There has been increasing emphasis on understanding endocrine changes that occur in wild and hatchery salmon during downstream migration and early ocean entry (Björnsson et al., 2011). Fig. 5.6 presents a compilation of data from two studies on endocrine changes during migration of Atlantic salmon in river, coastal, and ocean environments (Stefansson et al., 2003; McCormick et al., 2013b). The results indicate that salmon smolt plasma GH levels increase in the river and after initial SW entry but in the long term (after several weeks) in the ocean are relatively low. Initial increases in plasma GH levels may be due to the metabolic demands of migration and the response to SW exposure. Since levels of plasma GH are low and IGF-I high in rapidly growing animals, the lower levels of plasma GH and higher IGF-I in the open ocean probably reflect the high growth rate of postsmolts (Stefansson et al., 2003).

6.2. Cortisol

Early histological studies showed hyperactivity of the interrenal cells during smolt development (see Specker, 1982, for review), and circulating cortisol was first shown to increase in Atlantic salmon smolts by Fontaine and Hatey (1954) and later in coho salmon by Specker and Schreck (1982). These results have since been confirmed by a large number of studies, although information in early smolting species, such as chum and pink salmon, is absent. Plasma levels of cortisol remain low and constant in Atlantic salmon parr in spring, but increase 10-fold in smolts held under the same conditions (McCormick et al., 2007). Similarly, plasma cortisol increases during smolting in anadromous Atlantic salmon, but spring increases are absent in a landlocked strain (Nilsen et al., 2008).

There is still an incomplete understanding of the factors that regulate the hypothalamic–pituitary–interrenal axis during smolting. The number of CRH neurons increases during smolt development of Atlantic salmon, at least in part due to upregulation by thyroid hormones (Ebbesson et al., 2011). Surprisingly, there has been no measure of the major cortisol secretagogue, adrenocorticotropic hormone (ACTH), either in circulation or in the pituitary, during smolt development. A marked increase in the sensitivity of the interrenal to produce cortisol in response to ACTH was observed in April in smolting coho salmon, which occurred several weeks

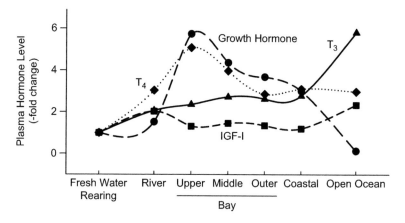

Fig. 5.6. Changes in plasma growth hormone, insulin-like growth factor-1 (IGF-I), thyroxine (T₄), and triiodothyronine (T₃) in Atlantic salmon smolts during freshwater (FW) rearing, river migration, and ocean entry. The graph is a composite from two studies: McCormick et al. (2013b) for river, bay, and coastal results, and Stefansson et al. (2012) for open ocean. Initial FW levels (river levels in the case of Stefansson et al., 2012) were normalized to 1.0 and other values expressed as a fold-change relative to these values.

before the peak in circulating cortisol (Young, 1989). This enhanced ACTH sensitivity is likely to be due to a direct effect of GH on the interrenal (Young, 1988), which can also explain how GH treatment leads to increased circulating cortisol levels (Quinn et al., 2003). Plasma clearance rate of cortisol increases in late spring and declines in summer in coho salmon smolts in FW (Patino et al., 1985) and after SW transfer (Redding et al., 1984), indicative of increased utilization, metabolism, or clearance of cortisol.

Using classical binding studies, the number of gill cortisol receptors was shown to decrease during smolt development of coho salmon (Shrimpton et al., 1994). This may in part be due to the effect of increasing endogenous cortisol, which decreases the number of ligand-free receptors (Shrimpton and Randall, 1994). In Atlantic salmon, both parr and smolt exhibited large increases in gill cortisol receptors in the spring (Shrimpton and McCormick, 1998b). Immunocytochemical and *in situ* hybridization approaches have shown higher levels of cortisol receptors in ionocytes compared to other cell types in the gill (Uchida et al., 1998). Recent molecular data indicate that there are one or two "glucocorticoid" receptors (GRs) and one "mineralocorticoid" receptor (MR) present in most teleosts; this nomenclature is based on their similarity to mammalian receptors that have distinct regulatory pathways for metabolic and ion regulatory effects of corticosteroids, although such distinctions do not apply to fish (Takei and McCormick, 2013, Chapter 3, this volume). Increased gill GR transcription has been observed in smolting masu and Atlantic salmon, whereas gill MR transcription does not change (Mizuno et al., 2001; Nilsen et al., 2008). Based on the current weight of evidence, it seems likely that one or both GRs signal most of the actions of cortisol to promote salt secretion, and that MR may also play a more limited role in osmoregulation in most species. The high level of MR transcription in the teleost brain suggests a role for this receptor in regulating behavior and/or neuroendocrine responses. Since cortisol appears to have actions in both ion uptake and salt secretion, some of the differential actions of GR and MR may relate to these dual actions of cortisol, although evidence for a role in MR in salmonid osmoregulation is still lacking.

Exogenous treatment with cortisol stimulates salinity tolerance and many of the underlying mechanisms involved in SW osmoregulation in the gill and the gut (McCormick, 2001). The major transport proteins involved in salt secretion, NKAα1b, NKCC1, and CFTR, can be increased by cortisol treatment, as can the abundance of ionocytes (Pelis and McCormick, 2001; Kiilerich et al., 2007a; McCormick et al., 2008). Using NKA activity as an endpoint, *in vitro* responsiveness of coho salmon gill to cortisol was low in winter, increased in spring just before endogenous increases in gill NKA activity, and subsequently declined in late spring and early summer (McCormick et al., 1991). Cortisol also increases transcription of gill claudin 28e, a tight junction protein that increases after SW acclimation of Atlantic salmon (Tipsmark et al., 2009). Cortisol treatment increases the intestinal water uptake capacity and NKA activity in juvenile Atlantic and Chinook salmon (Veillette et al., 1995; Veillette and Young, 2004).

There is an important interaction between the GH–IGF-I axis and cortisol that underlies the development of hypoosmoregulatory capacity of smolts. As noted above, GH increases the responsiveness of the interrenal to ACTH, increasing the amount of cortisol released at any given level of ACTH. GH also upregulates the abundance of corticosteroid binding sites and transcription of GR in the gill (Shrimpton et al., 1995; Kiilerich et al., 2007b). In turn, cortisol increases the transcription of gill GH and IGF-I receptors (Tipsmark and Madsen, 2009). In Atlantic salmon juveniles, cortisol treatment increases both the FW and SW isoforms of NKA (α1a and α1b), but when coinjected with GH cortisol decreases NKAα1a and increases NKAα1b, and results in high salinity tolerance similar to that of smolts (S.D. McCormick, unpublished results). Thus, GH appears to be acting as a switch for the actions of cortisol, moving it away from ion uptake and promoting salt secretion. While this interaction is well established for the gill, there is little evidence for its importance in the gut or kidney. Intestinal claudin 25b transcription is upregulated after SW exposure and GH treatment, but there is no apparent preparatory increase in transcription of this gene during smolting (Tipsmark et al., 2010b).

It seems likely that the hypothalamic–pituitary–interrenal (HPI) axis plays a role in behavioral changes during smolting. Corticotropin-releasing factor (CRF) increased downstream migratory behavior in juvenile coho and chum salmon (Clements and Schreck, 2004; Ojima and Iwata, 2009, 2010). The rapid response to this central administration suggests a direct response to this peptide rather than an action through cortisol, although the latter cannot be ruled out. Cortisol has been shown to increase the salinity preference of juvenile coho salmon, although to a lesser extent than GH or T_4 (Iwata et al., 1990). The HPI axis may also play a role in some of the morphological changes that occur during smolting, since ACTH (but not cortisol) has been found to increase fin darkening in Atlantic salmon (Langdon et al., 1984), a characteristic smolt-related change in this species.

6.3. Prolactin

Histological studies of the pituitary gave the first indication that prolactin played a role in smolt development, with greater secretory "activity" present in FW smolts than in parr or SW smolts (Nishioka et al., 1982). In both Atlantic and coho salmon, plasma prolactin is elevated in winter and early spring but decreases in April and May when smolting peaks (Fig. 5.2C) (Prunet et al., 1989; Young et al., 1989). Given the negative impacts of prolactin on salt secretory capacity (see below), this decrease can be viewed as removing an inhibitory factor that allows normal progression of smolt development. Although seasonal changes in plasma prolactin were observed in amago salmon (*O. rhodurus*), there were no obvious increases in silvering and salinity tolerance that occur in autumn in this species (Yada et al., 1991). Plasma prolactin is lower in downstream migrating Arctic char than in non-migrants (Hogasen and Prunet, 1997). As in most other teleosts, plasma prolactin levels are strongly reduced after SW exposure, further supporting the idea that prolactin promotes ion uptake and is inhibitory to salt secretion (Prunet et al., 1989; Young et al., 1989; Yada et al., 1991).

Information on the factors involved in prolactin regulation during smolting is limited. A hypothalamic prolactin-releasing peptide has recently been found in vertebrates including trout and salmon (Moriyama et al., 2002), but has not been examined in smolts. Unlike the euryhaline tilapia in which the direct response of prolactin-producing cells in the pituitary to osmolality is well established, the prolactin cells of coho salmon do not alter synthesis or secretion in response to physiological changes in osmolality (Kelley et al., 1990), but do appear to be responsive to calcium (MacDonald and McKeown, 1983). Cortisol directly inhibits prolactin synthesis and release, which may provide a mechanism for decreased prolactin during the peak of smolting and after SW exposure (Kelley et al., 1990). As noted above, cortisol has a role in ion uptake in fish, and there is some indication that prolactin and cortisol interact to promote ion uptake. The mechanism for this interaction has not been established, however, and there is little known of the interaction of cortisol and prolactin during smolt development.

In the euryhaline tilapia (*Oreochromis mossambicus*), prolactin receptors are found at high levels in osmoregulatory organs such as gill and kidney (Dauder et al., 1990). Transcription of gill prolactin receptors remains stable or decreases during smolting of Atlantic salmon, and decreases further after 1 month in SW (Kiilerich et al., 2007b; Nilsen et al., 2008). Prolactin receptor abundances in gill, gut, or kidney during smolting have not been examined.

Exogenous treatment with prolactin has been shown to decrease salinity tolerance of smolts (T.O. Nilsen, personal communication) and brown trout in autumn (Seidelin and Madsen, 1997). Prolactin also inhibits some of the mechanisms associated with developmental increases in salt secretion (Fig. 5.5), including GH- and cortisol-induced elevations of gill NKA activity (Seidelin and Madsen, 1997) and SW levels of NKA activity and NKAα1b transcription (Tipsmark and Madsen, 2009). This effect may occur primarily through the action of prolactin to decrease the capacity of the gill to respond to IGF-I (Seidelin and Madsen, 1999). Given the observed impact of prolactin on "water drive" behavior in amphibians, it is certainly possible that prolactin has behavioral effects in smolts, but this has yet to be examined.

6.4. Thyroid Hormones

Using a histological approach, Hoar (1939) found evidence for increased "activity" of the thyroid gland during smolt development of Atlantic salmon. The capacity to measure circulating levels of thyroid hormone by radioimmunoassay in salmon, first developed by Dickhoff et al. (1978), has led to a large number of studies that indicate increased circulating levels of T_4 and to a lesser extent triiodo-l-thyronine (T_3) during smolting of Atlantic and Pacific salmon (Fig. 5.2C) (Dickhoff and Sullivan, 1987). Large increases in plasma T_4 have been observed in hatchery smolts after release and in wild smolts during migration, and are possibly related to the stimulation of the thyroid axis by the act of migration itself (Iwata et al., 2003; McCormick et al., 2003). Increased plasma T_4 also occurs after exposure to "novel water" (water with a different chemical composition) (Hoffnagle and Fivizzani, 1990) and may play a role in imprinting, as discussed in detail below. Lunar rhythms in plasma T_4 have also been observed in coho salmon smolts (Grau et al., 1981) and are associated with lunar periodicity of movement into estuaries that has been observed in some populations. There is recent evidence that thyroid hormones increase during entry of smolts into estuarine and near-coastal environments (Fig. 5.6).

In their review of the role of thyroid hormones in smoltification, Dickhoff and Sullivan (1987) emphasized the metabolic and morphological impacts of this endocrine axis. Exogenous thyroid hormones have been shown to increase silvering, but not fin darkening, in several salmonid species. Hemoglobin isoforms change during smolt development and similar changes can be induced by thyroid hormone treatment. The increased lipolysis and reduced lipogenesis that occur during smolting can also be induced by thyroid hormone treatment, although as noted above GH and cortisol appear also to be involved in these metabolic changes.

There is both direct and indirect evidence that thyroid hormones play a role in imprinting. Treatment of presmolt coho salmon with T_4 resulted in the establishment of long-term odor "memories", whereas untreated fish did not imprint on odors (Hasler and Scholz, 1983). Increased responsiveness of olfactory tissue to amino acids has been found in Atlantic salmon in association with increased plasma T_4 levels (Morin et al., 1994), although paradoxically, T_4 treatment reduces olfactory sensitivity to alanine (Morin

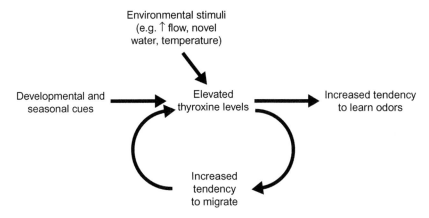

Fig. 5.7. Schematic of impacts smolt development, environmental stimuli, and migration on plasma thyroxine levels and the tendency to learn odors and imprint. From Dittman and Quinn (1996) with permission.

et al., 1995). Thyroid hormone receptors have been found in the olfactory neurons and olfactory bulb of masu salmon (Kudo et al., 2002). Increased cell proliferation in olfactory rosettes was detected in coho salmon smolts treated with T_3 (Lema and Nevitt, 2004). Dittman and Quinn (1996) suggested that thyroid hormone elevations before and during smolt development cue the imprinting of odor memories, and that both the experience of novel water and the act of migration feed back on one another to increase thyroid hormones and promote odor learning (Fig. 5.7). The observed increase in plasma thyroid hormones during estuarine and near-coastal migration (Fig. 5.6) suggests that imprinting (olfactory and/or spatial) may occur not only during downstream migration but also during early ocean entry (Iwata et al., 2003; McCormick et al., 2012).

Thyroid hormones have been shown to have effects on the behavior of fish, and there are several links between thyroid hormones and downstream migratory behavior. Aggressive behavior decreases during smolt development (presumably in order to promote schooling), and thyroid hormone treatment has been shown to decrease aggressive behavior in anadromous, but not in non-anadromous, salmonids (Hutchison and Iwata, 1998). In his review of the role of thyroid hormones and cortisol on smolt behavior, Iwata (1995) concluded that thyroid hormones had clear effects on the salinity preference of smolts but that effects on downstream migration were less certain. In experiments on chum salmon fry, T_4 treatment induced higher levels of circulating T_4 but had no effect on downstream movement (Ojima and Iwata, 2007b). Intracerebroventricular injections of gonado-tropin-releasing hormone (GnRH) and CRH in juvenile coho salmon resulted in both downstream movement and increased plasma T_4 levels, leading the authors to suggest that increases in thyroid hormones during migration are an *outcome* of other neuroendocrine hormones rather than a driving factor for migration (Ojima and Iwata, 2010).

Long-term dietary or injection treatments with thyroid hormones have only a limited impact on salinity tolerance of salmonids (McCormick, 2001). This contrasts with cortisol and GH, which as noted above can individually increase salt secretory capacity and together have an additive or synergistic effect. It should be noted that the repressor nature of thyroid hormone receptors may make it difficult to utilize exogenous hormone treatments, and negative effects should be viewed with some skepticism. However, the fact that some aspects of smolt development (e.g. morphology and metabolism) can be altered by exogenous thyroid hormone treatment indicates that the development of salinity tolerance is less likely to be under direct control of thyroid hormones. Rather, thyroid hormones are more likely to have a permissive role or to work indirectly through the GH–IGF-I and cortisol axes (Takei and McCormick, 2013, Chapter 3, this volume). For instance, exogenous T_3 treatment increases gill cortisol receptors and synergizes with GH to increase them even further (Shrimpton and McCormick, 1998a).

The hypothalamic control of thyroid hormones during smolt development has not been examined. Larsen et al. (1998) have provided convincing evidence that CRF, and not thyroid-releasing hormone or GHRH, is the most potent stimulator of thyroid-stimulating hormone (TSH) release in coho salmon parr. Thus, the increased CRF neurogenesis in Atlantic salmon smolts noted above may play a role in thyroid hormone regulation. In coho salmon smolts, pituitary TSH mRNA levels decrease in spring, whereas pituitary and plasma levels of TSH do not change and plasma thyroid hormones increase (Larsen et al., 2011). The capacity of mammalian thyroid stimulating hormone to increase plasma T_4 changes during smolt development, peaking at or near the time of downstream migration (Specker and Schreck, 1984). Thus, the seasonal changes in plasma T_4 may owe more to changes in sensitivity to TSH than to changes in TSH itself. Metabolic clearance rate of thyroid hormones is greatest early in smolt development when plasma T_4 levels are low, but subsequently decreases (Ojima and Iwata, 2007a). These authors suggest that the thyroid surge seen in some species near the peak of smolt development is due to lower tissue utilization of T_4 and T_3, consistent with observations on thyroid kinetics during smolt development (Specker et al., 1984). The importance of intracellular levels of thyroid hormones (and thus deiodinases) during smolt development has been emphasized by Specker et al. (1992), although a clear picture of their regulation and relation to thyroid hormone action has yet to emerge.

7. DEVELOPMENTAL AND ENVIRONMENTAL REGULATION

7.1. Development and Heterochrony

Rounsefell (1958) was the first to quantify the relative reliance on FW and SW (the "scale of anadromy") among salmonid species and genera, with a general increase in life history spent in SW among the genus *Oncorhynchus* (Fig. 5.8A). The increased proportion of time spent in SW is primarily driven by an earlier migration from FW to SW, which not surprisingly is also accompanied by earlier development of SW tolerance. This has been pointed out to be a heterochrony, in which the timing of developmental events (smolting and downstream migration) has been shifted earlier in more recently evolved species (McCormick, 1994) (Fig. 5.8B). Thus, species such as Atlantic salmon and steelhead trout spend one to several years in FW before undergoing a size-dependent transformation at 12–15 cm, whereas pink and chum salmon may spend as little as a few weeks after hatching in FW and begin their seaward migration at a much smaller size of 3–5 cm. Based on a consensus phylogeny in which *Salvelinus* is basal and *Salmo* and *Oncorhynchus* are sister genera, this suggests an evolutionary pattern in which salinity tolerance has been shifted to earlier developmental stages, especially within the most recently derived species of *Oncorhynchus* (Fig. 5.8). In heterochronic terms, this is a peramorphosis caused by predisplacement, the earlier onset of a developmental event.

In addition to the shift in timing, a change in the environmental factors that regulate smolt development has occurred (McCormick, 1994; Gallagher et al., 2013). In facultatively anadromous species such as brook trout and lake trout, smolt development is poorly developed or absent, and salinity itself is the primary inducer of mechanisms for salt tolerance. In smolting salmonids that spend at least a year in FW (Atlantic salmon, steelhead trout, coho salmon), a minimum size must be achieved, after which the animals alter their response to photoperiod. Thus, in Atlantic salmon a threshold size of 12–13 cm must be achieved by late January when daylength begins to increase (McCormick et al., 2007). In coho and Chinook salmon a growth threshold may also be involved in the initial "decision" to become smolts (Dickhoff et al., 1997; Beckman et al., 2007). If fish have not achieved this minimum size they will wait another year (or more) before transforming to smolts. In Atlantic salmon the attainment of smolt size as 1-year olds (1+) has been related to a bimodal growth pattern that develops several months before, in their first autumn (Kristinsson et al., 1985). How fish determine their size and/or growth rate at the appropriate time of year and translate this to increased responsiveness to daylength is an important area that has not received substantial attention. It has been suggested that proxies for size

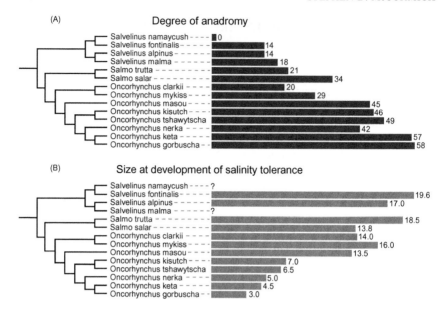

(A) Degree of anadromy

(B) Size at development of salinity tolerance

Fig. 5.8. Phylogeny of the subfamily Salmoninae mapped with the degree of anadromy (A) and the size at which salinity tolerance develops (B). The figures demonstrate that there is increased anadromy in more recently evolved salmonines, and that there has been a heterochrony in the size (and age) of smolt development with a clear trend towards smaller and earlier development of salinity tolerance and accompanying seaward migration. Phylogeny is a consensus phylogeny incorporating genetic, morphological, and life history data from 38 separate analyses in 34 published studies (Wilson and Williams, 2010). Data for the degree of anadromy are from Rounsefell (1958), with the exception of masu salmon, which was based on life history characteristics presented by Kato (1991). The degree of anadromy is based on six criteria, each with a maximum 10-point score: (1) extent of migration in the sea; (2) duration of stay in the sea; (3) state of maturity attained at sea; (4) spawning habits and habitat; (5) mortality after spawning; and (6) occurrence of freshwater (FW) forms. Size at development of salinity tolerance was taken from literature values in which fish were directly transferred from FW to at least 30 ppt seawater and survival was monitored for at least 14 days and was greater than 75% (with the exception of *S. fontinalis* in which a gradual acclimation was employed, and *O. nerka* in which survival was only monitored for 1 day); *S. fontinalis* (McCormick and Naiman, 1984), *S. alpinus* (Arnesen et al., 1992), *S. salar* (Johnston and Saunders, 1981), *S. trutta* (Parry, 1960), *O. mykiss* (Conte and Wagner, 1965), *O. clarkii* (Yeoh et al., 1991), *O. masou* (Ban et al., 1988), *O. tshawytscha* (Wagner et al., 1969), *O. kisutch* (Conte et al., 1966), *O. nerka* (Heifetz et al., 1989), *O. keta* (Black, 1951), *O. gorbuscha* (Gallagher et al., 2013). Phylogeny was created with software developed by Letunic and Bork (2011).

and growth such as energy reserves or fat content signaled through metabolic or neuroendocrine pathways may be involved (McCormick, 1994), but these have yet to be established.

In pink and chum salmon the early development of salinity tolerance, occurring soon after hatching, appears to be largely independent of environmental factors and to be hardwired to occur (McCormick, 1994; Gallagher et al., 2013). In addition to species difference, there appears to be diversity in the size and time of year at which smolt development and migration occur within the species, especially in Chinook salmon, and diversity can occur within and between populations (Taylor, 1990). Thus, for the following sections on environmental factors that regulate physiological and behavioral aspects of smolt development it is important to point out that the critical environmental cues are likely to differ among species and among populations within species. Extensive reviews on environmental factors controlling smolt development have been published by Hoar (1988) and Björnsson et al. (2011), the latter focusing on those factors most relevant to fish in the wild.

7.2. Photoperiod

For those smolting salmonids that migrate at least a year after hatching, photoperiod has been shown to be the most important environmental determinant for the timing of virtually all physiological and behavioral aspects of smolt development (Hoar, 1988). In Atlantic salmon, advanced increases in daylength or out-of-phase photoperiod can advance smolting by many months (Duston and Saunders, 1995). Although there is some evidence for an endogenous rhythm of smolting (Eriksson and Lundqvist, 1982), increased daylength is probably necessary for normal smolting to occur, and is certainly required for its normal timing. The capacity of photoperiod to affect smolting may be restricted at low temperatures (McCormick et al., 2000), indicating that temperature may be a controlling factor in smolting (Fry, 1971), at least in some species.

The actions of photoperiod on the timing of smolt development are cued by its impact on the neuroendocrine axis (Björnsson et al., 2011). Circulating levels of GH, cortisol, and thyroid hormones have all been shown to be responsive to changes in photoperiod. GH seems to be especially responsive, increasing within several days of an advanced photoperiod, and potentially causing a cascade of other endocrine events (McCormick et al., 1995). Atlantic salmon parr too small to undergo smolting do not change plasma GH or cortisol substantially in spring, nor do they respond to an advanced photoperiod; in contrast, fish that have reached the critical size for smolting in this species increased plasma GH and cortisol in the spring, the timing of which can be advanced by increased daylength (McCormick et al., 2007). Thus, developmental difference in the capacity to respond to photoperiod is a key component of smolting in this species. This developmental difference may be due in part to changes in retinal and pineal innervation of the hypothalamus. Atlantic salmon smolts develop robust retinal innervation of the preoptic nucleus, an area known to regulate pituitary function (Ebbesson et al., 2007). Furthermore, this innervation is blocked by exposure to continuous light, which also inhibits normal smolt development.

The mechanism by which photoperiodic information is translated to a neuroendocrine response in teleosts has not been established. Melatonin secretion by the pineal of salmonids can be directly entrained by photoperiod, but the relationship of melatonin rhythm to photoperiodism in fish is unclear (Falcon et al., 2007). The major clock genes found in other vertebrates (Clock, Bmal1, Per2, and Cry2) are expressed in the brains of juvenile Atlantic salmon and can be altered by changes in daylength (Davie et al., 2009). Diel rhythms of plasma melatonin and cortisol and expression patterns of brain and pineal Clock, Cry2, and Per1 were present in presmolt Atlantic salmon exposed to short days but were abolished after exposure to increased daylength (Huang et al., 2010). The presence of latitudinal variation in the OtsClock1b gene in four populations of Pacific salmon is hypothesized to play a role in the timing of spawning, although it is possible that it also plays a role in other seasonal life history events such as smolting and downstream migration. Further work is clearly needed to determine the mechanistic link among changes in daylength, neuroendocrine activation, and physiological and behavioral responses.

As noted above, photoperiod is the seasonal "zeitgeber" that triggers a complex series of developmental events that precede and prepare salmon for migration. Although photoperiod influences the timing of migratory behavior of salmon smolts (Hoar, 1976), it is not the only environmental mediator. Several authors have proposed that photoperiod may act to increase activity (migratory restlessness or Zugunruhe) and make fish sensitive to other environmental cues (releasing factors) that initiate migration (McCormick et al., 1998). Photoperiod will therefore determine the range of days or weeks during which migration may occur, but other factors such as water temperature, flow, and turbidity may act as releasing factors within that range.

7.3. Temperature

Studies using differing levels of constant temperature during smolt rearing have found that increased temperature results in earlier smolt development in Atlantic salmon and steelhead trout (although at temperatures of 15°C and higher, smolt development of steelhead is inhibited) (Adams et al., 1973; Johnston and Saunders, 1981; Solbakken et al., 1994). Exposure of Atlantic salmon smolts to several rates of temperature increase in spring confirms this effect of temperature and indicates a strong relationship between the cumulative temperature experience (degree days) and smolt development (Sigholt et al., 1998), indicating that temperature is acting to control the rate of development. Effects of temperature on smolt development may not be universal, as Larsen et al. (2001) found no difference in the timing or magnitude of changes in gill NKA activity of coho salmon smolts reared at 2.5 and 10°C in winter and spring. In reviewing the influence of temperature on smolt development, McCormick et al. (1997) concluded that although increased temperature was not necessary for the completion of smolting, it acted with photoperiod to advance smolting and could act as a directive factor in the absence of photoperiod cues. However, more recent studies indicate that temperature acts primarily on the rate of development and has only a limited ability to act as a directive factor. Under normal daylength conditions, an advanced temperature regimen resulted in more rapid increases in gill NKA activity, but the date at which increases began and peak levels occurred was the same for the two temperature regimens (McCormick et al., 2002). The same study found that smolting remained incomplete under a regimen of short days and an advanced temperature regimen, indicating that temperature increase alone cannot act to advance smolting in the absence of photoperiod cues.

Developmental events and circannual cues, working through the neuroendocrine axis, set the stage for both physiological development and migratory readiness. Environmental factors can then act as "releasers" of migratory behavior (Baggerman, 1960), as outlined in Fig. 5.7. Which hormones are most important for determining migratory readiness has not been examined. As noted above, temperature has a role in determining the timing of smolt development (and thus migratory readiness), but is also likely to have a role as a releasing factor. However, determining the independent effects of temperature on these two processes may prove to be difficult. A large number of field studies have found a correlation between migration timing and temperature, and it seems clear that water temperature is an important factor, perhaps the most important factor, initiating downstream migrations in salmon (Jonsson et al., 1991; McCormick et al., 1998). Several authors have reported that downstream migration of smolts of Atlantic salmon and brown trout occurs at about 10°C or slightly above, and similar aspects of threshold temperatures have been proposed for smolt migration in other species (Jonsson et al., 1991). There are several examples, however, of populations of Pacific salmon in northern latitudes beginning or even completing migration before these thresholds are reached (J. M. Shrimpton, personal communication). Thus, there is some reason to question whether a temperature threshold is a major driver of smolt migration. Zydlewski et al. (2005) found that the effect of an advanced temperature regimen on both the onset and termination of migratory behavior in captive Atlantic salmon smolts could best be explained by accumulated thermal units (ATUs, also known as cumulative degree days) rather than a threshold temperature. Similarly, the effect of temperature manipulation on migration of captive Chinook salmon smolts was related to ATUs. Analysis of a 12 year dataset of wild Chinook salmon smolt migration timing found that ATUs had a greater explanatory power than a threshold temperature (Sykes et al., 2009). In addition to cumulative degree days there may be other aspects of temperature such as the magnitude of diel changes or population-specific thresholds that play a role in migration.

7.4. Flow and Turbidity

High water flow in rivers may stimulate downstream movement in a large number of fish species (Jonsson et al., 1991), and the downstream migration of smolts has been linked to increased water flow (see references in McCormick et al., 1998). As pointed out above, it is often difficult to separate the impact of flow on the initiation and the speed of migration in field studies. The migration of Atlantic salmon smolts in the River Orkla, Norway, was coincident with the first spring peak in water discharge when the water temperature was above 3°C (Hesthagen and Garnas, 1986). It has been suggested that smolt migration is a result of a passive displacement by high water flow (Thorpe, 1984), although this hypothesis has not received substantial empirical support (Bourgeois and O'Connell 1988). It has been clearly demonstrated that smolts migrate more quickly at high water velocities than at low flow (Youngson et al., 1989). Using an artificial tank system to examine downstream migration, Sykes and Shrimpton (2010) found that temperature rather than flow most strongly influenced the initiation of migration. However, the presence of a strong, directional current resulted in a period of more defined movement, indicating an influence of current once migration is underway. Atlantic salmon smolts actively move out into the main current of the river, apparently to avoid being caught in backwaters and sloughs (Hansen and Jonsson, 1985). Smolts migrating through lakes move faster than the downstream current velocity (Bourgeois and O'Connell 1988), further indicating that the seaward migration is not purely passive.

There is strong evidence that the survival of hatchery-reared smolts improves significantly when they are released at high water discharge within the normal period of migration (Hosmer et al., 1979; Hvidsten and Hansen, 1988). Wild coho salmon survival (as measured by adult return rates) is highest in years with high flow rates during spring emigration (Lawson et al., 2004), further suggesting an important link between flow and survival.

Several factors may contribute to this greater survival associated with high flow, including higher migration speeds and reduced visibility due to turbidity, both of which can reduce predation (Gregory and Levings, 1998). Greater migration speeds may also minimize the chance that fish may enter the ocean after they have reached the peak of SW adaptability.

7.5. Physiological and Ecological Smolt Windows

Salinity tolerance, migratory behavior, and several other aspects of development that occur during smolting are lost if fish remain in FW (Fig. 5.2A,B). This process has been called "desmoltification", although this term should be used with some caution, as it implies that all aspects of smolting are reversed, which is not the case. Nonetheless, many of the reversible aspects of smolting such as salinity tolerance are critical for marine survival, and indicate that there is a limited period during which the fish are at peak preparedness for SW entry, known as the physiological smolt window. The boundaries of the physiological smolt window will be determined by the environmental factors that control the stimulation and loss of smolt characters. As noted above, photoperiod and temperature are the primary drivers of smolt development in coho, steelhead, and Atlantic salmon, whereas photoperiod may be less important in pink and chum salmon that develop salinity tolerance soon after hatching. In coho, steelhead, and Atlantic salmon, it appears that the loss of salinity tolerance and other aspects of smolt development are driven primarily by temperature and time (Zaugg and McLain, 1976; Duston et al., 1991). This relationship has been investigated in detail in Atlantic salmon, where there is a dome-shaped relationship between cumulative degree days and the development and loss of salinity tolerance and/or gill NKA activity (McCormick et al., 1997). Thus, after peak smolt development is reached, fish that experience an additional 500 ATU will have reverted back to parr levels of salinity tolerance. It appears that the regulation of smolt behavior is similar to that relating to salinity tolerance, in that higher temperatures lead to an earlier cessation of downstream migratory behavior (Wagner, 1974; Zydlewski et al., 2005).

Adult return rates of hatchery-reared salmon smolts, perhaps the most important measure of SW performance, are strongly dependent on the timing of release. Hatchery-reared Atlantic salmon released in late May during the normal migratory period of wild fish have higher adult return rates than fish released in April and June (Staurnes et al., 1993; Lundqvist et al., 1994). Similarly, Bilton et al. (1982) found substantially higher return rates in hatchery-reared coho salmon released in April and May than in June and July. In studies where it is measured, the peak of physiological smolt development corresponds with the timing of release that yields the highest adult return rates (Virtanen et al., 1991). This correspondence is likely to be dependent on both the physiological smolt window and coincidence of release and subsequent migration with optimum environmental conditions for smolt survival (the "ecological smolt window") (McCormick et al., 1998). It seems likely that natural selection will have shaped the physiological smolt window to coincide with the ecological smolt window. Stock-specific differences in the timing of downstream migration (Orciari and Leonard, 1996; Mcginnity et al., 2007) or increases in salinity tolerance (Handeland et al., 2004) are likely examples of selection acting on the timing of smolt development.

In addition to the timing of release, FW rearing conditions in hatcheries will affect the overall SW performance of smolts. Beckman et al. (1999) found that the same strain of spring Chinook salmon reared at three different hatcheries had different smolt-to-adult return rates, and that these were most strongly correlated with spring growth, plasma IGF-I levels, and gill NKA activity. Fish density and raceway flows experienced by juveniles in FW affect the adult return rates of spring Chinook salmon (Banks, 1994). Several studies have found a positive correlation between size of released smolts, and smolt survival and adult return rates (Miyakoshi et al., 2001; Connor et al., 2004). The effect of size appears to be more pronounced in years with poor overall marine survival (Saloniemi et al., 2004). The effects of rearing conditions are generally smaller than the effects of time of release, but clearly indicate that growth conditions in FW affect smolt survival. While these studies have focused on hatcheries, they underscore the principle that variations in environmental conditions in streams and rivers impact smolt development and SW performance, and it would be of interest to determine whether variations in winter and spring growth conditions affect survival in wild fish.

How the physiological smolt window affects marine survival is likely to be more complex than smolts simply suffering acute osmoregulatory failure and dying following movement into high-salinity waters. It is more likely that there will be a more subtle compromise in their ability to migrate, forage, and avoid predators, and an increase in time spent in estuaries where predation pressures are high. Minor compromises of physiological and/or behavioral preparation which have minimal effects in the laboratory may have dire consequences in the wild. Mortality rates of up to 34% due to predation during the first 2 days of SW entry have been reported in brown trout smolts (Dieperink et al., 2002). Atlantic salmon smolts directly transferred to SW suffered greater experimental predation in SW than smolts acclimated to SW for 30 days (Jarvi, 1990). Similarly, the greatest predation rates on smolts occurred during the first days of exposure to SW when plasma ion perturbations were at their highest levels (Handeland et al., 1996). Chinook salmon smolts subjected to a handling stress spent more time in the upper, FW layers of a simulated estuary than did unstressed fish (Price and Schreck, 2003), which the authors suggest would expose them to greater avian predation. Furthermore, lower levels of gill NKA activity are associated with greater levels of avian predation in estuaries during downstream migration of Chinook salmon (Schreck et al., 2006). Osmotic and other stressors may also result in increased susceptibility to diseases that reduce estuarine and marine survival (Loge et al., 2005). Adult return rates of coho salmon subjected to transportation stress just before release were reduced compared to those of fish released directly from the hatchery or those allowed 6 weeks to recover from transportation stress (Schreck et al., 1989). A variety of contaminants have been shown to affect smolt development, some of which specifically act on salinity tolerance and the capacity for marine survival (McCormick et al., 2009a). All of these studies indicate that factors compromising the physiological and behavioral capacity of smolts to make the transition from FW to SW are likely to result in an overall decrease in marine survival, even if the ultimate source of mortality is not directly due to osmoregulatory failure.

8. CONCLUSIONS AND PERSPECTIVES

The migration of smolts takes them from headwater streams, through large rivers, into estuaries, and out into the expansive ocean. Smolts can therefore be affected by natural and anthropogenic changes that affect any one of these aquatic environments. Sedimentation and pollution can degrade upland aquatic habitats. Dams and turbines can be directly lethal, but also alter migration patterns and affect the thermal environment experienced by smolts. Invasion of estuarine habitats by a new competitor, predator, or disease agent can increase mortality of smolts at a stage when mortality is already high. Natural variations in ocean conditions are also known to alter smolt survival and long-term alterations induced by climate change may cause a shift to less favorable conditions.

To understand the impacts of environmental change on salmon smolts, it will be critical to understand both their basic biology and their susceptibility to threats. There are large variations in the responses of individual species to environmental change. As this chapter has shown, smolts undergo a variety of preparatory developmental changes that allow them to migrate and live in dramatically different habitats. The control of migration timing appears to be critical for smolt survival (Thorstad et al., 2012), and knowledge of how climate change will affect the ecological and physiological smolt windows will be critical to predicting the future of salmon populations (McCormick et al., 1998). The incredible plasticity of smolts has some tradeoffs, and in some cases appears to make them more susceptible to external stressors. For instance, smolts are more sensitive than other life stages to the impacts of acidification (Rosseland et al., 2001). Exposure to contaminants in streams and rivers can affect not only FW survival, but the ability to make the transition to the marine environment (McCormick et al., 2009a). Climate change is likely to interact with other habitat alterations to compound negative effects on smolt survival and salmon persistence. Continued examination of the biology of smolts and their response to environmental changes will be critical to understanding their habitat requirements and conservation needs for ensuring salmon survival.

ACKNOWLEDGMENTS

I thank all of the members of my lab and my collaborators over the years for their willingness to share ideas, some of which have probably appeared here unknowingly without proper attribution. Thanks to Eric Schultz, Arne Christensen, Amy Regish, Michael O'Dea, and Tara Duffy for their help in composing several of the figures. I dedicate this chapter to Howard Bern, mentor, friend, and forever a source of inspiration.

REFERENCES

Adams, B. L., Zaugg, W. S. and McLain, L. R. (1973). Temperature effect on parr-smolt tranformation in steelhead trout (*Salmo gairdneri*) as measured by gill sodium-potassium stimulated ATPase. *Comp. Biochem. Physiol. [A]* 44, 1333–1339.

Agustsson, T., Sundell, K., Sakamoto, T., Johansson, V., Ando, M. and Bjornsson, BT. (2001). Growth hormone endocrinology of Atlantic salmon (*Salmo salar*): pituitary gene expression, hormone storage, secretion and plasma levels during parr-smolt transformation. *J. Endocr.* 170, 227–234.

Arnesen, A. M., Halvorsen, M. and Nilssen, K. J. (1992). Development of hypoosmoregulatory capacity in Arctic char (*Salvelinus alpinus*) reared under either continuous light or natural photoperiod. *Can. J. Fish. Aquat. Sci.* 49, 229–237.

Baggerman, B. (1960). Factors in the diadromous migrations of fish. *Zool. Soc. Symp., London I* 33–60.

Ban, M., Kanno, H., Izumi, T. and Yamauchi, K. (1988). Body size and seawater tolerance in the hatchery-reared underyearling masu salmon (*Oncorhynchus masou*). *Tohoku Reg. Fish. Res. Lab. Bull.* 50, 117–123.

Bandoh, H., Kida, I. and Ueda, H. (2011). Olfactory responses to natal stream water in sockeye salmon by BOLD fMRI. *PLOS ONE* 6, e16051.

Banks, J. L. (1994). Raceway density and water flow as factors affecting spring chinook salmon (*Oncorhynchus tshawytscha*) during rearing and after release. *Aquaculture* 119, 201–217.

Beckman, B. R. and Dickhoff, W. W. (1998). Plasticity of smolting in spring chinook salmon: relation to growth and insulin-like growth factor-I. *J. Fish Biol.* 53, 808–826.

Beckman, B. R., Dickhoff, W. W., Zaugg, W. S., Sharpe, C., Hirtzel, S., Schrock, R., Larsen, D. A., Ewing, R. D., Palmisano, A., Schreck, C. B. and Mahnken, C. W. (1999). Growth, smoltification, and smolt-to-adult return of spring chinook salmon from hatcheries on the Deschutes River, Oregon [Review]. *Trans. Am. Fish. Soc.* 128, 1125–1150.

Beckman, B. R., Gadberry, B., Parkins, P., Cooper, K. A. and Arkush, K. D. (2007). State-dependent life history plasticity in Sacramento River winter-run chinook salmon (*Oncorhynchus tshawytscha*): interactions among photoperiod and growth modulate smolting and early male maturation. *Can. J. Fish. Aquat. Sci.* 64, 256–271.

Beeman, J. W., Rondorf, D. W. and Tilson, M. E. (1994). Assessing smoltification of juvenile spring chinook salmon (*Oncorhynchus tshawytscha*) using changes in body morphology. *Can. J. Fish. Aquat. Sci.* 51, 836–844.

Bern, H. A. (1978). Endocrinological studies on normal and abnormal salmon smoltification. In *Comparative Endocrinology* (eds. P. J. Gaillard and H. H. Boer), pp. 77–100. Amsterdam: Elsevier/North Holland Biomedical Press.

Bertmar, G. (1983). Monthly development of Baltic salmon olfactory organs before and during the imprinting period, and comparisons with precocious males of the same stock. *Aquillo Ser. Zool.* 22, 37–44.

Bilton, H. T., Alderdice, D. F. and Schnute, J. T. (1982). Influence of time and size at release of juvenile coho salmon (*Oncorhynchus kisutch*) on returns at maturity. *Can. J. Fish. Aquat. Sci.* 39, 426–447.

Björnsson, B. Th., Stefansson, S. O. and McCormick, S. D. (2011). Environmental endocrinology of salmon smoltification. *Gen. Comp. Endocrinol.* 170, 290–298.

Black, V. S. (1951). Changes in body chloride, density and water content of chum (*Oncorhynchus keta*) and coho (*O. kisutch*) salmon fry when transferred from fresh water to sea water. *J. Fish. Res. Board Can.* 8, 164–177.

Boeuf, G. (1993). Salmonid smolting: a pre-adaptation to the oceanic environment. In *Fish ecophysiology* (eds. J. C. Rankin and F. B. Jensen), pp. 105–135. London: Chapman and Hall.

Bourgeois, C. E. and O'Connell, M. F. (1988). Observations on the seaward migration of Atlantic salmon (*Salmo salar* L.) smolts through a large lake as determined by radiotelemetry and Carlin tagging studies. *Can. J. Zool.* 66, 685–691.

Brannon, E. L. and Quinn, T. P. (1990). Field test of pheromone hypothesis for homing of pacific salmon. *J. Chem. Ecol.* 16, 603–610.

Brauner, C. J., Shrimpton, J. M. and Randall, D. J. (1992). Effect of short-duration seawater exposure on plasma ion concentrations and swimming performance in coho salmon (*Oncorhynchus kisutch*) parr. *Can. J. Fish. Aquat. Sci.* 49, 2399–2405.

Carey, J. B. and McCormick, S. D. (1998). Atlantic salmon smolts are more responsive to an acute handling and confinement stress than parr. *Aquaculture* 168, 237–253.

Clarke, W. C., Saunders, R. L. and McCormick, S. D. (1996). Smolt Production. In *Principles of Salmonid Culture*, vol. 29 (eds. W. Pennel and B. A. Barton), pp. 517–567. Amsterdam: Elsevier.

Clements, S. and Schreck, C. B. (2004). Central administration of corticotropin-releasing hormone alters downstream movement in an artificial stream in juvenile chinook salmon (*Oncorhynchus tshawytscha*). *Gen. Comp. Endocrinol.* 137, 1–8.

Collie, N. L. and Bern, H. A. (1982). Changes in intestinal fluid transport associated with smoltification and seawater adaptation in coho salmon, *Oncorhynchus kisutch* (Walbaum). *J. Fish Biol.* 21, 337–348.

Connor, W. P., Smith, S. G., Andersen, T., Bradbury, S. M., Burum, D. C., Hockersmith, E. E., Schuck, M. L., Mendel, G. W. and Bugert, R. M. (2004). Postrelease performance of hatchery yearling and subyearling fall chinook salmon released into the Snake River. *N. Am. J. Fish. Manag.* 24, 545–560.

Conte, F. P. and Wagner, H. H. (1965). Development of osmotic and ionic regulation in juvenile steelhead trout *Salmo gairdneri. Comp. Biochem. Physiol.* 14, 603–620.

Conte, F. P., Wagner, H. H., Fessler, J. and Gnose, C. (1966). Development of osmotic and ionic regulation in juvenile coho salmon (*Oncorhynchus kisutch*). *Comp. Biochem. Physiol.* 18, 1–15.

Dauder, S., Young, G., Hass, L. and Bern, H. A. (1990). Prolactin receptors in liver, kidney, and gill of the tilapia (*Oreochromis mossambicus*): characterization and effect of salinity on specific binding of iodinated ovine prolactin. *Gen. Comp. Endocrinol.* 77, 368–377.

Davie, A., Minghetti, M. and Migaud, H. (2009). Seasonal variations in clock-gene expression in Atlantic salmon (*Salmo salar*). *Chronobiol. Int.* 26, 379–395.

Dickhoff, W. W., Beckman, B. R., Larsen, D. A., Duan, C. and Moriyama, S. (1997). The role of growth in endocrine regulation of salmon smoltification. *Fish Physiol. Biochem.* 17, 231–236.

Dickhoff, W. W., Folmar, L. C. and Gorbman, A. (1978). Changes in plasma thyroxine during smoltification of coho salmon, *Oncorhynchus kisutch. Gen. Comp. Endocrinol.* 36, 229–232.

Dickhoff, W. W. and Sullivan, C. V. (1987). Involvement of the thyroid gland in smoltification, with special reference to metabolic and developmental processes. *Am. Fish. Soc. Symp.* 1, 197–210.

Dieperink, C., Bak, B. D., Pedersen, L. F., Pedersen, M. I. and Pedersen, S. (2002). Predation on Atlantic salmon and sea trout during their first days as postsmolts. *J. Fish Biol.* 61, 848–852.

Dittman, A. H. and Quinn, T. P. (1996). Homing in Pacific salmon: Mechanisms and ecological basis. *J. Exp. Biol.* 199, 83–91.

Dittman, A. H., Quinn, T. P. and Nevitt, G. A. (1996). Timing of imprinting to natural and artificial odors by coho salmon (*Oncorhynchus kisutch*). *Can. J. Fish. Aquat. Sci.* 53, 434–442.

Dittman, A. H., Quinn, T. P., Nevitt, G. A., Hacker, B. and Storm, D. R. (1997). Sensitization of olfactory guanylyl cyclase to a specific imprinted odorant in coho salmon. *Neuron* 19, 381–389.

Duan, C. M., Plisetskaya, E. M. and Dickhoff, W. W. (1995). Expression of insulin-like growth factor I in normally and abnormally developing coho salmon (*Oncorhynchus kisutch*). *Endocrinol.* 136, 446–452.

Dukes, J. P., Deaville, R., Bruford, M. W., Youngson, A. F. and Jordan, W. C. (2004). Odorant receptor gene expression changes during the parr-smolt transformation in Atlantic salmon. *Molec. Ecol.* 13, 2851–2857.

Duston, J. and Saunders, R. L. (1995). Advancing smolting to autumn in age 0+ Atlantic salmon by photoperiod, and long-term performance in sea water. *Aquaculture* 135, 295–309.

Duston, J., Saunders, R. L. and Knox, D. E. (1991). Effects of increases in freshwater temperature on loss of smolt characteristics in Atlantic salmon (*Salmo salar*). *Can. J. Fish. Aquat. Sci.* 48, 164–169.

Ebbesson, L. O. E., Ebbesson, S. O. E., Nilsen, T. O., Stefansson, S. O. and Holmqvist, B. (2007). Exposure to continuous light disrupts retinal innervation of the preoptic nucleus during parr-smolt transformation in Atlantic salmon. *Aquaculture* 273, 345–349.

Ebbesson, L. O. E., Nilsen, T. O., Helvik, J. V., Tronci, V. and Stefansson, S. O. (2011). Corticotropin-releasing factor neurogenesis during midlife development in salmon: genetic, environmental and thyroid hormone regulation. *J. Neuroendocrinol.* 23, 733–741.

Eddy, F. B. and Talbot, C. (1985). Urine production in smolting Atlantic salmon, *Salmo salar* L. *Aquaculture* 45, 67–72.

Edwards, S. L. and Marshall, W. S. (2013). Principles and patterns of osmoregulation and euryhalinity in fishes. In *Fish Physiology*, Vol. 32, *Euryhaline Fishes* (eds. S. D. McCormick, C. J. Brauner and A. P. Farrell). Amsterdam: Academic Press, pp. 1–44.

Eriksson, L.-O. and Lundqvist, H. (1982). Circannual rhythms and photoperiod regulation of growth and smolting in Baltic salmon (*Salmo salar* L.). *Aquaculture* 28, 113–121.

Evans, D. H., Piermarini, P. M. and Choe, K. P. (2005). The multifunctional fish gill: Dominant site of gas exchange, osmoregulation, acid-base regulation, and excretion of nitrogenous waste. *Physiolog. Rev.* 85, 97–177.

Falcon, J., Besseau, L., Sauzet, S. and Boeuf, G. (2007). Melatonin effects on the hypothalamo-pituitary axis in fish. *Trends Endocrinol. Metab.* 18, 81–88.

Fontaine, M. and Hatey, J. (1954). Sur la teneur en 17-hydroxycorticosteroides du plasm de saumon (*Salmo salar* L.). *Compt. Rend. Acad. Sci.* 239, 319–321.

Fry, F. E. J. (1971). The effect of environmental factors on the physiology of fish. In *Fish Physiology*, Volume VI, *Environmental Relations and Behavior* (eds. W. S. Hoar and D. J. Randall), pp. 1–98. New York: Academic Press.

Gallagher, Z., Bystriansky, J.S., Farrell, A.P. and Brauner, C.J. (2013). A novel pattern of smoltification in the most anadromous salmonid: Pink salmon (*Oncorhynchus gorbuscha*). *Can. J. Fish. Aquat. Sci.* (in press).

Gibson, R. J. (1983). Water velocity as a factor in the change from aggressive to schooling behaviour and subsequent migration of Atlantic salmon smolt (*Salmo salar*). *Le Naturaliste Canadien* 110, 143–148.

Gorbman, A., Dickhoff, W. W., Mighell, J. L., Prentice, E. F. and Waknitz, F. W. (1982). Morphological indices of developmental progress in the parr-smolt coho salmon, *Oncorhynchus kisutch*. *Aquaculture* 28, 1–19.

Grau, E. G., Dickhoff, W. W., Nishioka, R. S., Bern, H. A. and Folmar, L. C. (1981). Lunar phasing of the thyroxine surge preparatory to seaward migration of salmonid fish. *Science* 211, 607–609.

Gregory, R. S. and Levings, C. D. (1998). Turbidity reduces predation on migrating juvenile Pacific salmon. *Trans. Am. Fish. Soc.* 127, 275–285.

Grossell, M. (2011). In *The role of the gastrointestinal tract in salt and water balance* (eds. M. Grosell, A. P. Farrell and C. J. Brauner), pp. 135–164. London: Elsevier.

Handeland, S. O., Jarvi, T., Ferno, A. and Stefansson, S. O. (1996). Osmotic stress, antipredator behaviour, and mortality of Atlantic salmon (*Salmo salar*) smolts. *Can. J. Fish. Aquat. Sci.* 53, 2673–2680.

Handeland, S. O., Wilkinson, E., Sveinsbo, B., McCormick, S. D. and Stefansson, S. O. (2004). Temperature influence on the development and loss of seawater tolerance in two fast-growing strains of Atlantic salmon. *Aquaculture* 233, 513–529.

Hansen, L. P. and Jonsson, B. (1985). Downstream migration of hatchery-reared smolts of Atlantic salmon (*Salmo salar* L.) in the River Imsa. *Aquaculture* 45, 237–248.

Hansen, L. P. and Quinn, T. R. (1998). The marine phase of the Atlantic salmon (*Salmo salar*) life cycle, with comparisons to Pacific salmon. *Can. J. Fish. Aquat. Sci.* 55, 104–118.

Hara, T. J. (1994). Olfaction and gustation in fish: An overview. *Acta Physiol. Scand.* 152, 207–217.

Hasler, A. D. and Scholz, A. T. (1983). *Olfactory Imprinting and Homing in Salmon*. New York: Springer-Verlag.

Heifetz, J., Johnson, S. W., Koski, K. V. and Murphy, M. L. (1989). Migration timing, size, and salinity tolerance of sea-type sockeye salmon (*Oncorhynchus nerka*) in an Alaska estuary. *Can. J. Fish. Aquat. Sci.* 46, 633–637.

Hesthagen, T. and Garnas, E. (1986). Migration of Atlantic salmon smolts in River Orkla, central Norway in relation to management of a hydroelectric station. *N. Am. J. Fish. Manag.* 6, 376–382.

Hoar, W. S. (1939). The thyroid gland of the Atlantic salmon. *J. Morphol.* 65, 257–295.

Hoar, W. S. (1976). Smolt transformation: evolution, behavior, and physiology. *J. Fish. Res. Board Can.* 33, 1234–1252.

Hoar, W. S. (1988). The physiology of smolting salmonids. In *Fish Physiology*, Vol. XIB (eds. W. S. Hoar and D. Randall), pp. 275–343. New York: Academic Press.

Hoffnagle, T. L. and Fivizzani, A. J. (1990). Stimulation of plasma thyroxine levels by novel water chemistry during smoltification in chinook salmon (*Oncorhynchus tshawytscha*). *Can. J. Fish. Aquat. Sci.* 47, 1513–1517.

Hogasen, H. R. and Prunet, P. (1997). Plasma levels of thyroxine, prolactin, and cortisol in migrating and resident wild arctic char, *Salvelinus alpinus*. *Can. J. Fish. Aquat. Sci.* 54, 2947–2954.

Hosmer, M. J., Stanley, J. G. and Hatch, R. W. (1979). Effects of hatchery procedures on later return of Atlantic salmon to rivers. *Prog. Fish-Culturist* 41, 115–119.

Huang, T. S., Ruoff, P. and Fjelldal, P. G. (2010). Diurnal expression of clock genes in pineal gland and brain and plasma levels of melatonin and cortisol in Atlantic salmon parr and smolts. *Chronobiol. Int.* 27, 1697–1714.

Hutchison, M. J. and Iwata, M. (1998). Effect of thyroxine on the decrease of aggressive behaviour of four salmonids during the parr-smolt transformation. *Aquaculture* 168, 169–175.

Hvidsten, N. A. and Hansen, L. P. (1988). Increased recapture rate of adult Atlantic salmon, *Salmo salar* L., stocked as smolts at high water discharge. *J. Fish Biol.* 32, 153–154.

Iwata, M. (1995). Downstream migratory behavior of salmonids and its relationship with cortisol and thyroid hormones: A review. *Aquaculture* 135, 131–139.

Iwata, M., Tsuboi, H., Yamashita, T., Amemiya, A., Yamada, H. and Chiba, H. (2003). Function and trigger of thyroxine surge in migrating chum salmon *Oncorhynchus keta* fry. *Aquaculture* 222, 315–329.

Iwata, M., Yamauchi, K., Nishioka, R. S., Lin, R. and Bern, H. A. (1990). Effects of thyroxine, growth hormone and cortisol on salinity preference of juvenile coho salmon (*Oncorhynchus kisutch*). *Mar. Behav. Physiol.* 17, 191–201.

Jarrard, H. E. (1997). Postembryonic changes in the structure of the olfactory bulb of the chinook salmon (*Oncorhynchus tshawytscha*) across its life history. *Brain, Behav. Evol.* 49, 249–260.

Jarvi, T. (1990). Cumulative acute physiological stress in Atlantic salmon smolts: the effect of osmotic imbalance and the presence of predators. *Aquaculture* 89, 337–350.

Johnston, C. E. and Eales, J. G. (1967). Purines in the integument of the Atlantic salmon (*Salmo salar*) during parr-smolt transformation. *J. Fish. Res. Board Can.* 24, 955–964.

Johnston, C. E. and Saunders, R. L. (1981). Parr-smolt transformation of yearling Atlantic salmon (*Salmo salar*) at several rearing temperatures. *Can. J. Fish. Aquat. Sci.* 38, 1189–1198.

Johnstone, K. A., Lubieniekci, K. P., Koop, B. F. and Davidson, W. S. (2011). Expression of olfactory receptors in different life stages and life histories of wild Altantic salmon (*Salmo salar*). *Molec. Ecol.* 20, 4059–4069.

Jonsson, B., Jonsson, N. and Hansen, L. P. (1991). Differences in life history and migratory behaviour between wild and hatchery-reared Atlantic salmon in nature. *Aquaculture* 98, 69–78.

Kato, F. (1991). Life histories of masu and amago salmon (*Oncorhynchus masou* and *Oncorhynchus rhodurus*). In *Pacific salmon life histories* (eds. C. Groot and L. Margolis), pp. 447–520. Vancouver: UBC Press.

Kelley, K. M., Nishioka, R. S. and Bern, H. A. (1990). In vitro effect of osmotic pressure and cortisol on prolactin cell physiology in the coho salmon (*Oncorhynchus kisutch*) during the parr-smolt transformation. *J. Exp. Zool.* 254, 72–82.

Kiilerich, P., Kristiansen, K. and Madsen, S. S. (2007a). Cortisol regulation of ion transporter mRNA in Atlantic salmon gill and the effect of salinity on the signaling pathway. *J. Endocrinol.* 194, 417–427.

Kiilerich, P., Kristiansen, K. and Madsen, S. S. (2007b). Hormone receptors in gills of smolting Atlantic salmon, *Salmo salar*: Expression of growth hormone, prolactin, mineralocorticoid and glucocorticold receptors and 11 beta-hydroxysteroid dehydrogenase type 2. *Gen. Comp. Endocrinol.* 152, 295–303.

Klemetsen, A., Amundsen, P.-A., Dempson, J. B., Jonsson, B., Jonsson, N., O'Connell, M. F. and Mortensen, E. (2012). Atlantic salmon *Salmo salar* L., brown trout *Salmo trutta* L. and Arctic charr *Salvelinus alpinus*: a review of aspects of their life histories. *Ecol. Freshw. Fish* 12, 1–59.

Kristinsson, J. B., Saunders, R. L. and Wiggs, A. J. (1985). Growth dynamics during the development of bimodal length-frequency distribution in juvenile Atlantic salmon (*Salmo salar* L.). *Aquaculture* 45, 1–20.

Kudo, H., Tsuneyoshi, Y., Nagae, M., Adachi, S., Yamauchi, K., Ueda, H. and Kawamura, H. (2002). Detection of thyroid hormone receptors in the olfactory system and brain of wild masu salmon, *Oncorhynchus masou* (Brevoort), during smolting by *in vitro* autoradiography. *Aquacult. Fish. Mngmnt* 25 (**s2**), 171–182.

Langdon, J. S., Thorpe, J. E. and Roberts, R. J. (1984). Effects of cortisol and ACTH on gill Na$^+$/K$^+$-ATPase, SDH and chloride cells in juvenile Atlantic salmon, *Salmo salar*. *Comp. Biochem. Physiol. [A]* 77, 9–12.

Larsen, D. A., Beckman, B. R. and Dickhoff, W. W. (2001). The effect of low temperature and fasting during the winter on growth and smoltification of coho salmon. *N. Am. J. Aquacult.* 63, 1–10.

Larsen, D. A., Swanson, P., Dickey, J. T., Rivier, J. and Dickhoff, W. W. (1998). In vitro thyrotropin-releasing activity of corticotropin-releasing hormone-family peptides in coho salmon, *Oncorhynchus kisutch*. *Gen. Comp. Endocrinol.* 109, 276–285.

Larsen, D. A., Swanson, P. and Dickhoff, W. W. (2011). The pituitary-thyroid axis during the parr-smolt transformation of Coho salmon, *Oncorhynchus kisutch*: Quantification of TSH beta mRNA, TSH, and thyroid hormones. *Gen. Comp. Endocrinol.* 171, 367–372.

Lawson, P. W., Logerwell, E. A., Mantua, N. J., Francis, R. C. and Agostini, V. N. (2004). Environmental factors influencing freshwater survival and smolt production in Pacific Northwest coho salmon (*Oncorhynchus kisutch*). *Can. J. Fish. Aquat. Sci.* 61, 360–373.

Lema, S. C. and Nevitt, G. A. (2004). Evidence that thyroid hormone induces olfactory cellular proliferation in salmon during a sensitive period period for imprinting. *J. Exp. Biol.* 207, 3317–3327.

Leonard, J. B. K. and McCormick, S. D. (2001). Metabolic enzyme activity during smolting in stream- and hatchery-reared Atlantic salmon (*Salmo salar*). *Can. J. Fish. Aquat. Sci.* 58, 1585–1593.

Letunic, I. and Bork, P. (2011). Interactive Tree of Life v2: online annotation and display of phylogenetic trees made easy. *Nuc. Acid Res.* 39, W475–W478.

Levings, C. D. (1994). Feeding behaviour of juvenile salmon and significance of habitat during estuary and early sea phase. *Nord. J. Fresh. Res.* 69, 7–16.

Li, W. M., Sorensen, P. W. and Gallaher, D. D. (1995). The olfactory system of migratory adult sea lamprey (*Petromyzon marinus*) is specifically and acutely sensitive to unique bile acids released by conspecific larvae. *J. Gen. Physiol.* 105, 569–587.

Loge, F. J., Arkoosh, M. R., Ginn, T. R., Johnson, L. L. and Collier, T. K. (2005). Impact of environmental stressors on the dynamics of disease transmission. *Environ. Sci. Technol.* 39, 7329–7336.

Loretz, C. A., Collie, N. L., Richman, N. H. and Bern, H. A. (1982). Osmoregulatory changes accompanying smoltification in coho salmon. *Aquaculture* 28, 67–74.

Lorz, H. W. and McPherson, B. P. (1976). Effects of copper or zinc in fresh water on the adaptation to sea water and ATPase activity, and the effects of copper on migratory disposition of coho salmon (*Oncorhynchus kisutch*). *J. Fish. Res. Board Can.* 33, 2023–2030.

Lubin, R. T., Rourke, A. W. and Bradley, T. M. (1989). Ultrasturctural alteration in branchial chloride cells of Atlantic salmon, *Salmo salar*, during the parr-smolt transformation and early development in sea water. *J. Fish Biol.* 34, 259–272.

Lundqvist, H., Mckinnell, S., Fangstam, H. and Berglund, I. (1994). The effect of time, size and sex on recapture rates and yield after river releases of *Salmo salar* smolts. *Aquaculture* 121, 245–257.

MacDonald, D. J. and McKeown, B. A. (1983). The effect of Ca^{2+} levels on *in vitro* prolactin release from the rostral pars distalis of coho salmon (*Oncorhynchus kisutch*). *Can. J. Zool.* 61, 682–684.

McCormick, S. D. (1994). Ontogeny and evolution of salinity tolerance in anadromous salmonids: Hormones and heterochrony. *Estuaries* 17, 26–33.

McCormick, S. D. (2001). Endocrine control of osmoregulation in teleost fish. *Am. Zool.* 41, 781–794.

McCormick, S. D., Björnsson, B. Th., Sheridan, M., Eilertson, C., Carey, J. B. and O'Dea, M. (1995). Increased daylength stimulates plasma growth hormone and gill Na^+,K^+-ATPase in Atlantic salmon (*Salmo salar*). *J. Comp. Physiol.* 165, 245–254.

McCormick, S. D., Dickhoff, W. W., Duston, J., Nishioka, R. S. and Bern, H. A. (1991). Developmental differences in the responsiveness of gill Na^+,K^+-ATPase to cortisol in salmonids. *Gen. Comp. Endocrinol.* 84, 308–317.

McCormick, S. D., Hansen, L. P., Quinn, T. P. and Saunders, R. L. (1998). Movement, migration, and smolting of Atlantic salmon (*Salmo salar*). *Can. J. Fish. Aquat. Sci.* 55, 77–92.

McCormick, S. D., Lerner, D. T., Monette, M. Y., Nieves-Puigdoller, K., Kelly, J. T. and Björnsson, B. Th. (2009a). Taking it with you when you go: How perturbations to the freshwater environment, including temperature, dams, and contaminants, affect marine survival of salmon. *Am. Fish. Soc. Symp.* 69, 195–214.

McCormick, S. D., Moriyama, S. and Björnsson, B. Th. (2000). Low temperature limits photoperiod control of smolting in Atlantic salmon through endocrine mechanisms. *Am. J. Physiol. – Reg. Integr. Comp. Physiol.* 278, R1352–R1361.

McCormick, S. D. and Naiman, R. J. (1984). Osmoregulation in the brook trout, *Salvelinus fontinalis*. II. Effects of size, age and photoperiod on seawater survival and ionic regulation. *Comp. Biochem. Physiol.* 79A, 17–28.

McCormick, S. D., O'Dea, M. F., Moeckel, A. M. and Björnsson, B. Th. (2003). Endocrine and physiological changes in Atlantic salmon smolts following hatchery release. *Aquaculture* 222, 45–57.

McCormick, S. D., Regish, A., O'Dea, M. F. and Shrimpton, J. M. (2008). Are we missing a mineralocorticoid in teleost fish? Effects of cortisol, deoxycorticosterone and aldosterone on osmoregulation, gill Na+,K+-ATPase activity and isoform mRNA levels in Atlantic salmon. *Gen. Comp. Endocrinol.* 157, 35–40.

McCormick, S. D., Regish, A. M. and Christensen, A. K. (2009b). Distinct freshwater and seawater isoforms of Na+/K+-ATPase in gill chloride cells of Atlantic salmon. *J. Exp. Biol.* 212, 3994–4001.

McCormick, S. D., Regish, A. M., Christensen, A. K., and Björnsson, B. Th. (2013a). Differential regulation of sodium-potassium pump isoforms during smolt development and seawater exposure of Atlantic salmon. *J. Exp. Biol.* in press.

McCormick, S. D. and Saunders, R. L. (1987). Preparatory physiological adaptations for marine life in salmonids: osmoregulation, growth and metabolism. *Am. Fish. Soc. Symp.* 1, 211–229.

McCormick, S. D., Saunders, R. L. and MacIntyre, A. D. (1989). Mitochondrial enzyme activity and ion regulation during parr-smolt transformation of Atlantic salmon (*Salmo salar*). *Fish Physiol. Biochem.* 6, 231–241.

McCormick, S. D., Sheehan, T., Björnsson, B. Th., Lipsky, C., Kocik, J., Regish, A. M. and O'Dea, M. F. (2013b). Physiological and endocrine changes in Atlantic salmon smolts during hatchery rearing, downstream migration and ocean entry. *Can. J. Fish. Aquat. Sci.* in press.

McCormick, S. D., Shrimpton, J. M., Moriyama, S. and Björnsson, B. Th. (2002). Effects of an advanced temperature cycle on smolt development and endocrinology indicate that temperature is not a zeitgeber for smolting in Atlantic salmon. *J. Exp. Biol.* 205, 3553–3560.

McCormick, S. D., Shrimpton, J. M., Moriyama, S. and Björnsson, B. Th. (2007). Differential hormonal responses of Atlantic salmon parr and smolt to increased daylength: A possible developmental basis for smolting. *Aquaculture* 273, 337–344.

McCormick, S. D., Shrimpton, J. M. and Zydlewski, J. D. (1997). Temperature effects on osmoregulatory physiology of juvenile anadromous fish. In *Global Warming: Implications for Freshwater and Marine Fish* (eds. C. M. Wood and D. G. Mcdonald), pp. 279–301. Cambridge: Cambridge University Press.

Mcginnity, P., de Eyto, E., Cross, T. F., Coughlan, J., Whelan, K. and Ferguson, A. (2007). Population specific smolt development, migration and maturity schedules in Atlantic salmon in a natural river environment. *Aquaculture* 273, 257–268.

Miyakoshi, Y., Nagata, M. and Kitada, S. (2001). Effect of smolt size on postrelease survival of hatchery-reared masu salmon *Oncorhynchus masou. Fish. Sci.* 67, 134–137.

Mizuno, S., Misaka, N., Ando, D. and Kitamura, T. (2004). Quantitative changes of black pigmentation in the dorsal fin margin during smoltification in masu salmon, *Oncorhynchus masou. Aquaculture* 229, 433–450.

Mizuno, S., Misaka, N., Ando, D., Torao, M., Urabe, H. and Kitamura, T. (2007). Effects of diets supplemented with iron citrate on some physiological parameters and on burst swimming velocity in smoltifying hatchery-reared masu salmon (*Oncorhynchus masou*). *Aquaculture* 273, 284–297.

Mizuno, S., Ura, K., Onodera, Y., Fukada, H., Misaka, N., Hara, A., Adachi, S. and Yamauchi, K. (2001). Changes in transcript levels of gill cortisol receptor during smoltification in wild masu salmon, *Oncorhynchus masou. Zool. Sci.* 18, 853–860.

Monette, M. Y. and McCormick, S. D. (2008). Impacts of short-term acid and aluminum exposure on Atlantic salmon (*Salmo salar*) physiology: a direct comparison of parr and smolts. *Aquat. Toxicol.* 86, 216–226.

Morin, P. P., Andersen, O., Haug, E. and Doving, K. B. (1994). Changes in serum free thyroxine, prolactin, and olfactory activity during induced smoltification in Atlantic salmon (*Salmo salar*). *Can. J. Fish. Aquat. Sci.* 51, 1985–1992.

Morin, P. P. and Doving, K. B. (1992). Changes in the olfactory function of Atlantic salmon, *Salmo salar*, in the course of smoltification. *Can. J. Fish. Aquat. Sci.* 49, 1704–1713.

Morin, P. P., Hara, T. J. and Eales, J. G. (1995). T-4 depresses olfactory responses to L-alanine and plasma T-3 and T-3 production in smoltifying Atlantic salmon. *Am. J. Physiol – Regul. Integr. C* 38, R1434–R1440.

Moriyama, S., Ito, T., Takahashi, A., Amano, M., Sower, S. A., Hirano, T., Yamamori, K. and Kawauchi, H. (2002). A homolog of mammalian PRL-releasing peptide (fish arginyl-phenylalanyl-amide peptide) is a major hypothalamic peptide of PRL release in teleost fish. *Endocrinol.* 143, 2071–2079.

Nevitt, G. A., Dittman, A. H., Quinn, T. P. and Moody, W. J. (1994). Evidence for a peripheral olfactory memory in imprinted salmon. *Proc. Natl Acad. Sci. U. S. A.* 91, 4288–4292.

Nielsen, C., Madsen, S. S. and Björnsson, B. Th. (1999). Changes in branchial and intestinal osmoregulatory mechanisms and growth hormone levels during smolting in hatchery-reared and wild brown trout. *J. Fish Biol.* 54, 799–818.

Nilsen, T. O., Ebbesson, L. E., Stefansson, S. O., Madsen, S. S., McCormick, S. D., Björnsson, B. Th. and Prunet, P. (2007). Differential expression of gill Na$^+$,K$^+$-ATPase α- and β-subunits, Na$^+$,K$^+$,2Cl$^-$ cotransporter and CFTR anion channel in juvenile anadromous and landlocked Atlantic salmon *Salmo salar. J. Exp. Biol.* 210, 2885–2896.

Nilsen, T. O., Ebbesson, L. O. E., Kiilerich, P., Björnsson, B. Th., Madsen, S. S., McCormick, S. D. and Stefansson, S. O. (2008). Endocrine systems in juvenile anadromous and landlocked Atlantic salmon (*Salmo salar*): Seasonal development and seawater acclimation. *Gen. Comp. Endocrinol.* 155, 762–772.

Nishioka, R. S., Bern, H. A., Lai, K. V., Nagahama, Y. and Grau, E. G. (1982). Changes in the endocrine organs of coho salmon during normal and abnormal smoltification – an electron-microscopy study. *Aquaculture* 28, 21–38.

Nordeng, H. (1977). A pheromone hypothesis for homeward migration in anadromous salmonids. *Oikos* 28, 155–159.

Ogata, H. Y., Konno, S. and Silverstein, J. T. (1998). Muscular buffering capacity of the parr and smolts in *Oncorhynchus masou. Aquaculture* 168, 303–310.

Ojima, D. and Iwata, M. (2007a). Seasonal changes in plasma thyroxine kinetics in coho salmon *Oncorhynchus kisutch* during smoltification. *Aquaculture* 273, 329–336.

Ojima, D. and Iwata, M. (2007b). The relationship between thyroxine surge and onset of downstream migration in chum salmon *Oncorhynchus keta* fry. *Aquaculture* 273, 185–193.

Ojima, D. and Iwata, M. (2009). Central administration of growth hormone-releasing hormone triggers downstream movement and schooling behavior of chum salmon (*Oncorhynchus keta*) fry in an artificial stream. *Comp. Biochem. Physiol. A – Molecul. Integr. Physiol.* 152, 293–298.

Ojima, D. and Iwata, M. (2010). Central administration of growth hormone-releasing hormone and corticotropin-releasing hormone stimulate downstream movement and thyroxine secretion in fall-smolting coho salmon (*Oncorhynchus kisutch*). *Gen. Comp. Endocrinol.* 168, 82–87.

Orciari, R. D. and Leonard, G. H. (1996). Length characteristics of smolts and timing of downstream migration among three strains of Atlantic salmon in a southern New England stream. *N. Am. J. Fish. Manag.* 16, 851–860.

Parhar, I. S. and Iwata, M. (1996). Intracerebral expression of gonadotropin-releasing hormone and growth hormone-releasing hormone is delayed until smoltification in the salmon. *Neurosci. Res.* 26, 299–308.

Parry, G. (1960). The development of salinity tolerance in the salmon, *Salmo salar* (L.) and some related species. *J. Exp. Biol.* 37, 425–434.

Patino, R., Schreck, C. B. and Redding, J. M. (1985). Clearance of plasma corticostreroids during smoltification of coho salmon, *Oncorhynchus kisutch. Comp. Biochem. Physiol. [A]* 82A, 531–535.

Peake, S. and McKinley, R. S. (1998). A re-evaluation of swimming performance in juvenile salmonids relative to downstream migration. *Can. J. Fish. Aquat. Sci.* 55, 682–687.

Pelis, R. M. and McCormick, S. D. (2001). Effects of growth hormone and cortisol on Na$^+$K$^+$2Cl$^-$ cotransporter localization and abundance in the gills of Atlantic salmon. *Gen. Comp. Endocrinol.* 124, 134–143.

Pelis, R. M. and McCormick, S. D. (2003). Fin development in stream- and hatchery-reared Atlantic salmon. *Aquaculture* 220, 525–536.

Pelis, R. M., Zydlewski, J. and McCormick, S. D. (2001). Gill $Na^+K^+2Cl^-$ cotransporter abundance and location in Atlantic salmon: effects of seawater and smolting. *Am. J. Physiol. – Reg. Integr. & Comp. Physiol.* 280, R1844–R1852.

Pisam, M., Prunet, P., Boeuf, G. and Rambourg, A. (1988). Ultrastructural features of chloride cells in the gill epithelium of the Atlantic salmon, *Salmo salar*, and their modifications during smoltification. *Am. J. Anat.* 183, 235–244.

Price, C. S. and Schreck, C. B. (2003). Stress and saltwater-entry behavior of juvenile chinook salmon (*Oncorhynchus tshawytscha*): conflicts in physiological motivation. *Can. J. Fish. Aquat. Sci.* 60, 910–918.

Primmett, D. R. N., Eddy, F. B., Miles, M. S., Talbot, C. and Thorpe, J. E. (1988). Transepithelial ion exhange in smolting Atlantic salmon (*Salmo salar* L.). *Fish Physiol. Biochem.* 5, 181–186.

Prunet, P., Boeuf, G., Bolton, J. P. and Young, G. (1989). Smoltification and seawater adaptation in Atlantic salmon (*Salmo salar*): plasma prolactin, growth hormone, and thyroid hormones. *Gen. Comp. Endocrinol.* 74, 355–364.

Quinn, M. C. J., Veillette, P. A. and Young, G. (2003). Pseudobranch and gill Na+, K +-ATPase activity in juvenile chinook salmon, *Oncorhynchus tshawytscha*: developmental changes and effects of growth hormone, cortisol and seawater transfer. *Comp. Biochem. Physiol. A, Mol. Integr. Physiol.* 135, 249–262.

Quinn, T. P., Brannon, E. L. and Dittman, A. H. (1989). Spatial aspects of imprinting and homing in coho salmon. *Fish. Bull.* 87, 769–774.

Rechisky, E. L., Welch, D. W., Porter, A. D., Jacobs, M. C. and Ladouceur, A. (2009). Experimental measurement of hydrosystem-induced delayed mortality in juvenile Snake River spring Chinook salmon (*Oncorhynchus tshawytscha*) using a large-scale acoustic array. *Can. J. Fish. Aquat. Sci.* 66, 1019–1024.

Redding, J. M., Patino, R. and Schreck, C. B. (1984). Clearance of corticosteroids in yearling coho salmon, *Oncorhynchus kisutch*, in fresh water and seawater and after stress. *Gen. Comp. Endocrinol.* 54, 433–443.

Richards, J. G., Semple, J. W., Bystriansky, J. S. and Schulte, P. M. (2003). Na+/K+-ATPase α-isoform switching in gills of rainbow trout (*Oncorhynchus mykiss*) during salinity transfer. *J. Exp. Biol.* 206, 4475–4486.

Richman, N. H., Tai de Diaz, S., Nishioka, R. S., Prunet, P. and Bern, H. A. (1987). Osmoregulatory and endocrine relationships with chloride cell morphology and density during smoltification in coho salmon (*Oncorhynchus kisutch*). *Aquaculture* 60, 265–285.

Rimmer, D. M., Paim, U. and Saunders, R. L. (1983). Autumnal habitat shift in juvenile Atlantic salmon (*Salmo salar*) in a small river. *Can. J. Fish. Aquat. Sci.* 40, 671–680.

Robertson, L. S. and McCormick, S. D. (2012). Transcriptional profiling of the parr-smolt transformation in Atlantic salmon. *Comp. Biochem. Physiol. A – Molecul. Integr. Physiol.* 7, 351–360.

Rosseland, B. O., Kroglund, F., Staurnes, M., Hindar, K. and Kvellestad, A. (2001). Tolerance to acid water among strains and life stages of Atlantic salmon (*Salmo salar* L.). *Water Air Soil. Pollut.* 130, 899–904.

Rounsefell, G. A. (1958). Anadromy in North American salmonidae. *Fish. Bull.* 58, 171–185.

Sakamoto, T., Hirano, T., Madsen, S. S., Nishioka, R. S. and Bern, H. A. (1995). Insulin-like growth factor I gene expression during parr-smolt transformation of coho salmon. *Zool. Sci.* 12, 249–252.

Saloniemi, I., Jokikokko, E., Kallio-Nyberg, I., Jutila, E. and Pasanen, P. (2004). Survival of reared and wild Atlantic salmon smolts: size matters more in bad years. *ICES J. Mar. Sci.* 61, 782–787.

Schreck, C. B., Solazzi, M. F., Johnson, S. L. and Nickelson, T. E. (1989). Transportation stress affects performance of coho salmon, *Oncorhynchus kisutch. Aquaculture* 82, 15–20.

Schreck, C. B., Stahl, T. P., Davis, L. E., Roby, D. D. and Clemens, B. J. (2006). Mortality estimates of juvenile spring-summer Chinook salmon in the Lower Columbia River and estuary, 1992–1998: Evidence for delayed mortality? *Trans. Am. Fish. Soc.* 135, 457–475.

Seidelin, M. and Madsen, S. S. (1997). Prolactin antagonizes the seawater-adaptive effect of cortisol and growth hormone in anadromous brown trout (*Salmo trutta*). *Zool. Sci.* 14, 249–256.

Seidelin, M. and Madsen, S. S. (1999). Endocrine control of Na+,K+-ATPase and chloride cell development in brown trout (*Salmo trutta*): interaction of insulin-like growth factor-I with prolactin and growth hormone. *J. Endocr.* 162, 127–135.

Sheridan, M. (1986). Effects of thyroxine, cortisol, growth hormone, and prolactin on lipid metabolism of coho salmon, *Oncorhynchus kisutch*, during smoltification. *Gen. Comp. Endocrinol.* 64, 220–238.

Shimizu, M., Kudo, H., Ueda, H., Hara, A., Shimazaki, K. and Yamauchi, K. (1993). Identification and immunological properties of an olfactory system-specific protein in kokanee salmon (*Oncorhynchus nerka*). *Zool. Sci.* 10, 287–294.

Shoji, T., Ueda, H., Ohgami, T., Sakamoto, T., Katsuragi, Y., Yamauchi, K. and Kurihara, K. (2000). Amino acids dissolved in stream water as possible home stream odorants for masu salmon. *Chem. Sens.* 25, 533–540.

Shoji, T., Yamamoto, Y., Nishikawa, D., Kurihara, K. and Ueda, H. (2003). Amino acids in stream water are essential for salmon homing and migration. *Fish Physiol. Biochem.* 28, 249–251.

Shrimpton, J. M., Bernier, N. J., Iwama, G. K. and Randall, D. J. (1994). Differences in measurements of smolt development between wild and hatchery-reared juvenile coho salmon (*Oncorhynchus kisutch*) before and after saltwater exposure. *Can. J. Fish. Aquat. Sci.* 51, 2170–2178.

Shrimpton, J. M., Devlin, R. H., Mclean, E., Byatt, J. C., Donaldson, E. M. and Randall, D. J. (1995). Increases in gill corticosteroid receptor abundance and saltwater tolerance in juvenile coho salmon (*Oncorhynchus kisutch*) treated with growth hormone and placental lactogen. *Gen. Comp. Endocrinol.* 98, 1–15.

Shrimpton, J. M. and McCormick, S. D. (1998a). Regulation of gill cytosolic corticosteroid receptors in juvenile Atlantic salmon: Interaction effects of growth hormone with prolactin and triiodothyronine. *Gen. Comp. Endocrinol.* 112, 262–274.

Shrimpton, J. M. and McCormick, S. D. (1998b). Seasonal differences in plasma cortisol and gill corticosteroid receptors in upper and lower mode juvenile Atlantic salmon. *Aquaculture* 168, 205–219.

Shrimpton, J. M. and Randall, D. J. (1994). Downregulation of corticosteroid receptors in gills of coho salmon due to stress and cortisol treatment. *Am. J. Physiol.* 267, R432–R438.

Sigholt, T., Asgard, T. and Staurnes, M. (1998). Timing of parr-smolt transformation in Atlantic salmon (*Salmo salar*): effects of changes in temperature and photoperiod. *Aquaculture* 160, 129–144.

Singer, T. D., Clements, K. M., Semple, J. W., Schulte, P. M., Bystriansky, J. S., Finstad, B., Fleming, I. A. and McKinley, R. S. (2002). Seawater tolerance and gene expression in two strains of Atlantic salmon smolts. *Can. J. Fish. Aquat. Sci.* 59, 125–135.

Solbakken, V. A., Hansen, T. and Stefansson, S. O. (1994). Effects of photoperiod and temperature on growth and parr-smolt transformation in Atlantic salmon (*Salmo salar* L) and subsequent performance in seawater. *Aquaculture* 121, 13–27.

Specker, J. L. (1982). Interrenal function and smoltification. *Aquaculture* 28, 59–66.

Specker, J. L., Brown, C. L. and Bern, H. A. (1992). Asynchrony of changes in tissue and plasma thyroid hormones during the parr-smolt transformation of coho salmon. *Gen. Comp. Endocrinol.* 88, 397–405.

Specker, J. L., Distefano, J. J., Grau, E. G., Nishioka, R. S. and Bern, H. A. (1984). Development-associated changes in thyroxine kinetics in juvenile salmon. *Endocrinol.* 115, 399–406.

Specker, J. L. and Schreck, C. B. (1982). Changes in plasma corticosteroids during smoltification of coho salmon, *Oncorhynchus kisutch. Gen. Comp. Endocrinol.* 46, 53–58.

Specker, J. L. and Schreck, C. B. (1984). Thyroidal response to mammalian thyrotropin during smoltification of juvenile coho salmon (*Oncorhynchus kisutch*). *Comp. Biochem. Physiol.* 78, 441–444.

Staurnes, M., Lysfjord, G., Hansen, L. P. and Heggberget, T. G. (1993). Recapture rates of hatchery-reared Atlantic salmon (*Salmo salar*) related to smolt development and time of release. *Aquaculture* 118, 327–337.

Stefansson, S. O., Björnsson, B. Th., Ebbesson, L. O. E. and McCormick, S. D. (2008). Smoltification. In *Fish Larval Physiology.* (eds. R. N. Finn and B. G. Kapoon), pp. 639–681. Enfield, NH, USA: Science Publishers.

Stefansson, S. O., Björnsson, B. Th., Sundell, K., Nyhammer, G. and McCormick, S. D. (2003). Physiological characteristics of wild Atlantic salmon post-smolts during estuarine and coastal migration. *J. Fish Biol.* 63, 942–955.

Stefansson, S. O., Haugland, M., Björnsson, B. Th., McCormick, S. D., Holm, M., Ebbesson, L. O. E., Holst, J. C. and Nilsen, T. O. (2012). Growth, osmoregulation and endocrine changes in wild Atlantic salmon smolts and post-smolts during marine migration. *Aquaculture* 363, 127–136.

Stefansson, S. O., Nilsen, T. O., Ebbesson, L. O. E., Wargelius, A., Madsen, S. S., Björnsson, B. Th. and McCormick, S. D. (2007). Molecular mechanisms of continuous light inhibition of Atlantic salmon parr-smolt transformation. *Aquaculture* 273, 235–245.

Sundell, K., Jutfelt, F., Agustsson, T., Olsen, R. E., Sandblom, E., Hansen, T. and Björnsson, B. Th. (2003). Intestinal transport mechanisms and plasma cortisol levels during normal and out-of-season parr-smolt transformation of Atlantic salmon, *Salmo salar. Aquaculture* 222, 265–285.

Sykes, G. E., Johnson, C. J. and Shrimpton, J. M. (2009). Temperature and flow effects on migration timing of chinook salmon smolts. *Trans. Am. Fish. Soc.* 138, 1252–1265.

Sykes, G. E. and Shrimpton, J. M. (2010). Effect of temperature and current manipulation on smolting in Chinook salmon (*Oncorhynchus tshawytscha*): the relationship between migratory behaviour and physiological development. *Can. J. Fish. Aquat. Sci.* 67, 191–201.

Takei, Y. and McCormick, S. D. (2013). Hormonal control of fish euryhalinity. In *Fish Physiology,* Vol. 32, *Euryhaline Fishes* (eds. S. D. McCormick, C. J. Brauner and A. P. Farrell), pp. 69–124. Amsterdam: Academic Press.

Taylor, E. B. (1990). Phenotypic correlates of life-history variation in juvenile chinook salmon, *Oncorhynchus tshawytscha. J. Anim. Ecol.* 59, 455–468.

Thorpe, J. E. (1984). Downstream movements of juvenile salmonids: a forward speculative view. In *Mechanisms of migration in fishes* (eds. J. D. McCleave, G. P. Arnold, J. J. Dodson and W. H. Neill), pp. 387–396. New York: Plenum Press.

Thorpe, J. E. (1994a). An alternative view of smolting in salmonids. *Aquaculture* 121, 105–113.

Thorpe, J. E. (1994b). Salmonid fishes and the estuarine environment. *Estuaries* 17, 76–93.

Thorpe, J. E. and Morgan, R. I. G. (1978). Periodicity in Atlantic salmon, *Salmo salar*, smolt migration. *J. Fish Biol.* 12, 541–548.

Thorstad, E. B., Whoriskey, F., Uglem, I., Moore, A., Rikardsen, A. H. and Finstad, B. (2012). A critical life stage of the Atlantic salmon *Salmo salar*: behaviour and survival during the smolt and initial post-smolt migration. *J. Fish Biol.* 81, 500–542.

Tipsmark, C. K., Jorgensen, C., Brande-Lavridsen, N., Engelund, M., Olesen, J. H. and Madsen, S. S. (2009). Effects of cortisol, growth hormone and prolactin on gill claudin expression in Atlantic salmon. *Gen. Comp. Endocrinol.* 163, 270–277.

Tipsmark, C. K., Kiilerich, P., Nilsen, T. O., Ebbesson, L. O. E., Stefansson, S. O. and Madsen, S. S. (2008). Branchial expression patterns of claudin isoforms in Atlantic salmon during seawater acclimation and smoltification. *Am. J. Physiol. – Reg. Integr. Comp. Physiol.* 294, R1563–R1574.

Tipsmark, C. K. and Madsen, S. S. (2009). Distinct hormonal regulation of Na+,K+-ATPase genes in the gill of Atlantic salmon (*Salmo salar* L.). *J. Endocrinol.* 203, 301–310.

Tipsmark, C. K., Mahmmoud, Y. A., Borski, R. J. and Madsen, S. S. (2010a). FXYD-11 associates with Na+K+ATPase in the gill of Atlantic salmon: regulation and localization in relation to changed ion-regulatory status. *Am. J. Physiol. – Reg. Integr. Comp. Physiol.* 299, R1212–R1223.

Tipsmark, C. K., Sorensen, K. J., Hulgard, K. and Madsen, S. S. (2010b). Claudin-15 and -25b expression in the intestinal tract of Atlantic salmon in response to seawater acclimation, smoltification and hormone treatment. *Comp. Biochem. Physiol. A – Mol. Integr. Physiol.* 155, 361–370.

Tipsmark, C. K., Sorensen, K. J. and Madsen, S. S. (2010c). Aquaporin expression dynamics in osmoregulatory tissues of Atlantic salmon during smoltification and seawater acclimation. *J. Exp. Biol.* 213, 368–379.

Tucker, S., Trudel, M. and Welch, D. W. (2012). Annual coastal migration of juvenile Chinook salmon: static stock-specific patterns in a highly dynamic ocean. *Mar. Ecol. Prog. Ser.* 449, 245–284.

Uchida, K., Kaneko, T., Tagawa, M. and Hirano, T. (1998). Localization of cortisol receptor in branchial chloride cells in chum salmon fry. *Gen. Comp. Endocrinol.* 109, 175–185.

Ueda, H. (2011). Physiological mechanism of homing migration in Pacific salmon from behavioral to molecular biological approaches. *Gen. Comp. Endocrinol.* 170, 222–232.

Veillette, P. A., Sundell, K. and Specker, J. L. (1995). Cortisol mediates the increase in intestinal fluid absorption in Atlantic salmon during parr smolt transformation. *Gen. Comp. Endocrinol.* 97, 250–258.

Veillette, P. A., White, R. J. and Specker, J. L. (1993). Changes in intestinal fluid transport in Atlantic salmon (*Salmo salar* L) during parr-smolt transformation. *Fish. Physiol. Biochem.* 12, 193–202.

Veillette, P. A. and Young, G. (2004). Temporal changes in intestinal Na+, K+-ATPase activity and in vitro responsiveness to cortisol in juvenile chinook salmon. *Comp. Biochem. Physiol. A – Mol. Integr. Physiol.* 138, 297–303.

Virtanen, E., Soderholmtana, L., Soivio, A., Forsman, L. and Muona, M. (1991). Effect of physiological condition and smoltification status at smolt release on subsequent catches of adult salmon. *Aquaculture* 97, 231–257.

Wagner, H. H. (1974). Photoperiod and temperature regulation of smolting in steelhead trout. *Can. J. Zool.* 52, 219–234.

Wagner, H. H., Conte, F. P. and Fessler, J. L. (1969). Development of osmotic and ionic regulation in two races of chinook salmon *O. tshawytscha*. *Comp. Biochem. Physiol.* 29, 325–341.

Welch, D. W., Melnychuk, M. C., Payne, J. C., Rechisky, E. L., Porter, A. D., Jackson, G. D., Ward, B. R., Vincent, S. P., Wood, C. C. and Semmens, J. (2011). In situ measurement of coastal ocean movements and survival of juvenile Pacific salmon. *PNAS* 108, 8708–8713.

Wilson, M. V. H. and Williams, R. R. G. (2010). Salmoniform fishes: key fossils, supertree, and possible morphological synapomorphies. In *Origin and Phylogenetic Interrelationships of Teleosts* (eds. J. S. Nelson, H.-P. Schultze and M. V.H. Wilson), pp. 379–409. Munchen: Verlag Dr. Friedrich Pfeil.

Wood, A. W., Duan, C. M. and Bern, H. A. (2005). Insulin-like growth factor signaling in fish. *Int. Rev. Cytol.* 243, 215–285.

Yada, T., Takahashi, K. and Hirano, T. (1991). Seasonal changes in seawater adaptability and plasma levels of prolactin and growth hormone in landlocked sockeye salmon (*Oncorhynchus nerka*) and amago salmon (*O. rhodurus*). *Gen. Comp. Endocrinol.* 82, 33–44.

Yamamoto, Y., Hino, H. and Ueda, H. (2010). Olfactory imprinting of amino acids in lacustrine sockeye salmon. *PLOS ONE* 5, e8633.

Yeoh, C.-G., Kerstetter, T. H. and Loudenslager, E. J. (1991). Twenty-four hour seawater challenge test for coastal cutthroat trout. *Progr. Fish-Culturist* 53, 173–176.

Young, G. (1988). Enhanced response of the interrenal of coho salmon (*Oncorhynchus kisutch*) to ACTH after growth hormone treatment in vivo and in vitro. *Gen. Comp. Endocrinol.* 71, 85–92.

Young, G. (1989). Cortisol secretion in vitro by the interrenal of coho salmon (*Oncorhynchus kisutch*) during smoltification: relationship with plasma thyroxine and plasma cortisol. *Gen. Comp. Endocrinol.* 63, 191–200.

Young, G., Björnsson, B. Th., Prunet, P., Lin, R. J. and Bern, H. A. (1989). Smoltification and seawater adaptation in coho salmon (*Oncorhynchus kisutch*): plasma prolactin, growth hormone, thyroid hormones, and cortisol. *Gen. Comp. Endocrinol.* 74, 335–345.

Youngson, A. F., Hansen, L. P., Jonsson, B. and Naesje, T. F. (1989). Effects of exogenous thyroxine or prior exposure to raised water-flow on the downstream movement of hatchery-reared Atlantic salmon smolts. *J. Fish Biol.* 34, 791–797.

Zaugg, W. S. and McLain, L. R. (1970). Adenosine triphosphatase activity in gills of salmonids: seasonal variations and salt water influences in coho salmon, *Oncorhynchus kisutch. Comp. Biochem. Physiol.* 35, 587–596.

Zaugg, W. S. and McLain, L. R. (1976). Influence of water temperature on gill sodium, potassium-stimulated ATPase activity in juvenile coho salmon (*Oncorhynchus kisutch*). *Comp. Biochem. Physiol.* 54A, 419–421.

Zydlewski, G. B., Haro, A. and McCormick, S. D. (2005). Evidence for cumulative temperature as an initiating and terminating factor in downstream migratory behavior of Atlantic salmon (*Salmo salar*) smolts. *Can. J. Fish. Aquat. Sci.* 62, 68–78.

6

FRESHWATER TO SEAWATER TRANSITIONS IN MIGRATORY FISHES

JOSEPH ZYDLEWSKI
MICHAEL P. WILKIE

The transition from freshwater to seawater is integral to the life history of many fishes. Diverse migratory fishes express anadromous, catadromous, and amphidromous life histories, while others make incomplete transits between freshwater and seawater. The physiological mechanisms of osmoregulation are widely conserved among phylogenetically diverse species. Diadromous fishes moving between freshwater and seawater develop osmoregulatory mechanisms for different environmental salinities. Freshwater to seawater transition involves hormonally mediated changes in gill ionocytes and the transport proteins associated with hypoosmoregulation, increased seawater ingestion and water absorption in the intestine,

Euryhaline Fishes: Volume 32
FISH PHYSIOLOGY
Copyright © 2013 Elsevier Inc. All rights reserved
DOI: http://dx.doi.org/10.1016/B978-0-12-396951-4.00006-2

and reduced urinary water losses. Fishes attain salinity tolerance through early development, gradual acclimation, or environmentally or developmentally cued adaptations. This chapter describes adaptations in diverse taxa and the effects of salinity on growth. Identifying common strategies in diadromous fishes moving between freshwater and seawater will reveal the ecological and physiological basis for maintaining homeostasis in different salinities, and inform efforts to conserve and manage migratory euryhaline fishes.

1. INTRODUCTION

The understanding of euryhalinity in migratory fishes has been investigated over the past century in numerous studies, but the focus has mainly been on salmonid fishes (see McCormick, 2013, Chapter 5, this volume). In recent decades, however, more research has centered on other families including the lampreys, sturgeons, anguillid eels, herrings, and tilapias. Functional approaches to defining life history patterns (e.g. Elliott and Dewailly, 1995; Elliott et al., 2007; Franco et al., 2008) have effectively defined estuarine guilds based on use patterns to inform estuarine ecology. Estuaries provide critical habitat for many migratory species (McLusky and Elliott, 2004; Rountree and Able, 2007) and are relied upon for rearing, feeding, spawning, or simply serving as a corridor of migration (Elliott and Hemingway, 2002). This chapter will focus on the transition from freshwater (FW) to seawater (SW), which is an integral part of the life history of many migratory fishes. For most migratory fishes, such a transition generally punctuates longer residence times at more stable salinities.

Migrants between FW and SW are collectively termed "diadromous" and geographic patterns of diadromy have been well described (McDowall, 1987). General trends in occurrence are linked to productivity differences between FW and SW habitats (Gross, 1987; see also Shrimpton, 2013, Chapter 7, this volume). The success or failure of these migrants is largely dependent upon the timing of migration in the context of biotic and abiotic environments (McCormick et al., 1998; Limburg, 2001), size at migration (Saloniemi et al., 2004), and physiological preparation for changing environment (e.g. Zydlewski et al., 2003). While the estuary serves as a migratory corridor between the ocean and inland waters (Lobry et al., 2003), it can also serve as an important staging area for physiological acclimation (e.g. McCormick and Saunders, 1987), growth opportunity, and predator avoidance (Klemetsen et al., 2003; Lepage et al., 2005).

The idealized descriptions of anadromy, catadromy, and amphidromy remain useful and defining archetypes for most euryhaline migrations. These terms have considerable history (Myers, 1949; McDowall, 1987, 1988, 2007) and are widely applied (although the use of the term "amphidromy" continues to evolve). While they are useful heuristic constructs, there is a growing appreciation for the complexity of life histories. Many species exhibit divergent behaviors within a population – or an individual. These divergent movements can be, and have been argued to be, trophic movements rather than "true migrations" (Myers, 1949; Dingle, 1996) but some of the distinctions are difficult to determine empirically. The overview that follows is not intended to be exhaustive, or to define terms (as such contributions have been prominently noted). Rather, what follows is a sampling of the diversity of migratory patterns, with some consideration of those patterns that do not fit neatly into the three general patterns of diadromy. The physiological mechanistic and behavioral patterns that allow these fish to exploit the estuarine environment as part of their life history as they move from FW to SW are also discussed.

2. LIFE HISTORY PATTERNS

2.1. Anadromy

Anadromous fishes spawn in FW and the young (usually) remain in this dilute environment for some period before making a directed seaward migration. As adults, these fish return to FW to reproduce (Fig. 6.1). The period in SW is generally associated with accelerated growth opportunities and, as a result, greater fecundity. This group is exemplified by salmonines (McDowall, 1988; McCormick, 1994) but it is a conspicuous strategy across many species. Although phylogenetically distant, sea lampreys (*Petromyzon marinus*) have a distinct FW phase (ammocoete) that typically lasts for 3–7 years followed by a parasitic phase at sea, which is characterized by rapid growth as they ingest large quantities of blood from (mainly) fish (Hardisty and Potter, 1971; Beamish and Potter, 1975). Juvenile downstream migration is linked to a metamorphosis as the eyeless, suspension feeding, sediment-dwelling ammocoete transforms into a silvered, metamorph with distinct eyes and an adult-like oral disc (Youson and Potter, 1979). Seaward migration is bimodal, occurring in spring and fall (autumn), associated with flow events (Beamish and Potter, 1975; McCormick et al., 1997). Some sturgeon species have well-defined anadromous migrations, while others make multiple movements into the estuary habitat through their life

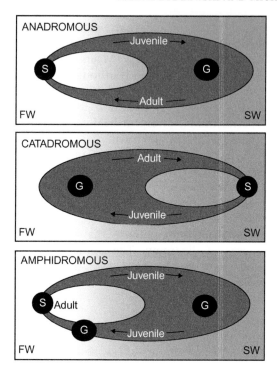

Fig. 6.1. Idealized patterns of diadromous migration. Life histories proceed in a clockwise fashion. Width of dark gray band represents variation in the salinity to which a given life stage may be exposed. Shading on right side of panels indicates increased salinity. G: periods of growth; S: spawning.

history (Doroshov, 1985). Atlantic sturgeons (*Acipenser oxyrinchus*) migrate from the ocean into riverine habitat to spawn, and juvenile anadromous sturgeons gradually enter SW within several years (Rochard et al., 2001; Wilson and McKinley, 2004). Green sturgeons (*Acipenser medirostris*) enter SW earlier than other sturgeon species, within 1–3 years (Allen and Cech, 2007). Their eggs are relatively large and juveniles grow rapidly (Deng et al., 2002); these may be adaptations for early entry into the estuary.

Anadromous clupeids such as the American shad (*Alosa sapidissima*) undergo migrations from coastal rivers of North America (Leggett and Carscadden, 1978). Adults enter FW when river temperatures are between 14 and 20°C (Leggett and Whitney, 1972). Spawning occurs in open water beyond tidal influence, and the young generally remain in FW until autumn.

The downstream migration of juvenile American shad is linked to declining river temperatures (O'Leary and Kynard, 1986). The use of otolith microchemistry has greatly informed the complexity of life history diversity within this and other alosine species to demonstrate varied use of the estuary by juveniles (Limburg, 1998).

Moronids demonstrate great flexibility in their anadromous life history. For striped bass (*Morone saxatilis*), there are fully FW populations at the northern and southern extents of their range (Rulifson and Dadswell, 1995; Haeseker et al., 1996; Carmichael et al., 1998), although anadromous spawning migrations that ensure early development occurs in low salinities are the norm (Gemperline et al., 2002; Wingate and Secor, 2008). Similarly, white perch (*Mornone americana*) have considerable diversity in migratory patterns linked to the estuary (Hanks and Secor, 2011). These fish spawn in FW before returning to the estuary where they reside through the winter (Setzler-Hamilton, 1991; Kerr and Secor, 2009). Larvae are associated with the low-salinity zone of the estuarine turbidity maximum and young-of-the-year juveniles are subsequently found rearing in both FW and SW (North and Houde, 2001). Members of the same population can be either resident or migratory (Kerr et al., 2009), diverging in behavior only after transitioning into the juvenile stage (Kraus and Secor, 2004).

2.2. Catadromy

Catadromous fishes spawn in SW and move into FW for a period of growth (Fig. 6.1). While catadromy is more commonly associated with tropical latitudes, American (*Anguilla rostrata*), European (*A. anguilla*), and Japanese eels (*A. japonica*) exemplify the temperate pattern of catadromy (Lecomte-Finiger, 1983; Sorensen, 1984). These fish are spawned in tropical ocean waters and leptocephalus larvae are carried passively towards coastal areas (Beumer and Harrington, 1980; Tsukamoto et al., 2003) before invading the estuary as "glass eels". These juveniles acquire pigmentation and migrate upstream to varying extents as "yellow eels". This growth phase can be as long as 20 years before seaward migration as an adult "silver eel" (Hourdry, 1995).

Although conventional wisdom has assumed that the majority of eel juveniles pass through the estuary (Tesch, 1977; Feunteun et al., 2003; Fontaine et al., 1995), recent work demonstrates that many eels rear in elevated salinities (Morrison and Secour, 2003; Arai et al., 2006; Thibault et al., 2007; Jessop et al., 2008), making them facultative in their catadromy. Facultative FW entry is related to energy status (Edeline, 2007; Bureau du

Columbier et al., 2011), where slower growing eels migrate farther upstream into FW, perhaps to avoid competition with conspecifics. These FW migrants may comprise a large proportion of migrating adults (12–25%) (Jessop et al., 2004; Morrison and Secour, 2003). Juveniles that rear in the estuary for an extended period before recruiting into FW may also move back and forth from FW to the estuary as subadults (Jessop et al., 2002, 2006, 2008). Recent work using telemetry and Sr:Ca ratios (Daverat et al., 2006; Thibeault et al., 2007; Arai et al., 2009; Shrimpton, 2013, Chapter 7, this volume) has provided convincing evidence for this variation (Fig. 6.2).

Although less well characterized, non-anguillid fishes also express catadromous life histories. Mullets (Mugilidae) (Anderson, 1957; Nordlie, 2000; Cardona, 2006) and a few species of the Galaxiidae (Pollard, 1971) are considered catadromous. Some species of tropical and subtropical gobies and some sculpins may also be considered catadromous (McDowall, 2007), although they have also been classified as amphidromous.

2.3. Amphidromy

The use and application of "amphidromy" has been less precise than either "anadromy" or "catadromy". For anadromous and catadromous migrations, patterns are tightly linked with reproduction (Klemetsen et al.,

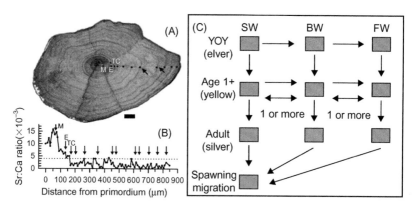

Fig. 6.2. Facultative catadromy in American eel. Using electron microprobe analysis of otoliths (A) Sr:Ca ratios along transects can be measured (B) to retrospectively infer movements between freshwater (FW) (low ratio) and seawater (SW) (high ratio). Arrows indicate onset of metamorphosis (M), elver phase (E), and habitat transition (TC). Based on such inferred histories (C) facultatively catadromous eels may employ a variety of contingent behaviors which exploit FW and estuarine habitats to different extents. BW: brackish water. Reproduced with permission from Jessop et al. (2008).

2003; Quinn and Myers, 2004). In contrast, amphidromy has been characterized by trophic movements between FW and SW such that there is a growth phase in each biome. "Freshwater" and "seawater" amphidromous patterns have been described (Myers, 1949; McDowall, 1987, 1988, 1992) and attributed to species such as shortnose sturgeon (*Acipenser brevirostrum*) (Bain, 1997; Bemis and Kynard, 1997) and tarpon (*Megalops* sp.) (Myers, 1949) based on early use of the term. As terminology has been sharpened, however, these species fit the definition more poorly. Here, the more precise definition of McDowall (2007) with specific criteria is employed. Amphidromous fishes migrate seaward as larvae, experience a brief growth phase in SW (or brackish water, BW), and return to FW as small juveniles (usually less than 50 mm). This definition restricts the term to what has previously been defined as "FW amphidromy" (the use of "FW" and "SW" amphidromy, as in McDowall, 1988, 1992, has been dropped). There is a clear distinction between amphidromy and other diadromous fishes. In general anadromous and catadromous fishes reduce or cease feeding altogether upon entry of the habitat in which they will spawn (McDowall, 2004), whereas amphidromous species have an extended secondary growth phase in FW which is followed by spawning (Keith, 2003) (Fig. 6.1).

The majority of amphidromous fishes are found in the southern hemisphere in the intertropical area. Particularly in island habitats, amphidromy can be the dominant life history form (McDowall, 1999), as exemplified by galaxiids (e.g. genus *Galaxias*). Spawning of amphidromous galaxiids takes place in FW, followed by the larval growth phase in SW (Waters et al., 2001) and then the return to rivers for the extended growth period (McDowall, 1990). Although this is the general pattern, larvae may be retained in the BW of the estuary before returning into FW river systems (David et al., 2004). Similarly, the red-tailed goby (indigenous to islands of the Indian Ocean and Pacific Ocean) spawns in FW followed by a short pelagic growth period in the ocean (Keith and Marquet, 2002; Keith, 2003). Amphidromous Sicydiine gobies enter into SW environment hours after hatching, whereupon they enter a growth phase lasting 3–6 months (Keith et al., 2008; Lord et al., 2010).

Migration back to FW habitats is generally associated with a striking metamorphosis (Nishimoto, 1996; Nishimoto and Kuamo'o, 1997; Balon, 1990; Schoenfuss et al., 1997) that includes changes in skull morphology including a shift from a subterminal to terminal mouth. Many gobies develop fused pelvic fins during the period of FW entry, allowing them to hang onto rocks and invade upstream habitat above waterfalls (Fitzsimons et al., 2002; Taillebois et al., 2011). *Lentipes concolor*, found in the Hawaiian Islands, is known to make impressive inland movements up waterfalls greater than 600 m in height (Nishimoto and Fitzsimons, 1986; Englund and Filbert, 1997).

In bullies (*Gobiomorphus* spp.) and galaxiids, the completion of an amphidromous life history can be facultative (Hicks et al., 2010). Larvae may successfully develop in FW lakes even where ocean access is possible (Closs et al., 2003; David et al., 2004; Hicks, 2012). The degree to which this indicates plasticity within an individual rather than variability within a population is unclear. Amphidromous species seem to be marked by lower maternal investment than resident relatives. These fish have smaller eggs and a less developed state at hatching, presumably adaptations for passive downstream drift. Congeneric species that are completely fluvial in their life histories hatch as larger, precocious young that can hold position in stream flows. Such differences are observed among gobies of the genus *Rhinogobius* and Japanese FW sculpin of the genus *Cottus* (Goto, 1990; Goto and Andoh, 1990; Iguchi and Mizuno, 1999).

2.4. Freshwater-Linked and Seawater-Linked Estuarine Movements

Rather than being linked to a distinct growth or reproductive phase, FW-linked estuarine migrants are generally opportunistic in their use of the estuary. These fish generally experience salinity transitions that are transient and usually low in salinity. Such patterns are best described as trophic movements rather than migrations between environments. There is great variation in osmotic tolerance among FW species. Ictalurid catfish and yellow perch (*Perca flavescens*) have extremely low tolerances to salinities above 1 ppt (Lutz, 1972; Furspan et al., 1984), yet other "stenohaline" FW species are commonly observed in estuaries (Lowe et al., 2009). Juvenile centrarchids are commonly found in the upper parts of estuaries (Hackney and de la Cruz, 1981; Rozas and Hackney, 1983; Rogers et al., 1984) and it is likely that the salinity gradient defines the extent of downstream abundance. Limits to survival in BW may be restricted by limited osmoregulatory capacity (Peterson, 1988; Meador and Kelso, 1990), by reduced growth (Meador and Kelso, 1990; Peterson, 1991), or by preventing successful spawning (Tebo and McCoy, 1964).

Local populations of FW-linked fishes may be adapted to tolerate increased or fluctuating salinity. Some largemouth bass (*Micropterus salmoides*) have developed greater tolerance to ion perturbation, rather than increasing osmoregulatory capacity in elevated salinities (Meador and Kelso, 1990; Lowe et al., 2009). Other centrarchids may likewise tolerate modest salinities typical of an upper estuary. Peterson (1988) demonstrated that red-ear sunfish (*Lepomis microlophus*) osmoregulated effectively up to

8 ppt. Local adaptation to salinity exposure has also been observed in western mosquitofish (*Gambusia affinis*) (Purcell et al., 2008).

While FW-linked movements are generally opportunistic, many SW species are dependent upon the estuary for the early part of their life history (Boesch and Turner, 1984; Kneib, 1997; Beck et al., 2001; Able, 2005). Because of the clear link of movement into the estuary to these fishes' life histories, such movements can be considered migrations. The importance of the estuary is often demonstrated by the strong link between estuarine conditions and SW recruitment (Elliott et al., 1990; Thiel and Potter, 2001). In general, these fish experience stable SW salinity for most of their life cycle, and exploit the near-shore or estuary habitat at the larval and juvenile stages. The facultative use of lower salinities confounds the use of diadromous definitions. These SW fish lay pelagic eggs, often near the coast or in the lower estuary, and parental investment is defined by selection of spawning site (Wootton, 1999; Elliott and Hemingway, 2002; DeMartini and Sikkel, 2006; Elliott et al., 2007).

Many benthic-oriented SW fishes have distinct settling patterns [e.g. seaboard goby (*Gobiosoma ginsburgi*) (Duval and Able, 1998) and window-pane flounder (*Scophthalmus aquosus*) (Neuman and Able, 2002)], such that larvae develop and take up residence in estuaries or coastal habitats (Able et al., 2006). SW-linked halibut (*Paralichthys californicus*), summer flounder (*P. dentatus*), and turbot (*Pleuronichthys guttulatus* and *P. ritteri*) spawn near the coast and larvae metamorphose and settle as juveniles near or in the estuary and remain there through the early part of their life history (Moser, 1996; Gibson, 1997; Love, 1996; Sackett et al., 2007, 2008; Herzka et al., 2009). Juvenile flatfish (e.g. *Solea solea* and *Solea senegalensis*) select areas based largely on abiotic conditions such as structure, temperature, and salinity (Vinagre et al., 2007, 2009). Although variable, many juvenile flatfish are likely to experience only narrow salinity fluctuations at or near full-strength SW (Herzka et al., 2009; Fairchild et al., 2008).

Other SW-linked fishes have pelagic larvae (e.g. gilthead sea bream, *Sparus aurata*) which hatch in the ocean and are recruited into coastal waters. Juveniles migrate further into estuaries, where they experience a wide range of salinities, up to 60 ppt (Ben-Tuvia, 1979; Tandler et al., 1995). Distribution of these fish is strongly influenced by prey availability (Timmons, 1995) and recruited young of the year experience rapid growth (Rountree and Able, 1992; Szedlmayer et al., 1992). Pinfish (*Lagodon rhomboides*) (Sparadae) spawn in the ocean (Muncy, 1984) and larvae are recruited into the estuary (Warlen and Burke, 1990). Juveniles subsequently exploit benthic habitats for several months (King and Sheridan, 2008).

European anchovy (*Engraulis encrasicolus*) is a conspicuous example of a SW-linked species that facultatively exploits estuarine habitat (Suzuki et al.,

2008; Hibino et al., 2006). Some fish complete their life history in a saline environment, while others use low-salinity reaches of the estuary as a rearing habitat (Morais et al., 2010). For these fish, spawning occurs in the lower estuary and larvae develop and move into the upper reaches of the estuary before migrating downstream after this period of growth (Chícharo et al., 2001). Similarly, sea bass (*Lateolabrax japonicas*) spawn offshore and the young migrate inshore populating both estuaries and surf zones as juveniles (Fuji et al., 2010). Conspecifics can therefore have very different osmotic experiences.

Many SW-linked fishes can dominate the upper reaches of the estuary where salinity approaches FW, necessitating tolerance to low salinities in these fish (Weisberg et al., 1996; Whitfield, 1998; Maes et al., 2005; Hoeksema and Potter, 2006). Juveniles remain in estuarine environments over wide temporal and spatial periods. Winter flounder (*Pseudopleuronectes americanus*) remain in the estuary for 2 years before moving into the ocean (Pereira et al., 1999). Ladyfish (*Elops saurus*) are found over a wide range of salinities, although rarely in FW (McBride et al., 2001). Spotted grunter (*Pomadasys commersonnii*) spawn in the ocean and developing juveniles recruit into estuaries, where they remain for a period of 1–3 years (Wallace and Van der Elst, 1975; Heemstra and Heemstra, 2004) and display strong site fidelity (Childs et al., 2008).

2.5. Estuarine Fishes

Estuarine fishes reside in the zone of fluctuating salinity through all periods of their life history, but individuals within a given cohort may experience substantial variability in salinity. These fish often gravitate to BW for spawning and are considered in detail in Marshall (2013, Chapter 8, this volume). Estuarine species are exemplified by black sea bream (*Acanthopagrus butcheri*), which can be found in salinities from 0 to 60 ppt (Hoeksema et al., 2006; Hindell et al., 2008; Sakkabe and Lyle, 2010) and generally complete their life cycle within the upper and middle part of an estuary (Butcher and Ling, 1958; Hindell, 2007; Hindell et al., 2008). For this species, moderate salinities are required as poor recruitment is correlated with high FW delivery into the estuary (Sakabe, 2009). The mummichog (*Fundulus heteroclitus*) may best exemplify estuarine species. This fish completes its life history within salt marshes (Taylor et al., 1979) and undergoes limited movements (Lotrich, 1975; Teo and Able, 2003). These remarkable fish may undergo daily fluctuations in salinity, from near FW to full-strength SW as the tide ebbs (Griffith, 1974). Estuarine fishes are not considered further in this chapter but are discussed by Marshall (2013, Chapter 8, this volume).

3. MOVEMENT PATTERNS

3.1. Control of Migration

The choice of "when and where" to spawn can have profound effects on the life history strategy assumed by the progeny. White perch spawn over a protracted period (North and Houde, 2001) and a greater proportion of resident (versus anadromous) fish occur in faster growing late-spawned cohorts (Kerr and Secor, 2010). Amphidromous ayu (*Plecoglossus altivelis*; Osmeriformes) move downstream in anticipation of spawning (Tsukamoto et al., 1987; Iguchi et al., 1998), which may be advantageous for young in that it reduces their seaward migratory distance. Similarly, *Galaxias maculatus* adults move downstream to spawn at the head of tide (McDowall, 2008). For amphidromous fishes, this does not appear to be a widespread pattern. Others, such as the Amur goby (*Rhinogobius brunneus*), show no indication of such prespawn movement patterns.

The internal condition of the organism influences its behavior during migration. Sea lamprey metamorphosis and subsequent survival during the prefeeding (parasitic phase), downstream migration is dependent upon the availability of sufficient lipid energy stores during these non-trophic periods (Lowe et al., 1973; Beamish et al., 1979; Youson, 1997). Similarly, the propensity for anguillid glass eels to migrate is dependent upon energy stores (Bureau du Colombier et al., 2007, 2009; Bolliet and Labonne, 2008). It has been hypothesized that the facultative diadromy observed in anguillids is linked to an individual's response to its own metabolic status (Edeline, 2007). While downstream migration of juvenile American shad may be more concerted in the fall (O'Leary and Kynard, 1986; Zydlewski and McCormick, 1997a,b), retrospective otolith analysis reveals a more protracted migration that is influenced by size (Limburg and Ross, 1995; O'Donnell and Letcher, 2008). Similar variability and correlation with size have been reported for other clupeid species, e.g. allis shad (*Alosa alosa*) and twaite shad (*Alosa fallax*), in terms of both the timing of migration and extent of residence in the estuary (Taverny, 1991; Lochet et al., 2008).

Endogenous rhythms are ubiquitous and are the underpinning of activity patterns and migrations in all animal clades (Dingle, 1996, 2006). For salmonines, timing of the development of SW tolerance is species specific and cued by photoperiod and modified by other environmental factors (Boeuf, 1993; Hoar, 1988; McCormick, 1994). Diurnal cycles are linked to redistributions of fish within estuaries (Rountree and Able, 1993; Gray et al., 1998; Miller and Skilleter, 2006; Hagan and Able, 2008). Larval and juvenile green sturgeon exhibit a diel pattern of migration that peaks at night

(Kynard et al., 2005) and photoperiod has been suggested as the major driver of preparatory adaptation in this fish (Allen et al., 2011).

Many estuarine fishes exhibit a demonstrable pattern of activity linked to tidal stage, e.g. toadfish (*Halobatrachus didactylus*) (Campos et al., 2008), and with demonstrated circatidal rhythms such as those seen with juvenile plaice (*Pleuronectes platessa*) (Burrows, 2001) and the juvenile intertidal blenny (*Zoarces viviparous*) (Cummings and Morgan, 2001). Vertical migration of glass eels in response to tidal currents remains one of the most conspicuous examples of circatidal patterns (McCleave and Wippelhauser, 1987; Wippelhauser and McCleave, 1988). In the absence of a tidal *zeitgeber*, the pattern dampens (Bardonnet et al., 2003), even with photoperiod information (Dou and Tsukamoto, 2003).

Behavioral patterns established by endogenous rhythms are sculpted by other cues for migration. Eel movements occur preferentially under low-light or high-turbidity conditions, e.g. European eel (Bardonnet et al., 2005). Migration into the estuary is linked to spring tides and therefore moon phase (Jellyman, 1979; McKinnon and Gooley, 1998; Tsukamoto et al., 2003). Patterns of upper estuary use by flatfish juveniles (e.g. *Plecoglossus altivenils* and *Lateolabrax japonicas*) may be facultative and linked to the diurnal flood tides (Ohmi, 2002). Olfactory cues such as those resulting from decaying organic material (e.g. decomposition of plants/animals) may serve to attract coastal juvenile anguillids (Sorensen, 1986; Sola, 1995), indirectly indicating lower salinity (Sola and Tongiorgi, 1996).

Temperature is considered a driving factor in the timing of spawning and migrations (Livingston, 1976; Marshall and Elliott, 1998; Witting et al., 1999). The relative roles of other factors (e.g. salinity, dissolved oxygen, FW input) can be masked through covariance (Morin et al., 1992; Potter et al., 1986; Valiela, 1995; Fraser, 1997). Temperature, however, may limit the use of FW habitat and estuaries for many fishes (Attrill and Power, 2002, 2004). White perch and striped bass spawning is cued by the initial rise in temperature during spring (Rutherford and Houde, 1995; Secor and Houde, 1995). Temperature is thought to provide a threshold for the initiation of upstream migration in anguillid eels, e.g. shortfinned eel (*A. australis*) (Kearney et al., 2009) and European eel (Creutzberg, 1961; Tesch, 2003).

For American shad juveniles, temperature in FW may serve as a migratory cue through its influence on hyperosmoregulatory ability. Anadromous American shad are fully competent to enter into SW at larval juvenile transition, months before autumnal migration. As temperatures decline, these fish cease feeding and downstream migration is hastened (Backman and Ross, 1990; Zydlewski and McCormick, 1997b). Below 10°C shad juveniles rapidly decline in osmoregulatory ability (Chittenden, 1972; Zydlewski and McCormick, 1997b). Late SW entry at low temperatures also comes with a physiological cost because it may impact survival (Zydlewski et al., 2003).

Downstream movements of young amphidromous fishes (e.g. gobiids and eleotrids) appear to be cued by seasonal and short-term changes in stream flow (Fitzsimons et al., 2002). Similarly, migration back into FW is also associated with rain events (Delacroix and Champeau, 1992). Flow may also clear the stream mouth, thus facilitating passage (Fitzsimons and Ogorman, 1996). Postlarvae congregate near the shore and initiate diurnal migrations, indicating a visual component of upstream searching (Miller, 1984; Fievet et al., 1999; Lim et al., 2002). Actively migrating amphidromous gobies select FW rather than SW and preferentially select stream-origin water over well water (Fitzsimons et al., 2002).

3.2. Passive and Active Movement in the Estuary

Migratory fishes can use estuary habitats for protracted periods, ranging from weeks to years (Able et al., 2005). For structure-oriented species such site fidelity can be strong (Able et al., 1995; Tupper and Boutilier, 1995), as seen in juvenile winter flounder (*Pseudopleuronectes americanus*) in which movements were limited to a 100 m zone for up to 3 weeks (Saucerman and Deegan, 1991). Similar observations were made in juvenile plaice (*Pleuronectes platessa*) over several weeks (Burrows et al., 2004). These fish also exhibited strong homing if displaced up to 3.5 km offshore (Riley, 1973). Resource ranging (Dingle, 1996) and site fidelity were observed in young of the year summer flounder (*Paralichthys dentatus*). These fish exhibited site fidelity, but made daily movements of up to 1 km (Szedlmayer and Able, 1993). Flounder move onto the mudflats with high tide to feed and return to deeper water at low tide (Wirjoatmodjo and Pitcher, 1984; Raffaelli et al., 1990).

The importance of shallow littoral zones for small fish is well established in estuarine ecology (Boesch and Turner, 1984; Loneragan et al., 1986; Manderson et al., 2004) because it affords some protection from predators (Paterson and Whitfield, 2000; Ruiz et al., 1993) and provides abundant feeding opportunities (Orth et al., 1984; Rozas and Odum, 1988). Movements synchronous with twilight periods into estuarine intertidal mudflats (Vinagre et al., 2007) may also reduce risk of predation (Cowley and Whitfield, 2001; Steinmetz et al., 2003; Zydelis and Kontautas, 2008).

Passive forces can be largely attributed to the volitional retention within or movement through the estuary by migrating fishes. Larval fish have limited mobility due to both size and morphology. Initial movements are therefore influenced by the selection of spawning sites (Boehlert and Mundy, 1988). Tidal currents can transport and retain inorganic material (Postma,

1961) and passively drifting organisms alike (de Wolf, 1973; Jager and Mulder, 1999). Such advective tidal transport occurs as larval fish sink during low-current velocities and are lifted in the water column by turbulence on an incoming tide. Thus, tidal currents are responsible for egg and early larval movement inshore (Power, 1984; Able and Fahay, 1998; Werner et al., 1999).

Rapid seaward migration may be critical for the larvae of some amphidromous species so that SW entry is synchronous with the transition from endogenous to exogenous feeding. Iguchi and Mizuno (1999) estimated that more than half of the Japanese goby (*Rhinogobius brunneus*) may starve during downstream migration in long river systems owing to a lack of appropriately sized prey in streams (Tsukamoto, 1991). Goby larvae passively drift downstream during the night (Iguchi and Mizuno, 1990, 1991; Moriyama et al., 1998) and actively remain in the water column by alternating movement towards the surface and sinking during rest periods (Kinzie, 1993; Balon and Bruton, 1994; Keith et al., 1999). These larvae accomplish this migration at the diminutive size of 1–4 mm in length (Han et al., 1998; Keith et al., 1999). For amphidromous fishes that re-enter FW at a small size, inshore and estuarine movements can also be dependent upon tidal currents, e.g. *Awaous guamensis* (Keith et al., 2000) and *L. concolor* (Nishimoto and Kuamo'o, 1997). Larvae are transported mostly by currents, as has been frequently observed for other fish (Borkin, 1991; Hare and Cowen, 1996), but larvae may also actively swim (Balon and Bruton, 1994; Cowen et al., 1993; Leis and Carson-Ewart, 1997; Stobutzki and Bellwood, 1997; Fisher et al., 2000).

Many SW-linked fishes move into temperate estuaries during the postflexion larval stage (Miskiewicz, 1986; Strydom et al., 2003) or during the juvenile stage (Wasserman and Strydom, 2011). This period is marked by physiological, behavioral, and morphological shifts that can occur synchronously with transition to nursery habitat (Balon, 1984; Kaufman et al., 1992; Able et al., 2006). Migration to these "critical zones" in the estuary favors successful recruitment (Dovel, 1971). Some pelagic juvenile species such as menhaden (*Brevoortia tyrannus*) rely on the strength of a tide to enter the estuary (Joyeaux, 1999). Passive forces may be sufficient to explain the invasion of the estuary by pufferfish (*Takifugu rubripes* and *T. xanthopterus*). These fish are moved inshore to the upper estuary by residual currents in the lower vertical stratum of the estuary (Yamaguchi and Kume, 2008).

Some larval fish exploit currents in order to move inland (Jenkins and Black, 1994; Jenkins et al., 1999). Oliveira et al. (2006) demonstrated through hydraulic modeling that passive individual recruitment to the estuary is dependent on FW flow. Vertical movements, however, can effectively move and retain organisms even under high flow conditions. Such a mechanism of selective tidal transport exploits the vertical velocity profile that approaches zero at the bottom during a period of opposing flow. Thus, a fish can control directional movement in areas of reversing flows while minimizing the energetic cost of swimming (Fortier and Leggett, 1983; Miller, 1988). The ability to maintain position in the estuary by selectively populating vertical habitat is a commonality in larval fish retained in the estuary (Creutzberg, 1961; Hobbs et al., 2006), developing as both behavioral and sensory abilities increase (Forward et al., 1999; Tolimieri et al., 2000). The degree to which these fish have active control of position is logically correlated with an increase in swimming ability (Clark et al., 2005; Leis et al., 2006). Many fish exhibit a clear pattern to estuarine entry linked to size and development. Bay anchovy (*Anchoa mitchilli*) in the Chesapeake Bay are spawned in the lower estuary (Zastrow et al., 1991; Rilling and Houde, 1999; Schultz et al., 2003) and the juveniles are recruited into low-salinity waters at the head of tide as they grow (Dovel, 1971; Kimura et al., 2000). The prevalence of vertical patterns is also correlated with tidal magnitude (Graham and Sampson, 1982).

Tidal transport is not limited to larval fish, and this mechanism of movement into and through the estuary is shared by many catadromous anguillid species (Jellyman, 1979; Sheldon and McCleave, 1985; Sugeha et al., 2001; Dou and Tsukamoto, 2003; Tesch, 2003). Selective vertical movement of glass eels drives upstream progress (McCleave and Kleckner, 1982), resulting in an accumulation of these juveniles at the head of tide (Gascuel, 1986; McCleave and Wipplehauser, 1987; De Casamajor et al., 1999). At this point, upstream progress necessitates the initiation of active swimming (Creutzberg, 1961).

4. OSMOREGULATORY COMPETENCE

As euryhaline fishes move between SW and FW, an obvious requirement is the maintenance of plasma osmotic concentration (Blaber, 1974; Mehl, 1974; Martin, 1990). Steady-state plasma ions are generally higher in SW-acclimated fish than in FW (Holmes and Donaldson, 1969; Allen and Cech, 2007; He et al., 2009) but regulated within a relatively narrow range (McDonald and Milligan, 1992; Evans et al., 2005). Even larval fishes tightly regulate internal osmolality with respect to external salinities (280–360 mOsm; Varsamos et al., 2005). In FW, fish use hyperosmoregulatory strategies to offset the osmotic influx of water and to counter passive ion losses. On the other hand, hypoosmoregulatory strategies are used to offset ion influx and dehydration in more saline waters (Evans, 1999; Varsamos et al., 2005). Fishes in SW must also offset the uptake of divalents such as Mg^{2+}, SO_4^{2-}, and Ca^{2+}, which can occur passively across the gills, or through the ingestion of food and water.

The gills, digestive tract, and kidneys play important roles in osmoregulation in both saline and FW environments. Gill ionocytes (also known as mitochondrion-rich cells and chloride cells), first noted by Keys and Wilmer (1932) in American eel, play a critical role in ion excretion, as definitively demonstrated by Foskett and Scheffey (1982). The excretion of Na^+ and Cl^- excretion by these SW ionocytes relies upon oubain-sensitive, basolateral Na^+/K^+-ATPase (NKA) pumps, which maintain the low intracellular Na^+ concentrations required to promote the excretion of Cl^- and Na^+ across the gills (Karnaky et al., 1976, 1977; Silva et al., 1977). Ion extrusion also involves the transport of Cl^- and Na^+ into the ionocyte via a basolateral $Na^+,K^+,2Cl^-$ cotransporter (NKCC), resulting in sufficiently high intracellular Cl^- concentrations to generate the electrochemical gradient needed to promote Cl^- excretion via an apical cystic fibrosis transmembrane conductance regulator (CFTR) channel normally found within the small apical pit of SW ionocytes (reviewed by Marshall, 2002; Evans et al., 2005; Marshall and Grosell, 2006; Edwards and Marshall, 2013, Chapter 1, this volume). The resulting local accumulation of negative charge then promotes paracellular Na^+ excretion between adjacent ionocytes and accessory cells or other ionocytes linked with relatively shallow tight junctions (Sardet et al., 1979). FW fishes actively take up Na^+ and Cl^- via FW ionocytes on the gill filament and lamellar epithelium (reviewed by Perry, 1997; Wilson and Laurent, 2002; Marshall, 2002; Evans et al., 2005; Edwards and Marshall, 2013, Chapter 1, this volume). The large bulbous, apical surface of the FW ionocyte is studded with numerous microvilli, increasing the surface area for ion uptake. Like the SW ionocytes, the basolateral surface of FW ionocytes is enriched with NKA transporters, but it is apical H^+-ATPase pumps and the basolateral NKA that establish the inward electrochemical gradient that promotes Na^+ influx via apical Na^+ channels. Chloride is taken up via apical Cl^-/HCO_3^- exchange and a presumed basolateral Cl^- channel (Perry, 1997; Marshall, 2002; Evans et al., 2005 for reviews).

To counteract osmotic water losses across the gills, SW-acclimated fish drink SW, relying on gill-mediated ion excretion to counter the resulting salt load (Smith, 1932; Keys, 1933). In general, Na^+ and Cl^- are actively taken up in the esophagus, which lowers the osmolality of the ingested SW prior to its reaching the intestine (Hirano and Mayer-Gostan, 1976), where the majority of water uptake takes place (Grosell, 2011). The excess Na^+ and Cl^- is then removed from the blood via the gill SW ionocytes. Water uptake is further driven by the alkalinization of the gut contents in the posterior intestine, resulting in the generation of CO_3^{2-} and the formation of $CaCO_3$ precipitates in the gut lumen, which further lowers the osmolality of the intestinal fluids, promoting water uptake (Grosell, 2006, 2011; Wilson et al., 2009). The gut also plays an important role, along with the kidneys, in the excretion of divalents such as SO_4^{2-}, Mg^{2+}, and Ca^{2+}. Glomerular filtration rate (GFR) and urinary flow rate (UFR) are much lower in SW fishes. In FW fishes, on the other hand, the kidneys reabsorb Na^+ and Cl^- and act as a "bilge-pump" that excretes copious amounts of dilute urine (Beyenbach, 2004; Marshall and Grosell, 2006; Evans et al., 2005; Edwards and Marshall, 2013, Chapter 1, this volume). Although it had been widely accepted that fish do not drink in FW, it now appears that there is some water ingestion, but at rates far below those observed in SW fishes (Marshall and Grosell, 2006).

4.1. Patterns of Osmoregulatory Competence

The fundamental mechanisms of transitioning from FW to SW are similar across all groups of euryhaline fishes. In general, there is a replacement of FW ionocytes with SW ionocytes, usually accompanied by an increase in gill NKA activity, increased drinking, and decreased urine output (Folmar and Dickhoff, 1980; Hoar, 1976, 1988; McCormick, 2001; Evans et al., 2005). While increasing NKA is generally observed during SW acclimation, an increase in this NKA activity can also be observed in fish transferred from SW to FW (Bystriansky and Schulte, 2011), underscoring the shared reliance of osmoregulatory mechanisms on this transporter in either environment. In addition to these homeostatic adjustments, factors such as life stage, behavior, and body size are key factors affecting survival following FW to SW transfer (Hoar, 1988; Folmar and Dickhoff, 1980). While there appears to be great commonality in the mechanisms of osmoregulation across clades of migratory fishes (Hoar, 1988; Evans, 1993; Marshall and Grosell, 2006), there is great variation in the developmental patterns of osmoregulatory competence for life in SW.

In assessing patterns of salinity tolerance, McCormick (1994) described three general pathways by which SW tolerance might be developed: (1) development at an early age; (2) gradual acclimation associated with increasing size; and (3) environmentally cued development as a juvenile "preparatory adaptation". Such processes are well characterized for salmonines, which represent only one group of migratory fishes. Development is a major driver for salmon that enter SW soon after hatching [e.g. fall Chinook (*Oncorhynchus tshawytscha*), chum (*O. keta*), and pink salmon (*O. gorbuscha*)], but the underlying mechanisms are not yet clearly worked out. On the other hand, the role of the preparatory adaptation of smolting is very well characterized and extensively described for Atlantic salmon (*Salmo salar*), coho salmon (*O. kisutch*), and steelhead trout (*O. mykiss*). This preparatory adaptation minimizes osmotic perturbations as these fish transit the estuary (McCormick and Saunders, 1987; Hoar, 1988; McCormick, 2013, Chapter 5, this volume). In brook trout (*Salvelinus fontinalis*) there are no apparent physiological changes and survival in SW depends upon gradual acclimation and body size (McCormick and Naiman, 1984). While these patterns are not mutually exclusive, this construct presents testable hypotheses that can be systematically applied to explain the patterns observed in other clades (as in Allen et al., 2011; see below).

4.1.1. DEVELOPMENT AT AN EARLY AGE

Osmoregulation begins with embryonic development (Alderdice, 1988), but the focus of this section will be on the mechanisms seen posthatch. Prior to the development of the gills, larval fishes have to counter diffusional ion losses that mainly take place across the body surface using ionocytes located in the integument (Tytler and Bell, 1989; Tytler et al., 1993; Rombough, 2007). The ionoregulatory capacity appears to be somewhat lower in these early life stages compared to later in development, as internal osmolality can vary widely in larval fishes, from 250 to 540 mOsm (Varsamos et al., 2005). SW-linked larval fishes may survive in BW, and be tolerant to very dilute salinities (Yin and Blaxter, 1987). Even as yolk-sac larvae, Atlantic herring (*Clupea harengus*) maintain low blood osmolality with respect to a saline environment (Holliday and Blaxter, 1960), as does the lumpsucker (*Cyclopterus lumpus*) (Kjörsvik et al., 1984). Plaice (*Pleuronectes platessa*) yolk-sac larvae display an impressive pattern of regulation from low salinities to full-strength SW with minimal perturbation (Holliday, 1965; Holliday and Jones, 1967).

There are many other examples of early ontogenic changes in capacity for salinity (see excellent review by Varsamos et al., 2005). In many SW-linked fishes there is an ontogenic shift in the capacity to osmoregulate in the early postembryonic stages, e.g. in European sea bass (*Morone labrax*) and starry flounder (*Platichthys stellatus*) (Hickman, 1959). The amphidromous goby (*Chaenogobius urotaenia*) has the ability to enter higher salinity immediately posthatch, but performs poorly in FW. Larvae moved into 50% SW survived for more than 30 days, while those in FW lived less than a week (Katsura and Hamada, 1986). A similar trend was observed in the goby (*Awaous guamensis*) when larvae were held in either 34 ppt SW or FW (Ego, 1956).

Striped mullet (*Mugil cephalus*) display osmotic perturbations during FW and SW exposure as young larvae, but osmoregulatory ability increases quickly through development (Nordlie et al., 1982). For these fish, early development of hypoosmoregulatory capacity may reflect an adaptive "spread the risk" strategy. Remaining in the estuary during early development may be just as likely an outcome as being recruited into full salinity and these fish are prepared for either. Indeed, these fish are capable of survival in full-strength SW very early in development, when they are still quite small (just over 40 mm) (Nordlie et al., 1982). In many cases, osmoregulatory competence in early life stages involves developmentally dependent restructuring and/or modifications to osmoregulatory organs including the gill, gastrointestinal tract, and kidneys (Varsamos et al., 2005; Rombough, 2007). For instance, in gilthead sea bream (*Sparus aurata*) osmoregulatory capacity is increased during development as ionocytes shift from the integument to the gills (Bodinier et al., 2010).

4.1.2. ACCLIMATION AND SIZE

Many teleosts can be gradually acclimated to increased salinity and either correct or establish new steady-state internal ion concentrations (Holmes and Donaldson, 1969; Jacob and Taylor, 1983; Evans, 1984). The ability to acclimate gradually to increasing ion concentrations, or in response to abrupt transfer to SW, can be profoundly influenced by size. For example, in tilapia (*Oreochromis aureus* and *O. niloticus*) the ontogeny of salinity tolerance is positively correlated with body size although there is no apparent developmental stage associated with this increase (Watanabe et al., 1985). Similarly, increased salinity tolerance has been linked to size in gilthead sea bream (*S. aurata*) (Bodinier et al., 2010) and European sea bass (Varsamos et al., 2001). Such patterns have been postulated to reflect the relatively reduced shift in diffusional surface with respect to increased volume achieved during growth, thereby reducing the burden of osmoregulation in larger fish (Allen et al., 2009).

With the notable exception of green sturgeon (Allen et al., 2009, 2011), the pattern of late SW entry of most sturgeon juveniles is consistent with the hypothesis that these fish do not have a discrete preparatory adaptation for SW entry, but slowly acquire the ability through an increase in size. The general behavioral paradigm has been that juvenile sturgeon migrate seaward over a period of years (Vladykov and Greeley, 1963), remaining in the estuary, and perhaps exploiting the lower metabolic costs associated with a lesser osmotic differential. Size-mediated increase in osmoregulatory ability has been documented in several sturgeon species, e.g. Gulf sturgeon (*A. oxyrinchus*) (Altinok et al., 1998), white sturgeon (*A. transmontanus*) (Amiri et al., 2009; McEnroe and Cech, 1985) and shortnose sturgeon (Ziegeweid et al., 2008). The behavior of Atlantic sturgeon indicates an avoidance of high-salinity environments at small sizes (Brundage and Meadows, 1982; Dovel and Berggren, 1983; Bain, 1997). There is also some evidence of staging in the estuary by European sturgeon (*A. sturio*) juveniles. This may indicate a period of acclimation prior to SW entry (Rochard et al., 2001).

Fish may actively occupy regions of moderate salinity in order to acclimate. Contact with increased salinities can elicit greater salinity tolerance in several sturgeon species, e.g. Gulf sturgeon (Altinok et al., 1998), white sturgeon (McEnroe and Cech, 1985), and Adriatic sturgeon (McKenzie et al., 2001). Anguillid glass eels may stratify in SW low in the estuary, perhaps to gradually make the transition into FW. Further up in the estuary these fish are more evenly distributed through the water column in FW (Adam et al., 2008). Similarly, the return to FW in amphidromous gobies may be linked to body size, based on the consistent sizes of species upon return (Keith et al., 2002). These fish may persist in the estuary for several weeks (Font and Tate, 1994), although it is unclear whether this is linked to the ability to osmoregulate or whether it reflects a size-dependent metamorphosis, or both.

4.1.3. DEVELOPMENTAL STAGE

Direct SW transfer is often used to characterize the ontogeny of salinity tolerance (Varsamos et al., 2001; Watanabe et al., 1985; Zydlewski and McCormick, 1997a; Allen et al., 2009, 2011). In many species SW tolerance is, after all, correlated with migration and seasonal movements (Varsamos et al., 2005). For some species, this connection is relatively clear. The parr–smolt transformation in salmonines is a synchronized shift in behavior, morphology, and physiology linked to seaward migration resulting in a rapid transition through the estuary (Schreck et al., 2006; McCormick, 2013, Chapter 5, this volume) that limits vulnerability to predators (Kennedy et al., 2007). Yet in other species, the ontogeny of salinity tolerance and of preparatory adaptation is the backdrop for a wide array of life history contingencies. For instance, in anguillids, glass eel metamorphosis and acclimation to FW are associated with irreversible changes to the gut structure (Ciccotti et al., 1993; Rodriguez et al., 2005). Body size and life stage are important determinants of osmoregulatory competency in saltwater. As illustrated below, however, in most cases these variables are part of a suite of processes associated with other preparatory adaptations that determine osmotic scope through a species's life history.

5. PREPARATORY ADAPTATION AND MECHANISTIC TRENDS

5.1. Anadromous Fishes

5.1.1. LAMPREY

All of the anadromous lampreys have an SW parasitic phase, in which they attach themselves to potential prey/hosts, which include not only large-bodied teleost fishes (Farmer, 1980; Renaud et al., 2009) but also elasmobranchs (Jensen and Schwartz, 1994; Wilkie et al., 2004; Gallant et al., 2006) and even cetaceans (Nichols and Tscherter, 2011). Before this parasitic phase, anadromous lampreys spend the first several years of their lives burrowed in the substrate of FW streams as functionally blind suspension feeding larvae, known as ammocoetes (Youson, 1980; Rovainen, 1996). Upon metamorphosis these relatively sedentary ammocoetes undergo marked structural and physiological changes that include a reorganization of the feeding apparatus in which the oral hood of the ammocoete is replaced by an oral disc and rasping tongue that is used to penetrate the hide and/or tear pieces of flesh from their hosts/prey in the parasitic stage (Youson, 1980, 2003; Renaud et al., 2009). The gills also switch from a unidirectional flow through gill to a tidally ventilated gill in which water is actively pumped in and out of gill pouches in order to continuously use the mouth for attachment to its parasitic host (Rovainen, 1996; Wilson and Laurent, 2002). Metamorphosis is a size-dependent process, reflecting the large stores of lipid needed to sustain the animal during this non-trophic life phase (Lowe et al., 1973; O'Boyle and Beamish, 1977; Holmes et al., 1994; Youson, 1997).

Development of SW tolerance is tightly linked to metamorphosis (Beamish, 1980a,b; Morris, 1980; Richards and Beamish, 1981) and upregulation of osmoregulatory capacity (Reis-Santos et al., 2008). The ammocoetes are exclusively FW, and incapable of surviving in even dilute SW for more than a few days (10–15 ppt ; Beamish et al., 1978; Reis-Santos et al., 2008). As in other fishes, the lamprey gill epithelium is comprised of pavement cells and lower numbers of mucus cells (Bartels and Potter, 2004; Evans et al., 2005). Differences in ionocyte structure differentiate lamprey gills from those of other fishes. The ammocoete gill has two types of ionocytes: ammocoete ionocytes (referred to as ammocoete MR cells in Bartels and Potter, 2004) and a population of intercalated ionocytes (Bartels and Potter, 2004), which are solitary and found between adjacent pavement cells in this life stage (Reis-Santos et al., 2008). The ammocoete ionocyte is distinct from the typical teleost ionocyte and the intercalated ionocyte in that it lacks an extensive tubular network, and has only minimal apical exposure and microvilli.

During metamorphosis, the ammocoete ionocytes lying between lamellae of the gills are replaced by SW ionocytes (Peek and Youson, 1979), which are subsequently lost during the upstream spawning migration (Bartels and Potter, 2004). The intercalated ionocytes are retained (or reappear) only in FW, implying a role in ion uptake. The emergence of SW ionocytes at metamorphosis is accompanied by increased activity and abundance of gill NKA (Reis-Santos et al., 2008) (Fig. 6.3A). Like teleost SW ionocytes, lamprey SW ionocytes have an extensive tubular network, but they lack the

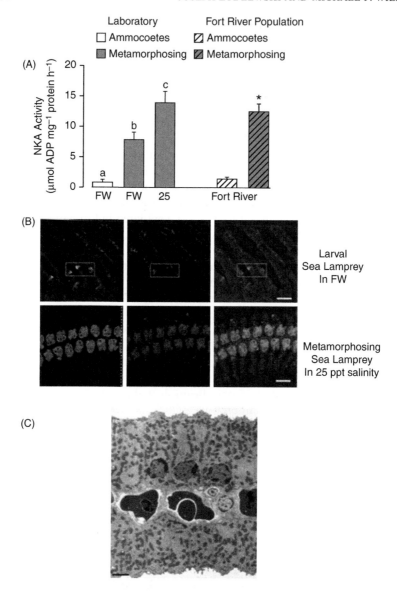

Fig. 6.3. Preparative physiological adaptation and seawater (SW) ionocytes in anadromous sea lampreys. (A) In both laboratory-held and river-captured sea lampreys, metamorphosis was accompanied by a marked increase in Na^+/K^+-ATPase (NKA) activity in freshwater (FW), and further increases occurred in 25 ppt SW (25). (B) NKA protein (red) and H^+-ATPase protein (green) were immunolocalized in ammocoete ionocytes on the lamellae in FW (upper panel), but immunofluorescence was restricted to ionocytes in the interlamellar spaces in SW-acclimated (metamorphosing) animals (lower panel). Differential interference contrast images of merged

apical crypt that is characteristic of analogous teleost SW ionocytes (Bartels et al., 1998). Characteristic features of the SW ionocyte are their plate-like appearance and propensity to be distributed in rows at the base of the filament (Fig. 6.3B,C). These SW ionocytes extend into the interlamellar region, probably with leaky paracellular junctions between adjacent cells (Bartels and Potter, 2004). These changes in gill ultrastructure are remarkably reminiscent of those that characterize smoltification in euryhaline salmonid fishes, which also normally includes an upregulation of NKA capacity and reliance on SW ionocytes (see McCormick, 2001; Evans et al., 2005 for excellent reviews). Like smolting salmonids, postmetamorphic lampreys take on a silvery sheen owing to the deposition of guanine in the epithelium (Youson, 1980). Despite the subtle but significant differences in gill ultrastructure, the mechanism of Na^+ and Cl^- extrusion in lamprey also appears similar to that of SW teleosts. Evidence for the NKCC is limited in lampreys, but using the T4 antibody to this protein (which cannot distinguish between NKCC1 and NKCC2) this cotransporter has been identified in sea lamprey ionocytes (S. Edwards, S. Blair and M. P. Wilkie, unpublished observations). While a CFTR-like protein has not yet been described in lampreys, its presence also seems likely.

Like other SW fishes, lampreys drink while in SW (Pickering and Morris, 1970; Rankin et al., 2001), with water absorption probably taking place across the anterior intestine down osmotic gradients generated by Na^+ and Cl^- uptake by enterocytes (Pickering and Morris, 1973). The divalent ions (Mg^{2+} and SO_4^{2-}) are not taken up by the intestine (Pickering and Morris, 1973) and are instead excreted in the very low amounts of urine and via defecation (Pickering and Morris, 1973; Rankin et al., 2001).

The role of lamprey kidneys in SW has been described in upstream migrant river lamprey (*Lampetra fluviatilis*) reacclimated to 50% SW. Pickering and Morris (1973) noted that kidney UFRs were extremely low after exposure to these hyperosmotic conditions. Further studies by Logan et al. (1980) demonstrated that it was mainly marked reductions in GFR (through a reduction in the number of renal corpuscles) that resulted in the 90% reduction in UFR observed. As in many SW teleosts, lamprey urine is either hypoosmotic or isosmotic to SW, and mainly concentrated with Mg^{2+} and SO_4^{2-}, and Cl^- (Logan et al., 1980).

---

images of NKA, H^+-ATPase protein, and DAPI (nuclear marker) to illustrate colocalization of the two transport proteins. (C) Electron micrograph illustrating that the SW ionocytes of lamprey (pouched lamprey, *Geotria australis* shown) are arranged in columns or rows, containing extensive tubular invaginations, but unlike teleost ionocytes, lack an apical crypt. Scale bar: 3 µm. Data and images in panel A and B from Reis-Santos et al. (2008); electron micrograph in panel C adapted from Bartels and Potter (2004) with permission. **SEE COLOR PLATE SECTION**

The contribution to lamprey ion and osmoregulation made by the ingestion of isosmotic fluids from their teleost hosts has yet to be investigated. Such physiological parasitism could explain why NKA activities tend to decrease with age and with body size in SW-acclimated sea lamprey (Beamish, 1980b). Fossil evidence suggests an SW origin for lampreys (Shu et al., 1999; Gess et al., 2006), implying that ancestral lampreys were likely to be anadromous (Gill et al., 2003). However, the conspicuous presence of the FW population of *P. marinus* in the Laurentian Great Lakes indicates that anadromy may be facultative in this species, or that rapid adaptation is possible (Lawrie, 1970; Eshenroder, 2009). Whether or not the several other FW populations of parasitic lampreys, including members of the *Ichthymzyon*, *Entosphenus*, and *Lampetra* genera (Potter and Gill, 2003), have retained some of the features associated with anadromy remains an open but intriguing question. Comparative work considering endocrine control of ionocyte structure and function in relation to SW tolerance in postmetamorphic juveniles may inform the evolutionary radiation of lamprey species.

Little is known about the hormonal factors involved in the preparative changes involved in the FW to SW transition in lampreys, but metamorphosis is well described and depends upon gradual increases in the thyroid hormones thyroxine (T_4) and triiodothyronine (T_3) in the larval phase. These increases are followed by a precipitous drop in both hormones that is thought to trigger metamorphosis (Youson, 1994, 2003; Youson and Manzon, 2012). It is notable that these "low" levels are near the peak concentrations known to initiate metamorphosis in amphibians (Youson, 2003), and could therefore be at physiologically relevant levels that are sufficient to play a role in the development of SW tolerance. In fact, thyroid hormones play an important role in preparatory adaptation of salmonines to SW by initiating the upregulation of corticosteroid receptors, which is critical for ensuring that the cortisol-induced upregulation of SW ionocytes takes place in these fishes (Evans et al., 2005). The recent finding that 11-deoxycortisol is the functional mineralocorticoid in lampreys (Close et al., 2010) could represent a turning point in our understanding of preparatory adaptation in these basal vertebrates. Indeed, 11-deoxycortisol administration to lampreys results in a marked increase in gill NKA activity, a prerequisite for downstream migration and SW acclimation (Close et al., 2010).

5.1.2. STURGEON

Salinity tolerance is related to ontogeny in the anadromous sturgeons, with body size being directly proportional to the ability of the fish to hypoosmoregulate in SW environments (Altinok, 1998; Allen and Cech,

2007; Allen et al., 2009, 2011). Amiri et al. (2009) noted that larger juvenile white sturgeon experienced lower mortality and a lower onset of osmotic disturbances when transferred to SW. Similar findings were reported by Altinok et al. (1998) for Gulf sturgeon (*A. oxyrinchus*) using fish of identical age (13 month posthatch) but grouped by size (110–170, 230–270, and 460–700 g). Smaller fish suffered greater osmotic stress (increases in plasma osmolality, Na^+, and K^+) and mortality following direct transfer from FW to SW (25 ppt). Following 96 h exposure to SW, however, plasma ions returned towards FW values in the larger fish, but Na^+ remained elevated in the smaller fish. Many sturgeon species require prior exposure to BW to make the transition from FW to SW (McKenzie et al., 1999). White and green sturgeons readily tolerate acute transfer from BW to the full-strength SW, but are less tolerant of acute transfer to FW (Potts and Rudy, 1972). Blood ion concentrations and patterns of Na^+ uptake are similar to those of teleosts. Transfer to SW also leads to decreased water permeability, with reduced urinary flow being mainly for the elimination of divalents such as Mg^{2+} and SO_4^{2-} (Potts and Rudy, 1972).

Size alone is not sufficient to explain patterns of osmotic competence in green sturgeon. There is a clear link between an ability to control plasma osmolality and age. Allen and Cech (2007) acclimated three different life stages of green sturgeon (100, 170, and 533 days posthatch age) to FW (<3 ppt), BW (10 ppt), and SW (33 ppt). There was significant mortality (23%), lower growth, and osmotic perturbations in the youngest group, but older fish experienced no change to osmolality or Na^+ concentration, and only minor changes in plasma Cl^-. Thus, the green sturgeon has the ability to completely acclimate to SW after 1.5 years, and life-stage dependent preparatory adaptation for SW residence is probably essential. Indeed, there was no relationship between body mass and plasma osmolality when green sturgeons were acclimated to SW over 7 weeks (Allen et al., 2009). Green sturgeon survived short-term transfer to SW after a gradual acclimation in BW at 4.5 months, but by 7 months they were capable of withstanding direct transfer with no mortality, and minimal osmotic perturbations (Fig. 6.4A) (Allen et al., 2011).

The development of increased osmoregulatory ability coincides with a peak in cortisol that was followed by peaks in thyroid hormones and the upregulation of gill and pyloric ceca NKA activity and abundance (Fig. 6.4A) (Allen et al., 2009). Cortisol, with growth hormone (GH), drives the proliferation of gill SW ionocytes in euryhaline teleosts. Thyroid hormone is thought to have permissive effects that include the upregulation of corticosteroid receptors in the gill (Takei and McCormick, 2013, Chapter 3, this volume). These changes occur at the time when decreased swimming performance and other behavioral shifts associated with downstream

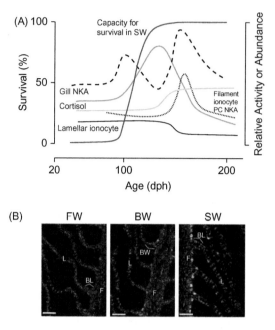

Fig. 6.4. Preparatory adaptation in the anadromous green sturgeon. Seawater (SW) tolerance in green sturgeon is linked to ontogeny (A), with full SW tolerance developing in less than 5 months (red line). Tolerance is linked to a steady increases in plasma cortisol (orange line), and increased gill (dashed line) and pyloric ceca (dotted line) Na$^+$/K$^+$-ATPase (NKA) activity. Marked decreases in ionocyte abundance on the lamellae (black) are coincident with increases in filamental ionocyte abundance (blue). dph: days posthatch. (B) Gill ionocytes rich in NKA (green) on the lamellae (L) are more numerous in freshwater (FW) than brackish water (BW) and SW. The intensity of the immunostaining on the filament located ionocytes (F), particularly at the lamellar base (BL), was also greater in SW. Scale bar: 10 μm. Drawing based on Allen et al. (2011); micrographs were obtained, with permission, from Allen et al. (2009); micrographs obtained with permission from Allen et al. (2009). **SEE COLOR PLATE SECTION**

migration occur (Kynard et al., 2005). This is the first evidence of a preparatory developmental stage in sturgeon and may reflect a preparatory adaptation of this species to enter SW at a relatively young age (Brown, 2007). Such preparatory adaptation may be triggered by external factors such as photoperiod (Allen et al., 2011), but whether this or other factors are involved requires further investigation. It remains to be determined whether similar preparatory strategies for SW acclimation are used by other migratory sturgeon, but this certainly represents an exciting and important direction of future study.

There has been very limited work on the underlying molecular mechanisms of ionocyte function in the sturgeon, but the general structure and function of sturgeon ionocytes are similar to those of teleosts. As in teleosts, sturgeon switch from an FW ionocyte to an SW ionocyte as they transition from FW to SW (Fig. 6.4B) (Altinok et al., 1998; McKenzie et al., 1999; Martinez-Alvarez, 2005; Allen et al., 2009, 2011; Sardella and Kültz, 2009; Zhao et al., 2010). The most detailed microscopic examination, using transmission electron microscopy, suggests that the SW ionocyte of at least one species, the Adriatic sturgeon (*A. naccarii*), has an apical crypt (Martinez-Alvarez et al., 2005), implying the presence of an apical Cl^- channel. An accompanying NKCC has been localized to the ionocyte in green sturgeon, which is upregulated during SW acclimation (Sardella and Kültz, 2009). Similarly, immunohistochemical localization indicates that the basolateral membrane of SW ionocytes is enriched with NKA (McKenzie et al., 1999; Allen et al., 2009; Sardella and Kültz, 2009; Zhao et al., 2010). Sardella and Kültz (2009) also noted a downregulation of the V-type H^+-ATPase with SW acclimation in green sturgeon, consistent with a decreased role for H^+-ATPase coupling to drive Na^+ uptake in saltwater environments.

There is some direct evidence that sturgeon drink during SW acclimation, as demonstrated by exposing FW-acclimated Siberian sturgeon (*A. baerri*) to elevated salinity (Taylor and Grosell, 2006). In the green sturgeon, the concentrations of Na^+ and Cl^- are markedly lower in the stomach compared to ambient SW, suggesting that the water is desalinated in the esophagus en route to the stomach (Allen et al., 2009). The ion concentrations decline further in the anterior–mid intestine, and in the rectum, which probably generates more favorable lumen–blood osmotic gradients in this region (Allen et al., 2009). As in SW teleosts there is circumstantial evidence for high rates of luminal Cl^-/HCO_3^- exchange in the intestine, based on lower Cl^- concentrations, more alkaline pH, and the presence of solid mucus tubes associated with precipitates comprised mainly of Ca^{2+}, CO_3^{2-}, and HCO_3^- (Allen et al., 2009). Thus, like its teleostean counterparts, it appears that the sturgeon maximizes water uptake through a combination of ion uptake in the esophagus and intestine, and Cl^-/HCO_3^--mediated base extrusion in the intestine to cause precipitation of solutes such as Ca^{2+} in the form of $CaCO_3$ (Grosell, 2011). Although increased NKA, NKCC, and aquaporin in the gut are likely to be involved, this has yet to be examined in sturgeons.

5.1.3. ALOSINE FISHES

Several anadromous clupeid species spawned in FW develop SW tolerance at the larval juvenile transition, e.g. allis shad (Leguen et al., 2007; Bardonnet and Jatteau, 2008) and American shad (Zydlewski and McCormick, 1997a). Development of larval American shad can occur at salinities greater than isosmotic (Limburg and Ross, 1995) but development of the gills at the larval–juvenile transition is necessary for survival in full SW (Zydlewski and McCormick, 1997a). This ontogeny is likely to afford some success of eggs and larvae displaced to the estuary before migration occurs. In temperate rivers, migration can be protracted into the fall; thus a wide window of SW entry is allowed. Size does affect migration. Larger juveniles are recruited into migration earlier in the season (Limburg, 1996; O'Donnell and Letcher, 2008) but there appears to be no increase in SW tolerance at the time of migration (Zydlewski and McCormick, 1997a,b). In allis shad, however, additional development of SW tolerance has been linked to size of the juvenile (Leguen et al., 2007), which corresponds with the timing of SW entry (Lochet et al., 2009).

In spite of being competent to enter SW early in development, migratory juvenile American shad in FW have markedly higher gill NKA activities than their non-migrant counterparts (Zydlewki and McCormick, 1997b). Remarkably, this increase in NKA is linked to marked reductions in hyperosmoregulatory ability in FW, as indicated by declines in plasma Cl^-. When held under FW conditions in the laboratory past the period of migration, dramatic 70% reductions in plasma Cl^- and increased mortality occur. This decline is delayed, but not prevented, in juvenile shad held constant at 24°C (Zydlewski and McCormick, 2001). Thus, declining hyperosmoregulating capacity probably represents a developmental shift, a "preparatory adaptation", associated with the FW to SW migratory period, although its adaptive significance is perplexing. As this reduction in ability to osmoregulate in FW is hastened by declining temperature, this development probably defines a window for successful ocean migration in temperate American shad populations (Zydlewski et al., 2003) (Fig. 6.5). The increased gill NKA in migrant American shad in FW is likely to be related to upregulation of ion uptake mechanisms. Increased NKA corresponds to a significant increase in ionocyte abundance in the gills, particularly on the lamellae (Zydlewski and McCormick, 2001; Zydlewski et al., 2003) (Fig. 6.5). In FW, these cells have a large surface and appear similar in morphology to FW ionocytes implicated in ion uptake in teleosts. As ionocytes on the lamellae proliferate, they begin to cover an increasing proportion of the respiratory surface (Zydlewski and McCormick, 2001). The proliferation and enlargement of ionocytes during seaward migration may present a considerable energetic challenge to late migrant shad or may directly interfere with respiration and other gill functions. In both American shad and alewife (*A. pseudoharengus*), acclimation to SW is associated with a marked increase in gill NKA activity, loss of ionocytes on the gill lamellae, and increased size of filamental ionocytes (Fig. 6.5) (Zydlewski et al., 2001; Christensen et al., 2012). Detailed immunohistochemical examination of the alewife gill has demonstrated that

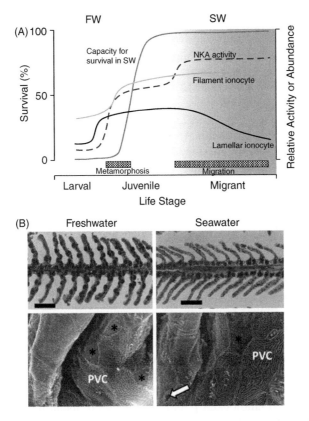

Fig. 6.5. Preparatory adaptation in American shad. (A) Seawater (SW) tolerance develops at the larval–juvenile transition in American shad, accompanied by parallel increases in gill Na^+/K^+-ATPase (NKA) and filamental and lamellar ionocytes as the gills are formed. FW=freshwater. (B) As shown by light (scale bar: 10 μm) and scanning electron microscopy, freshwater ionocytes are numerous on both the gill lamellae and filament with large apical surfaces (as indicated by *). During SW acclimation, total ionocyte numbers decrease and cells are restricted to the filament. Apical crypts characteristic of SW ionocytes are formed (arrow). Data modified from Zydlewski and McCormick (2001) and Zydlewski et al. (2003). Light micrographs reproduced with permission from Zydlewski and McCormick (2001). **SEE COLOR PLATE SECTION**

the FW and SW ionocytes are indistinct from those of other teleosts (Christensen et al., 2012). SW ionocytes have basolaterally located NKA and NKCC1, and an apical CFTR channel embedded in the apical crypt of the cells. The CFTR appears to be colocalized with the Na^+/H^+ exchange protein 3 (NHE3), which is likely to be important for acid–base regulation in SW (Evans et al., 2005). SW ionocytes are arranged in diads or triads, which probably facilitates paracellular Na^+ extrusion through the formation of leaky tight junctions (Christensen et al., 2012; Edwards and Marshall, 2013, Chapter 1, this volume).

The FW to SW transition in alewife leads to increased gill NKA activity and abundance on the basolateral membrane. NKCC abundance also increases (Christensen et al., 2012). Based on similar increases in gill NKA activity and changes in SW ionocyte distribution following SW acclimation (Zydlewski and McCormick, 1997a, 2001), the fundamental mechanisms can be presumed to be the same in American shad.

The decline in osmoregulatory competence seen in American shad with prolonged residence in FW suggests that it is an obligate anadromous species (Zydlewski et al., 2001). However, alewives appear to be facultative in the expression of anadromy, as demonstrated by their ability to regulate internal Na^+ (Stanley and Colby, 1961) and Cl^- (Christensen et al., 2012) during prolonged or even permanent residence in FW (Scott and Crossman, 1973). American shad residence in FW is apparently curtailed by the temporal loss of hyperosmoregulatory ability in FW (Zydlewski and McCormick, 1997b, 2001; Zydlewski et al., 2003) and these fish are rarely successful if landlocked (von Geldern, 1965; Lambert et al., 1980). Like American shad, similar reductions in ionoregulatory competence have also been reported in alewives exposed to cold temperatures following prolonged periods in FW, which may have contributed to the sudden die-offs of alewives that frequently occurred in the Great Lakes during the 1960s and 1970s (Stanley and Colby, 1971). There is a population of landlocked American shad in California (von Geldern, 1965; Lambert et al., 1980), not to mention several landlocked populations of shad, including the allis shad (*A. alosa*) and twaite shad (*A. fallax*), in Europe (Bianco, 2002; Bagliniere et al., 2003; Jolly et al., 2012). Direct comparisons between the anadromous and landlocked Alosines could help to tease out the relative importance of physiological preparation in these fishes.

5.1.4. Temperate Basses

The approximately 50 species of temperate basses inhabit FW, estuaries, and marine environments. Some, including the striped bass (*Morone saxitilis*) and white perch (*Morone americana*), are anadromous, spawning in FW and typically migrating towards the sea as juveniles (Scott and Scott, 1988). However, schools of juveniles often remain in estuaries, with occasional forays into FW that may last for several months or longer (Scott and Scott, 1988). The period of preparatory adaptation for SW appears to be minimal in the striped bass following hatching. While tolerance to increased salinity is limited during the yolk-sac and larval stages, juvenile striped bass are able to tolerate up to 10–15 ppt salinity only a few months after hatching (Tagatz, 1961; Otwell and Merriner, 1975; Kane et al., 1990; Grizzle and Mauldin, 1994; Winger and Lasier, 1994; Secor et al., 2000; Cook et al., 2010). Survival at higher salinities is positively correlated with warmer water temperatures in juvenile white perch and striped bass, perhaps due to limitations in ATP supply at lower temperatures. Reduced power for NKA pumps may therefore reduce gill-mediated Na^+ and Cl^- extrusion (Hurst and Conover, 2002; Hanks and Secor, 2011). This may explain why these fishes tend to overwinter in lower salinity regions of the estuary, when the metabolic demands of osmoregulation would be less than at higher salinities (Hurst and Conover, 2002). Indeed, marked reductions in energy stores have been reported in young-of-the-year white perch following acclimation to 16 ppt salinity compared to more dilute waters (Hanks and Secor, 2011).

Studies on striped bass, the most studied of the temperate basses, indicate that transfer from FW to SW is associated with transient increases in plasma Na^+ and osmolality that generally recover to pretransfer levels within 24 h (Madsen et al., 1994; Tipsmark et al., 2004). In both SW and FW, NKA and CFTR appears to be localized to large cells, which are presumably ionocytes, found at the base of the lamellae and in the interlamellar spaces (Madsen et al., 2007). Ionocytes appear early in the ontogeny of the striped bass, when they are mainly restricted to the gill filaments in larvae, followed by a shift to the lamellae in the juvenile stages (43 days posthatch) (Hirai et al., 2002). As suggested earlier, such findings imply that there may be two populations of ionocytes, a filamental population involved in osmoregulation at higher salinities and an FW population needed for hypoosmoregulation in more dilute waters. King and Hossler (1991) proposed that the rapid acclimation of FW-acclimated striped bass to SW was related to the restructuring of the ionocytes characterized by extensions of the apical membranes of ionocytes, which were associated with increased Cl^- efflux. These ionocytes were replaced within 7 days by cells displaying the distinct apical crypts characteristic of SW ionocytes, but ionocytes similar to those seen in FW were still retained. These early findings therefore suggest that the striped bass retains some of the physiological machinery required for osmoregulation in both SW and FW.

Unlike many other anadromous migrating fishes, SW acclimation is associated with only minor changes in gill ionocyte number or distribution (King and Hossler, 1991; Madsen et al., 1994), although the surface area of individual cells increases slightly (Madsen et al., 1994, 2007). Neither NKA activity nor protein abundance (α1 subunit) changes greatly following FW to SW (Madsen et al., 1994, 2007; Tipsmark et al., 2004) or SW to FW transfer (Tipsmark et al., 2004). Nor does CFTR messenger RNA (mRNA) expression or protein abundance appreciably change in response to greater salinity (Madsen et al., 2007). On the other hand, NKCC mRNA expression and protein abundance markedly increase following FW to SW transfer, consistent with a switch from ion uptake in FW to ion extrusion in SW (Tipsmark et al., 2004).

Like the estuarine killifish (*Fundulus heteroclitus*) (see Marshall, 2013, Chapter 8, this volume), the striped bass appears to be in a "state of readiness" to control osmotic balance in the event of sudden changes in salinity (Madsen et al., 2007). Madsen et al. (2007) have speculated that the striped bass uses an "ionocyte shift-reuse" strategy in which the CFTR is translocated to the basolateral membrane from its typical apical location, while the NKCC1 is internalized (or removed) from the basolateral membrane. Such reorganization would rapidly convert the ionocyte from a salt-secreting to salt-absorbing cell, in which Cl^- uptake would presumably take place via apical Cl^-/HCO_3^- exchange followed by basolateral uptake via the CFTR, and Na^+ uptake would take place via epithelial Na^+ channels coupled to H^+-ATPase-mediated proton extrusion. However, H^+-ATPase activity is quite low in striped bass compared to other FW fishes such as the brown trout (*Salmo trutta*), suggesting that higher basal rates of NKA activity or abundance may be needed to help generate the low intracellular Na^+ concentrations needed to generate favorable inwardly directed Na^+ electrochemical gradients in FW (Tipsmark et al., 2004). Indeed, there may also be a functional switch in NKA indicated by differential mRNA expression of the NKA $\alpha1$ subunit, shifting from the "FW" $\alpha1a$ to the "SW" $\alpha1b$ isoform, as described in rainbow trout (Richards et al., 2002). While intriguing, evidence for the ionocyte shift-reuse model is lacking. Approaches using *in vitro*, immunohistochemical, and immunoblotting approaches may be productive ways to elucidate intracellular signaling pathways that control ion uptake and extrusion in the gills of striped bass transitioning between FW and SW. However, it would also be advisable to combine such approaches with direct measurements of unidirectional movements of Na^+ and Cl^- across the gills to better relate the changes in ion uptake and diffusive loss to events occurring at the cellular, subcellular, and molecular level of the gills.

While physiological adjustments to changes in salinity are rapid, endocrine regulation is likely to play an important but slightly different role in mediating the FW to SW acclimation in the striped bass. Consistent with its role in SW acclimation in other species (Marshall, 2002), cortisol increases following FW to SW transfer in striped bass (Madsen et al., 1994). However, the importance of cortisol remains unresolved (Tipsmark et al., 2007). Exogenous cortisol administration has no effect on the expression of gill NKA or CFTR mRNA levels *in vivo*, and actually suppressed CFTR expression in isolated gill tissue *in vitro* (Madsen et al., 2007). Kiilerich et al. (2011) recently demonstrated that cortisol and 11-deoxycorticosterone activated the mitogen-activated protein (MAP) kinase signaling cascade in striped bass. Thus, even subtle, transient increases in hormone levels could activate or inactivate CFTR, NKCC1, and NKA (or other critical proteins) via phosphorylation pathways. As in other fishes, ionocytes are not regulated by corticosteroids alone. In combination with cortisol, epidermal growth factor (EGF) may play a key role in striped bass osmoregulation by regulating CFTR function via the MAP kinase pathway (Madsen et al., 2007). However, the role of cortisol plus EGF in the acquisition of SW tolerance remains unresolved as experiments with isolated gill cells have demonstrated a downregulation of CFTR abundance, rather than the expected increase following FW to SW transfer (Madsen et al., 2007). Insulin-like growth factor-1 (IGF-I) may also be critical because IGF-I receptor mRNA expression is upregulated in ionocytes following FW to SW transfer (Tipsmark et al., 2007). Increased plasma IGF-I has been shown to play an important role in smoltification in salmonines (Sakamoto et al., 1995). However, exogenously administered IGF-I impairs the hypoosmor-egulatory ability of striped bass, suggesting that its fundamental actions differ from those reported in salmonines (Tipsmark et al., 2007). It is more likely that IGF-I is involved in FW acclimation as it promotes prolactin release in striped bass pituitaries (Fruchtman et al., 2000). Work on the molecular signaling pathways controlling osmoregulation in striped bass, as well as other species including estuarine killifish (Marshall, 2013), is still in its early stages, but should fill in many gaps in our understanding of the processes controlling salt excretion by marine fishes.

Based on the rapid corrections to osmotic balance in striped bass with SW acclimation, changes in gut ion and water transport capacity are likely to occur in parallel to those taking place in the gill. Madsen et al. (1994) reported that the water transport capacity of the mid-intestine doubled with a 35% increase in intestinal NKA activity following FW to SW transfer (Madsen et al., 1994). Increased drinking rate occurs in 1-month-old striped bass following transfer from FW to BW (5 ppt) (Grizzle and Cummins, 1999). Like the gills, the intestine and the kidneys are probably also in a state of "readiness" to allow striped bass to rapidly adjust to changes in external salinity. However, most of the work thus far has focused on physiological adjustments occurring in the gills. Given the importance of the gut and kidneys in SW osmoregulation, studies focusing upon the temporal plasticity of these organ systems in response to changes in salinity are clearly warranted.

5.2. Catadromous Fishes

5.2.1. ANGUILLIDS

As described earlier, newly hatched eel larvae (leptocephali) are transported on ocean currents from ocean spawning grounds to intertidal areas where they transform into glass eels. These juveniles accumulate at the head of tide before initiating active swimming upstream. This delay in migration may be associated with morphological and physiological preparation for the riverine environment as they transition to pigmented "yellow eels" (Jellyman, 1977; McCleave and Wippelhauser, 1987; Pease et al., 2003). There is, however, no obvious shift in osmoregulatory capacity at this stage. Glass eels make the transition from SW to FW without notable changes in water content (Wilson et al., 2007) or plasma osmolality (Seo et al., 2009). Glass eels captured in the lower tidal area are already competent to osmoregulate in FW (Wilson et al., 2004).

Recent findings using telemetry and Sr:Ca ratios in otoliths suggest that northern temperate eels are facultatively catadromous (Tsukamoto et al., 2001; Daverat et al., 2006; Thibeault et al., 2007; Arai et al., 2009), with some animals persisting in the estuary and making sorties into FW (Fig. 6.2) (Jessop et al., 2008; see also Shrimpton, 2013, Chapter 7, this volume). Yellow eels must therefore be robust to the changes in salinity that occur through the tidal cycle and the gills play a critical role. Gill ionocytes are distributed on both the lamellar and filamental epithelium of the yellow eel (Thomson and Sargent, 1977; Sasai et al., 1998). In Japanese eel, two types of ionocytes, acidophilic type A and weakly acidophilic type B ionocytes, are present (Shirai and Utida, 1970). The type B ionocytes (referred to as CCs by the authors) were differentiated from type A based on their smaller size, smaller mitochondria with less distinct cristae, and a less elaborate tubular network (Shirai and Utida, 1970; Utida et al., 1971). More recent work revealed that these basolateral invaginations are rich in NKA (Marshall, 2002; Evans et al., 2005).

Abrupt transfer of yellow eels from FW to SW is accompanied by an initial increase in ionocyte abundance and size (Keys and Wilmer, 1932; Olivereau, 1970; Utida et al., 1971; Thomson and Sargent, 1977), and parallel increases in gill NKA activity (Kamiya and Utida, 1968; Utida et al., 1971; Thomson and Sargent, 1977; Rankin, 2009) and abundance (Cutler et al., 1995a,b). In Japanese eel it is the type A ionocytes that increase in number and size, followed by the gradual elimination of the type B ionocytes after 2 weeks in SW (Shirai and Utida, 1970).

While such plasticity is of inherent value when fish are faced the uncertain salinity of the estuary, an increase in active Na^+ and Cl^- extrusion capacity in full-strength SW probably results in greater energetic costs. However, otolith Sr:Ca ratio data also indicate that growth is much less in FW-dwelling yellow eels and those spending prolonged periods in BW or SW (Jessop et al., 2008; Cairns et al., 2009; Lamson et al., 2009). Therefore it remains unclear what the relationship is between energy expenditure and osmoregulatory capacity in the yellow eel. It would be informative to determine the energetic costs that transient excursions into SW from FW or

BW have on yellow eels, compared to silver eels that are undergoing physiological preparation for their seaward migration.

The transition from the yellow to silver stage ("silvering") in anguillid eels is perhaps the most obvious change associated with the initiation of seaward migration and reproductive maturation (e.g. Utida et al., 1967; Thomson and Sargent, 1977; Fontaine et al., 1995; Lignot et al., 2002; Tesch, 2003; Acou et al., 2005; Kalujnaia et al., 2007; van Ginneken et al., 2007b). Silvering also marks a period of physiological preparedness for the eel's extended, terminal stay in SW (see Tesch, 2003; Rankin, 2009 for recent reviews). However, there appears to be no difference in the SW tolerance of yellow and silver eels (Rankin, 2009).

Gill ionocyte abundance changes little during silvering (Thomson and Sargent, 1977; Sasai et al., 1998): ionocytes on the lamellae of the gill are lost, while ionocyte size and number on the gill filament increase (Fontaine et al., 1995; Sasai et al., 1998). These shifts result in modest increases in NKA activity in FW, which further increases after exposure to SW (Thomson and Sargent, 1977; Sasai et al., 1998; Rankin, 2009).

As in the clupeids (see above), the loss of lamellar ionocytes implicates these cells in ion uptake, while the filamental ionocytes are probably for salt extrusion (Sasai et al., 1998; Sakamoto et al., 2001). Detailed ultrastructure analysis also noted a more extensive tubular network and greater numbers of mitochondria in filamental ionocytes (Doyle and Epstein, 1972; Fontaine et al., 1995). Accessory cells were also observed in close association with the ionocytes in both yellow and silver eels, which would be consistent with a greater capacity to excrete Na^+ and Cl^- in SW (Fontaine et al., 1995).

"Silvering" may be associated with reduced FW osmoregulatory capacity, as evidenced by the loss of lamellar ionocytes; the silver eel may in fact be at a point of no return. The decrease in branchial aquaporin 3 (AQP3) mRNA (Cutler and Cramb, 2002a; Tse et al., 2006) and protein (Lignot et al., 2002) reported in SW-acclimated eels could also compromise FW osmoregulatory capacity by impairing cell volume control and other physiological processes in ionocytes (Cutler et al., 2007). Indeed, disturbances to ion homeostasis (demineralization) may be an important trigger of FW to SW migration in silver eels (Dutil et al., 1987; Durif et al., 2009), but this hypothesis requires further investigation.

There is ample evidence that FW to SW transfer by eels is accompanied by a necessary increase in ingestion of SW. Smith (1932) and Keys (1933) demonstrated this by preventing water ingestion by blocking the esophagus with a surgically implanted balloon. This led to uncontrolled water loss in SW. Maetz and Skadhague (1968) later demonstrated that FW-acclimated eels ingested water as well, but at lower rates than in SW. As in other teleosts, ingested SW is desalinated, followed by further dilution in the stomach before the water reaches the intestine. In the intestine the bulk of water uptake is driven by the lower osmolality of the fluid, further facilitated by Cl^- uptake via intestinal NKCC and CFTR (Cutler and Cramb, 2002b; Ando et al., 2003). Although mRNA AQP3 is located in the esophagus and gut, expression does not change between FW and SW (Cutler et al., 2007) and is not immunolocalized to enterocytes as would be expected if it were involved in water uptake (Lignot et al., 2002). It is more likely that water is taken up paracellularly, between adjacent enterocytes, as it is in mammals (Cutler et al., 2007). While Cutler and colleagues have examined the underlying endocrine basis of AQP regulation (e.g. Martinez-Alvarez et al., 2005; Cutler et al., 2007), control of drinking behavior and osmoregulatory capacity of the digestive tract remain poorly characterized.

Silvering is probably cued by changes in photoperiod and lunar cycles (Tsukamoto et al., 2003), as mediated by changes in endocrine status. The endocrine control of silvering is somewhat analogous to smoltification in salmonids (van Ginneken et al., 2007a; also see McCormick, 2013, Chapter 5, this volume). In yellow European eel, silvering may be initiated by T_4, which peaks in the spring and rises modestly in late summer prior to silvering (van Ginneken et al., 2007a). Similar findings, along with parallel increases in the β-subunit of thyroid-stimulating hormone (TSH) mRNA, were reported in Japanese eel (Han et al., 2004). However, Aroua et al. (2005) did not observe similar variation in T_4 profiles, and chronic T_4 administration to yellow eels did not induce silvering.

Elevated plasma cortisol and GH act in a dual manner to increase NKA activity and SW ionocyte number in smolting salmonids (McCormick, 2001), but the actions of these hormones in silvering eels is less clear-cut. Plasma GH concentrations, at least in female eels, show no distinct temporal variation in the months preceding silvering (van Ginneken et al., 2007a). It also seems unlikely that GH plays a significant role in osmoregulation because hypophysectomy has no effect on FW or SW tolerance in eels (Olivereau and Ball, 1970). However, there is a two-fold increase in plasma cortisol in silver compared to yellow European eel (van Ginneken et al., 2007a). This increase may be causal to increases in ionocyte size and number as well as increased NKA activity in the gill (Epstein et al., 1971; Wong and Chan, 2001) and intestine (Epstein et al., 1971). These shifts during silvering result in the higher salt excretory capacity (Mayer et al., 1967). Increased cortisol may also promote the mobilization of energy stores needed for migration (van Ginneken et al., 2007b).

Cortisol also produces an upregulation of AQP1 in the esophagus and intestine of SW-acclimating European eel (Martinez-Alvarez et al., 2005). This suggests that in addition to salt extrusion, cortisol plays an important role in water desalination and water uptake from ingested SW (Hirano and

Utida, 1968, 1971). It is unlikely that changes in cortisol alone are sufficient to trigger increased water uptake by the eel digestive tract, and it is still unclear whether such changes precede SW entry as preparatory adaptations. Other hormones have been implicated in osmoregulatory processes, including atrial natriuretic peptide and somatostatin (reviewed by Ando et al., 2003; Rankin, 2009), but their roles are unclear. The advent of high-throughput genomic techniques such as microarray analysis (Kalujnaia et al., 2007) should yield additional clues about the physiological and hormonal processes that prepare the temperate eels for seaward migration.

5.2.2. MUGILIDS

The catadromous mullets (Mugilidae) spawn in offshore SW waters, and the newly hatched larvae drift shoreward into saltmarshes and estuaries where they develop into juveniles (Moore, 1974; McDowall, 1988; Nordlie, 2000). Striped mullet (*Mugil cephalus*) larvae and small juveniles are not capable of osmoregulating in FW until they are least 40 mm in length (Nordlie et al., 1982; Ciccotti et al., 1995), but survive and grow in 17 ppt to full SW. While larger juveniles are tolerant of salinities from FW to SW, they generally remain in elevated salinities (Nordlie, 2000; Cardona, 2006), where growth is enhanced (Nordlie, 2000; Cardona, 2006). This may be because the energetic costs of osmoregulation are lowered when the animals are in oligomesohaline (brackish) waters nearer the osmolality of their own tissues (Murashige et al., 1991; Cardona, 2006).

The mullets also apparently maintain a state of physiological "prepared-ness" for a wide range of salinities. FW ionocytes are found on the gill filament, the interlamellar space, and the lamellae, and increase in density in response to low salinity (Cicotti et al., 1994; Khodabandeh et al., 2009). Entry into FW is marked by pronounced increases in gill NKA activity (Gallis and Bourdichon, 1976; Cicotti et al., 1994), which is localized to gill ionocytes (Cicotti et al., 1994; Khodabandeh et al., 2009).

SW ionocytes dominate the interlamellar space in SW-acclimated animals. Like the ionocytes of other euryhaline fishes, these cells are characterized by an elongated shape and a pronounced apical crypt (Cicotti et al., 1994). Given the increasing scarcity of temperate eels, and the ability to rear large numbers of mullet in hatcheries (Lee and Ostrowski, 2001), it is conceivable that this fish could serve as an important model to further our understanding of the physiological drivers and basis for catadromy.

5.3. Amphidromous Fishes

Because amphidromous fishes migrate seaward soon after hatching, early development of SW tolerance is necessitated. The fundamental mechanisms of osmoregulation in the amphidromous fishes are probably the same as in their diadromous counterparts (e.g. McCormick et al., 2003). FW ionocytes play a more dominant role in FW adults, but SW ionocytes are likely to be more important in the larvae as they passively drift from FW, through estuaries, and on to SW. Studies on the amphidromous ayu (*Plecoglossus altivelis*) of Japan suggest that increased temperature and salinity lead to more rapid yolk depletion and impaired growth in larvae (Iguchi and Takeshima, 2011). The gills probably take on added importance after the yolk sac is resorbed (Rombough, 1988), as described in the juvenile Hawaiian goby (*Stenogobius hawaiiensis*) (McCormick et al., 2003). These metamorphosed juveniles captured in FW have ionocytes found on both the gill filament and the lamellae. Following acclimation to SW (20 and 30 ppt), the ionocytes increase in size and number. This increase is accompanied by a modest increase in the abundance of basolateral NKCC and NKA proteins, and NKA activity. Immunopositive staining for CFTR increases markedly in SW, and the protein is restricted to the apical crypts of the SW ionocytes. Thus, these amphidromous gobies retain their ability to osmoregulate in SW for some time after FW entry. This is similar to the pattern noted earlier in the striped bass, but also the estuarine mummichog (Marshall, 2013, Chapter 8, this volume).

Preparatory adaptation is probably essential for larval amphidromous fishes. In the Antillian rock-climbing goby (*Sicydium punctatum*), early larvae (0–5 d posthatch) select salinities less than 10 ppt, but within a week volitionally occupy increasing salinities. Exposure to elevated salinities is associated with an early cessation of migratory behaviors (Bell and Brown, 1995). Yada et al. (2010) used a salinity gradient that allowed larval ayu to spontaneously move from FW to SW. Their study revealed that a marked downregulation of prolactin mRNA takes place within 10 days of hatching, coinciding with movement into SW. Similar reductions in GH are also observed, although slightly delayed. Whole-body water and Na^+ content are similar in SW-selecting and FW-selecting fishes, but these metrics are perturbed in larvae abruptly transferred into SW. It is therefore tempting to speculate that reduced prolactin secretion precedes SW entry in these (and perhaps other) amphidromous fishes.

Metamorphosis in the amphidromous goby (*Sicyopterus lagocephalus*) appears to be triggered by a rise in both T_3 and T_4 (Taillebois et al., 2011). Given the importance that these hormones have in osmoregulation and development in other diadromous fishes, they are likely to influence osmoregulation following this species' return to FW. While this work begins to prove the endocrine control of FW entry, the role of T_3 and T_4 in the preparatory adaptation for SW entry as larvae remains a conspicuous gap in our knowledge. Very little is known of the relationship between osmotic tolerances and development in amphidromous fishes.

5.4. Freshwater-Linked and Seawater-Linked Fishes

FW-linked organisms face the challenge of possibly having to hypoosmoregulate in saline waters, and the current evidence suggests that this is achieved through the modulation of FW and SW gill ionocyte abundance and distribution, and corresponding increases in NKA activity (Morgan et al., 1997; Varsamos et al., 2005; Tseng and Hwang, 2008). Drinking also increases in FW-linked fishes such as the Mozambique tilapia following the FW to SW transition, along with decreases in urinary output (Varsamos et al., 2005). Although the Mozambique tilapia can withstand full-strength SW, many other FW-linked fishes that facultatively use the estuary have upper tolerance levels well below full-strength SW. For instance, Sr:Ca otolith data suggest that wild largemouth bass (*M. salmoides*) can withstand salinities approaching 10 ppt in estuaries along the US Atlantic coast, but tend to avoid higher salinities, probably due to physiological constraints (Lowe et al., 2009). The bulk of evidence generated so far would seem to suggest that the FW-linked fishes respond to more saline waters, rather than making any preparatory physiological adjustments before entering higher salinities. However, this working hypothesis could change as more work is done on these estuarine fishes.

Preparatory adaptation is also important in SW-linked fishes such as the flounder (*P. dentatus*) and gilthead sea bream (*S. aurata*), which hatch at sea in full-strength SW and drift into the estuary (Schreiber and Specker, 2000; Bodinier et al., 2010). In the summer flounder, thyroid hormone is critical for triggering metamorphosis, and it probably triggers a shift from larval ionocytes that function in the dilute salinities of the estuary to juvenile ionocytes better suited to ion excretion in SW (Schreiber and Specker, 2000). In the sea bream, osmotic tolerance is dependent upon increased ionocytes in the integument and the gills, reflecting the variable salinity these fish experience (Bodinier et al., 2010). More work is needed, however, to determine whether changes to the osmoregulatory apparatus are initiated before entry into SW.

The use of the estuary may also be facultative. For instance, SW-linked juvenile sea bass (*Dicentrarchus labrax*) reproduce and hatch in saltwater, with some juveniles entering more dilute BW (Nebel et al., 2005). However, survival is highly variable in the more dilute waters of the estuary and rivers up which some animals are known to migrate (Lemaire et al., 2000; Varsamos, 2002). The underlying physiological basis for the inability of some fish to cope with more dilute waters appears to be related to the lack of renal tubules, which makes it impossible for the fish to produce the copious amounts of dilute (hypoosmotic) urine needed to counteract water uptake in more dilute waters (Nebel et al., 2005). While these differences may be genetically predetermined (Nebel et al., 2005), these findings also suggest that sea bass may have a narrow window of opportunity to take advantage of the estuary, before ontogenetically determined changes more suited to life in SW environments become irreversible. Ultimately, a better understanding of the underlying physiological mechanisms, and the endogenous (endocrine signals, genetic preprogramming) and exogenous signals (e.g. photoperiod, temperature) that regulate these processes, is needed to more fully understand habitat use by FW-linked and SW-linked fishes.

6. GROWTH AND OSMOREGULATION

Osmoregulating fish in FW or SW must actively transport ions, necessitating energy expenditures associated with standard metabolic rate. Assuming hypoosmoregulation and hyperosmoregulation both require active energy inputs, an isosmotic environment of approximately 12 ppt should be least taxing (Potts, 1954; Watanabe et al., 1989; Wootton, 1990; Gaumet et al., 1995; Imsland et al., 2001; Rocha et al., 2005), as demonstrated for Mozambique tilapia (*Oreochromis mossambicus*) (Febry and Lutz, 1987; Morgan et al., 1997; Chang et al., 2007). Observed patterns diverge from this construct, varying with species, ontogeny, and season (Gutt, 1985; Morgan and Iwama, 1991; Lambert et al., 1994; Deacon and Hecht, 1999). The energy cost of osmoregulation in different salinities may be relatively modest in teleosts (Boeuf and Payan, 2001) but can be significant when salinity shifts rapidly (Du Preez et al., 1990; Morgan et al., 1997; Morgan and Iwama, 1999). Salinity effects can be acute. If unable to offset the mass actions of the surrounding environment, disruption of ion balance occurs in advance of metabolic failure and mortality (Woo and Fung, 1981). An expanded examination of the energetics of osmoregulation can be found in Chapter 9 of this volume (Brauner et al., 2013).

Salinity may impact metabolism directly (increasing standard metabolic rate) or indirectly through food conversion efficiency, endocrine shifts associated with acclimation, or changing feeding behavior (Boeuf and Payan, 2001). Direct metabolic costs of osmoregulation in SW have been reported to be as low as approximately 10% in recent studies (e.g. Kidder et al., 2006) but as much as 50% in others (Boeuf and Payan, 2001). Because endocrine factors that affect osmoregulation are also important for growth (e.g. IGF-I and GH) (McCormick, 1996; Mancera and McCormick, 1998), the impact of salinity may transcend a simple energetic cost. Because drinking is an active part of the osmoregulatory process in SW, the scope for growth in SW may be influenced by the increased ATP demands that SW ingestion could place on the enterocytes (absorptive cells) of the gastro-intestinal tract. As a result, energy partitioning could influence the external and internal milieu of the gastrointestinal tract, influencing digestion and food conversion efficiency (DeSilva and Perera, 1976; MacLeod, 1977; Ferraris et al., 1986). For many fish, salinity does influence growth and this optimum can be empirically assessed.

Many FW spawners (FW-linked, anadromous, and amphidromous fishes) have higher growth rates in salinities at or below isosmotic (0–12 ppt). Even some stenohaline FW species exhibit increased growth at low salinities, below 2 ppt, by increasing their food conversion rate (Boeuf and Payan, 2001), perhaps by offsetting the costs of ion uptake in FW (e.g. the FW catfish *Mystus vittatus*) (Arunachalam and Reddy, 1979). As the environment becomes more saline, growth scope declines (Britz and Hecht, 1989; Morgan and Iwama, 1991; Brown et al., 1992). For green sturgeon there is no difference in metabolic rate in FW or SW, but osmotic perturbations in SW may prevent entry into SW or confer a growth disadvantage (Allen and Cech, 2007). Lower growth rates in SW have been observed in other sturgeon juveniles (e.g. shortnose sturgeon) (Jarvis and Ballantyne, 2003). Some species have fairly defined optima, e.g. approximately 7 ppt for striped bass (Brown et al., 1992). For some amphidromous fishes, the early growth phase may be optimized in BW. For the ayu high salinity results in the acceleration of yolk depletion and reduced growth (Iguchi and Takeshima, 2011). The distribution of ayu larvae near shore through the yolk-sac and larval stage (Tago, 2002; Yagi et al., 2006) suggests the significance of low-salinity BW in their survival.

Conversely, many SW-linked and catadromous fished generally have increasing growth near or above isosmotic (usually 5–18 ppt; Boeuf and Payan, 2001; Morrison and Secour, 2003; Jessop et al., 2008; Acou et al., 2003; Melia et al., 2006). While the estuary may be viewed as an environment to which many SW fishes are not adapted, many species exhibit greater growth at lower salinities, e.g. European flounder (*Platichthys flesus*) (Gutt, 1985) and Atlantic cod (*Gadus morhua*) (Lambert et al., 1994). Increased growth rate may not necessarily indicate the lowest cost of osmoregulation, but an integration of processes. Cod grown at 7 ppt had higher growth than at 28 ppt owing to greater food conversion efficiency (Boeuf and Payan, 2001), and feeding can decrease in juvenile cod at higher salinities (Lambert et al., 1994). Changing conditions can change feeding dynamics (Le Bail and Boeuf, 1997). For turbot (*Scophthalmus maximus*), greater growth at or near isosmotic salinity was linked to increased food ingestion (Gaumet et al., 1995; Imsland et al., 2001).

It should be noted that assessment of metabolic costs and growth is usually accomplished in a static setting and does not incorporate costs of movements through varying salinity environments. Swimming studies on tilapia (*Oreochromis niloticus*) (Farmer and Beamish, 1969) and *O. mossambicus* (Febry and Lutz, 1987) indicated that oxygen consumption at 12 ppt was lower than in full FW or SW, yet growth studies at near isotonic conditions provide less clear results (Morgan and Iwama, 1991; Boeuf and Payan, 2001). Such apparent differences may be associated with the multiple functions of the gill for respiration and osmoregulation. Across a steep osmotic gradient, there is a theoretical diminishing return on increased gill perfusion due to the cost of ion transport (Boeuf and Payan, 2001) leading to an osmoregulatory compromise (Nilsson, 1986).

Growth and metabolism are effective physiological assessments not only of salinity influence, but also of the optimal environmental conditions of a fish (Cech, 1990). Modeling of environmental parameters in an ecological framework has provided valuable insight into the role of osmoregulation in bluegill (*Lepomis macrochirus*) (Neill et al., 2004), red drum (*Sciaenops ocellatus*) (Fontaine et al., 2007), and juvenile sole (*Solea solea*) (Fonseca et al., 2010). For many fish in the estuary, salinity in conjunction with temperature and oxygen can define suitable habitat (Imsland et al., 2001). For example, white perch have reduced feeding and increased metabolic costs associated with salinities greater than 16 ppt (Hanks and Secor, 2011) and these fish are seldom observed in salinities that exceed this level (Setzler-Hamilton, 1991; Nemerson and Able, 2004). Juvenile white perch exhibit the greatest growth rates in salinities from 4 to 8 ppt (Kerr and Secor, 2009), but this benefit is not independent of temperature and dissolved oxygen. The interaction of these environmental factors has been characterized as a "habitat squeeze" (Coutant and Benson, 1990; Niklitschek and Secor, 2005; Hanks and Secor, 2011) where temperature increases escalate basal metabolic demands (Guderley and Pörtner, 2010), further perturbing the osmorespiratory compromise. In addition to increased metabolic rates under hypoxic conditions, food consumption decreases (Hanks and Secor, 2011). These three environmental factors therefore limit rearing areas for juvenile white perch through their impact on growth and, probably, performance (Miltner et al., 1995; Lankford et al., 2001; Harrison and Whitfield, 2006). It is not unreasonable to assume that other fishes would be similarly limited.

7. CONCLUSIONS AND PERSPECTIVES

Over the past few decades advances in a suite of technologies have informed both the "where" and "how" of diadromous migrations.

Telemetry has long been used for direct tracking of large species (e.g. sturgeon) (Taverny et al., 2002; Kelly et al., 2007) but technological advances have allowed effective tracking even for moderately sized juveniles such as salmon smolts (e.g. Holbrook et al., 2011). The advent of Sr:Ca microchemical analysis of otoliths, and the correlation with salinity, changed the playing field in the study of migrating fishes (Kalish, 1990; Secor, 1992; Secor et al., 1995; de Pontual et al., 2003). This approach has advanced our understanding of migratory patterns on one hand while blurring the lines of stereotyped life history variants on the other. The use of stable isotopes has also revealed complexities in movement patterns. Isotope signatures can reflect integration of diet over several weeks, allowing spatial feeding patterns to be inferred over a brief timescale (Hesslein et al., 1993; Barnes and Jennings, 2007). In addition to using $\delta^{13}C$ and $\delta^{15}N$ measures, $\delta^{34}S$ has likewise emerged as an important tool for assessing movement in estuarine fishes (Hoffman et al., 2007; Fry and Chumchal, 2011). Less technical approaches have also been informative. Trnski (2002) physically followed the larvae of estuary-dependent coastal spawners using scuba gear. Together, these efforts have revealed the degree to which many fish that migrate between FW and SW are facultative in their use of the estuary (Gerking, 1994; Blaber, 1997; Wootton, 1999; Elliott and Hemingway, 2002; Elliott et al., 2007). The construct of "contingent behaviors" as described by Kerr and Secor (2010) is useful in describing variations in migratory patterns.

Understanding dynamic habitat shifts of the estuary and their importance to euryhaline fishes will be critical as estuarine waters become more highly impacted by human activity (Able, 1999; Hoss et al., 1999; Quinlan and Crowder, 1999). The recruitment of ecologically important species like the anchovy (*Engraulis encrasicolus*) is strongly linked to river inflow (Chícharo et al., 2001; Drake et al., 2007) such that impoundments would be likely to compromise anchovy success (Morais et al., 2009; Morais et al., 2010). Degradation of estuarine conditions has been linked to lower survival rates in salmonine species (Magnusson and Hilborn, 2003), and there is no reason to suspect that these effects would be restricted to salmonines in the reduction of life history and species diversity. Longer term stressors due to factors such as climate change may influence settlement habitats (Able et al., 2006). A northward creep of distributions of euryhaline fishes has already begun to reshape temperate estuaries (Nicolas et al., 2010). Flow regimens into estuarine systems impact the suitability of the estuary as a rearing habitat through their influence on salinity, temperature, and oxygen (Chícharo et al., 2006; Lassalle and Rochard, 2009). Hanks and Secor (2011) speculate that this "habitat squeeze" would be exacerbated by climate change scenarios (e.g. Najjar et al., 2010) as regions of hypoxia are more likely to persist during the summer months of temperate estuaries (Breitburg et al., 2009). Industrial, agricultural, and municipal contaminants and reduced FW input could also have additive or synergistic impacts on available habitat (Cooper and Brush, 1991; Najjar et al., 2000; Niklitschek and Secor, 2005).

The estuarine environment is also likely to be a frontline for many invasive species. Salinity tolerance will probably define the speed and extent of expansions of invasives (e.g. armored catfish) (Capps et al., 2011) which may opportunistically use fluctuations in river plumes as a means of dispersal (Brown and St. Pierre, 2001; Bringolf et al., 2005; Scott et al., 2008). Non-native piscivores such as the largemouth bass (*M. salmoides*) may threaten estuary-dependent SW species along the Atlantic seaboard (Weyl and Lewis, 2006; Wasserman and Strydom, 2011; Peer et al., 2006; Norris et al., 2010). Invasives may also influence native euryhaline fishes through direct interaction, such as through competition between juveniles of SW-spawned and non-native fish species (Wathen et al., 2011; Wathen et al., 2012) or through diet overlap (Skelton, 1993; Mansfield and McCardle, 1998).

These ecological challenges are inextricably linked to the physiological abilities and ontogenic requirements of migrating species. Efforts to characterize the physiological adjustments that precede or accompany the FW to SW transition will inform measures to counter the inevitable challenges that these euryhaline fishes face. Although the past decade has seen an explosion of research on osmoregulation in migratory euryhaline species, much of the field remains open territory. However, several themes are emerging. There are notable commonalities between salmonine smoltification and many of the preparatory adaptations of other species described in this chapter. Like smolting salmonines, sea lamprey, alosines, and catadromous eels exhibit increases in NKA abundance and activity, as well as in branchial ionocytes, prior to seaward migration. In advance of SW entry (and/or upon SW entry) there is a shift from predominantly FW ionocytes to SW ionocytes. While preparatory adaptation may be obligatory, some fishes are quite plastic. Other species (e.g. alewife and striped bass) or other life stages (e.g. yellow eels) exhibit an impressive capacity to move between FW and SW. These fishes remain in a state of readiness, perhaps mediated by rapid control of ionocyte function.

Cortisol is a critical endocrine factor that mediates the changes in gill ionocyte structure and function in anticipation of SW entry. As in salmonines, thyroid hormone probably plays an important role by promoting corticosteroid receptor proliferation. The critical role played by T_4 and T_3 in initiating metamorphosis (e.g. in lampreys, flounders, and at least one species of ampidromous goby) appears to be directly tied to the preparatory adaptation process. It also appears likely, as in salmonids, that cortisol acts in concert with GH to initiate SW ionocyte proliferation (McCormick, 2001; Evans et al., 2005). Our understanding of ion and water transport processes in the digestive tract has advanced substantially (Ando et al., 2003; Grosell, 2011) but research on the endocrine control of preparatory changes is limited. Less is known about the preparative changes of the kidneys, as they shift from the role of "bilge pump" to an organ of water conservation and divalent ion excretion. Although research is now beginning to shed light on cell-sensing and intracellular signaling pathways, particularly the interactions between IGF-I and EGF, and the MAP kinases in sensing cell volume, much more work is needed in this area.

A great deal more integrative research is therefore needed to learn how environmental, somatic, temporal, and physiological factors interact to allow fish to undergo the dramatic and critical transition from FW to SW. With the advent and development of more sophisticated techniques (e.g. cell isolation, morphilino, immunohistochemistry) and continual advances in genomics and proteomics, the tools are currently available to address the data gaps identified here. We are in a period of great discovery and of great potential. Integrative research approaches can not only inform our understanding of fish physiology, but also provide fisheries managers with the information they need to protect these fascinating and vulnerable species.

REFERENCES

Able, K. W. (1999). Measures of juvenile fish habitat quality: examples from a national estuarine research reserve. In *Fish Habitat: Essential Fish Habitat and Rehabilitation*, Vol. 22 (ed. L. Banaka), pp. 134–147. Bethesda, MD: American Fisheries Society.

Able, K. and Fahay, M. P. (1998). *The First Year in the Life of Estuarine Fishes in the Middle Atlantic Bight*. New Brunswick, NJ: Rutgers University Press.

Able, K. W., Fahay, M. P. and Shepherd, G. R. (1995). Early life history of black sea bass, *Centropristis striata* in the mid-Atlantic Bight and a New Jersey estuary. *Fish. Bull.* 93, 429–445.

Able, K. W., Hales, L. S., Jr. and Hagan, S. M. (2005). Movement and growth of juvenile (age 0 and 1+) tautog (*Tautoga onitis* [L.]) and cunner (*Tautogolabrus adspersus* [Walbaum]) in a southern New Jersey estuary. *J. Exp. Mar. Biol. Ecol.* 327, 22–35.

Able, K. W., Fahay, M. P., Witting, D. A., McBrided, R. S. and Hagan, S. M. (2006). Fish settlement in the ocean vs. estuary: comparison of pelagic larval and settled juvenile composition and abundance from southern New Jersey. *Estuar. Coast. Shelf Sci.* 66, 280–290.

Acou, A., Lefebvre, F., Contournet, P., Poizat, G., Panfili, J. and Crivelli, A. J. (2003). Silvering of female eels (*Anguilla anguilla*) in two sub-populations of the Rhône Delta. *Bull. Fr. Pêche. Piscic.* 368, 55–68.

Acou, A., Boury, P., Laffaille, P., Crivelli, A. J. and Feunteun, E. (2005). Towards a standardized characterization of the potentially migrating silver European eel (*Anguilla anguilla* L.). *Arch. Hydrobiol.* 164, 237–255.

Adam, G., Feunteun, E., Prouzet, P. and Rigaud, C. (2008). *L'Anguille Européenne: Indicateurs d'Abondance et de Colonisation*. Versailles: Editions Quae.

Alderdice, D. F. (1988). Osmotic and ionic regulation in teleost eggs and larvae. In *Fish Physiology*, Vol. 11A, *The Physiology of Developing Fish: Eggs and Larvae* (eds. W. S. Hoar and D. J. Randall), pp. 163–242. London: Academic Press.

Allen, P. J. and Cech, J. J. (2007). Age/size effects on juvenile green sturgeon, *Acipenser medirostris*, oxygen consumption, growth, and osmoregulation in saline environments. *Environ. Biol. Fish.* 79, 211–229.

Allen, P. J., Cech, J. J. and Kültz, D. (2009). Mechanisms of seawater acclimation in a primitive, anadromous fish, the green sturgeon. *J. Comp. Physiol. B* 179, 903–920.

Allen, P. J., McEnroe, M., Forostyan, T., Cole, S., Nicholl, M. M., Hodge, B. and Cech, J. J. (2011). Ontogeny of salinity tolerance and evidence for seawater-entry preparation in juvenile green sturgeon. *Acipenser medirostris. J. Comp. Physiol. B* 181, 1045–1062.

Altinok, I., Galli, S. M. and Chapman, F. A. (1998). Ionic and osmotic regulation capabilities of juvenile Gulf of Mexico sturgeon (*Acipenser oxyrinchus desotoi*). *Comp. Biochem. Physiol. A Mol. Integr. Physiol.* 120, 609–616.

Amiri, B. M., Baker, D. W., Morgan, J. D. and Brauner, C. J. (2009). Size dependent early salinity tolerance in two sizes of juvenile white sturgeon (*Acipenser transmontanus*). *Aquaculture* 286, 121–126.

Anderson, W. W. (1957). Larval forms of freshwater mullet (*Agonostomus monticola*) from the open ocean off the Bahamas and South Atlantic coast of the United States. *U. S. Fish. Wildl. Serv. Fish. Bull.* 120, 415–425.

Ando, M., Makuda, T. and Kozaka, T. (2003). Water metabolism in the eel acclimated to sea water: from mouth to intestine. *Comp. Biochem. Physiol. B Biochem. Mol. Biol.* 136, 621–633.

Arai, T., Kotake, A. and McCarthy, T. K. (2006). Habitat use by the European eel (*Anguilla anguilla*) in Irish waters. *Estuar. Coast. Shelf Sci.* 67, 569–578.

Arai, T., Chino, N. and Kotake, A. (2009). Occurrence of estuarine and sea eels (*Anguilla japonica*) and a migrating silver eel (*Anguilla anguilla*) in the Tokyo Bay area, Japan. *Fish. Sci. (Tokyo Jpn)* 75, 1197–1203.

Aroua, S., Schmitz, M., Baloche, S., Vidal, B., Rousseau, K. and Dufour, S. (2005). Endocrine evidence that silvering, a secondary metamorphosis in the eel, is a pubertal rather than a metamorphic event. *Neuroendocrinology* 82, 221–232.

Arunachalam, S. and Reddy, S. R. (1979). Food intake, growth, food conversion, and body composition of catfish exposed to different salinities. *Aquaculture* 16, 163–171.

Attrill, M. J. and Power, M. (2002). Climatic influence on a marine fish assemblage. *Nature* 417, 275–278.

Attrill, M. J. and Power, M. (2004). Partitioning of temperature resources amongst an estuarine fish assemblage. *Estuar. Coast. Shelf Sci.* 61, 725–738.

Backman, T. W. and Ross, M. (1990). Comparison of three techniques for the capture and transport of impounded subyearling American shad. *Prog. Fish.-Cult.* 52, 246–252.

Bagliniere, J. L., Sabatie, M. R., Rochard, E., Alexamdrino, P. and Aprahamian, M. W. (2003). The allis shad (*Alosa alosa*): biology, ecology, range, and status of populations. *Am. Fish. Soc. Symp.* 35, 85–102.

Bain, M. B. (1997). Atlantic and shortnose sturgeons of the Hudson River: common and divergent life history attributes. *Prog. Fish.-Cult.* 48, 347–358.

Balon, E. K. (1984). Reflections on some decisive events in the early life of fishes. *Trans. Am. Fish. Soc.* 133, 178–185.

Balon, E. K. (1990). Epigenesis of an epigeneticist: the development of some alternative concepts on the early ontogeny and evolution of fishes. *Guelph Ichthyol. Rev.* 1, 1–42.

Balon, E. K. and Bruton, M. N. (1994). Fishes of the Tatinga River, Comoros, with comments on freshwater amphidromy in the goby (*Sicyopterus lagocephalus*). *Ichthyol. Explor. Freshw.* 5, 25–40.

Bardonnet, A., Bolliet, V., and Belon, V. (2005). Recruitment abundance estimation: Role of glass eel (*Anguilla anguilla* L.) response to light. *J. Exp. Mar. Biol. Ecol.* 321(2), 181–190.

Bardonnet, A. and Jatteau, P. (2008). Salinity tolerance in young Allis shad larvae (*Alosa alosa* L.). *Ecol. Freshw. Fish.* 17, 193–197.

Bardonnet, A., Dasse, S., Parade, M. and Heland, M. (2003). Study of glass-eels movements in a flume in relation to nyctemeral changes. *Bull. Fr. Pêche. Piscic.* 368, 9–20.

Barnes, C. and Jennings, S. (2007). Effect of temperature, ration, body size and age on sulphur isotope fractionation in fish. *Rapid Commun. Mass Spectrom.* 21, 1461–1467.

Bartels, H. and Potter, I. C. (2004). Cellular composition and ultrastructure of the gill epithelium of larval and adult lampreys – implications for osmoregulation in fresh and seawater. *J. Exp. Biol.* 207, 3447–3462.

Bartels, H., Potter, I. C., Pirlich, K. and Mallatt, J. (1998). Categorization of the mitochondria-rich cells in the gill epithelium of the freshwater phases in the life cycle of lampreys. *Cell Tissue Res.* 291, 337–349.

Beamish, F. W. H. (1979). Migration and spawning energetics of the anadromous sea lamprey, *Petromyzon marinus. Env. Biol. Fish.* 4(1), 3–7.

Beamish, F. W. H. (1980a). Biology of the North American anadromous sea lamprey, *Petromyzon marinus. Can. J. Fish. Aquat. Sci.* 37, 1924–1943.

Beamish, F. W. H. (1980b). Osmoregulation in juvenile and adult lampreys. *Can. J. Fish. Aquat. Sci.* 37, 1739–1750.

Beamish, F. W. H. and Potter, I. C. (1975). Biology of anadromous sea lamprey, *Petromyzon marinus*, in New Brunswick. *Can. J. Zool.* 177, 57–72.

Beamish, F. W. H., Strachan, P. D. and Thomas, E. (1978). Osmotic and ionic performance of anadromous sea lamprey *Petromyzon marinus. Comp. Biochem. Physiol. A Mol. Integr. Physiol.* 60, 435–443.

Beck, M., Heck, K., Able, K., Childers, D., Egglestone, D., Gillanders, B., Halpern, B., Hays, C., Hoshino, K., Minello, T., Orth, R., Sheridan, P. and Weinstein, M. (2001). The identification, conservation and management of estuarine and marine nurseries for fish and invertebrates. *BioScience* 51, 633–641.

Bell, K. N. I. and Brown, J. A. (1995). Active salinity choice and enhanced swimming endurance in 0 to 8-d-old larvae of diadromous gobies including *Sicydium punctatum* (Pisces) in Dominica, West Indies. *Mar. Biol.* 121, 409–417.

Bemis, W. E. and Kynard, B. (1997). Sturgeon rivers: an introduction to Acipenseriform biogeography and life history. *Environ. Biol. of Fish.* 48, 167–183.

Ben-Tuvia, A. (1979). Studies of the population and fisheries of (*Sparus aurata*) in the Bardawil Lagoon, eastern Mediterranean. *Invest. Pesq.* 43, 43–68.

Beumer, J. P. and Harrington, D. J. (1980). Techniques for collecting glass eels and brown elvers. *Austr. Fish.* 39, 16–22.

Beyenbach, K. W. (2004). Kidneys sans glomeruli. *Am. J. Physiol. Renal Physiol.* 286, F811–F827.

Bianco, P. G. (2002). The status of the twaite shad, *Alosa agone*, in Italy and the western Balkans. *Mar. Ecol. Publ. Stn. Zool. Napoli.* 23, 51–64.

Blaber, S. J. M. (1974). Osmoregulation in juvenile *Rhabdosargus holubi* (Steindachner) (Teleostei: Sparidae). *J. Fish Biol.* 6, 797–800.

Blaber, S. J. M. (1997). *Fish and Fisheries of Tropical Estuaries*. London: Chapman & Hall.

Bodinier, C., Sucré, E., Lecurieux-Belfond, L., Blodeau-Bidet, E. and Charmantier, G. (2010). Ontogeny of osmoregulation and salinity tolerance in the gilthead sea bream (*Sparus auratus*). *Comp. Biochem. Physiol. A Mol. Integr. Physiol.* 157, 200–228.

Boehlert, G. W. and Mundy, B. C. (1988). Roles of behavioral and physical factors in larval and juvenile fish recruitment to estuarine nursery areas. *Am. Fish. Soc. Symp.* 3, 51–67.

Boesch, D. F. and Turner, R. E. (1984). Dependence of fishery species on salt marshes – the role of food and refuge. *Estuaries* 7, 460–468.

Boeuf, G. (1993). Salmonid smolting: a pre-adaptation to the oceanic environment. In *Fish Ecophysiology* (eds. J. C. Rankin and F. B. Jensen), pp. 105–135. London: Chapman & Hall.

Boeuf, G. and Payan, P. (2001). How should salinity influence fish growth? *Comp. Biochem. Physiol. C Comp. Pharmacol.* 130, 411–423.

Bolliet, V. and Labonne, J. (2008). Individual patterns of rhythmic swimming activity in *Anguilla anguilla* glass eels synchronised to water current reversal. *J. Exp. Mar. Biol. Ecol.* 362, 125–130.

Borkin, I. V. (1991). Ichthyoplankton of western Spitzbergen coastal waters. *J. Ichthyol.* 31, 680–685.

Brauner, C. J., Gonzalez, R. J. and Wilson, J. M. (2013). Extreme environments: hypersaline, alkaline, and ion-poor waters. In *Fish Physiology,* Vol. 32, *Euryhaline Fishes* (eds. S. D. McCormick, A. P. Farrell and C. J. Brauner), pp. 435–476. New York: Elsevier.

Breitburg, D. L., Craig, J. K., Fulford, R. S., Rose, K. A., Boynton, W. R., Brady, D., Ciotti, B. J., Diaz, R. J., Friedland, K. D., Hagy, J. D., III, Hart, D. R., Hines, A. H., Houde, E. D., Kolesar, S. E., Nixon, S. W., Rice, J. A., Secor, D. H. and Targett, T. E. (2009). Nutrient enrichment and fisheries exploitation: interactive effects on estuarine living resources and their management. *Hydrobiologia* 629, 31–47.

Bringolf, R. B., Kwak, T. J., Cope, W. G. and Larimore, M. S. (2005). Salinity tolerance of flathead catfish: implications for dispersal of introduced populations. *Trans. Am. Fish. Soc.* 134, 927–936.

Britz, P. J. and Hecht, T. (1989). Effects of salinity on growth and survival of African sharptooth catfish (*Clarias gariepinus*). *J. Appl. Ichthyol.* 5, 194–202.

Brown, J. J., St. and Pierre, R. A. (2001). Restoration of American shad (*Alosa sapidissima*) populations in the Susquehanna and Delaware rivers. *OCEANS, 2001. MTS/IEEE Conference and Exhibition* 1–4, 321–326.

Brown, K. (2007). Evidence of spawning by green sturgeon, *Acipenser medirostris*, in the upper Sacramento River, California. *Environ. Biol. Fish.* 79, 297–303.

Brown, M. L., Nematipour, G. R. and Gatlin, D. M. (1992). The dietary protein requirement of juvenile sunshine bass at different salinities. *Prog. Fish.-Cult.* 54, 148–156.

Brundage, H. M. and Meadows, R. E. (1982). The Atlantic sturgeon, *Acipenser oxyrhynchus*, in the Delaware River estuary. *Fish. Bull.* 80, 337–343.

Bureau du Colombier, S., Bolliet, V., Lambert, P. and Bardonnet, A. (2007). Energy and migratory behavior in glass eels (*Anguilla anguilla*). *Physiol. Behav.* 92, 684–690.

Bureau du Colombier, S., Bolliet, V. and Bardonnet, A. (2009). Swimming activity and behaviour of European *Anguilla anguilla* glass eels in response to photoperiod and flow reversal and the role of energy status. *J. Fish Biol.* 74, 2002–2013.

Bureau du Columbier, S. B., Lambert, P. and Bardonnet, A. (2011). Metabolic loss of mass in glass eels at different salinities according to their propensity to migrate. *Estuar. Coast. Shelf Sci.* 93, 1–6.

Burrows, M. T. (2001). Depth selection behaviour during activity cycles of juvenile plaice on a simulated beach slope. *J. Fish Biol.* 59, 116–125.

Burrows, M. T., Gibson, R. N., Robb, L. and MacLean, A. (2004). Alongshore dispersal and site fidelity of juvenile plaice from tagging and transplants. *J. Fish Biol.* **65**, 620–634.

Butcher, A. and Ling, J. (1958). *Bream Tagging Experiments in East Gippsland during April and May (1944)*. East Melbourne: Victoria Fisheries and Wildlife Department.

Bystriansky, J. S. and Schulte, P. M. (2011). Changes in gill H^+-ATPase and Na^+/K^+-ATPase expression and activity during freshwater acclimation of Atlantic salmon (*Salmo salar*). *J. Exp. Biol.* 214, 2435–2442.

Cairns, D. K., Secor, D. A., Morrison, W. E. and Hallett, J. A. (2009). Salinity-linked growth in anguillid eels and the paradox of temperate-zone catadromy. *J. Fish Biol.* 74, 2094–2114.

Campos, M. C., Costa, J. L., Quintella, B. R., Costa, M. J. and Almeida, P. R. (2008). Activity and movement patterns of the Lusitanian toadfish inferred from pressure-sensitive dataloggers in the Mira estuary (Portugal). *Fish. Manag. Ecol.* 15, 449–458.

Capps, K. A., Nico, L. G., Carranza, M. M., Areválo-Frías, W., Ropicki, A. J., Heilpern, S. A. and Rodiles-Hernández, R. (2011). Salinity tolerance of the exotic armored catfish (*Siluriformes loricariidae*) in southern Mexico: potential new pathways for invasion. *Aquat. Conserv. Mar. Freshw. Ecosyst.* 21, 528–540.

Cardona, L. (2006). Habitat selection by grey mullets (Osteichthyes: Mugilidae) in Mediterranean estuaries: the role of salinity. *Sci. Mar.* 70, 443–455.

Carmichael, J. T., Haeseker, S. L. and Hightower, J. E. (1998). Spawning migration of telemetered striped bass in the Roanoke River, North Carolina. *Trans. Am. Fish. Soc.* 127, 286–297.

Cech, J. J., Jr. (1990). Respirometry. In *Methods for Fish Biology* (eds. C. B. Schreck and P. B. Moyle), pp. 335–362. Bethesda, MD: American Fisheries Society.

Chang, J. C., Wu, S., Tseng, Y., Lee, Y., Baba, O. and Hwang, P. P. (2007). Regulation of glycogen metabolism in gills and liver of the euryhaline tilapia (*Oreochromis mossambicus*) during acclimation to seawater. *J. Exp. Biol.* 210, 3494–3504.

Chícharo, L., Chícharo, M. A., Esteves, E., Andrade, J. P. and Morais, P. (2001). Effects of alterations in fresh water supply on the abundance and distribution of *Engraulis encrasicolus* in the Guadiana estuary and adjacent coastal areas of south Portugal. *Ecohydrol. Hydrobiol.* 1, 341–345.

Chícharo, M. A., Chícharo, L. and Morais, P. (2006). Inter-annual differences of ichthyofauna structure of the Gaudiana estuary and adjacent coastal area (SE Portugal/ SW Spain): before and after Alqueva dam construction. *Estuar. Coast. Shelf Sci.* 70, 39–51.

Childs, A. R., Booth, A. J., Cowley, P. D., Potts, W. M., Næsje, T. F., Thorstad, E. B. and Okland, F. (2008). Home range of an estuarine-dependent fish species (*Pomadasys commersonnii*) in a South African estuary. *Fish. Manag. Ecol.* 15, 441–448.

Chittenden, M. E., Jr. (1972). Responses of young American shad, *Alosa sapidissima*, to low temperatures. *Trans. Am. Fish. Soc.* 101, 680–685.

Christensen, A. K., Hiroi, J., Schultz, E. T. and McCormick, S. D. (2012). Branchial ionocyte organization and ion-transport protein expression in juvenile alewives acclimated to freshwater or seawater. *J. Exp. Biol.* 215, 642–652.

Ciccotti, E., Macchi, E., Rossi, A., Cataldi, E. and Cataudella, S. (1993). Glass eel (*Anguilla anguilla*) acclimation to freshwater and seawater: morphological changes of the digestive tract. *J. Appl. Ichthyol.* 9, 74–81.

Ciccotti, E., Marino, G., Pucci, P., Cataldi, E. and Cataudella, S. (1995). Acclimation trial of *Mugil cephalus* juveniles to freshwater: morphological and biochemical aspects. *Environ. Biol. Fish.* 43, 163–170.

Clark, D. L., Leis, J. M., Hay, A. C. and Trnski, T. (2005). Swimming ontogeny of larvae of four temperate marine fishes. *Mar. Ecol. Prog. Ser.* 292, 287–300.

Close, D. A., Yun, S. S., McCormick, S. D., Wildbill, A. J. and Li, W. M. (2010). 11-Deoxycortisol is a corticosteroid hormone in the lamprey. *Proc. Natl Acad. Sci. U. S. A.* 107, 13942–13947.

Closs, G. P., Smith, M., Barry, B. and Markwitz, A. (2003). Non-diadromous recruitment in coastal populations of common bully (*Gobiomorphus cotidianus*). *N. Z. J. Mar. Freshw. Res.* 37, 301–313.

Cook, A. M., Duston, J. and Bradford, R. G. (2010). Temperature and salinity effects on survival and growth of early life stage shubenacadie river striped bass. *Trans. Am. Fish. Soc.* 139, 749–757.

Cooper, S. R. and Brush, G. S. (1991). Long-term history of Chesapeake Bay anoxia. *Science* 254, 992–996.

Coutant, C. C. and Benson, D. L. (1990). Summer habitat suitability for striped bass in Chesapeake Bay: reflections on a population decline. *Trans. Am. Fish. Soc.* 119, 757–778.

Cowen, R. K., Hare, J. A. and Fahay, M. P. (1993). Beyond hydrography: can physical processes explain larval fish assemblages within the Middle Atlantic Bight? *Bull. Mar. Sci.* 53, 567–587.

Cowley, P. D. (2001). Ichthyofaunal characteristics of a typical temporarily open/closed estuary on the south east coast of South Africa. *Ichthyol. Bull.* 71, 1–17.

Creutzberg, F. (1961). On the orientation of migrating elvers (*Anguilla vulgaris* Turt.) in a tidal area. *Neth. J. Sea Res.* 1, 257–338.

Cummings, S. M. and Morgan, E. (2001). Time-keeping system of the eel pout (*Zoarces viviparous*). *Chronobiol. Int.* 18, 1–9.

Cutler, C. P. and Cramb, G. (2002a). Branchial expression of an aquaporin 3 (AQP-3) homologue is downregulated in the European eel (*Anguilla anguilla*) following seawater acclimation. *J. Exp. Biol.* 205, 2643–2651.

Cutler, C. P. and Cramb, G. (2002b). Two isoforms of the $Na^+/K^+/2Cl^-$ cotransporter are expressed in the European eel (*Anguilla anguilla*). *Biochim. Biophys. Acta Biomembr.* 1566, 92–103.

Cutler, C. P., Sanders, I. L., Hazon, N. and Cramb, G. (1995a). Na^+,K^+-ATPase α1 subunit in the European eel (*Anguilla anguilla*). *Comp. Biochem. Physiol. B Biochem. Mol. Biol.* 111, 567–573.

Cutler, C. P., Sanders, I. L., Hazon, N. and Cramb, G. (1995b). Primary sequence, tissue specificity and expression of the Na^+, K^+-ATPase beta-1 subunit in the European eel (*Anguilla anguilla*). *Fish Physiol. Biochem.* 14, 423–429.

Cutler, C. P., Martinez, A. S. and Cramb, G. (2007). The role of aquaporin 3 in teleost fish. *Comp. Biochem. Physiol. A Mol. Integr. Physiol.* 148, 82–91.

Daverat, F., Limburg, K. E., Thibault, I., Shiao, J. C., Dodson, J. J., Caron, F. O., Tzeng, W. N., Lizuka, Y. and Wickstrom, H. (2006). Phenotypic plasticity of habitat use by three temperate eel species, *Anguilla anguilla*, *A. japonica* and *A. rostrata*. *Mar. Ecol. Prog. Ser.* 308, 231–241.

David, B. O., Chadderton, L., Closs, G., Barry, B. and Markwitz, A. (2004). Evidence of flexible recruitment strategies in coastal populations of giant kokopu (*Galaxias argenteus*). DOC Science Internal Series 160, Department of Conservation Wellington, New Zealand.

De Casamajor, M. N., Bru, N. and Prouzet, P. (1999). Influence de la luminosité nocturne et de la turbidité sur le comportement vertical de migration de la civelle d'anguille (*Anguilla anguilla* L.) dans l'estuaire de l'Adour. *Bull. Fr. Pêche. Piscic.* 355, 327–347.

de Wolf, P. (1973). Ecological observations on the mechanisms of dispersal of barnacle larvae during planktonic life and settling. *Neth. J. Sea Res.* 6, 1–29.

Deacon, N. and Hecht, T. (1999). The effect of reduced salinity on growth, food conversion and protein efficiency ratio in juvenile spotted grunter, *Pomadasys commersonii* (Lacepede) (Teleostei: Haemulidae). *Aquacult. Res.* 30, 13–20.

Delacroix, P. and Champeau, A. (1992). Ponte en eau douce de *Sicyopterus lagocephalus* (Pallas) poisson Gobiidaeamphi bionte des rivières de La Réunion. *Hydroecology* 14, 49–63.

DeMartini, E. E. and Sikkel, P. C. (2006). Reproduction. In The Ecology of Marine Fishes – California and Adjacent Waters (eds. L. G. Allen, D. J. Pondella, II and M. H. Horn), pp. 483–523. San Francisco, CA: University of California Press.

Deng, X., Van Eenennaam, J. P. and Doroshov, S. I. (2002). Comparison of early life stages and growth of green and white sturgeon. In Biology, Management, and Protection of North American Sturgeon, Symposium 28 (eds. W. Van Winkle, P. J. Anders, D. H. Secor and D. A. Dixon), pp. 237–248. Bethesda, MD: American Fisheries Society.

DeSilva, S. S. and Perera, P. A. B. (1976). Studies on young grey mullet, Mugil cephalus L. I. Effects of salinity on food intake, growth and food conversion. Aquaculture 7, 327–338.

de Pontual, H., Lagardere, F., Amara, R., Bohn, M. and Ogor, A. (2003). Influence of ontogenetic and environmental changes in the otolith microchemistry of juvenile sole (Solea solea). J. Sea Res. 50, 199–210.

Dingle, H. (1996). Migration: The Biology of Life on the Move. New York: Oxford University Press.

Dingle, H. (2006). Animal migration: is there a common migratory syndrome? J. Ornithol. 147, 212–220.

Doroshov, S. I. (1985). The biology and culture of sturgeon. In Recent Advances in Aquaculture, Vol. 2 (eds. J. Muir and R. Roberts), pp. 251–274. London: Croom Helm.

Dou, S. Z. and Tsukamoto, K. (2003). Observations on the nocturnal activity and feeding behavior of Anguilla japonica glass eels under laboratory conditions. Environ. Biol. Fish. 67, 389–395.

Dovel, W. L. (1971). Fish eggs and larvae of the Upper Chesapeake Bay. College Park, University of Maryland: National Research Institute, Special Report No. 4.

Dovel, W. L. and Berggren, T. J. (1983). Atlantic sturgeon of the Hudson estuary, New York. N. Y. Fish Game J. 30, 140–172.

Doyle, W. L. and Epstein, F. H. (1972). Effects of cortisol treatment and osmotic adaptation on chloride cells in eel, Anguilla rostrata. Cytobiologie 6, 58–73.

Drake, P., Borla, A., González-Ortegón, E., Baldó, F., Vilas, C. and Fernández-Delgado, C. (2007). Spatio-temporal distribution of early life stages of the European anchovy (Engraulis encrasicolus L.) within an European temperate estuary with regulated freshwater inflow: effects of environmental variables. J. Fish Biol. 70, 1689–1709.

Du Preez, H. H., McLachlan, A., Marais, J. F. K. and Cockcroft, A. C. (1990). Bioenergetics of fishes in a high-energy surf-zone. Mar. Biol. 106, 1–12.

Durif, C. M. F, van Ginneken, V., Dufour, S., Müller, T. and Elie, P. (2009). Seasonal evolution and individual differences in silvering eels from different locations. In Spawning Migration of the European Eel (eds. G. van den Thillart, S. Dufour and J. C. Rankin), pp. 13–38. New York: Springer.

Dutil, J. D., Besner, M. and McCormick, S. D. (1987). Osmoregulatory and ionoregulatory changes and associated mortalities during the transition of maturing American eels to a marine environment. Am. Fish. Soc. Symp. 1, 175–190.

Duval, E. J. and Able, K. W. (1998). Aspects of the life history of the seaboard goby (Gobiosoma ginsburgi) in estuarine and continental shelf waters. Bull. N.J. Acad. Sci. 43, 5–10.

Edeline, E. (2007). Adaptive phenotypic plasticity of eel diadromy. Mar. Ecol. Prog. Ser. 341, 229–232.

Edwards, S. L. and Marshall, W. S. (2013). Principles and patterns of osmoregulation and euryhalinity in fishes. In Fish Physiology, Vol. 32, Euryhaline Fishes (eds. S. D. McCormick, A. P. Farrell and C. J. Brauner), pp. 1–44. New York: Elsevier.

Ego, K. (1956). Life history of freshwater gobies. Proj. No. F-4-R, Freshwater Game Fish Management Research, Department of Land and Natural Resources, State of Hawaii, Honolulu. 24 pp.

Elliott, M. and Dewailly, F. (1995). The structure and components of European estuarine fish assemblages. *Neth. J. Aquat. Ecol.* 29, 397–417.

Elliott, M. and Hemingway, K. L. (2002). *Fishes in Estuaries.* London: Blackwell Science.

Elliott, M., O'Reilly, M. G. and Taylor, C. J. L. (1990). The Forth estuary: a nursery and overwintering area for North Sea fishes. *Hydrobiologia* 195, 89–103.

Elliott, M., Whitfield, A. K., Potter, I. C., Blaber, S. J. M., Cyrus, D. P., Nordlie, F. G. and Harrison, T. D. (2007). The guild approach to categorizing estuarine fish assemblages: a global review. *Fish Fish.* 8, 241–268.

Englund, R. A. and Filbert, R. B. (1997). *Native and exotic stream organisms study in the Kawainui, Alakahi, Koiawe, and Lalakea Streams, Lower Hamakua Ditch watershed project, County of Hawaii.* USDA-NRCS Contract No. 53-9251-6-275.

Epstein, F. H., Cynamon, M. and McKay, W. (1971). Endocrine control of Na–K-ATPase and seawater adaptation in *Anguilla rostrata. Gen. Comp. Endocrinol.* 16, 323–328.

Eshenroder, R. L. (2009). Comment: Mitochondrial DNA analysis indicates sea lampreys are indigenous to Lake Ontario. *Trans. Am. Fish. Soc.* 138, 1178–1189.

Evans, D. H. (1984). The roles of gill permeability and transport mechanisms in euryhalinity. In *Fish Physiology*, Vol. 10B (eds. W. S. Hoar and D. J.Z. Randall), pp. 239–283. New York: Academic Press.

Evans, D. H. (1993). Osmotic and ionic regulation. In *The Physiology of Fishes* (ed. D. H. Evans), pp. 315–341. Boca Raton, FL: CRC Press.

Evans, D. H. (1999). Ionic transport in the fish gill epithelium. *J. Exp. Zool.* 283, 641–652.

Evans, D. H., Piermarini, P. M. and Choe, K. P. (2005). The multifunctional fish gill: dominant site of gas exchange, osmoregulation, acid–base regulation, and excretion of nitrogenous waste. *Physiol. Rev.* 85, 97–177.

Fairchild, E. A., Sulikowski, J., Rennels, N., Howell, W. H. and Gurshin, C. W. D. (2008). Distribution of winter flounder (*Pseudopleuronectes americanus*) in the Hampton-Seabrook Estuary, New Hampshire: observations from a field study. *Estuar. Coasts* 31, 1158–1173.

Farmer, G. J. (1980). Biology and physiology of feeding in adult lampreys. *Can. J. Fish. Aquat. Sci.* 37, 1751–1761.

Farmer, G. J. and Beamish, F. W. H. (1969). Oxygen consumption of *Tilapia nilotica* in relation to swimming speed and salinity. *J. Fish. Res. Bd. Can.* 26, 2807–2821.

Febry, R. and Lutz, P. (1987). Energy partitioning in fish: the activity-related cost of osmoregulation in a euryhaline cichlid. *J. Exp. Biol.* 128, 63–85.

Ferraris, R. P., Catacutan, M. R., Mabelin, R. L. and Jazul, A. P. (1986). Digestibility in milkfish, *Chanos chanos* (Forsskal): effects of protein source, fish size and salinity. *Aquaculture* 59, 93–105.

Feunteun, E., Laffaille, P., Robinet, T., Briand, C., Baisez, A., Olivier, J. M. and Acou, A. (2003). A review of upstream migration and movements in inland waters by anguillid eels: toward a general theory. In *Eel Biology* (eds. K. Aida, K. Tsukamoto and K. Yamauchi), pp. 191–212. Tokyo: Springer.

Fievet, E. P., Bonnet-Arnaud, P. B. and Mallet, J. P. (1999). Efficiency and sampling bias of electrofishing for freshwater shrimp and fish in two Caribbean streams, Guadeloupe Island. *Fish. Res.* 44, 149–166.

Fisher, R., Bellwood, D. R. and Job, S. D. (2000). Development of swimming abilities in reef fish larvae. *Mar. Ecol. Prog. Ser.* 202, 163–173.

Fitzsimons, J. D. and Ogorman., R. (1996). Fecundity of hatchery lake trout in Lake Ontario. *J. Great Lakes Res.* 22, 304–309.

Fitzsimons, J. M., Parham, J. E. and Nishimoto, R. T. (2002). Similarities in behavioral ecology among amphidromous and catadromous fishes on the oceanic islands of Hawaii and Guam. *Environ. Biol. Fish.* 65, 123–129.

Folmar, L. C. and Dickhoff, W. W. (1980). The parr–smolt transformation (smoltification) and seawater adaptation in salmonids – a review of selected literature. *Aquaculture* 21, 1–37.

Fonseca, V. F., Neill, W. H., Miller, J. M. and Cabral, H. N. (2010). Fish perspectives on growth of juvenile soles, *Solea solea* and *Solea senegalensis*, in the Tagus estuary, Portugal. *J. Sea Res.* 64, 118–124.

Font, W. F. and Tate, D. C. (1994). Helminth-parasites of native Hawaiian fresh-water fishes – an example of extreme ecological isolation. *J. Parasitol.* 80, 682–688.

Fontaine, L. P., Whiteman, K. W., Li, P., Burr, G. S., Webb, K. A., Goff, J., Gatlin, D. M., Neill, W. H., Davis, K. B. and Vega, R. R. (2007). Effects of temperature and feed energy on the performance of juvenile red drum. *Trans. Am. Fish. Soc.* 136, 1193–1205.

Fontaine, Y. A., Pisam, M., LeMoal, C. and Rambourg, A. (1995). Silvering and gill "mitochondria-rich" cells in the eel, *Anguilla anguilla. Cell Tiss. Res.* 281, 465–471.

Fortier, L. and Leggett, W. C. (1983). Vertical migrations and transport of larval fish in a partially mixed estuary. *Can. J. Fish. Aquat. Sci.* 40, 1543–1555.

Forward, R. B., Jr., Reinsel, K. A., Peters, D. S., Tankersley, R. A., Churchill, J. H., Crowder, L. B., Hettler, W. F., Warlen, S. M. and Green, M. D. (1999). Transport of fish larvae through a tidal inlet. *Fish. Oceanogr.* 8, 153–172.

Foskett, J. K. and Scheffey, C. (1982). The chloride cell: definitive identification as the salt-secreting cell in teleosts. *Science* 215, 164–166.

Franco, A., Elliott, M., Franzoi, P. and Torricelli, P. (2008). Life strategies of fishes in European estuaries: the functional guild approach. *Mar. Ecol. Prog. Ser.* 354, 219–228.

Fraser, T. H. (1997). Abundance, seasonality, community indices, trends and relationships with physiochemical factors of trawled fish in upper Charlotte Harbor, Florida. *Bull. Mar. Sci.* 60, 739–763.

Fry, B. and Chumchal, M. (2011). Sulfur stable isotope indicators of residency in estuarine fish. *Limnol. Oceanogr.* 56, 1563–1576.

Fuji, T., Kasai, A., Suzuki, K. W., Ueno, M. and Yamashita, Y. (2010). Freshwater migration and feeding habits of juvenile temperate seabass (*Lateolabrax japonicus*) in the stratified Yura River estuary, the Sea of Japan. *Fish. Sci.* 76, 643–652.

Furspan, P., Prange, H. D. and Greenwald, L. (1984). Energetics and osmoregulation in the catfish *Ictalurus nebulosus* and *I. punctatus. Comp. Biochem. Physiol. A Mol. Integr. Physiol.* 77, 773–778.

Gallant, J., Harvey-Clark, C., Myers, R. A. and Stokesbury, M. J. W. (2006). Sea lamprey attached to a Greenland shark in the St. Lawrence Estuary, Canada. *Northeast. Nat.* 13, 35–38.

Gallis, J. L. and Bourdichon, M. (1976). Changes of (Na-K) dependent ATPase activity in gills and kidneys of two mullets *Chelon labrosus* (Risso) and *Liza ramada* (Risso) during fresh water adaptations. *Biochimie* 58, 627–635.

Gascuel, D. (1986). Flow carried and active swimming migration of the glass eel (*Anguilla anguilla*) in the tidal area of a small estuary on the French Atlantic coast. *Helgolander Meeresuntersuchungen* 40, 321–326.

Gaumet, F., Boeuf, G., Sévère, A., Le Roux, A. and Mayer-Gostan, N. (1995). Effects of salinity on the ionic balance and growth of juvenile turbot. *J. Fish Biol.* 47, 865–876.

Gemperline, P. J., Rulifson, R. A. and Paramore, L. (2002). Multiway analysis of trace elements in fish otoliths to track migratory patterns. *Chemom. Intell. Lab. Syst.* 60, 135–146.

Gerking, S. D. (1994). Feeding variability. In *Feeding Ecology of Fish* (ed. S. D. Gerking), pp. 41–53. San Diego, CA: Academic Press.

Gess, R. W., Coates, M. I. and Rubidge, B. S. (2006). A lamprey from the Devonian period of South Africa. *Nature* 443, 981–984.

Gibson, R. N. (1997). Behavior and the distribution of flatfishes. *J. Sea Res.* 37, 241–256.

Gill, H. S., Renaud, C. B., Chapleau, F., Mayden, R. L. and Potter, I. C. (2003). Phylogeny of living parasitic lampreys (*Petromyzontiformes*) based on morphological data. *Copeia* 2003, 687–703.

Goto, A. (1990). Alternative life-history styles of Japanese sculpins revisited. *Environ. Biol. Fish.* 28, 101–112.

Goto, A. and Andoh, T. (1990). Genetic divergence between the sibling species of river-sculpin, *Cottus amblystomopsis* and *C. nozawae*, with special reference to speciation. *Environ. Biol. Fish.* 28, 257–266.

Graham, B. and Sampson, D. B. (1982). An experiment on factors affecting depth distribution of larval herring, *Clupea harengus*, in coastal Maine. *NAFO Scient. Council Stud.* 3, 33–38.

Gray, C. A., Chick, R. C. and McElligott, D. J. (1998). Diel changes in assemblages of fishes associated with shallow seagrass and bare sand. *Estuar. Coast. Shelf Sci.* 46, 849–859.

Griffith, R. W. (1974). Environment and salinity tolerance in genus. *Funulus. Copeia* 1974, 319–331.

Grizzle, J. M. and Cummins, K. A. (1999). Drinking rates of stressed one-month-old striped bass: effects of calcium and low concentrations of sodium chloride. *Trans. Am. Fish. Soc.* 128, 528–531.

Grizzle, J. M. and Mauldin, A. C. (1994). Age-related-changes in survival of larval and juvenile striped bass in different concentrations of calcium and sodium. *Trans. Am. Fish. Soc.* 123, 1002–1005.

Grosell, M. (2006). Intestinal anion exchange in marine fish osmoregulation. *J. Exp. Biol.* 209, 2813–2827.

Grosell, M. (2011). Intestinal anion exchange in marine teleosts is involved in osmoregulation and contributes to the oceanic inorganic carbon cycle. *Acta Physiol.* 202, 421–434.

Gross, M. R. (1987). Evolution of diadromy in fishes. *Am. Fish. Soc. Symp.* 1, 14–25.

Guderley, H. and Pörtner, H. O. (2010). Metabolic power budgeting and adaptive strategies in zoology: examples from scallops and fish. *Can. J. Zool.* 88, 753–763.

Gutt, J. (1985). The growth of juvenile flounders (*Platyichthys flesus* L.) at salinities of 0, 5, 15 and 35 ppt. *J. Appl. Ichthyol.* 1, 17–26.

Hackney, C. T. and de la Cruz, A. A. (1981). Some notes on the macrofauna of an oligohaline tidal creek in Mississippi. *Bull. Mar. Sci.* 31, 658–661.

Haeseker, S. L., Carmichael, J. T. and Hightower, J. E. (1996). Summer distribution and condition of striped bass within Albemarle Sound, North Carolina. *Trans. Am. Fish. Soc.* **125**, 690–704.

Hagan, S. M. and Able, K. W. (2008). Diel variation in the pelagic fish assemblage in a temperate estuary. *Estuar. Coasts* 31, 33–42.

Han, K. H., Kim, Y. U. and Choe, K. J. (1998). Spawning behavior and development of eggs and larvae of the Korea freshwater goby (*Rhinogobius brunneus*) (Gobiidae: Perciformes). *J. Kor. Fish. Soc.* 31, 114–120.

Han, Y. S., Liao, I. C., Tzeng, W. N. and Yu, J. Y. L. (2004). Cloning of the cDNA for thyroid stimulating hormone β subunit and changes in activity of the pituitary–thyroid axis during silvering of the Japanese eel, *Anguilla japonica. J. Mol. Endocrinol.* 32, 179–194.

Hanks, D. M. and Secor, D. H. (2011). Bioenergetic responses of Chesapeake Bay white perch (*Morone americana*) to nursery conditions of temperature, dissolved oxygen, and salinity. *Mar. Biol.* 158, 805–815.

Hardisty, M. W. and Potter, I. C. (1971). The behaviour, ecology, and growth of larval lampreys. In *The Biology of Lampreys*, Vol. 1 (eds. M. W. Hardisty and I. C. Potter), pp. 82–125. New York: Academic Press.

Hare, J. A. and Cowen, R. K. (1996). Transport mechanisms of larval and pelagic juvenile bluefish (*Pomatomus saltatrix*) from South Atlantic Bight spawning grounds to Middle Atlantic Bight nursery habitats. *Limnol. Oceanogr.* 41, 1264–1280.

Harrison, T. D. and Whitfield, A. K. (2006). Temperature and salinity as primary determinants influencing the biogeography of fishes in South African estuaries. *Estuar. Coast. Shelf Sci.* 66, 335–345.

He, X., Zhuang, P., Zhang, L. and Xie, C. (2009). Osmoregulation in juvenile Chinese sturgeon (*Acipenser sinensis* Gray) during brackish water adaptation. *Fish Physiol. Biochem.* 35, 223–230.

Heemstra, P. and Heemstra, E. (2004). *Coastal Fishes of Southern Africa.* Grahamstown, South Africa: National Inquiry Service Centre (NISC) and South African Institute for Aquatic Biodiversity (SAIAB).

Herzka, S. Z., Griffiths, R., Fodrie, F. J. and McCarthy, I. D. (2009). Short-term size-specific distribution and movement patterns of juvenile flatfish in a Pacific estuary derived through length–frequency and mark–recapture data. *Cienc. Mar.* 35, 41–57.

Hesslein, R. H., Hallard, K. A and Ramlal, P. (1993). Replacement of sulfur, carbon, and nitrogen in tissue of growing broad whitefish (*Coregonus nasus*) in response to a change in diet traced by $\delta34S$, $\delta13C$, and $\delta15N$. *Can. J. Fish. Aquat. Sci.* 50, 2071–2076.

Hibino, M., Ohta, T., Isoda, T., Nakayama, K. and Tanaka, M. (2006). Diel and tidal changes in the distribution and feeding habits of Japanese temperate bass (*Lateolabrax japonicus*) juveniles in the surf zone of Ariake Bay. *Ichthyol. Res.* 53, 129–136.

Hickman, C. P. (1959). The osmoregulatory role of the thyroid gland in the starry flounder. *Platichthys stellatus. Can. J. Zool.* 37, 997–1006.

Hicks, A. S., Closs, G. P. and Swearer, S. E. (2010). Otolith microchemistry of two amphidromous galaxiids across an experimental salinity gradient: a multi-element approach for tracking diadromous migrations. *J. Exp. Mar. Biol. Ecol.* 394, 86–97.

Hindell, J. S. (2007). Determining patterns of use by black bream (*Acanthopagrus butcheri*) (Munro, 1949) of re-established habitat in a southeastern Australian estuary. *J. Fish Biol.* 71, 1331–1346.

Hindell, J. S., Jenkins, G. P. and Womersley, B. (2008). Habitat utilisation and movement of black bream (*Acanthopagrus butcheri*) (Sparidae) in an Australian estuary. *Mar. Ecol. Prog. Ser.* 366, 219–229.

Hirai, N., Tagawa, M., Kaneko, T., Secor, D. H. and Tanaka, M. (2002). Freshwater adaptation in Japanese sea bass and striped bass: a comparison of chloride cell distribution during their early life history. *Fish. Sci.* 68, 433–434.

Hirano, T. and Mayergostan, N. (1976). Eel esophagus as an osmoregulatory organ. *Proc. Natl Acad. Sci. U. S. A.* 73, 1348–1350.

Hirano, T. and Utida, S. (1968). Effects of ACTH and cortisol on water movement in isolated intestine of eel *Anguilla japonica. Gen. Comp. Endocrinol.* 11, 373–380.

Hirano, T. and Utida, S. (1971). Plasma cortisol concentration and rate of intestinal water absorption in eel, *Anguilla japonica. Endocrinol. Jpn.* 18, 47–52.

Hoar, W. S. (1976). Smolt transformation – evolution, behavior, and physiology. *J. Fish. Res. Bd. Can.* 33, 1233–1252.

Hoar, W. S. (1988). The physiology of smolting salmonids. In *Fish Physiology*, Vol. XI (eds. H. W. Hoar and D. J. Randall), pp. 275–343. San Diego, CA: Academic Press.

Hobb, J. A., Bennett, W. A. and Burton, J. E. (2006). Assessing nursery habitat quality for native smelts (Osmeridae) in the low-salinity zone of the San Francisco estuary. *J. Fish Biol.* 69, 907–922.

Hoeksema, S. D. and Potter, I. C. (2006). Diel, seasonal, regional and annual variations in the characteristics of the ichthyofauna of the upper reaches of a large Australian microtidal estuary. *Estuar. Coast. Shelf Sci.* 67, 503–520.

Hoeksema, S. D., Chuwen, B. M. and Potter, I. C. (2006). Massive mortalities of the black bream (*Acanthopagrus butcheri*) (Sparidae) in two normally-closed estuaries, following extreme increases in salinity. *J. Mar. Biol. Assoc. U. K.* 86, 893–897.

Hoffman, J. C., Bronk, D. A. and Olney, J. E. (2007). Tracking nursery habitat use in the York River estuary, Virginia, by young American shad using stable isotopes. *Trans. Am. Fish. Soc.* 136, 1285–1297.

Holbrook, C., Kinnison, M. and Zydlewski, J. (2011). Survival of migrating Atlantic salmon smolts through the Penobscot River, Maine, USA: a pre-restoration assessment. *Trans. Am. Fish. Soc.* 140, 1255–1268.

Holliday, F. G. T. (1965). Osmoregulation in marine teleost eggs and larvae. *Calif. Coop. Ocean. Fish. Investig. Rep.* 10, 89–95.

Holliday, F. G. T. and Blaxter, J. H. S. (1960). The effects of salinity on the developing eggs and larvae of the herring. *J. Mar. Biol.* 39, 591–603.

Holliday, F. G. T. and Jones, M. P. (1967). Some effects of salinity on the developing eggs and larvae of the plaice (*Pleuronectes platessa*). *J. Mar. Biol.* 47, 39–48.

Holmes, J. A., Beamish, F. W. H., Seelye, J. G., Sower, S. A. and Youson, J. H. (1994). Long-term influence of water temperature, photoperiod, and food-deprivation on metamorphosis of sea lamprey, *Petromyzon marinus*. *Can. J. Fish. Aquat. Sci.* 51, 2045–2051.

Holmes, W. N. and Donaldson, E. M. (1969). The body compartments and the distribution of electrolytes. In *Fish Physiology*, Vol. I (eds. W. S. Hoar and D. J. Randall), pp. 1–89. New York: Academic Press.

Hoss, D. E., Bath, G. E., Cross, F. A. and Merriner, J. V. (1999). US Atlantic coastal fisheries – how effective have management plans been in maintaining "sustainable fisheries"? *Limnol. Oceanogr.* 29, 227–232.

Hourdry, J. (1995). Fish and cyclostome migrations between fresh water and sea water – osmoregulatory modifications. *Boll. Zool.* 62, 97–108.

Hurst, T. P. and Conover, D. O. (2002). Effects of temperature and salinity on survival of young-of-the-year Hudson River striped bass (*Morone saxatilis*): implications for optimal overwintering habitats. *Can. J. Fish. Aquat. Sci.* 59, 787–795.

Iguchi, K. and Mizuno, N. (1990). Diel changes of larval drift among amphidromous gobies in Japan, especially *Rhinogobius brunneus*. *J. Fish Biol.* 37, 255–264.

Iguchi, K. and Mizuno, N. (1991). Mechanisms of embryonic drift in the amphidromous goby, *Rhinogobius brunneus*. *Env. Biol. Fish.* 31, 295–300.

Iguchi, K. and Takeshima, H. (2011). Effect of saline water on early success of amphidromous fish. *Ichthyol. Res.* 58, 33–37.

Iguchi, K., Ito, F., Yamaguchi, M. and Matsubara, N. (1998). Spawning downstream migration of ayu in the Chikuma River. *Bull. Nat. Res. Inst. Fish. Sci.* 11, 75–84.

Iguchi, K. and Mizuno, N. (1999). Early starvation limits survival in amphidromous fishes. *J. Fish. Biol.* 54, 705–712.

Imsland, A. K., Foss, A., Gunnarsson, S., Berntssen, M. H. G., FitzGerald, R., Bonga, S. W., Ham, E. V., Nævdal, G. and Stefansson, S. O. (2001). The interaction of temperature and salinity on growth and food conversion in juvenile turbot (*Scophthalmus maximus*). *Aquaculture* 198, 353–367.

Jacob, W. F. and Taylor, M. H. (1983). The time course of seawater acclimation in *Fundulus heteroclitus* L. *J. Exp. Zool.* 228, 33–39.

Jager, Z. and Mulder, H. P. J. (1999). Transport velocity of flounder larvae (*Platichthys flesus* L.) in the Dollard (Ems Estuary). *Estuar. Coast. Shelf Sci.* 49, 327–346.

Jarvis, P. L. and Ballantyne, J. S. (2003). Metabolic responses to salinity acclimation in juvenile shortnose sturgeon (*Acipenser brevirostrum*). *Aquaculture* 219, 891–909.

Jellyman, D. J. (1977). Invasion of a New Zealand freshwater stream by glass-eels of two *Anguilla* spp. *N. Z. J. Mar. Freshw. Res.* 11, 193–209.

Jellyman, D. J. (1979). Upstream migration of glass-eels (*Anguilla* spp.) in the Waikato River. *N. Z. J. Mar. Freshw. Res.* 13, 13–22.

Jenkins, G. P. and Black, K. P. (1994). Temporal variability in settlement of a coastal fish (*Sillaginodes punctata*) determined by low-frequency hydrodynamics. *Limnol. Oceanogr.* 39, 1744–1754.

Jenkins, G. P., Black, K. P. and Keough, M. J. (1999). The role of passive transport and the influence of vertical migration on the pre-settlement distribution of a temperate, demersal fish: numerical model predictions compared with field sampling. *Mar. Ecol. Prog. Ser.* 184, 259–271.

Jensen, C. and Schwartz, F. J. (1994). Atlantic Ocean occurrences of the sea lamprey, *Petromyzon marinus* (Petromyzontiformes: Petromyzontidae), parasitizing sandbar, *Carcharhinus plumbeus*, and dusky *C. obscurus* (Carcharhiniformes: Carcharhinidae) sharks off North and South Carolina. *Brimleyana* 21, 69–72.

Jessop, B. M., Shiao, J. C., Iizuka, Y. and Tzeng, W. N. (2002). Migratory behaviour and habitat use by American eels (*Anguilla rostrata*) as revealed by otolith microchemistry. *Mar. Ecol. Prog. Ser.* 233, 217–229.

Jessop, B. M., Shiao, J. C., Iizuka, Y. and Tzeng, W. N. (2004). Variation in the annual growth, by sex and migration history, of silver American eels (*Anguilla rostrata*). *Mar. Ecol. Prog. Ser.* 272, 231–244.

Jessop, B. M., Shiao, J. C., Iizuka, Y. and Tzeng, W. N. (2006). Migration of juvenile American eels (*Anguilla rostrata*) between freshwater and estuary, as revealed by otolith microchemistry. *Mar. Ecol. Prog. Ser.* 310, 219–233.

Jessop, B. M., Cairns, D. K., Thibault, I. and Tzeng, W. N. (2008). Life history of American eel (*Anguilla rostrata*): new insights from otolith microchemistry. *Aquat. Biol.* 1, 205–216.

Jolly, M. T., Aprahamian, M. W., Hawkins, S. J., Henderson, P. A., Hillman, R., O'Maoileidigh, N., Maitland, P. S., Piper, R. and Genner, M. J. (2012). Population genetic structure of protected allis shad (*Alosa alosa*) and twaite shad (*Alosa fallax*). *Mar. Biol.* 159, 675–687.

Joyeaux, J. C. (1999). The abundance of fish larvae in estuaries: within tide variability at inlet and migration. *Estuaries* 22, 889–904.

Kalish, J. M. (1990). Use of otolith microchemistry to distinguish progeny of sympatric anadromous and non-anadromous salmonids. *Fish. Bull.* 88, 657–666.

Kalujnaia, S., McWilliam, I. S., Zaguinaiko, V. A., Feilen, A. L., Nicholson, J., Hazon, N., Cutler, C. P., Balment, R. J., Cossins, A. R., Hughes, M. and Cramb, G. (2007). Salinity adaptation and gene profiling analysis in the European eel (*Anguilla anguilla*) using microarray technology. *Gen. Comp. Endocrinol.* 152, 274–280.

Kamiya, M. and Utida, S. (1968). Changes in activity of sodium–potassium-activated adenosine triphosphatase in gills during adaptation of Japanese eel to sea water. *Comp. Biochem. Physiol.* 26, 675–685.

Kane, A. S., Bennett, R. O. and May, E. B. (1990). Effect of hardness and salinity on survival of striped bass larvae. *N. Am. J. Fish. Manag.* 10, 67–71.

Karnaky, K. J., Kinter, L. B., Kinter, W. B. and Stirling, C. E. (1976). Teleost chloride cell. 2. Autoradiographic localization of gill Na^+,K^+-ATPase in killifish (*Fundulus heteroclitus*) adapted to low and high salinity environments. *J. Cell. Biol.* 70, 157–177.

Karnaky, K. J., Degnan, K. J. and Zadunaisky, J. A. (1977). Chloride transport across isolated opercular epithelium of killifish – a membrane rich in chloride cells. *Science* 195, 203–205.

Katsura, K. and Hamada, K. (1986). Appearance and disappearance of chloride cells throughout the embryonic and postembryonic development of the goby, *Rhinoogobius urotaenia*. *Bull. Fac. Fish. Hokkaido Univ.* 37, 95–100.

Kaufman, L., Ebersole, J., Beets, J. and McIvor, C. C. (1992). A key phase in the recruitment dynamics of coral reef fishes: post-settlement transition. *Environ. Biol. Fish.* 34, 109–118.

Kearney, M., Jeis, A. and Lee, P. (2009). Effects of salinity and temperature on the growth and survival of New Zealand shortfin, *Anguilla australis*, and longfin, *A. dieffenbachii*, glass eels. *Aquacult. Res.* 39, 1769–1777.

Keith, P. (2003). Biology and ecology of amphidromous Gobiidae of the Indo-Pacific and the Caribbean regions. *J. Fish Biol.* 63, 831–847.

Keith, P. V. E. and Marquet, G. (2002). Atlas des poissons et crustaces d'eau de Polynesie Francaise. *Patrim. Nat.* 55, 1–177.

Keith, P., Hoareau, T., Lord, C., Ah-Yane, O., Gimmoneau, G., Robinet, T. and Valade, P. (2008). Characterisation of post-larvae to juvenile stages, metamorphosis, and recruitment of an amphidromous goby, *Sicyopterus lagocephalus* (Pallas, 1767) (Teleostei: Gobiidae: Sicydiinae). *Mar. Freshw. Res.* 5, 876–889.

Keith, P., Vigneux, E. and Bosc, P. (1999). Atlas des poissons et des crustaces d'eau douce de la Reunion. *Patrim. Nat.* 39, 1–136.

Keith, P., Watson, R. E. and Marquet, G. (2000). Découverte d'*Awaous ocellaris* (Gobiidae) en Nouvelle Calédonie et au Vanuatu. *Cybium* 24, 395–400.

Keith, P., Watson, R. E. and Marquet, G. (2002). *Stenogobius insularigobius yateiensis*, a new species of freshwater goby from New Caledonia (Teleostei: Gobioidei). *Bull. Franc. Pêche. Pisc.* 364, 187–196.

Kelly, J. T., Klimleya, A. P. and Crocker, C. E. (2007). Movements of green sturgeon, *Acipenser medirostris*, in the San Francisco Bay estuary, California. *Environ. Biol. Fish.* 79, 281–295.

Kennedy, B. M., Gale, W. L. and Ostrand, K. G. (2007). Relationship between smolt gill Na^+, K^+-ATPase activity and migration timing to avian predation risk of steelhead trout (*Oncorhynchus mykiss*) in a large estuary. *Can. J. Fish. Aquat. Sci.* 64, 1506–1516.

Kerr, L. A., Secor, D. H. and Piccoli, P. M. (2009). Partial Migration of Fishes as Exemplified by the Estuarine-Dependent White Perch. *Fisheries* 34, 114–123.

Kerr, L. A. and Secor, D. H. (2010). Latent effects of early life history on partial migration for an estuarine-dependent fish. *Environ. Biol. Fish.* 89, 479–492.

Keys, A. (1933). The mechanism of adaptation to varying salinity in the common eel and the general problem of osmotic regulation in fishes. *Proc. R. Soc. Lond. Ser. B* 112, 184–199.

Keys, A. and Willmer, E. N. (1932). "Chloride secreting cells" in the gills of fishes, with special reference to the common eel. *J. Physiol.* 76, 368–378.

Khodabandeh, S., Moghaddam, M. S. and Abtahi, B. (2009). Changes in chloride cell abundance, Na^+, K^+-ATPase immunolocalization and activity in the gills of golden grey mullet, *Liza aurata*, fry during adaptation to different salinities. *Yakhteh Medical Journal* 11(1), 49–54.

Kidder, G. W., Petersen, C. W. and Preston, R. L. (2006). Energetics of osmoregulation: II. water flux and osmoregulatory work in the euryhaline fish *Fundulus heteroclitus*. *J. Exp. Zool. A Comp. Exp. Biol.* 305, 318–327.

Kiilerich, P., Tipsmark, C. K., Borski, R. J. and Madsen, S. S. (2011). Differential effects of cortisol and 11-deoxycorticosterone on ion transport protein mRNA levels in gills of two euryhaline teleosts, Mozambique tilapia (*Oreochromis mossambicus*) and striped bass (*Morone saxatilis*). *J. Endocrinol.* 209, 115–126.

Kimura, R., Secor, D. H., Houde, E. D. and Piccoli, P. M. (2000). Up-estuary dispersal of young-of-the-year bay anchovy (*Anchoa mitchilli*) in the Chesapeake Bay: inferences from microprobe analysis of strontium in otoliths. *Mar. Ecol. Prog. Ser.* 20, 217–227.

King, J. A. C. and Hossler, F. E. (1991). The gill arch of striped bass (*Morone saxatilis*). 4. Alterations in the ultrastructure of chloride cell apical crypts and chloride efflux following exposure to sea water. *J. Morphol.* 209, 165–176.

King, S. P. and Sheridan, P. (2008). Nekton of new seagrass habitats colonizing a subsided salt marsh in Galveston Bay, Texas. *Estuaries* 29, 286–296.

Kinzie, R. A., III (1993). Evolution and life history patterns in freshwater gobies. *Micronesica* 30, 27–40.

Kjörsvik, E., Davenport, J. and Lonning, S. (1984). Osmotic changes during the development of eggs and larvae of the lumpsucker, *Cyclopterus lumpus* L. *J. Fish Biol.* 24, 311–321.

Klemetsen, A., Amundsen, P. A., Dempson, J. B., Jonsson, B., Jonsson, N., O'Connell, M. F. and Mortensen, E. (2003). Atlantic salmon (*Salmo salar* L.), brown trout (*Salmo trutta* L.) and Arctic charr (*Salvelinus alpinus* L.): a review of aspects of their life histories. *Ecol. Freshw. Fish.* 12, 1–59.

Kneib, R. T. (1997). The role of tidal marshes in the ecology of estuarine nekton. *Oceanogr. Mar. Biol. Annu. Rev.* 35, 163–220.

Kraus, R. T. and Secor, D. H. (2004). Dynamics of white perch (*Morone americana*) population contingents in the Patuxent River estuary, Maryland, USA. *Mar. Ecol. Prog. Ser.* 279, 247–259.

Kynard, B., Parker, E. and Parker, T. (2005). Behavior of early life intervals of Klamath River green sturgeon, *Acipenser medirostris*, with a note on body color. *Environ. Biol. Fish.* 72, 85–97.

Lambert, T. R., Toole, C. L., Handley, J. M., Mitchell, D. F., Wang, J. C. S. and Koeneke, M. A. (1980). Environmental conditions associated with spawning of a landlocked American shad (*Alosa sapidissima*) population. *Am. Zool.* 20, 813.

Lambert, Y., Dutil, J. D. and Munro, J. (1994). Effect of intermediate and low salinity conditions on growth rate and food conversion of Atlantic cod, *Gadus morhua*. *Can. J. Fish. Aquat. Sci.* 51, 1569–1576.

Lamson, H. M., Cairns, D. K., Shiao, J. C., Iizuka, Y. and Tzeng, W. N. (2009). American eel, *Anguilla rostrata*, growth in fresh and salt water: implications for conservation and aquaculture. *Fish. Manag. Ecol.* 16, 306–314.

Lankford, T. E., Billerbeck, J. M. and Conover, D. O. (2001). Evolution of intrinsic growth and energy acquisition rates. II. Trade-offs with vulnerability to predation in (*Menidia menidia*). *Evolution* 55, 1873–1881.

Lassalle, G. and Rochard, E. (2009). Impact of twenty-first century climate change on diadromous fish spread over Europe, North Africa and the Middle East. *Global Change Biol.* 15, 1072–1089.

Lawrie, A. H. (1970). Sea lamprey in the Great Lakes. *Trans. Am. Fish. Soc.* **99**, 766–775.

Le Bail, P. Y. and Boeuf, G. (1997). What hormones may regulate appetite in fish? *Aquat. Living Resourc.* 10, 371–439.

Lecomte-Finiger, R. (1983). Régime alimentaire des Civelles et Anguillettes (*Anguilla anguilla*) dans trois étangs saumâtresdu Roussillon [Diet of elvers and small eels (*Anguilla anguilla*) in three brackish lagoons from Roussillon Province]. *Bull. Ecol.* 14, 297–306.

Lee, C. S. and Ostrowski, A. C. (2001). Current status of marine finfish larviculture in the United States. *Aquacult.* 200, 89–109.

Leggett, W. C. and Carscadden, J. E. (1978). Latitudinal variation in reproductive characteristics of American shad (*Alosa sapidissima*): evidence for population specific life-history strategies in fish. *J. Fish. Res. Bd. Can.* 35, 1469–1478.

Leggett, W. C. and Whitney, R. R. (1972). Water temperature and migrations of American shad. *Fish. Bull.* 70, 659–670.

Leguen, I., Véron, V., Sevellec, C., Azam, D., Sabatiè, M. R., Prunet, P. and Baglinière, J. L. (2007). Development of hypoosmoregulatory ability in allis shad (*Alosa alosa*). *J. Fish Biol.* 70, 630–637.

Leis, J. M., Hay, A. C. and Trnski, T. (2006). *In situ* ontogeny of behaviour in pelagic larvae of three temperate, marine, demersal fishes. *Mar. Biol.* 148, 655–669.

Leis, M. J. and Carson-Ewart, B. M. (1997). *In situ* swimming speeds of the late pelagic larvae of some Indo-Pacific coral-reef fishes. *Mar. Ecol. Prog. Ser.* 159, 165–174.

Lemaire, C., Allegrucci, G., Naciri, M., Bahri-Sfar, L., Kara, H. and Bonhomme, F. (2000). Do discrepancies between microsatellite and allozyme variation reveal differential selection between sea and lagoon in the sea bass (*Dicentrarchus labrax*). *Mol. Ecol.* 9, 457–467.

Lepage, M., Taverny, C., Piefort, S., Dumont, P., Rochard, E. and Brosse, L. (2005). Juvenile sturgeon (*Acipenser sturio*) habitat utilization in the Gironde estuary as determined by acoustic telemetry. In *Fifth Conference on Fish Telemetry held in Europe* (eds. M. T. Spedicato, G. Lembo and G. Marmulla), pp. 169–177. Ustica, Italy: FAO/COISPA.

Lignot, J. H., Cutler, C. P., Hazon, N. and Cramb, G. (2002). Immunolocalisation of aquaporin 3 in the gill and the gastrointestinal tract of the European eel (*Anguilla anguilla* L.). *J. Exp. Biol.* 205, 2653–2663.

Lim, P., Meunier, F., Keith, P. and Noël, P. (2002). Atlas des poissons d'eau douce de la Martinique. *Patrim. Nat.* 51, 1–120.

Limburg, K. E. (1996). Modelling the ecological constraints on growth and movement of juvenile American shad (*Alosa sapidissima*) in the Hudson River Estuary. *Estuaries* 19, 794–813.

Limburg, K. E. (1998). Anomalous migrations of anadromous herrings revealed with natural chemical tracers. *Can. J. Fish. Aquat. Sci.* 55, 431–437.

Limburg, K. E. (2001). Through the gauntlet again: demographic restructuring of American shad by migration. *Ecology* 82, 1584–1596.

Limburg, K. E. and Ross, R. M. (1995). Growth and mortality rates of larval American shad, *Alosa sapidissima*, at different salinities. *Estuaries* 18, 335–340.

Livingston, R. J. (1976). Diurnal and seasonal fluctuations of organisms in a north Florida estuary. *Estuar. Coast. Mar. Sci.* 4, 373–400.

Lobry, J., Mourand, L., Rochard, E. and Elie, P. (2003). Structure of the Gironde estuarine fish assemblages: a comparison of European estuaries perspective. *Aquat. Living Resourc.* 16, 47–58.

Lochet, A., Jatteau, P., Tomás, J. and Rochard, E. (2008). Retrospective approach to investigating the early life history of a diadromous fish: allis shad (*Alosa alosa* L.) in the Gironde–Garonne–Dordogne watershed. *J. Fish Biol.* 72, 946–960.

Lochet, A., Boutry, S. and Rochard, E. (2009). Estuarine phase during seaward migration for allis shad (*Alosa alosa*) and twaite shad (*Alosa fallax*) future spawners. *Ecol. Freshw. Fish.* 18, 323–335.

Logan, A. G., Morris, R. and Rankin, J. C. (1980). A micropuncture study of kidney-function in the river lamprey (*Lampetra fluviatilis*) adapted to sea water. *J. Exp. Biol.* 88, 239–247.

Loneragan, N. R., Potter, I. C., Lenanton, R. C. J. and Caputi, N. (1986). Spatial and seasonal differences in the fish fauna in the shallows of a large Australian estuary. *Mar. Biol.* 92, 575–586.

Lord, C., Brun, C., Hautecoeur, M. and Keith, P. (2010). Comparison of the duration of the marine larval phase estimated by otolith microstructural analysis of three amphidromous *Sicyopterus* species (Gobiidae: Sicydiinae) from Vanuatu and New Caledonia: insights on endemism. *Ecol. Freshw. Fish.* 19, 26–38.

Lotrich, V. A. (1975). Summer home range and movements of *Fundulus heteroclitus* in a tidal creek. *Ecology* 56, 191–198.

Love, M. (1996). *Probably More Than You Want to Know About the Fishes of the Pacific Coast*. Santa Barbara, CA: Really Big Press.

Lowe, D. R., Beamish, F. W. H. and Potter, I. C. (1973). Changes in proximate body composition of landlocked sea lamprey *Petromyzon marinus* (L.) during larval life and metamorphosis. *J. Fish Biol.* 5, 673–682.

Lowe, M. R., DeVries, D. R., Wright, R. A., Ludsin, S. A. and Fryer, B. J. (2009). Coastal largemouth bass (*Micropterus salmoides*) movement in response to changing salinity. *Can. J. Fish. Aquat. Sci.* 66, 2174–2188.

Lutz, B. L. (1972). Body composition and ion distribution in the teleost *Perca juviatilis*. *Comp. Biochem. Physiol. A* 41, 181–193.

MacLeod, M. G. (1977). Effects of salinity on fasted rainbow trout (*Salmo gairdneri*). *Mar. Biol.* 43, 103–108.

Madsen, S. S., McCormick, S. D., Young, G., Endersen, J. S., Nishioka, R. S. and Bern, H. A. (1994). Physiology of seawater acclimation in the striped bass, *Morone saxatilis* (Walbaum). *Fish Physiol. Biochem.* 13, 1–11.

Madsen, S. S., Jensen, L. N., Tipsmark, C. K., Kiilerich, P. and Borski, R. J. (2007). Differential regulation of cystic fibrosis transmembrane conductance regulator and Na$^+$, K$^+$-ATPase in gills of striped bass, *Morone saxatilis*: effect of salinity and hormones. *J. Endocrinol.* 192, 249–260.

Maes, J., Stevens, M. and Ollevier, F. (2005). The composition and community structure of the ichthyofauna of the upper Scheldt estuary: synthesis of a 10-year data collection (1991–2001). *J. Appl. Ichthyol.* 21, 86–93.

Maetz, J. and Skadhaug, E. (1968). Drinking rates and gill ionic turnover in relation to external salinities in eel. *Nature* 217, 371–373.

Magnusson, A. and Hilborn, R. (2003). Estuarine influence on survival rates of coho (*Oncorhynchus kisutch*) and Chinook salmon (*Oncorhynchus tshawytscha*) released from hatcheries on the US Pacific coast. *Estuaries* 26, 1094–1103.

Mancera, J. M. and McCormick, S. D. (1998). Osmoregulatory actions of the GH/IGF axis in non-salmonid teleosts. *Comp. Biochem. Physiol. B Biochem. Mol. Biol.* 121, 43–48.

Manderson, J. P., Pessutti, J., Hilbert, J. G. and Juanes, F. (2004). Shallow water predation risk for a juvenile flatfish (winter flounder; *Pseudopleuronectes americanus* Walbaum) in a northwest Atlantic estuary. *J. Exp. Mar. Biol. Ecol.* 304, 137–157.

Mansfield, S. and McCardle, B. H. (1998). Dietary composition of *Gambusia affinis* (Family Poeciliidae) populations in the northern Waikato region of New Zealand. *N. Z. J. Mar. Freshw. Res.* 32, 375–383.

Marshall, S. and Elliott, M. (1998). Environmental influences on the fish assemblage of the Humber Estuary, UK. *Estuar. Coast. Shelf Sci.* 46, 175–184.

Marshall, W. S. (2002). Na$^+$, Cl$^-$, Ca^{2+} and Zn^{2+} transport by fish gills: retrospective review and prospective synthesis. *J. Exp. Zool.* 293, 264–283.

Marshall, W. (2013). Osmoregulation in estuarine and intertidal fishes. In *Fish Physiology,* Vol. 32, *Euryhaline Fishes* (eds. S. D. McCormick, A. P. Farrell and C. J. Brauner), pp. 395–434. New York: Elsevier.

Marshall, W. S. and Grosell, M. (2006). Ion transport, osmoregulation, and acid–base balance. In *The Physiology of Fishes* (eds. D. H. Evans and J. B. Claiborne), 3rd edn, pp. 177–230. Boca Raton, FL: CRC Press.

Martin, T. J. (1990). Osmoregulatory in three species of Ambassidae (Osteichthyes: Perciformes) from estuaries in Natal, South Africa. *Can. J. Zool.* 25, 229–234.

Martinez-Alvarez, R. M., Sanz, A., Garcia-Gallego, M., Domezain, A., Domezain, J., Carmona, R., Ostos-Garrido, M. D. and Morales, A. E. (2005). Adaptive branchial mechanisms in the sturgeon (*Acipenser naccarii*) during acclimation to saltwater. *Comp. Biochem. Physiol. A Mol. Integr. Physiol.* 141, 183–190.

Mayer, N., Maetz, J., Chan, D. K. O., Forster, M. and Jones, I. C. (1967). Cortisol, a sodium excreting factor in eel (*Anguilla anguilla* L.) adapted to sea water. *Nature* 214, 1118–1120.

McBride, R. S., MacDonald, T. C., Matheson, R. E., Rydene, D. A. and Hood, P. B. (2001). Nursery habitats for ladyfish, *Elops saurus*, along salinity gradients in two Florida estuaries. *Fish. Bull.* 99, 443–458.

McCleave, J. D. and Kleckner, R. C. (1982). Selective tidal stream transport in the estuarine migration of glass eels of the American eel (*Anguilla rostrata*). *J. Cons. Int. Explor. Mer.* 40, 262–271.

McCleave, J. D. and Wippelhauser, G. S. (1987). Behavioral aspects of selective tidal stream transport in juvenile American eels (*Anguilla rostrata*). *Am. Fish. Soc. Symp.* 1, 138–150.

McCormick, S. D. (1994). Ontogeny and evolution of salinity tolerance in anadromous salmonids: hormones and heterochrony. *Estuaries* 17, 26–33.

McCormick, S. D. (1996). Effects of growth hormone and insulin-like growth factor I on salinity tolerance and gill Na^+,K^+-ATPase in Atlantic salmon (*Salmo salar*): interaction with cortisol. *Gen. Comp. Endocrinol.* 101, 3–11.

McCormick, S. D. (2001). Endocrine control of osmoregulation in teleost fish. *Am. Zool.* 41, 781–794.

McCormick, S. D. (2013). Smolt physiology and endocrinology. In *Fish Physiology*, Vol. 32, *Euryhaline Fishes* (eds. S. D. McCormick, A. P. Farrell and C. J. Brauner), pp. 199–251. New York: Elsevier.

McCormick, S. D. and Naiman, R. J. (1984). Osmoregulation in the brook trout, *Salvelinus fontinalis*. II. Effects of size, age and photoperiod on seawater survival and ionic regulation. *Comp. Biochem. Physiol. A* 79, 17–28.

McCormick, S. D. and Saunders, R. L. (1987). Preparatory physiological adaptations for marine life in salmonids: osmoregulation, growth and metabolism. Common strategies of anadromous and catadromous fishes. *Am. Fish. Soc. Symp.* 1, 211–229.

McCormick, S. D., Shrimpton, J. M. and Zydlewski, J. D. (1997). Temperature effects on osmoregulatory physiology of anadromous fish. In *Global Warming Implications for Freshwater and Marine Fish. Society for Experimental Biology, Seminar Series* (eds. C. M. Wood and D. G. McDonald), Vol. 61. Cambridge: Cambridge University Press.

McCormick, S. D., Hansen, L. P., Quinn, T. P. and Saunders, R. L. (1998). Movement, migration and smolting in Atlantic salmon. *Can. J. Fish. Aquat. Sci.* 55 (Suppl. 1), 77–92.

McCormick, S. D., Sundell, K., Bjornsson, B. T., III, Brown, C. L. and Hiroi, J. (2003). Influence of salinity on the localization of Na^+/K^+-ATPase, $Na^+/K^+/2Cl^-$ cotransporter (NKCC) and CFTR anion channel in chloride cells of the Hawaiian goby (*Stenogobius hawaiiensis*). *J. Exp. Biol.* 206, 4575–4583.

McDonald, D. G. and Milligan C. L (1992). Chemical properties of the blood. In *Fish Physiology*, Vol. 12B (eds. W. S. Hoar, D. J. Randall and A.P. Farrell), pp. 56–113. New York: Academic Press.

McDowall, R. M. (1987). The occurrence and distribution of diadromy among fishes, Vol. 1 In *Common Strategies of Anadromous and Catadromous Fishes* (eds. M. J. Dadswell, R. J. Klauda, C. M. Moffit, R. L. Saunders, R. A. Rulifson and J. E. Cooper), pp. 1-13, American Fisheries Society Symposium, Bathesda, MD.

McDowall, R. M. (1988). *Diadromy in Fishes: Migrations Between Freshwater and Marine Environments.* London: Croom Helm.

McDowall, R. M. (1990). When galaxiid and salmonid fishes meet – a family reunion in New Zealand. *J. Fish Biol.* 37, 35–43.

McDowall, R. M. (1992). Diadromy – origins and definitions of terminology. *Copeia* 1992, 248–251.

McDowall, R. M. (1999). Driven by diadromy: its role in the historical and ecological biogeography of the New Zealand freshwater fish fauna. *Ital. J. Zool.* 65, 73–85.

McDowall, R. M. (2004). Ancestry and amphidromy in island freshwater fish faunas. *Fish Fish.* 5, 75–85.

McDowall, R. M. (2007). On amphidromy, a distinct form of diadromy in aquatic organisms. *Fish Fish.* 8, 1–13.

McDowall, R. M. (2008). Early hatch: a strategy for safe downstream larval transport in amphidromous gobies. *Rev. Fish Biol. Fish.* 19, 1–8.

McEnroe, M. and Cech, J. J., Jr. (1985). Osmoregulation in juvenile and adult white sturgeon, *Acipenser transmontanus. Environ. Biol. Fish.* 14, 23–30.

McKenzie, D. J., Cataldi, E., Di Marco, P., Mandich, A., Romano, P., Ansferri, S., Bronzi, P. and Cataudella, S. (1999). Some aspects of osmotic and ionic regulation in Adriatic sturgeon (*Acipenser naccarii*). II: Morpho-physiological adjustments to hyperosmotic environments. *J. Appl. Ichthyol.* 15, 61–66.

McKenzie, D. J., Cataldi, E., Romano, P., Taylor, E. W., Cataudella, S. and Bronzi, P. (2001). Effects of acclimation to brackish water on tolerance of salinity challenge by young-of-the-year Adriatic sturgeon (*Acipenser naccarii*). *Can. J. Fish. Aquat. Sci.* 58, 1113–1121.

McKinnon, L. J. and Gooley, G. J. (1998). Key environmental criteria associated with the invasion of *Anguilla australis* glass eels into estuaries of southeastern Australia. *Bull. Fr. Pêche. Piscic.* 349, 117–128.

McLusky, D. S. and Elliott, M. (2004). *The Estuarine Ecosystem: Ecology, Threats and Management.* Oxford: Oxford University Press.

Meador, M. R. and Kelso, W. E. (1990). Physiological responses of largemouth bass, *Micropterus salmoides*, exposed to salinity. *Can. J. Fish. Aquat. Sci.* 47, 2358–2363.

Mehl, J. A. P. (1974). Ecology, osmoregulatory and reproductive biology of the white steenbras (*Lithognathus lithognathus*) (Teleostei: Sparidae). *Zool. Afr.* 8, 157–230.

Melia, P., Bevacqua, D., Crivelli, A. J., De Leo, G., Panfili, J. and Gatto, M. (2006). Age and growth of *Anguilla anguilla* in the Camargue lagoons. *J. Fish Biol.* 68, 876–890.

Miller, J. M. (1988). Physical processes and the mechanisms of coastal migrations of immature marine fishes. *Am. Fish. Soc. Symp.* 3, 68–76.

Miller, P. J. (1984). The tokology of gobioid fishes. In *Fish Reproduction: Strategies and Tactics* (eds. G. W. Potts and R. J. Wooton), pp. 119–152. London: Academic Press.

Miller, S. J. and Skilleter, G. A. (2006). Temporal variation in habitat use by nekton in a subtropical estuarine system. *J. Exp. Mar. Biol. Ecol.* 337, 82–95.

Miltner, R. J., Ross, S. W. and Posey, M. H. (1995). Influence of food and predation on the depth distribution of juvenile spot (*Leiostomus xanthurus*) in tidal nurseries. *Can. J. Fish. Aquat. Sci.* 52, 971–982.

Miskiewicz, A. G. (1986). The season and length of entry into a temperate Australian estuary of the larvae of *Acanthopagrus australis*, *Rhabdosargus sarba* and *Chrysophrys auratus* (Teleostei: Sparidae). In *Indo-Pacific Fish Biology* (eds. T. Uyeno, R. Arai, T. Taniuchi and K. Matsuura), pp. 740–747. Tokyo: Ichthyological Society of Japan.

Moore, R. H. (1974). General ecology, distribution and relative abundance of *Mugil cephalus* and *Mugil curema* on the south Texas coast. *Contrib. Mar. Sci.* 18, 241–255.

Morais, P., Faria, A., Chicharo, M. A. and Chicharo, L. (2009). The unexpected occurrence of late *Sardina pilchardus* (Walbaum, 1792) (Osteichthyes: Clupeidae) larvae in a temperate estuary. *Cah. Biol. Mar.* 50, 79–89.

Morais, P., Babaluk, P., Correia, A. T., Chícharo, M. A., Campbell, J. L. and Chícharo, L. (2010). Diversity of anchovy migration patterns in an European temperate estuary and in its adjacent coastal area: implications for fishery management. *J. Sea Res.* 64, 295–303.

Morgan, J. D. and Iwama, G. K. (1991). Effects of salinity on growth, metabolism, and ion regulation in juvenile rainbow and steelhead trout (*Oncorhynchus mykiss*) and fall Chinook salmon (*Oncorhynchus tshawytscha*). *Can. J. Fish. Aquat. Sci.* 48, 2083–2094.

Morgan, J. D. and Iwama, G. K. (1999). Energy cost of NaCl transport in isolated gills of cutthroat trout. *Am. J. Physiol.* 277, R631–R639.

Morgan, J. D., Sakamoto, T., Grau, E. G. and Iwama, G. K. (1997). Physiological and respiratory responses of the Mozambique tilapia (*Oreochromis mossambicus*) to salinity acclimation. *Comp. Biochem. Physiol. A* 117, 391–398.

Morin, B., Hudon, C. and Whoriskey, F. G. (1992). Environmental influences on seasonal distribution of coastal and estuarine fish assemblages at Wemindji, Eastern James Bay. *Environ. Biol. Fish.* 35, 219–229.

Moriyama, A., Yanagisawa, Y., Mizuno, N. and Omori, K. (1998). Starvation of drifting goby free embryos due to retention in the upper stream. *Environ. Biol. Fish.* 52, 321–329.

Morris, R. (1980). Blood composition and osmoregulation in ammocoete larvae. *Can. J. Fish. Aquat. Sci.* 37, 1665–1679.

Morrison, W. E. and Secor, D. H. (2003). Demographic attributes of yellow-phase American eels (*Anguilla rostrata*) in the Hudson Estuary. *Can. J. Fish. Aquat. Sci.* 60, 1487–1501.

Moser, H. G. (1996). *The Early Stages of Fishes in the California Current Region.* Lawrence: Allen Press, CalCOFI Atlas 33.

Muncy, R. J. (1984). *Species Profiles: Life Histories and Environmental Requirements of Coastal Fishes and Invertebrates (Gulf of Mexico) Pinfish.* US Army Corps of Engineers, TR EL. Technical Report FWS/OBS. 82 (11.26). US Fish and Wildlife Service, Division of Biological Survey.

Murashige, R., Bass, P., Wallace, L., Molnar, A., Eastham, B., Sato, V., Tamaru, C. and Lee, C. S. (1991). The effect of salinity on the survival and growth of striped mullet (*Mugil cephalus*) larvae in the hatchery. *Aquacult.* 96, 249–254.

Myers, G. S. (1949). Usage of anadromous, catadromous and allied terms for migratory fishes. *Copeia* 1949, 89–97.

Najjar, R. G., Walker, H. A., Anderson, P. J., Barron, E. J., Bord, R. J., Gibson, J. R., Kennedy, V. S., Knight, C. G., Megonigal, J. P., O'Connor, R. E., Polsky, C. D., Psuty, N. P., Richards, B. A., Sorenson, L. G., Steele, E. M. and Swanson, R. S. (2000). The potential impacts of climate change on the mid-Atlantic coastal region. *Clim. Res.* 14, 219–233.

Najjar, R. G., Pyke, C. R., Adams, M. B., Breitburg, D., Hershner, C., Kemp, M., Howarth, R., Mulholland, M. R., Paolisso, M., Secor, D., Sellner, K., Wardrop, D. and Wood, R. (2010). Potential climate-change impacts on the Chesapeake Bay. *Estuar. Coast. Shelf Sci.* 86, 1–20.

Nebel, C., Romesand, B., Nègre-Sadargues, G., Grousset, E., Ajoulat, F., Bacal, J., Bonhomme, F. and Charmantier, G. (2005). Differential freshwater adaptation in juvenile sea bass *Dicentrarchus labrax*: involvement of gills and urinary system. *J. Exp. Biol.* 208, 3859–3871.

Neill, W., Brandes, T., Burke, B., Craig, S., Dimichele, L., Duchon, K., Edwards, R., Fontaine, L., Gatlin, D., Hutchins, C., Miller, J. M., Ponwith, B., Stahl, C., Tomasso, J. and Vega, R. (2004). Fish: a simulation model of fish growth in time varying environmental regimes. *Rev. Fish. Sci.* 12, 233–288.

Nemerson, D. M. and Able, K. W. (2004). Spatial patterns in diet and distribution of juveniles of four fish species in Delaware Bay marsh creeks: factors influencing fish abundance. *Mar. Ecol. Prog. Ser.* 276, 249–262.

Neuman, M. J. and Able, K. W. (2002). Quantification of ontogenetic transitions during the early life of a flatfish, windowpane, *Scophthalmus aquosus* (Pleuronectiformes: Scophthalmidae). *Copeia* 2002, 597–609.

Nichols, O. C. and Tscherter, U. T. (2011). Feeding of sea lampreys (*Petromyzon marinus*) on minke whales (*Balaenoptera acutorostrata*) in the St Lawrence Estuary, Canada. *J. Fish Biol.* 78, 338–343.

Nicolas, D., Chaalali, A., Drouineau, H., Lobry, J., Uriarte, A., Borja, A. and Boët, P. (2010). Impact of global warming on European tidal estuaries: some evidence of northward migration of estuarine fish species. *Region. Environ. Change* 11, 639–649.

Niklitschek, E. J. and Secor, D. H. (2005). Modeling spatial and temporal variation of suitable nursery habitats for Atlantic sturgeon in the Chesapeake Bay. *Estuar. Coast. Shelf Sci.* 64, 135–148.

Nilsson, S. (1986). Control of gill blood flow. In *Fish Physiology: Recent Advances* (eds. S. Nilsson and S. Holmgren), pp. 87–101. London: Croom Helm.

Nishimoto, R. T. (1996). Recruitment of goby and crustacean postlarvae juveniles into Hakalau Stream with comments on recruitment into an outflow canal (Wailihi "Stream"). In *Will Stream Restoration Benefit Freshwater, Estuarine, and Marine Fisheries?* (ed. W.S. Devick), pp. 148–151. Proceedings of the 1994 Hawai'I Stream Restoration Symposium, State of Hawai'i, Department of Land and Natural Resources, Division of Aquatic Resources, Honolulu.

Nishimoto, R. T. and Fitzsimons, J. M. (1986). Courtship, territoriality, and coloration in the endemic Hawaiian freshwater goby, *Lentipes concolor*. In *Indo-Pacific Fish Biology: Proceedings of the Second International Conference on Indo-Pacific Fishes* (eds. T. Uyeno, R. Arai, T. Taniuchi and K. Matsuura), pp. 811–817. Tokyo: Ichthyological Society of Japan.

Nishimoto, R. T. and Kuamo'o, D. G. K. (1997). Recruitment of goby larvae into Hakalau Stream, Hawai'i Island. *Micronesia* 30, 41–49.

Nordlie, F. G. (2000). Patterns of reproduction and development of selected resident teleosts of Florida salt marshes. *Hydrobiologia* 434, 165–182.

Nordlie, F. G., Szelistowski, W. A. and Nordlie, W. C. (1982). Ontogenesis of osmotic regulation in the striped mullet, *Mugil cephalus* L. *J. Fish Biol.* 20, 79–86.

Norris, A. J., DeVries, D. R. and Wright, R. A. (2010). Coastal estuaries as habitat for a freshwater fish species: exploring population-level effects of salinity on largemouth bass. *Trans. Am. Fish. Soc.* 139, 610–625.

North, E. W. and Houde, E. D. (2001). Retention of white perch and striped bass larvae: biological–physical interactions in Chesapeake Bay estuarine turbidity maximum. *Estuaries* 24, 756–769.

O'Boyle, R. N. and Beamish, F. W. H. (1977). Growth and intermediary metabolism of larval and metamorphosing stages of the landlocked sea lamprey *Petromyzon marinus*. *Environ. Biol. Fish.* 2, 103–120.

O'Donnell, M. J. and Letcher, B. H. (2008). Size and age distributions of juvenile Connecticut River American shad above Hadley Falls: influence on outmigration representation and timing. *River Res. Applic.* 24, 929–940.

O'Leary, J. A. and Kynard, B. (1986). Behavior, length and sex ratio of seaward migrating juvenile American shad and blueback herring in the Connecticut River. *Trans. Am. Fish. Soc.* 115, 529–536.

Ohmi, H. (2002). Juvenile ecology in Yura river estuary in Wakasa Bay (in Japan). In *Temperate Bass and Biodiversity* (eds. M. Tanaka and I. Kinoshita), pp. 44–53. Tokyo: Koseisha-Koseikaku.

Oliveira, A., Fortunato, A. B. and Pinto, L. (2006). Modelling the hydrodynamics and the fate of passive and active organisms in the Guadiana estuary. *Estuar. Coast. Shelf Sci.* 70, 76–84.

Olivereau, M. (1970). Reaction of chloride cells in gills of European eel after transfer into sea water. *Compt. Rend. Seances Soc. Biol. Filiales* 164, 1951–1955.

Olivereau, M. and Ball, J. N. (1970). Pituitary influences on osmoregulation in teleosts. *Mem. Soc. Endocrinol.* 18, 57–85.

Orth, R. J., Heck, K. L. and Montfrans, J. (1984). Faunal composition in seagrass beds: a review of the influence of plant structure and prey characteristics on predator–prey relationships. *Estuaries* 4, 339–350.

Otwell, W. S. and Merriner, J. V. (1975). Survival and growth of juvenile striped bass, *Morone saxatilis*, in a factorial experiment with temperature, salinity and age. *Trans. Am. Fish. Soc.* 104, 560–566.

Paterson, A. W. and Whitfield, A. K. (2000). Do shallow-water habitats function as refugia for juvenile fishes? *Estuar. Coast. Shelf Sci.* 51, 359–364.

Pease, B. C., Silberschneider, V. and Walford, T. (2003). Upstream migration by glass eels of two (*Anguilla* species) in the Hacking River, New South Wales, Australia. *Trans. Am. Fish. Soc.* 33, 47–61.

Peek, W. D. and Youson, J. H. (1979). Ultrastructure of chloride cells in young adults of the anadromous sea lamprey, *Petromyzon marinus* L. in fresh water and during adaptation to seawater. *J. Morphol.* 160, 143–163.

Peer, A. C., DeVries, D. R. and Wright, R. A. (2006). First-year growth and recruitment of coastal largemouth bass (*Micropterus salmoides*): spatial patterns unresolved by critical periods along a salinity gradient. *Can. J. Fish. Aquat. Sci.* 63, 1911–1924.

Pereira, J. J., Goldberg, R., Ziskowski, J. J., Berrien, P. L., Morse, W. W. and Johnson, D. L. (1999). *Essential Fish Habitat Source Document: Winter Flounder,* Pseudopleuronectes americanus, *Life History and Habitat Characteristics.* Special Scientific Report – Fish. Technical Memorandum NMFS-NE-138, 1–39. National Oceanic and Atmospheric Administration USA.

Perry, S. F. (1997). The chloride cell: Structure and function in the gills of freshwater fishes. *Annu. Rev. Physiol.* 59, 325–347.

Peterson, M. S. (1988). Comparative physiological ecology of centrarchids in hyposaline environments. *Can. J. Fish. Aquat. Sci.* 45, 827–833.

Peterson, M. S. (1991). Differential length–weight relations among centrarchids (Pisces: Centrarchidae) from tidal freshwater and oligohaline wetland habitats. *Wetlands* 11, 325–332.

Pickering, A. D. and Morris, R. (1970). Osmoregulation of *Lampetra fluviatilis* L. and *Petromyzon marinus* (Cylcostomata) in hyperosmotic solutions. *J. Exp. Biol.* 53, 231–243.

Pickering, A. D. and Morris, R. (1973). Localization of ion-transport in intestine of migrating river lamprey, *Lampetra fluviatilis* L. *J. Exp. Biol.* 58, 165–176.

Pollard, D. A. (1971). The biology of a landlocked form of the normally catadromous salmoniform fish (*Galaxias maculatus*). Jenyns I. Life cycle and origin. *Aust. J. Mar. Freshw. Res.* 22, 91–123.

Postma, H. (1961). Transport and accumulation of suspended matter in the Dutch Wadden Sea. *Neth. J. Sea Res.* 1, 148–190.

Potter, I. C. and Gill, H. S. (2003). Adaptive radiation of lampreys. *J. Great Lakes Res.* 29, 95–112.

Potter, I. C., Claridge, P. N. and Warwick, R. M. (1986). Consistency of seasonal changes in an estuarine fish assemblage. *Mar. Ecol. Prog. Ser.* 32, 217–228.

Potts, W. T. W. (1954). The energetics of osmotic regulation in brackish- and fresh-water animals. *J. Exp. Zool.* 31, 618–630.

Potts, W. T. W. and Rudy, P. P. (1972). Aspects of osmotic and ionic regulation in sturgeon. *J. Exp. Biol.* 56, 703–715.

Power, J. H. (1984). Advection, diffusion, and drift migrations of larval fish. In *Mechanisms of Migration in Fishes* (eds. J. D. McCleave, G. P. Arnold, J. J. Dodson and W. H. Neill), pp. 27–37. New York: Plenum Press.

Purcell, K. M., Hitch, A. T., Klerks, P. L. and Leberg, P. L. (2008). Adaptation as a potential response to sea level rise: a genetic basis for salinity tolerance in populations of a coastal marsh fish. *Evol. Appl.* 1, 155–160.

Quinlan, J. A. and Crowder, L. B. (1999). Searching for sensitivity in the life history of Atlantic menhaden: inferences from a matrix model. *Fish. Oceanogr.* 8, 124–133.

Quinn, T. P. and Myers, K. W. (2004). Anadromy and the marine migrations of Pacific salmon and trout: Rounsefell revisited. *Rev. Fish Biol. Fish.* 14, 421–442.

Raffaelli, D., Richner, H., Summers, R. and Northcott, S. (1990). Tidal migrations in the flounder. *Mar. Behav. Physiol.* 16, 249–260.

Rankin, J. C. (2009). Acclimation to seawater in the European eel *Anguilla anguilla*: effects of silvering. In *Spawning Migration of the European Eel* (eds. G. van den Thillart, S. Dufour and J. C. Rankin), pp. 129–145. New York: Springer.

Rankin, J. C., Cobb, C. S., Frankling, S. C. and Brown, J. A. (2001). Circulating angiotensins in the river lamprey, *Lampetra fluviatilis*, acclimated to freshwater and seawater: possible involvement in the regulation of drinking. *Comp. Biochem. Physiol. B Biochem. Mol. Biol.* 129, 311–318.

Reis-Santos, P., McCormick, S. D. and Wilson, J. M. (2008).). Ionoregulatory changes during metamorphosis and salinity exposure of juvenile sea lamprey (*Petromyzon marinus* L.). *J. Exp. Biol.* 211, 978–988.

Renaud, C. B., Gill, H. S. and Potter, I. C. (2009). Relationships between the diets and characteristics of the dentition, buccal glands and velar tentacles of the adults of the parasitic species of lamprey. *Can. J. Zool.* 278, 231–242.

Richards, J. E. and Beamish, F. W. H. (1981). Initiation of feeding and salinity tolerance in the Pacific lamprey (*Lampetra tridentate*). *Mar. Biol.* 63, 73–77.

Richards, J. G., Semple, J. W., Bystriansky, J. S. and Schulte, P. M. (2002). Na$^+$/K$^+$-ATPase α-isoform switching in gills of rainbow trout (*Oncorhynchus mykiss*) during salinity transfer. *J. Exp. Biol.* 206, 4475–4486.

Riley, J. (1973). Movements of 0-group plaice (*Pleuronectes platessa* L.) as shown by latex tagging. *J. Fish Biol.* 5, 323–343.

Rilling, G. C. and Houde, E. D. (1999). Regional and temporal variability in distribution and abundance of bay anchovy (*Anchoa mitchilli*) eggs, larvae and adult biomass in the Chesapeake Bay. *Estuaries* 22, 1096–1109.

Rocha, A. J. D., Gomes, V., Van Ngan, P., Passos, M. J. D. and Furia, R. R. (2005). Metabolic demand and growth of juveniles of (*Centropomus parallelus*) as function of salinity. *J. Exp. Mar. Biol. Ecol.* 3162, 157–165.

Rochard, E., LePage, M., Dumont, P., Tremblay, S. and Gazeau, C. (2001). Downstream migration of juvenile European sturgeon (*Acipenser sturio* L.) in the Gironde estuary. *Estuaries* 24, 108–115.

Rodriguez, A., Gisbert, E. and Castello-Orvay, F. (2005). Nutritional condition of *Anguilla anguilla* starved at various salinities during the elver phase. *J. Fish Biol.* 67, 521–534.

Rogers, S. G., Targett, T. E. and Van Sant, S. B. (1984). Fish-nursery use in Georgia salt-marsh estuaries: the influence of springtime freshwater conditions. *Trans. Am. Fish. Soc.* 113, 595–606.

Rombough, P. J. (1988). Respiratory gas exchange, aerobic metabolism, and effects of hypoxia during early life. In *Fish Physiology*, Vol. 11A (eds. W. S. Hoar and D. J. Randall), pp. 59–161. New York: Academic Press.

Rombough, P. J. (2007). The functional ontogeny of the teleost gill: which comes first, gas or ion exchange? *Comp. Biochem. Physiol. A Mol. Integr. Physiol.* 148, 732–742.

Rountree, R. A. and Able, K. W. (1992). Foraging habits, growth and temporal patterns of salt-marsh creek habitat use by young-of-the-year summer flounder in New Jersey. *Trans. Am. Fish. Soc.* 121, 765–776.

Rountree, R. A. and Able, K. W. (1993). Diel variation in decapod crustacean and fish assemblages in New Jersey polyhaline marsh creeks. *Estuar. Coast. Shelf Sci.* 37, 181–201.

Rountree, R. A. and Able, K. W. (2007). Spatial and temporal habitat use patterns for salt marsh nekton: implications for ecological functions. *Aquat. Ecol.* 41, 25–45.

Rovainen, C. M. (1996). Feeding and breathing in lampreys. *Brain Behav. Evol.* 48, 297–305.

Rozas, L. P. and Hackney, C. T. (1983). The importance of oligohaline estuarine wetland habitats to fisheries resources. *Wetlands* 3, 77–89.

Rozas, L. P. and Odum, W. E. (1988). Occupation of submerged aquatic vegetation by fishes: testing the roles of food and refuge. *Oecologia* 77, 101–106.

Ruiz, G. M., Hines, A. H. and Posey, M. H. (1993). Shallow water as a refuge habitat for fish and crustaceans in non-vegetated estuaries: an example from Chesapeake Bay. *Mar. Ecol. Prog. Ser.* 99, 1–16.

Rulifson, R. A. and Dadswell, M. J. (1995). Life history and population characteristics of striped bass in Atlantic Canada. *Trans. Am. Fish. Soc.* 124, 477–507.

Rutherford, E. S. and Houde, E. D. (1995). The influence of temperature on cohort-specific growth, survival, and recruitment of striped bass, *Morone saxatilis* larvae in Chesapeake Bay. *Fish. Bull.* 93, 315–332.

Sackett, D. K., Able, K. W. and Grothues, T. H. (2007). Dynamics of summer flounder, *Paralichthys dentatus*, seasonal migrations based on ultrasonic telemetry. *Estuar. Coast. Shelf Sci.* 74, 119–130.

Sackett, D. K., Able, K. W. and Grothues, T. M. (2008). Habitat dynamics of summer flounder, *Paralichthys dentatus* within a shallow USA estuary, based on multiple approaches using acoustic telemetry. *Mar. Ecol. Prog. Ser.* 364, 199–212.

Sakabe, R. (2009). Ecology and life history characteristics of black bream, *Acanthopagrus butcheri*, in Tasmanian estuarine ecosystems. PhD Thesis, University of Tasmania, Hobart.

Sakamoto, T., Hirano, T., Madsen, S. S., Nishioka, R. S. and Bern, H. A. (1995). Insulin-like growth factor-1 gene expression during parr–smolt transformation of coho salmon. *Zool. Sci.* 12, 249–252.

Sakamoto, T., Uchida, K. and Yokota, S. (2001). Regulation of the ion-transporting mitochondrion-rich cell during adaptation of teleost fishes to different salinities. *Zool. Sci.* 18, 1163–1174.

Saloniemi, I., Jokikokko, E., Kallio-Nyberg, I., Jutila, E. and Pasanen, P. (2004). Survival of reared and wild Atlantic salmon smolts: size matters more in bad years. *ICES J. Mar. Sci.* 61, 782–787.

Sardella, B. A. and Kültz, D. (2009). Osmo- and ionoregulatory responses of green sturgeon (*Acipenser medirostris*) to salinity acclimation. *J. Comp. Physiol. B* 179, 383–390.

Sardet, C., Pisam, M. and Maetz, J. (1979). Surface epithelium of teleostean fish gills – cellular and junctional adaptations of chloride cell in relation to salt adaptation. *J. Cell. Biol.* 80, 96–117.

Sasai, S., Kaneko, T., Hasegawa, S. and Tsukamoto, K. (1998). Morphological alteration in two types of gill chloride cells in Japanese eels (*Anguilla japonica*) during catadromous migration. *Can. J. Zool. Rev. Can. Zool.* 76, 1480–1487.

Saucerman, S. F. and Deegan, L. A. (1991). Lateral and cross-channel movements of young-of-the-year flounder (*Pseudopleuronectes americanus*) in Waquoit Bay, Massachusetts. *Estuaries* 14, 440–446.

Schoenfuss, H. L., Blanchard, T. A. and Kuamo'o, D. G. (1997). Metamorphosis in the cranium of postlarval (*Sicyopterus stimpsoni*), an endemic Hawaiian stream goby. *Micronesica* 30, 93–104.

Scott, W. B. and Crossman, E. J. (1973). Freshwater Fishes of Canada. Fisheries Research Board of Canada; 184, Ottawa: Fisheries Research Board of Canada, pp. 966.

Schreck, C. B., Stahl, T. P., Davis, L. E., Roby, D. D. and Clemens, B. J. (2006). Mortality estimates of juvenile spring–summer Chinook salmon in the lower Columbia River estuary (1992)–(1998): evidence for delayed mortality? *Trans. Am. Fish. Soc.* 135, 457–475.

Schreiber, A. M. and Specker, J. L. (2000). Metamorphosis in the summer flounder (*Paralichthys dentatus*): thyroidal status influences gill mitochondria-rich cells. *Gen. Comp. Endocrinol.* 117, 238–250.

Schultz, E. T., Lwiza, K. M. M., Fencil, M. C. and Martin, J. M. (2003). Mechanisms promoting upriver transport of larvae of two fishes in the Hudson River Estuary (USA). *Mar. Ecol. Prog. Ser.* 251, 263–277.

Scott, G. R., Baker, D. W., Schulte, P. M. and Wood, C. M. (2008). Physiological and molecular mechanisms of osmoregulatory plasticity in killifish after seawater transfer. *J. Exp. Biol.* 211, 2450–2459.

Scott, W. B. and Scott, M. G. (1988). Atlantic fishes of Canada. *Can. Bull. Fish. Aquat. Sci.* 219, 1–731.

Secor, D. H. (1992). Application of otolith microchemistry analysis to investigate anadromy in Chesapeake Bay striped bass *Morone saxatilis. Fish. Bull.* 90, 798–806.

Secor, D. H. and Houde, E. D. (1995). Temperature effects on the timing of striped bass egg-production, larval viability, and recruitment potential in the Patuxent River (Chesapeake Bay). *Estuaries* 18, 527–544.

Secor, D. H., Hendersonarzapalo, A. and Piccoli, P. M. (1995). Can otolith microchemistry chart patterns of migration and habitat utilization in anadromous fishes? *J. Exp. Mar. Biol. Ecol.* 192, 15–33.

Secor, D. H., Gunderson, T. E. and Karlsson, K. (2000). Effect of temperature and salinity on growth performance in anadromous (Chesapeake Bay) and nonanadromous (Santee-Cooper) strains of striped bass *Morone saxatilis. Copeia* 2000, 291–296.

Seo, M. Y., Lee, K. M. and Kaneko, T. (2009). Morphological changes in gill mitochondria rich cells in cultured Japanese eel (*Anguilla japonica*) acclimated to a wide range of environmental salinity. *Fish. Sci.* 75, 1147–1156.

Setzler-Hamilton, E. M. (1991). White perch. In *Habitat Requirements for Chesapeake Bay Living Resources* (eds. S. L. Funderbank, J. A. Mihursky, S. J. Jordan and D. Riley), pp. 12.11–12.99. Annapolis, MD: Chesapeake Bay Research Consortium.

Sheldon, M. R. and McCleave, J. D. (1985). Abundance of glass eels of the American eel, *Anguilla rostrata*, in mid-channel and near shore during estuarine migration. *Nat. Can.* 112, 425–430.

Shirai, N. and Utida, S. (1970). Development and degeneration of chloride cell during seawater and freshwater adaptation of Japanese eel, *Anguilla japonica. Z. Zellforsch. Mikrosk. Anat.* 103, 247–264.

Shrimpton, J. M. (2013). Seawater to freshwater transitions in diadromous fish. In *Fish Physiology*, Vol. 32, *Euryhaline Fishes* (eds. S. D. McCormick, A. P. Farrell and C. J. Brauner), pp. 327–393. New York: Elsevier.

Shu, D. G., Luo, H. L., Morris, S. C., Zhang, X. L., Hu, S. X., Chen, L., Han, J., Zhu, M., Li, Y. and Chen, L. Z. (1999). Lower Cambrian vertebrates from south China. *Nature* 402, 42–46.

Silva, P., Solomon, R., Spokes, K. and Epstein, F. H. (1977). Oubain inhibition of gill Na^+K^+-ATPase – relationship to active chloride transport. *J. Exp. Zool.* 199, 419–426.

Skelton, P. H. (1993). *A Complete Guide to the Freshwater Fishes of Southern Africa.* South Africa: Macmillan, Halfway House.

Smith, H. W. (1932). Water regulation and its evolution in the fishes. *Q. Rev. Biol.* 7, 1–26.

Sola, C. (1995). Chemoattraction of upstream migrating glass eels *Anguilla anguilla* to earthy and green odorants. *Environ. Biol. Fish.* 43, 179–185.

Sola, C. and Tongiorgi, P. (1996). The effect of salinity on the chemotaxis of glass eels, *Anguilla anguilla*, to organic earthy and green odorants. *Environ. Biol. Fish.* 47, 213–218.

Sorensen, P. W. (1984). Juvenile eels rely on odor cues for migration. *Maritimes* 28, 8–9.

Sorensen, P. W. (1986). Origins of the freshwater attractant(s) of migrating elvers of the American eel *Anguilla rostrata. Environ. Biol. Fish.* 17, 185–200.

Stanley, J. G. and Colby, P. J. (1971). Effects of temperature on electrolyte balance and osmoregulation in the alewife (*Alosa pseudoharengus*) in fresh and sea water. *Trans. Am. Fish. Soc.* 100, 624–638.

Steinmetz, J., Kohler, S. L. and Soluk, D. A. (2003). Birds are overlooked top predators in aquatic food webs. *Ecology* 84, 1324–1328.

Sugeha, H. Y., Arai, T., Miller, M. J., Limbong, D. and Tsukamoto, K. (2001). Inshore migration of the tropical eels (*Anguilla spp.*) recruiting to the Poigar River estuary on north Sulawesi Island. *Mar. Ecol. Prog. Ser.* 221, 233–243.

Suzuki, K. W., Kasai, A., Ohta, T., Nakayama, K. and Tanaka, M. (2008). Migration of Japanese temperate bass (*Lateolabrax japonicas*) juveniles within the Chikugo River estuary revealed by δ13C analysis. *Mar. Ecol. Prog. Ser.* 358, 245–256.

Szedlmayer, S. T. and Able, K. W. (1993). Ultrasonic telemetry of age-0 summer flounder, *Paralichthys dentatus*, movements in a southern New Jersey estuary. *Copeia* 1993, 728–736.

Szedlmayer, S. T., Able, K. W. and Rountree, R. A. (1992). Growth and temperature-induced mortality of young-of-the-year summer flounder (*Paralichthys dentatus*), in southern New Jersey. *Copeia* 1992, 120–128.

Tagatz, M. E. (1961). *Tolerance of Striped Bass and American Shad to Changes of Temperature and Salinity. Special Scientific Report: Fisheries.* No. 388, pp. 1–8. US Fish and Wildlife Service.

Tago, Y. (2002). Larval distribution of ayu *Plecoglossus altivelis* in the surface layer of estuary regions in Toyama Bay. *Nippon Suisan Gakkaishi* 68, 61–71.

Taillebois, L., Keith, P., Valade, P., Torres, P., Baloche, S., Dufour, S. and Rousseau, K. (2011). Involvement of thyroid hormones in the control of larval metamorphosis in *Sicyopterus lagocephalus* (Teleostei: Gobioidei) at the time of river recruitment. *Gen. Comp. Endocrinol.* 173, 281–288.

Takei, Y. and McCormick, S. D. (2013). Hormonal control of fish euryhalinity. In *Fish Physiology, Vol. 32, Euryhaline Fishes* (eds. S. D. McCormick, A. P. Farrell and C. J. Brauner), pp. 69–123. New York: Elsevier.

Tandler, A., Anav, F. A. and Choshniak, I. (1995). The effect of salinity on growth rate, survival and swimbladder inflation in gilthead seabream, *Sparus aurata*, larvae. *Aquaculture* 135, 343–353.

Taverny, C., Lepage, M., Piefort, S., Dumont, P. and Rochard, E. (2002). Habitat selection by juvenile European sturgeon (*Acipenser sturio*) in the Gironde estuary (France). *J. Appl. Ichthyol.* 18, 536–541.

Taylor, J. R. and Grosell, M. (2006). Feeding and osmoregulation: dual function of the marine teleost intestine. *J. Exp. Biol.* 209, 2939–2951.

Taylor, M. H., Leach, G. J., DiMichele, L., Levitan, W. M. and Jacob, W. F. (1979). Lunar spawning cycle in the mummichog (*Fundulus heteroclitus*) (Pisces: Cyprinodontidae). *Copeia* 1979, 291–297.

Tebo, L. B., Jr. and McCoy, E. G. (1964). Effect of sea-water concentration on the reproduction and survival of largemouth bass and bluegills. *Prog. Fish.-Cult.* 26, 99–106.

Teo, S. L. H. and Able, K. W. (2003). Habitat use and movement of the mummichog (*Fundulus heteroclitus*) in a restored salt marsh. *Estuaries* 26, 720–730.

Tesch, F. W. (1977). *The Eel: Biology and Management of Anguillid Eels*. London: Chapman & Hall.

Tesch, F. W. (2003). *The Eel* (ed. J. E. Thorpe), p. 407. London: Blackwell.

Thibault, I., Dodson, J. J., Caron, F., Tzeng, W. N., Iizuka, Y. and Shiao, J. C. (2007). Facultative catadromy in American eels: testing the conditional strategy hypothesis. *Mar. Ecol. Prog. Ser.* 344, 219–229.

Thiel, R. and Potter, I. C. (2001). The ichthyofauna of the Elbe Estuary: an analysis in space and time. *Mar. Biol.* 138, 603–616.

Thomson, A. J. and Sargent, J. R. (1977). Changes in levels of chloride cells and Na^+–K^+-dependent ATPase in gills of yellow and silver eels adapting to seawater. *J. Exp. Zool.* 200, 33–40.

Timmons, T. J. (1995). Density and natural mortality of paddlefish, *Polyodon spathula*, in an unfished Cumberland River subimpoundment, South Cross Creek Reservoir, Tennessee. *J. Freshw. Ecol.* 10, 421–431.

Tipsmark, C. K., Madsen, S. S. and Borski, R. J. (2004). Effect of salinity on expression of branchial ion transporters in striped bass (*Morone saxatilis*). *J. Exp. Zool. A* 301, 979–991.

Tipsmark, C. K., Luckenbach, J. A., Madsen, S. S. and Borski, R. J. (2007). IGF-I and branchial IGF receptor expression and localization during salinity acclimation in striped bass. *Am. J. Physiol.* 292, R535–R543.

Tolimieri, N., Jeffs, A. and Montgomery, J. C. (2000). Ambient sound as a cue for navigation by the pelagic larvae of reef fishes. *Mar. Ecol. Prog. Ser.* 207, 219–224.

Trnski, T. (2002). Behaviour of settlement-stage larvae of fishes with an estuarine juvenile phase: in situ observations in a warm–temperate estuary. *Mar. Ecol. Prog. Ser.* 242, 205–214.

Tse, W. K. F., Au, D. W. T. and Wong, C. K. C. (2006). Characterization of ion channel and transporter mRNA expressions in isolated gill chloride and pavement cells of seawater acclimating eels. *Biochem. Biophys. Res. Commun.* 346, 1181–1190.

Tseng, Y. C. and Hwang, P. P. (2008). Some insights into energy metabolism for osmoregulation in fish. *Comp. Biochem. Physiol. C Toxicol. Pharmacol.* 148, 419–429.

Tsukamoto, K. (1991). Age and growth of ayu larvae (*Plecoglossus altivelis*) collected in the Nagara, Kiso and Tone rivers during the downstream migration. *Nippon Suisan Gakkaishi* 57, 2013–2022.

Tsukamoto, K., Ishida, R., Naka, K. and Kajihara, T. (1987). Switching of size and migratory pattern in successive generations of landlocked ayu. In *Common Strategies of Anadromous and Catadromous Fishes*, Vol. 1 (eds. M. J. Dadswell, R. J. Klauda, C. M. Moffitt, R. L. Saunders, R. A. Rulifson and J. E. Cooper), pp. 492–506. Bethesda, MD: American Fisheries Society.

Tsukamoto, K., Otake, T., Mochioka, N., Lee, T. W., Fricke, H., Inagaki, T., Aoyama, J., Ishikawa, S., Kimura, S., Miller, M. J., Hasumoto, H., Oya, M. and Suzuki, Y. (2003). Seamounts, new moon and eel spawning: the search for the spawning site of the Japanese eel. *Environ. Biol. Fish.* 66, 221–229.

Tupper, M. and Boutilier, R. G. (1995). Size and priority at settlement determine growth and competitive success of newly settled Atlantic cod. *Mar. Ecol. Prog. Ser.* 118, 295–300.

Tytler, P. and Bell, M. V. (1989). A study of diffusional permeability of water, sodium and chloride in yolk-sac larvae of cod (*Gadus morhua* L.). *J. Exp. Biol.* 147, 125–132.

Tytler, P., Bell, M. V. and Robinson, J. (1993). The ontogeny of osmoregulation in marine fish: effects of changes in salinity and temperature. In *Physiological and Biochemical Aspects of Fish Development* (eds. B. T. Walther and H. J. Fyhn), pp. 249–258. Bergen: University of Bergen.

Utida, S., Oide, M., Saishu, S. and Kamiya, M. (1967). Preestablishment of adaptation mechanism to salt water in intestine and gills isolated from silver eel during catadromous migration. *Compt. Rend. Seances Soc. Biol. Filiales* 161, 1201–1204.

Utida, S., Kamiya, M. and Shirai, N. (1971). Relationship between activity of Na^+K^+ activated adenosine triphosphatase and number of chloride cells in eel gills with special reference to seawater adaptation. *Comp. Biochem. Physiol.* 38, 443–446.

Valiela, I. (1995). *Marine Ecological Processes* (2nd edn.). Berlin: Springer.

van Ginneken, V., Dufour, S., Sbaihi, M., Balm, P., Noorlander, K., de Bakker, M., Doornbos, J., Palstra, A., Antonissen, E., Mayer, I. and van den Thillart, G. (2007a). Does a 5500-km swim trial stimulate early sexual maturation in the European eel (*Anguilla anguilla* L.)?. *Comp. Biochem. Physiol. A Mol. Integr. Physiol.* 147, 1095–1103.

van Ginneken, V., Durif, C., Dufour, S., Sbaihi, M., Boot, R., Noorlander, K., Doornbos, J., Murk, A. J. and van den Thillart, G. (2007b). Endocrine profiles during silvering of the European eel (*Anguilla anguilla* L.) living in saltwater. *Anim. Biol.* 57, 453–465.

Varsamos, S. (2002). Tolerance range and osmoregulation in hypersaline conditions in the European sea bass (*Dicentrarchus labrax*). *J. Mar. Biol. Assoc. U.K.* 82, 1047–1048.

Varsamos, S., Connes, R., Diaz, J. P., Barnabe, G. and Charmantier, G. (2001). Ontogeny of osmoregulation in the European sea bass (*Dicentrarchus labrax* L.). *Mar. Biol.* 138, 909–915.

Varsamos, S., Nebel, C. and Charmantier, G. (2005). Ontogeny of osmoregulation in postembryonic fish: a review. *Comp. Biochem. Physiol. A Mol. Integr. Physiol.* 141, 401–429.

Vinagre, C., França, S. and Cabral, H. N. (2007). Diel and semi-lunar patterns in the use of an intertidal mudflat by the juveniles of Senegal sole, *Solea senegalensis*. *Estuar. Coast. Shelf Sci.* 69, 246–254.

Vinagre, C., Santos, F. D., Cabral, H. N. and Costa, M. J. (2009). Impact of climate and hydrology on juvenile fish recruitment towards estuarine nursery grounds in the context of climate change. *Estuar. Coast. Shelf Sci.* 85, 479–486.

Vladykov, V. D. and Greeley, J. R. (1963). Order (Acipenseroidei). In *Fishes of the Western North Atlantic*, Vol. 1 (ed. Y. H. Olsen), pp. 24–60. Memoirs of the Sears Foundation for Marine Research, Yale University. Part III.

von Geldern, C. E. (1965). Evidence of American shad reproduction in a landlocked environment. *Calif. Fish Game* 51, 212–213.

Wallace, J. H. and Van der Elst, R. P. (1975). The estuarine fishes of the east coast of South Africa. IV. Occurrence of juveniles in estuaries. V. Ecology, estuarine dependence and status. *Investig. Rep. Oceanogr. Res. Inst.* 42, 1–63.

Warlen, S. M. and Burke, J. S. (1990). Immigration of larvae of fall/winter spawning marine fishes into a North Carolina estuary. *Estuaries* 13, 453–461.

Wasserman, R. J. and Strydom, N. A. (2011). The importance of estuary head waters as nursery areas for young estuary- and marine-spawned fishes in temperate South Africa. *Estuar. Coast. Shelf Sci.* 94, 56–67.

Watanabe, W. O., Kuo, C. M. and Huang, M. C. (1985). The ontogeny of salinity tolerance in the tilapias *Oreochromis aureus*, *O. niloticus*, and *O. mossambicus* × *O. niloticus* hybrid, spawned and reared in freshwater. *Aquaculture* 47, 353–368.

Watanabe, W. O., French, K. E. and Ernt, D. H. (1989). Salinity during early development influences growth and survival of Florida red tilapia in brackish and seawater. *J. World Aquacult. Soc.* 20, 134–142.

Waters, J. M., Dijkstra, L. H. and Wallis, G. P. (2001). Biogeography of a southern hemisphere freshwater fish: how important is marine dispersal? *Mol. Ecol.* 9, 1815–1821.

Wathen, G., Coghlan, S., Zydlewski, J. and Trial, J. (2011). Habitat selection and overlap of Atlantic salmon and smallmouth bass juveniles in nursery streams. *Trans. Am. Fish. Soc.* 140, 1145–1157.

Wathen, G., Coghlan, S., Zydlewski, J. and Trial, J. (2012). Effects of smallmouth bass on Atlantic salmon habitat use and diel movements in an artificial stream. *Trans. Am. Fish. Soc.* 141, 174–184.

Weisberg, S. B., Wilson, H. T., Himchak, P., Baum, T. and Allen, R. (1996). Temporal trends in the abundance of fish in the tidal Delaware River. *Estuaries* 19, 723–729.

Werner, F. E., Blanton, B. O., Quinlan, J. A. and Luettich, R. A., Jr. (1999). Physical oceanography of the North Carolina continental shelf during the fall and winter seasons: implications for the transport of larval menhaden. *Fish. Oceanogr.* 8 (Suppl. 2), 7–21.

Weyl, O. L. F. and Lewis, H. (2006). First record of predation by the alien invasive freshwater fish (*Micropterus salmoides*) on migrating estuarine fishes in South Africa. *Zool. Afr.* 41, 294–296.

Whitfield, A. K. (1998). Biology and ecology of fishes in southern African estuaries. *Ichthyol. Monogr. J.L.B. Smith Inst. Ichthyol.* 2, 1–223.

Wilkie, M. P., Turnbull, S., Bird, J., Wang, Y. S., Claude, J. F. and Youson, J. H. (2004). Lamprey parasitism of sharks and teleosts: high capacity urea excretion in an extant vertebrate relic. *Comp. Biochem. Physiol. A Mol. Integr. Physiol.* 138, 485–492.

Wilson, J. M. and Laurent, P. (2002). Fish gill morphology: inside out. *J. Exp. Zool.* 293(3), 192–213.

Wilson, J. A. and McKinley, R. S. (2004). Distribution, habitat and movements. In *Sturgeons and Paddlefish of North America* (eds. G. T.O. LeBreton, F. W.H. Beamish and R. S. McKinley), pp. 40–72. Dordrecht: Kluwer.

Wilson, J. M., Antunes, J. C., Bouça, P. D. and Coimbra, J. (2004). Osmoregulation plasticity of the glass eel of *Anguilla anguilla*: freshwater entry and changes in branchial ion-transport protein expression. *Can. J. Fish. Aquat. Sci.* 61, 432–442.

Wilson, J. M., Leitão, A., Gonçalves, A. F., Ferreira, C., Reis-Santos, P., Fonseca, A. V., Da Silva, J. M., Antunes, J. C., Pereira-Wilson, C. and Coimbra, J. (2007). Modulation of branchial ion transport protein expression by salinity in glass eels (*Anguilla anguilla* L.). *Mar. Biol.* 151, 1633–1645.

Wilson, R. W., Millero, F. J., Taylor, J. R., Walsh, P. J., Christensen, V., Jennings, S. and Grosell, M. (2009). Contribution of fish to the marine inorganic carbon cycle. *Science* 323, 359–362.

Wingate, R. L. and Secor, D. H. (2008). The effects of winter temperature and flow on a summer–fall nursery fish assemblage in the Chesapeake Bay, Maryland. *Trans. Am. Fish. Soc.* 137, 1147–1156.

Winger, P. V. and Lasier, P. J. (1994). Effects of salinity on striped bass eggs and larvae from the Savanna River, Georgia. *Trans. Am. Fish. Soc.* 123, 904–912.

Wippelhauser, G. S. and McCleave, J. D. (1988). Rhythmic activity of migrating juvenile American eels (*Anguilla rostrata*). *J. Mar. Biol.* 68, 81–91.

Wirjoatmodjo, S. and Pitcher, T. J. (1984). Flounders follow the tides to feed: evidence from ultrasonic tracking in an estuary. *Estuar. Coast. Shelf Sci.* 19, 231–241.

Witting, D. A., Able, K. W. and Fahay, M. P. (1999). Larval fishes of a Middle Atlantic Bight estuary: assemblage structure and temporal stability. *Can. J. Fish. Aquat. Sci.* 56, 222–230.

Wong, C. K. C. and Chan, D. K. O. (2001). Effects of cortisol on chloride cells in the gill epithelium of Japanese eel, *Anguilla japonica*. *J. Endocrinol.* 168, 185–192.

Woo, N. Y. S. and Fung, A. C. Y. (1981). Studies on the biology of red sea bream, *Chrysophrys major*. II. Salinity adaptation. Comp. Biochem. Physiol. A 69, 237–242.

Wootton, R. J. (1990). *Ecology of Teleost Fishes* (1st edn.). London: Chapman & Hall.

Wootton, R. J. (1999). *Ecology of Teleost Fishes* (2nd edn.). Dordrecht: Kluwer.

Yada, T., Tsuruta, T., Sakano, H., Yamamoto, S., Abe, N., Takasawa, T., Yogo, S., Suzuki, T., Iguchi, K., Uchida, K. and Hyodo, S. (2010). Changes in prolactin mRNA levels during downstream migration of the amphidromous teleost, ayu (*Plecoglossus altivelis*). *Gen. Comp. Endocrinol.* 167, 261–267.

Yagi, Y., Bito, C., Funakoshi, T., Kinoshita, I. and Takahashi, I. (2006). Distribution and feeding habits of ayu (*Plecoglossus altivelis altivelis*) larvae in coastal waters of Tosa Bay. *Nippon Suisan Gakkaishi* 72, 1057–1067.

Yamaguchi, A. and Kume, G. (2008). Evidence for up-estuary transport of puffer Takifugu larvae (Tetraodontidae) in Ariake Bay, Japan. *J. Appl. Ichthyol.* 24, 60–62.

Yin, M. C. and Blaxter, J. H. S. (1987). Temperature, salinity tolerance and buoyancy during development and starvation of Clyde and North Sea herring, cod and flounder larvae. *J. Exp. Mar. Biol. Ecol.* 107, 279–290.

Youson, J. H. (1980). Morphology and physiology of lamprey metamorphosis. *Can. J. Fish. Aquat. Sci.* 37, 1687–1710.

Youson, J. H. (1994). Environmental and hormonal cues and endocrine glands during lamprey metamorphosis. In *Perspectives in Comparative Endocrinology* (eds. K. G. Davey, R. E. Peter and S. S. Tobe), pp. 400–407. Ottawa: National Research Council of Canada Publications.

Youson, J. H. (1997). Is lamprey metamorphosis regulated by thyroid hormones? *Am. Zool.* 37, 441–460.

Youson, J. H. (2003). The biology of metamorphosis in sea lampreys: endocrine, environmental, and physiological cues and events, and their potential application to lamprey control. *J. Great Lakes Res.* 29, 26–49.

Youson, J. H. and Manzon, R. G. (2012). Lamprey metamorphosis. In *Metamorphosis in Fish* (eds. S. Doufour, K. Rousseau and B. G. Kapoor), pp. 12–75. Enfield, NH: Science Publishers.

Youson, J. H. and Potter, I. C. (1979).). Description of the stages in the metamorphosis of the anadromous sea lamprey (*Petromyzon marinus* L.). *Can. J. Zool.* 57, 1808–1817.

Zastrow, C. E., Houde, E. D. and Morin, L. G. (1991). Spawning, fecundity, hatch-date frequency and young-of-the-year growth of bay anchovy (*Anchoa mitchilli*) in mid-Chesapeake Bay. *Mar. Ecol. Prog. Ser.* 73, 161–171.

Zhao, F., Zhuang, P., Zhang, L. Z. and Hou, J. L. (2010). Changes in growth and osmoregulation during acclimation to saltwater in juvenile Amur sturgeon (*Acipenser schrenckii*). *Chin. J. Oceanol. Limnol.* 28, 603–608.

Ziegeweid, J. R., Jennings, C. A., Peterson, D. L. and Black, M. C. (2008). Effects of salinity, temperature, and weight on the survival of young-of-year shortnose sturgeon. *Trans. Am. Fish. Soc.* 137, 1490–1499.

Zydelis, R. and Kontautas, A. (2008). Piscivorous birds as top predators and fishery competitors in the lagoon ecosystem. *Hydrobiologia* 611, 45–54.

Zydlewski, J. and McCormick, S. D. (1997a). The ontogeny of salinity tolerance in the American shad, *Alosa sapidissima*. *Can. J. Fish. Aquat. Sci.* 54, 182–189.

Zydlewski, J. and McCormick, S. D. (1997b). The loss of hyperosmoregulatory ability in migrating juvenile American shad, *Alosa sapidissima*. *Can. J. Fish. Aquat. Sci.* 54, 2377–2387.

Zydlewski, J. and McCormick, S. D. (2001). Developmental and environmental regulation of chloride cells in young American shad, *Alosa sapidissima*. *J. Exp. Zool.* 290, 73–87.

Zydlewski, J., McCormick, S. D. and Kunkel, J. G. (2003). Late migration and seawater entry is physiologically disadvantageous for American shad juveniles. *J. Fish Biol.* 63, 1521–1537.

7

SEAWATER TO FRESHWATER TRANSITIONS IN DIADROMOUS FISHES

J. MARK SHRIMPTON

Diadromous fishes have evolved mechanisms to transition between seawater and freshwater environments. For anadromous species, and probably also amphidromous species, movement into freshwater is obligate for successful spawning. Landlocked populations that do not migrate to seawater indicate that movement into marine habitats for feeding and rearing is facultative in anadromous species, but some amphidromous species may be obligate. For catadromous species, movement into freshwater is facultative as fish seek areas of greater productivity, but some individuals may complete their life cycle without entering freshwater. Conditions for spawning, therefore, appear to be more strongly dictated

Euryhaline Fishes: Volume 32
FISH PHYSIOLOGY
Copyright © 2013 Elsevier Inc. All rights reserved
DOI: http://dx.doi.org/10.1016/B978-0-12-396951-4.00007-4

than habitats for feeding. This chapter describes the behavior and timing of movements from seawater to freshwater in diadromous fishes, the physiological and endocrine changes that enable successful transition between the two environments, mechanisms for selecting freshwater habitats, and the effect of movement between seawater and freshwater on genetic population structure of diadromous species.

1. INTRODUCTION

Migration is one of the most energetically demanding and physiologically challenging phases during the life history of an animal, as it represents an interaction among behavioral, morphological, and physiological changes (Hinch et al., 2006). This is particularly the case for animals that migrate between regions where the environments differ greatly, and migrations between inland freshwater (FW) systems and saline marine environments represent one of the most difficult transitions. Few taxa have developed strategies to successfully move between FW and seawater (SW) environments. Diadromy is a migratory strategy in fish that involves regular seasonally timed migration between FW and marine environments (McDowall, 2007). Even within the vast number of bony fishes (estimated at more than 25,000 species), fish that make diadromous migrations are relatively rare and occur in only approximately 250 different species (Nelson, 1994; McDowall, 2001; Schultz and McCormick, 2013, Chapter 10, this volume). Yet these fish represent ecologically, economically, and culturally important species beyond their species numbers owing to their movement patterns and redistribution of productivity.

Three forms of diadromy have been recognized (McDowall, 2001). Anadromy is certainly the most well-known form of diadromy as fish migrate from the ocean to FW to spawn. Juveniles rear in FW for variable lengths of time, but at some stage migrate to the sea where they feed, grow, and become sexually mature, and then return to FW to spawn. Anadromy is found in a diverse group of fish species, including the families Petromyzontidae (northern lamprey), Geotriidae and Mordacidae (southern lamprey), Acipenseridae (sturgeon), Clupeidae (shad), and Osmeridae (smelts), but is best known because of the family Salmonidae (salmon and trout). A second form of diadromy is catadromy, where fish migrate from FW to the ocean to spawn. Juveniles migrate to FW to feed, grow, and become sexually mature when the adults migrate back to the sea to spawn. Catadromy is often thought to be restricted to one family, Anguillidae (eels), but a few species within several other families are also known to be catadromous; notably the families Mugillidae (mullets) and Clupeidae (herring). Considerable variations in patterns exist within the broad definitions of anadromy and catadromy, but the essential element is that the return to their natal environment is associated with reproduction. The third form of diadromy is amphidromy, where fish migrate back to their natal environment as juveniles rather than mature adults. Amphidromous fishes typically spawn in FW, larvae being carried downstream to the ocean immediately after hatch, where they feed and grow for a relatively short period before returning to FW as small juveniles (McDowall, 2007). Consequently, most feeding and growth occur in FW systems, where the fish eventually become sexually mature and spawn. Amphidromy is found in a diverse number of families: Retropinnidae (southern smelts), Galaxiidae, Cottidae (sculpins), Gobiidae (gobies), and Eleotridae (sleeper gobies) (McDowall, 1987, 2007).

In amphidromous species, fish return to FW as small juveniles, usually less than 50 mm in length, and they undergo most of their somatic growth in FW, before maturing and spawning. In contrast, anadromous fishes returning to FW have completed their somatic growth. In terms of timing for FW entry, amphidromy is more similar to catadromy than to anadromy, as it is the juveniles that migrate from the ocean to FW initially just for growth opportunities. The geographic range of amphidromous species also sets this life history strategy apart from other forms of diadromy. Amphidromy is uncommon on the major continents, but is very common among the tropical and temperate oceanic islands, especially in the Sicydiinae gobies (McDowall, 2007).

For diadromous species, some of the most striking examples are where timing of movement from the marine to FW environments occurs in conjunction with significant developmental changes. For example, maturation in anadromous salmon is associated with a change from bright silver to red, brown, green, or black colors associated with reproduction (Fig. 7.1A) and males particularly undergo extensive morphological remodeling with the development of secondary sexual characteristics such as the kype and hump. Morphological changes in catadromous eels are arguably even more dramatic as leptocephali metamorphose into glass eels before recruiting to FW (Fig. 7.1B); the changes also continue with residence in estuaries and migration upriver as glass eels develop pigmentation and transform into elvers (Tesch, 2003). Morphological changes are no less dramatic in amphidromous gobies as they metamorphose from pelagic marine larvae to benthic FW forms (Fig. 7.1C) and develop pigmentation. It is not just morphology that changes, but there is a complex interplay between behavior

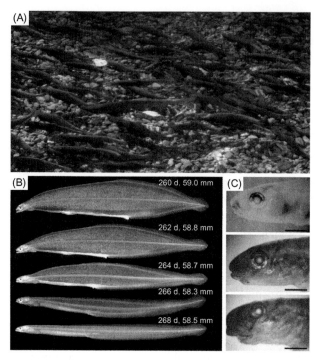

Fig. 7.1. Morphological changes that occur in diadromous fish during the transition from marine to freshwater habitats. (A) The bright nuptial colors associated with maturation are seen in mature *Oncorhynchus nerka*. Photograph courtesy of Marley C. Bassett, University of Northern British Columbia. (B) Metamorphosis of laboratory-reared *Anguilla japonica* from the leptocephalus stage (59.0 mm, 260 days old) to the glass eel stage (58.5 mm, 268 days old). Reprinted from Tsukamoto et al. (2009) with permission from Springer. (C) Changes in head structure of *Sicyopterus japonicas* showing the different mouth shapes and pigmentation with metamorphosis from pelagic larvae to benthic juvenile: top panel shows the terminal mouth, 34 mm total length (TL) postlarva; middle panel shows the intermediate phase, 33 mm TL juvenile; bottom panel show the subterminal mouth, 33 mm TL juvenile. Scale bars: 2 mm. Reprinted from Shen and Tzeng (2002) with permission from Inter-Research. **SEE COLOR PLATE SECTION**

and physiology that ensures successful transition between the ocean and FW environments.

Movement between marine and FW environments occurs at very different stages of development for anadromous, catadromous, and amphidromous species. Such differences are expected to affect fish at both the population and individual level. The behavior of fish migrating toward coastal regions and into FW also differs as some animals are seeking natal systems to spawn whereas others are dispersing to exploit new habitats for foraging. Population genetics of diadromous species has shown interesting patterns of gene flow and population structure that vary for the different forms of diadromy. Although all species migrating from the marine environment to FW experience a physiological challenge as they move from a hyperosmotic medium to a hypoosmotic medium, there are physiological differences among diadromous fish due to stage of development. Migration from the ocean and upriver also represents a considerable energetic cost for fish swimming against the current, a factor that will affect seasonal timing of movement into FW, but the energetics differs tremendously among forms of diadromous fishes. Catadromous and amphidromous fishes move into rivers to forage, whereas anadromous species generally cease feeding before leaving the ocean (e.g. Miller et al., 2009; Clemens et al., 2010). This chapter will examine differences in the behavior of diadromous fishes associated with directed migration, physiological changes associated with the osmoregulatory challenge, and the consequence of migration from the marine environment into FW on genetic population structure.

Although diadromous species are known from diverse taxonomic groups of fish (Schultz and McCormick, 2013, Chapter 10, this volume), surprisingly little is known about most of them. Even among the species and families that have been studied, knowledge gaps exist regarding their behavior, ecology, and physiology. This chapter will describe patterns that have been recognized for well-studied species. Consequently, specific genera and families will be used as examples; for anadromy the families Salmonidae and Petromyzontidae, for catadromy the family Anguillidae, and for amphidromy the family Gobiidae, but where knowledge gaps exist examples from other taxa are given. The family Salmonidae contains the salmon and is perhaps the best characterized of all diadromous fishes. The commercial, economic, and social importance of this group has motivated much research. For almost the opposite reasons, a considerable amount of work has been done on the family Petromyzontidae, particularly the landlocked sea lamprey found in the Great Lakes of North America. Commercial trout and whitefish fisheries collapsed following the invasion of sea lamprey to the Great Lakes and basic life history information has been gathered in efforts to eradicate them. A decline in populations of anadromous species, however, has led to recent conservation efforts and stimulated research (Clemens et al., 2010). Declines in abundance of economically important species of Anguillidae has also stimulated research (Tesch, 2003). Of the diadromous forms, the least is known about amphidromy (McDowall, 2007). The small size of most adult forms has likely also contributed to the paucity of information on amphidromous species, but fisheries for amphidromous species are generally small, although they can be of regional importance (Lord et al., 2010; Walter et al., 2012).

1.1. Why Migrate to Freshwater?

Aquatic productivity has been suggested as a factor favoring the evolution of diadromy (Gross et al., 1988). These authors argued that anadromy should evolve at high latitudes, because the productivity of oceans is greater than that of neighboring FW. Migration to the ocean, therefore, provides growth opportunities that exceed those of the FW environments. Differences in growth opportunities are readily seen between different life history forms within the same species; anadromous sockeye salmon (*Oncorhynchus nerka*) are approximately 10-fold larger by weight when mature than non-anadromous kokanee of the same age that remain in FW throughout their life cycle. Returning to nutrient-poor regions for spawning is also of benefit as such systems usually have few predators and competitors.

Differences in aquatic productivity would predict that there should be latitudinal variation in anadromy versus residency within a species. Direct observation and tagging studies have given us insight into patterns of movement for anadromous salmonids, but the use of elemental signatures in bones has provided a powerful tool to track movement of fish among habitats, particularly movements between marine and FW environments (Fig. 7.2A, B). For Pacific salmon, however, there is little evidence for latitudinal variation in degree of anadromy along the west coast of North America. In *O. nerka* the two life history forms, anadromous sockeye and FW resident kokanee, occur sympatrically. The kokanee life history is polyphyletic, with kokanee populations showing greater genetic relatedness to sockeye in the same river system than to other kokanee populations from adjacent systems (Taylor et al., 1996). Latitude appears to have no influence on the relative proportion of kokanee to sockeye populations across the geographic range of the species. Chum (*O. keta*) and pink (*O. gorbuscha*) salmon have the most abbreviated FW residence, but no natural non-anadromous populations exist (Quinn and Myers, 2004). Greater support for latitudinal variation in residency is seen in sea run trout populations. Brook trout (*Salvelinus fontinalis*) (McCormick, 1994), Arctic char (*S. alpinus*) (Klemetsen et al., 2003), Dolly Varden (*S. malma*) (Morita et al., 2005), and brown trout (*Salmo trutta*) (Klemetsen et al., 2003) populations from the northern parts of their ranges in eastern North America, Japan, and Europe migrate to the ocean; but there is little to no evidence of marine residence in more southerly populations.

In contrast, the aquatic productivity hypothesis predicts that catadromy should evolve at low latitudes, because FW productivity is greater than in marine environments in tropical regions. Migration into FW by catadromous species, therefore, represents movement along a gradient in food abundance between the ocean and FW in the tropics. Catadromy was assumed to be obligate in the anguillid eels, but some individuals never enter FW. Again, elemental signatures in otoliths have proven useful in assessing habitat use and the degree of catadromy (Fig. 7.2C). Non-catadromous eels have been found in temperate populations of American eel (*Anguilla rostrata*) (Lamson et al., 2006; Jessop et al., 2008), European eel (*A. anguilla*) (Harrod et al., 2005), Japanese eel (*A. japonica*) (Arai and Hirata, 2006), and the giant mottled eel (*A. marmorata*) (Chino and Arai 2010). Daverat et al. (2006) showed a strong relationship between FW residency and latitude for American, European, and Japanese eels. In the northern hemisphere at low latitudes (20°N) approximately 95% of eels were FW residents, but the proportion declines and at higher latitudes (55°N) only 25% of eels were FW residents. The aquatic productivity model of Gross et al. (1988) would also suggest that all tropical eels should be obligate catadromous. Chino and Arai (2010), however, using otolith microchemistry, provide evidence that in a tropical eel (*A. bicolor*) some individuals never enter FW. They also found that the age when FW eels moved to the estuary or into the ocean was highly variable. Flexible habitat use by *A. bicolor* suggests that FW residency is not obligate even in tropical regions such as Indonesia (Chino and Arai, 2010).

There is considerable evidence that growth differs between eels in FW and marine environments. In general, *A. rostrata* from brackish water (BW) have higher growth rates than those entering FW owing to the greater productivity in BW (Thibault et al., 2007) and American glass eels also grow faster in BW than in FW (Côté et al., 2009). Growth was similarly better in marine environments for *A. anguilla* in temperate regions (Edeline et al., 2005a; Harrod et al., 2005). Despite the poorer growth opportunity, the selection of low-productivity FW habitats by temperate eels may enhance survival owing to lower predation. Cairns et al. (2009) found no evidence that mortality rates are lower in FW than in SW, but use of diverse habitats by a population spreads risk and potentially decreases variation in fitness.

Our understanding of evolutionary factors favoring amphidromy is limited. The greatest number of amphidromous species has been described for southern temperate regions of the world (McDowall, 1987). Amphi-dromous species tend to occur on oceanic islands and it has been speculated that amphidromy is a life history strategy that facilitates colonization of newly available habitats (McDowall, 2007). Large numbers of planktonic marine larvae facilitate dispersal. For example, one species of goby, *Sicyopterus lagocephalus*, is found throughout the mid Pacific Ocean across a range of approximately 18,000 km (Keith et al., 2005). Amphidromy may also be advantageous for fish living in FW habitats of small island streams that are prone to rapid changes, making these systems unfavorable for short

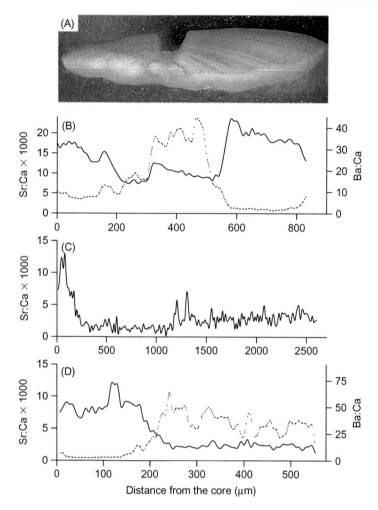

Fig. 7.2. (A) Otolith from *Oncorhynchus tshawytscha* showing laser ablation burn to determine elemental composition of the structure. Photograph courtesy of Nicole Laforge, University of Victoria. (B) Line scans of otolith Sr:Ca (solid line) and Ba:Ca (dotted line) elemental ratios from *O. kisutch* captured in the Coldwater River, British Columbia, Canada. Elemental signatures from the core to ~150 μm represent maternal signature with high Sr and low Ba characteristic of anadromous spawners. The lower Sr and higher Ba section of the line scan represent movements of juveniles among freshwater (FW) habitats. The increase in Sr and drop in Ba at ~550 μm indicates ocean entry at the smolt stage and the end of the trace indicates return to FW (J. M. Shrimpton, K. H. Telmer, G. J. Glova and N. L. Todd, unpublished data). (C) Line scans of otolith Sr:Ca elemental ratios for *Anguilla marmorata* captured in the Yatsuse River, Bonin Islands, Japan. The high Sr:Ca ratio near the core indicates the marine leptocephali phase, followed by low Sr:Ca values characteristic of residence in FW, and the intermediate Sr:Ca values suggest movement from FW to brackish water later in life. Adapted

periods, such that the marine environment serves as a refuge from which juveniles reinvade FW after disturbances (McDowall, 2007).

Elemental chemistry of otoliths has also proven useful for tracking movement among habitats in amphidromous gobies and galaxiids (Fig. 7.2D). Amphidromy appears to be obligate for the Sicydiinae gobies. Radtke and Kinzie (1996) found no evidence of landlocked gobies in Hawaii. All fish had a SW signature at the otolith core – even fish collected from upper sections of interrupted streams and above waterfalls, including above Akaka Falls with a free-falling drop of 130 m. Tsunagawa and Arai (2011) found 159 of 160 samples of FW goby from Japan belonging to the genus *Rhinogobius* to be amphidromous, with no marine signature in only one individual. Amphidromy, however, is facultative among many species of galaxiids. Humphries (1989), working on spotted galaxias (*Galaxias truttaceus*), identified both amphidromous and landlocked populations.

2. BEHAVIOR AND TIMING

Diadromous migrations cover thousands of kilometers through marine and FW environments and movement from the ocean and into river systems consists of several phases: initiation, approach to the coast, transition to FW, and migration upstream. There are tremendous differences and also common features among diadromous species for the four stages of migration.

Timing of migration to coastal waters is difficult to determine owing to the vast area over which fish migrate. We have a better idea of when fish approach the coast and return to FW as this has traditionally been the location where fish have been intercepted for harvest. For commercially important species it has been known for a long time that populations of fish annually migrate through specific areas at particular times. Fisheries have taken advantage of this information by targeting harvest efforts on the larger populations, while restricting the capture of smaller populations that are not abundant enough to sustain harvest pressure. The regular timing of

---

from Chino and Arai (2010) with permission from John Wiley & Sons. (D) Line scans showing changes in Sr:Ca (solid line) and Ba:Ca (dotted line) elemental ratios in *Sicyopterus lagocephalus* captured in the Barendeu River, New Caledonia. The high Sr:Ca and low Ba:Ca ratios represent the pelagic marine larval stage and the low Sr:Ca and higher Ba:Ca ratios represent FW signatures. Adapted from Lord et al. (2011) with permission from John Wiley & Sons. **SEE COLOR PLATE SECTION**

movement to FW as part of the life cycle of anadromous fishes, therefore, is well understood for many species, particularly where humans exploit these species as a food resource. The seasonality of these movements will be reviewed among different species of diadromous fishes; patterns that exist within anadromous, catadromous, and diadromous species and how such patterns are related to motivations for exploiting the FW environment will be examined.

2.1. Anadromous Salmon

The timing of migration in salmon differs among species, but also among populations within a single species that spawn in the same watershed (Fig. 7.3). Such differences have been used to argue that in salmon the timing of FW entry is under selective pressure and is therefore strongly controlled by genetics, e.g. in sockeye salmon (Cooke et al., 2004), Chinook salmon (*O. tshawytscha*) (Keefer et al., 2004), and Atlantic salmon (*Salmo salar*) (Jonsson et al., 2007). Selective pressure is not restricted to events immediately associated with FW entry as long-distance migrations to spawning areas and extended residence of early life stages separate the environmental conditions experienced by adults from those experienced by juveniles. Initiation of migration in maturing adults in the ocean, therefore, is likely to be determined by responses to photoperiod that are specific for each population (Hodgson et al., 2006).

The initial phase of migration is a directional movement from the open ocean to the coast and begins many months prior to arrival at the animal's natal estuary. Japanese chum salmon have been shown to leave feeding areas in the Gulf of Alaska in the spring en route to the Bering Sea (Onuma et al., 2009), then leave the Bering Sea in early summer, but still have 2–3 months of migration before reaching their natal rivers in Japan (Tanaka et al., 2005). In Atlantic salmon this initial phase of migration has been referred to as a crude directional movement towards the coastline. Hansen et al. (1993) released tagged Atlantic salmon smolts at three locations: into Norwegian rivers, along the coast of Norway, and at their feeding area north of the Faroe Islands during winter. Experience gained by out-migrating smolts was not required to ensure successful directional move-ment towards the coast of returning adult salmon (Hansen et al., 1993).

Timing of salmon migrations from the ocean to coastal waters shows interannual variation within populations that often correlates with temperature. Following a spring–summer period with warm sea surface temperatures, populations of sockeye salmon from south-western Alaska returned earlier, while populations from southern British Columbia returned to the Fraser River later (Hodgson et al., 2006). The authors argued that

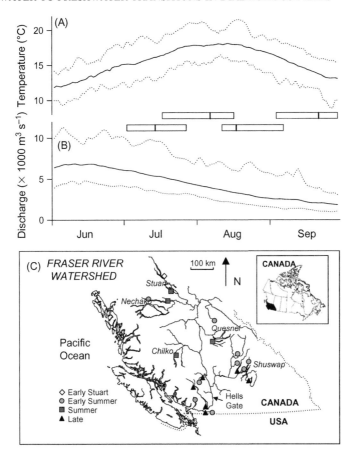

Fig. 7.3. Mean temperature (A) and discharge (B) of the Fraser River for June to September from 1976 to 2006. Dotted lines represent minimum and maximum temperature and flow. Boxes between the temperature and discharge graphs represent the historic peak period of the sockeye salmon run-timing groups at Hells Gate passage with the median dates from 1976 to 2006 shown by the vertical solid line. From left to right the run timing groups are Early Stuart, Early Summer, Summers, and Lates. The locations of spawning grounds for some of the major populations representing the four different run-timing groups in the Fraser River watershed are shown in (C). Data from Patterson et al. (2007).

ocean distributions of adult sockeye would be affected by temperature and in warmer years fish would be distributed further north. If the date when salmon commence homeward migration differs little among years and swimming speed is constant, Alaskan populations should arrive earlier because of the shorter distance and Fraser populations should arrive later owing to the longer time taken to cover the greater distance. Warmer oceanographic conditions also appear to delay arrival of chum salmon returning to natal rivers in Japan (Onuma et al., 2003a).

Migration rate in the ocean appears to vary somewhat with population, but distance travelled per day is fairly consistent within run-timing groups (Crossin et al., 2007). Late-run sockeye migrated at about 20 km day^{-1}, while summer run sockeye migrated faster, at about 33 km day^{-1}. Work on chum salmon has also shown that this species swims actively along the east coast of Japan and prefers surface waters (Tanaka et al., 2001), except in warmer months when fish swim deeper to avoid warmer sea surface temperatures and appear to behaviorally thermoregulate (Tanaka et al., 2000). Migration along the coast requires precise navigation in Atlantic salmon (Hansen et al., 1993) and experience gained during the outmigration is necessary for successful navigation along the coast and return to the river of release. Coastal migration is also a time of changing metabolic physiology. Using a functional genomics approach, Miller et al. (2009) showed that sockeye salmon returning to the Fraser River in British Columbia had stopped feeding when they were intercepted approximately 850 km from the river estuary. Enhanced protein turnover and reduced transcription of muscle and oxygen transport proteins were early indicators of starvation while fish were still in saltwater.

Arrival at the river mouth is not associated with immediate migration through the estuary and river entry. Holding times in the estuary vary among species and populations, but this period appears to be important for the successful transitioning from the marine environment to FW and in some cases timing the arrival at spawning areas. Movement of salmon in the estuary has often been described as appearing to be passive as fish pause before upstream migration. Directed movement towards the river mouth through an estuary, however, is related to tidal current. Sockeye salmon moved with the tidal current towards the river mouth, but during ebb tides showed relatively little movement (Groot et al., 1975). More recently, Levy and Cadenhead (1995) also showed that for early-run sockeye returning to the Fraser, migration through the estuary was synchronous with the tidal cycle, a behavior referred to as selective tidal stream transport (STST). STST has also been shown in Atlantic salmon (Aprahamian et al., 1998). In contrast to Pacific salmon, Atlantic salmon can reside in an estuary for several months before the start of upstream migration (Brawn, 1982). Despite differences in timing of river entry among populations, both early- and late-run Atlantic salmon can be present in the estuary at the same time (Brawn, 1982).

Date of river entry is linked to reproduction and will affect when fish will arrive at spawning grounds and, spawn and also determine the rate of larval and juvenile development (Burgner, 1991; Healey, 1991; Salo, 1991). Indicative of the strong genetic control for migration, variations in entry timing within a population are usually small. For example, in sockeye salmon interannual timing for each population rarely deviates by more than a week (Cooke et al., 2004). Examples of large deviations from expected river entry date are few, but it is expected that the consequences would be severe if this should occur. Recently, there has been an unprecedented change in entry timing for late-run sockeye salmon entering the Fraser River that has been associated with significant in-river mortality; in some years mortality rates have exceeded 90% of the fish observed to enter the Fraser River (Cooke et al., 2004).

There is enormous variation in the stage of maturity when salmon enter FW, even within a species. Chum salmon, for example, spawn over a wide geographic range along the Pacific and Arctic coasts of North America and Asia. The majority of chum salmon become sexually mature while still in the ocean, migrate short distances upriver, and spawn shortly after entering FW (Salo, 1991). Such individuals develop their breeding coloration while still in the ocean and are referred to as ocean-maturing. Some populations of chum, however, migrate over 2000 km to reach their spawning grounds. Long-distance migrating chum populations are found in the Yukon, Mackenzie, and Amur Rivers and are referred to as river-maturing. River-maturing chum are less mature when they enter the river and do not develop breeding coloration until late in the FW migration. Genetic analyses indicate that rivers with ocean-maturing and river-maturing chum were colonized by a single group of chum salmon which then diverged (J. D. Hamilton, J. M. Shrimpton, and D. D. Heath, unpublished data).

Examples of variation in state of maturity are also seen in other species of salmon. Some populations of Chinook enter FW in the fall (autumn) and spawn within a month. Longer river systems also have populations that enter in the spring and migrate most of the way to their spawning grounds, hold during the summer, and then spawn in the fall. These two life history variations are referred to as fall and spring Chinook (Healey, 1991). As with the chum populations that migrate different distances within a single watershed, most rivers were colonized by a single group of Chinook which then diverged into spring and fall runs; run types within a river are more closely related than they are to runs with similar timing from other rivers (Waples et al., 2004). Quinn and Myers (2004) suggest that the spring life history pattern evolved to enable Chinook to exploit areas suitable for fall spawning by adults and rearing by juveniles that are inaccessible before spawning because flow or temperature limit access. Consequently, adults leave the ocean early when access is possible and minimize energy losses until they spawn. A similar pattern of early entry into FW is also seen in sockeye. In southern coastal populations sockeye avoid high summer temperatures by migrating upriver earlier in the year, and then remain in lakes below the thermocline until the fall when they enter tributaries to spawn (Hodgson and Quinn, 2002).

Despite the strong genetic control for seasonal migration, variations in time of river entry are seen and appear to be influenced by environmental factors. Differences among studies exist and findings are not consistent within or among species. For example, Smith et al. (1994) found that Atlantic salmon entered the river in response to periods of elevated flow, whereas Jonsson et al. (2007) found that fish size and water temperature were also important variables. Other work on Atlantic salmon from a large river in northern Norway found that temperature and discharge had little effect on timing of river entry (Lilja and Romakkaniemi, 2003). In sockeye salmon returning to the Fraser River, British Columbia, early entry has been linked to weaker ocean currents and lower coastal salinities (Thomson and Hourston, 2011).

Populations of salmon within a species are also well adapted to migrate upstream and encounter river discharges and temperatures within a relatively narrow range (Fig. 7.3) (Eliason et al., 2011), but problems occur for migrating adults when environmental conditions fall outside this range (Farrell et al., 2008). It could be argued, therefore, that environmental factors experienced in-river should influence timing of upstream migration and supportive evidence exists. Data collected over 23 years for Atlantic salmon in the Connecticut River drainage found that migration has occurred earlier by about 0.5 days year^{-1} (Juanes et al., 2004). Similarly, spring warming in the Columbia River has progressively occurred earlier over the past 50 years and sockeye salmon are migrating upriver approximately 6 days earlier than they did 50 years ago (Quinn and Adams, 1996). The mean temperature that the fish experience during upstream migration, however, is approximately 2.5°C warmer over this period, so temperature does not influence migration timing as much as would be predicted. Quinn and Adams (1996) argue that the long migration and delayed spawning make salmon more strongly controlled by innate responses to photoperiod. Consequently, migration occurs at a time of year that enables maximal survival of incubating larvae and juveniles and is not based on conditions in the lower river when adults are migrating. In-river conditions, however, can influence the rate at which fish migrate upstream. Keefer et al. (2004) radiotagged Chinook salmon in the Columbia River and over 5 years found that run timing within a population was highly predictable, but during low-flow years fish migrated faster and reached upstream dams earlier.

2.2. Anadromous Lamprey

At the end of the marine parasitic phase, lamprey cease feeding and begin their spawning migration. Lamprey spawn in spring or early summer, but considerable interspecific variation exists for FW entry. Pacific lamprey (*Entosphenus tridentatus*) enter FW approximately 1 year before spawning (Clemens et al., 2010) and the pouched lamprey (*Geotria australis*) enters FW up to 16 months before spawning (Jellyman et al., 2002). Other species generally enter FW in the fall before spawning. In British Columbia, adult North American river lamprey (*Lampetra ayresii*) enter FW in September (Beamish and Youson, 1987). Species with relatively short distances to migrate in FW leave the ocean only months before spawning, such as the sea lamprey (*Petromyzon marinus*) (Larsen, 1980; Clemens et al., 2010) and the Caspian lamprey (*Caspiomyzon wagneri*) (Ahmadi et al., 2011).

Intraspecific variation also exists in most species and as in salmon photoperiod may be the primary stimulus for the initiation of lamprey migration given the general importance it plays in maturation and reproductive timing in most species. Consequently, there is a latitudinal cline in timing of FW entry ranging from September to March along the east coast of North American for sea lamprey (Beamish, 1980a). Clemens et al. (2010) suggest that southern populations of Pacific lamprey may migrate into FW earlier than northern populations. Within rivers, variation in FW entry has also been documented for the European river lamprey (*Lampetra fluviatilis*) (Abou-Seedo and Potter, 1979). In this species migration is predominantly in the fall, but fish migrating in the spring enter the river in a more advanced stage of sexual maturity (Abou-Seedo and Potter, 1979). In addition, upstream migration of the Caspian lamprey in the Shirud River, Iran, typically occurs in the spring shortly before spawning, but fall migration and overwintering in FW have also been observed (Ahmadi et al., 2011). In Pacific lamprey, most individuals enter FW the year before spawning and Robinson and Bayer (2005) estimate that overwinter holding locations are on average 87% of the migration distance to the spawning areas. For Pacific lamprey in the Willamette River, Oregon, some individuals enter FW in an advanced stage of maturation only weeks before spawning (Clemens et al., 2012). Two life history forms appear to exist for Pacific lamprey, therefore, that parallel patterns observed in salmon: a river-maturing form that enters FW to migrate long distances to spawn and an ocean-maturing form that migrates short distances upriver and spawns shortly after entering FW.

The reasons for such a prolonged FW residency prior to spawning in Pacific lamprey are unknown, but this life history strategy has been suggested to be a function of the large river systems on the west coast of

North America. Anadromous Pacific lamprey migrate variable distances in FW to reach spawning areas and in the Columbia River watershed have been recorded at locations over 500 km upriver (Keefer et al., 2009). It could be argued that timing of migration then is linked to reaching appropriate spawning locations that are well suited for embryonic and larval life stages and adult lamprey moving upstream are relatively insensitive to environmental factors. Lamprey tracked by radiotelemetry were observed to migrate during the spring and early summer before stopping when river temperatures peaked above 20°C (Clemens et al., 2012). Conversely, the early migration strategy and long period holding in FW should decouple migration timing from reproduction timing and may allow considerable flexibility in how lamprey respond to proximate environmental conditions during upstream migration (Keefer et al., 2009). Indeed, migration timing of Pacific lamprey as they reach Bonneville Dam in the Columbia River (235 km from river entry) does vary over time; migration was earliest in warm, low-discharge years and later in cold, high-flow years, and shifted approximately 13 days earlier between 1939 and 2007 (Keefer et al., 2009). The interval examined is similar to that of Quinn and Adams (1996) working on sockeye salmon (as discussed above in Section 2.1), but run timing was advanced by only approximately 6 days in the salmon. The difference could be due to greater responsiveness in lamprey than in salmon to proximate cues for adult migration, but observed change over time is far less than that for American shad (*Alosa sapidissima*). Shad migrating in the Columbia River have experienced the same environmental changes, but now ascend the river approximately 38 days earlier than they did in 1938 (Quinn and Adams, 1996). Shad spawn in the mainstem of rivers and do not hold before spawning; environmental conditions for larvae will be similar to those experienced by spawners. It has been suggested that shad have evolved a migratory pattern that allows for greater behavioral flexibility to respond to environmental fluctuations than salmon (e.g. sockeye) (Quinn and Adams, 1996), or even lamprey. It appears, therefore, that the long spawning migrations of Pacific lamprey are innately determined by seasonal changes such as photoperiod and that responses to environmental change may be modest.

Spawning migrations for many other species of lamprey are not as protracted as seen in Pacific lamprey, potentially owing to the shorter length of river systems; a characteristic of the rivers flowing into the Atlantic, for example (Clemens et al., 2010). In sea lamprey the shorter holding times in FW may indicate that this species is more responsive to environmental factors during spawning migration than Pacific lamprey. Binder et al. (2010) found that water temperature was the best predictor of upstream migratory activity. Increasing water temperature stimulated, while decreasing water temperature suppressed, migration. In a radiotelemetry experiment, Almeida et al. (2002) found that sea lamprey in Portugal were more active in response to increased discharge when water was released from a power station, but ground speed was actually reduced. Responses to short-term changes in flow are not surprising and may be common among species. Jellyman et al. (2002), working on pouched lamprey in New Zealand, also found that upstream movement was stimulated by increased flow but prevented by very high flows.

2.3. Catadromous Eels

Amphidromous and catadromous fishes seem to show much greater range in time of year for FW entry than anadromous fishes. The greater range may be due to passive transit of small larvae that are limited in directional movement compared to directional swimming movements of large adult anadromous fishes. It is also likely that greater flexibility can exist when FW entry does not have to be synchronized with reproduction. Following spawning, leptocephalus larvae of eels are carried by ocean currents towards the continents. For American and European eels that spawn in the Sargasso Sea, the leptocephali of both species are carried by the Gulf Stream; American eels transform more quickly, at a younger age and smaller size, and postmetamorphic glass eels migrate to the coast (Wang and Tzeng, 2000). European eels, in contrast, appear to grow more slowly and transform at an older age and larger size, and are carried further by the Gulf Stream to the coasts of Europe. Iceland is an intermediate location where both species occur: American eels colonizing Iceland metamorphose at an older age than those found in North America and European eels are younger than those found in Europe (Kuroki et al., 2008) (Table 7.1). Similarly, Japanese eel larvae are spawned near the Mariana Ridge and are transported westward in the North Equatorial Current and then are carried by the Kuroshioi current to recruit to East Asia (Shinoda et al., 2011). This pattern has also been described for tropical and southern temperate eels, but specific spawning sites for these species have not been definitively located (Tesch, 2003).

The body shape of leptocephalus larvae is well suited for drifting with ocean currents, unlike the more characteristically eel-like shape of glass eels, which are not as buoyant and must actively swim to leave the ocean currents and move into coastal waters (Chino and Arai, 2010). The cue for metamorphosis is unclear, but the continental shelf appears to be a clear dividing line between larvae and postmetamorphic animals (Tesch, 2003). The decrease in depth has been speculated to be a trigger (Tesch, 2003); however, age and size are also likely to determine whether leptocephali are

Table 7.1

Studies examining the size, age, and season when glass eels belonging to the genus *Anguilla* recruit to freshwater habitats among north temperate, tropical, and south temperate species.

	Location	Total length (mm)	Age (days)	Month	Source
North temperate species					
A. anguilla	Iceland	67.9 ± 3.9 (58.0–78.5)	337 ± 41.7 (260–416)	May to Jul[a]	Kuroki et al. (2008)
A. anguilla	Europe	(54–74)	448 ± 41.7 (420–468)	Apr to Sep	Wang and Tzeng (2000)
A. anguilla	France	69.0 ± 3.8 (55–82)		Jan to Apr	Laffaille et al. (2007)
A. anguilla	France	68.0 ± 2.2	(180–185)	Nov to May	Lecomte-Finiger (1992)
A. anguilla	Portugal	60.6 ± 3.2 (54.1–63.6)	249 ± 22.6 (220–281)	May[a]	Arai et al. (2000)
A. rostrata	Iceland	66.0 ± 4.4 (58.5–73.0)	319 ± 36.0 (248–365)	May to Jul[a]	Kuroki et al. (2008)
A. rostrata	N. America	(41–65)	255 ± 30.2 (220–284)	Dec to May	Wang and Tzeng (2000)
A. rostrata	USA	57.7 ± 1.5 (41.1–86.1)	206 ± 22.3 (171–252)	Feb to Oct	Overton and Rulifson (2009)
A. rostrata	Maine, USA	57.8 ± 2.2 (53.9–61)	218 ± 29.0 (151–276)	Apr[a]	Arai et al. (2000)
A. japonica	Japan	56.3 ± 2.3 (51–61)	182 ± 12.4	Dec to Apr	Tsukamoto (1990)
A. japonica	Japan	57.4 ± 2.3 (51.3–62.2)	157 ± 17.4	Jan	Cheng and Tzeng (1996)
A. japonica	Taiwan	56.1 ± 2.4 (51.3–60.1)		Dec to Mar	Cheng and Tzeng (1996)
Tropical species					
A. marmorata	Phillipines	49.9 ± 1.4 (47.2–51.6)	154 ± 13.5	Sep[a]	Arai et al. (1999a)
A. marmorata	Indonesia	(47.9–52.3)	155 ± 14.8 (144–182)	Jan to Dec	Arai et al. (2001)
A. marmorata	Indonesia	51.1 ± 2.0 (42.0–57.0)		Jan to Dec	Sugeha et al. (2001)
A. celebesensis	Phillipines	51.2 ± 1.7 (48.4–54.6)	157 ± 13.7	Sep[a]	Arai et al. (1999a)
A. celebesensis	Phillipines	51.2 ± 1.7 (48.4–54.6)	157 ± 13.7 (130–177)	Jul[a]	Arai et al. (2003)
A. celebesensis	Indonesia	(44.7–52.6)	109 ± 10.9 (104–118)	Jan to Dec	Arai et al. (2001)
A. celebesensis	Indonesia	49.1 ± 2.2 (40.0–55.8)		Jan to Dec	Sugeha et al. (2001)
A. celebesensis	Indonesia	52.8 ± 1.9 (50–56.2)	112 ± 14.2 (92–139)	Jul[a]	Arai et al. (2003)
A. bicolar bicolor	Indonesia	49.4 ± 2.4 (45.5–52.3)	177 ± 16.4	Sep[a]	Arai et al. (1999b)

A. bicolor bicolor	Indonesia	49.2 ± 2.2 (43.0–54.0)		Dec to Oct	Sugeha et al. (2001)
A. bicolor pacifica	Indonesia	48.9 ± 1.7 (45.7–51.2)	167 ± 19.3 (124–202)	Jun and Jul[a]	Arai et al. (1999a)
A. bicolor pacifica	Indonesia	(48.6–51.5)	173 ± 20.9 (158–201)	Dec to Oct	Arai et al. (2001)
South temperate species					
A. australis	Australia	50.3 ± 1.5 (48–53.9)	208 ± 17.4 (186–239)	July[a]	Arai et al. (1999c)
A. australis	New Zealand	60.3 ± 2.2 (57.0–64.4)	268 ± 31.3 (216–326)	Sep[a]	Marui et al. (2001)
A. australis	New Zealand	59.7 ± 1.5 (56.4–63.1)	237 ± 20.0 (208–266)	Sep[a]	Arai et al. (1999c)
A. dieffenbachii	New Zealand	65.3 ± 6.3 (62.8–71.6)	302 ± 28.2 (240–332)	Aug[a]	Marui et al. (2001)

Data are presented as mean ± SD (range).
[a] Limited sampling.

ready to transform to glass eels. Both American and European eels are carried by the Gulf Stream along the edge of the North American continental shelf and past many islands that recruit American eels to their streams, but not European eels. The slower development and smaller size of European eels may preclude metamorphosis despite appropriate proximate cues (Wang and Tzeng, 2000). Leptocephali are capable of active swimming movements (Tesch, 2003) and if not ready to transform to glass eels, they may actively swim to avoid being carried into shallower water.

Glass eels do not feed until entry into river estuaries, where they develop pigmentation (marking the transition to elvers). The age that eels recruit to FW varies considerably among and within species, but generally temperate eels recruit at an older age and larger size than tropical eel species (Marui et al., 2001) (Table 7.1). Owing to the long larval period and great distance from spawning sites, arrival of glass eels at estuaries occurs over many months with significant variation in size and age (Table 7.1). For example, Overton and Rulifson (2009) report that recruitment of American glass eels to rivers of the south-eastern USA was highly variable throughout the year and no consistent seasonal migration pattern was evident. In contrast, Haro and Kreuger (1988) showed a highly seasonal but protracted movement of American eels into a river in the north-eastern USA, occurring from February to May. Although accumulation of American glass eels and early stage elvers at lower estuary sites shows variability over time, the development into late-stage elvers and upriver movement is contingent on water temperature and also synchronous between estuaries (Sullivan et al., 2009). Tropical eel species, *A. marmorata*, *A. bicolor*, and *A. celebesensis*, exhibit year-round recruitment to rivers in Indonesia, but transform to glass eels at a similar age within a species (Arai et al., 2001). Size of glass eels is similar, but age at recruitment is younger at lower latitudes for Japanese (Tsukamoto, 1990) and American (Wang and Tzeng, 2000) eels, although size at recruitment appears more variable in European eels (Wang and Tzeng, 2000).

A preference for FW has been linked to greater swimming activity in European glass eels caught in estuaries, a behavior suggested to promote recruitment to rivers (Edeline et al., 2005a), whereas lower swimming activity appears to be associated with a preference for higher salinity water. There is large interannual variation in glass eel recruitment, but in rivers in France the greatest migration into rivers occurred during the month following peak discharge (Acou et al., 2009). Consequently, the higher flows may act as an attractant for eels to move towards FW habitats. On a shorter term basis, however, Overton and Rulifson (2009) found that glass eels recruiting to rivers were routinely caught when river discharge was low, and when discharge levels were high no glass eels were caught.

Although low salinities may stimulate locomotor activity and be an attractant for river recruitment, migration periods are also linked to the tidal cycle. Typically, the greatest numbers of glass eels are caught within a few hours of the peak of high tide, as seen in shortfin eels (*A. australis*) and longfin eels (*A. dieffenbachii*) in New Zealand (Jellyman et al., 2009). The link between timing of river entry and tidal cycle suggests that glass eels use STST, similar to that reported for salmon. European glass eels have also been shown to move with the current (Bureau du Colombier et al., 2007) and both tropical and temperate eels enter rivers at the beginning and middle of the flood tide (Sugeha et al., 2001; Laffaille et al., 2007). Other proximate cues for migration have been indicated, although variable results have been found. For example, Naismith and Knights (1988) indicate that upriver migration was mainly dependent on water temperature. Edeline et al. (2006) found that locomotor activity was reduced at low temperature in European glass eels, but Wuenshel and Able (2008) collected migrating American glass eels when river temperatures ranged from 4 to 21°C. Light intensity also plays a role in FW recruitment and glass eels are more active at night. Sugeha et al. (2001) found that *A. celebesensis*, *A. marmorata*, and *A. bicolor pacifica* glass eels migrated into Indonesian rivers beginning after sunset and peaked around midnight. Laffaille et al. (2007) also found higher capture rates in FW at night compared to the day for *A. anguilla* in France. The increase in migration at night appears to be a negative response to light. Glass eels show a strong light avoidance when placed in an experimental choice selection tank, particularly glass eels that have recently migrated into estuarine regions, as light sensitivity decreases with age (Bardonnet et al., 2005).

Based on size and age of eels in river systems, Ibbotson et al. (2002) suggest that movement into FW is composed of two phases: an initial rapid dispersal by young eels and then a slower dispersal by older eels. After FW entry, eels disperse variable distances upstream; some become resident soon after entering FW while others continue to disperse upstream (Imbert et al., 2010). Consequently, there is a consistent decline in abundance with distance upstream (Ibbotson et al., 2002), but the average age and size of eels increase with distance upstream (Naismith and Knights, 1988). FW eels are known to move within FW locations and some move downstream into estuarine areas or even back into the ocean (Jessop et al., 2008; Lamson et al., 2006; Kotake et al., 2004; Chino and Arai, 2010). Duration of FW residence, however, tends to increase with distance migrated upstream (Jessop et al., 2008).

2.4. Amphidromous Gobies

The return to FW of small amphidromous juveniles is functionally and strategically more similar to movement into FW by catadromous species than to the return by maturing anadromous adults. As with catadromous eels, FW recruitment of postlarvae has been documented to occur year round, but numbers vary seasonally and the season when the greatest numbers of juvenile gobies enter the mouths of rivers differs regionally (Keith, 2003). The duration of the amphidromous larval phase can be quite lengthy. In Hawaii, gobies spawn in the rainy season (January and February) and embryos are easily swept downstream and the marine phase lasts for more than 4 months, 135 ± 9.2 days for *Stenogobius genivittatus* and 161 ± 5.7 days for *Awaous stamineus* (Radtke et al., 1988). Within the Sicydiinae gobies there is considerable variation in the marine larval phase among species (Table 7.2). The variable length of time that larval gobies spend at sea has been suggested to favor dispersal as the duration of the larval phase is related to the breadth of a species geographic distribution (Hoareau et al., 2007; Lord et al., 2010). Widespread species such as *Sicyopterus lagochephalus* that are distributed over 18,000 km have a much longer larval stage than locally endemic goby species such as *S. aiensis* from Vanuatu, *S. sarasini* from New Caledonia (Lord et al., 2010), or *Cotylopus acutipinnis* from the Mascarene Archipelago in the south-west Indian Ocean (Hoareau et al., 2007). It has also been suggested that size at which goby recruit to FW systems does not vary much within a species (Keith et al., 2008). Among species with differing pelagic larval phase duration, species with longer larval stages are generally larger at the time of river recruitment (Table 7.2).

Amphidromous species of goby are found in small high-gradient coastal streams that discharge directly into the sea and do not have extensive estuaries or low-gradient reaches before entering the ocean. Theusen et al. (2011) found that the fauna of small coastal Australian streams was similar to that of distant Pacific Island streams dominated by amphidromous species. In contrast, obligate FW and catadromous species were characteristic of large low-gradient Australian rivers. Detection of flowing water in the ocean may be an important cue for gobies to select appropriate FW systems for colonizing. Recruitment of juvenile gobies to FW is associated with a strong attraction to flowing water (Smith and Smith, 1998). In addition, rapid increases in river flows following precipitation events may trigger inland migration (Keith, 2003). Working with *Sicyopterus stimpsoni* and *Awaous guamensis*, Smith and Smith (1998) demonstrated that gobies rapidly responded to directional cues and moved upstream against a current. When flow was stopped the fish continued to move in the same compass direction; potentially a mechanism for these fish to recruit to rivers that flow only intermittently. A strong selective pressure may exist for gobies to recruit to fast-flowing streams so that after hatch, embryos are rapidly transported to the ocean. Newly hatched larvae do not survive if maintained

Table 7.2

Studies examining the size, age, and season when amphidromous Sicydiinae gobies recruit to freshwater habitats for different species with small and larger geographic ranges.

	Location	Total length (mm)	Age (days)	Month	Source
Small geographic range					
Sicyopterus aiensis	Vanuatu		79.2 ± 4.6		Lord et al. (2010)
Sicyopterus sarasini	New Caledonia		76.5 ± 3.9		Lord et al. (2010)
Lentipes concolor	Hawaii	16.0 ± 0.7 (14.1–17.9)	86.2 ± 8.5 (63–106)	Jan to Dec	Radtke et al. (2001)
Stiphodon percnopterygionus	Okinawa	(12.7–13.6)	99 ± 16 (78–146)	Jan to Dec	Yamasaki et al. (2007)
Cotylopus acutipinnis	Mascarene archipelago	22.5 ± 1.2	101 ± 14 (78–150)		Hoareau et al. (2007)
Large geographic range					
Sicyopterus japonicus	Japan	26.3 ± 1.1 (23.5–30.0)	208 ± 22 (173–253)	Apr to Aug	Iida et al. (2008)
Sicyopterus japonicus	Taiwan	33.95 ± 1.31 (30.7–38.1)	163.7 ± 12.8 (130–198)	Sep to Jun	Shen and Tzeng (2008)
Sicyopterus lagocephalus	Vanuatu	26.81 ± 0.61	131.5 ± 7.8		Lord et al. (2010)
Sicyopterus lagocephalus	New Caledonia		132.2 ± 7.2		Lord et al. (2010)
Sicyopterus lagocephalus	Mascarene archipelago	32.7 ± 1.1	199 ± 33 (133–266)		Hoareau et al. (2007)

Data are presented as mean $\pm$ SD (range).

in FW (see Section 3.3 on ionoregulation, below). Slow-moving rivers that flow through lowland areas and have extensive estuaries therefore may not be selected, as insufficient attractive flow exists for recruitment.

Once in the river, juveniles migrate upstream towards adult habitat, but the timing and mechanisms for upstream movement are not well known. Juveniles undergo substantial metamorphic changes after entering FW, which are clearly important as they switch from a planktonic to a benthic feeding mode (Keith, 2003). The translucent postlarvae acquire pigmentation, the caudal fin transforms from forked to truncated (Keith et al., 2008), a pelvic sucking disk is formed by fusion of the pelvic fins (Schoenfuss and Blob, 2004), and the head remodels for grazing (Taillebois et al., 2011). In *S. stimpsoni*, the mouth also develops into a secondary sucking disk to aid locomotion; such modifications enable juveniles of this species less than 30 mm in length to climb waterfalls over 300 m high by "inching up" vertical surfaces (Schoenfuss and Blob, 2004). The presence of gobies with marine otolith signatures in intermittent streams and above waterfalls (Radtke and Kinzie, 1996) indicates the importance of high gradient systems for these species to complete their life cycle. Within FW systems, evidence has been provided by Lord et al. (2011) that different movement patterns exist in the adult goby; otolith microchemistry indicates three patterns: sedentary FW fish, migratory FW fish, and fish that stayed in the lower river reaches.

3. IONOREGULATION

Migration from the ocean to rivers creates a considerable osmotic challenge for fish; in SW fish are continually forced to defend against dehydration, while dilution of internal fluids is a problem for FW fish. Mechanisms for SW and FW ionoregulation in fish have been provided elsewhere (Edwards and Marshall, 2013, Chapter 1, this volume), but considerable remodeling of the gill, gut, and kidney occurs as fish transition from ion excretion to ion uptake as they leave the ocean. The preparatory changes that occur for fish migrating from FW to SW are fairly well understood (McCormick, 2013, Chapter 5, this volume; Zydlewski and Wilkie, 2013, Chapter 6, this volume) and the success of ocean entry is clearly related to preparatory changes that occur in FW that enable SW entry with minimal osmotic disturbance. Changes that occur in diadromous fish migrating into FW systems are less well understood and primarily limited to species of significant economic importance. Sampling migratory fishes in the ocean has presented considerable challenges, but with concerns associated with the potential collapse of the sockeye salmon commercial fisheries on the west coast of Canada and the USA, considerable effort has been expended to understand the biological challenges these fish experience during their return migration to spawn. The discussion below on ionoregulatory changes will highlight what has been learned for anadromous fish from these investigations and what is known for species of catadromous and amphidromous species.

3.1. Anadromous Fishes

As migration to FW systems is associated with maturation, the seasonality for river entry within a population tends to be quite specific for anadromous species with little variation from year to year. Given the strong seasonal nature of anadromous migrations to FW environments, it would be expected that fish prepare to hyperosmoregulate before moving into FW. Evidence of preparatory physiological changes has been observed for sockeye salmon intercepted approximately 200 km from the estuarine environment. The kinetic activity of the enzyme Na^+/K^+-ATPase (NKA) in the gills of juvenile salmon has long been used as a marker of SW tolerance (McCormick and Saunders, 1987) as this enzyme increases during the spring when salmon smolt. A corresponding decline in gill NKA has also been found in migrating adult sockeye off the coast of British Columbia. A comparison of gill NKA activity for fish captured approximately 400 km from river entry with that for fish captured 200 km from river entry found significant declines over multiple years: a change from 4.8 ± 0.6 to 1.9 ± 0.2 µmol ADP mg protein^{-1} h^{-1} (Shrimpton et al., 2005; Flores et al., 2012) (Fig. 7.4A). Adult sockeye salmon migrate approximately 25 km day^{-1} in coastal waters as they approach their natal estuary (Hanson et al., 2008), indicating that changes in the gill occur at least 1–2 weeks before they reach the estuary and fish are preparing for FW entry while still in the ocean. A comparison of chum salmon captured in the Bering Sea and also off the north-east coast of Hokkaido (~300 km from FW entry) showed a 25–50% reduction in gill NKA activity; however, a further reduction in NKA activity was not observed until after FW entry (Onuma et al., 2010a). The decline in NKA activity reported in these studies may promote a higher tolerance to the lower levels of salinity in FW, the converse of the parr–smolt transformation.

It is difficult to assess when physiological changes needed for FW entry are initiated, but it is possible that preparatory changes occurred even earlier than when fish were intercepted along the coasts of Canada and Japan reported in the above studies. Gill NKA activity was recently measured for migrating sockeye salmon smolts captured 1 month after ocean entry. Values were greater than 25 µmol ADP mg protein^{-1} h^{-1} (M. C. Bassett and

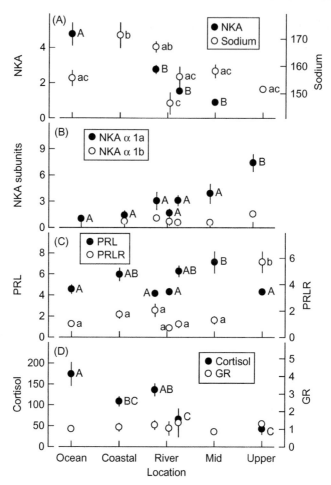

Fig. 7.4. (A) Gill Na$^+$/K$^+$-ATPase (NKA) activity (μmol ADP mg^{-1} of protein h^{-1}) and plasma sodium concentration (mmol L^{-1}); (B) gill NKA isoform α1a and α1b mRNA (expressed relative to β-actin); (C) plasma prolactin (PRL) concentration (ng mL^{-1}) and gill prolactin receptor (PRLR) mRNA (expressed relative to β-actin); (D) plasma cortisol concentration (ng mL^{-1}) and gill glucocorticoid receptor (GR) mRNA (expressed relative to β-actin) for maturing Late Shuswap sockeye salmon (*Oncorhynchus nerka*) captured in seawater and freshwater during their migration to spawn in the Fraser River watershed, British Columbia, Canada. *Ocean* indicates fish caught 350–850 km from the mouth of the Fraser River. *Coastal* indicates fish caught ~200 km from the mouth of the Fraser River. *River* indicates the mouth of the Fraser River; fish were caught in estuarine waters, in tidal portions of the river, and in the lower non-tidal part of the river. *Mid* indicates fish captured ~200 km upstream of the river mouth. *Upper* indicates fish captured over 400 km upstream, but before reaching the spawning grounds. Significant differences determined by Bonferroni test on all possible comparisons are shown for values that do not have a common letter ($p < 0.05$). Values are means ± 1 SE. Adapted from Flores et al. (2012) with permission from Springer.

J. M. Shrimpton, unpublished data), levels much higher than measured in studies examining migrating adults off the coast of British Columbia. It is also important to realize that measurement of gill NKA activity does not adequately reveal the biochemical changes that occur in the gill in the living animal. Two isoforms of the NKA α-subunit (α1a and α1b) have been identified to change in rainbow trout (*O. mykiss*) transferred to waters of different salinity (Richards et al., 2003). The biochemical method to quantify specific activity of NKA does not discriminate between the two isoforms. In FW, NKA α1a is the abundant isoform present in lamellar ionocytes in the gills, while NKA α1b becomes the dominant isoform after SW acclimation and is found primarily in filamental ionocytes (McCormick et al., 2009). Messenger RNA (mRNA) levels have recently been quantified in migrating adult sockeye salmon in the ocean and FW (Fig. 7.4B). In this study, no differences in mRNA levels for the NKA α1b subunit were found among sockeye sampled over 800 km from river entry, along their coastal migration, or as they approach estuarine waters (Flores et al., 2012). Such a finding suggests that downregulation of the α1b isoform may already have occurred when the fish were first intercepted over 800 km from FW entry. Conversely, gill NKA α1a mRNA, the FW isoform of NKA (Richards et al., 2003), increased before FW entry (Flores et al., 2012); this provides further evidence that the sockeye salmon gill appears to remodel for FW ionoregulation while the fish are still in SW.

Consequently, it has been suggested that low NKA activity may be a useful indicator that fish are prepared for FW entry, similar to high NKA activity as a marker for SW tolerance in salmon smolts. There is some evidence to support a link between low gill NKA activity and successful migration to spawning areas in FW. A number of studies have tagged sockeye salmon off the coast of British Columbia in Johnstone Strait (~200 km from the river mouth) and fish were tracked after FW entry up to 85 km upstream. Crossin et al. (2009) found that there was a relationship between gill NKA activity and holding time off the river mouth before FW entry and migration upstream to spawning areas; salmon with lower NKA activity delayed river entry before migrating upriver to spawn. These fish also had higher gross somatic energy levels than fish with low gill NKA activity that entered the river directly and ultimately failed to reach spawning areas. In contrast, using a similar radiotagging approach, Cooke et al. (2006a, b) found no difference in gill NKA activity for sockeye salmon that were detected to successfully enter FW and fish that failed to enter the river. It is likely that gill NKA activity is already downregulated by the time the fish are within 200 km of the river mouth. Bystriansky and Schulte (2011), working on adult Atlantic salmon, however, found an increase in gill NKA activity following rapid transfer from SW to FW. The higher kinetic activity was associated with a seven-fold increase in NKA α1a mRNA levels and a 60% decrease in NKA α1b mRNA.

With a decline in gill NKA activity, it would be expected that maturing salmon progressively lose their ability to hypoosmoregulate in SW and plasma ions should increase while fish are still in SW. Confounding the anticipated increase in plasma ions as fish enter coastal waters is a general decrease in salinity, decreasing the osmotic difference between the water and fish plasma, but even coastal waters remain hyperosmotic to the internal tissues of the fish. There is little evidence, however, that plasma ions increase as maturing salmon migrate in SW before river entry. Flores et al. (2012) found that plasma [Na$^+$] increased (Fig. 7.4A), but not plasma [Cl$^-$] or osmolality in sockeye. Onuma et al. (2010a) found significant declines in plasma [Na$^+$] and [Cl$^-$] in chum salmon for some years, with little change in others. Saito et al. (2001), also working on chum salmon, found that plasma ions changed with location as fish migrated from the ocean, through a bay and into the river, but the differences were not consistent among years. A consistent increase in plasma ions, therefore, does not appear to exist for maturing anadromous salmon as they migrate in SW, indicating that they defend plasma ionic homeostasis. A decline in plasma ions would be expected when anadromous salmon smolt while still in FW, but such a change has not consistently been shown (McCormick and Saunders, 1987).

There is also a fundamental difference between changes observed in adult salmon and the preparatory changes that occur in juvenile salmon smolts. Marine fishes are amphihaline and able to tolerate direct transfer to FW. For example, SW-acclimated Atlantic salmon directly transferred to FW showed a transient decline in plasma ions and a rebound within 4 days with no mortality (Bystriansky and Shulte, 2011). The reverse, however, is not the case and juvenile coho salmon (*O. kisutch*) transferred to SW that are not smolts show large perturbations in plasma ions and high mortality (Shrimpton et al., 1994). Juvenile Atlantic salmon smolts prevented from entering SW also survive and can smolt the following year (Shrimpton et al., 2000). In contrast, the preparatory changes for FW entry are irreversible and adult salmon retained in SW do not survive. Maturing chum salmon captured before river entry and held in SW showed a decline in gill NKA activity, an increase in plasma osmolality, and subsequent mortality (Uchida et al., 1997). Onuma et al. (2003b) found that chum salmon maintained in SW for 4 days after capture before they could enter FW did not survive. Holding sockeye in SW also resulted in higher rates of mortality than for fish transferred to FW (Cooperman et al., 2010). The impaired hypoosmoregulatory function is directly linked to sexual maturation in adult salmon (Clarke and Hirano, 1995) and is discussed in Section 4.2.1 on endocrine control. Returning adult salmon also cease feeding before entering FW and the cessation of feeding leads to atrophy of the gastrointestinal tract; sockeye returning to the Fraser River had already stopped feeding when intercepted 850 km from the estuary (Miller et al., 2009), a distance traveled in approximately 4 weeks (Crossin et al., 2007). Gastrointestinal atrophy may compromise drinking ability in adult salmon and be linked to the loss of SW tolerance (see Edwards and Marshall, 2013, Chapter 1, this volume).

Although preparatory physiological changes occur while anadromous fishes are still in the ocean, river entry is still accompanied by significant drops in plasma [Na$^+$] and [Cl$^-$] (Shrimpton et al., 2005; Onuma et al., 2010a; Flores et al., 2012). The drop in plasma ions indicates that the transition to FW requires further physiological adjustments. Flores et al. (2012) showed that plasma [Na$^+$] and [Cl$^-$] rebounded in adult sockeye as fish migrated upriver from an area still influenced by estuarine waters (river 11 km) to an area above tidal influence (river 50 km), a distance the fish would cover in approximately 2 days (Hanson et al., 2008). In a laboratory study, Bystriansky and Schulte (2011) directly transferred adult Atlantic salmon from SW to FW and also found a decrease in plasma ions and total osmolality with recovery after 4 days. It is not clear whether the temporal differences in plasma ion dynamics between the natural migrating salmon and rapid transfer experiments are due to preparatory or species differences.

Studies on sockeye salmon in British Columbia have provided an opportunity to track anadromous fishes during an extensive FW migration, in some cases up to 1000 km for fish to reach spawning grounds. Gill NKA activity, plasma osmolality, [Na$^+$], and [Cl$^-$] consistently decline with the time and distance fish migrate upriver (Shrimpton et al., 2005; Flores et al., 2012). The lowest values for each measurement after the initial perturbation following river entry were observed as fish arrived at the spawning area. Gill NKA α1a mRNA levels, however, increased in sockeye salmon following movement into the river, and while migrating upstream and onto the spawning grounds. Similar findings for gill NKA α1a were also reported for wild anadromous Arctic char migrating from SW into FW (Bystriansky et al., 2007). The NKA α1b isoform, however, did not change in response to movement from SW to FW. The lack of change in the α1b isoform again suggests preparation for FW entry while salmon are still in the ocean and differs from the decline in mRNA levels following the acute transfer experiments on Atlantic salmon of Bystriansky and Schulte (2011), but not those performed on SW-acclimated rainbow trout when transferred to FW (Richards et al., 2003).

Knowledge of the ionoregulatory changes that occur in adult anadromous lamprey is quite limited. The ultrastructure and distribution of cell types in the gill differ from those of teleosts (Bartels and Potter, 2004), making it unclear whether our understanding of teleost ionoregulation is transferable to lamprey, but the ultrastructure of ionocytes resembles that of cells found in ion-transporting epithelia of other vertebrates. Choe et al. (2004) provided evidence that the major transport proteins involved in both SW and FW ionoregulation in teleosts are present in the gill of adult pouched lamprey. Further, Reis-Santos et al. (2008) found that gill NKA and salinity tolerance increased in metamorphosing sea lamprey and gill H^+-ATPase was downregulated by salinity; H^+-ATPase immunoreactivity was localized to cells not expressing NKA, a finding consistent with models for teleost ionoregulation. Similar to findings for salmon, anadromous sea lamprey are amphihaline as feeding juveniles when in SW, but nearly mature adults are unable to regulate their internal osmolality at salinities of 16 ppt or higher (Beamish, 1980b). Beamish (1980b) also reports that gill NKA activity is high in feeding immature lamprey in SW, but is at very low levels in spawners sampled in FW. Given the exceptional euryhalinity observed in anadromous lamprey during the feeding stage, the decline in NKA and loss of SW tolerance suggest that preparatory changes occur. Atrophy of the gastrointestinal tract also occurs when lamprey cease feeding (Larsen, 1980) and may compromise the ability of the fish to ionoregulate in SW (see Edwards and Marshall, 2013, Chapter 1, this volume). Gastrointestinal atrophy is correlated with gonadal development. In European river lamprey captured soon after FW entry, gonadectomy prevented atrophy of the intestine. Treatment with sex hormones, however, increased the rate of atrophy (Pickering, 1976).

3.2. Catadromous Fishes

Catadromous species enter FW after larval juvenile metamorphosis and therefore individuals are smaller than the adult form. The small size of catadromous animals in itself does not account for the paucity of information on ionoregulatory changes that occur at this life history stage (compared to our understanding of smolting in salmonids). Juveniles in the marine environment are generally at quite low densities and it is often only feasible to collect animals once they have entered the estuary or are in close proximity. Consequently, ionoregulatory changes that occur in catadromous species associated with preparation for FW entry are not well characterized. Most of what is known about catadromous animals has been learned from studies conducted on eels.

For eels a number of studies have examined the ecological and behavioral aspects of riverine recruitment (reviewed in Section 2.3 on behavior and timing) although relatively little attention has been focused on osmoregulation. The change in environmental salinity for catadromous species requires a similar change in ionoregulation to anadromous species where ion transport processes in the gill realign from ion excretion to ion uptake and preacclimation to hyperosmotic regulation may occur while glass eels are still in SW. Compared to salmonids, glass eels are difficult to study because it is not possible to maintain the larvae indefinitely in culture conditions (Tanaka et al., 2003). The age of juvenile glass eels is also highly variable (Tesch, 2003) (Table 7.1). In addition, timing for FW entry does not show the same precise seasonality that is seen in anadromous species. Consequently, a seasonal cue that would enable preparation for entry to FW is difficult to ascertain. Adult eels appear to prepare for SW entry while still in FW (Sasai et al., 1998); however, there is no evidence of preacclimation in glass eels. Recent work on both *A. anguilla* and *A. japonica* has shown that glass eels are very euryhaline.

Using Japanese eels, Sasai et al. (2007) examined the development of ionocytes in leptocephalus larvae and glass eels. Ionocytes were first detected when the gill filaments developed and gradually increased as fish grew and advanced to the later larval stages. After metamorphosis to glass eels, lamellae were observed to develop on the gill filaments; ionocytes were abundant on the gill filaments but ionocytes were not apparent on the lamellae. These authors did report immunoreactivity for their NKA-specific antiserum on the lamellar epithelium. The NKA immunoreactivity was observed for glass eels caught in the ocean and also found for fish transferred to FW. The NKA immunoreactivity, however, disappeared in fish that were maintained in SW for 14 days after capture. Such a finding conforms to the model proposed for salmonids where distributions of ion-secretory ionocytes and ionocytes for ion uptake differ in the gills of SW and FW fish (see McCormick et al., 2009). The presence and subsequent loss of NKA immunoreactivity on the lamellar epithelium for fish maintained in SW suggest preparatory changes for FW entry may indeed occur in glass eels. Given the wide range of times that glass eels migrate to FW, it is difficult to predict what environmental cue might stimulate such preparatory changes. Conversely, if preparatory changes occur for FW entry, it would be expected that they could be detected biochemically. Wilson et al. (2007a) collected European glass eels monthly throughout the year and found no seasonal change in gill NKA activity. These authors also measured the activity of V-type H^+ ATPase, an enzyme linked to Na^+ uptake in the FW fish gill, and found no significant seasonal changes.

Independent of whether preparatory changes occur before glass eels recruit to FW habitats, eels show a remarkable euryhalinity at this life stage. *Anguilla anguilla* glass eels captured in an estuary and held in BW (24 ppt) were capable of adapting to FW following acute transfer with no associated mortality or significant changes in whole-body water content (Wilson et al., 2004). Gill NKA activity did not differ 24 h after transfer to FW, but was lower by 7 days. The drop in gill NKA activity following transfer to FW is similar to that observed for upriver migrating adult sockeye salmon (Shrimpton et al., 2005; Flores et al., 2012). In acute transfer laboratory experiments, it has been shown that acclimation to FW continues over a period of weeks with a downregulation of the branchial ion transporters [NKA, the $Na^+/K^+/2Cl^-$ cotransporter (NKCC), and the cystic fibrosis transmembrane receptor (CFTR) Cl^- channel] associated with active ion excretion (Wilson et al., 2007b). The time it takes for glass eels to fully acclimate to FW is not clear, but significant differences between glass eels and elvers have been documented. Gill NKA activity is lower in elvers than in glass eels, but no difference was found for the V-type H^+-ATPase activity (Wilson et al., 2007a). Mortality in glass eels following abrupt transfer has been reported only for fish in distilled water treatments or hypersaline treatments (150% SW; 48 ppt) (Wilson et al., 2007b). Even sublethal osmoregulatory indicators such as whole-body water content and Na^+ levels were only disturbed at the extreme changes in salinity. The glass eel life stage, therefore, appears to be well adapted for FW entry. Eels and particularly eels found in temperate waters, however, do not always migrate into FW habitats (see Section 2.3 on behavior and timing, above). The strong euryhalinity at this stage enables these fish to opportunistically exploit both FW and estuarine habitats.

3.3. Amphidromous Fishes

Return migration of small juveniles to FW systems for feeding and growth, where the fish eventually mature and spawn characterizes amphidromous species of fish. As with catadromous eels a number of studies have examined the ecological and behavioral aspects of riverine recruitment (reviewed in Section 2.4, above, on behavior and timing), but little is known about ionoregulatory changes that occur during this life history stage. Postmetamorphic juveniles that recruit to FW over an extended duration are at a similar stage of development to catadromous species. It is expected, therefore, that ionoregulatory preparations in SW and changes after FW entry would be similar to those described earlier for the catadromous eels. Ionoregulatory studies on amphidromous species, however, are extremely limited, but it is likely that amphidromous species are highly euryhaline and able to tolerate a wide range of salinities over a protracted period. Support for this assertion comes from work on the common galaxias (*Galaxias maculatus*). This species is found to inhabit salinities ranging from FW to 49 ppt (Chessman and Williams, 1975). In laboratory experiments where fish collected from FW were transferred to waters of different salinities, fish did not show a significant perturbation in plasma osmolality, except in waters that were higher than 49 ppt. It is probable, therefore, that larval fish swept downstream are euryhaline and maintain their high tolerance of a wide range of salinities throughout the larval dispersal phase.

Euryhalinity of larval amphidromous fishes may not be a characteristic of all species and families that have evolved this life history strategy. Work on a species of Japanese goby, *Sicyopterus japonicas*, found that salinity tolerance varied with life history stage (Iida et al., 2010). Larval premetamorphic goby do not survive if maintained in FW and adult goby do not survive in SW. This species is therefore dependent on both marine and FW environments, obligately amphidromous, and not euryhaline throughout its life. The oceanic larval stage for *S. japonicus* may be up to 8 months (Iida et al., 2010) (Table 7.2), but euryhalinity at this stage of development has not been examined. The long oceanic larval stage and variable timing of entry to FW for gobies suggest that euryhalinity would be an important attribute for successful recruitment to FW. Consequently, as amphidromous larval fish approach estuarine areas and metamorphose into the juvenile form that migrate upstream, they are probably euryhaline and already able to tolerate low salinities. Euryhalinity, however, does not persist and juvenile gobies from Hawaii (*Stenogobius genivittatus* and *Awaous stamineus*) will not survive if maintained in SW (Radtke et al., 1988), although McCormick et al. (2003) found no mortality in stream-captured *S. hawaiiensis* acclimated to salinities as high as 30 ppt for 10 days.

4. ENDOCRINE CONTROL

Many hormonal changes occur as fish migrate from marine to FW environments. Numerous studies have helped to reveal the endocrine factors that control ionoregulation (Takei and McCormick, 2013, Chapter 3, this volume), but the transition from a hyperosmotic to a hypoosmotic environment also occurs at a complex time during the life history of diadromous fishes. Migration to FW in diadromous fishes is accompanied by significant developmental changes. Anadromous species are returning to rivers to spawn and, depending on migration distance and holding patterns in FW, are at varying stages of maturation. Catadromous and amphidromous species undergo metamorphosis from the larval to juvenile form prior to movement into riverine systems and some of the hormonal regulation of metamorphosis in these animals is known. Motivation for diadromous migrations is also linked to endocrine changes. It is often assumed that diadromous fishes moving into river systems are seeking specific locations for successful rearing or reproduction and much work has been done to examine how fish select specific locations and elucidate the hormones that enhance this process. Migratory activity and movement are also influenced by endocrine cues. What is known of the endocrine changes in diadromous fishes as they migrate from the ocean to rivers will be examined in this section.

4.1. Ionoregulatory Hormones

4.1.1. PROLACTIN AND GROWTH HORMONE

Two closely related pituitary hormones, prolactin (PRL) and growth hormone (GH), have been shown to be important endocrine factors for controlling ionoregulation in anadromous fishes. It is generally accepted that PRL is important for FW ionoregulation, while GH is important for SW ionoregulation among a large and phylogenetically diverse number of teleosts (McCormick, 2001; Takei and McCormick, 2013, Chapter 3, this volume). During FW acclimation, pituitary and plasma PRL levels increase to regulate hydromineral balance by decreasing water uptake and increasing ion retention (Manzon, 2002). Changes in pituitary hormones have recently been characterized by Onuma et al. (2010a), who examined chum salmon captured at four locations during their migration from the Gulf of Alaska to their natal stream on Hokkaido, Japan. Pituitary mRNA levels for PRL and GH began to increase soon after the initiation of homeward migration; mRNA levels were 10-fold and five-fold higher for PRL and GH, respectively, when fish captured in the Bering Sea were compared to fish captured in the Gulf of Alaska (Onuma et al., 2010a). GH mRNA levels in the pituitaries of fish, however, declined as they approached the coastal waters off Hokkaido. A decrease in GH mRNA levels would be expected to lead to lower plasma GH levels and would be consistent with preparatory changes for FW entry. Benedet et al. (2010), however, caution the extrapolation of mRNA levels to protein concentration as they did not find a consistent relationship between pituitary GH mRNA and plasma GH levels. In contrast, levels of pituitary PRL mRNA for coastal fish did not differ from the maturing adults early in migration. Following FW entry and migration upriver, PRL mRNA showed a moderate but significant increase. Such changes in PRL mRNA would predict that plasma PRL concentrations should be high in migrating salmon long before they reach FW. Circulating levels of PRL have been shown to increase while adult Atlantic salmon were still in SW (Andersen et al., 1991). Flores et al. (2012) found that plasma PRL was also high in adult sockeye salmon captured approximately 400 km away from their natal estuary; plasma concentrations were greater than 4 ng mL^{-1} and comparable to values reported for FW fish (Fig. 7.4C). For example, PRL concentrations less than 0.5 ng mL^{-1} have been reported for SW fish, but increase to greater than 3 ng mL^{-1} after transfer to FW for adult coho salmon (Sakamoto et al., 1991). As Onuma et al. (2010a) found an increase in pituitary PRL mRNA following FW entry in chum, Flores et al. (2012) found that plasma PRL levels increased after river entry in sockeye. Further work on prespawning chum salmon has indicated that pituitary PRL mRNA increased during upstream migration in both long river systems (Taniyama et al., 1999; Onuma et al., 2010a) and short river systems (Onuma et al., 2003b). Increases in the PRL mRNA occurred when prespawning chum salmon were transferred to FW, but also for fish retained in SW (Onuma et al., 2003b), providing a strong indication that upregulation of PRL is preparatory for entry to the FW environment.

There is also evidence for a differential role of PRL and GH based on changes in mRNA for the receptors of these two hormones measured in the gills of migrating adult salmon. Gill prolactin receptor (PRLR) mRNA was shown to increase in adult sockeye salmon in SW as they approached the river mouth (Flores et al., 2012). Following FW entry, mRNA for PRLR (Fig. 7.4C) and growth hormone receptor (GHR) increased; the mRNA levels for both receptors were highest for fish arriving at the spawning grounds. PRL and GH differentially affect NKA α-subunit expression. Tipsmark and Madsen (2009) showed that PRL injection decreased gill NKA α1b mRNA, while GH injection increased NKA α1b mRNA. These authors, however, found no effect of PRL treatment on gill NKA α1a mRNA, which contrasts with the correlation between PRLR and NKA α1a mRNA found by Flores et al. (2012). In salinity challenge experiments with rainbow trout, gill NKA α1a mRNA and PRLR mRNA increased for fish transferred to lower salinity water while NKA α1b mRNA and GHR mRNA increased for fish exposed to higher salinity water (Flores and Shrimpton, 2012). High levels of PRLR mRNA in the gills of migrating adult sockeye salmon in SW and following FW entry further indicate that PRL plays an important role in ionoregulation throughout the FW migration. The increase in GHR mRNA in the gills is less clear, but higher mRNA levels of PRLR and GHR may have an ionoregulatory role and modify gill NKA to limit osmotic perturbations. Plasma GH has been shown to increase in maturing Atlantic salmon just before spawning (Benedet et al., 2010), an increase consistent with the gill GHR mRNA changes observed in wild sockeye salmon migrating upstream by Flores et al. (2012). The function of high GH levels in spawning salmon does not appear to be associated with an increase in insulin-like growth factor-1 (IGF-I); Onuma et al. (2010b) found that IGF-I levels in plasma of maturing chum salmon were lower in FW than in SW. Elevated circulating levels of

GH, however, may be linked to locomotor activity and homing in maturing salmon (see Section 4.3, below).

Hormonal regulation in anadromous lamprey is less well understood. Earlier work indicated the presence of GH in the pituitary of sea lamprey based on immunohistochemical evidence (Wright, 1984). More recently, Kawauchi et al. (2002) reported the identification of GH and a GH–IGF system in the sea lamprey. Sea lamprey GH is expressed in the pituitary and stimulates the expression of an IGF gene in the liver, as in other vertebrates (Kawauchi and Sower, 2006). No evidence of PRL has been found in sea lamprey, however, which suggests that GH may be the ancestral hormone in the molecular evolution of the GH/PRL family. Given that GH and PRL have opposing functions in ionoregulation in teleosts and only GH is present in the lamprey, it is not clear which hormone may stimulate ion uptake for fish in FW (but see Section 4.1.2 below on corticosteroid hormones). Immunoreactivity to GH antibodies has been shown to change with stage of development in sea lamprey (Nozaki et al., 2008). Few GH-like cells were present in the pituitary during the larval and metamorphic phases, but the number of cells increased markedly during the parasitic period and corresponded with the rapid growth, suggesting a somatropic rather than an ionoregulatory role for the hormone. The area occupied by GH-like cells, however, was greatest in prespawning adult lampreys. Whether the increase is associated with an ionoregulatory role as adult lamprey migrate from a marine to FW environment has not been elucidated.

Knowledge of endocrine factors that control ionoregulatory changes as catadromous fishes move from the ocean to FW is also limited, but the role of pituitary hormones appears to agree with that reported for anadromous teleosts. An immunocytochemical study detected PRL and GH cells in the pituitary of Japanese eels at all stages of development of the leptocephalus larvae (Ozaki et al., 2006). The cell area of the pituitary that exhibited PRL immunoreactivity tended to decrease with larval size, as did the GH-immunoreactive cell area. In glass eels caught just before upstream migration, the PRL cell area doubled, whereas there was no change in GH cell area (Ozaki et al., 2006). The increase in PRL cells in glass eels before FW entry suggests that PRL has a preparatory osmoregulatory role for FW acclimation. The large proportion of GH-immunoreactive cells supports a role for GH in control of ionoregulation in SW during early life stages of the leptocephalus larvae.

4.1.2. CORTICOSTEROIDS

It is well established that cortisol is an SW-acclimating hormone in teleosts and there is also considerable evidence that this hormone may play a role in FW ionoregulation. For example, cortisol treatment was found to stimulate ionocyte proliferation and increase ion uptake in FW rainbow trout (Laurent and Perry, 1990). Furthermore, cortisol has been shown to increase both NKA α1a and NKA α1b mRNA in the gills of juvenile Atlantic salmon held in FW (McCormick et al., 2008; Tipsmark and Madsen, 2009), but not for Atlantic salmon acclimated to SW (Tipsmark and Madsen 2009). Levels of cortisol in migrating adult salmon in SW were higher than for fish captured after FW entry, suggesting that cortisol stimulated physiological adjustments in SW fish to enable successful FW entry (Onuma et al., 2010a; Flores et al., 2012) (Fig. 7.4D). The high cortisol levels measured in fish caught in the ocean, however, could be due to stress and the difficulty of sampling fish in this location.

An alternative method to assess cortisol action is to measure changes in the hormone receptor as the biological effect of cortisol depends not only on the plasma concentration of the hormone, but also on the number and affinity of intracellular receptors in the gills (Shrimpton and McCormick, 1999). Cortisol receptor dynamics take several hours to change in response to cortisol treatment and are thus not as responsive to the acute effects of stress as are changes in circulating cortisol levels (Sathiyaa and Vijayan, 2003). It is likely that cortisol is the functional ligand for both mineralocorticoid receptors (MRs) and glucocorticoid receptors (GRs) in salmonids (McCormick et al., 2008). Given the work of Sathiyaa and Vijayan (2003) on autoregulation of GRs, the high plasma cortisol concentrations for marine salmon would result in high mRNA levels expressed for at least one of the cortisol receptors. Flores et al. (2012) found little change in mRNA levels for either MRs or GRs for sockeye salmon sampled at different locations in both SW and FW, and suggested that gene transcription also remains high throughout migration (Fig. 7.4D). The similar pattern and limited difference in mRNA expression for GRs and MRs indicate that cortisol may function through both an MR and a GR in FW ionoregulation during upstream migration of this species.

The highest levels of cortisol observed by Flores et al. (2012) for sockeye salmon were found for fish on the spawning grounds. High cortisol levels characterize sexual maturation in semelparous Pacific salmon (Donaldson and Fagerlund, 1972), but have also been measured in kokanee, the non-anadromous life history form of sockeye salmon (Carruth et al., 2000b). The increase in plasma cortisol is linked to enhanced olfactory ability and homing (see Section 4.4 below), but it is also generally accepted that high cortisol levels play a role in postspawning death of semelparous Pacific salmon. Gonadectomy blocks the increase in cortisol in prespawning salmon (Donaldson and Fagerlund, 1972) and evidence of a gonadal factor contributing to the cortisol excess has recently been shown by Barry et al. (2010). Normally, cortisol-metabolizing enzymes regulate the peripheral concentrations of cortisol; however, the process is compromised during sexual maturation in semelparous Pacific salmon. The steroid 17α,20β-dihydroxy-4-pregnen-3-one (17,20-P) is produced by the gonads and inhibits cortisol metabolism in semelparous coho salmon, but not in the iteroparous rainbow trout (Barry et al., 2010). Depression of metabolizing enzymes, therefore, leads to cortisol excess and contributes to postspawning mortality in semelparous Pacific salmon.

Although the role of corticosteroid hormones has been well documented in anadromous migrations of teleosts, the presence of a corticosteroid hormone in lamprey has only recently been shown. In sea lamprey, 11-deoxycortisol (DOC) is the corticosteroid hormone and a receptor for this hormone in the gill has been characterized (Close et al., 2010). The authors found that DOC responded to acute stress, indicating that the hormone is a glucocorticoid. Further, DOC upregulated gill NKA activity of adults in FW, indicating that it is also a mineralocorticoid. Whether DOC functions to stimulate ion uptake is not known, but given the lack of PRL-immunoreactive cells in lamprey pituitaries it is intriguing to speculate that corticosteroids may function in opposition to GH in these phylogenetically ancient fishes.

Investigations on the role of corticosteroids in FW ionoregulation of catadromous fishes have largely been limited to eels. An effect of cortisol on FW ionoregulation has been demonstrated by Perry et al. (1992), who injected European eel with cortisol and found that the surface area of ionocytes as well as the influx of Na^+ and Cl^- in the gill increased. Changes in circulating levels of cortisol also support a role for this hormone in ion uptake; transfer of European eel to lower salinity water has also been shown to result in a transient increase in plasma cortisol (Leloup-Hatey, 1974). It is not known whether an increase in cortisol occurs during the larval stage or in glass eels migrating into FW. In the non-anadromous conger eel (*Conger myriaster*), small leptocephalus larvae had whole-body cortisol levels greater than 200 ng g^{-1}, but concentrations were lower in larger premetamorphic larvae and remained low throughout the metamorphic period (Yamano et al., 1991). Whether cortisol levels are high in anguillid leptocephalus larvae during development has not been examined.

4.2. Hormones Associated with Development

4.2.1. MATURATION

The return to FW by anadromous fishes is coincident with sexual maturity. Reproductive hormones are upregulated while anadromous salmon are still in the ocean. Plasma levels of testosterone, 11-

ketotestosterone (11KT), and estradiol-17β (E2) were higher in maturing adult chum salmon than in immature fish sampled in the ocean (Onuma et al., 2009); in fact, endocrine signals were elevated in maturing adult chum captured in the winter while still far out at sea in the Gulf of Alaska. An increase in reproductive hormones for sockeye migrating back to the Fraser River, British Columbia, was also found before entry into FW (Cooke et al., 2006b). Circulating testosterone, however, differed between populations and reflected differences in migration distance to spawning locations (Crossin et al., 2009). The seasonal increases in the maturation hormones, therefore, appear to be inseparable from initiation of spawning migration of chum salmon (Onuma et al., 2009). For chum salmon, plasma levels of gonadotropins and plasma sex steroid hormones continued to increase throughout migration and peaked after migration upstream from the coast to the natal hatchery in Hokkaido (Onuma et al., 2009; Ueda, 2011). Similarly, sex steroids increased with distance migrated upstream from the ocean in sockeye salmon in the Fraser River, British Columbia (Hinch et al., 2006). Year-to-year variation in plasma levels of steroid hormones and sexual maturation has also been observed for chum salmon returning to Hokkaido. Changes in the plasma levels of testosterone and 11KT during prespawning migration differed from year to year in association with the sea surface temperature. Warmer oceanographic conditions appeared to affect migratory behavior, resulting in delayed arrival of fish at their natal river, so that the levels of testosterone and 11KT increased at an earlier period of upstream migration (Onuma et al., 2003a).

Treatment of juvenile Atlantic salmon with testosterone, 11KT (Lundqvist et al., 1989), and E2 (Madsen and Korsgaard, 1991) has been shown to impair SW tolerance and smolting. In the spring a slight but significant increase in plasma testosterone in precocious mature male Atlantic salmon was negatively correlated with gill NKA activity (Shrimpton and McCormick, 2002). 11-Ketotestosterone has also been shown to decrease basal levels of adrenocorticotropic hormone (ACTH) (Pottinger et al., 1996) and suppress interrenal activity (Young et al., 1996), although testosterone has been found to have less of an effect. Lower pituitary GH mRNA and gill NKA activity occurred coincidently with an increase in plasma levels of sex steroids, which were elevated in ocean-migrating chum salmon from the Bering Sea to the coast (Onuma et al., 2009). Exogenous treatment of adult sockeye salmon with gonadotropin-releasing hormone (GnRH) to accelerate maturation resulted in lower gill NKA activity, higher sex steroid concentration, and greater mortality of fish held in SW (Cooperman et al., 2010). Sexual maturation, therefore, appears to interfere with hypoosmoregulation in anadromous salmon, making FW entry obligatory. A relationship between reproductive state and the ability of fish to reach spawning areas has also been shown. Salmon that delayed river entry and successfully reached spawning areas had lower plasma testosterone levels than salmon that entered the river directly but ultimately failed to reach spawning areas (Crossin et al., 2009). Although steroid hormones associated with sexual maturation inhibit SW tolerance, there is no evidence that these hormones directly promote ion uptake.

Sex steroids have been examined in several lamprey species, but the majority of studies have focused on the sea lamprey. Estradiol-17β has been identified as the primary female sex steroid and androstenedione is the main male sex steroid; both have been demonstrated to have hormonal roles similar to those found in jawed vertebrates (Bryan et al., 2008). The maturation process also begins in parasitic lamprey before they enter FW for their spawning migration (Clemens et al., 2010). The maturation process may occur more slowly in the Pacific lamprey that migrate longer distances in the rivers of western North America than for the sea lamprey migrating to eastern North American streams; however, this has not been demonstrated. Differences in physiological changes associated with maturation in fall-migrating compared to spring-migrating lamprey have also not been well characterized. In a study comparing time of FW entry within a single species, Ahmadi et al. (2011) report that gonads were in the final stages of maturity for Caspian lamprey that entered FW in the fall or the spring. Mesa et al. (2010) indicate that endocrine changes in Pacific lamprey held in a laboratory occurred 6–8 months after the fish had moved into FW; levels of E2 and progesterone, but not 15α-hydroxytestosterone (15α-T), increased before final maturation.

4.3. Metamorphosis

For catadromous and amphidromous species movement into FW environments is accompanied by or follows metamorphosis from a pelagic larval form to a juvenile form capable of migration through estuaries and upriver. Of the hormones that change during the larval stage, the thyroid hormones are most directly related to metamorphosis. In Japanese eels, thyroid hormones were not detected in premetamorphic larvae, but could be measured during metamorphosis (Yamano et al., 2007). Maximum thyroid hormone levels were found at the end of metamorphosis and in juveniles just postmetamorphosis. Histology revealed that the thyroid gland was active in early metamorphosis, but was low by the end of metamorphosis. Ozaki et al. (2000) examined pituitary and thyroid glands in leptocephalus larvae of the Japanese eel and two other species, *A. obscura* and *A. bicolor pacifica*. Immunohistochemistry showed that small leptocephali did not contain thyroid-stimulating hormone (TSH)-immunoreactive cells, but these cells appeared in more developed leptocephali. Conversely, thyroxine-immunoreactive thyroid follicles were detected in all specimens, both leptocephalus larvae and glass eel. Thyroid hormone production, therefore, started before TSH production, but both TSH and thyroid hormone are secreted during the metamorphosis from leptocephalus to glass eel. TSH and thyroid hormones are involved in the metamorphosis from leptocephalus to glass eel, therefore, but not in the early growth from preleptocephalus to leptocephalus (Ozaki et al., 2000).

Endocrine control of metamorphosis in the amphidromous goby *Sicyopterus lagocephalus* was recently investigated and thyroid hormones also appear to play a role in metamorphosis in this species. Changes in thyroid hormone levels, thyroxine (T_4) and triiodothyronine (T_3), were followed during postlarval metamorphosis. There was an acute increase in whole-body T_4 levels 2 days after entry into FW with a peak on day 10, the end of the postlarval stage (Taillebois et al., 2011). T_4 was highest when morphological changes, such as the change in the position of the mouth, were most important and then decreased. Hormonal treatment with T_4 or thiourea (TU, a thyroid inhibitor) affected the rate of metamorphosis; the change in the position of the mouth was accelerated in the T_4-treated postlarvae, while it was delayed in the TU-treated postlarvae, compared to controls. Metamorphosis of *S. lagocephalus* postlarva, therefore, is under the control of thyroid hormones when this species recruits into rivers (Taillebois et al., 2011). Nothing is known, however, about how thyroid hormones might control ionoregulation at this stage in amphidromous species.

4.4. Hormones Associated with Migration

Endocrine changes that occur during FW entry are not limited to control of ionoregulation, maturation, or metamorphosis, but hormones are also involved in enhancing activity associated with movement upstream and enhancing mechanisms for orientation in the river environment. Although such endocrine changes have not been characterized for all diadromous species as they enter FW, there appears to be some commonality in endocrine changes among the diverse taxa of diadromous fish species.

4.4.1. MOVEMENT

Thyroid hormones have been shown to be elevated in migratory fishes at a number of different life stages, suggesting that the role of this hormone in migration may be general. T_4 surges that occur during the spring in smolts have been linked to morphological changes and salinity preference, but whether T_4 has a stimulatory role on downstream migration is not clear

(Iwata, 1995; McCormick, 2013, Chapter 5, this volume). A link between circulating T_4 and upstream migration has also been suggested for adult salmon. T_4 and T_3 levels have been shown to be elevated in wild adult Atlantic salmon migrating upstream (Youngson and Webb, 1992). These authors also found a relationship between river discharge and plasma thyroid hormone levels. Hamano et al. (1996) found that T_4 levels were high in coastal migrating chum salmon, but were lower in river fish migrating a short distance upriver to an enhancement facility. Given the work on juvenile salmon and the rapid changes that have been seen in T_4 in response to changes in flow and increases in turbidity (Ojima and Iwata, 2007), the relationship between upstream migration and thyroid hormones is likely to be complex. Support for thyroid hormones playing a role in enhancing migratory movements in anadromous species of fish is actually provided by work done on the catadromous eel.

European glass eels captured at the tidal limit and entering FW had higher whole-body thyroid hormone concentrations than glass eels collected sheltering on the bottom of the estuary (Edeline et al., 2004). Jegstrup and Rosenkilde (2003) also showed that benthic associated eels had lower T_4 levels compared to pelagic eels. Elvers moving upstream also had higher T_4 and T_3 levels compared to inactive elvers (Imbert et al., 2008). Separating the effects of thyroid hormones on metamorphosis from swimming behavior may be difficult given the close temporal association between these two events in catadromous eels. Examination of migratory eels at other life stages has shown that spontaneous locomotor activity is linked to thyroid hormones. Migratory juvenile yellow American eels were caught moving upstream and climbing waterfalls and found to have plasma thyroxine concentrations twice as high as in sedentary eels captured in the estuary (Castonguay et al., 1990). European glass eels have also been experimentally treated with T_4 and TU; locomotor activity was increased among T_4-treated eels and decreased among TU-treated eels (Edeline et al., 2005b). T_4, however, did not appear to affect rheotactic behavior as T_4 treatment increased both upstream and downstream migration.

Migratory behavior has also been linked to a number of other endocrine factors, particularly those associated with sexual maturation. GnRH treatment in prespawning sockeye salmon shortened the duration of upstream migration to the natal hatchery of the fish (Kitahashi et al., 1998; Ueda, 2011). Plate et al. (2003) found that sexually mature sockeye were more likely than immature fish to jump over an artificial waterfall. This behavior was linked to GnRH as fish injected with GnRH exhibited greater swimming activity and jumping behavior. Other hormones have also been shown to increase migratory behavior. Corticotropin-releasing hormone (CRH) injected intracerebroventricularly in juvenile Chinook salmon increased locomotor activity (Clements et al., 2002). The effect of CRH on locomotor activity, however, was not related to changes in plasma cortisol or T_4 as neither of these hormones was elevated. There is also evidence that GH has an effect on behavior in fish. Juvenile rainbow trout implanted with ovine GH increased swimming activity (Johannson et al., 2004). Levels of the dopamine metabolite 3,4-hydroxyphenylacetic acid (DOPAC) were also higher in the brains of GH-treated fish, suggesting that GH may function as a neuromodulator. The higher levels of GH reported by Benedet et al. (2010) in maturing Atlantic salmon, therefore, may enhance upstream migratory behavior.

4.4.2. DIRECTION

Upstream migrations in diadromous species are often directed as fish search for specific locations for spawning or feeding. Fish use a number of cues to orient movements; one of the senses used for navigation is olfaction and sensitivity is high at migratory stages. Olfactory cues are detected by peripheral olfactory receptors. In an experiment where olfactory receptor cells were removed and isolated for patch-clamp recordings, imprinted fish showed greater sensitivity than naïve fish (Nevitt et al., 1994). Thyroid hormones also appear to play a role in enhancing olfactory sensitivity as T_4 receptors have been found in the olfactory epithelium and the olfactory bulb of the brain in masu salmon, *O. masou* (Kudo et al., 1994). T_3 increases cell proliferation in the olfactory epithelium of juvenile coho salmon and natural fluctuations in T_4 stimulate the proliferation of neural progenitor cells in the salmon epithelium (Lema and Nevitt, 2004), establishing a link between the thyroid hormone axis and measurable anatomical changes in the peripheral olfactory system.

It has also been argued that increases in corticosteroids during the upstream migration of spawning salmon are adaptive for enhancing the ability of fish to recall imprinted memory. Plasma cortisol levels are high in salmon migrating through coastal waters, but are also high in fish migrating upstream (Donaldson and Fagerlund, 1972; Onuma et al., 2010a; Flores et al., 2012). There were significant differences between GR-immunoreactive neuronal cell body numbers in brains of sexually immature and spawning kokanee (Carruth et al., 2000a). Olfactory neurons in the brains of kokanee are GR immunoreactive; GR-immunoreactive staining was nuclear in kokanee spawners, but cytoplasmic in immature fish, suggesting that cortisol has an effect on olfactory neurons. Such findings indicate a role for cortisol in olfactory-mediated directional migration in salmon and may be consistent for diadromous species in general, but this has not been investigated.

5. MECHANISMS FOR SELECTION OF FRESHWATER HABITAT

Given the diversity of species and stages of development when diadromous fishes enter FW, the mechanisms that are used to orient diadromous migrations as they recruit to FW would be expected to differ. What is known about habitat selection in FW by diadromous species indicates that methods differ among and within the different forms of diadromy.

There is considerable evidence that olfaction guides the return migration of salmon to their natal streams and cues are composed of conspecific chemical signals (Solomon, 1973), but more importantly specific home stream odors (Scholz et al., 1976). The information for successful homing is thought to be acquired during the parr–smolt transformation (see McCormick, 2013, Chapter 5, this volume). Active migration downstream appears to be important for successful homing. Juvenile Chinook salmon and steelhead (*O. mykiss*) transported downstream in barges to reduce outmigration mortality exhibited higher rates of permanent straying into non-natal rivers as returning adults (Keefer et al., 2008a). Juvenile transport, therefore, may impair adult homing by disrupting the sequential imprinting process. Other life stages, however, must also be important for imprinting as species of Pacific salmon that reside in FW for a year before smolting move among different habitats. A proportion of juvenile Chinook salmon from large river systems migrate downstream the year before they smolt (Sykes et al., 2009), yet adults return to specific spawning locations in the upper parts of the watershed each year (Shrimpton and Heath, 2003). A study using otolith microchemistry in coho salmon showed that FW signatures vary considerably and indicate extensive movements of juveniles while in FW, and the adults return to their natal river (Shrimpton et al., unpublished data) (Fig. 7.5). Quinn et al. (1999) also used otolith microchemistry to demonstrate that sockeye returned to natal incubation sites, not to a specific lake. These findings indicate that imprinting or a mechanism to find appropriate spawning sites must occur at times other than the smolt stage. Intragravel electrical conductance, dissolved oxygen, temperature, and specific discharge have been found to play a prominent role in spawning site selection for coho salmon from the interior Fraser River watershed of British Columbia, Canada (McRae et al., 2012).

Olfaction is also important for other diadromous species to select rivers to enter, but the cues for species that are not returning to a specific spawning site or home stream may be quite different from those described for the anadromous salmonids. An established population of conspecifics could be a good predictor of suitable habitat and chemical cues associated with their

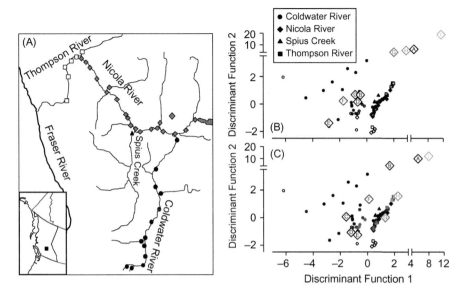

Fig. 7.5. (A) Map of water sampling sites throughout the mainstem and tributaries of the Thompson, Nicola, and Coldwater Rivers, British Columbia, Canada. Discriminant function analysis was used to provide a visualization of the geographic separation using Sr:Ca, Ba:Ca, and Mn:Ca ratios for the water samples collected. Relationships between water chemistry and elemental signatures from otoliths of juvenile coho salmon (*Oncorhynchus kisutch*) have been established (J. M. Shrimpton, K. H. Telmer, G. J. Glova and N. L. Todd, unpublished data). Using discriminant function equations, scores for the first two discriminant functions were calculated for changes in elemental signatures for each 50 μm section of otolith from coho salmon spawners caught in the Coldwater River (see Fig. 7.2B, C). The data were plotted with the water chemistry data to indicate putative locations of residence for the fish as juveniles in (A) and (B). (B) Putative juvenile residence locations of a coho salmon spawner caught in the Coldwater River showing residence in the Coldwater River as a juvenile. Closed symbols represent water chemical signatures from mainstem rivers and open symbols represent tributary chemical signatures. Large diamonds with letters are putative residence locations. A: elemental signature near the core; I: seawater entry signature; B–F: movement among habitats in the Coldwater River. (C) A plot similar to (B), but for a coho that spent the initial part of the freshwater phase in the Coldwater River, then the Nicola River, and probably also the Thompson River before migrating to the ocean. Symbols A–I as in (B) (Shrimpton et al., unpublished data).

presence are important olfactory signals. For lamprey the bile acids, petromyzonol sulfate, petromyzonamine disulfate, and petromyzosterol disulfate, are potent and specific stimulants of adult olfactory system and maturing adults use larval odor to select streams for spawning (Sorensen and Hoye, 2007; Yun et al., 2011). Migrating sea lamprey showed fine-scale movements in streams to avoid swimming in waters that lacked larval odor (Wagner et al., 2009). The bile acids are produced and released by larvae from diverse species of lamprey and appear to be evolutionarily conserved pheromones, although the relative amounts produced and the sensitivity appear to differ among species (Fine et al., 2004). The similar spawning and larval habitat requirements among species, therefore, may make the common pheromone a useful cue for any species. Fine and Sorensen (2005) have also measured petromyzonal sulfate in picomolar concentrations in streams with larval lamprey. Pheromones, however, are only a long-distance cue that helps lampreys to locate a spawning stream and other environmental cues are likely to be used to find specific locations for spawning as ammocoetes usually disperse downstream from spawning locations to find appropriate feeding habitats.

Glass eels show a strong attraction to FW streams, but odors originating from terrestrial and FW microbes appear to be important (Miles, 1968; Sorensen, 1986; Tosi and Sola, 1993). The odors of adult conspecifics, particularly amino acids and bile acids, may also serve as olfactory cues for migration into FW (Sola, 1995), but this factor appears to be limited to low-salinity waters (Sola and Tongiorgi, 1996). Olfaction continues to be important for adult eels to navigate to locations of original capture when moved in a displacement experiment (Barbin, 1988). Considerably less is known about the cues for selection of FW habitat in amphidromous species. Galaxiid juveniles from two species, *Galaxias maculatus* and *G. brevipinnis*, were attracted to odors of adults within the same genus, but not in the presence of other stream-dwelling fish (Baker and Hicks, 2003). Whether pheromones from adults may attract juvenile gobies to migrate upstream is not known, but has been suggested as a mechanism to recruit postlarval fish to FW systems (Fitzsimmons et al., 2002).

6. EFFECT OF DIADROMY ON GENETIC POPULATION STRUCTURE

Homing behavior and the physical nature of river networks have led to reproductive isolation among populations of anadromous fishes. This has been particularly well documented in anadromous salmon. In sockeye salmon, for example, genetic differentiation has been observed among populations indicating strong philopatry (Beacham et al., 2004). Genetic divergence among populations has been linked to the evolution of adaptive characters and the selection of locally adapted traits that influence fitness. Relationships between local adaptations and physiological function have been demonstrated in Pacific salmon (Heath et al., 2002) and the link to genetic differences is associated with strong isolation by distance (IBD) (Heath et al., 2006). The importance of locally adapted traits has been neatly demonstrated in a recent study on sockeye salmon from the Fraser River,

British Columbia. Cardiac and respiratory physiology differ among populations and are related to historic temperature conditions experienced by adult sockeye salmon as they migrate upstream. Fish from populations with more challenging migratory environments have greater aerobic scope, larger hearts, and better coronary supply. Consequently, thermal optima for aerobic, cardiac, and heart rate scopes are consistent with the historic river temperature ranges experienced by each population (Eliason et al., 2011).

Evidence of homing has also been shown in a number of non-salmonid anadromous species. Hatchery-reared American shad with otoliths marked by tetracycline immersion showed a high degree of homing to their natal hatchery (Hendricks et al., 2002). The use of geochemical signatures in the otoliths of American shad also suggests that most fish were homing to their natal river, but there was less fidelity to an individual tributary (Walther et al., 2008). Genetic variation and temporally stable genetic differentiation among drainages indicates that rivers support distinct populations of American shad with a strong pattern of IBD (Hasselman et al., 2010). Similarly, genetic analysis has revealed strong IBD suggestive of philopatry in Atlantic sturgeon (*Acipenser oxyrinchus*) from the east coast of North America (King et al., 2001). Genetic analysis for shortnose sturgeon (*A. brevirostrum*) also showed highly structured populations across their range in eastern North America (Grunwald et al., 2002). From the Pacific coast of North America, genetic population structure is also supported for work done on both green sturgeon (*A. medirostris*) (Israel et al., 2004) and white sturgeon (*A. transmontanus*) (Smith et al., 2002). Population structuring was also found in three-spined stickleback (*Gasterosteus aculeatus*) inhabiting the Saint Lawrence River estuary (McCairns and Bernatchez, 2008). Consistent with work on salmonids, relationships between local adaptations and physiological function have been demonstrated in a non-salmonid euryhaline species, the common killifish (*Fundulus heteroclitus*) (Fangue et al., 2006).

In contrast, the parasitic feeding strategy of lamprey has been argued to be problematic for homing as fish are dispersed by movements of the host animals they parasitize rather than directed migratory movements of salmon (Waldman et al., 2008). Consequently, lamprey do not appear to home to specific streams, but seek suitable rivers based on the presence of other lamprey. The lack of strong natal homing would promote gene flow among drainages (Goodman et al., 2008) and sea lamprey populations exhibit regional panmixia (Bryan et al., 2005; Waldman et al., 2008). Lin et al. (2008) sampled Pacific lamprey populations that ranged from Japan to Oregon and over this range found significant differences among the populations and an IBD relationship. Differences appear to exist, therefore, between Pacific and sea lamprey. Pacific lamprey probably travel shorter distances in the ocean than anadromous sea lamprey, perhaps owing to their smaller body size, potentially leading to geographic genetic structure even in the absence of natal homing (Spice et al., 2012).

Given the proximate cues used by catadromous eels and amphidromous species, it would be expected that populations are panmictic. For eels there is considerable controversy over panmixia. In earlier studies, American eels exhibited no genetic divergence along the east coast of North America (Avise et al., 1986), Japanese eel from Taiwan, Japan, and China showed no significant population structure (Sang et al., 1994), and European eel sampled from Iceland to Morocco also indicated panmixia (Dannewitz et al., 2005). The findings for panmictic anguillid eel populations in FW in many ways are not surprising. Selective pressures should be much greater for fidelity to spawning habitat than for juvenile and adult feeding habitats; fish in FW systems show seasonal and even year-to-year preferences for different habitats (Gowan et al., 1994). A number of other studies, however, provide evidence of weak population structure over large geographic ranges. Japanese eels are genetically differentiated into two management units: a low-latitude group and a high-latitude group (Tseng et al., 2006). Wirth and Bernatchez (2001) indicate even stronger patterns of genetic structure implying restricted gene flow in European eels sampled from Iceland to North Africa. Daemen et al. (2001) examined European glass eels from Ireland, Italy, Morocco, Sweden, and the UK, and found weak differentiation among populations. Kettle and Haines (2005) suggest that adult European eels may select spawning locations to preferentially target currents that will disperse larvae back to the same region that adults formerly inhabited. The conclusion that temperate eels show regional genetic distinction in FW, however, has been further questioned as glass eels collected throughout their inshore migration period showed no temporal genetic structure for European eels (Dannewitz et al., 2005) and Japanese eels (Ishikawa et al., 2001). Spawning time differences, variation in reproductive success among individuals, or even differences in spawning locations may cause the genetic differentiation observed in the earlier studies. Consequently, Als et al. (2011) conducted genetic analysis of larval European eels sampled throughout their spawning areas in the Sargasso Sea and provide strong evidence that the species is a single panmictic unit. Population structure, however, may differ for tropical eel species. The most widely distributed tropical species, the giant mottled eel (*A. marmorata*), appears to be fairly panmictic in the North Pacific, but metapopulation structure exists for the south Pacific and Indian Ocean populations based on genetic analysis (Minegishi et al., 2008) and total vertebral counts (Watanabe et al., 2011). The metapopulation structure for the giant mottled eel is suggested to result from at least four spawning locations (Minegishi et al., 2008).

Amphidromy is an important adaptation for colonization of FW of oceanic islands that are characterized by small size and high gradient, and that discharge directly into the ocean. The pelagic larval phase is the mechanism used for dispersal of amphidromous species between island streams (McDowall, 2001). As with eels, dispersal by oceanic currents would be expected to limit defined regional population structure of amphidromous fishes in FW. A number of studies show a lack of genetic population structure. Gobies in Hawaii exhibit no genetic population structure, with high gene flow among populations (McDowall, 2003). The larval marine stage also facilitates extensive gene flow and no genetic structure among populations in the Australian grayling (*Prototroctes maraena*) (Schmidt et al., 2011), an obligate amphidromous species (Crook et al., 2006). The goby *Sicyopterus lagochephalus* occurs over an extensive 18,000 km range from the western Indian Ocean to the eastern Pacific Ocean, but with little genetic divergence and potentially just two genetic groups indicating extensive dispersal capabilities (Keith et al., 2005). Unlike catadromous fishes, however, amphidromous species select FW habitats for juvenile growth opportunities and also for subsequent spawning habitat. Endemic populations of goby have been described with more limited distribution: *S. aiensis* from Vanuatu, *S. sarasini* from New Caledonia (Lord et al., 2010), and *Cotylopus acutipinnis* from the Mascarene Archipelago (Hoareau et al., 2007). Presumably the shorter pelagic larval stage has contributed to higher rates of recruitment to the natal river or adjacent rivers and driving speciation (Table 7.2) (Berrebi et al., 2005). A general absence of population structure was found for the widely distributed *S. lagocephalus*, but Berrebi et al. (2005) suggest that their results do not support panmixia owing to the lack of heterozygotes, potentially due to recruitment to the natal island stream, but also recruitment from other island streams.

Based on the weak genetic population structure and high rates of gene flow, diadromy is an excellent strategy for dispersal throughout FW systems, with the exception of the non-lamprey anadromous species which exhibit high fidelity for natal locations. Even among the anadromous species of salmonids, however, incredibly rapid rates of dispersal have been documented. Quinn et al. (2001) report that Chinook salmon stocked from 1901 to 1907 in New Zealand naturally strayed and colonized rivers up to 230 km away from point of introduction within 15 years. Chinook salmon have also established breeding populations in South America, expanding from the point of introduction more than 200 km in a similar time span (Correa and Gross, 2008). Within approximately 35 years, anadromous steelhead have dispersed downstream in the Saint Lawrence system of eastern Canada (Thibault et al., 2009), the probability of dispersal being greater for the anadromous steelhead life history form of *O. mykiss* than the FW-resident rainbow trout life history form (Thibault et al., 2010). Rapid expansion of geographic range has also been seen in non-salmonid anadromous species. American shad were introduced to the Sacramento River in 1871; there were reports of mature shad in the Columbia River by 1876 and shad had reached Alaska by 1904 (Welander, 1940). Even within their native range, dispersal rates are high. Anderson and Quinn (2007) found that coho salmon immediately exploited new habitat for spawning when 33 km of habitat lost for over a century was opened up by providing passage around a dam. These authors suggest that the immediate use of new habitat by colonists and widespread movements indicate that exploration is an innate component of salmon breeding. Habitat improvements have also led to recolonization of formerly inhospitable habitat (Perrier et al., 2010). Improvements in water quality of rivers in Belgium in the latter part of the twentieth century resulted in the natural recolonization of diadromous species, European eel, river lamprey (*Lampetra fluviatilis*) and smelt (*Osmerus eperlanus*), that were found as far upriver as 170 km (Buysse et al., 2008).

Given the evidence for rapid invasion of new habitat by diadromous species, continued high rates of gene flow would be expected for most species. This is clearly seen in species that select FW habitat by the presence of conspecifics as outlined above. The observed fidelity to natal sites would be expected to lead to breeding isolation and local adaptation and genetic evidence supports this pattern in anadromous salmon. Such a line of reasoning would suggest that salmon may stop straying into new habitat when conspecifics already exist in that location, but this is not the case. Candy and Beacham (2000) found that Chinook salmon tagged as juveniles strayed and spawned in streams from 6 to 480 km from their home stream, but most fish returned to near their home stream and were recovered within 30 km. Extensive straying over similar distances have also been shown in Atlantic salmon, with straying rates 2.5 times higher in hatchery-produced fish compared to their wild conspecifics (Jonsson et al., 2003). Using assignment tests based on genetic analysis for five populations of Chinook salmon, Walter et al. (2009) estimated a mean straying rate of 13% in the fish sampled.

Straying may be related to non-direct homing behaviors, as has been observed in radiotagged Chinook salmon that overshot natal tributaries by over 250 km; on average, 15% of fish temporarily used non-natal tributaries within a given year (Keefer et al., 2008b). Dittman et al. (2010) suggest that the instinct to home to a specific spawning location may be overridden by factors such as environmental and social factors. The benefits of such high rates of straying among populations seem counterproductive to the selection of locally adapted traits, but the majority of straying actually occurs among streams in close proximity. Low levels of straying in salmonids may have evolved to favor local adaptation, but gene flow among populations has recently been shown to elevate the effective population size and preserve genetic variability within populations (Walter et al., 2009).

Dispersal among amphidromous species enables colonization and recolonization of local populations across watersheds within and among oceanic islands (McDowall, 2001; Keith, 2003), both for endemic species (Lord et al., 2010) and for widely distributed species (Keith et al., 2005). Dispersal of leptocephali has resulted in extensive geographic distribution for many of the anguillid eels (Tesch, 2003). Therefore, it is likely that dispersal and straying are effective strategies among diadromous species for sustaining populations throughout their geographic ranges.

7. CONCLUSIONS AND PERSPECTIVES

Diadromy has enabled a relatively small number of species of fish to be immensely successful over a vast geographic range and to exploit different habitats within both marine and FW environments. The evolution of diadromy, particularly obligate movements between SW and FW, however, requires dependence on widely divergent habitat types, and the loss of any habitat that is specifically required for life cycle completion will place a species at risk. Amphidromous gobies are often endemic and many are bordering on extinction (Keith, 2003). Recovery programs and management efforts have been developed to limit further declines in catadromous eels (Tesch, 2003). Anadromous salmon and lamprey have experienced precipitous declines in population abundance over much of their geographic range (Nehlsen et al., 1991; Clemens et al., 2010). The reasons are diverse. Contamination of coastal waters compromises marine larval survival, creating a dispersal barrier that impedes exchange and recruits among stream populations for Hawaiian gobies (Walter et al., 2012). Blocked upstream passage and even inadequate flows from anthropogenic disturbance limit FW dispersal (Nehlsen et al., 1991; Buysse et al., 2008). Species richness is altered by land use; rivers in forested watersheds have more species than rivers in watersheds impacted by agricultural and urban development (Walter et al., 2012). Loss of specific habitat needed for critical stages of development may further threaten regional populations (McRae et al., 2012). Life history strategies that have been so successful but require existence in and movement between FW and marine environments may ultimately limit the continued success of diadromous species unless more effective methods are developed to limit and mitigate anthropogenic effects on the environment.

ACKNOWLEDGMENTS

Work presented in this chapter was funded by the Natural Sciences and Engineering Research Council of Canada through the Discovery and Strategic research grant programs, the Environmental Watch Program of Fisheries and Oceans Canada, and the Pacific Salmon Foundation. I thank the members of my lab and my collaborators who over the years have always been willing to share ideas. Thanks to Steve McCormick, Margaret Docker, and an anonymous reviewer for their comments and suggestions that greatly improved this chapter.

REFERENCES

Abou-Seedo, F. S. and Potter, I. C. (1979). Estuarine phase in the spawning run of the river lamprey, *Lampetra fluviatilis*. *J. Zool. Lond.* 188, 5–25.

Acou, A., Legault, A., Laffaille, P. and Feunteun, E. (2009). Environmental determinism of year-to-year recruitment variability of European eel *Anguilla anguilla* in a small coastal catchment, the Frémur River, north-west France. *J. Fish Biol.* 74, 1985–2001.

Ahmadi, M., Amiri, B. M., Abdoli, A., Fakharzade, S. M. E. and Hoseinifar, S. H. (2011). Sex steroids, gonadal histology and biological indices of fall and spring Caspian lamprey (*Caspiomyzon wagneri*) spawning migrants in the Shirud River, southern Caspian Sea. *Environ. Biol. Fish.* 92, 229–235.

Almeida, P. R., Quintella, B. R. and Dias, N. M. (2002). Movement of radio-tagged anadromous sea lamprey during the spawning migration in the River Mondego (Portugal). *Hydrobiologia* 483, 1–8.

Als, T. D., Hansen, M. M., Maes, G. E., Castonguay, M., Riemann, L., Aarestrup, K., Munk, P., Sparholt, H., Hanel, R. and Bernatchez, L. (2011). All roads lead to home: panmixia of European eel in the Sargasso Sea. *Mol. Ecol.* 20, 1333–1346.

Andersen, O., Skibeli, V., Haug, E. and Gautvik, K. M. (1991). Serum prolactin and sex steroids in Atlantic salmon (*Salmo salar*) during sexual maturation. *Aquaculture* 95, 169–178.

Anderson, J. H. and Quinn, T. P. (2007). Movements of adult coho salmon (*Oncorhynchus kisutch*) during colonization of newly accessible habitat. *Can. J. Fish. Aquat. Sci.* 64, 1143–1154.

Aprahamian, M. W., Jones, G. O. and Gough, P. J. (1998). Movement of adult Atlantic salmon in the Usk estuary, Wales. *J. Fish Biol.* 53, 221–225.

Arai, T. and Hirata, T. (2006). Differences in the trace element deposition in otoliths between marine- and freshwater-resident Japanese eels, *Anguilla japonica*, as determined by laser ablation ICPMS. *Environ. Biol. Fish.* 75, 173–182.

Arai, T., Limbong, D., Otake, T. and Tsukamoto, K. (1999a). Metamorphosis and inshore migration of tropical eels *Anguilla* spp. in the Indo-Pacific. *Mar. Ecol. Prog. Ser.* 182, 282–293.

Arai, T., Otake, T., Limbong, D. and Tsukamoto, K. (1999b). Early life history and recruitment of the tropical eel *Anguilla bicolor pacifica*, as revealed by otolith microstructure and microchemistry. *Mar. Biol.* 133, 319–326.

Arai, T., Otake, T., Jellyman, D. J. and Tsukamoto, K. (1999c). Differences in the early life history of the Australasian shortfinned eel *Anguilla australis* from Australia and New Zealand, as revealed by otolith microstructure and microchemistry. *Mar. Biol.* 135, 381–389.

Arai, T., Otake, T. and Tsukamoto, K. (2000). Timing of metamorphosis and larval segregation of the Atlantic eels *Anguilla rostrata* and *A. anguilla*, as revealed by otolith microstructure and microchemistry. *Mar. Biol.* 137, 39–45.

Arai, T., Limbong, D., Otake, T. and Tsukamoto, K. (2001). Recruitment mechanisms of tropical eels *Anguilla* spp. and implications for the evolution of oceanic migration in the genus *Anguilla*. *Mar. Ecol. Prog. Ser.* 216, 253–264.

Arai, T., Miller, M. J. and Tsukamoto, K. (2003). Larval duration of the tropical eel *Anguilla celebesensis* from Indonesian and Philippine coasts. *Mar. Ecol. Prog. Ser.* 251, 255–261.

Avise, J. C., Helfman, G. S., Saunders, N. C. and Hales, L. S. (1986). Mitochondrial DNA differentiation in North Atlantic eels: population genetic consequences of an unusual life history pattern. *Proc. Natl Acad. Sci. U. S. A.* 83, 4350–4354.

Baker, C. F. and Hicks, B. J. (2003). Attraction of migratory inanga (*Galaxias maculatus*) and koaro (*Galaxias brevipinnis*) juveniles to adult galaxiid odours. *N. Z. J. Mar. Freshw. Res.* 37, 291–299.

Barbin, G. P. (1988). The role of olfaction in homing and estuarine migratory behavior of yellow-phase American eels. *Can. J. Fish. Aquat. Sci.* 55, 564–575.

Bardonnet, A., Bolliet, V. and Belon, V. (2005). Recruitment abundance estimation: role of glass eel (*Anguilla anguilla* L.) response to light. *J. Exp. Mar. Biol. Ecol.* 321, 181–190.

Barry, T. P., Marwah, A. and Nunez, S. (2010). Inhibition of cortisol metabolism by 17α,20β-P: mechanism mediating semelparity in salmon? *Gen. Comp. Endocrinol.* 165, 53–59.

Bartels, H. and Potter, I. C. (2004). Cellular composition and ultrastructure of the gill epithelium of larval and adult lampreys: implications for osmoregulation in fresh and seawater. *J. Exp. Biol.* 207, 3447–3462.

Beacham, T. D., Lapointe, M., Candy, J. R., McIntosh, B., MacConnachie, C., Tabata, A., Kaukinen, K., Deng, L., Miller, K. M. and Withler, R. E. (2004). Stock identification of Fraser River sockeye salmon using microsatellites and major histocompatibility complex variation. *Trans. Am. Fish. Soc.* 133, 1117–1137.

Beamish, F. W. H. (1980a). Biology of the North American anadromous sea lamprey *Petromyzon marinus*. *Can. J. Fish. Aquat. Sci.* 37, 1924–1943.

Beamish, F. W. H. (1980b). Osmoregulation in juvenile and adult lampreys. *Can. J. Fish. Aquat. Sci.* 37, 1739–1750.

Beamish, R. J. and Youson, J. H. (1987). Life history and abundance of young adult *Lampetra ayresi* in the Fraser River and their possible impact on salmon and herring stocks in the Strait of Georgia. *Can. J. Fish. Aquat. Sci.* 44, 525–537.

Benedet, S., Andersson, E., Mittelholzer, C., Taranger, G. L. and Björnsson, B. Th. (2010). Pituitary and plasma growth hormone dynamics during sexual maturation of female Atlantic salmon. *Gen. Comp. Endocrinol.* 167, 77–85.

Berrebi, P., Cattaneo-Berrebi, G., Valade, P., Ricou, J.-F. and Hoareau, T. (2005). Genetic homogeneity in eight freshwater populations of *Sicyopterus lagocephalus*, an amphidromous gobiid of La Réunion Island. *Mar. Biol.* 148, 179–188.

Binder, T. R., McLaughlin, R. L. and McDonald, D. G. (2010). Relative importance of water temperature, water level, and lunar cycle to migratory activity in spawning-phase sea lampreys in Lake Ontario. *Trans. Am. Fish. Soc.* 139, 700–712.

Brawn, V. M. (1982). Behavior of Atlantic salmon (*Salmo salar*) during suspended migration in an estuary, Sheet Harbour, Nova Scotia, observed visually and by ultrasonic tagging. *Can. J. Fish. Aquat. Sci.* 39, 248–256.

Bryan, M. B., Zalinski, D., Filcek, K. B., Libants, S., Li, W. and Scribner, K. T. (2005). Patterns of invasion and colonization of the sea lamprey (*Petromyzon marinus*) in North America as revealed by microsatellite genotypes. *Mol. Ecol.* 14, 3757–3773.

Bryan, M. B., Scott, A. P. and Li, W. (2008). Sex steroids and their receptors in lampreys. *Steroids* 73, 1–12.

Bureau du Colombier, S., Bolliet, V., Lambert, P. and Bardonnet, A. (2007). Metabolic loss of mass in glass eels at different salinities according to their propensity to migrate. *Estuar. Coast. Shelf Sci.* 93, 1–6.

Burgner, R. L. (1991). The life history of sockeye salmon (*Oncorhynchus nerka*). In *Pacific Salmon Life Histories* (eds. C. Groot and L. Margolis), pp. 3–117. Vancouver: University of British Columbia Press.

Buysse, D., Coeck, J. and Maes, J. (2008). Potential re-establishment of diadromous fish species in the River Scheldt (Belgium). *Hydrobiologia* 602, 155–159.

Bystriansky, J. S. and Schulte, P. M. (2011). Changes in gill H^+-ATPase and Na^+/K^+-ATPase expression and activity during freshwater acclimation of Atlantic salmon (*Salmo salar*). *J. Exp. Biol.* 214, 2435–2442.

Bystriansky, J. S., Frick, N. T., Richards, J. G., Schulte, P. M. and Ballantye, J. S. (2007). Wild Arctic char (*Salvelinus alpinus*) upregulate gill Na^+/K^+-ATPase during freshwater migration. *Physiol. Biochem. Zool.* 80, 270–282.

Cairns, D. K., Secor, D. A., Morrison, W. E. and Hallett, J. A. (2009). Salinity-linked growth in anguillid eels and the paradox of temperate-zone catadromy. *J. Fish Biol.* 74, 2094–2114.

Candy, J. R. and Beacham, T. D. (2000). Patterns of homing and straying in southern British Columbia coded-wire tagged Chinook salmon (*Oncorhynchus tshawytscha*) populations. *Fish. Res.* 47, 41–56.

Carruth, L. L., Jones, R. E. and Norris, D. O. (2000a). Cell density and intracellular translocation of glucocorticoid receptor-immunoreactive neurons in the kokanee salmon (*Oncorhynchus nerka kennerlyi*) brain, with an emphasis on the olfactory system. *Gen. Comp. Endocrinol.* 117, 66–76.

Carruth, L. L., Dores, R. M., Maldonado, T. A., Norris, D. O., Ruth, T. and Jones, R. E. (2000b). Elevation of plasma cortisol during the spawning migration of landlocked kokanee salmon (*Oncorhynchus nerka kennerlyi*). *Comp. Biochem. Physiol. C Toxicol. Pharmacol.* 127, 123–131.

Castonguay, M., Dutil, J. D., Audet, C. and Miller, R. (1990). Locomotor-activity and concentration of thyroid-hormones in migratory and sedentary American eels. *Trans. Am. Fish. Soc.* 119, 946–956.

Cheng, P. W. and Tzeng, W. N. (1996). Timing of metamorphosis and estuarine arrival across the dispersal range of the Japanese eel *Anguilla japonica*. *Mar. Ecol. Prog. Ser.* 131, 87–96.

Chessman, B. C. and Williams, W. D. (1975). Salinity tolerance and osmoregulatory ability of *Galaxias maculatus* (Jenyns) (Pisces, Salmoniformes, Galaxiidae). *Freshw. Biol.* 5, 135–140.

Chino, N. and Arai, T. (2010). Migratory history of the giant mottled eel (*Anguilla marmorata*) in the Bonin Islands of Japan. *Ecol. Freshw. Fish* 19, 19–25.

Choe, K. P., O'Brien, S., Evans, D., Toop, T. and Edwards, S. (2004). Immunolocalization of Na^+/K^+-ATPase, carbonic anhydrase II, and vacuolar H^+- ATPase in the gills of freshwater adult lampreys *Geotria australis*. *J. Exp. Zool* A 301, 654–665.

Clarke, W. C. and Hirano, T. (1995). Osmoregulation. In *Physiological Ecology of Pacific Salmon* (eds. C. Groot, L. Margolis and W. C. Clarke), pp. 319–377. Vancouver: University of British Columbia Press.

Clemens, B. J., Binder, T. R., Docker, M. F., Moser, M. L. and Sower, S. A. (2010). Similarities, differences, and unknowns in biology and management of three parasitic lampreys of North America. *Fisheries* 35, 580–594.

Clemens, B. J., Mesa, M. G., Magie, R. J., Young, D. A. and Schreck, C. B. (2012). Pre-spawning migration of adult Pacific lamprey, *Entosphenus tridentatus*, in the Willamette River, Oregon, USA. *Environ. Biol. Fish.* 93, 245–254.

Clements, S., Schreck, C. B., Larsen, D. A. and Dickhoff, W. W. (2002). Central administration of corticotropin-releasing hormone stimulates locomotor activity in juvenile Chinook salmon (*Oncorhynchus tshawytscha*). *Gen. Comp. Endocrinol.* 125, 319–327.

Close, D. A., Yun, S.-S., McCormick, S. D., Wildbill, A. J. and Li, W. (2010). 11-Deoxycortisol is a corticosteroid hormone in the lamprey. *Proc. Natl Acad. Sci. U. S. A.* 107, 13942–13947.

Cooke, S. J., Hinch, S. G., Farrell, A. P., Lapointe, M., Healey, M., Patterson, D., Macdonald, S., Jones, S. and Van Der Kraak, G. (2004). Early-migration and abnormal mortality of late-run sockeye salmon in the Fraser River, British Columbia. *Fisheries* 29 (2), 22–33.

Cooke, S. J., Hinch, S. G., Crossin, G. T., Patterson, D. A., English, K. K., Healey, M. C., Shrimpton, J. M., Van der Kraak, G. and Farrell, A. P. (2006a). Mechanistic basis of individual mortality in Pacific salmon during spawning migrations. *Ecology* 87, 1575–1586.

Cooke, S. J., Hinch, S. G., Crossin, G. T., Patterson, D. A., English, K. K., Shrimpton, J. M., Van Der Kraak, G. and Farrell, A. P. (2006b). Physiology of individual late-run Fraser river sockeye salmon (*Oncorhyncus nerka*) sampled in the ocean correlates with fate during spawning migration. *Can. J. Fish. Aquat. Sci.* 63, 1469–1480.

Cooperman, M. C., Hinch, S. G., Crossin, G. T., Cooke, S. J., Patterson, D. A., Olsson, I., Lotto, A., Welch, D. W., Shrimpton, J. M., Van Der Kraak, G. and Farrell, A. P. (2010). Effects of experimental manipulations of salinity and maturation status on the physiological condition and mortality of homing adult sockeye salmon held in a laboratory. *Physiol. Biochem. Zool.* 83, 459–472.

Correa, C. and Gross, M. R. (2008). Chinook salmon invade southern South America. *Biol. Invasions* 10, 615–639.

Côté, C. L., Castonguay, M., Verreault, G. and Bernatchez, L. (2009). Differential effects of origin and salinity rearing conditions on growth of glass eels of the American eel *Anguilla rostrata*: implications for stocking programmes. *J. Fish Biol.* 74, 1934–1948.

Crook, D. A., Macdonald, J. I., O'Connor, J. P. and Barry, B. (2006). Use of otolith chemistry to examine patterns of diadromy in the threatened Australian grayling *Prototroctes maraena*. *J. Fish Biol.* 69, 1330–1344.

Crossin, G. T., Hinch, S. G., Cooke, S. J., Welch, D. W., Batten, S. D., Patterson, D. A., Van Der Kraak, G., Shrimpton, J. M. and Farrell, A. P. (2007). Behaviour and physiology of sockeye homing through coastal waters to a natal stream. *Mar. Biol.* 152, 905–918.

Crossin, G. T., Hinch, S. G., Cooke, S. J., Cooperman, M. S., Patterson, D. A., Welch, D. W., Hanson, K. C., Olsson, I., English, K. K. and Farrell, A. P. (2009). Mechanisms influencing the timing and success of reproductive migration in a capital breeding semelparous fish species, the sockeye salmon. *Physiol. Biochem. Zool.* 82, 635–652.

Daemen, E., Cross, T., Ollevier, F. and Volckaert, F. A. M. (2001). Analysis of the genetic structure of European eel (*Anguilla anguilla*) using microsatellite DNA and mtDNA markers. *Mar. Biol.* 139, 755–764.

Dannewitz, J., Maes, G. E., Johansson, L., Wickström, H., Volckaert, F. A. M. and Järvi, T. (2005). Panmixia in the European eel: a matter of time. *Proc. R. Soc. B* 272, 1129–1137.

Daverat, F., Limburg, K. E., Thibault, I., Shiao, J.-C., Dodson, J. J., Caron, F., Tzeng, W.-N. and Iizuka, Y. (2006). Phenotypic plasticity of habitat use by three temperate eel species, *Anguilla anguilla, A. japonica* and *A. rostrata*. *Mar. Ecol. Prog. Ser.* 308, 231–241.

Dittman, A. H., May, D., Larsen, D. A., Moser, M. L., Johnston, M. and Fast, D. (2010). Homing and spawning site selection by supplemented hatcher- and natural-origin Yakima River spring Chinook salmon. *Trans. Am. Fish. Soc.* 139, 1014–1028.

Donaldson, E. M. and Fagerlund, U. H. M. (1972). Corticosteroid dynamics in Pacific salmon. *Gen. Comp. Endocrinol.* 3 (Suppl), 254–265.

Edeline, E., Dufour, S., Briand, C., Fatin, D. and Elie, P. (2004). Thyroid status is related to migratory behavior in *Anguilla anguilla* glass eels. *Mar. Ecol. Prog. Ser.* 282, 261–270.

Edeline, E., Dufour, S. and Elie, P. (2005a). Role of glass eel salinity preference in the control of habitat selection and growth plasticity in *Anguilla anguilla*. *Mar. Ecol. Prog. Ser.* 304, 191–199.

Edeline, E., Bardonnet, A., Bolliet, V., Dufour, S. and Elie, P. (2005b). Endocrine control of *Anguilla anguilla* glass eel dispersal: effect of thyroid hormones on locomotor activity and rheotactic behavior. *Horm. Behav.* 48, 58–63.

Edeline, E., Lambert, P., Rigaud, C. and Elie, P. (2006). Effects of body condition and water temperature on *Anguilla anguilla* glass eel migratory behavior. *J. Exp. Mar. Biol. Ecol.* 331, 217–225.

Edwards, S. L. and Marshall, W. S. (2013). Principles and patterns of osmoregulation and euryhalinity in fishes. In *Fish Physiology*, Vol. 32, *Euryhaline Fishes* (eds. S. D. McCormick, A. P. Farrell and C. J. Brauner), pp. 1–44. New York: Elsevier.

Eliason, E. J., Clark, T. D., Hague, M. J., Hanson, L. M., Gallagher, Z. S., Jeffries, K. M., Gale, M. K., Patterson, D. A., Hinch, S. G. and Farrell, A. P. (2011). Differences in thermal tolerance among sockeye salmon populations. *Science* 332, 109–112.

Fangue, N. A., Hofmeister, M. and Schulte, P. M. (2006). Intraspecific variation in thermal tolerance and heat shock protein gene expression in common killifish *Fundulus heteroclitus*. *J. Exp. Biol.* 209, 2859–2872.

Farrell, A. P., Hinch, S. G., Cooke, S. J., Patterson, D. A., Crossin, G. T., Lapointe, M. and Mathes, M. T. (2008). Pacific salmon in hot water: applying aerobic scope models and biotelemetry to predict the success of spawning migrations. *Physiol. Biochem. Zool.* 81, 697–708.

Fine, J. M. and Sorensen, P. W. (2005). Biologically relevant concentrations of petromyzonol sulfate, a component of the sea lamprey migratory pheromone, measured in stream water. *J. Chem. Ecol.* 31, 2205–2210.

Fine, J. M., Vrieze, L. A. and Sorensen, P. W. (2004). Evidence that petromyzontid lampreys employ a common migratory pheromone that is partially comprised of bile acids. *J. Chem. Ecol.* 30, 2091–2110.

Fitzsimmons, J. M., Parhama, J. E. and Nishimotob, R. T. (2002). Similarities in behavioral ecology among amphidromous and catadromous fishes on the oceanic islands of Hawai'i and Guam. *Environ. Biol. Fish.* 65, 123–129.

Flores, A.-M. and Shrimpton, J. M. (2012). Differential physiological and endocrine responses of rainbow trout, *Oncorhynchus mykiss*, transferred from fresh water to ion-poor or salt water. *Gen. Comp. Endocrinol.* 175, 244–250.

Flores, A.-M., Shrimpton, J. M., Patterson, D. A., Hills, J. A., Cooke, S. J., Yada, T., Moriyama, S., Hinch, S. G. and Farrell, A. P. (2012). Physiological and molecular endocrine changes in maturing wild sockeye salmon, *Oncorhynchus nerka*, during ocean and river migration. *J. Comp. Physiol. B* 182, 77–90.

Goodman, D. H., Reid, S. B., Docker, M. F., Haas, G. R. and Kinziger, A. P. (2008). Mitochondrial DNA evidence for high levels of gene flow among populations of a widely distributed anadromous lamprey *Entosphenus tridentatus* (Petromyzontidae). *J. Fish Biol.* 72, 400–417.

Gowan, C., Young, M. K., Fausch, K. D. and Riley, S. C. (1994). Restricted movement in resident stream salmonids – a paradigm lost. *Can. J. Fish. Aquat. Sci.* 51, 2626–2637.

Groot, C., Simpson, K., Todd, I., Murray, P. D. and Buxton, G. A. (1975). Movements of sockeye salmon (*Oncorhynchus nerka*) in the Skeena River estuary as revealed by ultrasonic tagging. *J. Fish. Res. Bd. Can.* 32, 233–242.

Gross, M. R., Coleman, R. M. and McDowall, R. M. (1988). Aquatic productivity and the evolution of diadromous fish migration. *Science* 239, 1291–1293.

Grunwald, C., Stabile, J., Waldman, J. R., Gross, R. and Wirgin, I. (2002). Population genetics of shortnose sturgeon *Acipenser brevirostrum* based on mitochondrial DNA control region sequences. *Mol. Ecol.* 11, 1885–1898.

Hamano, K., Yosida, K., Suzuki, M. and Ashida, K. (1996). Changes of thyrotropin-releasing hormone concentration in the brain and levels of prolactin and thyroxin in the serum during spawning migration of the chum salmon *Oncorhynchus keta. Gen. Comp. Endocrinol.* 101, 275–281.

Hansen, L. P., Jonsson, N. and Jonsson, B. (1993). Oceanic migration in homing Atlantic salmon. *Anim. Behav.* 45, 927–941.

Hanson, K. C., Cooke, S. J., Hinch, S. G., Crossin, G. T., Patterson, D. A., English, K. K., Donaldson, M. R., Shrimpton, J. M., Van Der Kraak, G. and Farrell, A. P. (2008). Individual variation in migration speed of upriver-migrating sockeye salmon in the Fraser River in relation to their physiological and energetic status at marine approach. *Physiol. Biochem. Zool.* 81, 255–268.

Haro, A. J. and Krueger, W. H. (1988). Pigmentation, size, and migration of elvers (*Anguilla rostrata* (Lesueur)) in a coastal Rhode Island stream. *Can. J. Zool.* 66, 2528–2533.

Harrod, C., Grey, J., McCarthy, T. K. and Morrissey, M. (2005). Stable isotope analyses provide new insights into ecological plasticity in a mixohaline population of European eel. *Oecologia* 144, 673–683.

Hasselman, D. J., Bradford, R. G. and Bentzen, P. (2010). Taking stock: defining populations of American shad (*Alosa sapidissima*) in Canada using neutral genetic markers. *Can. J. Fish. Aquat. Sci.* 67, 1021–1039.

Healey, M. C. (1991). Life history of Chinook salmon (*Oncorhynchus tshawytscha*). In *Pacific Salmon Life Histories* (eds. C. Groot and L. Margolis), pp. 311–394. Vancouver: University of British Columbia Press.

Heath, D. D., Bryden, C. A., Shrimpton, J. M., Iwama, G. K., Kelly, J. and Heath, J. W. (2002). Relationships between heterozygosity, genetic distance (d^2), and reproductive traits in Chinook salmon, *Oncorhynchus tshawytscha. Can. J. Fish. Aquat. Sci.* 59, 77–84.

Heath, D. D., Shrimpton, J. M., Hepburn, R. I., Jamieson, S. K., Brode, S. K. and Docker, M. F. (2006). Population structure and divergence using microsatellite and gene locus markers in Chinook salmon (*Oncorhynchus tshawytscha*) populations. *Can. J. Fish. Aquat. Sci.* 63, 1370–1383.

Hendricks, M. L., Hoopes, R. L., Arnold, D. A. and Kaufman, M. L. (2002). Homing of hatchery-reared American shad to the Lehigh River, a tributary to the Delaware River. *N. Am. J. Fish. Manage.* 22, 243–248.

Hinch, S. G., Cooke, S., Healey, M. C. and Farrell, A. P. (2006). Behavioural physiology of fish migrations: salmon as a model approach. In *Fish Physiology,* Vol. 24, *Behaviour and Physiology of Fish* (eds. K. Sloman, S. Balshine and R. Wilson), pp. 239–295. San Diego: Academic Press.

Hoareau, T. B., Lecomte-Finiger, R., Grondin, H.-P., Conand, C. and Berrebi, P. (2007). Oceanic larval life of La Réunion "bichiques", amphidromous gobiid post-larvae. *Mar. Ecol. Prog. Ser.* 333, 303–308.

Hodgson, S. and Quinn, T. P. (2002). The timing of adult sockeye salmon migration into fresh water: adaptations by populations to prevailing thermal regimes. *Can. J. Zool.* 80, 542–555.

Hodgson, S., Quinn, T. P., Hilborn, R., Francis, R. C. and Rogers, D. E. (2006). Marine and freshwater climatic factors affecting interannual variation in the timing of return migration to fresh water of sockeye salmon (*Oncorhynchus nerka*). *Fish. Oceanogr.* 15, 1–24.

Humphries, P. (1989). Variation in the life history of diadromous and landlocked populations of the spotted galaxias, *Galaxias truttaceus* Valenciennes, in Tasmania. *Aust. J. Mar. Freshw. Res.* 40, 501–518.

Ibbotson, A., Smith, J., Scarlett, P. and Aprhamian, M. (2002). Colonisation of freshwater habitats by the European eel *Anguilla anguilla*. *Freshwater Biol.* 47, 1696–1706.

Iida, M., Watanabe, S., Shinoda, A. and Tsukamoto, K. (2008). Recruitment of the amphidromous goby *Sicyopterus japonicus* to the estuary of the Ota River, Wakayama, Japan. *Environ. Biol. Fish.* 83, 331–341.

Iida, M., Watanabe, S., Yamada, Y., Lord, C., Keith, P. and Tsukamoto, K. (2010). Survival and behavioral characteristics of amphidromous goby larvae of *Sicyopterus japonicus* (Tanaka, 1909) during their downstream migration. *J. Exp. Mar. Biol. Ecol.* 383, 17–22.

Imbert, H., Arrowsmith, R., Dufour, S. and Elie, P. (2008). Relationships between locomotor behavior, morphometric characters and thyroid hormone levels give evidence of stage-dependent mechanisms in European eel upstream migration. *Horm. Behav.* 53, 69–81.

Imbert, H., Labonne, J., Rigaud, C. and Lambert, P. (2010). Resident and migratory tactics in freshwater European eels are size-dependent. *Freshw. Biol.* 55, 1483–1493.

Ishikawa, S., Aoyama, J., Tsukamoto, K. and Nishida, M. (2001). Population structure of the Japanese eel *Anguilla japonica* as examined by mitochondrial DNA sequencing. *Fish. Sci.* 67, 246–253.

Israel, J. A., Cordes, J. F., Blumberg, M. A. and May, B. (2004). Geographic patterns of genetic differentiation among collections of green sturgeon. *N. Am. J. Fish. Manage.* 24, 922–931.

Iwata, M. (1995). Downstream migratory behavior of salmonids and its relationship with cortisol and thyroid hormones: a review. *Aquaculture* 135, 131–139.

Jegstrup, I. M. and Rosenkilde, P. (2003). Regulation of post-larval development in the European eel: thyroid hormone level, progress of pigmentation and change in behaviour. *J. Fish Biol.* 63, 168–175.

Jellyman, D. J., Glova, G. J. and Sykes, J. R. E. (2002). Movements and habitats of adult lamprey (*Geotria australis*) in two New Zealand waterways. *N. Z. J. Mar. Freshw. Res.* 36, 53–65.

Jellyman, D. J., Booker, D. J. and Watene, E. (2009). Recruitment of *Anguilla* spp. glass eels in the Waikato River, New Zealand. Evidence of declining migrations?. *J. Fish Biol.* 74, 2014–2033.

Jessop, B. M., Cairns, D. K., Thibault, I. and Tzeng, W. N. (2008). Life history of American eel *Anguilla rostrata*: new insights from otolith microchemistry. *Aquat. Biol.* 1, 205–216.

Johannson, V., Winberg, S., Jönsson, E., Hall, D. and Björnsson, B. Th. (2004). Peripherally administered growth hormone increases brain dopaminergic activity and swimming in rainbow trout. *Horm. Behav.* 46, 436–443.

Jonsson, B., Jonsson, N. and Hansen, L. P. (2003). Atlantic salmon straying from the River Imsa. *J. Fish Biol.* 62, 641–657.

Jonsson, B., Jonsson, N. and Hansen, L. P. (2007). Factors affecting river entry of adult Atlantic salmon in a small river. *J. Fish Biol.* 71, 943–956.

Juanes, F., Gephard, S. and Beland, K. F. (2004). Long-term changes in migration timing of adult Atlantic salmon (*Salmo salar*) at the southern edge of the species distribution. *Can. J. Fish. Aquat. Sci.* 61, 2392–2400.

Kawauchi, H. and Sower, S. A. (2006). The dawn and evolution of hormones in the adenohypophysis. *Gen. Comp. Endocrinol.* 148, 3–14.

Kawauchi, H., Suzuki, K., Yamazaki, T., Moriyama, S., Nozaki, M., Yamaduchi, K., Takahashi, A., Youson, J. and Sower, S. A. (2002). Identification of growth hormone in the sea lamprey, an extant representative of a group of the most ancient vertebrates. *Endocrinology* 143, 4916–4921.

Keefer, M. L., Peery, C. A., Jepson, M. A., Tolotti, K. R., Bjornn, T. C. and Stuehrenberg, L. C. (2004). Stock specific migration timing of adult spring–summer Chinook salmon in the Columbia River basin. *N. Am. J. Fish. Manage.* 24, 1145–1162.

Keefer, M. L., Caudill, C. C., Peery, C. A. and Lee, S. R. (2008a). Transporting juvenile salmonids around dams impairs adult migration. *Ecol. Appl.* 18, 1888–1900.

Keefer, M. L., Caudill, C. C., Peery, C. A. and Boggs, C. T. (2008b). Non-direct homing behaviours by adult Chinook salmon in a large, multi-stock river system. *J. Fish Biol.* 72, 27–44.

Keefer, M. L., Moser, M. L., Boggs, C. T., Daigle, W. R. and Peery, C. A. (2009). Variability in migration timing of adult Pacific lamprey (*Lampetra tridentata*) in the Columbia River, USA. *Environ. Biol. Fish.* 85, 253–264.

Keith, P. (2003). Biology and ecology of amphidromous Gobiidae of the Indo-Pacific and the Caribbean regions. *J. Fish Biol.* 63, 831–847.

Keith, P., Galewski, T., Cattaneo-Berrebi, G., Hoareau, T. and Berrebi, P. (2005). Ubiquity of *Sicyopterus lagocephalus* (Teleostei: Gobioidei) and phylogeography of the genus *Sicyopterus* in the Indo-Pacific area inferred from mitochondrial cytochrome b gene. *Mol. Phylogen. Evol.* 37, 721–732.

Keith, P., Hoareau, T. B., Lord, C., Ah-Yane, O., Gimonneau, G., Robinet, T. and Valade, P. (2008). Characterisation of post-larval to juvenile stages, metamorphosis and recruitment of an amphidromous goby, *Sicyopterus lagocephalus* (Pallas) (Teleostei: Gobiidae: Sicydiinae). *Mar. Freshw. Res.* 59, 876–889.

Kettle, A. J. and Haines, K. (2005). How does the European eel (*Anguilla anguilla*) retain its population structure during its larval migration across the North Atlantic Ocean? *Can. J. Fish. Aquat. Sci.* 63, 90–106.

King, T. L., Lubinski, B. A. and Spidle, A. P. (2001). Microsatellite DNA variation in Atlantic sturgeon (*Acipenser oxyrinchus oxyrinchus*) and cross-species amplification in the Acipenseridae. *Cons. Gen.* 2, 103–119.

Kitahashi, T., Sato, A., Alok, D., Kaeriyama, M., Zohar, Y., Yamauchi, K., Urano, A. and Ueda, H. (1998). Gonadotropin-releasing hormone analog and sex steroids shorten homing duration of sockeye salmon in Lake Shikotsu. *Zool. Sci.* 15, 767–771.

Klemetsen, A., Amundsen, P.-A., Dempson, J. B., Jonsson, B., Jonsson, N., O'Connell, M. F. and Mortensen, E. (2003). Atlantic salmon *Salmo salar* L., brown trout *Salmo trutta* L. and Arctic charr *Salvelinus alpinus* (L.): a review of aspects of their life histories. *Ecol. Freshw. Fish* 12, 1–59.

Kotake, A., Arai, T., Ohji, M., Yamane, S., Miyazaki, N. and Tsukamoto, K. (2004). Application of otolith microchemistry to estimate the migratory history of Japanese eel *Anguilla japonica* on the Sanriku Coast of Japan. *J. Appl. Ichthyol.* 20, 150–153.

Kudo, H., Tsuneyoshi, Y., Nagae, M., Adachi, S., Yamauchi, K., Ueda, H. and Kawamura, H. (1994). Detection of thyroid hormone receptors in the olfactory system and brain of wild masu salmon, *Oncorhynchus masou* (Brevoort), during smolting by *in vitro* autoradiography. *Aquacult. Fish. Manage.* 25 (Suppl. 2), 171–182.

Kuroki, M., Kawai, M., Jónsson, B., Aoyama, J., Miller, M. J., Noakes, D. L. and Tsukamoto, K. (2008). Inshore migration and otolith microstructure/microchemistry of anguillid glass eels recruited to Iceland. *Environ. Biol. Fish.* 83, 309–325.

Laffaille, P., Caraguel, J.-M. and Legault, A. (2007). Temporal patterns in the upstream migration of European glass eels (*Anguilla anguilla*) at the Couesnon estuarine dam. *Estuar. Coast. Shelf Sci.* 73, 81–90.

Lamson, H. M., Shiao, J.-C., Iizuka, Y., Tzeng, W.-N. and Cairns, D. K. (2006). Movement patterns of American eels (*Anguilla rostrata*) between salt- and freshwater in a coastal watershed, based on otolith microchemistry. *Mar. Biol.* 149, 1567–1576.

Larsen, L. O. (1980). Physiology of adult lampreys, with special regard to natural starvation, reproduction, and death after spawning. *Can. J. Fish. Aquat. Sci.* 37, 1762–1779.

Laurent, P. and Perry, S. F. (1990). Effects of cortisol on gill chloride cell morphology and ionic uptake in the freshwater trout, *Salmo gairdneri*. *Cell Tissue Res.* 259, 429–442.

Lecomte-Finiger, R. (1992). Growth history and age at recruitment of European glass eels (*Anguilla anguilla*) as revealed by otolith microstructure. *Mar. Biol.* 114, 205–210.

Leloup-Hatey, J. (1974). Influence de l'adaptation a l'eau de mer sur la function interrenalienne de l'Anguille (*Anguilla anguilla* L.). *Gen. Comp. Endocrinol.* 24, 28–37.

Lema, S. C. and Nevitt, G. A. (2004). Evidence that thyroid hormone induces olfactory cellular proliferation in salmon during a sensitive period for imprinting. *J. Exp. Biol.* 207, 3317–3327.

Levy, D. A. and Cadenhead, A. D. (1995). Selective tidal stream transport of adult sockeye salmon (*Oncorhynchus nerka*) in the Fraser River estuary. *Can. J. Fish. Aquat. Sci.* 52, 1–12.

Lilja, J. and Romakkaniemi, A. (2003). Early-season river entry of adult Atlantic salmon: its dependency on environmental factors. *J. Fish Biol.* 62, 41–50.

Lin, B., Zhang, Z., Wang, Y., Currens, K. P., Spidle, A., Yamazaki, Y. and Close, D. A. (2008). Amplified fragment length polymorphism assessment of genetic diversity in Pacific lampreys. *N. Am. J. Fish. Manage.* 28, 1182–1193.

Lord, C., Brun, C., Hautecoeur, M. and Keith, P. (2010). Insights on endemism: comparison of the duration of the marine larval phase estimated by otolith microstructural analysis of three amphidromous *Sicyopterus* species (Gobioidei: Sicydiinae) from Vanuatu and New Caledonia. *Ecol. Freshw. Fish* 19, 26–38.

Lord, C., Tabouret, H., Claverie, F., Pécheyran, C. and Keith, P. (2011). Femtosecond laser ablation ICP-MS measurement of otolith Sr:Ca and Ba:Ca composition reveal differential use of freshwater habitats for three amphidromous *Sicyopterus* (Teleostei: Gobioidei: Sicydiinae) species. *J. Fish Biol.* 79, 1304–1321.

Lundqvist, H., Borg, B. and Berglund, I. (1989). Androgens impair seawater adaptability in smolting Baltic salmon (*Salmo salar*). *Can. J. Zool.* 67, 1733–1736.

Madsen, S. S. and Korsgaard, B. (1991). Opposite effects of 17β-estradiol and combined growth hormone–cortisol treatment on hypo-osmoregulatory performance in sea trout presmolts, *Salmo trutta*. *Gen. Comp. Endocrinol.* 83, 276–282.

Manzon, L. (2002). The role of prolactin in fish osmoregulation: a review. *Gen. Comp. Endocrinol.* 125, 291–310.

Marui, M., Arai, T., Miller, M. J., Jellyman, D. J. and Tsukamoto, K. (2001). Comparison of early life history between New Zealand temperate eels and Pacific tropical eels revealed by otolith microstructure and microchemistry. *Mar. Ecol. Prog. Ser.* 213, 273–284.

McCairns, R. J. S. and Bernatchez, L. (2008). Landscape genetic analyses reveal cryptic population structure and putative selection gradients in a large-scale estuarine environment. *Mol. Ecol.* 17, 3901–3916.

McCormick, S. D. (1994). Ontogeny and evolution of salinity tolerance in anadromous salmonids: hormone and heterochrony. *Estuaries* 17, 26–33.

McCormick, S. D. (2001). Endocrine control of osmoregulation in teleost fish. *Am. Zool.* 41, 781–794.

McCormick, S. D. (2013). Smolt physiology and endocrinology. In *Fish Physiology*, Vol. 32, *Euryhaline Fishes* (eds. S. D. McCormick, A. P. Farrell and C. J. Brauner), pp. 199–251. New York: Elsevier.

McCormick, S. D. and Saunders, R. L. (1987). Preparatory physiological adaptations for marine life of salmonids: osmoregulation, growth, and metabolism. *Am. Fish. Soc. Symp.* 1, 211–229.

McCormick, S. D., Sundell, K., Björnsson, B. Th., Brown, C. L. and Hiroi, J. (2003). Influence of salinity on the localization of Na^+/K^+-ATPase, $Na^+/K^+/2Cl^-$ cotransporter (NKCC) and CFTR anion channel in chloride cells of the Hawaiian goby (*Stenogobius hawaiiensis*). *J. Exp. Biol.* 206, 4575–4583.

McCormick, S. D., Regish, A., O'Dea, M. F. and Shrimpton, J. M. (2008). Are we missing a mineralocorticoid in teleost fish? Effects of cortisol, deoxycorticosterone and aldosterone on osmoregulation, gill Na^+,K^+-ATPase activity and isoform mRNA levels in Atlantic salmon. *Gen. Comp. Endocrinol.* 157, 35–40.

McCormick, S. D., Regish, A. M. and Christensen, A. K. (2009). Distinct freshwater and seawater isoforms of Na^+/K^+-ATPase in gill chloride cells of Atlantic salmon. *J. Exp. Biol.* 212, 3994–4001.

McDowall, R. M. (1987). The occurrence and distribution of diadromy among fishes. *Am. Fish. Soc. Symp.* 1, 1–13.

McDowall, R. M. (2001). Diadromy, diversity and divergence: implications for speciation processes in fish. *Fish Fish.* 2, 278–285.

McDowall, R. M. (2003). Hawaiian biogeography and the islands' freshwater fish fauna. *J. Biogeogr.* 30, 703–710.

McDowall, R. M. (2007). On amphidromy, a distinct form of diadromy in aquatic organisms. *Fish Fish.* 8, 1–13.

McRae, C. J., Warren, K. D. and Shrimpton, J. M. (2012). Spawning site selection in Interior Fraser River coho salmon (*Oncorhynchus kisutch*): an imperiled population of anadromous salmon from an interior, snow-dominated watershed. *Endang. Species Res.* 16, 249–260.

Mesa, M. G., Bayer, J. M., Bryan, M. B. and Sower, S. A. (2010). Annual sex steroid and other physiological profiles of Pacific lampreys (*Entosphenus tridentatus*). *Comp. Biochem. Physiol. A Mol. Integr. Physiol.* 155, 56–63.

Miles, S. G. (1968). Rheotaxis of elvers of the American eel in the laboratory from different streams in Nova Scotia. *J. Fish. Res. Bd Can.* 25, 1591–1602.

Miller, K. M., Schulze, A. D., Ginther, N., Li, S., Patterson, D. A., Farrell, A. P. and Hinch, S. G. (2009). Salmon spawning migration: metabolic shifts and environmental triggers. *Comp. Biochem. Physiol. D Genom. Proteom.* 4, 75–89.

Minegishi, Y., Aoyama, J. and Tsukamoto, K. (2008). Multiple population structure of the giant mottled eel *Anguilla marmorata*. *Mol. Ecol.* 17, 3109–3122.

Morita, K., Arai, T., Kishi, D. and Tsuboi, J. (2005). Small anadromous *Salvelinus malma* at the southern limits of its distribution. *J. Fish Biol.* 66, 1187–1192.

Naismith, I. A. and Knights, B. (1988). Migrations of elvers and juvenile European eels, *Anguilla anguilla* L., in the River Thames. *J. Fish Biol.* 33 (Suppl. A), 161–175.

Nehlsen, W., Williams, J. E. and Lichatowich, J. A. (1991). Pacific salmon at the crossroads: stocks at risk from California, Oregon, Idaho, and Washington. *Fisheries* 16 (2), 4–21.

Nelson, J. S. (1994). *Fishes of the World*. New York: John Wiley & Sons.

Nevitt, G. A., Dittman, A. H., Quinn, T. P. and Moody, W. J., Jr. (1994). Evidence for a peripheral olfactory memory in imprinted salmon. *Proc. Natl Acad. Sci. U. S. A.* 91, 4288–4292.

Nozaki, M., Ominato, K., Shimotani, T., Kawauchi, H., Youson, J. H. and Sower, S. A. (2008). Identity and distribution of immunoreactive adenohypophysial cells in the pituitary during the life cycle of sea lampreys *Petromyzon marinus*. *Gen. Comp. Endocrinol.* 155, 403–412.

Ojima, D. and Iwata, M. (2007). The relationship between thyroxine surge and onset of downstream migration in chum salmon *Oncorhynchus keta* fry. *Aquaculture* 273, 185–193.

Onuma, T., Higashi, Y., Ando, H., Ban, M., Ueda, H. and Urano, A. (2003a). Year-to-year differences in plasma levels of steroid hormones in pre-spawning chum salmon. *Gen. Comp. Endocrinol.* 133, 199–215.

Onuma, T., Kitahashi, T., Taniyama, S., Saito, D., Ando, H. and Urano, A. (2003b). Changes in expression of genes encoding gonadotropin subunits and growth hormone/prolactin/somatolactin family hormones during final migration and freshwater adaptation in prespawning chum salmon. *Endocrine* 20, 23–33.

Onuma, T. A., Sato, S., Katsumata, H., Makino, K., Hu, W. W., Jodo, A., Davis, N. D., Dickey, J. T., Ban, M., Ando, H., Fukuwaka, M.-a., Azumaya, T., Swanson, P. and Urano,

A. (2009). Activity of the pituitary–gonadal axis is increased prior to the onset of spawning migration of chum salmon. *J. Exp. Biol.* 212, 56–70.

Onuma, T. A., Ban, M., Makino, K., Katsumata, H., Hu, W. W., Ando, H., Fukuwaka, M., Azumaya, T. and Urano, A. (2010a). Changes in gene expression for GH/PRL/SL family hormones in the pituitaries of homing chum salmon during ocean migration through upstream migration. *Gen. Comp. Endocrinol.* 166, 537–548.

Onuma, T. A., Makino, K., Katsumata, H., Beckman, B. R., Ban, M., Ando, H., Fukuwaka, M.-a., Swanson, P. and Urano, A. (2010b). Changes in the plasma levels of insulin-like growth factor-I from the onset of spawning migration through upstream migration in chum salmon. *Gen. Comp. Endocrinol.* 165, 237–243.

Overton, A. S. and Rulifson, R. A. (2009). Annual variability in upstream migration of glass eels in a southern USA coastal watershed. *Environ. Biol. Fish.* 84, 29–37.

Ozaki, Y., Okumura, H., Kazeto, Y., Ikeuchi, T., Ijiri, S., Nagae, M., Acachi, S. and Yamauchi, K. (2000). Developmental changes in pituitary–thyroid axis, and formation of gonads in leptocephali and glass eels of the *Anguilla* spp. *Fish. Sci.* 66, 1115–1122.

Ozaki, Y., Fukada, H., Tanaka, H., Kagawa, H., Ohta, H., Adachi, S., Hara, A. and Yamauchi, K. (2006). Expression of growth hormone family and growth hormone receptor during early development in the Japanese eel (*Anguilla japonica*). *Comp. Biochem. Physiol. B Biochem. Mol. Biol.* 145, 27–34.

Patterson, D. A., Macdonald, J. S., Skibo, K. M., Barnes, D. P., Guthrie, I. and Hills, J. (2007). Reconstructing the summer thermal history for the Lower Fraser River, 1941 to 2006, and implications for adult sockeye salmon (*Oncorhynchus nerka*) spawning migration. *Can. Tech. Report Fish. Aquat. Sci.* 2724, 1–43.

Perrier, C., Evanno, G., Belliard, J., Guyomard, R. and Balinière, J.-L. (2010). Natural recolonization of the Seine River by Atlantic salmon (*Salmo salar*) of multiple origins. *Can. J. Fish. Aquat. Sci.* 67, 1–4.

Perry, S. F., Goss, G. G. and Laurent, P. (1992). The interrelationships between gill chloride cell morphology and ionic uptake in four freshwater teleosts. *Can. J. Zool.* 90, 1775–1786.

Pickering, A. D. (1976). Stimulation of intestinal degeneration by oestradiol and testosterone implantation in the migrating river lamprey, *Lampetra fluviatilis* L. *Gen. Comp. Endocrinol.* 30, 340–346.

Plate, E. M., Wood, C. C. and Hawryshyn, C. W. (2003). GnRH affects activity and jumping frequency in adult sockeye salmon, *Oncorhynchus nerka*. *Fish Phys. Biochem.* 28, 245–248.

Pottinger, T. G., Carrick, T. R., Hughes, S. E. and Balm, P. H. M. (1996). Testosterone, 11-ketotestosterone, and estradiol-17 modify baseline and stress-induced interrenal and corticotropic activity in trout. *Gen. Comp. Endocrinol.* 104, 284–295.

Quinn, T. P. and Adams, D. J. (1996). Environmental changes affecting the migratory timing of American shad and sockeye salmon. *Ecology* 77, 1151–1162.

Quinn, T. P. and Myers, K. W. (2004). Anadromy and the marine migrations of Pacific salmon and trout: Rounsefell revisited. *Rev. Fish Biol. Fish.* 14, 421–442.

Quinn, T. P., Volk, E. C. and Hendry, A. P. (1999). Natural otolith microstructure patterns reveal precise homing to natal incubation sites by sockeye salmon (*Oncorhynchus nerka*). *Can. J. Zool.* 77, 766–775.

Quinn, T. P., Kinnison, M. T. and Unwin, M. J. (2001). Evolution of Chinook salmon (*Oncorhynchus tshawytscha*) populations in New Zealand: pattern, rate, and process. *Genetica* 112, 493–513.

Radtke, R. L. and Kinzie, R. A., III (1996). Evidence of a marine larval stage in endemic Hawaiian stream gobies from isolated high-elevation locations. *Trans. Am. Fish. Soc.* 125, 613–621.

Radtke, R. L., Kinzie, R. A., III and Folsom, S. D. (1988). Age at recruitment of Hawaiian freshwater gobies. *Environ. Biol. Fish.* 23, 205–213.

Radtke, R. L., Kinzie, R. A., III and Shafer, D. J. (2001). Temporal and spatial variation in length of larval life and size at settlement of the Hawaiian amphidromous goby *Lentipes concolor. J. Fish Biol.* 59, 928–938.

Reis-Santos, P., McCormick, S. D. and Wilson, J. M. (2008). Ionoregulatory changes during metamorphosis and salinity exposure of juvenile sea lamprey (*Petromyzon marinus* L.). *J. Exp. Biol.* 211, 978–988.

Richards, J. G., Semple, J. W., Bystriansky, J. S. and Schulte, P. M. (2003). Na^+/K^+-ATPase α-isoform switching in gills of rainbow trout (*Oncorhynchus mykiss*) during salinity transfer. *J. Exp. Biol.* 206, 4475–4486.

Robinson, T. C. and Bayer, J. M. (2005). Upstream migration of the Pacific lampreys in the John Day River, Oregon: behavior, timing, and habitat use. *Northwest. Sci.* 79, 106–119.

Saito, D., Ota, Y., Hiraoka, S., Hyodo, S., Ando, H. and Urano, A. (2001). Effect of oceanographic environments on sexual maturation, salinity tolerance, and vasotocin gene expression in homing chum salmon. *Zool. Sci.* 18, 389–396.

Sakamoto, T., Iwata, M. and Hirano, T. (1991). Kinetic studies of growth hormone and prolactin during adaptation of coho salmon, *Oncorhynchus kisutch*, to different salinities. *Gen. Comp. Endocrinol.* 82, 184–191.

Salo, E. O. (1991). Life history of chum salmon (*Oncorhynchus keta*). In *Pacific Salmon Life Histories* (eds. C. Groot and L. Margolis), pp. 231–309. Vancouver: University of British Columbia Press.

Sang, T.-K., Chang, H.-Y., Chen, C.-T. and Hui, C.-F. (1994). Population structure of the Japanese eel, *Anguilla japonica. Mar. Biol. Evol.* 11, 250–260.

Sasai, S., Kaneko, T., Hasegawa, S. and Tsukamoto, K. (1998). Morphological alteration in two types of gill chloride cells in Japanese eels (*Anguilla japonica*) during catadromous migration. *Can. J. Zool.* 76, 1480–1487.

Sasai, S., Katoh, F., Kaneko, T. and Tsukamoto, K. (2007). Ontogenetic change of gill chloride cells in leptocephalus and glass eel stages of the Japanese eel, *Anguilla japonica. Mar. Biol.* 150, 487–496.

Sathiyaa, R. and Vijayan, M. M. (2003). Autoregulation of glucocorticoid receptor by cortisol in rainbow trout hepatocytes. *Am. J. Physiol. Cell Physiol.* 284, C1508–C1515.

Schmidt, D. J., Crook, D. A., O'Connor, J. P. and Hughes, J. M. (2011). Genetic analysis of threatened Australian grayling *Prototroctes maraena* suggests recruitment to coastal rivers from an unstructured marine larval source population. *J. Fish Biol.* 78, 98–111.

Schoenfuss, H. L. and Blob, R. W. (2004). Kinematics of waterfall climbing in Hawaiian freshwater fishes (Goblidae): vertical propulsion at the aquatic–terrestrial interface. *J. Zool.* 261, 191–205.

Scholz, A. T., Horrall, R. M., Cooper, J. C. and Hasler, A. D. (1976). Imprinting to chemical cues: the basis for home stream selection in salmon. *Science* 192, 1247–1249.

Schultz, E. T. and McCormick, S. D. (2013). Euryhalinity in an evolutionary context. In *Fish Physiology*, Vol. 32, *Euryhaline Fishes* (eds. S. D. McCormick, A. P. Farrell and C. J. Brauner), pp. 69–124. New York: Elsevier,

Shen, K.-N. and Tzeng, W.-N. (2002). Formation of a metamorphosis check in otoliths of the amphidromous goby *Sicyopterus japonicas. Mar. Ecol. Prog. Ser.* 228, 205–211.

Shinoda, A., Aoyama, J., Miller, M. J., Otake, T., Mochioka, N., Watanabe, S., Minegishi, Y., Kuroki, M., Yoshinaga, T., Yokouchi, K., Fukuda, N., Sudo, R., Hagihara, S., Zenimoto, K., Suzuki, Y., Oya, M., Inagaki, T., Kimura, S., Fukui, A., Lee, T. W. and Tsukamoto, K. (2011). Evaluation of the larval distribution and migration of the Japanese eel in the western North Pacific. *Rev. Fish Biol. Fish.* 21, 591–611.

Shrimpton, J. M. and Heath, D. D. (2003). Temporal changes in genetic diversity and effective population size in Chinook salmon (*Oncorhynchus tshawytscha*) populations: large versus small scale perturbation effects. *Mol. Ecol.* 12, 2571–2583.

Shrimpton, J. M. and McCormick, S. D. (1999). Responsiveness of gill Na^+/K^+-ATPase to cortisol is related to gill corticosteroid receptor concentration in juvenile rainbow trout. *J. Exp. Biol.* 202, 987–995.

Shrimpton, J. M. and McCormick, S. D. (2002). Seasonal changes in androgen levels in Atlantic salmon parr and their relationship to smolting. *J. Fish Biol.* 61, 1294–1304.

Shrimpton, J. M., Bernier, N. J., Iwama, G. K. and Randall, D. J. (1994). Differences in measurements of smolt development between wild and hatchery-reared juvenile coho salmon (*Oncorhynchus kisutch*) before and after saltwater exposure. *Can. J. Fish. Aquat. Sci.* 51, 2170–2178.

Shrimpton, J. M., Björnsson, B. Th. and McCormick, S. D. (2000). Can Atlantic salmon smolt twice? Endocrine and biochemical changes during smolting. *Can. J. Fish. Aquat. Sci.* 57, 1969–1976.

Shrimpton, J. M., Patterson, D. A., Richards, J. G., Cooke, S. J., Schulte, P. M., Hinch, S. G. and Farrell, A. P. (2005). Ionoregulatory changes in different populations of maturing sockeye salmon *Oncorhynchus nerka* during ocean and river migration. *J. Exp. Biol.* 208, 4069–4078.

Smith, B. T., Nelson, R. J., Pollard, S., Rubidge, E., McKay, S. J., Rodzen, J., May, B. and Koop, B. (2002). Population genetic analysis of white sturgeon (*Acipenser transmontanus*) in the Fraser River. *J. Appl. Ichthyol.* 18, 307–312.

Smith, G. W., Smith., I. P. and Armstrong, S. M. (1994). The relationship between river flow and entry to the Aberdeenshire Dee by returning adult Atlantic salmon. *J. Fish Biol.* 45, 953–960.

Smith, J. F. and Smith, M. J. (1998). Rapid acquisition of directional preferences by migratory juveniles of two amphidromous Hawaiian gobies, *Awaous guamensis* and *Sicyopterus stimpsoni*. *Environ. Biol. Fish.* 53, 275–282.

Sola, C. (1995). Chemoattraction of upstream migrating glass eels *Anguilla anguilla* to earthy and green odorants. *Environ. Biol. Fish.* 43, 179–185.

Sola, C. and Tongiorgi, P. (1996). The effect of salinity on the chemotaxis of glass eels, *Anguilla anguilla* to organic earthy and green odorants. *Environ. Biol. Fish.* 47, 213–218.

Solomon, D. J. (1973). Evidence for pheromone-influenced homing by migrating Atlantic salmon, *Salmo salar* (L.). *Nature* 244, 231–232.

Sorensen, P. W. (1986). Origins of the freshwater attractant(s) of migration elvers of the American eel *Anguilla rostrata*. *Environ. Biol. Fish.* 17, 185–200.

Sorensen, P. W. and Hoye, T. R. (2007). A critical review of the discovery and application of a migratory pheromone in an invasive fish, the sea lamprey *Petromyzon marinus* L. *J. Fish Biol.* 71D, 100–114.

Spice, E. K., Goodman, D. H., Reid, S. B. and Docker, M. F. (2012). Neither philopatric nor panmictic: microsatellite and mtDNA evidence suggests lack of natal homing but limits to dispersal in Pacific lamprey. *Mol. Ecol.* 21, 2916–2930.

Sugeha, H. Y., Arai, T., Miller, M. J., Limbong, D. and Tsukamoto, K. (2001). Inshore migration of the tropical eels *Anguilla* spp. recruiting to the Poigar River estuary on north Sulawesi Island. *Mar. Ecol. Prog. Ser.* 221, 233–243.

Sullivan, M. C., Wuenschel, M. J. and Able, K. W. (2009). Inter and intra-estuary variability in ingress, condition and settlement of the American eel *Anguilla rostrata*: implications for estimating and understanding recruitment. *J. Fish Biol.* 74, 1949–1969.

Sykes, G. E., Johnson, C. J. and Shrimpton, J. M. (2009). Temperature and flow effects on migration timing of Chinook salmon smolts. *Trans. Am. Fish. Soc.* 138, 1252–1265.

Taillebois, L., Keith, P., Valade, P., Torres, P., Baloche, S., Dufour, S. and Rousseau, K. (2011). Involvement of thyroid hormones in the control of larval metamorphosis in *Sicyopterus lagocephalus* (Teleostei: Gobioidei) at the time of river recruitment. *Gen. Comp. Endocrinol.* 173, 281–288.

Takei, Y. and McCormick, S. D. (2013). Hormonal control of fish euryhalinity. In *Fish Physiology*, Vol. 32, *Euryhaline Fishes* (eds. S. D. McCormick, A. P. Farrell and C. J. Brauner), pp. 69–123. New York: Elsevier.

Tanaka, H., Takagi, Y. and Naito, Y. (2000). Behavioural thermoregulation of chum salmon during homing migration in coastal waters. *J. Exp. Biol.* 203, 1823–1833.

Tanaka, H., Takagi, Y. and Naito, Y. (2001). Swimming speeds and buoyancy compensation of migrating adult chum salmon *Oncorhynchus keta* revealed by speed/depth/acceleration data logger. *J. Exp. Biol.* 204, 3895–3904.

Tanaka, H., Kagawa, H., Ohta, H., Unuma, T. and Nomura, K. (2003). The first production of glass eel in captivity: fish reproductive physiology facilitates great progress in aquaculture. *Fish. Physiol. Biochem.* 28, 493–497.

Tanaka, H., Naito, Y., Davis, N. D., Urawa, S., Ueda, H. and Fukuwaka, M. (2005). First record of the at-sea swimming speed of a Pacific salmon during its oceanic migration. *Mar. Ecol. Prog. Ser.* 291, 307–312.

Taniyama, S., Kitahashi, T., Ando, H., Ban, M., Ueda, H. and Urano, A. (1999). Changes in the levels of mRNAs for GH/prolactin/somatolactin family and Pit-1/GHF-1 in the pituitaries of pre-spawning chum salmon. *J. Mol. Endocrinol.* 23, 189–198.

Taylor, E. B., Foote, C. J. and Wood, C. C. (1996). Molecular genetic evidence for parallel life-history evolution within a Pacific salmon (sockeye salmon and kokanee, *Oncorhynchus nerka*). *Evolution* 50, 401–416.

Tesch, F.-W. (2003). *The Eel.* London: Blackwell Science.

Theusen, P. A., Ebner, B. C., Larson, H., Keith, P., Silcock, R. M., Prince, J. and Russell, D. J. (2011). Amphidromy links a newly documented fish community of continental Australian streams, to oceanic islands of the West Pacific. *PLOS ONE* 6, e26685.

Thibault, I., Dodson, J. J., Caron, F., Tzeng, W.-N., Iizuka, Y. and Shiao, J.-C. (2007). Facultative catadromy in American eels: testing the conditional strategy hypothesis. *Mar. Ecol. Prog. Ser.* 344, 219–229.

Thibault, I., Bernatchez, L. and Dodson, J. J. (2009). The contribution of newly established populations to the dynamics of range expansion in a one-dimensional fluvial–estuarine system: rainbow trout (*Oncorhynchus mykiss*) in Eastern Quebec. *Divers. Distrib.* 15, 1060–1072.

Thibault, I., Hedger, R. D., Dodson, J. J., Shiao, J.-C., Iizuka, Y. and Tzeng, W.-N. (2010). Anadromy and the dispersal of an invasive fish species (*Oncorhynchus mykiss*) in Eastern Quebec, as revealed by otolith microchemistry. *Ecol. Freshw. Fish* 19, 348–360.

Thomson, R. E. and Hourston, R. A. S. (2011). A matter of timing: the role of ocean conditions in the initiation of spawning migration by late-run Fraser River sockeye salmon (*Oncorhynchus nerka*). *Fish. Oceanogr.* 20, 47–65.

Tipsmark, C. K. and Madsen, S. S. (2009).). Distinct hormonal regulation of Na^+,K^+-ATPase genes in the gill of Atlantic salmon (*Salmo salar* L.). *J. Endocrinol.* 203, 301–310.

Tosi, L. and Sola, C. (1993). Role of geosmin a typical inland water odour, in guiding glass eel *Anguilla anguilla* (L.) migration. *Ethology* 95, 177–185.

Tseng, M.-C., Tzeng, W.-N. and Lee, S.-C. (2006). Population genetic structure of the Japanese eel *Anguilla japonica* in the northwest Pacific Ocean: evidence of non-panmictic populations. *Mar. Ecol. Prog. Ser.* 308, 221–230.

Tsukamoto, K. (1990). Recruitment mechanism of the eel, *Anguilla japonica*, to the Japanese coast. *J. Fish Biol.* 36, 659–671.

Tsukamoto, K., Yamada, Y., Okamura, A., Tanaka, H., Miller, M. J., Kaneko, T., Horie, N., Utoh, T., Mikawa, N. and Tanaka, S. (2009). Positive buoyancy in eel leptocephali: an adaptation for life in the ocean surface layer. *Mar. Biol.* 156, 835–846.

Tsunagawa, T. and Arai, T. (2011). Migratory history of the freshwater goby *Rhinogobius* sp. CB in Japan. *Ecol. Freshw. Fish* 20, 33–41.

Uchida, K., Kaneko, T., Yamaguchi, A., Ogasawara, T. and Hirano, T. (1997). Reduced hypoosmoregulatory ability and alteration in gill chloride cell distribution in mature chum salmon (*Oncorhynchus keta*) migrating upstream for spawning. *Mar. Biol.* 129, 247–253.

Ueda, H. (2011). Physiological mechanism of homing migration in Pacific salmon from behavioral to molecular biological approaches. *Gen. Comp. Endocrinol.* 170, 222–232.

Wagner, C. M., Twohey, M. B. and Fine, J. M. (2009). Conspecific cueing in the sea lamprey: do reproductive migrations consistently follow the most intense larval odour? *Anim. Behav.* 78, 593–599.

Waldman, J., Grunwald, C. and Wirgin, I. (2008). Sea lamprey *Petromyzon marinus*: an exception to the rule of homing in anadromous fishes. *Biol. Lett.* 4, 659–662.

Walter, R. P., Aykanat, T., Kelly, D. W., Shrimpton, J. M. and Heath, D. D. (2009). Gene flow increases temporal stability of Chinook salmon (*Oncorhynchus tshawytscha*) populations in the Upper Fraser River, British Columbia, Canada. *Can. J. Fish. Aquat. Sci.* 66, 167–176.

Walter, R. P., Hogan, J. D., Blum, M. J., Gagne, R. B., Hain, E. F., Gilliam, J. F. and McIntyre, P. B. (2012). Climate change and conservation of endemic amphidromous fishes in Hawaiian streams. *Endang. Species Res.* 16, 261–272.

Walther, B. D., Thorrold, S. R. and Olney, J. E. (2008). Geochemical signatures in otoliths record natal origins of American shad. *Trans. Am. Fish. Soc.* 137, 57–69.

Wang, C. H. and Tzeng, W. N. (2000). The timing of metamorphosis and growth rates of American and European eel leptocephali: a mechanism of larval segregative migration. *Fish. Res.* 46, 191–205.

Waples, R. S., Teel, D. J., Myers, J. M. and Marshall, A. R. (2004). Life-history divergence in Chinook salmon: historic contingency and parallel evolution. *Evolution* 58, 386–403.

Watanabe, S., Miller, M. J., Aoyama, J. and Tsukamoto, K. (2011). Analysis of vertebral counts of the tropical anguillids, *Anguilla megastoma*, *A. obscura*, and *A. reinhardtii*, in the western South Pacific in relation to their possible population structure and phylogeny. *Environ. Biol. Fish.* 91, 353–360.

Welander, A. D. (1940). Notes on the dissemination of shad, *Alosa sapidissima* (Wilson), along the Pacific coast of North America. *Copeia* 1940, 221–223.

Wilson, J. M., Antunes, J. C., Bouça, P. D. and Coimbra, J. (2004). Osmoregulatory plasticity of the glass eel *Anguilla anguilla*: freshwater entry and changes in branchial ion-transport protein expression. *Can. J. Fish. Aquat. Sci.* 61, 432–442.

Wilson, J. M., Reis-Santos, P., Fonseca, A.-V., Antunes, J. C., Bouça, P. D. and Coimbra, J. (2007a). Seasonal changes in ionoregulatory variables of the glass eel *Anguilla anguilla* following estuarine entry: comparison with resident elvers. *J. Fish Biol.* 70, 1239–1253.

Wilson, J. M., Leitão, A., Gonçalves, A. F., Ferreira, C., Reis-Santos, P., Fonseca, A.-V., Moreira da Silva, J., Antunes, J. C., Pereira-Wilson, C. and Coimbra, J. (2007b). Modulation of branchial ion transport protein expression by salinity in glass eels (*Anguilla anguilla* L.). *Mar. Biol.* 151, 1633–1645.

Wirth, T. and Bernatchez, L. (2001). Genetic evidence against panmixia in the European eel. *Nature* 409, 1037–1040.

Wright, G. M. (1984). Immunohistochemical study of growth hormone, prolactin, and thyroid-stimulating hormone in the adenohypophysis of the sea lamprey, *Petromyzon marinus* L., during its upstream migration. *Gen. Comp. Endocrinol.* 55, 269–274.

Wuenshel, M. J. and Able, K. W. (2008). Swimming ability of eels (*Anguilla rostrata, Conger oceanicus*) at estuarine ingress: contrasting patterns of cross-shelf transport? *Mar. Biol.* 154, 775–786.

Yamano, K., Tagawa, M., de Jesus, E. G., Hirano, T., Miwa, S. and Inui, Y. (1991). Changes in whole body concentrations of thyroid hormones and cortisol in metamorphosing conger eel. *J. Comp. Phys. B* 161, 371–375.

Yamano, K., Nomura, K. and Tanaka, H. (2007). Development of thyroid gland and changes in thyroid hormone levels in Leptocephali of Japanese eel (*Anguilla japonica*). *Aquaculture* 270, 499–504.

Yamasaki, N., Maeda, K. and Tachihara, K. (2007). Pelagic larval duration and morphology at recruitment of *Stiphodon percnopterygionus* (Gobiidae: Sicydiinae). *Raffles Bull. Zool.* Suppl. 14, 209–214.

Young, G., Thorarensen, H. and Davie, P. S. (1996). 11-Ketotestosterone suppresses interrenal activity in rainbow trout (*Oncorhynchus mykiss*). *Gen. Comp. Endocrinol.* 103, 301–307.

Youngson, A. F. and Webb, J. H. (1992). The relationship between stream or river discharge and thyroid hormone levels in wild Atlantic salmon (*Salmo salar* L.). *Can. J. Zool.* 70, 140–144.

Yun, S.-S., Wildbill, A. J., Siefkes, M. J., Moser, M. L., Dittman, A. H., Corbett, S. C., Li, W. and Close, D. A. (2011). Identification of putative migratory pheromones from Pacific lamprey (*Lampetra tridentata*). *Can. J. Fish. Aquat. Sci.* 68, 2194–2203.

Zydlewski, J. and Wilkie, M. (2013). Freshwater to seawater transitions in migratory fishes. In *Fish Physiology,* Vol. 32, *Euryhaline Fishes* (eds. S. D. McCormick, A. P. Farrell and C. J. Brauner), pp. 253–326. New York: Elsevier.

8

OSMOREGULATION IN ESTUARINE AND INTERTIDAL FISHES

WILLIAM S. MARSHALL

Intertidal and estuarine fishes stand out among euryhaline fishes because of their physiological plasticity in response to frequent salinity changes and other environmental challenges, including polar ice and tropical heat. Examples are the northern killifish and tropical Nile tilapia. Estuarine fishes combine low water permeability of skin and gill epithelia with efficient NaCl secretion to live in seawater and hypersaline conditions, but have variable abilities to absorb NaCl from dilute environments, with some species requiring dietary salt intake for survival in freshwater. Changing salinity produces temporary shifts in plasma osmolality, and osmosensing ionocytes respond by respectively increasing or shutting off NaCl secretion if shrunken

Euryhaline Fishes: Volume 32
FISH PHYSIOLOGY
Copyright © 2013 Elsevier Inc. All rights reserved
DOI: http://dx.doi.org/10.1016/B978-0-12-396951-4.00008-6

or swollen osmotically. Aerial stranding and semi-terrestrial living stress respiratory, osmoregulatory, and acid–base systems, requiring large coping capabilities. Some estuarine fishes go through a tolerance phase before launching true acclimating mechanisms, so the response to cycling salinities is slight. This delayed response calls for more investigations through genomics and proteomics.

1. INTRODUCTION

The focus of this review is on estuarine and tide pool resident teleost fish species and includes some well-studied model species such as mummichog (*Fundulus heteroclitus*), Nile tilapia (*Oreochromis niloticus*), mudskippers (e.g. *Periopthalmodon modestus*), some euryhaline flounders (e.g. starry flounder, *Platichthys stellatus*), stickleback (e.g. three-spined stickleback, *Gasterosteus aculeatus*), silversides (Atlantic silverside, *Menidia menidia*), sculpins (e.g. the tide pool sculpin, *Oligocottus maculosus*, and coastal prickly sculpin, *Cottus asper*), intertidal blennies (e.g. *Blennius pholis*), and gobies (e.g. longjaw mudsucker, *Gillichthys mirabilis*). These species stand out because of their physiological plasticity in response to salinity changes and other environmental challenges and the wide use of a few of these species as models for osmoregulatory ability. This chapter excludes species that transiently pass through estuaries on migrations (eels, salmonids, bass, herring, alewife) or that use estuaries as a nursery habitat, covered by Chapters 5 and 6 in this volume (Takei and McCormick, 2013; Zydlewski and Wilkie, 2013). The environmental range includes estuaries, the intertidal zone, and tide pools, where there are many frequent and often large variations in salinity. In estuaries, haloclines and salinity gradients can develop, such as from rain and runoff that reduce salinity. In the intertidal zone some fish allow themselves to be stranded, a habit that can produce physiological challenges, including desiccation, gas exchange, and nitrogenous waste excretion. In tide pools, rain can reduce salinity, while evaporation can produce hypersaline conditions. This review centers on the special adaptations that have evolved in these animals to cope with their highly variable environment. Some of these euryhaline estuarine forms (e.g. brackish water mummichog) are essentially marine fish that tolerate freshwater (FW) well (Whitehead et al., 2011b), while others, such as Nile tilapia, tend to be FW-like forms that can, if challenged, develop salt secretory mechanisms (Guner et al., 2005; Inokuchi et al., 2009).

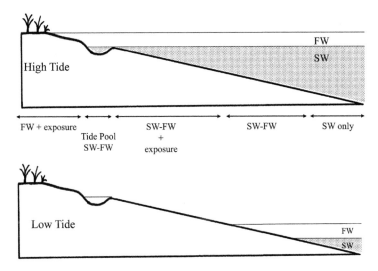

Fig. 8.1. Profile through estuary intertidal zone at high tide (upper panel) and at low tide (lower panel). Hatched underwater area is full-strength seawater (SW) below the halocline. Note that there are zones, indicated along the bottom of the figure, where a locus may be exposed to SW only (subtidal), SW alternating with freshwater (FW), SW – FW plus exposure and FW only, depending on the elevation above tide datum. The tide pool in the estuary may fill with SW on a large tide, or brackish water on a moderate tide, or be inundated by rain and fill with FW at low tide. During large tides and with strong winds, mixing can be higher amplitude and disrupt stratification, producing uniform brackish mixed salinity (not shown).

2. INTERTIDAL HABITATS: ESTUARIES AND TIDE POOLS

2.1. Physical Characteristics

Estuaries are characterized as brackish water (BW) intermediate zones between riverine habitat and the oceans. The size of an estuary depends on the shallowness of the slope and the size of the river(s) feeding the upstream end of the estuary. Thus, the largest estuaries are on the order of a million hectares and small streams may have less than a hectare. In most cases, salinity of the surface water is less than at the bottom and, because of the general roughly triangular shape with the narrow end pointing upstream, and the incursion of dense cold seawater (SW) underneath the brackish upper layers, these are called wedge estuaries (Fig. 8.1) (Acha et al., 2008). The halocline and thermocline can be stable over time and very sharp, such that salinity can rise from FW at the surface to essentially full-strength SW in a depth of 1–2 m. Wedge estuaries, where typically a wedge of cold (in temperate zones), high-density SW occupies the deeper parts of the estuary, are made more stable during neap tides where mixing and shear stresses are smaller, but during spring tides the incursion of the salt wedge progresses higher into the estuary and is more mixed, the result of the higher mixing energy of the larger tidal flux (Simons et al., 2010). In models of moderate-sized estuaries with large tidal flux, the salt wedge at the bottom of the estuary strengthens during neap tides and stratification decays during more energetic spring tides (Wang et al., 2011). Strong wind events also can reduce stratification and produce well-mixed conditions (Acha et al., 2008). In general, stratification of estuaries enhances the intensity of haloclines such that fish moving vertically in an estuary can encounter large changes in salinity over short distances.

Many estuaries have barrier dunes and islands that separate the main body of estuary from the ocean; in these cases the stratification and salt wedges are restricted to the occasional river-like openings to the ocean and the majority of the estuary is a weakly brackish lagoon. In these cases, fish moving into or out of the estuary would experience a change from BW to full-strength SW, sometimes accompanied by temperature shock. Thus, in most estuaries, resident fish species will experience salinity change with tidal flux, when they move through the estuary and on leaving and entering the estuary.

Tide pools are small bodies of SW captured as the tide ebbs that are exposed to warming from the sun, exposure to the cold in winter, but less wave action. Biological communities are highly variable and differ with wave exposure (open coast versus inner bays), vertical position in the intertidal zone, and tidal amplitude (Jordaan et al., 2011). The pools that include fish are typically isolated from the ocean for 6–10 h until the tide returns, as opposed to pools in the spray zone that can be isolated for weeks, but do not contain fish. Salinity in coastal marine tide pools thus begins as full-strength SW and if the pool is isolated from the ocean for several days there can be evaporative concentration often resulting in hypersaline conditions, warming, and hypoxia. There are reports of salinities up to 76 ppt ($2.4 \times$ SW) in salt marsh pools where the gulf killifish (*Fundulus grandis*) is found (Genz and Grosell, 2011). Alternatively, isolated tide pools can become diluted from rain and runoff and become BW or FW. These tide pools are suddenly cooled and the salinity is restored to full-strength SW when the tide returns. Estuarine tide pools are typically brackish. Fish that preferentially inhabit tide pools are generally small and camouflaged, and are benthic or suprabenthic in habit. Pelagic fish species are occasionally trapped in tide pools and thus technically are part of the community (Jordaan et al., 2011).

2.2. Climatic Extremes

2.2.1. POLAR

Estuaries in the Baltic and Gulf of St. Lawrence, and points northward, form thick pack-ice covers that minimize light and oxygen availability, while the haloclines remain intact. In winter, marine fishes such as flounder seek estuaries because the estuarine waters are warmer (1°C) than the ocean (−1.5°C) (Hanson and Courtenay, 1996). Different species react differently to the cold, some maintaining osmoregulation and activity, others becoming inactive. Intertidal zones in locations with seasonal ice cover endure severe cold and scouring of the shoreline, making year-round occupation by fish impossible. Only the lower littoral zone has significant vegetation and few animal species. In these zones, intertidal fishes generally do not migrate large distances, but rather seek coastal subtidal refuges for overwintering. One strategy is for the fish to occupy the deeper parts of estuaries in protected water. Alternatively, estuarine fishes can migrate moderately upstream where warmer FW streams run all year. The former strategy means that the animals must survive in cold, full-strength SW. The second strategy involves survival in cold FW. In either case, survival requires strong eurythermic and euryhaline capabilities.

2.2.2. TROPICAL

Tropical estuaries can stratify and produce haloclines and, unlike their temperate counterparts, can have inverted haloclines that trap warm saline water on the bottom (Stith et al., 2011). In tropical estuaries, mangrove swamps have extreme heat and low oxygen levels, conditions that force many species to evolve accessory breathing specializations and/or become semiterrestrial. There appear to be only slight osmoregulatory consequences of air breathing in teleosts and many species revert to surface breathing in hypoxic conditions. Terrestrial sorties can be for breeding, foraging, or predator avoidance, and the fish retain osmoregulatory abilities when they return to the water. Of the mangrove fish species, mangrove killifish and mudskippers are particularly well studied. These aerial species rely on cutaneous gas exchange and maintain acid–base balance by a suite of adaptations, including active excretion of NH_4^+, lowering of environmental pH, low NH_3 permeability of epithelial surfaces and, in a minority of species, volatilization of NH_3 (Ip et al., 2004; Chew et al., 2006), and all are subject to evaporative water loss during terrestrial sorties (see below).

3. OSMOREGULATORY STRATEGIES

3.1. Thermodynamics

An analysis of the ion and osmotic gradients at different salinities compared to blood plasma and to cytosol of ion transporting cells of the gill epithelium illustrates the gradients against which estuarine fishes must ion-regulate and osmoregulate. Several authors have presented this in different ways (Evans, 1984; Potts, 1984) but always with ion concentrations expressed on a linear scale. Here, the ion gradients across the gill are considered as fold gradients favoring inward (+) or outward (−) ion movement and as their Nernst equilibrium potentials (in millivolts), imagining that the membrane is conductive to one ion (Fig. 8.2A, B). The Nernst equilibrium potential for a monovalent ion is: $\Delta\psi = (RT/F) \ln[C_o/C_i]$, where R is 8.3143 coul. V/K mol., T is in K, F is 96,494 coul./mol. equiv., and C is the ion concentration on the inside, i, and outside, o, of the membrane, respectively. The log scale helps to expand the dilute end of the environmental concentration, thus allowing visual distinction between normal FW, at approximately 1 mM NaCl, and ion-poor environments. This distinction becomes important because few estuarine fishes can survive in ion-poor environments. Figure 8.2(A) shows a graphical interpretation of blood plasma ion concentrations versus environmental concentrations in a log-linear plot that keeps the equilibrium potentials linear, while the "isoionic" line, now steeply curved, delineates the border above which the fish hyperosmoregulates and below which the animal hypoosmoregulates.

In Fig. 8.2B are expressed the transmembrane concentrations (approximated, because many have not been measured) and the local apical membrane concentration gradients and equivalent electrical potentials in millivolts. It can be seen that in FW the gradients can exceed 100-fold and the equivalent voltages rise above 120 mV, well above the normal physiological range of transmembrane potentials. The Na^+ transmembrane gradient is smaller than the equilibrium potential in FW in the range of 1.0–10.0 mM Na^+, implying that a simple apical membrane Na^+ conductance by itself could not produce Na^+ influx (uptake). In BW above 10 mM NaCl, estuarine fishes use a neutral ion uptake mechanism, Na^+/Cl^- cotransporter (NCC), or ion exchange to uptake Na^+ and Cl^- (see Section 5 and review by Evans et al., 2005; Evans, 2011). In ion-poor conditions, the higher affinity NaCl uptake mechanisms involving V-type H^+-ATPase are necessary, yet most estuarine teleosts do not utilize these pathways (see Section 5), thus limiting their access to some FW environments. On the SW side, at the apical membrane the presumed intracellular ion activities are lower than in the environment, thus producing positively directed gradients of 11–31-fold and a Nernst equilibrium potential for Na (E_{Na}) of +87 mV (far from the estimated transmembrane voltage of −60 to −80 mV), but well within the normal range for E_{Cl}, thus showing that Cl^- can be secreted easily into SW simply via an apical membrane conductance, limited in marine teleosts to the cystic fibrosis transmembrane conductance regulator (CFTR) anion channel (see also Section 5). The apical membrane voltage has not been accurately measured (see below), but in chloride-secreting systems that are more amenable to intracellular impalement, such as corneal epithelium of rabbit, the intracellular potential at the apical membrane is −50.9 ± 1.0 mV n=121, the voltage drop through the tissue is an additional −6.0 mV (for a total of −56.9 mV) and the basolateral voltage is −78.2 ± 0.6, with the transepithelial voltage being the difference, +21.4 ± 0.8 mV (Marshall and Klyce, 1983). Thus, there is sufficient transepithelial potential (TEP) to drive Na^+ secretion paracellularly (see below) as well as transcellular Cl^- secretion across the apical membrane. In airway epithelium, $Na^+/K^+/2Cl^-$ cotransporter (NKCC) operation on the basolateral side causes intracellular Cl^- to accumulate to 39 ± 3 mM, well above the Nernst equilibrium potential of 17 ± 2 mM; the apical membrane voltage is −60 ± 2 and decreases to −46 ± 2 mV in resting versus secreting airway epithelial monolayers (reviewed by Welsh, 1986).

Na^+ efflux follows a paracellular pathway and requires a transjunctional E_{Na} of +32 mV, less than the measured TEP seen in some marine systems (+37 to +40 mV) (Guggino, 1980a; Pequeux et al., 1988). Thus, there is sufficient electrochemical driving force to secrete Cl^- transcellularly and Na^+ paracellularly into SW. The secretion of Cl^- across the apical membrane into hypersaline solutions has not been examined energetically before, to the author's knowledge, and it does present some mechanistic challenges. Either intracellular Cl^- needs to rise and/or the intracellular potential has to deepen to larger negative values. The former would be produced by more NKCC activity and the latter would be produced by higher basolateral K^+ conductance, and each requires higher Na^+/K^+-ATPase (NKA) activity. In hypersaline conditions the same pattern could be sustained simply if basolateral K^+ conductance were increased with a concomitant increase in NKA expression and activity. If K^+ conductance were to increase by 50%, the transmembrane potential would hyperpolarize to −85 to −90 mV and the maximum salinity into which the animal could secrete Cl^- would be almost 2 M Cl^-, a level readily obtained by teleosts and invertebrates such as brine shrimp (Griffith, 1974).

The log-linear presentation helps to distinguish the ion gradients in dilute media, such that transposed data for a group of related intertidal blenniid species show clearly the divergence of osmoregulatory ability in dilute media (Fig. 8.3). Mummichogs, the most hardy of killifish, can sustain

(A)   **Na⁺ and Cl⁻ equilibrium PD for strong osmoregulators**
      (*Fundulus heteroclitus* and *Oreochromis mossambicus*)

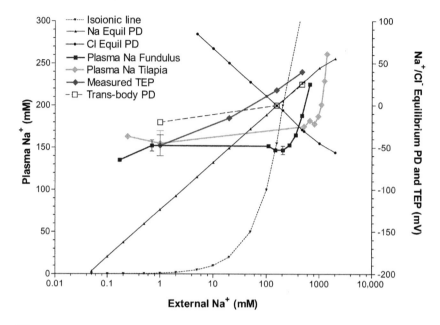

(B)

	Freshwater		**Seawater**	
Environment Ions (mM)	Na⁺ 0.1–1.0 Cl⁻ 0.1–1.0		Na⁺ 469 Cl⁻ 546	
Intracellular Ions (mM)	Na⁺ 15	Cl⁻ 10	Na⁺15	Cl⁻ 50
Fold-Gradient	⁻15 –⁻150	⁻10 –⁻100	⁺31	+11
As Equil. PD (mV)	⁻68–⁻126	⁺58–⁺116	⁺87	⁻60
Extracellular ions(mM)	Na⁺ 140	Cl⁻ 130	Na⁺ 155	Cl⁻ 145

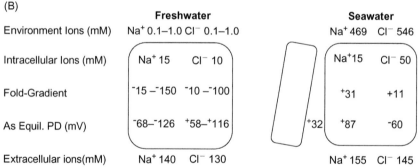

Fig. 8.2. (A) Presentation of theoretical Nernst potentials $\{E_x=(RT/z_iF)\ \ln(C_o/C_i)\}$ for x=Na⁺ and Cl⁻ plotted against the isosmotic/isoionic conformational line, which, in this log-linear plot, is a curve asymptotic to the abscissa. The two E_{Na} and E_{Cl} lines and the isoionic line cross 0 mV at an external Na⁺ of approximately 155 mM, isoionic to the blood plasma. Note that the measured transepithelial potential (TEP) in isoionic to hyperionic (one-third seawater and above) salinities approximates the slope of the Na⁺ equilibrium potential line but is above the line by 10–20 mV (the result of active Cl⁻ secretion), producing a consistent voltage gradient favoring cation (Na⁺) secretion via a cation-selective paracellular pathway. The transbody potential measured *in vivo* approximates E_{Na} in isotonic and hypertonic saline. In more dilute environments, the TEP deviates from the Na⁺ equilibrium potential (presumably because of less

osmoregulation in salinity equivalent to 1966 mM Cl^- (Griffith, 1974), so an extreme voltage gradient would be necessary to secrete Cl^- into that medium. Importantly, the transmembrane voltages at the apical and basolateral membranes of ionocytes have never been measured accurately. A serious attempt to impale ionocytes obtained rather small potentials of only -18 mV (Zadunaisky et al., 1988) but it is likely that the complex tubular system of ionocytes would be damaged by microelectrode impalement, thus degrading the measured potential. Perhaps non-invasive methods could obtain the necessary values for these potentials. The transepithelial (transbody) potentials, however, are easily obtained and are often measured.

3.2. Electrical Potentials

Transbody electrical potentials can be measured easily *in vivo* and demonstrate ion selectivity of the gill epithelium, the organ with the largest surface area and greatest overall permeability. In SW, estuarine teleosts, including European flounder (*Platichthys flesus*), blenny (*Blennius pholis*), longjaw mudsucker, Mozambique tilapia (*Oreochromis mossambicus*) and mummichog generally have inside positive potentials in the range of +10 to

---

cation selectivity) and has a polarity that would aid cation uptake, but is insufficient alone to produce cation uptake. The TEP is even farther from the Cl^- equilibrium potential, demonstrating that anion uptake occurs against a steep electrochemical gradient. Whereas some estuarine teleosts (e.g. *Oreochromis niloticus*) have effective NaCl uptake, others (e.g. *Fundulus heteroclitus*) simply fail to generate anion-active uptake at the gill and substitute Cl^- uptake at the gill by anion absorption via the intestine. In hypersaline conditions, both *Fundulus* (Marshall et al., 1999; Scott et al., 2006; Genz and Grosell, 2011) and hybrid tilapia (Garcia-Santos et al., 2006; Wang et al., 2009; Gonzalez, 2012) sharply increase plasma Na^+ in parallel to the isoionic line, as they reach their physiological limit. (B) Depiction of the transmembrane concentration gradients in freshwater (FW) and seawater (SW) ionocytes and the re-expression of the fold-gradients as Nernst equilibrium potentials. Intracellular ion activities for ionocytes have not been measured and are estimates based on other NaCl-secreting epithelia (corneal epithelium and airway). In two cases, Na^+ in FW and Cl^- in SW, the equilibrium potential for the apical membrane approximates a normal negative-inside cellular electrical potential; thus in these two cases, Na^+ uptake and Cl^- secretion across the apical membrane could be via ion channels (driven by the action of other active transporters). In SW, the apical cystic fibrosis transmembrane conductance regulator (CFTR) anion channel is the pathway for Cl^- secretion (Marshall et al., 1995; Singer et al., 2008) and in FW the recently discovered acid-sensing ion channels (Chen et al., 2007) are Na^+ channels that, by knockdown and pharmacological experiments with zebrafish (Dymowska, personal communication), are possible candidates for the Na^+ uptake channel.

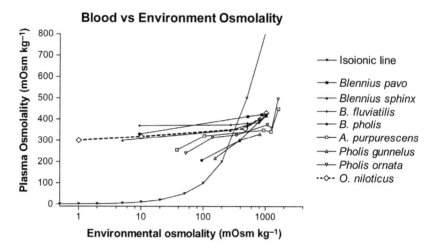

Fig. 8.3. Plot of blood versus environmental variables for osmoregulation of a series of blenniid species of intertidal teleosts, two of which survive in hypersaline conditions (>1100 mOsm kg^{-1}) but none of which survives in extremely dilute (<10 mOsm kg^{-1}) freshwater. Superimposed are data from the Nile tilapia (*Oreochromis niloticus*), which can survive in ion-poor environments, even in deionized water (Fall et al., 2011; Velan et al., 2011). In hypersaline conditions, plasma osmolality rises sharply in parallel to the isosmotic line as the fish approach their upper physiological limit (see also Fig. 8.2A).

+35 mV (Potts, 1984; Potts et al., 1991) and their approximation to the E_{Na} strongly suggests that the Na$^+$ conductance of the gill is very large relative to anions such as Cl$^-$. In European flounder, the TEP comprises a large diffusional component and a smaller electrogenic active component (Potts et al., 1991). In a study of Mozambique tilapia the measured transbody potential actually exceeded the E_{Na}, demonstrating that Na$^+$ may be secreted passively by a favorable electrochemical gradient (Dharmamba et al., 1975). In larval mummichog, animals where the ionocytes are in the yolk-sac membrane and the gill has not yet developed, the measured transbody potential was consistently higher than the E_{Na}, +37 to +40 mV (Guggino, 1980a). In isolated mummichog opercular membranes, without the shunting effect of the gill lamellae (Wood and Grosell, 2008, 2009), and in asymmetrical conditions with SW on the mucosal side, the TEP is again higher than E_{Na} (+28.8 mV), averaging +36.8 mV (Pequeux et al., 1988), illustrating that there is sufficient driving force for Na$^+$ to be secreted passively down its electrochemical gradient, if it traverses a paracellular pathway. Although it is technically difficult, it would be ideal to measure the local electrical potential adjacent to the apical crypts of ionocytes to demonstrate that Na$^+$ exit from SW fish is via passive diffusion down the Na$^+$ electrochemical gradient. The calculated E_{Cl} is universally negative in polarity (Fig. 8.2) and a few species of fully marine teleosts have measured transbody potentials with this polarity (reviewed by Potts, 1984; Evans et al., 1999). The negative inside potentials in seahorse (*Hippocampus erectus*) and toadfish (*Opsanus beta*) still call for further investigation as to how these species secrete NaCl into SW. The transbody potential does demonstrate indirectly that there is a predominant cation (Na^+) permeability in marine fish and that there is much less tendency to allow high Na^+ or Cl^- conductance in dilute environments. For instance, transbody potentials of fish first exposed to FW are strongly negative, but within 24 h these potentials decrease to within a few millivolts of zero (Wood and Grosell, 2008, 2009). The small potentials measured in acclimated fish in dilute environments suggest approximately equal and low Na^+ and Cl^- permeabilities (Kirschner, 1997).

3.3. Water Permeability

The osmotic permeability coefficient (P_{os}) is measured by the net rate of water gain by the whole animal or isolated gill in the presence of a transmural osmotic gradient. P_{os} represents the rate of water intake that must be balanced by fluid excretion in order to maintain an osmotic steady state; a more formal treatment appears in Alderdice (1988). Most studies measure P_{os} gravimetrically as mass changes, but dye dilution techniques are also used. P_{os} is distinguishable from the diffusional permeability coefficient (P_{diff}) of water movement measured by radioactive tracers in the absence of an osmotic gradient. Although the two coefficients are distinct, they are coupled, and a high P_{os} would require a high P_{diff}. A low P_{os} is a critically important characteristic to isolate the animal from its osmotic environment and minimize costs of osmoregulation. Even though the gills have a low P_{os}, the large surface area that must be available for gas exchange means that the gill accounts for the majority of the total osmotic water flow in the animal (Evans et al., 2005). The primary barrier to water permeation is the lipid bilayer and Mozambique tilapia and rainbow trout have approximately similar P_{os} in isolated gill arches in FW (Robertson and Hazel, 1999). Understandably, increasing temperature also increases P_{os} (Robertson and Hazel, 1999), suggesting that lower membrane viscosity and/or higher molecular activity decreases the barrier function of the lipid bilayer. P_{os} of the whole animal, however, is apparently lower than that accountable by the apical membrane lipid bilayer alone, based on studies of winter flounder membrane vesicles *in vitro* (Hill et al., 2004), thus opening speculation as to other mechanisms that could contribute. Unstirred layers and tight intercellular junctions are supplementary mechanisms that could contribute to the lower P_{os} *in vivo*. Inasmuch as the apical membranes lack water channels, the basolateral membranes, particularly of ionocytes, are rich in aquaporin 3 (AQP3), a water channel (Agre et al., 2002; Watanabe et al., 2005); thus volume regulatory responses of ionocytes may occur via these water channels. AQP3 is also one of the rapidly upregulated genes in several races of mummichog when these animals are challenged with 0.1 ppt FW (Whitehead et al., 2011b). Tight junctions that are impermeant to water have many strands of membrane-spanning proteins and function in part to isolate apical membrane from basolateral membrane, thus excluding the AQP3 from the apical membrane. In tilapia, the tight junction protein claudin 4 and in southern flounder (*Paralichthys lethostigma*) claudins 3 and 4 are localized predominantly in the tight junctions of gill filament outer epithelial layer, and FW fish have more claudin mRNA and protein present than do SW fish, as measured by western blots and immunofluorescence (Tipsmark et al., 2008a, b). Claudins 8 and 27 are expressed in the skin and gill epithelia of the euryhaline pufferfish *Tetraodon nigroviridis* and moving to SW increases claudin 8c and 27b,c expression in skin and gill (Bagherie-Lachidan et al., 2009), while claudin 3 expression in gill is reduced by SW and a hypersaline environment and in skin claudin 3 expression increases (Bagherie-Lachidan et al., 2008). Older research on whole-body P_{os} suggested that FW fish have lower P_{os} (Isaia, 1984), but there needs to be a re-examination with modern techniques. The low osmotic permeability in fish gills is clearly connected with elaboration of intercellular tight junctions in the gill. It will be exciting to see the plasticity of tight junction structure revealed in the near future.

The digestive tract also alters osmotic permeability with salinity acclimation. Whereas the esophagus epithelium is a simple columnar, osmotically permeable tissue in SW-acclimated mudskipper *Periophthalmodon modestus*, stimulated by cortisol, operating via glucocorticoid-type receptors, acclimation to FW causes epithelial proliferation and a decline in osmotic permeability, stimulated by prolactin (Takahashi et al., 2006b). Mozambique tilapia has a similar esophageal response to cortisol and prolactin (Takahashi et al., 2006a, 2007). In mummichog intestine where fluid absorption normally occurs, pharmacological stimulation of the epithelium via calcium and cyclic adenosine monophosphate (cAMP) reverses fluid flow through stimulation of CFTR anion channels in the apical membrane, thus producing NaCl and fluid secretion (similar to "secretory diarrhea" in mammals typical of cholera), thus demonstrating CFTR function in intestinal NaCl transport (Marshall et al., 2002a). A careful examination of polyethylene glycol (PEG) and water fluxes across intestinal epithelium of mummichogs has demonstrated that water (absorptive) flux is transcellular and not paracellular (Wood and Grosell, 2012). Transcellular water transport involves AQP1 in European eel (*Anguilla anguilla*) intestine (Martinez et al., 2005), rather than AQP3 (Cutler et al., 2007).

Osmotic permeability of eggs and embryos is much lower than that of yolk-sac larvae, juveniles, and adults; the large area of yolk sac and gill membranes clearly contributes to osmotic permeability (Alderdice, 1988). In intertidal spawners, where terrestrial egg-laying is common, the eggs are extremely resistant to desiccation (Taylor, 1999). Embryos of the Australian killifish (*Austrofundulus limnaeus*) have osmotic permeability 2% that of FW zebrafish (*Danio rerio*) embryos at similar stages (Machado and Podrabsky, 2007). Embryos of mummichog also have low water permeability, measured as P_{diff}. Early prehatch embryos have P_{diff} of 0.4×10^{-6} cm s^{-1}, which is low relative to other animals but in spite of the low P_{diff} there is drinking of perivitelline fluid, apparently to maintain osmotic balance (Guggino, 1980b). The low P_{diff} is consistent with a barrier function of the chorionic membrane which, in many estuarine species, has evolved to stick to vegetation and dry during development (Taylor, 1999).

4. OSMOREGULATORY STRESSES

4.1. Responses to Salinity Changes

4.1.1. STATIC CHANGES

Transfer experiments from SW to FW and vice versa have classically examined the changes in ion and water transport and the transition from short-term (minutes to hours) coping mechanisms to long-term (weeks) full acclimation responses. Estuarine fishes have evolved both and respond depending on circumstances. Many experimenters have induced euryhaline animals to acclimate to salinity change and, through these experiments, the induction of SW- and FW-adaptive mechanisms has been revealed, along with the hormones that affect the acclimation processes (see Takei and McCormick, 2013, Chapter 3, this volume). Estuarine fishes experience a large rise in blood osmolality (60–100 mOsm kg^{-1}) when transferred directly to SW; this is true for mummichog (Zadunaisky et al., 1995; Marshall et al., 1999), flounder (*Paralichthys lethostigma*) (Tipsmark et al., 2008b), the longjaw mudsucker (*Gillichthys mirabilis*), and Mozambique and Nile tilapia (Velan et al., 2011), and transfer of mummichog to FW produces a decrease of similar magnitude (Marshall et al., 2000), yet these imbalances are largely corrected by 24–72 h post-transfer in the new medium.

CFTR anion channel. A low conductance cAMP-activated anion channel has been identified in the apical membrane of SW ionocytes (Marshall et al., 1995) and the mummichog CFTR has been cloned and sequenced and expressed in *Xenopus* oocytes (Singer et al., 1998). CFTR is a member of the ATP binding cassette (ABC) protein family and, unlike a typical ligand-gated channel, its gating to the open state is connected to adenosine triphosphate

(ATP) hydrolysis (Kirk and Wang, 2011). ATP hydrolysis exposes certain amino acids, Q98 in transmembrane domain one and I344 in transmembrane domain 6, in the pore of the channel (Wang and Linsdell, 2012). Acclimation of euryhaline teleosts to SW is tightly associated with increased expression of CFTR in ionocytes and onset of Cl^- secretion by the opercular membrane (Marshall et al., 1999). Blockade of glucocorticoid-type cortisol receptors by RU486 impaired SW acclimation and blocked the increase in CFTR messenger RNA (mRNA) expression normally associated with acclimation to higher salinity (Marshall et al., 2005a; Shaw et al., 2007), whereas blockade of mineralocorticoid type receptors by spironolactone was without effect (Shaw et al., 2007), indicating that the teleost glucocorticoid-type cortisol receptor acts as an osmoregulatory mediator. Intracellular trafficking of CFTR from a subapical location to the apical membrane over the first 24 h after transfer to SW is an important means of increasing the NaCl secretion rate (Marshall et al., 2002a). A glucocorticoid-inducible kinase (SGK1) that enhances trafficking of CFTR to the cell surface is unaffected by RU486 blockade of these receptors, suggesting that the cortisol activation pathway of SGK1 is not via the glucocorticoid receptor (Shaw et al., 2008).

Na^+/K^+-ATPase (NKA). NKA is pivotal to ion uptake and secretion by the gill epithelium and is present in the basolateral membrane of all ionocytes. The enzyme has been cloned and sequenced from several teleost species including mummichog (Semple et al., 2002) and the gill isoform is similar to the $\alpha 1$ isoform of humans. Acclimation to SW is also associated with increased expression of NKA in gill tissues of Mozambique tilapia (Hwang et al., 1998; Hwang and Lee, 2007), mudskipper (*Periophthalmus cantonesis*) (Hwang et al., 1998; Hwang and Lee, 2007), and mummichog (Scott and Schulte, 2005).

Transfer to hypersaline conditions from SW produces a magnification of the NaCl secretion by the gill and opercular membrane and the fluid reabsorption function of the intestine (reviewed by Gonzalez, 2012) and produces upregulation of NKA in the gill and intestine and morphological changes to ionocytes in the opercular membrane. In the Mozambique tilapia, there are separate FW and SW isoforms of NKA and the $\alpha 1a$ isoform, supported by prolactin, decreases expression, while the $\alpha 1b$ isoform increases expression during SW acclimation (Tipsmark et al., 2011). In SW, the intestine of mummichog normally absorbs NaCl and water (Marshall et al., 2002a), although the epithelium can be induced to secrete NaCl and fluid in a response similar to secretory diarrhea. The posterior intestine of mummichog, as in most marine species, has active bicarbonate ion secretion that causes Ca and Mg carbonate precipitation in the posterior intestinal lumen and the precipitation of these solutes reduces the osmolality in the intestinal lumen, thus enhancing the continued resorption of fluid (Wilson and Grosell, 2003; Marshall and Grosell, 2006). Transfer to a hypersaline environment accentuated the bicarbonate secretion in mummichog, but otherwise, there were few changes in overall function (Genz and Grosell, 2011).

Estuarine species, on initial transfer to FW, have a large decrease in plasma Na^+ and Cl^- and osmolality, approximately 60 mOsm, a decrease that is slowed down by the cessation of NaCl secretion by ionocytes in opercular membranes subjected to a decrease in osmolality (Marshall et al., 2000; Daborn et al., 2001; Marshall et al., 2005b; Hoffmann et al., 2007). These ion-sparing responses can be observed also as a reduction in unidirectional efflux in whole mummichog transferred to FW (Wood, 2011). Estuarine fishes generally survive in FW, but the salinity decrease is not necessarily connected with the development of high-affinity ion-uptake pumps that are typical of FW-resident fish. For instance, the mummichog has a K_m of 2.0 mM for Na^+ uptake across the whole body (Potts and Evans, 1967) after acclimation to FW, while the rainbow trout has a K_m of 0.11 mM (Evans et al., 2005). Without a high-affinity Na^+ uptake mechanism, many estuarine teleosts cannot live in ion-poor environments. For instance, the gulf killifish (*Fundulus grandis*) has difficulty acclimating to salinity below 1.0 ppt (Patterson et al., 2012).

4.1.2. CYCLING SALINITY STRESSES

Recent experiments have examined the effects of cycling of temperature and salinity on estuarine species to mimic tidal cycles. These realistic experiments reveal some novel combinations of osmoregulatory responses with stress responses. When mummichogs were exposed to static or cycling salinity changes, while having their overall ionic permeability assessed through measurement of transbody electrical potential, these animals manifested an apparent selective permeability for cations, primarily Na^+, over anions, as they had TEP more positive than +15 mV in 100% SW, decreasing to 0 mV at 20–40% SW, and more negative than −30 mV in FW (Wood and Grosell, 2009). However, mummichogs that were fully acclimated to FW showed a lower slope to the TEP change with salinity, crossed 0 mV at about 20% SW, and had smaller positive values in SW (+10 mV), suggesting that FW acclimation produced a physiological commitment to life in a dilute environment and a lower overall ionic permeability (Wood and Grosell, 2009). In mummichogs, upregulation of important regulatory genes (14-3-3 protein) (Kültz et al., 2001) and some transport associated genes, such as cell-junction related genes (OCLN, CX32, CLDN3, and CLDN4), apparently lags by more than 24 h after salinity transfer (Whitehead et al., 2011b); thus one would expect that major overhaul of ion transporting cells cannot occur in a transient salinity change. The core transporters CFTR, NKA, and AQP3, however, appear to upregulate mRNA expression and protein function more rapidly. CFTR in apical membrane of ionocytes is significantly upregulated (mRNA) within 8 h post-transfer and Cl^- secretion is elevated by 24 h post-transfer (Singer et al., 1998; Marshall et al., 1999). In the case of NKA there is a rapid and transitory increase in gill NKA activity in the first few hours after the transfer of mummichogs to SW, a response that is independent of translational and transcriptional processes (Mancera and McCormick, 2000; McCormick et al., 2009). The basolateral aquaporin water channel, AQP3, is rapidly upregulated by 6 h (Whitehead et al., 2011b). Thus, mummichogs can rapidly modify ion and water transport function in ionocytes within hours of experiencing a new salinity, but, in some way, there is a delay in the major changes in gene expression that could lead to replacement of ion transporting cells with a new cell population. Transfer from SW to hypersaline conditions is accompanied by only slight modifications of gill (Hossler et al., 1985) and intestinal epithelia (Genz and Grosell, 2011), and electrophysiology of opercular membranes of Mozambique tilapia shows increased ionocyte density and decreased leak conductance (Kültz and Onken, 1993).

Transfer in the opposite direction, from SW to FW, induces in some estuarine species a novel reaction of rapid withdrawal of salt-secreting cells from the surface, as has been observed independently in mudskippers, detected by concanavalin-A fluorescence (Sakamoto et al., 2000) and in mummichogs, detected by scanning electron microscopy (Daborn et al., 2001). The retraction of the ionocytes is accompanied by a rapid reduction in salt secretion rate and a reduction in epithelial ion conductance (Daborn et al., 2001), and may be driven by a ring-like band of actin at the apical crypt opening (Daborn et al., 2001). In this way, the animals minimize ion loss on transfer to FW, even though this species does not apparently develop efficient ion uptake pathways in FW (Patrick et al., 1997; Wood and Grosell, 2009). The retraction of ionocytes to minimize ion loss in the first few hours of exposure to FW and the reversal of the transepithelial electrical potential are the likely explanation for the classically observed "exchange diffusion effect" seen when euryhaline fishes (European flounder and mummichog) were transferred to FW and showed rapid (20 min) reductions in Na^+ efflux across the gill (Motais et al., 1966).

4.2. Aerial Exposure

In sun-exposed tide pools, resident animals can become heat stressed and hypoxic, while still needing to maintain osmoregulation. Estuarine species have evolved to cope with short periods of osmotic and thermal stress that generally is relieved when the tide returns. Intertidal species such as blennies and mudskippers do not retreat with the ebbing tide and instead take refuge in cracks and holes and beneath rocks, where they remain exposed but, relatively inactive until the return of the tide. Mummichogs also survive well during aerial exposure and do not resort to anaerobic metabolism, but, rather, use cutaneous respiration at approximately half the usual rate of aquatic respiration in well-aerated water (Halpin and Martin, 1999). This aerial exposure can result in desiccation sufficient to cause significant concentration of the blood and tissues. There is thus a tradeoff between the need to maintain efficient cutaneous gas exchange, mostly across the wetted buccal epithelium (Randall et al., 2004), and the need to reduce the rate of water loss. The mudskipper *Periophthalmus cantonensis* loses between 6 and 8% of body weight per hour of aerial exposure (Gordon et al., 1978) and members of this genus die after water loss approaching 22% (Gordon et al., 1969). Four species – blennies *Istiblennius lineatus* and *Paralticus amboinensis* and mudskippers *Periophthalmus argentilineatus* and *Periophthalmus kalolo* – all lose water at approximately the same rate, 2.6–5.6% h^{-1}, as inanimate aqueous gel (Dabruzzi et al., 2011), implying that these fish have not evolved significant defenses against desiccation. Mudskippers do not have a dry keratinized skin or waxy secretion, but there are structural modifications of the skin including increased mucus secretion and a porous corky layer that may protect against ultraviolet light (Suzuki, 1992). In response to high ammonia exposure, mudskipper (*Periophthalmodon schlosseri*) skin increases levels of cholesterol and saturated fats, suggestive of a protective response to reduce NH_3 permeability with the side-effect of reduced evaporative water loss (Randall et al., 2004). Aerial exposure of mangrove killifish (*Kryptolebias marmoratus*) causes skin ionocytes to retract and become covered over, an effect reversible with immersion in SW (LeBlanc et al., 2010). During terrestrial sorties, interlamellar cell masses (ILCMs) accumulate in the gills, an effect that may reduce evaporation on land, but also reduces the effective aquatic respiratory surface area on return to the aquatic medium (Turko et al., 2011). Behaviorally, rockskippers (*Hypsoblennius gilberti*) typically emerge at night when evaporation rates are lower (Luck and Martin, 1999) and clingfish curl up near rocks to reduce the surface area available for evaporation. Mudskippers and other gobies keep their dorsal sides wet by rolling periodically (Lee et al., 2005) and restrict desiccation by spending time in their burrows (Ikebe and Oishi, 1997).

Aerial exposure also evokes changes in gill transporter distribution, an apparent evolutionary adaptation for effective excretion of ammonium ions. In mudskipper *Periophthalmodon schlosseri*, branchial ionocytes express NKA and NKCC on the basolateral surface, but have Na^+/H^+ exchangers NHE2, NHE3, V-type H^+-ATPase, and CFTR anion channels in the apical membrane (Wilson et al., 2000). NH_4^+ active secretion during aerial exposure results in astonishingly high accumulation of NH_4^+ in gill fluid to 90 mM in a few hours (Chew et al., 2007). In this case, the elimination of NH_4^+ may be aided by anions (HCO_3^-) permeating the CFTR pathway. Although the terrestrial sorties in mudskippers elevate plasma cortisol above resting values and cortisol actually evokes terrestrial behavior, expression of the heat shock protein HSP90 is not elevated (Sakamoto et al., 2002, 2011).

4.3. Osmosensitivity

Osmosensitivity is reviewed in Chapter 2 of this volume (Kültz, 2013), so only a few example mechanisms that control NaCl secretion rates in euryhaline teleost fishes are considered here. Estuarine teleosts experience large increases in plasma ions and osmolality when exposed to SW and equally large reductions in ions and osmolality when transferred to FW; these changes range to approximately 60 mOsm. Ionocytes in the gills and opercular epithelium are exquisitely sensitive to small changes in osmolality, as small as $5 \, mOsm \, kg^{-1}$, and slight swelling of ionocytes shuts off NaCl secretion (Marshall et al., 2000), while hypertonic shock stimulates NaCl secretion (Zadunaisky et al., 1995). The effect appears to be mediated directly at the transporter level, activating NKCC in the basolateral membrane (Marshall et al., 2008; Zadunaisky et al., 1995) and CFTR in the apical membrane (Marshall et al., 2009). The activation pathway involves α/ βintegrin as the volume sensor, and several intermediate kinase steps, probably involving p38 mitogen-activated protein kinase (MAPK), c-Jun N-terminal kinase (JNK), protein kinase C, osmotic stress response kinase (OSR1), ste20-proline rich kinase (SPAK), and focal adhesion kinase (FAK) (Marshall et al., 2005b). Of particular interest is FAK, which is phosphorylated at tyrosine 407 (FAKpY407) exclusively when the cells are perturbed osmotically, whereas other phosphorylation sites (Y397, Y576, Y861) are osmotically insensitive (Marshall et al., 2005b, 2008). FAKpY576 appears in the apical membrane colocalized with CFTR of ionocytes, while FAKpY861 is localized in the intercellular tight junctions between epithelial cells in the opercular epithelium (Marshall et al., 2008). Mummichog ionocytes are not the only fish epithelia to respond to osmotic shock by changing transport rates, as the European eel intestinal epithelia also regulate transport rate using cell volume (reviewed by Hoffmann et al., 2007). Whereas the osmotic responses have no effect on cAMP and are not mediated via protein kinase A (PKA), FAK is dephosphorylated by inhibitors and rephosphorylated at position 407 by adrenergic agonists (Marshall et al., 2009). There appears to be a multimolecular complex involving the volume sensor β1 integrin, the tyrosine kinase FAKpY407, and the transporter target NKCC, colocalized by immunocytochemistry and immunogold transmission electron microscopy, and confirmed by coimmunoprecipitation (Marshall et al., 2008). In the apical membrane, FAKpY407 also operates, but in connection with the anion channel CFTR, as FAKpY407 colocalizes with CFTR at the light and electron microscope levels by immunomicroscopy (Marshall et al., 2009). These reactions are effectively instantaneous, changing ion transport rates within a few minutes, and yet the effect is long lasting, as long as the osmotic shock is applied (Marshall et al., 2005b, 2008). The extreme sensitivity of this transport regulation system makes it ideal to operate in aid of estuarine fishes moving between salinities for short durations, and would help them to cope with salinity changes without large, metabolically expensive, acclimation processes being necessary (Marshall, 2003).

5. ESTUARINE FISHES AS PHYSIOLOGICAL MODELS

5.1. Fundulus heteroclitus

The mummichog is selected as a representative of an estuarine euryhaline species that fundamentally is a marine form that copes well with FW and has been an important experimental model species for over 50 years. This physiological grouping of marine-like estuarine teleost fishes includes mudskippers, gobies, blennies, and sculpins. Mummichogs are well known for their euryhaline capabilities and can be readily acclimated to environments ranging from FW to hypersaline conditions as high as 120 ppt (Griffith, 1974). Based upon this attribute, the mummichog has been and continues to be an important model organism for understanding mechanisms of teleost osmoregulation, as documented in two major reviews examining ionocyte structure and function (Karnaky, 1986; Wood and Marshall, 1994). Ion transport and acid–base models for teleost ion transport by gills (Evans et al., 2005; Evans, 2011) and other major osmoregulatory organs (Marshall and Grosell, 2006) also rely extensively on research with this species.

5.1.1. OSMOREGULATION IN SEAWATER

Philpott and Copeland (1963) recognized numerous ionocytes in mummichog gills, skin, and buccal epithelium, and described a curious ultrastructure, with an elaborated basolateral membrane surface in serpentine tubules that ramified among the well-organized mitochondria. These putative ion transporting cells were similar to those seen previously in American eel (*Anguilla rostrata*) gill (Keys and Willmer, 1932). NKA, localized specifically on the basolateral membrane of these "chloride cells" (Karnaky et al., 1976), displayed higher activity in the gills of mummichogs acclimated to SW than to FW, and higher in both conditions than in fish acclimated to isotonic one-third SW (Epstein et al., 1967; Towle et al., 1977). The link between NKA activity and Cl^- exit from the animal into SW was an Na^+/Cl^- cotransporter (Silva et al., 1977) now recognized as NKCC1, a member of a larger transporter family that is rapidly phosphorylated and activated and increases gene expression during acclimation of mummichogs to SW (Flemmer et al., 2010). NKCC mediates $Na^+/K^+/2Cl^-$ transport that accumulates Cl^- above its electrochemical equilibrium inside the cell, a secondarily active transport, while K^+ recycles across the basolateral membrane via Ba^{2+}-sensitive potassium channels. An explanation for the movement of Cl^- into SW against both electrical and concentration gradients, the "pump-leak" pathway, relied on an additional discovery: an anion channel in the apical membrane that would allow the Cl^- to leak out into the environment, as summarized in Chapter 1 of this volume (Edwards and Marshall, 2013) The mummichog chloride cell has a low conductance apical membrane channel that is similar to the mammalian CFTR and activated by cAMP and PKA (Marshall et al., 1995). The CFTR homologue in *F. heteroclitus* was cloned from mummichog tissues (Singer et al., 1998), showing that its expression increased after transfer to SW, and produces a cAMP-activated anion conductance when expressed in amphibian oocytes. Acclimation to SW augments ion secretion in parallel with increased CFTR expression (Marshall et al., 1999) and there is mobilization of both the CFTR and NKCC in chloride cells, placing more of the former in the apical membrane and more of the latter in the basolateral membrane (Marshall et al., 2002b).

NKCC was also characterized pharmacologically. NKCC is inhibited by the loop diuretics bumetanide and furosemide in isolated opercular membranes of mummichog (Degnan et al., 1977; Eriksson and Wistrand, 1986) and there is dependence of NKCC on basolateral K^+ (Marshall, 2002) consistent with the $Na^+/K^+/2Cl^-$ stoichiometry. The Cl^- secretion pathway relies indirectly on the recycling of K^+ across the basolateral membrane, thus serosal Ba^{2+} that blocks K^+ channels strongly inhibits Cl^- secretion by the opercular membrane (Degnan, 1985). Also, expression of NKCC increases during acclimation to SW (Scott and Schulte, 2005). NKCC activation involves phosphorylation (Flemmer et al., 2002) and, discovered in shark NKCC1, at three specific conserved threonine residues, T184, T189, and T202 (Darman and Forbush, 2002). NKCC activation involves a suite of kinases, including SPAK (Marshall et al., 2005b; Flemmer et al., 2010), OSR1 (Marshall et al., 2005b), and a tyrosine kinase FAK (Marshall et al., 2005b, 2008). Another kinase, with-no-lysine (WNK1), regulates NKCC in mammalian cells and in tissues from *Caenorhabdites elegans* (Delpire and Gagnon, 2008; Kahle et al., 2010), but this needs to be confirmed for teleost fish systems. Inducible cyclooxygenase, COX-2, regulates prostaglandin metabolism and mediates eicosanoid synthesis which, in turn, regulates NaCl secretion by the mummichog opercular epithelium (Van Praag et al., 1987; Evans et al., 2004).

The NKA, which is responsible for producing the Na^+ transmembrane gradient that drives NKCC operation, has also been cloned from mummichog (Semple et al., 2002). Current evidence indicates a complicated layering of regulation at the protein and gene expression levels associated with acclimation to SW. There is evidence for rapid activation of NKA within minutes, clearly derived from the enzyme already in place (Mancera and McCormick, 2000). At the RNA level, NKA expression increases after transfer to SW, but is delayed by 2–3 days (Scott and Schulte, 2005). The Na^+/H^+ exchangers (NHEs), which are usually associated with acid secretion in FW ion and acid–base balance (Evans et al., 2005), also appear to play a role in the regulation of acid transport in marine conditions (Edwards et al., 2005). Mummichogs drink more in SW than in FW, based on sulfate and inulin radioactive markers (Potts and Evans, 1967), findings that were among the early confirmations of Homer Smith's paradigm of marine fish drinking SW to offset osmotic water loss. Whereas the intestine is mostly absorptive of ions and water (Marshall and Grosell, 2006; Marshall et al., 2002b), there is active transport of a carbonate rich secretion by the posterior intestine that aids acid–base balance and water absorption (Genz and Grosell, 2011).

5.1.2. OSMOREGULATION IN FRESHWATER

Most of the osmoregulatory research with the mummichog has been directed towards mechanisms accommodating increased salinities, yet the converse, adjustment to lower salinities, is just as challenging and has resulted in diverse strategies (Evans et al., 2005; Marshall and Grosell, 2006). Because FW habitats are geologically transient, with inland lakes and streams forming and disappearing with geological changes, there have been innumerable opportunities for different FW osmoregulatory mechanisms to evolve (see Schultz and McCormick, 2013, Chapter 10, this volume). As a result, the paradigm followed by a major model species, rainbow trout, in which acid-secreting and base-secreting cells operate in parallel to link acid–base regulation with NaCl uptake (Evans et al., 2005), is not present in many teleosts. For instance, mummichogs express the pivotal enzyme V-type H^+-ATPase in the basolateral membrane (Katoh et al., 2003), rather than the apical membrane as in trout (Lin et al., 1994), implying a less direct involvement of this enzyme with ion uptake. In SW the opercular membrane secretes NaCl, as does the gill. However, in FW the gills actively absorb Na^+

but not Cl^- (Patrick et al., 1997; Patrick and Wood, 1999), while the opercular epithelium actively absorbs Ca^{2+} (Verbost et al., 1997; Marshall et al., 1997; Marshall, 2002) and Cl^- but not Na^+ (Marshall et al., 1997). Acclimation to FW relies on complex dynamics of Na^+/H^+-exchanger isoforms, NKA, and carbonic anhydrase (Edwards et al., 2005; Scott et al., 2005) and probably reflects both a rapid turnover of transporting cells (Daborn et al., 2001; Laurent et al., 2006) and an Na^+ uptake system that could instead rely on Na^+/H^+ exchange or neutral NaCl cotransport. As a result, Na^+ uptake is much less efficient in mummichog than in the trout, yet it is a mechanism that still operates sufficiently to allow mummichogs to occupy even ion-poor FW habitats, so long as the animals are actively feeding. Indeed, the ionic contribution of food is essential for mummichogs to survive in ion-poor FW because Cl^- uptake is limited in the opercular membrane and is absent from the gill (Marshall et al., 1997; Laurent et al., 2006).

5.1.3. KIDNEY AND INTESTINE

In the marine environment, mummichogs drink to offset osmotic water loss to the environment (Potts and Evans, 1967) in a reflex stimulated by angiotensins (Malvin et al., 1980). The intestine absorbs water and ions from the SW ingested by the animal and in the posterior portion there is also some ion secretion, specifically of bicarbonate (Marshall and Grosell, 2006). The digestive tract of the mummichog is unusual, as it lacks a stomach and is relatively short. The intestine is normally absorptive, involving NaCl absorption parallel to water channels, presumably AQP1, as discovered in European eel (*A. anguilla*) (Martinez et al., 2005), that impart hydraulic conductivity and allow fluid reabsorption. The kidney of teleosts is an important means for excretion of water by FW-acclimated teleosts and in marine teleosts for the secretion of divalent ions, especially Mg^{2+}, Ca^{2+}, and SO_4^{2-} (Marshall and Grosell, 2006). The mummichog kidney is small and has no urinary bladder, where ureteral urine might accumulate and be altered before release. In SW, tubular secretion of Na^+ and Cl^- occurs as secondary active Cl^- transport with electrically coupled paracellular Na^+ transport. Electrochemical gradients for K^+ and Na^+ allow Cl^- entry across the basolateral membrane, presumably via NKCC, and result in cytosolic Cl^- concentrations above electrochemical equilibrium. Apical Cl^- secretion is via a cAMP-stimulated conductive pathway, by an as yet unidentified anion channel (Marshall and Grosell, 2006). In both SW and FW killifish, NKCC immunoreactivity was detected in the apical membrane of the distal and collecting tubules, and in the basolateral membrane of the second segment of the proximal tubules, the major part of killifish renal tubules

(Katoh et al., 2008). These results support the secretory function of the early and absorptive function of the more distal portions of renal tubules of these euryhaline fishes. In FW, the killifish proximal tubule is also secretory (Beyenbach and Liu, 1996), being balanced by reabsorptive NaCl transport in the distal portions of the tubule for the net excretion of a dilute urine. Study of the spatial and temporal expression of transporters in the kidney would reveal the dynamics of this tissue during salinity changes. Development of miniaturized techniques and molecular approaches that use small tissue samples will enable future studies of transporter and channel expression in the kidney, a wide-open and particularly fascinating area for study.

Although the important discovery of CFTR in fish occurred in mummichog, and other transport protein genes are now known (NKA, NHE, carbonic anhydrase, H^+-ATPase, NKCC), many other transporter homologues (e.g. K^+ channels, AE1, NCC, Ca-ATPase, and Cl^- channels) and their regulatory proteins remain enigmatic. To date, four potential regulatory proteins have been sequenced: CFTR itself (Singer et al., 1998), the glucocorticoid receptor (Mommsen et al., 1999), COX-2 (Choe et al., 2006), and the 14-3-3 protein (Kültz et al., 2001).

5.2. Oreochromis niloticus

The Nile tilapia (*O. niloticus*) is a rapidly growing, omnivorous FW and estuarine species, closely related to an FW euryhaline teleost model species, Mozambique tilapia (*O. mossambicus*). Nile tilapia is subject to worldwide aquaculture exceeding 3.5 million tonnes as of 2012 (Table 8.1) and occupies brackish lagoons of the Nile delta (Oczkowski and Nixon, 2008) and estuaries of Africa, Indonesia, China, the Caribbean, and South America. It is FW tolerant and can even acclimate to deionized FW (Pisam et al., 1995) and to salinities up to 60 ppt (Guner et al., 2005; Schofield et al., 2011). Because the species is tropical, salinity acclimation ability is better in warm (30°C) water, and colder water temperatures limit survivability in temperate SW (Schofield et al., 2011). The gill epithelial structure indicates pavement cells interrupted by exposed ionocytes, mucus and pillar cells (Monteiro et al., 2010, see Fig. 8.4).

5.2.1. FRESHWATER

Acclimation to deionized water evokes elaboration of the microvilli (Velan et al., 2011) on the apical membrane of ionocytes, suggestive of increased area for ion uptake proteins (Pisam et al., 1995). The ion uptake mechanism in the Nile and Mozambique tilapias involves the Na^+/Cl^-

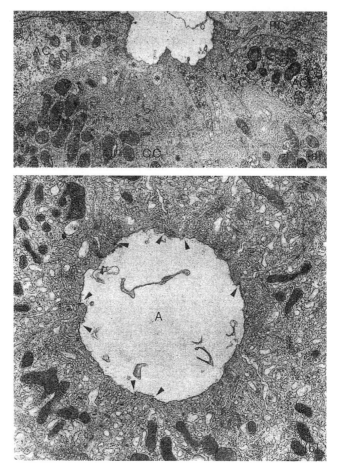

Fig. 8.4. Multicellular ionocyte complexes that develop in the gill epithelia of tilapias in full-strength seawater when the fish are slowly acclimated to increasing salinity. (a) side view, (*Oreochromis niloticus*) 12,000x, (b) transverse view (*O. mossambicus*) 17,800x. These complexes have numerous ionocytes (CC) and accessory cells (AC) congregated about a single large apical crypt. (A) surrounded by pavement cells (P); arrows indicate ionocyte-accessory cell shallow tight junctions, *apical interdigitations. From: Cioni et al. (1991), with permission.

cotransporter (NCC) expressed in the apical membrane of gill ionocytes (Velan et al., 2011) as well as basolateral NKA and NKCC1. If the relatively low-affinity (a few millimolar) NCC is the main NaCl uptake mechanism, it is not clear how these animals survive in deionized water, unless this environment invokes an as yet undiscovered high-affinity Na^+ uptake mechanism. It may be that dietary ions act as a supplement. In any case, the

NCC and low ion permeability are supported by high expression of PRL1, the gene for prolactin synthesis (Breves et al., 2011).

5.2.2. SEAWATER

Because of the critical importance of Nile tilapia aquaculture utilizing more estuarine habitats, salinity tolerance of various tilapia species and hybrids among tilapia species (Wang et al., 2000; Fall et al., 2011) has attracted much interest (Suresh and Lin, 1992; Guner et al., 2005; Rengmark et al., 2007). These fish have a U-shaped NKA activity relationship with salinity, in that NKA activity is high in very dilute environments, low in brackish conditions, and high again in SW and hypersaline conditions up to 60 ppt (Guner et al., 2005). Transfer from FW to 15 ppt BW downregulates NCC and prolactin (PRL1) expression in Nile tilapia, while Mozambique tilapia had this reaction plus a strong immediate upregulation of NKCC and NKA, yet growth hormone (GH) expression was apparently unchanged in both species (Velan et al., 2011). During acclimation to hypersalinity, Nile and Mossambique tilapia develop unusually large multicellular ionocyte complexes in the gills (Fig. 8.4) (Cioni et al., 1991; Guner et al., 2005). The multicellular complexes appear to comprise multiple accessory cells and one large ionocyte (Fontainhas-Fernandes et al., 2003). Four genes that have been identified as important in SW acclimation and growth in Nile tilapia are hemoglobin, Ca^{2+}-ATPase, pro-opiomelanocortin (POMC), and actin (Rengmark et al., 2007). Hyperosmotic transfers of Nile tilapia also increase expression of osmotic stress factor 1 (OSTF1), PRL1, GH, and branchial GH receptor, while branchial expression of the ion uptake transporter NCC is downregulated and NKCC expression is unchanged after 6 h (Breves et al., 2010). The slow response of Nile tilapia to increasing salinity may be part of a larger adaptive pattern for estuarine species to cope with short-term changes and to react to permanent changes some 24 h later.

5.3. Tolerance Versus Acclimation

Full acclimation to new salinity regimens is achieved rapidly in estuarine animals and they usually tolerate direct FW to SW and SW to FW transfers, unlike diadromous species that may only tolerate these changes during certain life stages. Estuarine fishes combine the short-term coping strategies with long-term acclimation to effect a full plasticity in osmoregulation (Marshall, 2003; Wood, 2011). It may be that euryhaline estuarine fishes have evolved a means of delaying expression of stress-responsive proteins such as 14-3-3 that are expressed in killifish transferred to FW only after 24 h (Kültz et al., 2001). NKA expression also is delayed by 2–3 days after transfer to SW (Scott and Schulte, 2005). By delaying the onset of large, energetically expensive remodeling of tissues, the tolerance strategy would save metabolic energy. Similarly, the SW type of cation permeability that produces the large negative-inside transbody potential in mummichogs is not lost for 24 h in FW, an effect thought to contribute to their ability to return to SW quickly (Wood and Grosell, 2008). When challenged by 0.1 ppt salinity, mummichog populations native to brackish or marine habitats lose osmotic homeostasis more severely and take longer to recover compared to FW populations, and, although AQP3 expression responds quickly, many other genes respond more slowly (Whitehead et al., 2011a). In a group with FW, estuarine, and marine members, the sculpin genus *Cottus*, FW and marine-resident species transferred to SW immediately initiate a strong drinking response, whereas the estuarine species has a much lower drinking rate even 24 h after transfer (Foster, 1969). Rapid small changes in osmoregulation, involving the reversible retraction of ionocytes from the surface on exposure to hypotonic media or hypotonic shock, have been observed in mudskippers (Sakamoto et al., 2000) and in killifish (Daborn et al., 2001). In mangrove killifish (*Kryptolebias marmoratus*) ionocytes retract during aerial exposed sessions (LeBlanc et al., 2010). The emergence of these cells is triggered in mudskippers by external SW or high calcium (Sakamoto and Ando, 2002). Transfer of climbing perch to SW initially evokes high blood sodium, to 170 and 164 mM in one-third and full-strength SW respectively, which is corrected by 7 days in SW and accompanied by a rise in gill NKA activity (Chang et al., 2007). Nile tilapia transferred to SW for 6 h show no increase in NKCC expression (Breves et al., 2010), yet this species can acclimate to twice SW salinity, if salinity is increased gradually (Guner et al., 2005). Transfer of mummichogs to SW also elicits rapid (~8 h) increases in CFTR (Marshall et al., 1999) and NKCC expression (Flemmer et al., 2010), and trafficking of CFTR to the apical crypt, where NaCl secretion is increased by 24 h (Marshall et al., 1999). There is also evidence for rapid activation of NKA and corresponding ion fluxes apparently without the need for protein synthesis (Mancera and McCormick, 2000; Wood, 2011). Therefore, rapid upregulation and activation of a small cadre of genes seems to be the only major change that is necessary to initiate full SW level NaCl secretion by ionocytes of the gill epithelium, rather than proliferation and differentiation of ionocytes that characterize long-term acclimation.

Similarly, tide pool and estuarine species may be more tolerant of temperature fluctuations and tolerate change rather than react to it. Tide pool fishes are exposed to large environmental fluctuations; for example, tide pool sculpins have evolved higher tolerance to temperature shifts and their heat shock protein (HSP) responses are activated at higher temperatures than their subtidal equivalent species (Nakano and Iwama,

2002). This may result from the animals in fluctuating conditions having higher resting levels of HSPs. Whereas the means by which estuarine fishes delay the large tissue turnover acclimation steps, from a variety of perspectives, the tolerance strategy seems to fit the combined results of many transfer experiments.

Hemoconcentration during terrestrial sorties by tropical intertidal mudskippers is minimized behaviorally by these animals. Mudskippers roll in the mud to reduce evaporative water loss (Ikebe and Oishi, 1997). They also will return to their water-filled burrows, a behavior driven hormonally by 11-deoxycorticosterone, apparently operating through mineralocorticoid type receptors, an effect partially blocked by the steroid receptor blocker RU-486 (Sakamoto et al., 2011). Behavioral solutions to osmoregulatory responses are economical and do not require large physiological adjustments.

Exposure to hypotonic conditions of intertidal blennies causes behavioral changes consistent with a tolerance response. Upper intertidal black pricklebacks (*Xiphister atropurpureus*) consume oxygen at a significantly lower rate in dilute SW, compared to full-strength SW, whereas there was no significant difference found in oxygen consumption by penpoint gunnels (*Apodichthys flavidus*) that frequent lower intertidal tide pools (Haynes et al., 2009), suggesting a hypometabolic tolerance response to salinity change, rather than acclimation, in animals exposed to more environmental variability.

5.4. Genomics and Proteomics

The estuarine fish genomes available are expanding and the genomes of several euryhaline species – fugu (*Takifugu nigroviridis*), three-spined stickleback (*Gasterosteus aculeatus*), and mummichog (*Fundulus heteroclitus*) – are complete or almost so, so that there will be more access to full genomic models in the near future (Table 8.1). Given the huge commercial importance and deep expressed sequence tag (EST) file, it is surprising that Nile tilapia has attracted less scientific interest than its cousin Mozambique tilapia (355 ESTs submitted). Scientists could do well to shift attention to Nile tilapia, the most aquacultured euryhaline teleost on the planet, and a critical protein source for many developing nations. These genomic advances will allow full molecular manipulation experiments to be performed, thus opening new and exciting possibilities for understanding the acclimation processes to various combinations of stressors often faced by estuarine species. The studies will be made all the more interesting because of the ancient total genomic duplication event in the actinopterygian fish lineage (Cutler and Cramb, 2001; Larhammar et al., 2009). The

Table 8.1

Genomes of osmoregulatory model species and related commercial importance**.

Trivial name	Scientific name	Source (NCBI)	Total EST[a]	Complete genome	Habitat+halinity	Aquaculture[b]	Fishery[b]
Zebrafish	*Danio rerio*	RefSeq	1,488,275	Sanger Institute (2002)***	Freshwater stenohaline	0	0
Japanese pufferfish	*Fugu rubripes*	Ensembl		Aparicio et al. (2002)	Marine stenohaline	?	?
Greenspotted pufferfish	*Tetraodon nigroviridis*	GenBank		Jaillon et al. (2004)	Estuarine euryhaline	?	?
Japanese medaka	*Oryzias latipes*	dbEST	666,891	Hubbard et al. (2007)	Freshwater euryhaline	0	0
Atlantic salmon	*Salmo salar*	dbEST	498,212		Diadromous	1.4 Mt	(Incl.)
Channel catfish	*Ictalurus punctatus*	dbEST	354,516		Freshwater stenohaline	1.5 Mt	1,000 t
Rainbow trout	*Oncorhynchus mykiss*	dbEST	287,967		Freshwater euryhaline	750,000 t	(Incl.)
Three-spined stickleback	*Gasterosteus aculeatus*	dbEST	276,992	Hubbard et al. (2007)	Estuarine euryhaline	0	0
Fathead minnow	*Pimephales promelas*	dbEST	258,504		Freshwater stenohaline	0	0
Atlantic cod	*Gadus morhua*	dbEST	257,217		Marine stenohaline	12,000 t	900,000 t
Blue catfish	*Ictalurus furcatus*	dbEST	139,475		Freshwater stenohaline	(Incl. in channel catfish)	
Nile tilapia	*Oreochromis niloticus*	dbEST	120,991	Unpublished	Estuarine euryhaline	3.5 Mt	200,000 t
Mummichog	*Fundulus heteroclitus*	dbEST	90,441	Unpublished	Estuarine euryhaline	0	0
Gilthead seabream	*Sparus aurata*	dbEST	7216		Marine euryhaline	140,000 t	?

[a]Total expressed sequence tags filed to the National Center for Biotechnology Information (NCBI) as of November 2012.
[b]Food and Agriculture Organization of the United Nations (FAO) (2012), http://www.fao.org/fishery/publications/yearbooks/en.
**Fishery tonnage (if any) included with aquaculture.
***Wellcome Trust Sanger Institute, http://www.sanger.ac.uk/resources/zebrafish.

biomonitoring function of the estuarine genomic fish model *Fundulus heteroclitus* has been demonstrated in a recent study examining salinity gradients and the effects of salinity variation in the habitat with genetic variation in the important clusters of osmoregulatory genes (Whitehead, 2010; Whitehead et al., 2011b). With careful and ongoing documentation of the genomic database (Paschall et al., 2004), comparative transcriptomics can reveal essential functional genes that allow the northern subspecies of the mummichog to be stronger hypoosmoregulators. Key to effective hypoosmoregulation in FW is reduction in passive permeability through the development of effective tight junctions in skin and gill epithelia, supportive calcium metabolism and rapidly responsive (6 h) osmotic stress factors (OSTF1), and aquaporin3 (AQP3) genes (Whitehead et al., 2011b) that help to maintain junctional integrity and enhance basolateral hydraulic conductivity in aid of cell volume regulation. A few well-placed transcriptomic studies have provided many physiological questions that can be answered using genomic techniques, such as selective knockout and knockdown approaches to revealing the functional importance of certain genes. For instance, morpholino knockdown techniques have been developed in mummichogs to test cytochrome P450-1A (CYP-1A) in examining the role of this enzyme in detoxification (Matson et al., 2008), and in zebrafish to discern the function of Rhcg1 in ammonium and Na^+ transport (Kumai and Perry, 2011). Importantly, multiple genes of unknown function will be highlighted and subsequently their functions can be revealed.

6. CONCLUSIONS AND PERSPECTIVES

From this discussion emerges a trend towards a tolerance strategy used by some estuarine fishes, wherein the animals cope with short-term salinity (and other) changes without launching full salinity-acclimating mechanisms and only invoke long-term acclimation mechanisms after a delay of a day or more. The mechanisms controlling this delay are currently unknown. This tolerance strategy is shared by semiterrestrial estuarine fishes that cope with hemoconcentration from evaporative water loss during terrestrial sorties. Estuarine fishes, in coping with frequent salinity change, use osmosensitive ion transport cells that respond to slight changes in environment and blood to shut off ion secretion and to retract the cells below the epithelial surface to protect from excessive ion loss in FW. Estuarine fishes use many adaptive behaviors and salinity preferences to minimize their exposure to large salinity changes. Estuarine fishes will continue to be important model species that are representative of most osmoregulatory strategies of the teleosts. The completion of important genomes will allow more sophisticated and probing questions to be asked. Remaining unresolved mechanisms include the structure and function of paracellular pathways in FW and SW, the interactions between the various transporters, especially post-translational variations and regulation by kinases, the role of NHE and Rh proteins, and further resolution of the Na^+ uptake channel. Soon there will be RNA-sequencing (RNA-seq) studies to discover important regulatory factors for ion transporters and water channels. Future studies may also focus on behavioral osmoregulation at the population extreme and molecular regulatory responses at the reductionist extreme, to complete the picture of adaptation to a continually changing environment.

REFERENCES

Acha, E. M., Mianzan, H., Guerrero, R., Carreto, J., Giberto, D., Montoya, N. and Carignan, M. (2008). An overview of physical and ecological processes in the Rio de la Plata Estuary. *Cont. Shelf Res.* 28, 1579–1588.

Agre, P., King, L., Yasui, M., Guggino, W., Ottersen, O., Fujiyoshi, Y., Engel, A. and Nielsen, S. (2002). Aquaporin water channels – from atomic structure to clinical medicine. *J. Physiol. Lond.* 542, 3–16.

Alderdice, D. F. (1988). Osmotic and ionic regulation in teleost eggs and larvae. In *Fish Physiology*, Vol. XIA (eds. W. S. Hoar and D. J. Randall), pp. 163–251. New York: Academic Press.

Aparicio, S, Chapman, J, Stupka, E, Putnam, N, Chia, J. M., Dehal, P., Christoffels, A., et al. (2002). Whole-genome shotgun assembly and analysis of the genome of *Fugu rubripes*. *Science* 297, 1301–1310.

Bagherie-Lachidan, M., Wright, S. I. and Kelly, S. P. (2008). Claudin-3 tight junction proteins in *Tetraodon nigroviridis*: cloning, tissue-specific expression, and a role in hydromineral balance. *Am. J. Physiol. Reg. Integr. Comp. Physiol.* 294, R1638–R1647.

Bagherie-Lachidan, M., Wright, S. I. and Kelly, S. P. (2009). Claudin-8 and -27 tight junction proteins in puffer fish *Tetraodon nigroviridis* acclimated to freshwater and seawater. *J Comp. Physiol. B Syst. Environ. Physiol.* 179, 419–431.

Beyenbach, K. and Liu, P. (1996). Mechanism of fluid secretion common to aglomerular and glomerular kidneys. *Kidney Int.* 49, 1543–1548.

Breves, J. P., Hasegawa, S., Yoshioka, M., Fox, B. K., Davis, L. K., Lerner, D. T., Takei, Y., Hirano, T. and Grau, E. G. (2010). Acute salinity challenges in Mozambique and Nile tilapia: differential responses of plasma prolactin, growth hormone and branchial expression of ion transporters. *Gen. Comp. Endocrinol.* 167, 135–142.

Breves, J. P., Seale, A. P., Helms, R. E., Tipsmark, C. K., Hirano, T. and Grau, E. G. (2011). Dynamic gene expression of GH/PRL-family hormone receptors in gill and kidney during freshwater-acclimation of Mozambique tilapia. *Comp. Biochem. Physiol. A Mol. Integr. Physiol.* 158, 194–200.

Chang, E. W. Y., Loong, A. M., Wong, W. P., Chew, S. F., Wilson, J. M. and Ip, Y. K. (2007). Changes in tissue free amino acid contents, branchial Na^+/K^+-ATPase activity and bimodal breathing pattern in the freshwater climbing perch, *Anabas testudineus* (Bloch), during seawater acclimation. *J. Exp. Zool. A* 307, 708–723.

Chen, X., Polleichtner, G., Kadurin, I. and Gruender, S. (2007). Zebrafish acid-sensing ion channel (ASIC) 4, characterization of homo- and heteromeric channels, and identification of regions important for activation by H^+. *J. Biol. Chem.* 282, 30406–30413.

Chew, S. F., Wilson, J. M., Ip, Y. K. and Randall, D. J. (2006). Nitrogen excretion and defense against ammonia toxicity. In *Fish Physiology*, Vol. 21, *The Physiology of Tropical Fishes* (eds. A. L. Val, V. M. F. de Almeida-Val and D. J. Randall), pp. 307–395. San Diego: Academic Press.

Chew, S. F., Sim, M. Y., Phua, Z. C., Wong, W. P. and Ip, Y. K. (2007). Active ammonia excretion in the giant mudskipper, *Periophthalmodon schlosseri* (Pallas), during emersion. *J. Exp. Zool. A* 307, 357–369.

Choe, K. P., Havird, J., Rose, R., Hyndman, K., Piermarini, P. and Evans, D. H. (2006). COX2 in a euryhaline teleost, *Fundulus heteroclitus*: primary sequence, distribution, localization, and potential function in gills during salinity acclimation. *J. Exp. Biol.* 209, 1696–1708.

Cioni, C., Demerich, D., Cataldi, E. and Cataudella, S. (1991). Fine structure of chloride cells in fresh water adapted and seawater adapted *Oreochromis niloticus* (Linnaeus) and *Oreochromis mossambicus* (Peters). *J. Fish Biol.* 39, 197–209.

Cutler, C. P. and Cramb, G. (2001). Molecular physiology of osmoregulation in eels and other teleosts: the role of transporter isoforms and gene duplication. *Comp. Biochem. Physiol. A Mol. Integr. Physiol.* 130, 551–564.

Cutler, C. P., Martinez, A. and Cramb, G. (2007). The role of aquaporin 3 in teleost fish. *Comp. Biochem. Physiol. A Mol. Integr. Physiol.* 148, 82–91.

Daborn, K., Cozzi, R. R. F. and Marshall, W. S. (2001). Dynamics of pavement cell–chloride cell interactions during abrupt salinity change in *Fundulus heteroclitus*. *J. Exp. Biol.* 204, 1889–1899.

Dabruzzi, T. F., Wygoda, M. L., Wright, J. E., Eme, J. and Bennett, W. A. (2011). Direct evidence of cutaneous resistance to evaporative water loss in amphibious mudskipper (family Gobiidae) and rockskipper (family Blenniidae) fishes from Pulau Hoga, southeast Sulawesi, Indonesia. *J. Exp. Mar. Biol. Ecol.* 406, 125–129.

Darman, R. and Forbush, B. (2002). A regulatory locus of phosphorylation in the N terminus of the NaKCl cotransporter, NKCC1. *J. Biol. Chem.* 277, 37542–37550.

Degnan, K. J. (1985). The role of K^+ and Cl^- conductances in chloride secretion by the opercular epithelium. *J. Exp. Zool.* 236, 19–25.

Degnan, K. J., Karnaky, K. J. and Zadunaisky, J. A. (1977). Active chloride transport in the *in vitro* opercular skin of a teleost (*Fundulus heteroclitus*), a gill-like epithelium rich in chloride cells. *J. Physiol. Lond.* 271, 155–191.

Delpire, E. and Gagnon, K. B. E. (2008). SPAK and OSR1: STE20 kinases involved in the regulation of ion homoeostasis and volume control in mammalian cells. *Biochem. J.* 409, 321–331.

Dharmamba, M., Bornancin, M. and Maetz, J. (1975). Environmental salinity and sodium and chloride exchanges across the gill of *Tilapia mossambica*. *J. Physiol. (Paris)* 70, 627–635.

Edwards, S. L. and Marshall, W. S. (2013). Principles and patterns of osmoregulation and euryhalinity in fishes. In *Fish Physiology*, Vol. 32, *Euryhaline Fishes* (eds. S. D. McCormick, A. P. Farrell and C. J. Brauner), pp. 1–44. New York: Elsevier.

Edwards, S., Wall, B., Morrison-Shetlar, A., Sligh, S., Weakley, J. and Claiborne, J. (2005). The effect of environmental hypercapnia and salinity on the expression of NHE-like isoforms in the gills of a euryhaline fish (*Fundulus heteroclitus*). *J. Exp. Zool. A* 303, 464–475.

Epstein, F. H., Katz, A. I. and Pickford, G. E. (1967). Sodium- and potassium-activated adenosine triphosphatase of gills: role in adaptation of teleosts to salt water. *Science* 156, 1245–1247.

Eriksson, Ö and Wistrand, P. J. (1986). Chloride transport inhibition by various types of "loop" diuretics in fish opercular epithelium. *Acta Physiol. Scand.* 126, 93–101.

Evans, D. H. (1984). The roles of gill permeability and transport mechanisms in euryhalinity. In *Fish Physiology,* Vol. XB, *Gills Part B Ion and Water Transfer* (eds. W. S. Hoar and D. J. Randall), pp. 239–283. New York: Academic Press.

Evans, D. H. (2011). Freshwater fish gill ion transport: August Krogh to morpholinos and microprobes. *Acta Physiol.* 202, 349–359.

Evans, D. H., Piermarini, P. M. and Potts, W. T. W. (1999). Ionic transport in the fish gill epithelium. *J. Exp. Zool.* 283, 641–652.

Evans, D. H., Rose, R. E., Roeser, J. M. and Stidham, J. D. (2004). NaCl transport across the opercular epithelium of *Fundulus heteroclitus* is inhibited by an endothelin to NO, superoxide, and prostanoid signaling axis. *Am. J. Physiol. Reg. Integr. Comp. Physiol.* 286, R560–R568.

Evans, D. H., Piermarini, P. M. and Choe, K. P. (2005). The multifunctional fish gill: dominant site of gas exchange, osmoregulation, acid–base regulation, and excretion of nitrogenous waste. *Physiol. Rev.* 85, 97–177.

Fall, J., Yi-ThengTseng, Ndong, D. and Sheen, S. (2011). The effects of different protein sources on the growth of hybrid tilapia (*Oreochromis niloticus x O. aureus*) reared under fresh water and brackish water. *Afr. J. Agric. Res.* 6, 5024–5029.

Flemmer, A. W., Gimenez, I., Dowd, B. F. X., Darman, R. B. and Forbush, B. (2002). Activation of the Na-K-Cl cotransporter NKCC1 detected with a phospho-specific antibody. *J. Biol. Chem.* 277, 37551–37558.

Flemmer, A. W., Monette, M. Y., Djurisic, M., Dowd, B., Darman, R., Gimenez, I. and Forbush, B. (2010). Phosphorylation state of the $Na^+K^+2Cl^-$ cotransporter (NKCC1) in the gills of Atlantic killifish (*Fundulus heteroclitus*) during acclimation to water of varying salinity. *J. Exp. Biol.* 213, 1558–1566.

Fontainhas-Fernandes, A., Gomes, E., Reis-Henriques, M. and Coimbra, J. (2003). Effect of cortisol on some osmoregulatory parameters of the teleost, *Oreochromis niloticus* L., after transference from freshwater to seawater. *Arq. Bras. Med. Vet. Zootec.* 55, 562–567.

Foster, M. A. (1969). Ionic and osmotic regulation in three species of *Cottus* (Cottidae, Teleost). *Comp. Biochem. Physiol.* 30, 751–759.

Garcia-Santos, S., Fontainhas-Fernandes, A. and Wilson, J. (2006). Cadmium tolerance in the Nile tilapia (*Oreochromis niloticus*) following acute exposure: assessment of some ionoregulatory parameters. *Environ. Toxicol.* 21, 33–46.

Genz, J. and Grosell, M. (2011). *Fundulus heteroclitus* acutely transferred from seawater to high salinity require few adjustments to intestinal transport associated with osmoregulation. *Comp. Biochem. Physiol. A Mol. Integr. Physiol.* 160, 156–165.

Gonzalez, R. J. (2012). The physiology of hyper-salinity tolerance in teleost fish: a review. *J. Comp. Physiol. B Biochem. Syst. Environ. Physiol.* 182, 321–329.

Gordon, M. S., Boetius, J., Boetius, I., Evans, D. H., McCarthy, R. and Oglesby, L. C. (1969). Aspects of the physiology of the terrestrial life in amphibious fish. I. The mudskipper *Periophthalmus sobrinus. J. Exp. Biol.* 50, 141–149.

Gordon, M. S., Ng, W. and Yip, A. Y. (1978). Aspects of the physiology of terrestrial life in amphibious fishes. III. The Chinese mudskipper *Periophthalmus cantoniensis. J. Exp. Biol.* 72, 57–77.

Griffith, R. W. (1974). Environment and salinity tolerance in the genus *Fundulus. Copeia* 1974, 319–331.

Guggino, W. B. (1980a). Salt balance in embryos of *Fundulus heteroclitus* and *F. bermudae* adapted to seawater. *Am. J. Physiol. Reg. Integr. Comp. Physiol.* 238, R42–R49.

Guggino, W. B. (1980b). Water balance in embryos of *Fundulus heteroclitus* and *F. bermudae* in seawater. *Am. J. Physiol. Reg. Integr. Comp. Physiol.* 238, R36–R41.

Guner, Y., Ozden, O., Cagirgan, H., Altunok, M. and Kizak, V. (2005). Effects of salinity on the osmoregulatory functions of the gills in Nile tilapia (*Oreochromis niloticus*). *Turk. J. Vet. Anim. Sci.* 29, 1259–1266.

Halpin, P. and Martin, K. (1999). Aerial respiration in the salt marsh fish *Fundulus heteroclitus* (Fundulidae). *Copeia* 1999, 743–748.

Hanson, J. and Courtenay, S. (1996). Seasonal use of estuaries by winter flounder in the southern Gulf of St Lawrence. *Trans. Am. Fish. Soc.* 125, 705–718.

Haynes, T. B., Phillips-Mentzos, E. and Facey, D. E. (2009). A comparison of the hyposaline tolerances of black prickleback (*Xiphister atropurpureus*) and penpoint gunnel (*Apodichthys flavidus*). *Northwest. Sci.* 83, 361–366.

Hill, W., Mathai, J., Gensure, R., Zeidel, J., Apodaca, G., Saenz, J., Kinne-Saffran, E., Kinne, R. and Zeidel, M. (2004). Permeabilities of teleost and elasmobranch gill apical membranes: evidence that lipid bilayers alone do not account for barrier function. *Am. J. Physiol. Cell Physiol.* 287, C235–C242.

Hoffmann, E. K., Schettino, T. and Marshall, W. S. (2007). The role of volume-sensitive ion transport systems in regulation of epithelial transport. *Comp. Biochem. Physiol. A Mol. Integr. Physiol.* 148, 29–43.

Hossler, F. E., Musil, G., Karnaky, K. J. J. and Epstein, F. H. (1985). Surface ultrastructure of the gill arch of the killifish, *Fundulus heteroclitus*, from seawater and freshwater, with special reference to the morphology of apical crypts of chloride cells. *J. Morphol.* 185, 377–386.

Hubbard, T. J. P., Aken, B. L., Beal, K., Ballester, B., Caccamo, M., Chen, Y., Clarke, L., et al. (2007). Ensembl 2007. *Nuc. Acids Res. Spec. Iss.* 35, D610–D617.

Hwang, P. P. and Lee, T. H. (2007). New insights into fish ion regulation and mitochondrion-rich cells. *Comp. Biochem. Physiol. A Mol. Integr. Physiol.* 148, 479–497.

Hwang, P. P., Fang, M. J., Tsai, J. C., Huang, C. J. and Chen, S. T. (1998). Expression of mRNA and protein of Na^+K^+-ATPase alpha subunit in gills of tilapia (*Oreochromis mossambicus*). *Fish Physiol. Biochem.* 18, 363–373.

Ikebe, Y. and Oishi, T. (1997). Relationships between environmental factors and diel and annual changes of the behaviors during low tides in *Periophthalmus modestus*. *Zool. Sci.* 14, 49–55.

Inokuchi, M., Hiroi, J., Watanabe, S., Hwang, P. and Kaneko, T. (2009). Morphological and functional classification of ion-absorbing mitochondria-rich cells in the gills of Mozambique tilapia. *J. Exp. Biol.* 212, 1003–1010.

Ip, Y., Chew, S., Wilson, J. and Randall, D. (2004). Defences against ammonia toxicity in tropical air-breathing fishes exposed to high concentrations of environmental ammonia: a review. *J. Comp. Physiol. B Biochem. Syst. Environ. Physiol.* 174, 565–575.

Isaia, J. (1984). Water and nonelectrolyte permeation. In *Fish Physiology*, Vol. XB, *Gills Part B Ion and Water Transfer* (eds. W. S. Hoar and D. Randall), pp. 1–38. New York: Academic Press.

Jaillon, O., Aury, J. M., Brunet, F., Petit, J. L., Stange-Thomann, M., Mauceli, E., Bouneau, L., et al. (2004). Genome duplication in the teleost fish *Tetraodon nigroviridis* reveals the early vertebrate proto-karyotype. *Nature* 431, 946–957.

Jordaan, A., Crocker, J. and Chen, Y. (2011). Linkages among physical and biological properties in tidepools on the Maine Coast. *Environ. Biol. Fish.* 92, 13–23.

Kahle, K. T., Rinehart, J. and Lifton, R. P. (2010). Phosphoregulation of the Na–K–2Cl and K–Cl cotransporters by the WNK kinases. *Biochim. Biophys. Acta Mol. Basis Dis.* 1802, 1150–1158.

Karnaky, K. J. Jr. (1986). Structure and function of the chloride cell of *Fundulus heteroclitus* and other teleosts. *Am. Zool.* 26, 209–224.

Karnaky, K. J., Kinter, L. B., Kinter, W. B. and Stirling, C. E. (1976). Teleost chloride cell. II Autoradiographic localization of gill Na^+,K^+-ATPase in killifish (*Fundulus heteroclitus* adapted to low and high salinity environments. *J. Cell Biol.* 70, 157–177.

Katoh, F., Hyodo, S. and Kaneko, T. (2003). Vacuolar-type proton pump in the basolateral plasma membrane energizes ion uptake in branchial mitochondria-rich cells of killifish *Fundulus heteroclitus*, adapted to a low ion environment. *J. Exp. Biol.* 206, 793–803.

Katoh, F., Cozzi, R. R. F., Marshall, W. S. and Goss, G. G. (2008). Distinct $Na^+/K^+/2Cl^-$ cotransporter localization in kidneys and gills of two euryhaline species, rainbow trout and killifish. *Cell Tissue Res.* 334, 265–281.

Keys, A. and Willmer, E. N. (1932). "Chloride secreting cells" in the gills of fishes, with special reference to the common eel. *J. Physiol. Lond.* 76, 368–378.

Kirk, K. L. and Wang, W. (2011). A unified view of cystic fibrosis transmembrane conductance regulator (CFTR) gating: combining the allosterism of a ligand-gated channel with the enzymatic activity of an ATP-binding cassette (ABC) transporter. *J. Biol. Chem.* 286, 12813–12819.

Kirschner, L. B. (1997). Extrarenal mechanisms in hydromineral and acid–base regulation in aquatic vertebrates. In *Handbook of Physiology*, Section 13, *Comparative Physiology*, Vol. 1 (ed. W. H. Dantzler), pp. 577–622. New York: Oxford University Press.

Kültz, D. (2013). Osmosensing. In *Fish Physiology*, Vol. 32, *Euryhaline Fishes* (eds. S. D. McCormick, A. P. Farrell and C. J. Brauner), pp. 45–68. New York: Elsevier.

Kültz, D. and Onken, H. (1993). Long-term acclimation of the teleost *Oreochromis mossambicus* to various salinities – two different strategies in mastering hypertonic stress. *Mar. Biol.* 117, 527–533.

Kültz, D., Chakravarty, D. and Adilakshmi, T. (2001). A novel 14-3-3 gene is osmoregulated in gill epithelium of the euryhaline teleost *Fundulus heteroclitus*. *J. Exp. Biol.* 204, 2975–2985.

Kumai, Y. and Perry, S. F. (2011). Ammonia excretion via Rhcg1 facilitates Na^+ uptake in larval zebrafish, *Danio rerio*, in acidic water. *Am. J. Physiol. Regul. Integr. Comp. Physiol.* 301, R1517–R1528.

Larhammar, D., Sundstrom, G., Dreborg, S., Daza, D. O. and Larsson, T. A. (2009). Major genomic events and their consequences for vertebrate evolution and endocrinology. *Ann. N. Y. Acad. Sci.* 1163, 201–208.

Laurent, P., Chevalier, C. and Wood, C. M. (2006). Appearance of cuboidal cells in relation to salinity in gills of *Fundulus heteroclitus*, a species exhibiting branchial Na^+ but not Cl^- uptake in freshwater. *Cell Tissue Res.* 325, 481–492.

LeBlanc, D. M., Wood, C. M., Fudge, D. S. and Wright, P. A. (2010). A fish out of water: gill and skin remodeling promotes osmo- and ionoregulation in the mangrove killifish *Kryptolebias marmoratus*. *Physiol. Biochem. Zool.* 83, 932–949.

Lee, H. J., Martinez, C. A., Hertzberg, K. J., Hamilton, A. L. and Graham, J. B. (2005). Burrow air phase maintenance and respiration by the mudskipper *Scartelaos histophorus* (Gobiidae: Oxudercinae). *J. Exp. Biol.* 208, 169–177.

Lin, H., Pfeiffer, D. C., Vogl, A. W., Pan, J. and Randall, D. J. (1994). Immunolocalization of H^+-ATPase in the gill epithelia of rainbow trout. *J. Exp. Biol.* 195, 169–183.

Luck, A. S. and Martin, K. L. M. (1999). Tolerance of forced air emergence by a fish with a broad vertical distribution, the rockpool blenny *Hypsoblennius gilberti* (Blenniidae). *Environ. Biol. Fish.* 54, 295–301.

Machado, B. E. and Podrabsky, J. E. (2007). Salinity tolerance in diapausing embryos of the annual killifish *Austrofundulus limnaeus* is supported by exceptionally low water and ion permeability. *J. Comp. Physiol. B Biochem. Syst. Environ. Physiol.* 177, 809–820.

Malvin, R. L., Schiff, D. and Eiger, S. (1980). Angiotensin and drinking rates in the euryhaline killifish. *Am. J. Physiol. Reg. Integr. Comp. Physiol.* 239, R31–R34.

Mancera, J. M. and McCormick, S. D. (2000). Rapid activation of gill Na^+,K^+-ATPase in the euryhaline teleost *Fundulus heteroclitus. J. Exp. Zool.* 287, 263–274.

Marshall, W. S. (2002). Na^+, Cl^-,Ca^{2+} and Zn^{2+} transport by fish gills: retrospective review and prospective synthesis. *J. Exp. Zool.* 293, 264–283.

Marshall, W. S. (2003). Rapid regulation of NaCl secretion by estuarine teleost fish: coping strategies for short-duration freshwater exposures. *Biochim. Biophys. Acta Biomembr.* 1618, 95–105.

Marshall, W. S. and Grosell, M. (2006). Ion transport, osmoregulation, and acid–base balance. In *The Physiology of Fishes*, 3rd Edition (eds. D. H. Evans and J. B. Claiborne), pp. 177–230. Boca Raton, FL: CRC Press.

Marshall, W. S. and Klyce, S. D. (1983). Cellular and paracellular pathway resistances in the "tight" Cl^--secreting epithelium of rabbit cornea. *J. Membr. Biol.* 73, 275–282.

Marshall, W. S., Bryson, S. E., Midelfart, A. and Hamilton, W. F. (1995). Low-conductance anion channel activated by cAMP in teleost Cl^--secreting cells. *Am. J. Physiol. Regul. Integr. Comp. Physiol.* 268, R963–R969.

Marshall, W. S., Bryson, S. E., Darling, P., Whitten, C., Patrick, M., Wilkie, M., Wood, C. M. and Buckland Nicks, J. (1997). NaCl transport and ultrastructure of opercular epithelium from a freshwater-adapted euryhaline teleost, *Fundulus heteroclitus. J. Exp. Zool.* 277, 23–37.

Marshall, W. S., Emberley, T. R., Singer, T. D., Bryson, S. E. and McCormick, S. D. (1999). Time course of salinity adaptation in a strongly euryhaline estuarine teleost, *Fundulus heteroclitus*: a multivariable approach. *J. Exp. Biol.* 202, 1535–1544.

Marshall, W. S., Bryson, S. E. and Luby, T. (2000). Control of epithelial Cl^- secretion by basolateral osmolality in the euryhaline teleost *Fundulus heteroclitus. J. Exp. Biol.* 203, 1897–1905.

Marshall, W. S., Howard, J. A., Cozzi, R. R. F. and Lynch, E. M. (2002a). NaCl and fluid secretion by the intestine of the teleost *Fundulus heteroclitus*: involvement of CFTR. *J. Exp. Biol.* 205, 745–758.

Marshall, W. S., Lynch, E. A. and Cozzi, R. R. F. (2002b). Redistribution of immuno-fluorescence of CFTR anion channel and NKCC cotransporter in chloride cells during adaptation of the killifish *Fundulus heteroclitus* to sea water. *J. Exp. Biol.* 205, 1265–1273.

Marshall, W. S., Cozzi, R. R. F., Pelis, R. M. and McCormick, S. D. (2005a). Cortisol receptor blockade and seawater adaptation in the euryhaline teleost *Fundulus heteroclitus. J. Exp. Zool. A Comp. Exp. Biol.* 303, 132–142.

Marshall, W., Ossum, C. and Hoffmann, E. (2005b). Hypotonic shock mediation by p38 MAPK, JNK, PKC, FAK, OSR1 and SPAK in osmosensing chloride secreting cells of killifish opercular epithelium. *J. Exp. Biol.* 208, 1063–1077.

Marshall, W. S., Katoh, F., Main, H. P., Sers, N. and Cozzi, R. R. F. (2008). Focal adhesion kinase and beta 1 integrin regulation of $Na^+,K^+,2Cl^-$ cotransporter in osmosensing ion transporting cells of killifish, *Fundulus heteroclitus. Comp. Biochem. Physiol. A Mol. Integr. Physiol.* 150, 288–300.

Marshall, W. S., Watters, K. D., Hovdestad, L. R., Cozzi, R. R. F. and Katoh, F. (2009). CFTR Cl^- channel functional regulation by phosphorylation of focal adhesion kinase at tyrosine 407 in osmosensitive ion transporting mitochondria rich cells of euryhaline killifish. *J. Exp. Biol.* 212, 2365–2377.

Martinez, A., Cutler, C., Wilson, G., Phillips, C., Hazon, N. and Cramb, G. (2005). Regulation of expression of two aquaporin homologs in the intestine of the European eel: effects of seawater acclimation and cortisol treatment. *Am. J. Physiol. Regul. Integr. Comp. Physiol.* 288, R1733–R1743.

Matson, C. W., Clark, B. W., Jenny, M. J., Fleming, C. R., Hahn, M. E. and Di Giulio, R. T. (2008). Development of the morpholino gene knockdown technique in *Fundulus heteroclitus*: a tool for studying molecular mechanisms in an established environmental model. *Aquat. Toxicol.* 87, 289–295.

McCormick, S. D., Regish, A. M. and Christensen, A. K. (2009). Distinct freshwater and seawater isoforms of Na^+/K^+-ATPase in gill chloride cells of Atlantic salmon. *J. Exp. Biol.* 212, 3994–4001.

Mommsen, T. P., Vijayan, M. M. and Moon, T. W. (1999). Cortisol in teleosts: dynamics, mechanisms of action, and metabolic regulation. *Rev. Fish Biol. Fish.* 9, 211–268.

Monteiro, S. M., Oliveira, E., Fontainhas-Fernandes, A. and Sousa, M. (2010). Fine structure of the branchial epithelium in the teleost *Oreochromis niloticus*. *J. Morphol.* 271, 621–633.

Motais, R., Garcia-Romeu, F. and Maetz, J. (1966). Exchange diffusion effect and euryhalinity in teleosts. *J. Gen. Physiol.* 50, 391–422.

Nakano, K. and Iwama, G. (2002). The 70-kDa heat shock protein response in two intertidal sculpins, *Oligocottus maculosus* and *O. snyderi*: relationship of hsp70 and thermal tolerance. *Comp. Biochem. Physiol. A Mol. Integr. Physiol.* 133, 79–94.

Oczkowski, A. and Nixon, S. (2008). Increasing nutrient concentrations and the rise and fall of a coastal fishery; a review of data from the Nile Delta, Egypt. *Estuar. Coast. Shelf Sci.* 77, 309–319.

Paschall, J., Oleksiak, M., VanWye, J., Roach, J., Whitehead, J., Wyckoff, G., Kolell, K. and Crawford, D. (2004). FunnyBase: a systems level functional annotation of *Fundulus* ESTs for the analysis of gene expression. *BMC Genomics* 5, 96.

Patrick, M. L. and Wood, C. M. (1999). Ion and acid–base regulation in the freshwater mummichog (*Fundulus heteroclitus*): a departure from the standard model for freshwater teleosts. *Comp. Biochem. Physiol. A Mol. Integr. Physiol.* 122, 445–456.

Patrick, M. L., Part, P., Marshall, W. S. and Wood, C. M. (1997). Characterization of ion and acid–base transport in the fresh water adapted mummichog (*Fundulus heteroclitus*). *J. Exp. Zool.* 279, 208–219.

Patterson, J., Bodinier, C. and Green, C. (2012). Effects of low salinity media on growth, condition, and gill ion transporter expression in juvenile Gulf killifish, *Fundulus grandis*. *Comp. Biochem. Physiol. A Mol. Integr. Physiol.* 161, 415–421.

Pequeux, A., Gilles, R. and Marshall, W. S. (1988). NaCl Transport in Gills and Related Structures. (ed. R. Greger) *Advances in Comparative & Environmental Physiology 1. NaCl Transport in Epithelia*, pp. 1–73. New York: Springer.

Philpott, C. W. and Copeland, D. E. (1963). Fine structure of chloride cells from three species of *Fundulus*. *J. Cell. Biol.* 18, 389–404.

Pisam, M., LeMoal, C., Auperin, B., Prunet, P. and Rambourg, A. (1995). Apical structures of mitochondria-rich alpha and beta cells in euryhaline fish gill – their behavior in various living conditions. *Anat. Rec.* 241, 13–24.

Potts, W. T. W. (1984). Transepithelial potentials in fish gills Part B Ion and Water Transfer. In *Fish Physiology*, vol. XB, *Gills Part B Ion and Water Transfer* (eds. W. S. Hoar and D. J. Randall), pp. 105–128. New York: Academic Press.

Potts, W. T. and Evans, D. H. (1967). Sodium and chloride balance in the killifish *Fundulus heteroclitus*. *Biol. Bull.* 133, 411–425.

Potts, W. T. W., Fletcher, C. R. and Hedges, A. J. (1991). The *in vivo* transepithelial potential in a marine teleost. *J. Comp. Physiol. B Biochem. Syst. Environ. Physiol.* 161, 393–400.

Randall, D. J., Ip, Y. K., Chew, S. F. and Wilson, J. M. (2004). Air breathing and ammonia excretion in the giant mudskipper, *Periophthalmodon schlosseri*. *Physiol. Biochem. Zool.* 77, 783–788.

Rengmark, A. H., Slettan, A., Lee, W. J., Lie, O. and Lingaas, F. (2007). Identification and mapping of genes associated with salt tolerance in tilapia. *J. Fish Biol.* 71, 409–422.

Robertson, J. C. and Hazel, J. R. (1999). Influence of temperature and membrane lipid composition on the osmotic water permeability of teleost gills. *Physiol. Biochem. Zool.* 72, 623–632.

Sakamoto, T. and Ando, M. (2002). Calcium ion triggers rapid morphological oscillation of chloride cells in the mudskipper, *Periophthalmus modestus*. *J. Comp. Physiol. B Biochem. Syst. Environ. Physiol.* 172, 435–439.

Sakamoto, T., Yokota, S. and Ando, M. (2000). Rapid morphological oscillation of mitochondrion-rich cell in estuarine mudskipper following salinity changes. *J. Exp. Zool.* 286, 666–669.

Sakamoto, T., Yasunaga, H., Yokota, S. and Ando, M. (2002). Differential display of skin mRNAs regulated under varying environmental conditions in a mudskipper. *J. Comp. Physiol. B Biochem. Syst. Environ. Physiol.* 172, 447–453.

Sakamoto, T., Mori, C., Minami, S., Takahashi, H., Abe, T., Ojima, D., Ogoshi, M. and Sakamoto, H. (2011). Corticosteroids stimulate the amphibious behavior in mudskipper: potential role of mineralocorticoid receptors in teleost fish. *Physiol. Behav.* 104, 923–928.

Schofield, P. J., Peterson, M. S., Lowe, M. R., Brown-Peterson, N. J. and Slack, W. T. (2011). Survival, growth and reproduction of non-indigenous Nile tilapia, *Oreochromis niloticus* (Linnaeus 1758). I. Physiological capabilities in various temperatures and salinities. *Mar. Freshw. Res.* 62, 439–449.

Schultz, E. T. and McCormick, S. D. (2013). Euryhalinity in an evolutionary context. In *Fish Physiology, Vol. 32, Euryhaline Fishes* (eds. S. D. McCormick, A. P. Farrell and C. J. Brauner), pp. 475–533. New York: Elsevier.

Scott, G. R. and Schulte, P. M. (2005). Intraspecific variation in gene expression after seawater transfer in gills of the euryhaline killifish *Fundulus heteroclitus*. *Comp. Biochem. Physiol. A Mol. Integr. Physiol.* 141, 176–182.

Scott, G. R., Claiborne, J. B., Edwards, S. L., Schulte, P. M. and Wood, C. M. (2005). Gene expression after freshwater transfer in gills and opercular epithelia of killifish: insight into divergent mechanisms of ion transport. *J. Exp. Biol.* 208, 2719–2729.

Scott, G. R., Schulte, P. M. and Wood, C. M. (2006). Plasticity of osmoregulatory function in the killifish intestine: drinking rates, salt and water transport, and gene expression after freshwater transfer. *J. Exp. Biol.* 209, 4040–4050.

Semple, J. W., Green, H. J. and Schulte, P. M. (2002). Molecular cloning and characterization of two Na/K-ATPase isoforms in *Fundulus heteroclitus*. *Mar. Biotechnol.* 4, 512–519.

Shaw, J. R., Gabor, K., Hand, E., Lankowski, A., Durant, L., Thibodeau, R., Stanton, C. R., Barnaby, R., Coutermarsh, B., Karlson, K. H., Sato, J. D., Hamilton, J. W. and Stanton, B. A. (2007). Role of glucocorticoid receptor in acclimation of killifish (*Fundulus heteroclitus*) to seawater and effects of arsenic. *Am. J. Physiol. Reg. Integr. Comp. Physiol.* 292, R1052–R1060.

Shaw, J. R., Sato, J. D., VanderHeide, J., LaCasse, T., Stanton, C. R., Lankowski, A., Stanton, S. E., Chapline, C., Coutermarsh, B., Barnaby, R., Karlson, K. and Stanton, B. A. (2008). The role of SGK and CFTR in acute adaptation to seawater in *Fundulus heteroclitus*. *Cell. Physiol. Biochem.* 22, 69–78.

Silva, P., Stoff, J., Field, M., Fine, L., Forrest, J. N. and Epstein, F. H. (1977). Mechanism of active chloride secretion by shark rectal gland: role of Na$^+$K$^+$-ATPase in chloride transport. *Am. J. Physiol.* 233, F298–F306.

Simons, R. D., Monismith, S. G., Saucier, F. J., Johnson, L. E. and Winkler, G. (2010). Modelling stratification and baroclinic flow in the estuarine transition zone of the St. Lawrence Estuary. *Atmos. Ocean* 48, 132–146.

Singer, T. D., Tucker, S. J., Marshall, W. S. and Higgins, C. F. (1998). A divergent CFTR homologue: highly regulated salt transport in the euryhaline teleost *F. heteroclitus. Am. J. Physiol. Cell Physiol.* 274, C715–C723.

Singer, T. D., Keir, K. R., Hinton, M., Scott, G. R., McKinley, R. S. and Schulte, P. M. (2008). Structure and regulation of the cystic fibrosis transmembrane conductance regulator (CFTR) gene in killifish: a comparative genomics approach. *Comp. Biochem. Physiol. D Genomics Proteomics* 3, 172–185.

Stith, B. M., Reid, J. P., Langtimm, C. A., Swain, E. D., Doyle, T. J., Slone, D. H., Decker, J. D. and Soderqvist, L. E. (2011). Temperature inverted haloclines provide winter warmwater refugia for manatees in Southwest Florida. *Estuar. Coasts* 34, 106–119.

Suresh, A. and Lin, C. (1992). Tilapia culture in saline waters – a review. *Aquaculture* 106, 201–226.

Suzuki, N. (1992). Fine-structure of the epidermis of the mudskipper, *Periophthalmus modestus* (Gobiidae). *Jpn. J. Ichthyol.* 38, 379–396.

Takahashi, H., Sakamoto, T., Hyodo, S., Shepherd, B., Kaneko, T. and Grau, E. (2006a). Expression of glucocorticoid receptor in the intestine of a euryhaline teleost, the Mozambique tilapia (*Oreochromis mossambicus*): effect of seawater exposure and cortisol treatment. *Life Sci.* 78, 2329–2335.

Takahashi, H., Takahashi, A. and Sakamoto, T. (2006b). *In vivo* effects of thyroid hormone, corticosteroids and prolactin on cell proliferation and apoptosis in the anterior intestine of the euryhaline mudskipper (*Periophthalmus modestus*). *Life Sci.* 79, 1873–1880.

Takahashi, H., Prunet, P., Kitahashi, T., Kajimura, S., Hirano, T., Grau, E. G. and Sakamoto, T. (2007). Prolactin receptor and proliferating/apoptotic cells in esophagus of the Mozambique tilapia (*Oreochromis mossambicus*) in fresh water and in seawater. *Gen. Comp. Endocrinol.* 152, 326–331.

Takei, Y. and McCormick, S. D. (2013). Hormonal control of fish euryhalinity. In *Fish Physiology*, Vol. 32, *Euryhaline Fishes* (eds. S. D. McCormick, A. P. Farrell and C. J. Brauner), pp. 69–123. New York: Elsevier.

Taylor, M. H. (1999). A suite of adaptations for intertidal spawning. *Am. Zool.* 39, 313–320.

Tipsmark, C. K., Baltzegar, D. A., Ozden, O., Grubb, B. J. and Borski, R. J. (2008a). Salinity regulates claudin mRNA and protein expression in the teleost gill. *Am. J. Physiol. Regul. Integr. Comp. Physiol.* 294, R1004–R1014.

Tipsmark, C. K., Luckenbach, J. A., Madsen, S. S., Kiilerich, P. and Borski, R. J. (2008b). Osmoregulation and expression of ion transport proteins and putative claudins in the gill of southern flounder (*Paralichthys lethostigma*). *Comp. Biochem. Physiol. A Mol. Integr. Physiol.* 150, 265–273.

Tipsmark, C. K., Breves, J. P., Seale, A. P., Lerner, D. T., Hirano, T. and Grau, E. G. (2011). Switching of Na$^+$,K$^+$-ATPase isoforms by salinity and prolactin in the gill of a cichlid fish. *J. Endocrinol.* 209, 237–244.

Towle, D. W., Gilman, M. E. and Hempel, J. D. (1977). Rapid modulation of gill Na$^+$+K$^+$-dependent ATPase activity during acclimation of the killifish *Fundulus heteroclitus* to salinity change. *J. Exp. Zool.* 202, 179–185.

Turko, A. J., Earley, R. L. and Wright, P. A. (2011). Behaviour drives morphology: voluntary emersion patterns shape gill structure in genetically identical mangrove rivulus. *Anim. Behav.* 82, 39–47.

Van Praag, D., Farber, S. J., Minkin, E. and Primor, N. (1987). Production of eicosanoids by the killifish gills and opercular epithelia and their effect on active transport of ions. *Gen. Comp. Endocrinol.* 67, 50–57.

Velan, A., Hulata, G., Ron, M. and Cnaani, A. (2011). Comparative time-course study on pituitary and branchial response to salinity challenge in Mozambique tilapia (*Oreochromis mossambicus*) and Nile tilapia (*O. niloticus*). *Fish Physiol. Biochem.* 37, 863–873.

Verbost, P. M., Bryson, S. E., Bonga, S. E. W. and Marshall, W. S. (1997). Na^+-dependent Ca^{2+} uptake in isolated opercular epithelium of *Fundulus heteroclitus*. *J. Comp. Physiol. B Syst. Environ. Physiol.* 167, 205–212.

Wang, B., Giddings, S. N., Fringer, O. B., Gross, E. S., Fong, D. A. and Monismith, S. G. (2011). Modeling and understanding turbulent mixing in a macrotidal salt wedge estuary. *J. Geophys. Res. Oceans* 116, C02036.

Wang, P. J., Lin, C. H., Hwang, L. Y., Huang, C. L., Lee, T. H. and Hwang, P. P. (2009). Differential responses in gills of euryhaline tilapia, *Oreochromis mossambicus*, to various hyperosmotic shocks. *Comp. Biochem. Physiol. A Mol. Integr. Physiol.* 152, 544–551.

Wang, W. and Linsdell, P. (2012). Conformational change opening the CFTR chloride channel pore coupled to ATP-dependent gating. *Biochim. Biophys. Acta Biomembr.* 1818, 851–860.

Wang, Y., Cui, Y. B., Yang, Y. X. and Cai, F. S. (2000). Compensatory growth in hybrid tilapia, *Oreochromis mossambicus* × *O. niloticus*, reared in seawater. *Aquaculture* 189, 101–108.

Watanabe, S., Kaneko, T. and Aida, K. (2005). Aquaporin-3 expressed in the basolateral membrane of gill chloride cells in Mozambique tilapia *Oreochromis mossambicus* adapted to freshwater and seawater. *J. Exp. Biol.* 208, 2673–2682.

Welsh, M. (1986). The respiratory epithelium. In *Physiology of Membrane Disorders* (eds. T. E. Andreoli, J. F. Hoffman, D. D. Fanestil and S. G. Schultz), pp. 751–766. New York: Plenum Medical.

Whitehead, A. (2010). The evolutionary radiation of diverse osmotolerant physiologies in killifish (*Fundulus* sp.). *Evolution* 64, 2070–2085.

Whitehead, A., Galvez, F., Zhang, S., Williams, L. M. and Oleksiak, M. F. (2011a). Functional genomics of physiological plasticity and local adaptation in killifish. *J. Hered.* 102, 499–511.

Whitehead, A., Roach, J. L., Zhang, S. and Galvez, F. (2011b). Genomic mechanisms of evolved physiological plasticity in killifish distributed along an environmental salinity gradient. *Proc. Natl Acad. Sci. U. S. A.* 108, 6193–6198.

Wilson, J. M., Randall, D. J., Donowitz, M., Vogl, A. W. and Ip, A. K. Y. (2000). Immunolocalization of ion-transport proteins to branchial epithelium mitochondria-rich cells in the mudskipper (*Periophthalmodon schlosseri*). *J. Exp. Biol.* 203, 2297–2310.

Wilson, R. W. and Grosell, M. (2003). Intestinal bicarbonate secretion in marine teleost fish-source of bicarbonate, pH sensitivity, and consequences for whole animal acid–base and calcium homeostasis. *Biochim. Biophys. Acta Biomembr.* 1618, 163–174.

Wood, C. M. (2011). Rapid regulation of Na^+ and Cl^- flux rates in killifish after acute salinity challenge. *J. Exp. Mar. Biol. Ecol.* 409, 62–69.

Wood, C. M. and Grosell, M. (2008). A critical analysis of transepithelial potential in intact killifish (*Fundulus heteroclitus*) subjected to acute and chronic changes in salinity. *J. Comp. Physiol. B Biochem. Syst. Environ. Physiol.* 178, 713–727.

Wood, C. M. and Grosell, M. (2009). TEP on the tide in killifish (*Fundulus heteroclitus*): effects of progressively changing salinity and prior acclimation to intermediate or cycling salinity. *J. Comp. Physiol. B Biochem. Syst. Environ. Physiol.* 179, 459–467.

Wood, C. M. and Grosell, M. (2012). Independence of net water flux from paracellular permeability in the intestine of *Fundulus heteroclitus*, a euryhaline teleost. *J. Exp. Biol.* 215, 508–517.

Wood, C. M. and Marshall, W. S. (1994). Ion balance, acid–base regulation, and chloride cell function in the common killifish, *Fundulus heteroclitus*: a euryhaline estuarine teleost. *Estuaries* 17, 34–52.

Zadunaisky, J. A., Curci, S., Schettino, T. and Scheide, J. I. (1988). Intracellular voltage recordings in the opercular epithelium of *Fundulus heteroclitus*. *J. Exp. Zool.* 247, 126–130.

Zadunaisky, J. A., Cardona, S., Au, L., Roberts, D. M., Fisher, E., Lowenstein, B., Cragoe, E. J. and Spring, K. R. (1995). Chloride transport activation by plasma osmolarity during rapid adaptation to high salinity of *Fundulus heteroclitus*. *J. Membr. Biol.* 143, 207–217.

Zydlewski, J. and Wilkie, M. (2013). Freshwater to seawater transitions in migratory fishes. In *Fish Physiology*, Vol. 32, *Euryhaline Fishes* (eds. S. D. McCormick, A. P. Farrell and C. J. Brauner), pp. 253–326. New York: Elsevier.

9

EXTREME ENVIRONMENTS: HYPERSALINE, ALKALINE, AND ION-POOR WATERS

COLIN J. BRAUNER

RICHARD J. GONZALEZ

JONATHAN M. WILSON

The physiological mechanisms required for fish to live in freshwater and seawater are well described for some species and are the same mechanisms exploited in amphihaline species that migrate between freshwater and seawater. Many fish not only tolerate, but can acclimate and adapt to conditions outside conventional freshwater and seawater conditions, specifically salinities greater than seawater, alkaline waters (up to pH 10), and ion-poor waters. These environments exist around the globe and in many instances can support recreational and commercial fisheries. This chapter will describe the chemical characteristics of these water types, the physiological challenges associated with living in these extreme

Euryhaline Fishes: Volume 32
FISH PHYSIOLOGY
Copyright © 2013 Elsevier Inc. All rights reserved
DOI: http://dx.doi.org/10.1016/B978-0-12-396951-4.00009-8

environments, and the physiological solutions that permit fish not only to survive, but in some cases to thrive, in hypersaline, alkaline, and ion-poor waters.

1. INTRODUCTION

A great deal is known about the physiological mechanisms that permit fish to live in freshwater (FW) and seawater (SW; typically 33–35 ppt), and in fish that migrate between, which is the focus of this volume. However, fish can also acclimate or have adapted to environments that lie outside the range of salinities considered "normal" for either FW or SW, specifically salinities greater than those of SW (referred to as hypersaline waters from this point forward), alkaline waters, and low ionic content dilute FW (referred to as ion-poor waters from this point forward). These water types exist throughout the world, supporting a large biomass and biodiversity of fish, and in some cases recreational and even commercial fisheries. Much less is known about the mechanisms that permit fish to acclimate to or adapt and live in these waters and the focus of this chapter will be to review what is known.

Hypersaline waters are commonly found in inland saline lakes, coastal lagoons, embayments, inverted or closed estuaries, and tidal flats and pools. The main physiological challenge for fish in these environments is to minimize ion gain from and water loss to the hypersaline water, which is not different from that for SW fish; however, hypersaline water creates greater osmotic and ionic gradients (potentially leading to faster and greater exchanges), as well as an altered ionic composition relative to SW and a variable nature of both the abiotic and biotic environment.

In general, hypersaline waters are associated with low fish biodiversity and productivity and tend to have dynamic populations. Yet, some of these systems have relatively stable fish populations. Many saline lakes are also alkaline, with water pH reaching values as high as 9–10. This imposes an additional physiological challenge to these fish, primarily related to nitrogenous waste excretion. But again, certain fish species can inhabit and, in some cases, thrive in extreme environments up to pH 10. Finally, ion-poor waters create ionoregulatory challenges for fish owing to low ion availability for uptake and the potential for increased diffusional ion loss. Remarkably, this physiological challenge has been overcome by many fish in the ion-poor waters of the Amazon, which is characterized by its tremendous fish biodiversity.

This chapter will describe the environments and the fish fauna that survive and thrive in hypersaline, alkaline, and ion-poor waters. Furthermore, it will describe the physiological challenges associated with living in these environments as well as the physiological solutions and limitations that ultimately dictate species abundance in these systems.

2. HYPERSALINE WATERS

Hypersaline waters are commonly found in saline lakes, coastal lagoons, embayments, inverted and closed estuaries, and tidal flats and pools. Saline lakes by definition have an ionic content above 3 ppt and in some cases exceed 400 ppt (Hammer, 1986). Saline lakes are found on every continent and the total volume has been estimated at 104,000 km^3, which is just less than that of the world's freshwaters, illustrating that they are much more widespread than commonly appreciated (Hammer, 1986). Some of these saline lakes are very large, such as the Caspian and Aral Seas and Lake Balkhash in Central Asia, but, numerically, most are much smaller and in many cases ephemeral. Saline lakes are generally shallow. Only about 25% of saline lakes exceed 10 m in depth and only 25% of saline lakes have a surface area greater than 200 km^2 (Hammer, 1986).

The accumulation of salts defines these systems and thus specific climatic and geographic characteristics are required for their existence. In particular, saline lakes are generally, but not exclusively, found in drainage basins with no outflow (endorheic basins) where evaporation matches or exceeds precipitation and incoming salts accumulate. This tends to occur in relatively arid, semi-arid, and subhumid regions. While most of these lakes lie below 1500 m, some even below sea level, there are also saline lakes that occur at elevations of 3500–5000 m on the Tibetan plateau and South American Altiplano (Hammer, 1986).

Although there are areas of the oceans (tropical seas and the Mediterranean Sea) and open coastal zones (e.g. Great Barrier Reef and Australian Bight) that are experiencing increases in salinity of 0.4–1.5 ppt (Curry et al., 2003; Andutta et al., 2011), these increases are relatively small in magnitude and likely to have relatively minor physiological consequence for fish. However, when exchange with open waters is limited, such as in coastal lagoons or lakes, embayments, and estuaries (Bayly, 1972; Potter et al., 2010), water may become substantially more hypersaline (40–164 ppt) as evaporative water losses exceed FW inputs from surface or groundwater flow or precipitation. (An estuary is defined by Potter et al., 2010, as "a partially enclosed coastal body of water which is either permanently or periodically open to the sea and within which there is a measurable variation of salinity due to the mixture of seawater with freshwater derived from land drainage".) These conditions tend to be met in arid to semi-arid regions and may only occur seasonally; however, an extreme example of this is the salt evaporation ponds of San Francisco Bay which range in salinity from SW to over 300 ppt, with ponds of different salinities sustaining different communities.

In the case of estuaries, hypersalinization can result during natural and anthropogenic, seasonal or perennial droughts when FW input is insignificant, evaporation high, and exchange with the sea limited. Inversion of the salinity gradient (inverted estuary) where salinity is greatest at the head of the estuary can occur, as well as complete closure of the estuary from sandbar formation across the mouth of the estuary (Potter et al., 2010). Seasonally closed estuaries include the Wellstead and Beaufort estuaries in Australia, and the Sine-Saloum estuary in West Africa is an example of an open inverted estuary adversely affected by decades of drought. Some well-studied hypersaline coastal lagoons include the Coorong Lagoon (Australia), Largo de Araruama (Brazil), Sivash Sea (Ukraine), Laguna Ojo de Liebre (Mexico), and Laguna Madre (US–Mexico), as well as large embayments such as Shark Bay and Spencer Gulf in western and southern Australia, respectively (Javor, 1989).

2.1. Environmental Characteristics of Hypersaline Waters

The ionic composition of saline lakes is largely determined by that of the incoming water (influent water and rainfall), dissolution of materials from the rocks, soils, and sediments of the drainage and lake basin, and the differential precipitation or solution of salts as the lake water becomes more concentrated or dilute through subsequent water loss (usually by evaporation) or addition. In general, the ionic composition of saline lakes varies dramatically globally, but tends to be consistent regionally. For a detailed description of the ionic composition of the main saline lakes of the world, see Hammer (1986). Of 167 lakes where total salinity was reported, 84 had salinities greater than that of SW (33 ppt), indicating that hypersaline environments are relatively common among saline lakes. In SW, the dominant ions are Na^+ (followed by $Mg^{2+} > Ca^{2+}$, K^+) and Cl^- (followed by $SO_4^{2-} > HCO_3^- CO_3$) (Table 9.1). Of the 154 lakes summarized, the dominant cation is Na^+ (136 lakes), with Mg^{2+} the second most prominent cation and varying levels of Ca^{2+} and K^+. The dominant anion is Cl^- (80 lakes); however, there is a large number that are $HCO_3^- CO_3$

Table 9.1

Physicochemical water parameters of select saline and alkaline lakes, and ion-poor waters.

Water body	pH	Alkalinity (titratable mM)	Salinity (ppt)	Na^+ (mM)	Mg^{2+} (mM)	Ca^{2+} (mM)	Cl^- (mM)	SO_4^{2-} (mM)	HCO_3^- (mM)	References
Seawater	8.1	2.4	35	469	52.8	10.3	546	28	1.8	Department of Energy (1994)
Saline/alkaline lakes										
Lake Magadi (Kenya); Fish Spring Lagoon	9.9	290	21.3	356	0.04	0.65	112	0.8	67	Wood et al. (2012); Jones et al. (1977)
Lake Qinghai (China)	9.4	30	9	200	18.7 36	0.3 0.23	173			Wang et al. (2003) Wood et al. (2007)
Lake Van (Turkey)	9.8	151	22.7	337.9	4.4	0.1	160.6	24.3		Danulat (1995); Oguz (personal communication) (Yuzuncu Ÿil University)
Pyramid Lake (USA)	9.4	23	4.4	58	7.3	0.2	60	1.7		Wright et al. (1993); Iwama et al. (1997)
Walker Lake (USA)	9.5	33	12	287.1	8.6	0.2	121.3	45.8	26.2	Bigelow et al. (2010)
Salton Sea (USA)	8.2	2	44	538	58	24	486	109	4	Holdren and Montano (2002)
Aral Sea (Uzbekistan) 1952	8		10	95.3	21.4	11.5	97.3	32.4	2.5	Zavialov et al. (2009)
South Aral Sea Western basin (Uzbekistan) 2007	8.1		127	1364.5	324.0	19.0	1400.7	360.0	10.4	Gertman and Zavialov (2011)
Ion-poor waters										
Rio Negro (Brazil)				16.5		5.3	47.9			Furch (1984)
World rivers				270		370	220			Wetzel (1983)

dominated (31 lakes), both of which have varying amounts of SO_4^{2-} (Table 9.1). There are many other minor ions not mentioned here. The implications of these different ionic compositions relative to that of SW are not well studied in fish.

The ionic composition of saline lakes can vary spatially and temporally, the magnitude of which is system dependent. The former is largely a result of the location and magnitude of the incoming FW source that may result in a horizontal salinity gradient and in some cases may act as a salinity refuge for fish. Temporal variation is dependent upon the relative magnitudes of lake water dilution, predominantly through water addition from rain or snow melt, and lake water concentration, largely due to evaporative water loss or extensive ice formation, both of which exclude salts and reduce the water volume (Hammer, 1986).

The climate of saline lakes is mostly influenced by temperature, insolation, evaporation, precipitation, humidity, and winds of the region, and further influenced by altitude, among other factors. Given that most saline lakes tend to be shallow and in relatively windy areas, water temperatures tend to parallel air temperatures (Hammer, 1986). Depending upon the season, there may be thermal stratification, the degree of which generally increases with lake depth (Hammer, 1986). However, there is a tendency for turnover and destratification, at least seasonally, which may dramatically alter the water chemistry, resulting in either hypoxia or anoxia and elevated sulfides, in some cases leading to large-scale mortality of fish that live in these systems (see Section 2.5).

In the case of hypersaline SW bodies, the dominant ions are Na^+ and Cl^-, as is the case in SW. With increasing salinity due to evaporation the proportions of the major ions remain relatively constant, with the exception of Ca^{2+}, HCO_3^-, and SO_4^{2-}, which precipitate out as $CaCO_3$ (at a salinity of 70 ppt or with 50% evaporation) and as $CaSO_4$ (at salinities above 90 ppt) (Fernandez et al., 1982; Marion et al., 2009). In hypersaline coastal waters, pelagic primary production (phytoplankton) tends to be low and, instead, is dependent on benthic and microphytobenthic production (e.g. seagrass, Laguna Madre; or algal mats, Hamelin Pool stromatolites). Changes in salinity are common with seasonal differences in rainfall, river flow, and evaporation rates, as well as storms that can inundate coastal areas.

2.2. Fishes that Inhabit Hypersaline Waters

Saline lakes are often characterized by a high density of bacteria, algae, and invertebrates, but fish are not as common. In general, the fish fauna in saline lakes is limited by the fact that many lakes are ephemeral and/or occupy endorheic drainage basins. Thus, there are no direct water links to other basins for fish migrations. Mechanisms of fish dispersal are much more limited relative to invertebrates and consequently fish fauna in many saline lakes reflect human introductions to establish new or to enhance existing fisheries. Adequate records of introduction are often lacking. Thus, it is often unknown whether the native populations in permanent lakes are natural or stocked (Bayly, 1972; Hammer, 1986).

While there is a general trend for decreasing species numbers with increasing salinity (Hammer, 1986), it is generally thought that a given saline lake can support more fish species than it does. Species diversity is limited by dispersal opportunities. Saline lakes with a salinity of 3–5 ppt tend to have the most fish species, generally around 10, regardless of the lake's geographic location. Lakes with salinities greater than 20 ppt may have a single species or no species at all. Hammer (1986) provides detailed descriptions of documented fish species in many of the saline lakes, including introduced species. At extreme hypersalinities (80–120 ppt) generally only cichlid (tilapias), mugilid, and antherinid species are successful.

As described above, the ionic composition of saline lakes varies dramatically. Based upon species composition and biomass of fish living in saline lakes, whether intentionally introduced or not, the greatest fish successes appear to be in waters where Na^+ and Cl^- dominate relative to SO_4^{2-}-rich or alkaline waters (Hammer, 1986). Regardless, several species have adapted to the latter waters (see Section 3.2).

In coastal hypersaline waters, euryhaline species are typically better represented than SW or FW species (Bayly, 1972; Potter et al., 1990; Zampatti et al., 2010). However, these waters can act as spawning or nursery grounds for SW and FW species. The antherinids are particularly well adapted to these conditions (*Allanetta mugiloides*, *Atherinosoma microstoma*, *A. wallacei*, and *A. elongata*) (Potter et al., 1986; Wedderburn et al., 2008), as well as the Mugilidae (*Mugil cephalus*). Potter and Hyndes (1999) propose that closure of estuaries has led to the selection of SW species that are capable of completing their life cycles within the estuary. Their review on ichthyofaunas in permanently open, intermittently open, and seasonally closed estuaries of south-western Australia shows that estuarine-spawning species are generally more prevalent in the latter two (>95% vs. <34% total fish numbers in permanently open estuaries). Atherinids and gobies, with their short life cycles, contribute significantly to these numbers. In contrast, in the permanently saline Largo de Araruama (Brazil; 46–56 ppt), only an artisanal mullet fishery exists (Kjerfve et al., 1996).

2.3. Physiological Challenges of Hypersaline Waters

2.3.1. Ion and Water Balance Relative to Seawater

The fundamental physiological mechanisms associated with living in hypersaline waters are those used by SW fishes, but at higher salinities these mechanisms, which are detailed in Chapter 1 of this volume (Edwards and Marshall, 2013), are upregulated and modified as described below.

In SW, teleost fishes regulate osmotic pressure of internal body fluids below that of the surrounding medium. Therefore, water is lost across the large, permeable surface area of the gill epithelium. To compensate for water loss, teleosts drink SW. However, to create an osmotic gradient favorable for water absorption across the gut, they must first take up most of the Na^+ and Cl^- in the imbibed water. Secretion of HCO_3^- into the gut lumen raises the pH of the gut fluid and precipitates most Ca^{2+} and some Mg^{2+} (Wilson et al., 1996, 2002), which prevents absorption of these divalent salts as well as significantly reducing the osmotic pressure of the gut fluid, facilitating water absorption. It is estimated that over 90% of Na^+, Cl^-, and K^+ of imbibed water in the gut is transferred into the blood, which then drives water uptake (Hickman, 1968; Shehadeh and Gordon, 1969; Kirschner, 1997). Much of the Na^+ and Cl^- may enter the blood passively in the esophagus owing to a favorable ionic gradient, but some, if not most, must be actively transported across the intestinal epithelium, a process driven by basolateral Na^+/K^+-ATPase (NKA) and facilitated by apical $Na^+/K^+/2Cl^-$ (NKCC) and Na^+/Cl^- (NCC) cotransporters (Grosell, 2010). In addition to the Na^+ and Cl^- taken up across the gut, fish in SW experience large concentration gradients which drive diffusive uptake of ions across the gill epithelium. It seems likely that Cl^- enters across the gills, but because the transepithelial potential (TEP) across the gills is slightly greater than the equilibrium potential for Na^+ (around +25 mV, plasma relative to SW), gill Na^+ entry may be minor (Evans, 1993; Marshall, 2002).

In Gulf toadfish (*Opsanus beta*), intestinal Na^+ concentrations in hypersaline-acclimated fish (50 and 70 ppt) are lower than in SW-control fish while Cl^- levels are higher (McDonald and Grosell, 2006). Luminal Mg^{2+}, the dominant cation, was significantly elevated, although only at 70 ppt. All salts entering the body via the gut and gill must be excreted to prevent overall body salt accumulation. The small amounts of divalent salts that enter (primarily Mg^{2+} and SO_4^{2-}) are excreted in small volumes of isosmotic urine (Hickman, 1968; Genz et al., 2011), while monovalent salts, primarily Na^+ and Cl^-, are excreted across the gill (see review by Evans et al., 2005; Edwards and Marshall, 2013, Chapter 1, this volume). In brief, ionocytes in the branchial epithelium and opercular epithelium utilize NKA and NKCC on the basolateral membrane to excrete Cl^- through apical chloride channels, which are homologous to the human cystic fibrosis transmembrane conductance regulator (CFTR) (Singer et al., 1998; Hiroi et al., 2005, 2008). Excretion of Cl^- generates the positive TEP that drives Na^+ out through leaky paracellular channels.

In hypersaline water, both water and salt balance using the above mechanisms become more challenging. At 70 ppt, for instance, the osmotic gradient across the gills more than doubles. If the rate of water loss also doubles then fish must double their drinking rate to keep pace. Furthermore, to absorb the ingested water fish must absorb more Na^+ and Cl^- per unit volume of water imbibed. So a doubling of the drinking rate at 70 ppt could quadruple salt absorbed across the gut. Furthermore, at 105 ppt ($3 \times$ SW) if drinking rate triples then salt absorption across the gut must increase nine-fold. All this additional Na^+ and Cl^- taken in must be excreted against a much greater concentration gradient. In addition, elevated salinities require higher rates of HCO_3^- secretion in the gut to precipitate the additional Ca^{2+}; the renal excretion of extra divalent salts, Mg^{2+} in particular, is another challenge. To date, for fish the consequences of hypersaline waters with altered ion composition relative to SW have been poorly studied.

2.4. The Physiology of Hypersaline-Tolerant Fishes

2.4.1. EARLY LIFE STAGES

Early life stages in fish are often very sensitive to environmental challenges, both natural and anthropogenic (Alderdice, 1988); hypersaline waters are no exception. While some information on the mechanisms and thresholds of hypersaline tolerance in juvenile and adult fish exists, little is known for developing fish. Sensitivity to high salinity arises from the complex and dynamic processes that occur during development, which if disrupted lead to deformity or mortality.

In general, increased salinity tolerance starts between 45 and 70 days posthatch in a number of tilapia species. In fact, 39 days posthatch is the minimum recommended age to begin successful SW acclimation after hatching in FW (Watanabe et al., 1997). Similarly, in the mangrove red snapper (*Lutjanus argentimaculatus*), salinity tolerance is also lowest during the first 7 days following fertilization, and then salinity progressively increases over the next several weeks (Estudillo et al., 2000). Salinity exposure (brackish waters rather than full-strength SW or higher salinities) during spawning and hatching can increase salinity tolerance of larval fish (Watanabe et al., 1997). In the mudskipper (*Boleophthalmus pectinirostris*), prelarval development can occur in 10–40 ppt, but the optimal salinity is

25 ppt, and early juveniles prefer a lower salinity in behavioral choice experiments (Chen et al., 2008).

Understanding the physiological limitations of fishes in the Salton Sea, which has a current salinity of 47 ppt, has generated considerable attention because of a recreational fishery as well as general scientific interest (Hurlbert et al., 2007). Most fish in the Salton Sea are of SW origin and have been transplanted into that system. The upper lethal salinity tolerance for embryos and larvae of two such species, bairdiella (*Bairdiella icistia*) and sargo (*Anisotremus davidsoni*), is 40 ppt (Brocksen and Cole, 1972; Lasker et al., 1972; May, 1975), and large free-breeding populations of these species existed in the Salton Sea until a decade ago when salinity was 43 ppt. The most dominant fish in the Salton Sea is the "California" Mozambique tilapia (*Oreochromis mossambicus* × *O. urolepis hornorum*) hybrid. As adults, they have an acute salinity tolerance in excess of 95 ppt (Sardella et al., 2004a); however, salinities well below this have a large effect on larval growth and development. In F_2 larval tilapia spawned from Salton Sea, a more or less dose-dependent effect of salinity on growth exists, where 9 weeks postrearing at 35–55 ppt halved body mass compared with rearing at 5 ppt (Sardella et al., 2007). The mechanism for this stunting of growth is unknown.

Desert pupfish (*Cyprinodon macularius*) are extremely saline tolerant and embryos can be transferred from FW to salinities of 35, 45, 55, 70, and 85 ppt within just 4 h postfertilization. However, developmental rate, eye development, and pigmentation were most rapid in FW and progressively decreased with salinity, and no hatch occurred at 70 or 85 ppt (Kinne and Kinne, 1962). Clearly, high salinities can be tolerated, but not without a pronounced effect on development.

Newly hatched larval California killifish (*Fundulus parvipinnis*) tolerate salinities up to 70 ppt, depending on the egg incubation salinity (Rao, 1975), while adults tolerate salinities up to 128 ppt (Feldmeth and Waggoner, 1972). The killifish *Austrofundulus limnaeus*, which inhabits ephemeral lakes in Venezuela that probably become hypersaline, have diapausing embryos that are able to defend internal osmolality up to a salinity of 50 ppt, primarily by reducing ion and water permeability. The water permeability of *A. limnaeus* embryos is 1000 times less than that of zebrafish embryos (Machado and Podrabsky, 2007). However, other developmental stages of *A. limnaeus* are fairly sensitive to elevated salinity, indicating that if the water salinity has not been reduced at hatch, these fish are likely to die.

2.4.2. JUVENILE AND ADULT FISH

When saline-tolerant fishes are incrementally exposed to hypersaline waters (e.g. a 5 ppt increase every 5–7 days), ionoregulatory ability is better compared with an abrupt transfer to a high salinity (Nordlie, 1985; Hotos

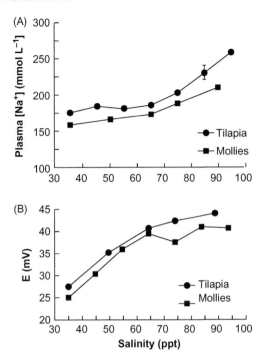

Fig. 9.1. Effect of salinity on (A) plasma Na^+ concentration and (B) equilibrium potential (E) for Na^+ across the branchial epithelium for "California" Mozambique tilapia hybrids (data from Sardella et al., 2004b) and sailfin mollies (data from Gonzalez et al., 2005a). Values for E were calculated from plasma Na^+ and water Na^+ concentrations from individual fish. Values are means $\pm$ SE. Error bars that are not visible are contained within the symbols.

and Vlahos, 1998; Sardella et al., 2004a, b; Gonzalez et al., 2005a). Gradual salinity transfer up to about 70–75 ppt has either no effect or only a modest effect on plasma osmolality or salt concentrations; however, above 75 ppt, values increase considerably (Figs. 9.1 and 9.2) (Valentine and Miller, 1969; Lotan, 1971; Griffith, 1974; Nordlie, 1985; Nordlie and Walsh, 1989; Nordlie et al., 1992; Jordan et al., 1993; Sardella et al., 2004b; Gonzalez et al., 2005a). Muscle water content, when measured as an indicator of overall water balance status (Sardella et al., 2004a, b; Gonzalez et al., 2005a), decreases only slightly (5%) or not at all, even at 85–95 ppt, indicating that despite the elevations in plasma ion concentrations hypersaline-tolerant fishes avoid problematic internal fluid shifts. One study found elevated levels of the osmolyte glycine in muscle tissue and *myo*-inositol in brain tissue of Mozambique tilapia at 70 ppt (Fiess et al., 2007), which is an area clearly worthy of further investigation given the possibility of a physiological mechanism to maintain the water volume of these tissues.

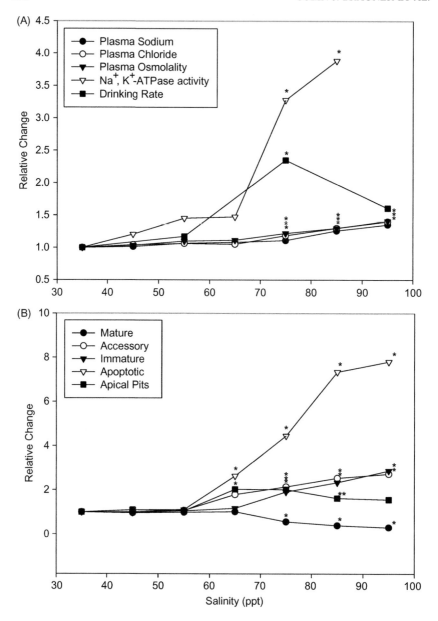

Fig. 9.2. Effect of salinity on relative changes (standardized to 35 ppt values) in (A) plasma [Na$^+$], [Cl$^-$], osmolality, gill Na$^+$/K$^+$-ATPase activity and drinking rate, and (B) gill ionoregulatory cell developmental stages (mature, immature, and apoptotic) and characteristics (apical pits, accessory cells) in "California" Mozambique tilapia hybrids. Reproduced from Sardella et al. (2007) with permission from Taylor and Francis.

Hypersaline exposure is initially associated with water loss. Thus, an important adaptation of hypersaline fishes is a reduction in branchial water permeability. This can generally be inferred from drinking rate, which is a proxy for the rate of water loss. Drinking rate increases with salinity, but not in proportion to the osmotic gradient to which the fish are exposed (Fig. 9.2) (Skadhauge and Lotan, 1974; Sardella et al., 2004b; Gonzalez et al., 2005a; Genz et al., 2008). For example, sailfin mollies (*Poecilia latipinna*) exposed to 60 ppt, elevated drinking rate by only 35%, despite a doubling of the osmotic gradient across the gills (Gonzalez et al., 2005a). The lower than expected increase in drinking rate at 60 ppt indicates that water permeability has probably been reduced. However, the 35% increase in drinking rate at 60 ppt, where salt concentrations are 70% higher than SW, may double salt entry into the gut. Water uptake from the gut is accomplished by an elevation in gut NKA activity (Sardella et al., 2004b; Gonzalez et al., 2005a) which drives NaCl uptake into the blood along with osmotically obliged water. Furthermore, there is an increased secretion of HCO_3^- into the gut, resulting in Ca^{2+} and Mg^{2+} precipitation, reducing osmolality, and further enhancing water uptake from the gut (Genz et al., 2008).

The plasma ion load incurred by drinking hypersaline water must be excreted across the branchial epithelium. An increase in ionocyte density and/or size (Kültz and Onken, 1993; Kültz et al., 1995; Uchida et al., 2000; Sardella et al., 2004b; Ouattara et al., 2009), although not always (Kültz et al., 1992; Wilson et al., 2007), is likely to facilitate this, and there appears to be a greater turnover of ionocytes as indicated by a progressive increase in apoptotic cells with salinity, in conjunction with an increase in immature and decrease in mature ionocytes with salinity (Fig. 9.2). Branchial NKA can increase two- to five-fold with a doubling of salinity (Fig. 9.2) (Karnaky et al., 1976; Kültz et al., 1995; Uchida et al., 2000; Sardella et al., 2004b; Gonzalez et al., 2005a; Fiess et al., 2007). In addition, branchial expression of NKCC and CFTR can increase with hypersaline exposure (Wilson et al., 2007; Ouattara et al., 2009). Tine et al. (2010), studying tilapia (*Saratherodon melanotheron*) in the Sine-Soloum inverted estuary, observed a higher branchial heat shock protein 70 (hsp70) messenger RNA (mRNA) and NKA (ATP1A1) expression in hypersaline environments.

Increasing ionocyte density and size, as well as elevating gill NKA activity, are necessary not only to excrete more salt in a hypersaline environment, but to do so against an elevated gradient. This is evident from the calculated equilibrium potential (*E*) for Na^+ across the gills, as illustrated for sailfin mollies and "California" Mozambique tilapia hybrids (Fig. 9.1). Using plasma and water Na^+ concentration data in the Nernst equation, *E* rises markedly between 35 and 70 ppt because external Na^+ levels rise while internal Na^+ concentrations do not. At salinities ≥ 70 ppt, *E* plateaus (particularly in tilapia) owing to the rising internal Na^+ concentrations. It appears that a larger TEP cannot be generated at salinities above 70 ppt and plasma levels rise until E falls below the TEP and Na^+ excretion resumes. This observation may explain the "biphasic" response of plasma ion concentrations to increasing salinities that has been observed (Sardella et al., 2004a). *In vitro* measurements across opercular epithelia of *Oreochromis mossambicus* showed that TEP rose linearly in fish acclimated from 35 to 60 ppt (Kültz and Onken, 1993).

Little is known about renal function in the most hypersaline-tolerant species. A few studies have examined renal function of Gulf toadfish (McDonald and Grosell, 2006; Genz et al., 2008, 2011), discovering that urine production predictably decreases and divalent cation concentrations increase. Similar observations were made in European flounder (*Platichthys flesus*) (R. W. Wilson, personal communication). However, neither species is very salt tolerant and the toadfish possesses atypical aglomerular nephrons. It would be interesting to examine kidney function in salt-tolerant species such as the cichlids and cyprinodonts.

2.4.3. THE ENERGETIC COSTS OF IONO-OSMOREGULATION

An obvious metabolic implication of life in different salinities has to do with the cost of iono-osmoregulation, which has been surprisingly difficult to determine. As discussed above, the regulation of ionoregulatory and osmoregulatory status at a given salinity is dependent upon a host of energy-consuming transporters at the gills, gut, and kidney, depending upon the salinity inhabited. In a comprehensive and elegant analysis, Kirschner (1993, 1995) determined the theoretical thermodynamic cost of ionoregulation in the FW fish gill to be about 1.6% of resting metabolic rate, similar to that calculated by Eddy (1982). In an isolated perfused FW gill preparation, Morgan and Iwama (1999) determined that the metabolic cost of gill NaCl uptake was surprisingly close to these theoretical values (1.8% of estimated resting metabolic rate). Estimates for the metabolic cost of ionoregulation across the SW gill were slightly higher (5.7%) (Kirschner, 1993), but in general, these estimates reveal fairly low values.

In contrast, many studies have used oxygen consumption rates ($\dot{M}O_2$) in fish transferred to different salinities to calculate the whole animal cost of iono-osmoregulation. The assumption of these studies is that $\dot{M}O_2$ will be lowest in an isosmotic environment (where the cost of maintaining osmoregulatory status would be minimal), and an elevation in $\dot{M}O_2$ at salinities above or below isosmotic is indicative of additional costs associated with iono-osmoregulation. However, iono-osmoregulation costs estimated in this manner range from 20 to 68% of resting metabolic rate, depending upon species, life history stage, acclimation duration, and experimental design, and therefore must be interpreted with caution (see Boeuf and Payan, 2001; Soengas et al., 2007, for reviews). Some studies have even found costs estimated in this manner close to zero (Morgan and Iwama, 1991; Pérez-Robles et al., 2012).

It is not unreasonable to expect that resting $\dot{M}O_2$ would increase as salinity is elevated above SW values to deal with the metabolic costs associated with increased drinking and ion exchanges, to maintain water and ion balance as described above. Gonzalez and McDonald (1992) have hypothesized that maintaining osmotic balance places limitations on metabolic scope (difference between maximal and resting metabolic rate), which will thus affect activity levels. However, hypersaline fishes commonly have a metabolic rate up to 40% lower than resting values in SW (see Sardella and Brauner, 2007a, for a review). For example, resting metabolic rate of the euryhaline milkfish (*Chanos chanos*) (Swanson, 1998) was reduced by 25% after acclimation from SW to 55 ppt. Swimming activity was also reduced, an observation consistent with the hypothesis of Gonzalez and McDonald (1992) that minimization of osmoregulatory disturbance takes priority over activity. In the "California" Mozambique tilapia hybrid, 2 week acclimation to salinities from 60 to 95 ppt reduced resting metabolic rate by about 40%, which was associated with an increase in plasma osmolality and gill NKA activity. A reduction in brain NKA and liver total ATPases suggests tissue-specific metabolic suppression during hypersaline exposure (Sardella et al., 2004a; Sardella and Brauner, 2008). Thus, there appears to be at least some control over metabolic rate, at least during initial exposure to elevated salinity, which may be beneficial during acclimation to these high salinities (Sardella and Brauner, 2008). However, this possibility needs further investigation.

2.4.4. The Osmorespiratory Compromise in Hypersaline Waters

The fish gill is a multipurpose organ. It is responsible for gas exchange, acid–base balance, and nitrogenous waste excretion, as well as ionoregulation, as described above. Therefore, there is a tradeoff among gill functions, which in relation to ionoregulation and gas exchange has been referred to as the osmorespiratory compromise (Randall et al., 1972; Nilsson, 1986). Conditions that are beneficial to gas exchange, such as a large gill surface area, thin diffusion distance, and high water and blood flows, are the very characteristics that are detrimental to maintaining ion and water balance and must be actively counteracted (Sardella and Brauner, 2007a). Conversely, alterations in gill morphology to enhance ion excretion, such as increased ionocyte density and increased gill diffusion distance, may directly impair oxygen uptake. Life in hypersaline waters may have implications for the osmorespiratory compromise through salinity effects on metabolic rate (and thus the cost of iono-osmoregulation and metabolic suppression; see above) and gill morphology, and through seasonally variable water temperature, which directly affects metabolic rate and oxygen solubility of the water, all of which will briefly be discussed here.

Most shallow saline lakes and hypersaline coastal water bodies are greatly influenced by atmospheric temperatures. Furthermore, they are often located in semi-arid or arid locations where air temperatures, and thus water temperatures, are high. An increase in temperature of 10°C generally doubles or triples a fish's metabolic rate ($Q_{10} = 2$–3). Both temperature and salinity compound matters by decreasing the oxygen solubility of water and hence oxygen availability. So, for a given oxygen extraction from water, ventilation volume must be greater in warmer, hypersaline water, which then has implications for osmoregulation, but few studies have investigated this interaction. The "California" Mozambique tilapia hybrid acclimated to 35 ppt and 25°C and transferred directly to 43, 51, and 60 ppt for 24 h was able to maintain plasma osmolality constant over this salinity range. However, when conducted at 35°C, which presumably elevated metabolic rate and reduced oxygen solubility, plasma osmolarity progressively increased with salinity, reaching values at 60 ppt that were 50% higher than those at 35 ppt (Sardella and Brauner, 2008). Altered membrane fluidity and permeability could easily contribute to these clear and large effects of elevated temperature on osmoregulation in hypersaline waters; however, this remains to be specifically investigated.

2.5. Life on the Edge: Fish Population Dynamics in Hypersaline Environments

Although hypersaline waters exist on every continent and comprise a relatively large volume even in comparison with the Earth's total FW volume, only some of these sustain fish populations. The following are some specific examples of systems that currently sustain, or have historically sustained, fish populations in hypersaline waters.

2.5.1. SALTON SEA

In North America, there are many saline lakes, many of which are located in the western USA in the rain shadows of the Great Basin, Mojave, Sonoran, or Chihuauan deserts. One system that has been quite extensively studied in terms of its fish population dynamics over the past century is that of the Salton Sea, which is the largest lake in California (980 km^2) and resides below sea level in the Imperial Valley. Its most recent formation (it has historically existed and dried up repeatedly) occurred in 1905 when the

Colorado River accidentally breached a diversion structure, filling the Salton Sea basin over the following 16 months before the river was diverted back to its regular channel. Starting with Colorado River water, the Salton Sea rapidly increased salinity to SW values over a relatively short 15-year period owing to the combination of dissolution of salts from the sea floor, a high evaporative water loss associated with this arid region, and inflow of mildly saline (2–4 ppt) agricultural waters (Hurlbert et al., 2007). Currently, the salinity is 47 ppt and is increasing at 0.3 ppt annually. Coupled with dramatic changes in oxygen, temperature, and hydrogen sulfide (Watts et al., 2001), the Salton Sea presents very challenging conditions for fish inhabitants.

Fish such as the common carp (*Cyprinus carpio*), striped mullet (*Mugil cephalus*), humpback sucker (*Xyrauchen texanus*), and rainbow trout (*Oncorhynchus mykiss*) entered the Salton Sea with the Colorado River water. However, with salinity rapidly increasing, most of these species perished and fish biomass was greatly reduced (Fig. 9.3). The California Department of Fish and Game introduced as many as 35 species of SW fish from 1929 to 1956 (Whitney, 1967). The most successful stocking effort occurred in 1950–1951 when fish introduced from the Gulf of California successfully established relatively large populations of bairdiella, sargo, and orange mouth corvine (*Cynoscion xanthulus*). These species formed the basis of a successful recreational fishery beginning in the 1960s (Hurlbert et al., 2007). Some time in the 1960s it is thought that two species of tilapia entered the Salton Sea and by the early 1980s, one of these, the "California" Mozambique tilapia hybrid, became the most abundant fish species.

Pronounced fish mortality events have occurred in the Salton Sea throughout the past century owing to its dynamic environment. Those from the late 1980s and onwards (Hurlbert et al., 2007) are well documented and show a large reduction in fish biomass followed by a recovery in the mid-1990s. In the early 2000s a series of mortalities, referred to as the "millennium crash", resulted in the majority of fish perishing. Some recovery of tilapia has occurred, but other species, if present, are caught only in very low numbers. The basis for these high mortalities is not well understood, but they probably stem from a combination of mixing events and associated anoxia and high sulfide levels which tend to occur in the summer (Hurlbert et al., 2007), and winter kills of tilapia which are probably a result of increasing salinity and low temperature (Sardella and Brauner, 2007b). With the latter, the ability of tilapia to regulate plasma ion levels at 15°C or lower is greatly affected through the direct effect of temperature on gill NKA, as described above (Sardella et al., 2004b).

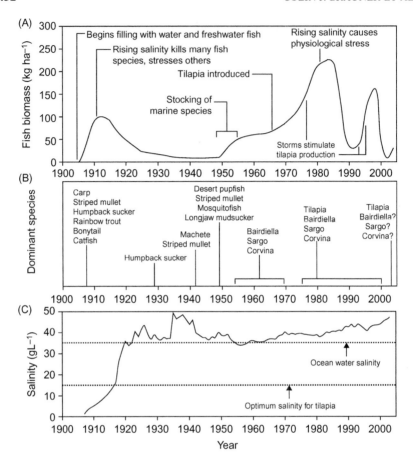

Fig. 9.3. Estimated changes in fish biomass, dominant fish species, and salinity of the Salton Sea in the twentieth century. Reproduced from Hurlbert et al. (2007) with permission from Taylor and Francis.

2.5.2. ARAL SEA

The Aral Sea in central Asia (Kazakhstan, Uzbekistan), once the fourth largest lake in the world (surface area of 66,100 km² and volume of 1064 km³) with a salinity of 8–10 ppt, is endorheic and has been isolated since the last ice age (10,000 years ago). Although characterized by low biodiversity and productivity it supported 20 endemic fish species, 12 of commercial importance including the Fringebarbel sturgeon (*Acipenser nudiventris*) and Aral barbell (*Barbus brachycephalus brachycephalus*) (Aladin and Potts, 1992). Since 1927, a total of 21 species have been introduced either intentionally to improve fisheries or accidentally. However, of these only the Baltic herring (*Clupea harengus membras*), silverside (*Atherina boyeri caspia*), bubyr (*Pomatoschistus caucasicus*), monkey goby (*Neogobius fluviatilis*), and round goby (*Neogobius melanostomus*) established populations. Commercial fisheries were also developed for the introduced grass carp (*Ctenopharyngodon idella*), silver carp (*Hypophthalmichthys molifrix*), black carp (*Mylopharyngodon piceus*) and, later, the European flounder (Aladin and Potts, 1992). Despite these introductions, overall productivity did not increase and native fauna were negatively impacted (Aladin et al., 2004).

The Aral Sea is most noted as an anthropogenic, ecological disaster (Mickle, 2007). In this arid region, the Amudarya (North) and Syrdarya (South) Rivers that drain into the Aral Sea were diverted for large-scale, Soviet-era (1960s), irrigation projects, which ultimately reduced the Aral Sea to less than 10% of its original volume (Mickle, 2007). Desiccation increased salinity and dramatically changed the ichthyofauna, causing extinction of species of FW origin by the 1970s (12–14 ppt) and the majority of brackish water species by the 1980s (22–24 ppt), resulting in the collapse of once important fisheries (Aladin and Potts, 1992). In 1989, falling water levels resulted in a division of what is now the North and South Aral Seas. At this time seven fish species were present [including native species: Ukrainian stickleback (*Pungitius platygaster*) and introduced species: Baltic herring, European flounder, silverside (*Atherina mochon*), monkey goby, and round goby] in both seas, when they were the same salinity (28–30 ppt). The North Aral Sea still receives water from the Amudarya River and has been in positive water balance, and construction of a dam has stabilized water levels, resulting in a decrease in salinity to 11 ppt. The European flounder fishery has been revived and FW species have recolonized the Amudarya River. In contrast, the South Aral Sea continues to receive little water from the Syrdarya River or overflow from the North Aral Sea, and continues its rapid decline with reported salinities of 82–150 ppt due to a strong negative water balance (evaporative loss exceeds input) (Aladin et al., 2004). The dropping water level divided the South Aral Sea into the Western and Eastern basins, with the latter completely drying up in 2009 (Table 9.1). No surviving species were reported by Aladin et al. (2004) when salinities had exceeded 70 ppt and no efforts are being made to save the South Sea. Instead, the dry lakebed is being explored for oil and gas.

2.5.3. SHARK BAY

Shark Bay in Western Australia is situated in the transition zone between temperate and tropical systems, which in combination with extensive seagrass beds and different salinity zones (ranging from 35 to over 65 ppt) with steep inner gulf salinity gradients, has resulted in an area with a rich biodiversity. Protective islands at the edge of the bay limit water mixing with the open sea and the development of the Fauré Sill 42,000 years ago across the south-eastern portion of Shark Bay has limited tidal mixing. The presence of persistent winds preventing stratification, high evaporative water loss, and an arid climate with little runoff from land have led to hypersalinization of Hamelin Pool and L'Haridon Bight at the bayhead. The hypersaline environment of the Hamelin Pool is famous as the home to stromatolites formed by mats of cyanobacteria.

In contrast to areas in adjacent seagrass beds and sandflats in the Eastern Gulf of Shark Bay, where 58 fish species have been identified (Black et al., 1990), the fish fauna of the hypersaline Hamelin Pool is limited to six species (Lenanton, 1977). Species included in Hamelin Pool are from the Mugilidae (flathead mullet, *Mugil cephalus*) and Antherinidae (few-ray hardyhead, *Craterocephalus pauciradiatus*), which are notably present in other hypersaline environments, as well as Theraponidae (yellowtail trumpeter, *Amphitherapon caudavittatus*), Sparidae (yellowfin seabream, *Acanthopagrus latus*) and Sillaginidae (yellowfin whiting, *Silago schomburgkii*, and golden-lined whiting, *S. analis*). The presence of some of these species in Hamelin Pool appears to be opportunistic as they are more abundant in other lower salinity regions of Shark Bay. However, the presence of both juvenile and adult stages of the few-ray hardyhead and yellowtail trumpeters indicates that these species may complete their life cycle within the pool (Lenanton, 1977). Indeed, the few-ray hardyhead is found in some of the most hypersaline environments in Australia (Potter et al., 1986). Also, in the bay as a whole, a total of 16 species has been recorded at salinities greater than 40 ppt (Bayly, 1972).

2.5.4. HYPERSALINE LAGOONS

The Laguna Madre, Texas, is an interesting example of a highly productive seagrass-based hypersaline lagoon that has been modified by human intervention over the past 70 years (Tunnell and Judd, 2002). Historically, the lagoon experienced cycles of boom-and-bust fisheries that were tied to flushing of the system by wet hurricanes, followed by a few years of high productivity as salinity increased, only to decrease again once the lagoon became extremely hypersaline (80–110 ppt). However, in the 1940s, the US Gulf Intracoastal Waterway (GIWW) was constructed, improving water circulation in the lagoon and eliminating the periods of extremely high salinity but also the posthurricane boom years (Tunnell and Judd, 2002). Subsequently, salinity has rarely exceeded 70 ppt and while stable fish populations exist, fish biodiversity is lower than in nearby non-hypersaline areas such as Corpus Christi Bay (Tunnell and Judd, 2002). The Sciaenidae (drums) were the most diverse family in Laguna Madre, with 10–13 species present. In the lagoon, 15 species have been recorded at salinities greater than 60 ppt, and an additional 10 are known to be able to tolerate even higher salinities (Bayly, 1972). However, at higher salinities few species dominate and only larger individual fish are found, indicating a lack of recruitment (Tunnell and Judd, 2002).

Another hypersaline lagoon, the Sivash Sea, which borders the Sea of Azov (Ukraine) and is the largest lagoon system in Europe (2500 km^2), was once a large hypersaline ecosystem (Siokhin et al., 2000). However, the Eastern and Central Sivash lagoons have been experiencing desalinization from agricultural (rice paddy) runoff and fish ponds, reducing salinity to one-tenth of its original level. The Eastern lagoon salinity ranges from 18 to 20 ppt and is now considered brackish. The decrease in salinity has improved fish productivity (glossa flounder, *Platichthys flesus luscus*, and the introduced grey mullet, *Mugil soiuy*), but suitable fish habitat has been lost owing to the spread of reed beds. The Western Sivash avoided desalinization but was maintained as an industrial reservoir for chemical extraction and has been subjected to industrial pollution (Siokhin et al., 2000). Estimates of ichthyofauna in the Sivash range from 30 to 45 species, with the three-spined stickleback (*Gasterosteus aculeatus*), the European anchovy (*Engraulis encrasicholus*), European flounder, grass goby (*Zostericola ophiocephalus*), and flathead mullet all reported to penetrate into historically hypersaline waters of the Sivash (Hedgpeth, 1959, cited in Bayly, 1972).

2.5.5. INVERTED ESTUARIES

The Sine Saloum estuary in Senegal has been an inverted estuary since the 1970s, when a persistent sub-Saharan drought eliminated the inflowing river and resulted in hypersaline conditions in the upper estuary where salinity is generally greater than 60 ppt, reaching values as high as 130 ppt. The influence of salinity on the life history traits of the hypersaline tolerant black-chinned tilapia (*S. melanotheron*), bongo shad (*Ethmalosa fimbriata*), and several mugilid species has been studied along this inverted salinity gradient. Panfili et al. (2004a) compared chinned tilapia populations in the Saloum with the nearby Gambia estuary, which has a normal head to mouth salinity gradient (0–38 ppt). They found a reduction in growth rate as well as size and age at maturity with increasing salinity in the Saloum, and an increase in fecundity during the wet season when salinity was reduced in the Saloum estuary. The bongo shad is a widely distributed brackish water species that supports important small-scale inland fisheries along the west coast of Africa. It tolerates a wide range of salinities from the FWs of coastal rivers to the hypersaline waters (90 ppt) of the Saloum and Casamance estuaries in Senegal. A comparison between populations from the Saloum (<60 ppt) and Gambia estuaries indicated higher fecundity and egg size in the former, although growth rate and size at maturity were inversely related to salinity, as in chinned tilapia (Panfili et al., 2004b). Among the six mugilid species found in the Saloum, young-of-the-year exhibited the lowest abundance in the uppermost region of the estuary, although recruitments were observed at salinities as high as 78 ppt (Trape et al., 2009). Salinity had no effect on size at maturity, unlike for chinned tilapia and bongo shad, perhaps indicating an early acquisition of osmoregulatory capacity in Mugilidae, which would be an area worthy of further study.

3. ALKALINE LAKES

Saline lakes tend to be alkaline to some degree, mostly with a pH ranging from 7.5 to 10 (in some cases up to 11) and titratable alkalinity ranging from 2 to 400 (Table 9.1). Of the 108 saline lakes where pH was reported by Hammer (1986), 45 had a pH greater than 9 and 13 had a pH exceeding 10. Perhaps the most studied alkaline lake with endemic fish populations at the extreme of alkalinity is Lake Magadi (Kenya). However, there are examples of other highly alkaline lakes (pH 9.4–10) supporting endemic fish populations, including the other East African rift lakes Natron and Manyara (Tazania) (Talling and Talling, 1965), as well as Lake Van (Turkey) (Danulat, 1995), Lake Qinghai (China) (Wood et al., 2007), and Pyramid and Walker Lakes (USA) (Galat et al., 1981, Beutel et al., 2001).

3.1. Environmental Characteristics of Alkaline Lakes

Lake Magadi and the other saline–alkaline lakes of the east African rift valley not only have high pH (9.5–10), but also have a $Na\text{-}HCO_3^-+CO_3^-$ dominated chemistry (Na^+ 356 mM, Cl^- 112 mM) that results in high alkalinity (>300 mM titratable alkalinity) (e.g. Wood et al., 2012) (Table 9.1). Wilson et al. (2004) have also shown that the physicochemical conditions within the different lagoons of Lake Magadi that support fish populations can vary widely (pH 9.13–10.05; titratable alkalinity 184–1625 mM; Na^+ 183–978 mM; Cl^- 46–693 mM; osmolality 278–1465 mOsm). Talling and Talling (1965) categorize these as Class III saline lakes (conductivity 6000–160,000 $\mu S\ cm^{-1}$; alkalinity >60 mM), which include the more saline lakes such as Lakes Magadi, Natron, Nakuru, Elmenteita, and Manyara. Lake Van has a similar chemistry (22 ppt, pH 9.8, Na^+ 337 mM, Cl^- 154 mM) (Table 9.1) (Danulat, 1995) to the rift lakes although less extreme alkalinity (150 mM titratable alkalinity), while Lake

Qinghai (9 ppt, Na^+, 200 mM, Cl^- 173 mM) (Wood et al., 2007) and Pyramid Lake (4.4 ppt, pH 9.4, Na^+ 58 mM, Cl^- 60 mM) (Galat et al., 1981) are Na^+/Cl^- dominant. All these lakes are also low in the divalent cation Ca^{2+} (<1 mM), although in Lakes Qinghai and Pyramid Mg^{2+} concentrations are high (18–36 mM and 7 mM, respectively). In addition, in Lake Magadi, over 90% of the surface is covered by a thick layer of trona, the floating precipitate of lake water consisting mainly of $NaHCO_3$ and Na_2CO_3, restricting open water to lagoons fed by volcanic saline hot springs along the lake edge. The lagoon waters can reach temperatures of 42°C and abundant cyanobacteria populations result in diurnal fluctuations of dissolved oxygen from 16 to 450 mmHg, during night and day, respectively (Narahara et al., 1996).

3.2. Fish Species that Reside in Alkaline Lakes

Alkaline lakes generally have low species diversity and occasionally only a single species is found. In the most studied example, Lake Magadi, only the cichlid *Alcolapia grahami* (formerly *Oreochromis alcalicus grahami* and *Tilapia grahami*) exists, although different populations of *A. grahami* are found within the isolated lagoons of Lake Magadi and Little Magadi (Wilson et al., 2004). *Alcolapia grahami* has also been transplanted and established in Lakes Nakuru (Vareschi, 1979) and Elmenteita (Owino et al., 2001). In neighboring Lake Natron, which was once (10,000–12,000 years ago) part of the larger paleolake Orolongo encompassing Lake Magadi (and Little Magadi), the closely related cichlids *Alcolapia alcalicus*, *A. latilabris*, and *A. ndalalani* are endemic (Seegers et al., 1999). Lake Manyara, 80 km further south, has an endemic population of cichlid, *Oreochromis amphimelas* (Beadle, 1962). For this species, Nagl et al. (2000) concluded on the basis of phylogenetic analyses that adaptations to saline–alkaline conditions evolved separately from the *Alcolapia* species of Lake Natron and Magadi.

The high-altitude saline–alkaline Lakes Van and Qinghai support endemic cyprinid species, the pearl mullet (*Chalcalburnus tarichi*) (Danulat and Kempe, 1992) and the scaleless carp (*Gymnocypris przewalskii*) (Wang et al., 2003), respectively. Both of these species are unique to their respective lakes and are stream spawners, migrating into FW rivers for breeding and returning to the lake for feeding and development. Food fisheries have existed for both species although both are threatened by population declines due to overfishing.

Pyramid and Walker Lakes are the remnant water bodies of the once larger paleolake Lahontan. They support more diversified ichthyofauna than the previously described lakes. Ten fish species have been reported and the four most abundant are the tui chub (*Gila bicolor*), Tahoe sucker (*Catostomus tahoensis*), Lahontan cutthroat trout (*Oncorhynchus clarki henshawi*), and cui-ui (*Chasmistes cujus*), which are endemic to these lakes (Galat et al., 1981). Notably, Lahontan cutthroat trout supports an important recreational fishery; they grow to trophy size by feeding on tui chub. Endemic lake breeding populations have been eliminated by damming of the spawning rivers of these obligate stream spawners and the lakes are stocked from hatcheries (Galat et al., 1981; Beutel et al., 2001). Lahontan cutthroat trout has also been successfully introduced into other alkaline lakes outside its native range in the USA (e.g. Omak and Grimes Lakes) (Galat et al., 1985).

3.3. Physiological Challenges of Alkaline Lakes

3.3.1. Ion and Water Balance

The Magadi tilapia *A. grahami* in many respects shows similarities to a SW teleost, given that its environment is hyperosmotic (580 mOsm) to blood. Although *A. grahami* is ureogenic, urea has only a minor role as an osmolyte (2–8% of internal osmolality) (Wood et al., 1994). *Alcolapia grahami* drinks the saline–alkaline water at a high rate (8 $\mu l\,g^{-1}\,h^{-1}$) even when exposed to dilute lake water (Wood et al., 2002a; Bergman et al., 2003). Of the intestinal water absorption, 70% is coupled with Na^+ and HCO_3^- uptake (consistent with the high imbibed Na^+ and HCO_3^-) and 30% is coupled with Na^+ and Cl^-, the latter of which is typical of SW fishes (Bergman et al., 2003) as described above. The Na^+, HCO_3^-, and Cl^- load (Bergman et al., 2003) is then excreted across the gills (Eddy et al., 1981). At the cellular level the gills of *A. grahami* possess SW-type ionocytes with neighboring accessory cells and leaky tight junctions (Laurent et al., 1995; Wood et al., 2002a). Wood et al. (2012) proposed that the electrogenic component of TEP in *A. grahami* is due to active HCO_3^- excretion and they found that the gills have an unusually low HCO_3^- permeability. A hypothetical model proposed by Laurent et al. (1995) is very similar to the classic chloride cell model proposed by Silva et al. (1977), in which electrogenic, secondarily active, HCO_3^- excretion is accomplished transcellularly in the gills, resulting in passive Na^+ efflux through the paracellular "shunt" pathway. The essential components in this excretory pathway (NKA, NKCC, and CFTR) have been demonstrated using immunohistochemistry (Fig. 9.4), indicating that they possess an apical NCC that may facilitate Cl^- uptake, similarly to FW-adapted Mozambique tilapia (*Oreochromis mossambicus*) (Hiroi et al., 2008).

One unique characteristic of *A. grahami* is that it has an exceptionally high blood pH (8.6–9.0) and intracellular pH (7.6) (Wood et al., 2002a), and

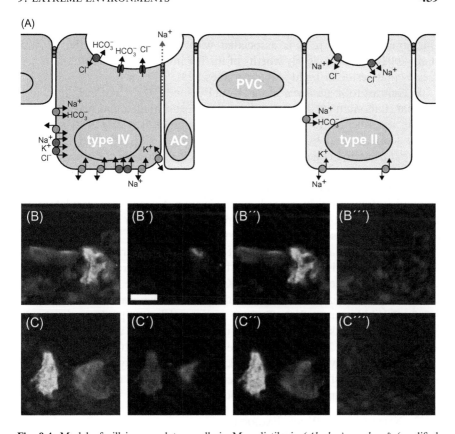

Fig. 9.4. Model of gill ionoregulatory cells in Magadi tilapia (*Alcolapia grahami*) (modified from Laurent et al., 1995). A seawater-like ionocyte (type IV*) with apical cystic fibrosis transmembrane regulator (CFTR) anion channel and basolateral Na^+/K^+-ATPase (NKA) and $Na^+/K^+/2Cl^-$ cotransporter (NKCC) is identified, but rather than functioning in NaCl secretion as in marine teleosts, it facilitates $NaHCO_3^-$ secretion with the CFTR operating as an HCO_3^- channel. A freshwater ionocyte (type II*) with apical Na^+/Cl^- cotransporter (NCC) and basolateral NKA uses the inward Na^+ gradient to drive Cl^- uptake. An apical Cl^-/HCO_3^- exchanger and a basolateral Na^+/HCO_3^- cotransporter (NBC) are also proposed for the type IV cell but their presence has not yet been confirmed. Immunofluorescent co-localization of (B, B″, C, C″) NKA (green) with either (B, B′) CFTR (red) or (C, C′) NKCC/NCC (red) in the gill cryosections of *A. grahami*. Sections were counterstained (B, B‴, C, C‴) with DAPI (blue) to label nuclei and differential interference contrast images also overlain for orientation. PVC pavement cell, AC accessory cell. Scale bar: 10 μm. From J. M. Wilson, P. Laurent, C. M. Wood, C. Chevalier, H. L. Bergman, A. Bianchini, J. N. Maina, O. E. Johannsson, L. F. Bianchini, G. D. Kavembe, M. B. Papah and R. O. Ojoo (unpublished data). *Ionocyte subtype categorized by Hiroi et al. (2008) for *Oreochromis mossambicus*. **SEE COLOR PLATE SECTION**

it has been estimated that up to 50% of this species' high metabolic rate ($34.5\,\mu mol\ O_2\ kg^{-1}\ h^{-1}$) is associated with acid–base regulation (Wood et al., 2002b), a topic clearly worthy of further investigation in this and other species that reside in highly alkaline lakes.

In contrast to *A. grahami*, species from Pyramid Lake, Lake Van, and Lake Qinghai (Lahontan cutthroat trout, tui chub, pearl mullet, and scaleless carp) are far less tolerant to either salinity or alkalinity. Taylor (1972) reported 16 ppt (concentrated by evaporating lake water) to be lethal to Lahontan cutthroat trout and concluded that this species lives close to its current limit in Walker Lake (Beutel et al., 2001). Wilkie et al. (1993) found that exposure to pH 10 was unsustainable and would lead to ionoregulatory failure and ammonia intoxication. In general, cyprinids are not strong osmoregulators. In Lake Van (22 ppt), the pearl mullet maintains very high plasma osmolality (472 mOsm) that is only 15% lower than that of lake water (551 mOsm) (Danulat, 1995). Plasma Na^+ levels (184 mM) are in the range of SW fishes' and given that Cl^- is lower (116 mM), the strong ion difference must be made up by HCO_3^-. Unexpectedly, gill NKA-immunoreactive cells decreased in distribution and size following acclimation to lake water (A. R. Oğuz, Yüzüncü Ýil University, Van, Turkey). In the scaleless carp, plasma ion levels increase towards lake levels following transfer from river water (plasma Na^+ levels were 188 and 222 mM, and Cl^- levels were 130 and 208 mM in river and lake water, respectively), and gill and kidney NKA activity decreases in lake-water acclimated fish, which is associated with up to a 40% reduction in resting metabolic rate (Wang et al., 2003; Wood et al., 2007).

3.3.2. NITROGENOUS WASTE EXCRETION

Teleosts are almost exclusively ammoniotelic, excreting the majority of their nitrogenous wastes as ammonia using an outwardly directed NH_3 partial pressure (PNH_3) gradient that can be enhanced through gill boundary layer acid trapping of NH_3 to NH_4^+ (Wright and Wood, 2009). At the molecular level, ammonia flux is facilitated by Rh glycoprotein ammonia transporters and acidification by either a proton pump or sodium proton exchanger (Wright and Wood, 2009). Urea makes up the remaining smaller portion of nitrogen flux and is generated by arginolysis and/or uricolysis rather than the ornithine urea cycle (OUC), which operates only in early life history stages and in a few adult fish (Wilkie and Wood, 1996). Ammoniotely is usually ideal in an aquatic environment because it is energetically less expensive than conversion to a less toxic molecule such as urea (OUC requires 5 ATP per urea molecule) and may diffuse directly into the water. The high pH of alkaline lakes, however, creates challenges for ammonia excretion by diffusion. The dissociation constant (pKa) of NH_4^+ is 9.25 and 9.35 in FW and SW, respectively (25°C) (Khoo et al., 1977), so an increase in water pH from 7 to 10 will increase the percentage of total ammonia that exists as NH_3 from <1 to 85%, greatly impairing NH_3 diffusion across the gills. Furthermore, in highly buffered lake water, boundary layer acidification becomes ineffective, preventing acid trapping of ammonia.

Fish from Pyramid Lake, Lake Van, and Lake Qinghai all show consistent adaptations in nitrogenous waste handling (Wilkie et al., 1993; McGeer et al., 1994; Wang et al., 2003; Danulat, 1995). Positive PNH_3 gradients are accomplished through higher plasma ammonia and/or pH levels (Wilkie et al., 1993; Danulat, 1995; Wood et al., 2007), ammonia excretion rates are reduced, and the relative contribution of urea (through uricolysis and/or arginolysis) to total nitrogen excretion is elevated (up to 35%). In Lake Van, which is markedly more saline (22 ppt), NH_4^+ diffusion may be of importance, although the molecular mechanisms of ammonia excretion in pearl mullet have yet to be studied.

In contrast to the other fishes described above that have adapted to alkaline lake conditions by modifying existing mechanisms of nitrogen excretion, the Magadi tilapia (*Alcolapia grahami*) is remarkable. It is the only teleost that is 100% ureotelic, and it uses the OUC (Randall et al., 1989) to produce urea at very high rates (7771 ± 849 µmol urea kg^{-1} h^{-1}) (Wood et al., 1989). The OUC enzymes function not only in the liver (Randall et al., 1989), but also in the white muscle (Lindley et al., 1999). The Magadi tilapia, which feeds almost continuously on nitrogen-rich cyanobacteria, has a very high metabolic rate ($\dot{M}O_2 \sim 34.5$ µmol g^{-1} h^{-1} at 36–42.5°C) (Narahara et al., 1996), which contrasts with the hypometabolic responses of the Lahontan trout and scaleless carp under alkaline conditions (Wilkie et al., 1993; Wood et al., 2007). Associated with this high metabolic rate, *A. grahami* has a very high affinity blood oxygen (Narahara et al., 1996) and a large gill diffusing capacity (Maina et al., 1996), as well as a capacity for facultative air-breathing via a physostomous air bladder (Narahara et al., 1996). *Alcolapia alcalicus* from the neighboring Lake Natron has also been shown to be ureotelic and to possess the complete OUC (Wilson et al., 2004). However, it remains to be determined whether the more distantly related *Oreochromis amphimelas*, which adapted independently in Lake Manyara, has also evolved ureotely to survive under alkaline conditions.

4. ION-POOR WATERS

4.1. Environmental Characteristics of Ion-Poor Waters

Ion-poor waters are found on virtually every continent. A few examples are waters of the Atlantic Coastal Plain of North America, much of the

Canadian Shield, peat bogs of northern Europe and much of Scandinavia, some rivers of Queensland, Australia, and the Rio Negro, a major tributary of the Amazon. They occur typically wherever soils are largely silicate sands, which bind minerals loosely. These soils have long been stripped and the waters draining this region are extremely ion poor, with Na^+, Cl^-, and Ca^{2+} concentrations well below global averages for rivers (Table 9.1). It is not unusual for ion-poor waters to contain large quantities of dissolved organic acids from partially decayed plant matter (Leenheer, 1980; Ertel et al., 1986; Walker and Henderson, 1996), and since the water has a very low buffer capacity the result is very low pH levels as well. Adaptations to ion-poor, acidic waters have been reviewed recently in this series (Gonzalez et al., 2005b) and therefore the discussion will focus on mechanisms of acclimation to ion-poor waters with some comparisons to species native to ion-poor environments.

4.2. Physiological Challenges in Ion-Poor Waters

4.2.1. ION AND WATER BALANCE RELATIVE TO FRESHWATER

In FW, fish regulate plasma Na^+ and Cl^- concentrations at levels well above those of the surrounding water. Consequently, they experience diffusive salt loss across the branchial epithelium, largely thought to be through paracellular tight junctions, although recent evidence indicates that transcellular pathways may also be involved (Wood et al., 2009; Iftikar et al., 2010; Matey et al., 2011). A number of studies that correlate the magnitude of ion efflux with water Ca^{2+} concentration indicate that permeability of gill epithelia is governed, at least in part, by the binding of Ca^{2+} to tight junction proteins (Hunn, 1985; Freda and McDonald, 1988; Gonzalez and Dunson, 1989) and thus, the lower the water Ca^{2+} the greater the gill permeability. To maintain internal Na^+ and Cl^- levels, both ions are actively taken up from the water across the gills. It is generally believed that Cl^- is taken up in exchange for HCO_3^-, but the mechanisms involved in Na^+ uptake are less clear. There is evidence for the presence of a range of transport systems on the apical membrane of the gills, including Na^+/H^+ exchangers, H^+-ATPase/Na^+ channel arrangements, and even Na^+/Cl^- cotransport (see Hwang, 2009, for review), and it is likely that there is no one mechanism common to all fish.

Ion-poor waters pose a range of challenges for ion regulation in teleost fishes. First, low ion concentrations may generate a "substrate limitation" for active Na^+ and Cl^- uptake. Both Na^+ and Cl^- transport, regardless of the mechanism involved, exhibit Michaelis–Menten-type saturation kinetics, and thus uptake tends to be low in ion-poor waters simply owing to the scarcity of these ions in the bulk medium (Potts, 1994). Potentially more important, however, is that low Ca^{2+} levels increase branchial ion permeability, presumably by removing Ca^{2+} from tight junctions, and stimulate diffusive loss of Na^+ and Cl^- (Hunn, 1985; Freda and McDonald, 1988; Gonzalez and Dunson, 1989). Together, the limitation of uptake and stimulation of efflux results in a net loss of Na^+ and Cl^- and if losses are too great (30–50%) then serious, potentially fatal, internal ionic and osmotic disturbances result (Milligan and Wood, 1982).

4.2.2. Acclimation to Ion-Poor Waters

For many FW fishes, migration or transfer into ion-poor water disrupts the ability to regulate internal salt levels and they cannot recover (McDonald and Rogano, 1986). However, some species can re-establish ion balance in ion-poor water through upregulation of ion uptake, a reduction in diffusive ion loss, or some combination. Upregulation of ion transport is typically associated with proliferation of ionocytes in the branchial epithelium (Bindon et al., 1994; Greco et al., 1995, 1996; Perry, 1998). FW ionocytes in general are much larger than surrounding pavement cells and are primarily found on the gill filaments and in intralamellar spaces, comprising less than 10% of the total gill surface area. As ionocytes proliferate during exposure to ion-poor water, they appear on the lamellar epithelium and can comprise up to 30% of the total gill surface area (Perry, 1998). There is also some indication of changes occurring in the transporters themselves. For instance, studies of Na^+ transport in zebrafish (*Danio rerio*) during acclimation to ion-poor water suggest that there is a switch from an Na^+/H^+ exchanger in ion-rich water to an H^+-ATPase/Na^+-channel arrangement in dilute water (Boisen et al., 2003; Yan et al., 2007). At the same time, fish may reduce branchial permeability and thus diffusive ion loss (McDonald, 1983; Audet and Wood, 1988; Gonzalez and Dunson, 1987; 1989). The mechanism for reduction in branchial permeability is not known but may involve adjustments in cell volume and/or tight junctions, the latter of which may involve hormones such as prolactin or cortisol. Given enough time, the degree to which efflux is reduced and influx is increased may allow fish to migrate into ion-poor waters; however, this is likely to be species specific and has only been examined in a very limited number of species.

4.2.3. Acclimation Versus Adaptation

Given the challenges to osmoregulation posed by ion-poor waters it is not surprising that most naturally occurring ion-poor environments are characterized by low piscine diversity. For example, along the Atlantic Coastal Plain of North America relatively few species inhabit these waters, with the exception of a few families such as the Centrarchidae, Ictaluridae, and Esocidae (Hastings, 1979; Graham and Hastings, 1984). Notably missing are cyprinids, the most species-rich family of fish in North America. In stark contrast to this general pattern is the rich and varied diversity of the Rio Negro. It is estimated that over 1000 species of fish from over 40 different families inhabit the Rio Negro, including a lungfish (*Lepidosiren*), two Osteoglossiformes (*Arapaima gigas* and *Osteoglussum bicirrhosum*), several dozen gymnotids, and many cichlids. The order Characiformes is particularly diverse, represented by 12 families, including the Characidae with almost 800 species (Val and Almeida-Val, 1995).

Species inhabiting ion-poor environments tend to possess specializations for ion regulation that sometimes differ from the acclimatory adjustments described above. The few measurements of plasma ion levels in these species reveal levels typical of FW fishes (Wood et al., 1998, 2002c; Brauner et al., 2004; Gonzalez et al., 2010), indicating that they do not reduce levels to mitigate the physiological challenge to ionoregulation in such a dilute environment. Gonzalez et al. (2002) described two basic patterns of ion regulation in species from the Rio Negro. One group displays high-capacity, high-affinity transport mechanisms that maintain high rates of uptake even in very dilute media. Despite these high rates of transport there is no proliferation of ionocytes on the lamellae as seen in non-native fish. It is not clear what specific transporters are involved in ion uptake in these species, but they possess some of the highest affinities reported and are able to function unhindered at pH 3.0–3.25 (Gonzalez and Preest, 1999; Gonzalez and Wilson, 2001; Preest et al., 2005). The second group has low-capacity, low-affinity transporters that produce low rates of transport in ion-poor water. These fish rely on equally low rates of diffusive loss to maintain ion balance. In both groups, regardless of their rate of uptake, diffusive ion loss is very resilient to ion-poor (or low pH) water. The waters of the Rio Negro have incredibly low Ca^{2+} concentrations (1–10 μmol L^{-1}), yet this does not seem to interfere with their ability to limit efflux (Gonzalez and Preest, 1999; Preest et al., 2005). Furthermore, there are some indications that dissolved organic carbon material in the waters of the aptly named Rio Negro can substitute for Ca^{2+} at the gills of fish in ion-poor waters, reducing passive ion efflux (Gonzalez et al., 2002; Wood et al., 2003). Such independence from water Ca^{2+} concentration has not been observed in non-native species. For further exploration of this topic see the review by Gonzalez et al. (2005b).

4.2.4. The Osmorespiratory Compromise in Ion-Poor Waters

As described above, changes in gill morphology associated with exposure to a changing environment may have implications for gas exchange through the osmorespiratory compromise. As described above, exposure to ion-poor water results in proliferation of ionocytes to enhance ion uptake; however, this is associated with a thickening of the blood–water diffusion distance of the lamellae in rainbow trout. Within 4 weeks of exposure to soft water, the blood–water diffusion distance was shown to double as a result of ionocyte proliferation (Greco et al., 1996), which negatively affected both oxygen uptake and carbon dioxide excretion. During exposure to progressive hypoxia, fish exhibiting lamellar thickening exhibited lower arterial PO_2 values than control fish exposed to the same water PO_2 (Perry, 1998). To compensate, ventilation rate was greatly elevated in rainbow trout exhibiting lamellar thickening (Greco et al., 1995). In rainbow trout acclimated to soft water, maximal swimming performance was reduced by 14% relative to control fish (Dussault et al., 2008), indicating that the impairment of gas exchange associated with lamellar thickening comes at a cost to whole-animal performance. Clearly, more research is required to understand the nature of the tradeoff between gas exchange and ionoregulation during acclimation to ion-poor waters. In fish that have adapted to ion-poor waters, the response may be quite different. In the only study where this has been investigated, exposure to hypoxia in the Amazonian oscar (*Astronotus ocellatus*) endemic to ion-poor waters was associated with a reduction in branchial ion efflux and water permeability, indicating a rapid and dramatic overall reduction in gill permeability. This was associated with changes in gill ultrastructure that were proposed to result in rapid closure of transcellular channels, thus reducing permeability, without compromising gas exchange capacity and mitigating the impact of the osmorespiratory compromise (Wood et al., 2009). Whether this is a common trait in fish native to ion-poor waters remains to be investigated, but may represent a more general adaptation among fishes to these conditions.

5. CONCLUSIONS AND PERSPECTIVES

Aquatic environments that deviate from "typical" FW or SW values are abundant throughout the world and in many cases are inhabited by fish populations that support recreational and even commercial fisheries. Fish have acclimated or adapted to the physiological challenges associated with these environments; however, we are only just starting to understand the physiological mechanisms involved. In some cases, the mechanisms represent fine-tuning or upregulation of processes that are well described in FW and SW fishes; however, in other cases more unique solutions have been discovered. A great deal remains to be learned, and the following are some areas raised in this chapter that are timely for further investigation.

In hypersaline lakes, upregulation of the mechanisms that are well described in SW fishes are important; however, little is known about the renal function and the role of the kidney in the most hypersaline-tolerant species. Furthermore, many species of hypersaline-tolerant fish exhibit a reduction in whole-animal $\dot{M}O_2$ with an increase in salinity. Is this associated with metabolic suppression (as some data imply) and, if so, how is this controlled? The ionic composition of hypersaline lakes often varies dramatically from SW ratios and little is known about the physiological implications of this. As discussed above, fish have more successfully inhabited lakes where Na^+ and Cl^- dominate over SO_4^{2-}- or HCO_3^--rich waters. Is there a physiological basis for this and, if so, what are the mechanisms that allow those fish to live in SO_4^{2-}- or HCO_3^--rich waters?

In alkaline lakes, water pH limits ammonia excretion and in the few species investigated, ammonia excretion is dependent upon an elevation in PNH_3 gradients through higher plasma ammonia and/or pH levels, and an elevation in the relative contribution of urea to total nitrogen excretion, in some cases up to 100%. How widespread is the elevation in blood pH in fishes from alkaline lakes? How widespread is ureotely among fishes that have independently adapted to alkaline lakes? In some fishes, NH_4^+ diffusion may be of importance to nitrogenous waste excretion: what are the mechanisms involved? In one species, *A. graham* (Wood et al., 2002b), it has been proposed that 50% of resting metabolic rate is associated with acid–base regulation. What is the cost of acid–base regulation in fishes in general, and is it higher in fishes adapted to alkaline lakes?

Ion-poor waters in some cases contain among the highest FW biodiversity in the world (such as in the Amazon). However, very little is known about how these fish deal with limited ion availability and the potential for diffusive ion efflux due to low environmental Ca^{2+}. In fishes known to possess high ion uptake capacity and affinity, what transporters are involved and how do they function at low pH values of 3–3.25? In other fishes that exhibit low efflux, what is the basis for low branchial permeability and how can this be accomplished in the face of low environmental Ca^{2+}? How can dissolved organic carbon substitute for Ca^{2+} at the gills to reduce efflux? In the Amazonian Oscar, which is adapted to ion-poor waters, branchial ion and water permeability can be rapidly reduced during exposure to hypoxia to minimize ionoregulatory disturbances without impairing gas exchange, apparently eliminating the osmorespiratory compromise. Is this a general characteristic of fishes adapted to ion-poor waters?

Clearly, there is a great deal to learn about how fish acclimate and adapt to hypersaline, alkaline, and ion-poor waters. Elucidation of novel physiological mechanisms that may be involved, and the degree to which described mechanisms can be modified, will give insight into the physiological control of euryhalinity and further our understanding of how fish have been able to inhabit almost every aquatic environment on the planet, and in some cases even those that have been recently drastically altered.

ACKNOWLEDGMENTS

Katelyn Tovey is thanked for bibliographic assistance. CJB was supported by a Natural Sciences and Engineering Research Council Discovery grant and JMW was supported by a Foundation for Science and Technology grant (PTDC/MAR/098035).

REFERENCES

Aladin, N. V. and Potts, W. T. W. (1992). Changes in the Aral Sea ecosystems during the period 1960–1990. *Hydrobiologia* 237, 67–79.

Aladin, N. V., Plotnikov, I. S. and Letolle, R. (2004). Hydrobiology of the Aral Sea. In *Dying and Dead Seas: Climatic Versus Anthropic Causes. NATO Science Series. IV: Earth and Environmental Sciences*, Vol. 36 (eds. J. C.J. Nihoul, P. O. Zavialov and P. E. Micklin), pp. 125–157. Dordrecht: Kluwer.

Alderdice, D. F. (1988). Osmotic and ionic regulation in teleost eggs and larvae. In *Fish Physiology*, Vol. 11A, *The Physiology of Developing Fish: Eggs and Larvae* (eds. W. S. Hoar and D. J. Randall), pp. 163–251. New York: Academic Press.

Andutta, F. P., Ridd, P. V. and Wolanski, E. (2011). Dynamics of hypersaline coastal waters in the Great Barrier Reef. *Estuar. Coast. Shelf Sci.* 94, 299–305.

Audet, C. and Wood, C. M. (1988). Do rainbow trout (*Salmo gairdneri*) acclimate to low pH? *Can. J. Fish. Aquat. Sci.* 45, 1399–1405.

Bayly, I. A. E. (1972). Salinity tolerance and osmotic behavior of animals in athalassic saline and marine hypersaline waters. *Annu. Rev. Ecol. Syst.* 3, 233–268.

Beadle, L. C. (1962). The evolution of species in lakes of East Africa. *Uganda J.* 26, 44–54.

Bergman, A. N., Laurent, P., Otiang'a-Owiti, G., Bergman, H. L., Walsh, P. J., Wilson, P. and Wood, C. M. (2003). Physiological adaptations of the gut in the Lake Magadi tilapia, *Alcolapia grahami*, and alkaline- and saline-adapted fish. *Comp. Biochem. Physiol. A Mol. Integr. Physiol.* 136, 701–715.

Beutel, M. W., Horne, A. J., Roth, J. C. and Barratt, N. J. (2001). Limnological effects of anthropogenic desiccation of a large, saline lake, Walker Lake, Nevada. *Hydrobiologia* 466, 91–105.

Bigelow, J. P., Rauw, W. M. and Gomez-Raya, L. (2010). Acclimation in simulated lake water increases survival of Lahontan cutthroat trout challenged with saline, alkaline water from Walker lake, Nevada. *Trans. Am. Fish. Soc.* 139, 876–887.

Bindon, S. D., Gilmour, K. M., Fenwick, J. C. and Perry, S. F. (1994). The effects of branchial chloride cell proliferation on respiratory function in the rainbow trout (*Oncorhynchus mykiss*). *J. Exp. Biol.* 197, 47–63.

Black, R., Robertson, A. I., Peterson, C. H. and Peterson, N. M. (1990). Fishes and benthos of near-shore seagrass and sand flat habitats at Monkey Mia, Shark Bay, Western Australia.

In *Research in Shark Bay. Report of the France–Australe Bicentenary Expedition Committee* (eds. P. F. Berry, S. D. Bradshaw and B. R. Wilson), pp. 245–261. Perth: WA Museum.

Boeuf, G. and Payan, P. (2001). How should salinity influence fish growth? *Comp. Biochem. Physiol. C Toxicol. Pharmacol.* 130, 411–423.

Boisen, A. M. Z., Amstrup, J., Novak, I. and Grosell., M. (2003). Sodium and chloride transport in soft water and hard water acclimated zebrafish (*Danio rerio*). *Biochim. Biophys. Acta* 1618, 207–218.

Brauner, C. J., Wang, T., Wang, Y., Richards, J. G., Gonzalez, R. J., Bernier, N. J., Xi, W., Patrick, M. L. and Val., A. L. (2004). Limited extracellular but complete intracellular acid–base regulation during short-term environmental hypercapnia in the armoured catfish, *Liposarcus pardalis. J. Exp. Biol.* 207, 3381–3390.

Brocksen, R. W. and Cole, R. E. (1972). Physiological responses of three species of fishes to various salinities. *J. Fish. Res. Bd. Can.* 29, 399–405.

Chen, S. X., Hong, W. S., Su, Y. Q. and Zhang, Q. Y. (2008). Microhabitat selection in the early juvenile mudskipper *Boleophthalmus pectinirostris* (L.). *J. Fish Biol.* 72, 585–593.

Curry, R., Dickson, B. and Yashayaev, I. (2003). A change in the freshwater balance of the Atlantic Ocean over the past four decades. *Nature* 426, 826–829.

Danulat, E. (1995). Biochemical–physiological adaptations of teleosts to highly alkaline, saline lakes. *Biochem. Mol. Biol. Fish.* 5, 229–249.

Danulat, E. and Kempe, S. (1992). Nitrogenous waste excretion and accumulation of urea and ammonia in *Chalcalburnus tarichi* (Cyprinidae) endemic to Lake Van (eastern Turkey). *Fish Physiol. Biochem.* 9, 377–386.

Department of Energy (1994). *Handbook of Methods for the Analysis of the Various Parameters of the Carbon Dioxide System in Sea Water*, Version 2 (eds. A. G. Dickson and C. Goyet). Oak Ridge, Tennessee, USA: ORNL/CDIAC-74.

Dussault, E., Playle, R., Dixon, D. and McKinley, R. (2008). Effects of soft-water acclimation on the physiology, swimming performance, and cardiac parameters of the rainbow trout, *Oncorhynchus mykiss. Fish Physiol. Biochem.* 34, 313–322.

Eddy, F. B. (1982). Osmotic and ionic regulation in captive fish with particular reference to salmonids. *Comp. Biochem. Physiol. B* 73, 125–141.

Eddy, F. B., Bamford, O. S. and Maloiy, G. M. O. (1981). Na and Cl effluxes and ionic regulation in *Tilapia grahami*, a fish living in conditions of extreme alkalinity. *J. Exp. Biol.* 91, 35–349.

Edwards, S. L. and Marshall, W. S. (2013). Principles and patterns of osmoregulation and euryhalinity in fishes. In *Fish Physiology*, Vol. 32, *Euryhaline Fishes* (eds. S. D. McCormick, A. P. Farrell and C. J. Brauner), pp. 1–44. New York: Elsevier.

Ertel, J. R., Hedges, J. I., Devol, A. H., Richey, J. E. and de Nazare Goes Ribeiro, M. (1986). Dissolved humic substances of the Amazon River system. *Limn. Oceanogr.* 31, 739–754.

Estudillo, C. B., Duray, M. N., Marasigan, E. T. and Emata, A. C. (2000). Salinity tolerance of larvae of the mangrove red snapper (*Lutjanus argentimaculatus*) during ontogeny. *Aquaculture* 190, 155–167.

Evans, D. H. (1993). Osmotic and ionic regulation. In *The Physiology of Fishes* (ed. D. H. Evans), pp. 315–342. Boca Raton, FL: CRC Press.

Evans, D. H., Piermarini, P. M. and Choe, K. P. (2005). The multifunctional fish gill: dominant site of gas exchange, osmoregulation, acid–base regulation, and excretion of nitrogenous waste. *Physiol. Rev.* 85, 97–177.

Feldmeth, C. R. and Waggoner, J. P. (1972). Field Measurements of tolerance to extreme hypersalinity in the California killifish, *Fundulus parvipinnis. Copeia* 1972, 592–594.

Fernendez, H., Vazquez, F. and Millero, F. J. (1982). The density and composition of hypersaline waters of a Mexican lagoon. *Limnol. Oceanogr.* 27, 315–321.

Fiess, J. C., Kunkel-Patterson, A., Mathias, L., Riley, L. G., Yancey, P. H., Hirano, T. and Grau, E. G. (2007). Effects of environmental salinity and temperature on osmoregulatory ability, organic osmolytes, and plasma hormone profiles in the Mozambique tilapia (*Oreochromis mossambicus*). *Comp. Biochem. Physiol. A Mol. Integr. Physiol.* 146, 252–264.

Freda, J. and McDonald, D. G. (1988). Physiological correlates of interspecific variation in acid tolerance in fish. *J. Exp. Biol.* 136, 243–258.

Furch, K. (1984). Water chemistry of the Amazon basin: the distribution of chemical elements among freshwaters. In *The Amazon. Limnology and Landscape Ecology of a Mighty Tropical River and its Basin* (ed. H. Sioli), pp. 167–199. Dordrecht: Dr. W. Junk.

Galat, D. L., Lider, E. L., Vigg, S. and Robertson, S. R. (1981). Limnology of a large, deep, North American terminal lake, Lake Pyramid, Nevada. *Hydrobiologia* 82, 281–317.

Galat, D. L., Post, G., Keefe, T. J. and Boucks, G. R. (1985). Histological changes in the gill, kidney and liver of Lahontan cutthroat trout, *Salmo clarki henshawi*, living in lakes of different salinity–alkalinity. *J. Fish Biol.* 27, 533–552.

Genz, J., Taylor, J. and Grosell, M. (2008). Effects of salinity on intestinal bicarbonate secretion and compensatory regulation of acid–base balance in *Opsanus beta*. *J. Exp. Biol.* 211, 2327–2335.

Genz, J., McDonald, M. D. and Grosell, M. (2011). Concentration of $MgSO_4$ in the intestinal lumen of *Opsanus beta* limits osmoregulation in response to acute hypersalinity stress. *Am. J. Physiol. R. I.* 300, R895–R909.

Gertman, I. and Zavialov, P. O. (2011). New equation of state for Aral Sea water. *Oceanology* 51, 367–369.

Gonzalez, R. J. and Dunson, W. A. (1987). Adaptations of sodium balance to low pH in a sunfish (*Enneacanthus obesus*) from naturally acidic waters. *J. Comp. Physiol.* 157, 555–566.

Gonzalez, R. J. and Dunson, W. A. (1989). Acclimation of sodium regulation to low pH and the role of calcium in the acid-tolerant sunfish *Enneacanthus obesus*. *Physiol. Zool.* 62, 977–992.

Gonzalez, R. J. and McDonald, D. G. (1992). The relationship between oxygen consumption and ion loss in a freshwater fish. *J. Exp. Biol.* 163, 317–332.

Gonzalez, R. J. and Preest, M. R. (1999). Ionoregulatory specializations for exceptional tolerance of ion-poor, acidic waters in the neon tetra (*Paracheirodon innesi*). *Physiol. Biochem. Zool.* 72, 156–163.

Gonzalez, R. J. and Wilson, R. W. (2001). Patterns of ion regulation in acidophilic fish native to the ion-poor, acidic Rio Negro. *J. Fish Biol.* 28, 1680–1690.

Gonzalez, R. J., Wilson, R. W., Wood, C. M., Patrick, M. L. and Val, A. L. (2002). Diverse strategies for ion regulation in fish collected from the ion-poor, acidic Rio Negro. *Physiol. Biochem. Zool.* 75, 37–47.

Gonzalez, R. J., Cooper, J. and Head, D. (2005a). Physiological responses to hyper-saline waters in sailfin mollies (*Poecilia latipinna*). *Comp. Biochem. Physiol. A Mol. Integr. Physiol.* 142, 397–403.

Gonzalez, R. J., Wilson, R. W. and Wood, C. M. (2005b). Ionoregulation in tropical fish from ion-poor, acidic blackwaters. In *Fish Physiology, Vol. 22, The Physiology of Tropical Fishes* (eds. A. L. Val, V. M.F.A. Val and D. J. Randall), pp. 397–442. San Diego: Academic Press.

Gonzalez, R. J., Brauner, C. J., Wang, Y. X., Richards, J. G., Patrick, M. L., Xi, W., Matey, V. and Val, A. L. (2010). Impact of ontogenetic changes in branchial morphology on gill function in *Arapaima gigas*. *Physiol. Biochem. Zool.* 83, 322–332.

Graham, J. H. and Hastings, R. W. (1984). Distributional patterns of sunfishes in the New Jersey coastal plain. *Environ. Biol. Fish.* 10, 137–148.

Greco, A. M., Gilmour, K., Fenwick, J. C. and Perry, S. F. (1995). The effects of soft-water acclimation on respiratory gas transfer in the rainbow trout (*Oncorhynchus mykiss*). *J. Exp. Biol.* 198, 2557–2567.

Greco, A. M., Fenwick, J. C. and Perry, S. F. (1996). The effects of soft water acclimation on gill structure in the rainbow trout (*Oncorhynchus mykiss*). *Cell Tissue Res.* 285, 75–82.

Griffith, R. W. (1974). Environment and salinity tolerance in the genus *Fundulus*. *Copeia* 1974, 319–331.

Grosell, M. (2010). The role of the gastrointestinal tract in salt and water balance. In *Fish Physiology*, Vol. 30, *The Multifunctional Gut in Fish* (eds. M. Grosell, A. P. Farrell and C. J. Brauner), pp. 136–165. Amsterdam: Academic Press.

Hammer, T. U. (1986). *Saline Lake Ecosystems of the World*. Boston, MA: Dr. W. Junk.

Hastings, R. W. (1979). Fish of the pine barrens. In *Pine Barrens: Ecosystem and Landscape* (ed. R. T.T. Foreman), pp. 489–504. New York: Academic Press.

Hickman, C. P. (1968). Ingestion, intestinal absorption and elimination of sea water and salts in the southern flounder, *Paralichthys lethostigma*. *Can. J. Zool.* 46, 457–466.

Hiroi, J., McCormick, S. D., Ohtani-Kaneko, R. and Kaneko, T. (2005). Functional classification of mitochondrion-rich cells in euryhaline Mozambique tilapia (*Oreochromis mossambicus*) embryos, by means of triple immunofluorescence staining for Na^+/K^+-ATPase, $Na^+/K^+/2Cl^-$ cotransporter and CFTR anion channel. *J. Exp. Biol.* 208, 2023–2036.

Hiroi, J., Yasumasu, S., McCormick, S. D., Hwang, P. P. and Kaneko, T. (2008). Evidence for an apical Na^+-Cl^- cotransporter involved in ion uptake in a teleost fish. *J. Exp. Biol.* 211, 2584–2599.

Holdren, G. C. and Montano, A. (2002). Chemical and physical characteristics of the Salton Sea, California. *Hydrobiologia* 473, 1–21.

Hotos, G. N. and Vlahos, N. (1998). Salinity tolerance of *Mugil cephalus* and *Chelon labrosus* Pisces: Mugilidae fry in experimental conditions. *Aquaculture* 167, 329–338.

Hunn, J. B. (1985). Role of calcium in gill function in freshwater fishes. *Comp. Biochem. Physiol. A* 82, 543–547.

Hurlbert, A. H., Anderson, T. W., Sturm, K. K. and Hurlbert, S. H. (2007). Fish and fish-eating birds at the Salton Sea: a century of boom and bust. *Lake Reserv. Manage.* 23, 469–499.

Hwang, P-P. (2009). Ion uptake and acid secretion in zebrafish (*Danio rerio*). *J. Exp. Biol.* 212, 1745–1752.

Iftikar, F. I., Matey, V. and Wood, C. M. (2010). The ionoregulatory responses to hypoxia in the freshwater rainbow trout *Oncorhynchus mykiss*. *Physiol. Biochem. Zool.* 83, 343–355.

Iwama, G. K., Takemura, A. and Takano, K. (1997). Oxygen consumption rates of tilapia in fresh water, sea water and hypersaline sea water. *J. Fish Biol.* 51, 886–894.

Javor, B. (1989). *Hypersaline Environments*. New York: Springer.

Jones, B. F., Eugster, H. P. and Rettig, S. L. (1977). Hydrochemistry of the Lake Magadi basin, Kenya. *Geochim. Cosmochim. Acta* 41, 53–72.

Jordan, F., Haney, D. C. and Nordlie, F. G. (1993). Plasma osmotic regulation and routine metabolism in the eustis pupfish, *Cyprinodon variegatus hubbsi* (Teleostei: Cyprinodontidae). *Copeia* 1993, 784–789.

Karnaky, K. J., Jr., Ernst, S. A. and Philpott, C. W. (1976). Killifish opercular skin: a flat epithelium with a high density of chloride cells. *J. Exp. Biol.* 199, 355–364.

Khoo, K. H., Culberson, C. H. and Bates, R. G. (1977). Thermodynamics of the dissociation of ammonium ion in seawater from 5 to 40°C. *J. Solut. Chem.* 6, 281–290.

Kinne, O. and Kinne, E. M. (1962). Rates of development in embryos of a cyprinodont fish exposed to different temperature–salinity–oxygen combinations. *Can. J. Zool.* 40, 231–253.

Kirschner, L. B. (1993). The energetics of osmotic regulation in ureotelic and hypoosmotic fishes. *J. Exp. Zool.* 267, 19–26.

Kirschner, L. B. (1995). Energetics of osmoregulation in fresh water vertebrates. *J. Exp. Zool.* 271, 243–252.

Kirschner, L. B. (1997). Extrarenal mechanisms of hydromineral and acid–base regulation in aquatic vertebrates. In *Handbook of Physiology – Comparative Physiology* (ed. W. H. Dantzer), pp. 577–622. New York: Oxford University Press.

Kjerfve, B., Schettini, C. A. F., Knoppers, B., Lessa, G. and Ferreira, H. O. (1996). Hydrology and salt balance in a large, hypersaline coastal lagoon: Lagoa de Araruama, Brazil. *Hydrobiol. Estuar. Coast. Shelf Sci.* 42, 701–725.

Kültz, D. and Onken, H. (1993). Long-term acclimation of the teleost *Oreochromis mossambicus* to various salinities: two different strategies in mastering hypertonic stress. *Mar. Biol.* 117, 527–533.

Kültz, D., Bastrop, R., Jurss, K. and Siebers, D. (1992). Mitochondria-rich (MR) cells and the activities of the Na$^+$/K$^+$-ATPase and carbonic anhydrase in the gill and opercular epithelium of *Oreochromis mossambicus* adapted to various salinities. *Comp. Biochem. Physiol. B* 102, 293–301.

Kültz, D., Jurss, K. and Jonas, L. (1995). Cellular and epithelial adjustments to altered salinity in the gill and opercular epithelium of a cichlid fish *Oreochromis mossambicus* adapted to various salinities. *Cell Tissue Res.* 279, 65–73.

Lasker, R., Tenaza, R. H. and Chamberlain, L. L. (1972). The response of the Salton Sea fish eggs and larvae to salinity stress. *Calif. Fish Game* 58, 58–66.

Laurent, P., Maina, J. N., Bergman, H. L., Narahara, A. N., Walsh, P. J. and Wood, C. M. (1995). Gill structure of a fish from an alkaline lake: effect of short-term exposure to neutral conditions. *Can. J. Zool.* 73, 1170–1181.

Leenheer, J. A. (1980). Origin and nature of humic substances in the waters of the Amazon River Basin. *Acta Amazon.* 10, 513–526.

Lenanton, R. C. J. (1977). Fishes from the hypersaline waters of the stromatolite zone of Shark Bay, WA. *Copeia* 1977, 387–390.

Lindley, T. E., Scheiderer, C. L., Walsh, P. J., Wood, C. M., Bergman., H. L., Bergman, A. N., Laurent, P., Wilson, P. and Anderson, P. M. (1999). Muscle as a primary site of urea cycle enzyme activity in an alkaline lake-adapted tilapia, *Oreochromis alcalicus grahami*. *J. Biol. Chem.* 274, 29858–29861.

Lotan, R. (1971). Osmotic adjustment in the euryhaline teleost *Aphanius dispar* (Cyprinodonti-dae). *Z. Vergl. Physiol.* 75, 383–387.

Machado, B. E. and Podrabsky, J. E. (2007). Salinity tolerance in diapausing embryos of the annual killifish *Austrofundulus limnaeus* is supported by exceptionally low water and ion permeability. *J. Comp. Physiol. B* 177, 809–820.

Maina, J. N., Kisia, S. M., Wood, C. M., Narahara, A., Bergman, H. L., Laurent, P. and Walsh, P. J. (1996). A comparative allometry study of the morphology of the gills of an alkaline adapted cichlid fish, *Oreochromis alcalicus grahami* of Lake Magadi, Kenya. *Int. J. Salt Lake Res.* 5, 131–156.

Marion, G. M., Millero, F. J. and Feistel, R. (2009). Precipitation of solid phase calcium carbonates and their effect on application of seawater SA-T-P models. *Ocean Sci.* 5, 285–291.

Marshall, W. S. (2002). Na$^+$, Cl$^-$, Ca^{2+}, and Zn^{2+} transport by fish gills: retrospective, review and prospective synthesis. *J. Exp. Biol.* 293, 264–283.

Matey, V., Iftikar, F. I., De Boeck, G., Scott, G. R., Sloman, K. A., Almeida-Val, V. M. F., Val, A. L. and Wood, C. M. (2011). Gill morphology and acute hypoxia: responses of mitochondria-rich, pavement, and mucous cells in two species with very different approaches to the osmo-respiratory compromise, the Amazonian oscar (*Astronotus ocellatus*) and the rainbow trout. *Can. J. Zool.* 89, 307–324.

May, R. C. (1975). Effects of temperature and salinity on fertilization, embryonic development, and hatching in *Dairiella icistia* (Pisces: Sciaenidae), and the effect of parental salinity acclimation on embryonic and larval salinity tolerance. *Fish. Bull.* 73, 1–22.

McDonald, D. G. (1983). The effects of H^+ upon the gills of freshwater fish. *Can. J. Zool.* 61, 691–703.

McDonald, M. D. and Grosell, M. (2006). Maintaining osmotic balance with an aglomerular kidney. *Comp. Biochem. Physiol. A Mol. Integr. Physiol.* 143, 447–458.

McDonald, M. D. and Rogano, M. S. (1986). Ion regulation by the rainbow trout, *Salmo gairdneri*, in ion-poor water. *Physiol. Zool.* 59, 318–331.

McGeer, J. C., Wright, P. A., Wood, C. M., Wilkie, M. P., Mazur, C. F. and Iwama, G. K. (1994). Nitrogen excretion in four species of fish from an alkaline lake. *Trans. Am. Fish. Soc.* 123, 824–829.

Mickle, P. P. (2007). The Aral Sea. *Annu. Rev. Earth Planet Sci.* 35, 47–72.

Milligan, C. L. and Wood, C. M. (1982). Disturbances in haematology, fluid volume distribution and circulatory function associated with low environmental pH in the rainbow trout, *Salmo gairdneri*. *J. Exp. Biol.* 99, 397–415.

Morgan, J. D. and Iwama, G. K. (1991). Effects of salinity on growth, metabolism, and ion regulation in juvenile rainbow and steelhead trout (*Oncorhynchus mykiss*) and fall Chinook salmon (*Onchorhynchus tshawytscha*). *Can. J. Fish. Aquat. Sci.* 48, 2083–2094.

Morgan, J. D. and Iwama, G. K. (1999). Energy cost of NaCl transport in isolated gills of cutthroat trout. *Am. J. Physiol.* 277, R631–R639.

Nagl, S., Tichy, H., Mayer, W. E., Takezaki, N., Takahata, N. and Klein, J. (2000). The origin and age of haplochromine fishes in Lake Victoria, East Africa. *Proc. R. Soc. Lond. B* 267, 1049–1061.

Narahara, A., Bergman, H. L., Laurent, P., Maina, J. N., Walsh, P. J. and Wood, C. M. (1996). Respiratory physiology of the Lake Magadi Tilapia (*Oreochromis alcalicus grahami*), a fish adapted to a hot, alkaline, and frequently hypoxic environment. *Physiol. Zool.* 69, 1114–1136.

Nilsson, S. (1986). Control of gill blood flow. In *Fish Physiology* (eds. S. Nilsson and S. Holmgren), pp. 87–101. London: Croom Helm.

Nordlie, F. G. (1985). Osmotic regulation in the sheepshead minnow *Cyprinodon variegatus* Lacepede. *J. Fish Biol.* 26, 161–170.

Nordlie, F. G. and Walsh, S. J. (1989). Adaptive radiation in osmotic regulatory patterns among three species of cyprinodontids (Teleostei: Atherniomorpha). *Physiol. Zool.* 62, 1203–1218.

Nordlie, F. G., Haney, D. C. and Walsh, S. J. (1992). Comparisons of salinity tolerance and osmotic regulatory capabilities in populations of sailfin mollies (*Poecilia latipinna*) from brackish and freshwaters. *Copeia* 1992, 741–746.

Ouattara, N., Bodinier, C., Negre-Sadargues, G., D'Cotta, H., Messad, S., Charmanteir, G., Panfili, J. and Baroiller, J. F. (2009). Changes in gill ionocyte morphology and function following transfer from fresh to hypersaline waters in the tilapia *Sarotherodon melanotheron*. *Aquaculture* 290, 155–164.

Owino, A. O., Oyugi, J. O., Nasirwa, O. O. and Bennun, L. A. (2001). Patterns of variation in waterbird numbers on four Rift Valley lakes in Kenya, 1991–1999. *Hydrobiologia* 458, 45–53.

Panfili, J., Mbow, A., Durand, J.-D., Diop, K., Diouf, K., Thior, D., Ndiaye, P. and Laë, R. (2004a). Influence of salinity on the life-history traits of the West African black chinned tilapia (*Sarotherodon melanotheron*): comparison between the Gambia and Saloum estuaries. *Aquat. Living Resour.* 17, 65–74.

Panfili, J., Durand, J.-D., Mbow, A., Guinand, B., Diop, K., Kantoussan, J., Thior, D., Thiaw, O. T., Albaret, J.-J. and Laë, R. (2004b). Influence of salinity on life history traits of the bonga shad *Ethmalosa fimbriata* (Pisces, Clupeidae): comparison between the Gambia and Saloum estuaries. *Mar. Ecol. Prog. Ser.* 270, 241–257.

Pérez-Robles, J., Ana Denisse, R., Ivone, G.-M. and Fernando, D. (2012). Interactive effects of salinity on oxygen consumption, ammonium excretion, osmoregulation and Na^+/K^+-ATPase expression in the bullseye puffer (*Sphoeroides annulatus*, Jenyns 1842). *Aquacult. Res.* 43, 1372–1383.

Perry, S. F. (1998). Relationships between branchial chloride cells and gas transfer in freshwater fish. *Comp. Biochem. Physiol. A* 119, 9–16.

Potter, I. C. and Hyndes, G. A. (1999). Characteristics of the ichthyofaunas of southwestern Australian estuaries, including comparisons with holarctic estuaries and estuaries elsewhere in temperate Australia: a review. *Austral. J. Ecol.* 24, 395–421.

Potter, I. C., Ivantsoff, W., Cameron, R. and Minnard, J. (1986). Life cycles and distribution of atherinids in the marine and estuarine waters of southern Australia. *Hydrobiologia* 139, 23–40.

Potter, I. C., Beckley, L. E., Whitfield, A. K. and Lenanton, R. C. J. (1990). Comparisons between the roles played by estuaries in the life cycles of fishes in temperate Western Australia and Southern Africa. *Environ. Biol. Fish.* 28, 143–178.

Potter, I. C., Benjamin, M., Chuwen, B. M., Hoeksema, S. D. and Elliott, M. (2010). The concept of an estuary: a definition that incorporates systems which can become closed to the ocean and hypersaline. *Estuar. Coast. Shelf Sci.* 87, 497–500.

Potts, W. T. W. (1994). Kinetics of sodium uptake in freshwater animals: a comparison of ion exchange and proton pump hypotheses. *Am. J. Physiol.* 266, R315–R320.

Preest, M. R., Gonzalez, R. J. and Wilson, R. W. (2005). A pharmacological examination of Na^+ and Cl^- transport in two species of freshwater fish. *Physiol. Biochem. Zool.* 78, 259–272.

Randall, D. J., Baumgarten, D. and Malyusz, M. (1972). The relationship between gas and ion transfer across the gills of fishes. *Comp. Biochem. Physiol. A Mol. Integr. Physiol.* 41, 629–637.

Randall, D. J., Wood, C. M., Perry, S. F., Bergman, H. L., Maloiy, G. M. O., Mommsen, T. P. and Wright, P. A. (1989). Urea excretion as a strategy for survival in a fish living in a very alkaline environment. *Nature Lond.* 337, 165–166.

Rao, T. R. (1975). Salinity tolerance of laboratory-reared larvae of the California killifish, *Fundulus parvipinnis* Girard. *J. Fish Biol.* 7, 783–790.

Sardella, B. A. and Brauner, C. J. (2007a). The osmo-respiratory compromise in fish: the effects of physiological state and the environment. In *Fish Respiration and Environment* (eds. M. N. Fernandes, F. T. Rantin, M. L. Glass and B. G. Kapoor), pp. 147–165. Enfield, NH: Science Publishers.

Sardella, B. A. and Brauner, C. J. (2007b). Cold temperature-induced osmoregulatory failure: the physiological basis for tilapia winter mortality in the Salton Sea. *Calif. Fish Game* 93, 200–213.

Sardella, B. A. and Brauner, C. J. (2008). The effect of elevated salinity on "California" Mozambique tilapia (*Oreochromis mossambicus × O. urolepis hornorum*) metabolism. *Comp. Biochem. Physiol. C Toxicol. Pharmacol.* 148, 430–436.

Sardella, B., Cooper, J., Gonzalez, R. and Brauner, C. J. (2004a). The effect of temperature on juvenile Mozambique tilapia hybrids (*Oreochromis mossambicus × O. urolepis hornorum*) exposed to full-strength and hypersaline seawater. *Comp. Biochem. Physiol. A Mol. Integr. Physiol.* 137, 621–629.

Sardella, B., Matey, V., Cooper, J., Gonzalez, R. and Brauner, C. J. (2004b). Physiological, biochemical, and morphological indicators of osmoregulatory stress in "California"

Mozambique tilapia (*Oreochromis mossambicus* × *O. urolepis hornorum*) exposed to hypersaline water. *J. Exp. Biol.* 207, 1399–1413.

Sardella, B. A., Matey, V. and Brauner, C. J. (2007). Coping with multiple stressors: physiological mechanisms and strategies in fishes of the Salton Sea. *Lake Reserv. Manage.* 23, 518–527.

Seegers, L., Sonnenberg, R. and Yamamoto, R. (1999). Molecular analysis of the Alcolapia flock from lakes Natron and Magadi, Tanzania and Kenya (Teleosti: Cichlidae), and implications for their systematics and evolution. *Ichthyol. Explor. Freshw.* 10, 175–199.

Shehadeh, Z. H. and Gordon, M. S. (1969). The role of the intestine in salinity adaptation of the rainbow trout, *Salmo gairdneri*. *Comp. Biochem. Physiol.* 30, 397–418.

Singer, T. D., Tucker, S. J., Marshall, W. S. and Higgins, C. G. (1998). A divergent CFTR homologue: highly regulated salt transport in the euryhaline teleost *F. heteroclitus*. *Am. J. Physiol. Cell Physiol.* 274, C715–C723.

Silva, P., Solomon, R., Spokes, K. and Epstein, F. (1977). Ouabain inhibition of gill Na-K-ATPase: relationship to active chloride transport. *J. Exp. Zool.* 199, 419–426.

Siokhin, V., Chernichko, I., Kostyushyn, V., Krylov, N., Andrushchenko, Y., Andrienko, T., Didukh, Y., Kolomijchuk, V., Parkhisenko, L., Chernichko, R. and Kirikova, T. (2000). *Sivash – The Lagoon Between Two Seas.* Kiev: Wetlands International–AEME.

Skadhauge, E. and Lotan, R. (1974). Drinking rate and oxygen consumption in the euryhaline teleost *Aphanius dispar* in waters of high salinity. *J. Exp. Biol.* 60, 547–556.

Soengas, J. L., Sangiao-Alvarellos, S., Laiz-Carrion, R. and Mancera, J. M. (2007). Energy metabolism and osmotic acclimation in teleost fish. In *Fish Osmoregulation* (eds. B. Baldisserotto, J. M. Mancera and B. G. Kapoor), pp. 277–307. Enfield, NH: Science Publishers.

Swanson, C. (1998). Interactive effects of salinity on metabolic rate, activity, growth osmoregulation in the euryhaline milkfish (*Chanos chanos*). *J. Exp. Biol.* 201, 3355–3366.

Talling, J. F. and Talling, I. B. (1965). The chemical composition of African lake waters. *Int. Revue Ges. Hydrobiol. Hydrogr.* 50, 421–463.

Taylor, R. E. (1972). *The Effects of Increasing Salinity of the Pyramid Lake Fishery.* Reno, NV: University of Nevada.

Tine, M., Bonhomme, F., McKenzie, D. J. and Durand, J. D. (2010). Differential expression of the heat shock protein Hsp70 in natural populations of the tilapia, *Sarotherodon melanotheron*, acclimatised to a range of environmental salinities. *BMC Ecol.* 10, 11.

Trape, S., Durand, J. D., Guilhauon, F., Vigliola, L. and Panfili, J. (2009). Recruitment patterns of young-of-the-year mugilid fishes in a West African estuary impacted by climate change. *Estuar. Coast. Shelf Sci.* 85, 357–367.

Tunnell, J. W. and Judd, F. W. (2002). *The Laguna Madre of Texas and Tamulipas.* Corpus Christi, TX: Texas A&M University Press.

Uchida, K., Kaneko, T., Miyazaki, H., Hasegawa, S. and Hirano, T. (2000). Excellent salinity tolerance of Mozambique tilapia (*Oreochromis mossambicus*): elevated chloride cell activity in the branchial and opercular epithelia of the fish adapted to concentrated seawater. *Zool. Sci.* 17, 149–160.

Val, A. L. and de Almeida-Val, V. M. F. (1995). *Fishes of the Amazon and Their Environment.* Berlin: Springer.

Valentine, D. W. and Miller, R. (1969). Osmoregulation in the California killifish, *Fundulus parvipinnis*. *Calif. Game Fish* 58, 20–25.

Vareschi, E. (1979). The ecology of Lake Nakuru (Kenya). II. Biomass and spatial distribution of fish (*Tilapia grahami Boulenger=Sarotherodon alcalicum grahami* Boulenger). *Oecologia* 37, 321–335.

Walker, I. and Henderson, P. A. (1996). Ecophysiological aspects of Amazonian blackwater litterbank fish communities. In *Physiology and Biochemistry of Fishes of the Amazon* (eds.

A. L. Randall, V. M.F. de Almeida-Val and D. J. Randall), pp. 7–22. Manaus: Instituto Nacional de Pesquisas da Amazonia.

Wang, Y. S., Gonzalez, R. J., Patrick, M. L., Grosell, M., Zhang, C., Feng, Q., Du, J. Z., Walsh, P. J. and Wood, C. M. (2003). Unusual physiology of scaleless carp, *Gymnocypris przewalskii*, in Lake Qinghai: a high altitude saline lake. *Comp. Biochem. Physiol. A Mol. Integr. Physiol.* 134, 409–421.

Watanabe, W. O., Olla, B. L., Wicklund, R. I. and Head, W. D. (1997). Saltwater culture of the Florida red tilapia and other saline-tolerant tilapias: a review. In *Tilapia Aquaculture in the Americas* (eds. B. A. Coasta-Pierce and J. E. Rakocy), pp. 55–141. Baton Rouge, LA: World Aquaculture Society.

Watts, J. M., Swan, B. K., Tiffany, M. A. and Hurlbert, S. H. (2001). Thermal mixing and oxygen regimes in the Salton Sea, California, 1997–1999. *Hydrobiologia* 466, 159–176.

Wedderburn, S. D., Walker, K. F. and Zampatti, B. P. (2008). Salinity may cause fragmentation of hardyhead (Teleostei: Atherinidae) populations in the River Murray, Australia. *Mar. Freshw. Res.* 59, 254–258.

Wetzel, R. G. (1983). *Limnology*. Philadelphia. PA: Saunders College Publishing.

Whitney, R. R. (1967). Introduction of commercially important species into inland mineral waters a review. *Contrib. Mar. Sci.* 12, 262–280.

Wilkie, M. P. and Wood, C. M. (1996). The adaptations of fish to extremely alkaline environments. *Comp. Biochem. Physiol. B Biochem. Mol. Biol.* 113, 665–673.

Wilkie, M. P., Wright, P. A., Iwama, G. K. and Wood, C. M. (1993). The physiological responses of the Lahontan cutthroat trout (*Oncorhynchus clarki henshawi*), a resident of highly alkaline Pyramid Lake (pH 9.4), to challenge at pH 10. *J. Exp. Biol.* 175, 173–194.

Wilson, J. M., Leitão, A., Gonçalves, A., Ferreira, C., Reis-Santos, P. N., Fonseca, A.-V., Antunes, J. C., Pereira, C. M. and Coimbra, J. (2007). Modulation of branchial ion transport protein expression by salinity in the glass eels (*Anguilla anguilla*). *Mar. Biol.* 151, 1633–1645.

Wilson, R. W., Gilmour, K., Henry, R. and Wood, C. M. (1996). Intestinal base excretion in the seawater-adapted rainbow trout: a role in acid–base balance? *J. Exp. Biol.* 199, 2331–2343.

Wilson, R. W., Wilson, J. M. and Grosell, M. (2002). Intestinal bicarbonate secretion by marine teleost fish – why and how? *Biochim. Biophys. Acta* 1566, 182–193.

Wilson, P. J., Wood, C. M., Walsh, P. J., Bergman, A. N., Bergman, H. L., Laurent, P. and White, B. N. (2004). Discordance between genetic structure and morphological, ecological, and physiological adaptation in Lake Magadi tilapia. *Physiol. Biochem. Zool.* 77, 537–555.

Wood, C. M., Perry, S. F., Wright, P. A., Bergman, H. L. and Randall, D. J. (1989). Ammonia and urea dynamics in the Lake Magadi tilapia, a ureotelic teleost fish adapted to an extremely alkaline environment. *Respir. Physiol.* 77, 1–20.

Wood, C. M., Bergman, H. I., Laurent, P., Maina, J. N., Narahara, A. and Walsh, P. J. (1994). Urea production, acid–base regulation and their interactions in the Lake Magadi tilapia, a unique teleost adapted to a highly alkaline environment. *J. Exp. Biol.* 189, 13–36.

Wood, C. M., Wilson, R. W., Gonzalez, R. J., Patrick, M. L., Bergman, H. L., Narahara, A. and Val, A. L. (1998). Responses of an Amazonian teleost, the tambaqui (*Colossoma macropomum*), to low pH in extremely soft water. *Physiol. Zool.* 71, 658–670.

Wood, C. M., Wilson, P. W., Bergman, H. L., Bergman, A. N., Laurent, P., Otiang'a-Owiti, G. and Walsh, P. J. (2002a). Obligatory urea production and the cost of living in the Magadi tilapia revealed by acclimation to reduced salinity and alkalinity. *Physiol. Biochem. Zool.* 75, 111–122.

Wood, C. M., Wilson, P. W., Bergman, H. L., Bergman, A. N., Laurent, P., Otiang'a-Owiti, G. and Walsh, P. J. (2002b). Ionoregulatory strategies and the role of urea in the Magadi tilapia (*Alcolapia grahami*). *Can. J. Zool.* 80, 503–515.

Wood, C. M., Matsuo, A. Y. O., Gonzalez, R. J., Wilson, R. W., Patrick, M. L. and Val, A. L. (2002c). Mechanisms of ion transport in *Potamotrygon*, a stenohaline freshwater elasmobranch native to the ion-poor blackwaters of the Rio Negro. *J. Exp. Biol.* 205, 3039–3054.

Wood, C. M., Matsuo, A. Y. O., Wilson, R. W., Gonzalez, R. J., Patrick, M. L., Playle, R. C. and Val, A. L. (2003). Protection by natural blackwater against disturbances in ion fluxes caused by low pH exposure in freshwater stingrays endemic to the Rio Negro. *Physiol. Biochem. Zool.* 76, 12–27.

Wood, C. M., Du, J. Z., Rogers, J., Brauner, C. J., Richards, J. G., Semple, J. W., Murray, B. W., Chen, X. Q. and Wang, X. (2007). Przewalski's naked carp (*Gymnocypris przewalskii*): an endangered species taking a metabolic holiday in Lake Qinghai, China. *Physiol. Biochem. Zool.* 80, 59–77.

Wood, C. M., Iftikar, F. I., Scott, G. R., De Boeck, G., Sloman, K. A., Matey, V., Valdez Domingos, F. A., Mendoza Duarte, R., Almeida-Val, V. M. F. and Val, A. L. (2009). Regulation of gill transcellular permeability and renal function during acute hypoxia in the Amazonian oscar (*Astronotus ocellatus*): new angles to the osmo-respiratory compromise. *J. Exp. Biol.* 212, 1949–1964.

Wood, C. M., Bergman, H. L., Bianchini, A., Laurent, P., Maina, J., Johannsson, E. O., Bianchini, L. F., Chevalier, C., Kavembe, G. D., Papah, M. B. and Ojoo, R. O. (2012). Transepithelial potential in the Magadi tilapia, a fish living in extreme alkalinity. *J. Comp. Physiol.* B 182, 247–258.

Wright, P. A., Iwama, G. K. and Wood, C. M. (1993). Ammonia and urea excretion in Lahontan cutthroat trout (*Oncorhynchus clarki henshawi*) adapted to the highly alkaline Pyramid Lake (pH 9.4). *J. Exp. Biol.* 175, 153–172.

Wright, P. A. and Wood, C. M. (2009). A new paradigm for ammonia excretion in aquatic animals: role of Rhesus (Rh) glycoproteins. *J. Exp. Biol.* 212, 2303–2312.

Yan, J. J., Chou, M. Y., Kaneko, T. and Hwang, P. P. (2007). Gene expression of Na^+/H^+ exchanger in zebrafish H^+-rich cells during acclimation to low-Na^+ and acidic environments. *Am. J. Physiol. Cell Physiol.* 293, C1814–C1823.

Zampatti, B. P., Bice, C. M. and Jennings, P. R. (2010). Temporal variability in fish assemblage structure and recruitment in a freshwater-deprived estuary: the Coorong, Australia. *Mar. Freshw. Res.* 61, 1298–1312.

Zavialov, P. O., Ni, A. A., Kudyshkin, T. V., Ishniyazov, D. P., Tomashevskaya, I. G. and Mukhamedzhanova, D. (2009). Ongoing changes in salt composition and dissolved gases in the Aral Sea. *Aquat. Geochem.* 15, 263–275.

10

EURYHALINITY IN AN EVOLUTIONARY CONTEXT

ERIC T. SCHULTZ
STEPHEN D. McCORMICK

This chapter focuses on the evolutionary importance and taxonomic distribution of euryhalinity. Euryhalinity refers to broad halotolerance (capability of surviving in both freshwater and seawater) and broad halohabitat distribution. Species vary widely in their range of tolerable salinity levels. Halotolerance breadth varies with species' evolutionary history and halohabitat. With respect to halohabitat distribution, a minority of species are euryhaline, but they are potent sources of evolutionary diversity. Euryhalinity is a key innovation trait whose evolution enables exploitation of new adaptive zones, triggering cladogenesis. This chapter reviews phylogenetically informed studies that demonstrate freshwater species diversifying from euryhaline ancestors through processes such as landlocking. Some euryhaline taxa are particularly susceptible to changes in halohabitat and subsequent diversification, and some geographic regions have been hotspots for transitions to freshwater. Comparative studies on

Euryhaline Fishes: Volume 32
FISH PHYSIOLOGY
Copyright © 2013 Elsevier Inc. All rights reserved
DOI: http://dx.doi.org/10.1016/B978-0-12-396951-4.00010-4

mechanisms among multiple taxa and at multiple levels of biological integration are needed to clarify evolutionary pathways to and from euryhalinity.

1. INTRODUCTION

In the living world, transitions beget diversification. Classic cases of adaptive radiation began with colonization of a new patch of ground such as a relatively unoccupied island or lake. Changes in morphology and physiology permitting exploitation of new habitats ushered in ascendance of major groups such as tetrapods and birds. In macroevolutionary history, taxa that endured mass extinction events often expanded into newly vacated ecospace. These homilies on diversification have a common moral, one that is close to a truism: the generalist is more likely to leave an evolutionary legacy than the specialist. In this chapter the authors endeavor to support this vague but lofty position for one group of generalists, the euryhaline fishes.

In this chapter both physiological and ecological meanings of euryhalinity are employed. Physiological euryhalinity focuses on halotolerance: it is defined as the capability of surviving in a wide range of salinity levels, potentially from freshwater (FW, ≤ 0.5 ppt) to seawater (SW, 30–40 ppt) and higher. Ecological euryhalinity focuses on halohabitat: it is defined as occurrence in both FW and SW [and brackish water (BW, 0.5–30 ppt)]. Ecological euryhalinity implies amphihalinity (the tolerance of both FW and SW), and more broadly, physiological euryhalinity; halohabitat can include both FW and SW only if halotolerance is sufficiently broad. However, the converse is not necessarily true, because a species may have a broad halotolerance but a restricted halohabitat. The distinction between the physiological and ecological facets of tolerance is the distinction between the fundamental niche, reflecting physiological capacity, and the realized niche, reflecting other ecological and historical factors (Whitehead, 2010). The next section considers how halotolerance is characterized through empirical work, and how it is distributed among the fishes across taxa, halohabitat, and ontogenetic stage. Using the halotolerance data, groups of species with similar tolerance limits are designated. Then the distribution of euryhalinity is examined in terms of halohabitat among the fishes, both in deep evolutionary time at the origin of the vertebrates, and among the major groups of extant fishes. Finally, the evolutionary potential of euryhalinity is reviewed, through cases of diversification arising within taxa that had the physiological capability of handling a broad range of salinity levels, occurring in habitats prone to subdivision.

2. DIVERSITY OF HALOTOLERANCE

This section reviews how halotolerance is empirically determined and examines how halotolerance is distributed among the ray-finned fishes. Halotolerance is tested and quantified in a variety of ways, and this part of the chapter is intended to improve comparability among future studies. The second part of the section compares halotolerance limits across 141 species of ray-finned fishes, assesses variability in halotolerance limits and halotolerance breadth with respect to higher taxa and habitat groups, and resolves ray-finned fishes into groups with similar halotolerances. Additional data on the halotolerance of fishes inhabiting extreme environments are presented by Brauner et al. (2013, Chapter 9, this volume).

2.1. Empirical Issues in Halotolerance Analysis

To test halotolerance limits, experimental subjects are exposed to altered salinity levels in several ways. One approach is to rear subjects from fertilization at constant salinity, and record the effect of salinity level on hatching and subsequent endpoints such as survival (e.g. Bohlen, 1999). This design rarely appears in the literature, presumably because few investigators begin work with subjects before hatching. A second approach (hereafter referred to as the "direct" design) entails altering environmental salinity rather instantaneously. Endpoints of different groups of subjects exposed to different salinity levels are compared for a prescribed period. A third approach (hereafter referred to as the "gradual" design) entails an incremental change in salinity on a prescribed schedule. Endpoints are monitored as salinity changes. The direct and gradual design approaches are represented in Fig. 10.1.

The direct and gradual methods both have virtues. The direct method focuses on the capacity of acute responses to cope with environmental change. For some ecological inquiries, such as the effort to link halotolerance of transient changes in salinity to distribution of FW fishes in estuaries, direct transfers among salinity levels may be more appropriate than gradual alterations of salinity. The simplicity of the experimental treatment in the direct method maximizes comparability among studies. The gradual design evidently permits a better assessment of halotolerance to chronic exposure, and requires fewer fish. Differences in the magnitude of salinity change and time at a given salinity can limit comparisons among studies. Because these designs are complementary rather than duplicative, the authors suggest that when possible investigators should use both in assessing halotolerance limits.

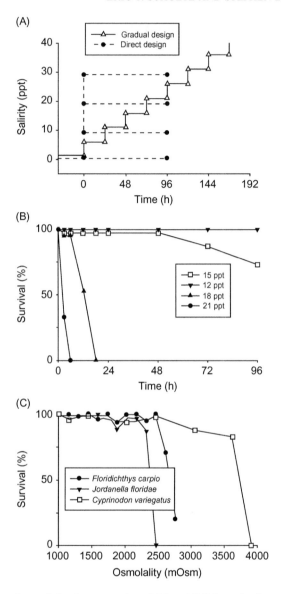

Fig. 10.1. Approaches to halotolerance testing of fishes. (A) Schematic of experimental testing via direct and gradual changes in salinity. In direct testing, subjects are transferred instantaneously at time *t*0 from the salinity of acclimation (here represented as 0.5 ppt) to one of several levels of treatment salinity (0.5 ppt as control, 10, 20, and 30 ppt). In gradual testing, salinity is changed in small increments, here represented as 5 ppt increases every 24 h. Some points are offset for clarity. (B) Results from direct testing: a typical example of survivorship curves for each of several salinity treatments. Points at 100% survival are offset slightly for clarity. Data from Guo et al. (1995). (C) Results from gradual testing: a typical example of change in survival rate at each salinity as salinity is progressively altered. Data from Nordlie and Walsh (1989).

Death appears to be the only reliable endpoint for determination of halotolerance when a species or life stage is examined for the first time. Some studies (Peterson, 1988; Scott et al., 2007) interpret the change (or constancy) of plasma osmolality over a range of salinity levels as indicative of intolerance (or tolerance). Plasma osmolality is a valuable metric of osmoregulatory performance, but interpreting it as a metric of halotolerance makes the assumption that departure from the plasma osmolality norm is tantamount to loss of function. This assumption is unwarranted without prior empirical demonstration for the species; some euryhaline species exhibit significant changes in plasma osmolality over the range of tolerated salinity, at least temporarily (Lotan, 1971; Nordlie, 1985, 2009; Shrimpton, 2013, Chapter 7, this volume; Marshall, 2013, Chapter 8, this volume). However, once a species has been examined and thresholds for mortality have been determined, then plasma osmolality can become an acceptable substitute (e.g. in the SW challenge test widely used in studies on salmon smolts) (Blackburn and Clarke, 1987). Loss of equilibrium has been used in some studies (Young and Cech, 1996) with benefits of minimizing destructive use of subjects and/or permitting their use at the endpoint for determinations that require living subjects, such as plasma osmolality. However, in the authors' experience subjects do not always demonstrate a loss of equilibrium before death due to high or low salinity exposure.

Tolerance is conventionally quantified as the central tendency of the distribution of stressor levels at which subjects succumb. It is unfortunate that many, if not most, studies investigating halotolerance do not provide statistics that summarize salinity limits. When provided, the most commonly used halotolerance statistic is referred to as the LC_{50} or LD_{50}, the concentration or dose at which half of the subjects are expected to die at a prescribed time-point.

Quantifying the LC_{50} requires an estimation procedure. In many cases, the procedure is arithmetic or graphical, such as linear interpolation between two dose–mortality points to estimate the dose at which mortality was 50% (Kendall and Schwartz, 1968; Kilambi and Zdinak, 1980; Watanabe et al., 1985; Britz and Hecht, 1989; Hotos and Vlahos, 1998; Garcia et al., 1999; Fashina-Bombata and Busari, 2003). A weakness of this approach is its possible reliance on a subset of the survival data. A statistical model relating the probability of survival to salinity is a better approach. Several regression models that are employed in environmental toxicology studies also appear in the halotolerance literature. The proportion surviving at a prescribed time has been modeled by linear or multiple linear regression (de March, 1989; Lemarie et al., 2004); however, probabilities rarely are distributed so that linear regression would be appropriate. More common approaches to estimating LC_{50} involve logit models (logistic regression) and probit models: examples of probit modeling include Cataldi et al. (1999) and Mellor and Fotedar (2005); examples of logistic regression in salinity tolerance studies include Ostrand and Wilde (2001) and Faulk and Holt (2006). Hamilton et al. (1977) identify several shortcomings of these methods and describe the Spearman–Karber method for calculating LC_{50}, which has been used in at least one salinity tolerance study (Bringolf et al., 2005).

Methods for deriving time-independent LC_{50} estimates have not been widely used in the halotolerance literature. In most studies, particularly when the direct method is used, additional exposure time at any salinity level would result in additional mortality. Hence, most LC_{50} estimates in the halotolerance literature are time dependent; extending the prescribed time at which the effect of salinity on mortality is assessed has the effect of moderating the LC_{50} (i.e. it increases the low limit and decreases the high limit). The range of a parameter such as salinity or temperature over which the extent of mortality is time exposure dependent is known as the lower or upper "zone of resistance" and lies just beyond the "zone of tolerance" within which the parameter level does not affect or induce mortality (Brett, 1956). The boundary between the zone of tolerance and the lower or upper zone of resistance is referred to as the "incipient lethal level", representing the most extreme value that can be tolerated for an indefinite period. Using line-fitting methods apparently first suggested by Doudoroff (1945) and modified by Green (1965), incipient lethal salinity limits have been determined by relatively few investigators (Reynolds and Thomson, 1974; Reynolds et al., 1976; Pfeiler, 1981). Incipient lethal estimates of LC_{50} are especially valuable, because they are time independent and are therefore most comparable among studies. The authors recommend that incipient lethal salinity limits be incorporated into direct design experiments. With few exceptions, the LC_{50} halotolerance limits compiled in this review are time dependent.

2.2. Interspecific Variability in Halotolerance

The authors accumulated a dataset on halotolerance by surveying four decades of salinity exposure experiments. The Aquatic Sciences and Fisheries Abstracts database was used for references from 1971 to 2012. An initial search using the terms "salinity tolerance" or "salt tolerance" and fish or fishes for all available years yielded 995 references. References were harvested from this list that presented salinity challenge experiments and quantified tolerance endpoints, which were mortality rates except on a few occasions reporting loss of equilibrium (Young and Cech, 1996). The search revealed surprisingly few references concerning salinity tolerance in elasmobranchs (Sulikowski and Maginniss, 2001) and none on sarcopterygians, and therefore the analysis was confined to studies on Actinopterygii. This analysis is based on a set of 108 studies, reporting results published as early as 1968, on 141 species (Table 10.S1).

Experimental results were divided into groups according to life stage of the experimental subjects and according to the method used to determine tolerance limits. Life stage was categorized as larva or juvenile and adult, because analyses of larvae often demonstrate pronounced changes in tolerance with development (Varsamos et al., 2001; Varsamos et al., 2005; Zydlewski and Wilkie, 2013, Chapter 6, this volume). Studies examining tolerance through metamorphosis (Hirashima and Takahashi, 2008) were placed among studies on larvae. Most studies involving field-collected individuals reported the size of experimental subjects if not the life stage, but in a few cases the life stage was inferred based on method of capture or other details (and in every case was identified as juvenile or adult). A small number of studies (Reynolds and Thomson, 1974; Reynolds et al., 1976) included experiments on both larvae and subsequent life stages. The experimental method was categorized as direct or gradual (Fig. 10.1). A few cases in which salinity was changed over a brief interval (less than 24 h, e.g. Chervinski, 1977b; Tsuzuki et al., 2000) relative to the time-course of response were categorized as direct, and studies that quantified the tolerance of individuals reared at different salinity levels from early life stages (e.g. Perschbacher et al., 1990; Bohlen, 1999) were categorized as gradual. When subjects were tested at multiple temperatures, results were used from temperatures that imposed the lowest level of mortality. The aggregation by species and stage yielded 168 estimates of lower and/or upper halotolerance limits. Determination of halotolerance limits was often not possible from the results, because subjects tolerated the most extreme salinity treatments used. As was frequently the case, when survival was high in FW a lower halotolerance limit of 0 ppt was imputed. Having imputed lower limits in this way, estimates of halotolerance breadth (the range of salinity levels that can be endured) were possible in most cases; lower and upper tolerance limits could not be determined for seven and 32, respectively, of the 168 records.

Most species tested by the direct or gradual method tolerated FW (Fig. 10.2A, B). The mean lower salinity limit among direct-method experiments was 1.2 ppt (SD = 2.5) and among gradual-method experiments it was 0.19 (SD = 0.90). The most common lower tolerance limit was 0.5 ppt or below (70 of 98 species tested by the direct method, 61 of 63 tested by the gradual method). The highest value for lower LC_{50} estimated by the direct method was 16 ppt, observed for *Scophthalmus maximus*. The highest

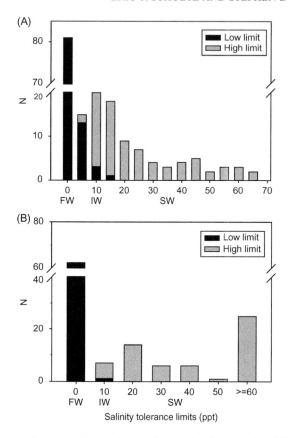

Fig. 10.2. Lower and upper salinity tolerance limits of actinopterygian fishes. Histograms represent the frequency of estimated tolerance limits by species. Typical salinity values for freshwater (FW), salinity at which fish are isotonic (IW), and seawater (SW), are indicated on each x-axis. Note that the x-axis scales of the two panels differ. (A) Tolerance limits, in classes of 5 ppt, determined via direct method. (B) Tolerance limits determined by gradual method.

value for lower LC_{50} estimated by the gradual method was 7 ppt, observed for *Parablennius sanguinolentus*.

Upper tolerance limits were broadly distributed among species (Fig. 10.2A, B). The mean upper salinity limit among direct-method experiments was 25 ppt (SD = 16) and among gradual-method experiments it was 52 ppt (SD = 36). The upper limit was distributed in a skewed or multimodal fashion in both datasets. Among the direct-method upper tolerance limits there was a clear mode close to isotonic salinity levels, around 10–15 ppt (Fig. 10.2A). The lowest values for upper LC_{50}

determined by the direct method were 6.7 and 6.8 ppt; both of these limits were observed for catfishes (*Hoplosternum thoracatum* and *Heterobranchus longifilis*, respectively). The highest value for upper LC_{50} estimated by the direct method was 65 ppt, observed for *Cyprinodon dearborni*. Among the gradual-method limits there was a clear mode around 20 ppt (Fig. 10.2B). The lowest value for upper LC_{50} determined by the gradual method was 6.6 ppt, observed for larval *Cobitis taena*. The highest values for upper LC_{50} estimated by the gradual method were 125 ppt and 126 ppt, for *Cyprinodon variegatus* and *Mugil cephalus*.

Halotolerance breadth varied by an order of magnitude or more among species. Estimates of breadth determined by the direct method (Fig. 10.3A) varied from 6.7 ppt (*H. thoracatum* and *H. longifilis*) to 59 ppt (*Leuresthes sardina* larvae; mean = 23, SD = 14). The values for breadth determined via direct challenge were distributed around a pronounced single mode at 10–15 ppt; the distribution for larval-stage subjects was comparable to that for later ontogenetic stages. Estimates of tolerance breadth determined by the gradual method (Fig. 10.3B) were about twice as long as estimates determined by the direct method, varying from 6 ppt (*C. taena* larvae) to 125 ppt (*C. variegatus*; mean = 50, SD = 34). The values for breadth determined via gradual salinity increases had a lower mode centered around 20 ppt.

Halotolerance breadth varied by order. Species in orders within the Otophysi all exhibited low values for breadth. The median value for breadth determined via the direct method for fishes in the Cypriniformes and Siluriformes was 13 ppt and 10 ppt, respectively (Fig. 10.4A). Breadth values for species in other well-represented orders were variable and the breadth distributions were comparable to each other (median values 20–30 for the direct method).

Halotolerance was aligned to halohabitat for each species. There have been few efforts to determine whether laboratory-determined salinity tolerance correlates with field limits, i.e. whether the fundamental niche and realized niche correspond. Kefford et al. (2004) found that direct-transfer experiments underestimated halohabitat breadth; early life stage and adult fish were often field-collected in salinity levels higher than direct-transfer experiments indicated they could tolerate. Gradual-method determinations of tolerance were better predictors of field distribution among the Australian fishes examined by Kefford et al. (2004). To test for correspondence of fundamental and realized haloniche among the species in this review, data on halohabitat were downloaded from FishBase (download 22 February 2012); every species in the database is listed as present or absent in FW, SW, and BW. Species in the halotolerance dataset were encoded as FW if they were present only in FW, SW if they were present only in SW, and BW if their halohabitat included BW; some of these latter species are diadromous and some are non-migratory.

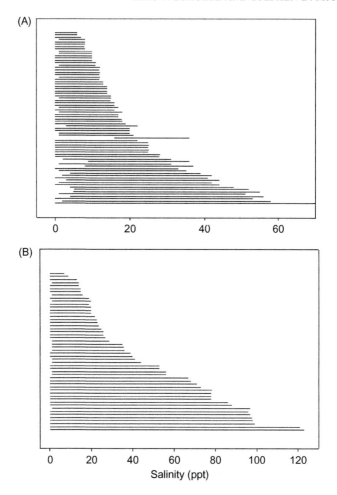

Fig. 10.3. Halotolerance breadth. Each species is represented by a line which extends between the lower and upper tolerance limit along the scale on the *x*-axis. In each plot species are sorted by tolerance breadth. (A) Tolerance breadth determined via direct method. (B) Tolerance breadth determined by gradual method. Note difference in range of *x*-axes.

Halotolerance limits and breadth varied among FW, BW, and SW fishes, but the experimentally determined fundamental haloniche was typically broader than the realized haloniche. Lower and upper halotolerance limits were lowest in FW species, intermediate in species whose halohabitat included BW, and highest in SW species (Table 10.1). On average, BW fishes tolerated salinity ranging from FW to nearly full-strength SW when subjected to direct testing, and up to about two times SW when subjected to

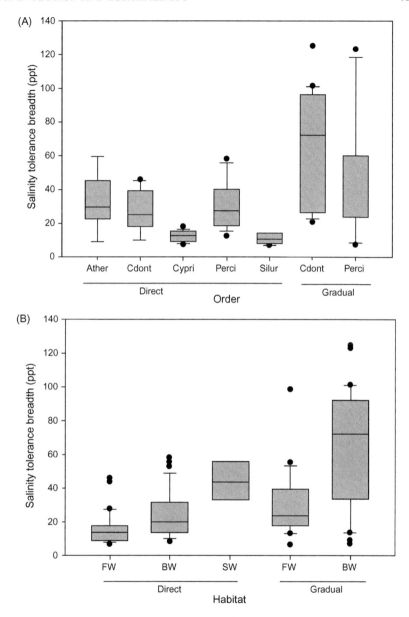

Fig. 10.4. Halotolerance breadth of selected groups. Lower, middle, and upper lines of boxes represent the quartiles of each distribution, whiskers represent the 10th and 90th percentiles, and points represent observations outside the 5th and 95th percentiles. Results were plotted if the sample size for a group was 9 or greater. (A) Well-represented orders in the dataset. The group of distributions on the left was estimated in direct experiments, the group on the right in gradual experiments. Ather: Atheriniformes; Cdont: Cyprinodontiformes; Cypri: Cyprinodontiformes; Perci: Perciformes; Silur: Siluriformes. (B) Grouping species by habitat. The group of distributions on the left was estimated in direct experiments, the group on the right in gradual experiments. FW: freshwater; SW: marine; BW: estuaries.

Table 10.1

Upper and lower halotolerance limits (ppt) of actinopterygian fishes by halohabitat.

Habitat	Lower limit				Upper limit			
	Mean	Range	SE	N	Mean	Range	SE	N
Direct method								
BW	1.11	(0–16)	0.37	53	27	(8–60)	2.4	37
FW	0.1	(0–1)	0.04	32	18	(7–65)	2.2	33
SW	4.5	(1–9)	0.79	10	49	(38–64)	4.9	5
Gradual method								
BW	0	(0–0)	0	38	68	(7–126)	6.4	32
FW	0.1	(0–1)	0.1	21	33	(7–109)	5	23
SW	7.0	(7–7)	–	1	–	–	–	0

Data are shown as the mean, range of values, standard error and number of species for the lower and upper tolerance limits in each of three halohabitat categories: freshwater (FW), brackish water (BW), and saltwater (SW). Tolerance limits determined using the direct method and those determined using the gradual method are presented separately.

gradual testing. The mean upper tolerance limit for FW fishes was about half-strength SW when determined by the direct method. However, FW fishes were able to tolerate SW when subjected to gradual salinity increases. Halotolerance studies therefore indicate that FW fishes generally have the capacity to survive in BW or SW. The mean lower tolerance limit for SW fishes was higher than for FW but well below the salinity of isotonicity, indicating that SW fishes also have the capacity to survive in BW. Halotolerance breadth also varied as expected by the habitat occupied (Fig. 10.4B). Quartile values of breadth distribution were lower among FW fishes than BW and SW fishes, but some FW fishes had breadth values as high as those of fishes in the other groups. Breadth values for BW and SW fishes were generally from 20 ppt to 50 ppt when determined by the direct method and were more than 70 ppt for BW fishes when determined by the gradual method. Hence, in contrast to previous findings (Kefford et al., 2004), the empirically determined fundamental haloniche is broader than the realized haloniche.

A comparable amount of variability in halotolerance breadth was explained by taxon and halohabitat, while less was explained by ontogenetic stage. Analyses of variance including all three effects (representing taxon by order), explained one-half to two-thirds of the variance in halotolerance breadth determined by both the direct and

Table 10.2
Predictors of halotolerance.

	R^2 of full model	Reduction in R^2 when dropped	R^2 of one-way
		Direct method	
	0.55		
Stage		0.020	0.000053
Order		0.25	0.43*
Habitat		0.12	0.29*
		Gradual method	
	0.64		
Stage		0.019	0.085*
Order		0.32	0.35*
Habitat		0.26	0.26*

Results are shown of analyses of variance on halotolerance breadth estimates derived from experiments using the direct method and the gradual method, testing the effect of ontogenetic stage (larva or juvenile+adult), taxonomic order, and halohabitat (five levels: freshwater, seawater, or brackish water plus freshwater and/or seawater).
Results of multiple models: R^2 of the full model including all three effects, decrease in R^2 when each effect is dropped from the full model, and R^2 of the model including each effect by itself.
* R^2 values of significant single-effects models.

gradual methods (Table 10.2). Taxon and habitat were significant ($p < 0.05$) in both full models and stage was not. To compare the contributions of the three effects to variance in halotolerance breadth, we examined changes in R^2 when each effect was eliminated from the full model, and the value of R^2 when each effect was by itself in a one-way model. Taxon explained more variance than halohabitat, from one-third more to twice as much. Ontogenetic stage was a weak predictor in both datasets; there is a significant stage effect only in analysis of the gradual dataset in which it is the sole predictor (mean breadth for larvae and juveniles+adults = 16 [$N = 5$] and 53 [$N = 48$], respectively). It was concluded that the degree of euryhalinity is predicted both by the present habitat of the species and by the evolutionary history of the species (i.e. the ancestral halohabitat) represented by the taxon.

Cluster analyses were conducted to define groups of species with similar halotolerances. The goal was to define a range of halotolerances that distinguish euryhaline from stenohaline fishes. In principle, stenohaline SW species should be intolerant of salinity substantially below isosmotic levels (9–10 ppt) and stenohaline FW species should be intolerant of salinity substantially above isosmotic levels. Euryhaline species should have the lower halotolerance limits of stenohaline FW species and the upper halotolerance limits of stenohaline SW species. The variables used for clustering were the upper halotolerance limit, and in the direct method dataset, halotolerance breadth (in the gradual method dataset there was perfect collinearity between upper tolerance limit and tolerance breadth). Clustering was conducted by the centroid method because the clusters were expected to be of unequal size (variable number of species per group) and dispersion (variable range of tolerance breadth). Because results for larvae were different from those for juveniles and adults, the cluster analysis was restricted to experiments using only juveniles and adults. If there were multiple determinations for a species they were not averaged. Inferential tools are not well established in cluster analysis and no attempt was made to assess the significance of cluster groupings. In the direct method dataset, two disparate groups were identified that were designated as empirically stenohaline and euryhaline; the tolerance limits of species by group are listed in Table 10.S2 (stenohaline tolerance breadth 7–35 ppt, euryhaline tolerance breadth 43–58 ppt). The groups are clearly separated based on centroid distance: the distance between clusters when the dataset is divided in two (standardized distance = 1.8) is large relative to the distance separating clusters at the next split in the tree (standardized distance = 0.7). In the gradual method dataset, three groups were identified that were designated as stenohaline FW and two levels of euryhaline: euryhaline–FW and euryhaline. The tolerance limits for each species are listed by group in Table 10.S3 (stenohaline FW tolerance breadth 9–46 ppt, euryhaline–FW tolerance breadth 55–80 ppt, euryhaline tolerance breadth 99–125 ppt). The division into groups is more subtle in the gradual dataset: the centroid distance between clusters does not change as dramatically as the number of clusters increases from two (standardized distance = 1.25) to three (standardized distance = 0.75).

These analyses and conclusions are unavoidably biased by the selection of species that have been subjected to tolerance tests. Tolerance tests such as these are often directed at revealing limits in broadly tolerant species; indeed, many of the studies in this chapter were motivated in some way to discern the limits of species known to be euryhaline, because of an interest in the culture or the ecology of the species. Most marine fishes that have been tested, even those that are not regarded as estuary dependent, can be regarded as tolerance-euryhaline: they have halotolerance limits well below isotonic salinity levels and a broad tolerance breadth. Only a few studies were identified that suggest that an SW species is limited to salinity levels above that at which it is expected to be isotonic, and it is hoped that more studies on SW fishes will be designed to test whether this limit is more prevalent than the existing literature suggests.

3. EVOLUTIONARY TRANSITIONS IN EURYHALINITY

Is euryhalinity a basal condition in fishes? How is it distributed phylogenetically – is there an evident phylogenetic signal among higher taxa, suggesting that gain or loss of broad tolerance occurred in deep nodes of the "fish tree", or alternatively is broad tolerance distributed uniformly among major fish groups, suggesting that lineages routinely switch from broadly to narrowly tolerant and back again? To develop answers to these questions, this section examines the debate over the environment in which the earliest fishes evolved, and analyzes how habitat-euryhalinity is distributed among broad taxonomic groups of extant fishes. Recent studies that have used phylogenetically informed analysis techniques to map salinity tolerance or halohabitat as a character are also reviewed.

3.1. Euryhalinity and Halohabitat Transitions in Early Fishes

Consideration of how euryhalinity was temporally and phylogenetically distributed among the earliest vertebrates must begin with the question of the halohabitat in which the first vertebrates evolved. Overall evidence supports the hypothesis that the earliest fishes were SW and stenohaline, followed by euryhalinity in some lineages and diversification in FW as well as SW (Evans et al., 2005). Early discussions (e.g. Smith, 1932; see also Vize, 2004) favored an FW origin, based on the predominance of a glomerular kidney in extant vertebrates and the intermediate concentration of inorganic ions in body fluids. Neither of these functional characters has proven to be decisive evidence for habitat of origin. Filtration by the glomerulus drives ionoregulatory functions of the kidney (particularly of divalent ions) in SW as well as FW habitats. Furthermore, a lower ionic concentration of plasma can plausibly be a derived rather than an ancestral condition, given the selective advantages of a more precisely tuned system of reactive tissues relying on membrane potentials (Ballantyne et al., 1987). Recent papers propose alternative scenarios in which the earliest vertebrates were estuarine or euryhaline. Ditrich (2007) suggests that vertebrates originated as osmoconformers in BW. According to his argument, protovertebrate kidney tubules functioned to maintain ion homeostasis and to recover metabolically important solutes but would not have been capable of the high-rate ionic exchange necessary for osmoregulation or urea retention. Ditrich's proposal has the substantial difficulty that it confers a requirement for stenohalinity on an organism in an estuary, which is likely to have highly variable salinity. Griffith (1987) proposes an anadromous life history for the protovertebrate, citing ancestral features of the kidney that he regards as evidence for hyperosmoregulation, and adaptive explanations for virtually all features shared by basal and derived fishes in terms of the advantages these features confer during migration. Molecular phylogenetic analysis also provides support for the euryhaline origin hypothesis. In contrast to morphologically based phylogenies, which identify stenohaline SW hagfish as basal to all other fishes, molecular analyses resolve jawless fishes as a monophyletic group (Heimberg et al., 2010). This placement implies that stenohalinity in the hagfish may be a derived condition, in which case the ancestral condition could be euryhalinity. An ecological difficulty of the euryhaline origin scenario is the harshness of fluvial habitats during the Cambrian; in the absence of banks stabilized by terrestrial or aquatic plants, waters would have been turbid, would have carried high sediment loads, and would have been completely unproductive. An additional count against the alternative scenarios is that the recent fossils illuminating the earliest emergence of Cambrian vertebrates or their precursors have been found in coastal SW deposits.

Transitions among halohabitats were frequent during the Paleozoic diversification of fishes, suggesting that physiological and ecological barriers were not difficult to surmount. Halstead (1985) discerned a proliferation of endemic genera and species upon colonization of BW and FW habitats in several major groups (e.g. thelodonts, cephalaspids). Diversification was less clearly associated with paleohabitat transition in other groups (e.g. Janvier et al., 1985, on osteostracans). Friedman and Blom (2006) assessed the paleoenvironment of basal actinopterygians using cladistic methods. They, like others, cautioned that paleoenvironmental reconstruction is subject to many uncertainties, especially for Paleozoic fossils for which there are no extant phyletic analogues. They proposed an SW origin for the clade based on earliest upper Silurian deposits in Sweden and China, and early Devonian SW diversification. Middle Devonian deposits record the appearance of actinopterygians in FW. Their evidence suggested four separate penetrations of FW, leading them to conclude that "the assembly of the earliest freshwater ecosystems was dominated not by unique, isolated 'seedings' of these novel environments by primitively marine clades, but instead by iterative and relatively frequent colonization events". Other analyses indicated that there were many transitions to FW, supported by multiple instances of genera that occurred in both FW and SW water deposits (Schultze and Cloutier, 1996); similarly, 53 trace fossil Paleozoic genera occurring in both marine and nonmarine deposits have been charted (Maples and Archer, 1989). Finally, ancestral-state reconstruction based on a molecular phylogeny of ray-finned fishes indicates that all extant ray-finned fishes are descended from an FW or a BW ancestor (Vega and Wiens, 2012), indicating that a complex history of transitions between SW and FW is embedded in the evolutionary history of this diverse group. To summarize, the halohabitat of the most recent common ancestor of all vertebrates was probably SW or BW, and that of the most recent common ancestor of ray-finned fishes was probably FW or BW. Euryhalinity may have played a significant role in Paleozoic diversification of fishes.

3.2. Euryhalinity Among Extant Fishes

Halohabitat use is distributed heterogeneously among broad taxa of fish, as is the case for other aquatic Metazoa. Hutchinson (1960), commenting on animal phyla that have FW and SW representatives, noted that "the distribution [of freshwater species] in the taxonomic system is highly irregular, suggesting a great degree of superdispersion of the physiological characters that preadapt marine organisms to entrance into freshwaters [sic]." Similarly, Nelson (2006) documented that the FW fishes are concentrated in certain orders. The likelihood of diadromy or euryhalinity is also known to vary taxonomically and phylogenetically. Diadromy is more prevalent among basal fishes (McDowall, 1988; but see Dodson, 1997, for a critique of McDowall's assignment of diadromy to taxa). Gunter (1967) suggested that euryhalinity is more pronounced in basal fishes, without quantifying the heterogeneity.

This section summarizes data on the phylogenetic distribution of halohabitat use among broad taxa of ray-finned fishes. Ballantyne and Fraser (2013, Chapter 4, this volume) demonstrate that euryhalinity and FW tolerance has evolved multiple times in the Elasmobranchii. To the authors' knowledge, no detailed description of the phylogenetic distribution of halohabitat in the Actinopterygii has been previously published; however, there have been several efforts to characterize halohabitat into distinct estuarine zones (Bulger et al., 1993) or to define euryhaline fish functional groups (Elliott et al., 2007), and the predominant halohabitat of fish families has been described by Evans (1984). The focus is on the Actinopterygii because it contains the vast majority of extant fish species and has arguably a greater heterogeneity in halohabitat use than the Chondrichthyes or the Sarcopterygii, and because this confines the analysis to an osmoregulatory physiology strategy. As described in Section 2.2, data on halohabitat use were downloaded from FishBase (download 22 February 2012). Any species that is found in BW is referred to here as halohabitat-euryhaline. Within this set there are subsets of habitat-euryhalinity: there are species that are found in SW and BW, species that are found in FW and BW, species that are found in all three halohabitats, and species that are found only in BW. Species occurring in both SW and FW are termed here as "halohabitat-amphihaline". The original application of "amphihaline" to a species that migrates between FW and SW (Fontaine, 1975) has been trumped by the common usage and more precise etymology of "diadromous".

FishBase currently recognizes 30,972 separate species or subspecies. Subspecies ($N = 397$) are recognized in 153 species, within 24% of which halohabitat varies among subspecies. For this analysis all subspecies were treated as if they were species and will henceforth be referred to them as such. Additional information downloaded from FishBase comprised entries on migratory behavior (e.g. amphidromous, oceanodromous). To date, migratory behavior has been recorded for about 3818 species of Actinopterygii, of which about 50% are listed as non-migratory. Because the taxonomic distribution of species for which migratory behavior has been recorded is uneven, any association between euryhalinity and migratory behavior should be regarded as tentative.

Phylogenetic relationships of major taxa followed Nelson (2006) for the placement of orders basal to the teleosts, and Wiley and Johnson's (2010) analysis of teleost clades. Relationships among derived Acanthopterygii are poorly resolved, and 30 orders (most of which are monophyletic but some of which are not monophyletic yet are widely regarded as taxa, e.g. "Perciformes") were aggregated into division Percomorphacea. Several polytomies (e.g., Ateleopodiformes+Stomiatiformes+Eurypterygia [not shown, consisting of Aulopiformes and more derived orders] and Percopsiformes+Gadiformes+Acanthopterygii) were retained because further aggregation would have obscured substantial phylogenetic detail. In addition, Hiodontiformes and Osteoglossiformes were aggregated into Osteoglossomorpha because the former has only two species. Assignment of species to each major taxon was done as follows: placement in family was done according to FishBase; family placement in higher taxa was done if possible according to Wiley and Johnson (2010) or according to Nelson (2006).

A minority of species is habitat-euryhaline. There are 2844 species (about 9% of the total) that include BW in their halohabitat (Table 10.3). The largest category of euryhaline species is found in BW and SW but not FW (4.2% of all Actinopterygii). Roughly one-quarter of these species may be diadromous; most species for which there are migration behavior entries in FishBase are listed as non-migratory, oceanodromous (migrating in SW only), or oceano-estuarine (migrating between SW and BW). About 2% of all Actinopterygii are amphihaline, and these species are almost exclusively diadromous. Another 2% of actinopterygians use BW and FW but not SW, and about 80% of these species for which there are migration behavior records are listed as non-migratory or potamodromous. Remarkably few species are found in only BW (0.3% of Actinopterygii). The apparently high percentage of species in the BW-only category that are diadromous must be viewed with caution as the number of migration behavior records is low.

Table 10.3
Halohabitat use of Actinopterygii.

Halohabitat	N (spp.)	Diadromous
+BW+FW+SW	732	93% (527)
+BW+FW−SW	727	20% (212)
+BW−FW+SW	1293	28% (288)
+BW−FW−SW	92	67% (9)
−BW+FW+SW	0	
−BW+FW−SW	14391	4% (1330)
−BW−FW+SW	13737	1% (1452)

Data are shown as the number of species [N (spp.)] and an estimate of the percentage of those species that are diadromous (with the number of species on which this estimate is based in parentheses), for each halohabitat category. Halohabitat categories are encoded according to whether a species is found (+) or is absent (−) in brackish water (BW), freshwater (FW), and/or saltwater (SW), such that a species occurring in all waters is encoded as +BW+FW+SW.

Over all categories, 60% of species that are halohabitat-euryhaline may be diadromous.

Habitat-euryhalinity appears primarily among the most basal and the most derived taxa in the Actinopterygii. At least half of the species are euryhaline in the basal clades Acipenseriformes, Lepisosteiformes, Elopiformes, Albuliformes, and Clupeiformes (Fig. 10.5). No more derived clade has a similarly high proportion of species that are euryhaline; nonetheless, most of the species that are euryhaline ($n = 2030$) are in the derived and speciose clade Percomorphacea. Although comprising only 12% of the clade, the halohabitat-euryhaline percomorphs are 71% of all the halohabitat-euryhaline fishes. There is significant heterogeneity among orders in the proportion of species that are euryhaline (test of independence of euryhalinity and order, chi-square $= 4360$, df $= 59$, $p < 0.0001$).

Among the habitat-euryhaline species, the representation of euryhalinity subsets varies among the clades. Euryhaline species are most commonly FW +BW in predominantly FW clades, such as Cypriniformes and Characiformes (Fig. 10.6). Conversely, euryhaline species are most commonly SW +BW in SW clades such as Clupeiformes. The predominant type of euryhalinity in a clade mirrors the predominant type of stenohalinity in the clade: the correlation between the proportion of the clade's euryhaline species that inhabit both FW and BW and the proportion of species in a clade that inhabit only FW is significant ($n = 12$ clades that have species occurring in FW and BW; $r = 0.78$, $p = 0.003$). Similarly, the correlation between the proportion of the clade's euryhaline species that inhabit both SW and BW and the proportion of species in a clade that inhabit only SW is

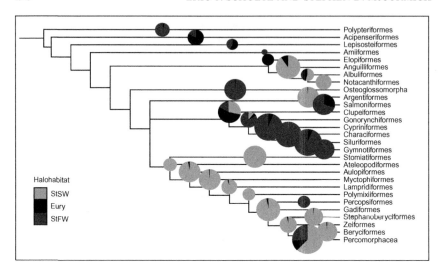

Fig. 10.5. Phylogenetic relationships and halohabitat use of Actinopterygian fishes. The pie chart for each terminal taxon in the phylogeny represents the proportion of species in the taxon that occur in saltwater only (stenohaline–saltwater: StSW), freshwater only (stenohaline–freshwater: StFW), or brackish water (Eury). The area of each pie chart is scaled to represent the number of species in the taxon (log₁₀ scale); the smallest pie, for Amiiformes, represents one species, whereas the largest pie, for Percomorphacea, represents 17,020 species. This figure was developed with the web-based tool Interactive Tree of Life (http://itol.embl.de: Letunic and Bork, 2011).

significant ($n = 16$ clades that have species occurring in SW and BW; $r = 0.86$, $p < 0.0001$). Two clades deviate notably from the strong association between predominant stenohalinity habitat and predominant euryhalinity habitat. In Salmoniformes, 70% of the species are stenohaline–FW but only 12% of the euryhaline species are confined to FW and BW, reflecting the high proportion in this group that occurs in all waters. Conversely, in Lepisosteiformes 42% of the species are stenohaline FW but all of the remaining species are confined to FW and BW, i.e. species in this clade do not inhabit SW.

Habitat-euryhalinity varies among taxa within the most derived clade, currently recognized as the Percomorphacea. For each major taxon in the Percomorphacea, the percentage of species within each of the habitat use categories was estimated (Table 10.4). With the exception of Elassomatiformes, orders within series Smegmamorpharia (also comprising Mugiliformes, Synbranchiformes, Gasterosteiformes, Atheriniformes, Beloniformes, and Cyprinodontiformes) are relatively euryhaline: by order the species that inhabit BW ranges from 7% (Cyprinodontiformes) to 76% (Mugiliformes),

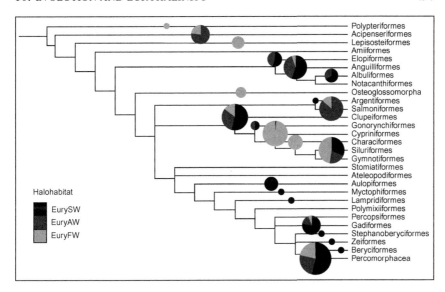

Fig. 10.6. Phylogenetic relationships and halohabitat use of euryhaline Actinopterygian fishes. Terminal branches are labeled with pie charts if the taxon has species occurring in brackish water. The pie chart represents the proportion of species in the taxon that occur in saltwater and brackish water (euryhaline–saltwater: EurySW), saltwater, brackish water, and freshwater (euryhaline–all waters: EuryAW), and freshwater and brackish water (euryhaline–freshwater: EuryFW). For clarity, species occurring in brackish water only are represented as EuryAW. The area of each pie chart is scaled to represent the number of euryhaline species in the taxon ($\log_{10}$ scale); the smallest pies represent one species and the largest pie for Percomorphacea represents 2030 species. This figure was developed with the web-based tool Interactive Tree of Life (http:// itol.embl.de: Letunic and Bork, 2011).

and overall 16% of smegmamorph species occur in BW, compared with 11% among species in the remaining orders. For the most part, a comparable percentage of smegmamorph fishes are euryhaline–SW, euryhaline–FW, and euryhaline–all waters. The most euryhaline among the remaining orders are Carangiformes, Gobiiformes, Scombriformes, and Batrachoidiformes. Relative to smegmamorphs, these euryhaline species are more likely to be euryhaline–SW.

Family-level variability in the tendency to be halohabitat-euryhaline is strong among Percomorphacea. Family-characteristic halohabitat use has long been recognized (Myers, 1938; Gunter, 1967) but has not been quantitatively assessed. To test the degree to which family is predictive of halohabitat use, the proportion of species that are euryhaline within each percomorph genus was estimated. The identity of higher taxonomic levels (order, and family nested within order) accounted for 35% of the variance in arcsine-transformed proportion of species that are euryhaline, whereas

Table 10.4
Halohabitat use of Percomorphacea.

Order	N (spp.)	% EuryFW	% EurySW	% EuryAW	% EuryBW	% StenoFW	% StenoSW
Elassomatiformes	7	0	0	0	0	100	0
Mugiliformes	82	8.5	20	45	2.4	3.7	21
Synbranchiformes	120	11	0	0	0	89	0
Gasterosteiformes	351	3.7	10	5.7	0.28	8.3	72
Atheriniformes	334	6.3	8.4	5.7	1.2	64	14
Beloniformes	273	8.8	9.2	8.8	0.73	25	47
Cyprinodontiformes	1231	5.7	0.57	0.89	0.16	93	0.081
Acanthuriformes	116	0	15	2.6	0	0	83
Anabantiformes	195	4.1	0	0	0	96	0
Batrachoidiformes	81	0	15	3.7	0	6.2	75
Blenniiformes	906	0.33	3.6	0.22	0.55	0.22	95
"Caproiformes"	18	0	0	0	0	0	100
Carangiformes	160	0	38	2.5	0	0	59
Cottiformes	1191	0.34	1.9	0.76	0.084	7.0	90
Dactylopteriformes	7	0	14	0	0	0	86
Gobiesociformes	362	0.28	3.9	0.28	0.28	2.8	93
Gobiiformes	1943	6.3	10	10	2.7	21	50
Icosteiformes	1	0	0	0	0	0	100
Labriformes	2688	2.2	1.3	0.11	0.074	59	37
Lophiiformes	347	0	0.29	0.29	0	0	99
Nototheniiformes	148	0	0	0	0	0	100
"Ophidiiformes"	517	0.39	0.97	0	0.19	0.97	97
"Perciformes"	2889	2.0	11	3.7	0.035	15	69
Pholidichthyiformes	2	0	0	0	0	0	100
Pleuronectiformes	782	1.3	9.2	3.5	0.13	3.7	82
Scombriformes	162	0	19	0.62	0	0	80
Scorpaeniformes	1314	0.15	4.6	0.38	0	0.61	94
Stromateiformes	66	0	9.1	0	0	0	91
Tetraodontiformes	432	1.9	9.3	1.9	0.46	6.7	80
"Trachiniformes"	294	0	3.1	0.68	0	0.34	96

For each order, data are shown as the number of species [N (spp.)] and an estimate of the percentage of those species that occur in: brackish and freshwater (% EuryFW), brackish and saltwater (% EurySW), brackish, freshwater, and saltwater (% EuryAW), brackish water only (% EuryBW), freshwater only (% StenoFW), and saltwater only (% StenoSW). Orders are arranged to reflect phylogeny insofar as it can be currently resolved (Wiley and Johnson, 2010); names in quotation marks indicate groups for which there is no evidence of monophyly.

order alone explained only 5.5%. It was concluded that the pronounced variability among percomorph higher taxa in patterns of halohabitat use is largely the result of shared ecology and physiology among species at an intermediate familial level of evolutionary relationship. Diversification at this level would have arisen primarily in the aftermath of the Cretaceous–Palaeogene extinction, when there was a sharp expansion in the number of extant fish families and a burst of morphological diversification among the percomorphs (Friedman, 2010). A satisfying concordance is suggested here between physiological, ecological, and morphological diversification.

3.3. Evolutionary Diversification upon Transitions in Halohabitat

Low prevalence notwithstanding, euryhaline species are potent sources of evolutionary diversity. A broadly tolerant physiology and wide range of occupied habitats heighten the likelihood of a transition to a new habitat and a more specialized regimen, potentially giving rise to new species, i.e. cladogenesis. In particular, euryhaline species are subject to landlocking, wherein a population becomes restricted to FW. This section reviews studies that provide conceptual or empirical insights into the cladogenetic potential of euryhalinity in fishes. Taxa and regions are identified that are well represented in recent literature on transitions and discuss the evolutionary processes associated with transitions.

As is often the case in evolutionary science, early contemplation on the diversifying potential of broad salinity tolerance can be found in the publications of Charles Darwin. Considering the puzzling distributions of some FW fish groups whose distribution includes multiple continents, he wrote (Darwin, 1876), "Salt-water fish can with care be slowly accustomed to live in fresh water; and, according to Valenciennes, there is hardly a single group of which all the members are confined to fresh water, so that a marine species belonging to a fresh-water group might travel far along the shores of the sea, and could, it is probable, become adapted without much difficulty to the fresh waters of a distant land." There is a clear connection between this thought and subsequent dispersalist explanations for the distribution of "secondary FW species" that may occasionally enter SW such as gar, synbranchids, cichlids, and cyprinodontids (Myers, 1938). Although widely adopted, the distinction between primary FW species, which spend their entire lives in FW, and secondary FW species has been criticized on the grounds that it is circular (i.e. if a taxon is widely distributed it must be capable of coastal or marine dispersal) (Rosen, 1974), and is not in fact predictive of a group's dispersal abilities (e.g. Sparks and Smith, 2005).

In more recent years, a series of insightful reviews have commented on the diversifying potential of either diadromous or estuarine life cycle or habitat. Lee and Bell (1999) briefly reviewed literature on postglacial (Pleistocene and recent) transitions to FW in invertebrates and diadromous fishes, emphasizing how recent invasions provide the opportunity to examine mechanisms involved in habitat transitions. McDowall (2001) described the paradoxically homogenizing and diversifying role of diadromous migration, on the one hand promoting gene flow and on the other hand yielding landlocking, isolation, and cladogenesis. Other authors have considered the diversifying potential of estuarine fishes. Bamber and Henderson (1988) hypothesize that "selection for plasticity has preadapted estuarine and lagoonal teleosts with the ability to invade fresh waters. The evolutionary history of fish has included repeated invasions from the estuary to fresh waters, followed by adaptive radiation." Bilton et al. (2002) generalize on this perspective both taxonomically (i.e. extend their review to all estuarine animals) and dynamically; they note that the estuarine habitat is itself spatially subdivided, potentially restricting gene flow and enhancing spatial differentiation in population genetic structure. Features that are explicitly or implicitly common to these discussions are adaptive change associated with shifts in halohabitat, speciation by allopatric, parapatric, or sympatric mechanisms, and repetition over space and/or time promoting adaptive radiation. Furthermore, virtually all studies on diversification in euryhaline fishes and their descendants (Table 10.5) allude to the role that changes in sea level have played in altering the habitat configuration of fishes living on the continental margin.

Some anadromous fishes and their landlocked derivatives furnish several model systems of diversification in evolutionary biology. Salmonids show high fidelity and local adaptation to natal sites (Hendry et al., 2003b), whereas other anadromous species show little tendency for homing and have weak geographic population structure (Shrimpton, 2013, Chapter 7, this volume). Modifications in landlocked populations of salmon and three-spined stickleback have illustrated the nature and pace of adaptive change, and the predictability of adaptive change has been highlighted in stickleback. There have been extensive recent reviews of diversification in salmonids and stickleback (McKinnon and Rundle, 2002; Kinnison and Hendry, 2003), and this subject will not be considered here in comparable detail.

Physiological and/or behavioral characteristics make some euryhaline taxa particularly susceptible to changes in halohabitat and subsequent differentiation (Table 10.5), such as silversides (Atheriniformes). The New World has multiple examples of atherinid species flocks or adaptive radiations arising from habitat transitions (Barbour, 1973; Beheregaray and Levy, 2000; Beheregaray and Sunnucks, 2001; Beheregaray et al., 2002; Bloom et al., 2009; Heras and Roldan, 2011) and species pairs in overlapping halohabitats (Fluker et al., 2011). In the Old World, the cosmopolitan species *Atherina boyeri* is known to be differentiated according to halohabitat (Klossa-Kilia et al., 2007). Australian coast atherinids have also diversified in halohabitat (Potter et al., 1986). As indicated above, Bamber and Henderson (1988) suggest that underlying this readiness to transition to FW habitat is a high intrinsic level of phenotypic plasticity in the family.

Table 10.5
Evolutionary transitions in euryhalinity.

Order	Family	Taxon	Ancestral halohabitat	Derived halohabitat	Timing of transition	Taxonomic level of diversification	References
Anguilliformes	Angullidae	*Anguilla*	SW	EuryAW (D)		Interspecific	Inoue et al. (2010)
Atheriniformes	Atherinidae	*Atherina boyeri*	EurySW	BW, FW		Intraspecific	Francisco et al. (2006); Klossa-Kilia et al. (2007)
Atheriniformes	Atherinidae	*Chirostoma*, *Poblana* spp.	EuryAW	FW	Plio-Pleistocene	Interspecific and intergeneric	Barbour (1973); Bloom et al. (2009)
Atheriniformes	Atherinidae	*Menidia beryllina*	BW	FW		Intraspecific	Fluker et al. (2011)
Atheriniformes	Atherinidae	*Menidia clarkhubbsi*	BW	BW		Intraspecific (clonal)	Echelle et al. (1989)
Atheriniformes	Atheriniopsidae	*Odontesthes argentinensis*	SW	BW		Intraspecific	Beheregaray and Levy (2000); Beheregaray and Sunnucks (2001); Heras and Roldan (2011)
Atheriniformes	Atheriniopsidae	*Odontesthes perugiae* complex	BW	FW	Pleistocene	Interspecific	Beheregaray et al. (2002)
Atheriniformes	Atheriniopsidae	*Odontesthes* spp.	SW, EurySW	FW	Pleistocene	Interspecific	Heras and Roldan (2011)
Atheriniformes	Cyprinodontidae	*Aphanius* spp.	EurySW	EuryFW, FW	Miocene	Interspecific	Kosswig (1967)

(Continued)

Table 10.5 (Continued)

Order	Family	Taxon	Ancestral halohabitat	Derived halohabitat	Timing of transition	Taxonomic level of diversification	References
Beloniformes	Belonidae	*Belonion, Potamorrhaphis, Pseudotylosurus, Xenontodon*	EurySW	FW		Interspecific and intergeneric	Lovejoy and Collette (2001)
Blenniiformes	Blenniidae	*Salaria fluviatilis*	EurySW	FW	Miocene	Interspecific	Kosswig (1967); Zander (1974); Plaut (1998)
Clupeiformes	Clupeidae	*Alosa pseudoharengus*	EuryAW (D)	FW	Recent	Intraspecific	Palkovacs et al. (2008); Post et al. (2008); Palkovacs and Post (2009)
Atheriniformes	Atherinidae	*Menidia clarkhubbsi*	BW	BW		Intraspecific (clonal)	Echelle et al. (1989)
Clupeiformes	Clupeidae	*Alosa* spp.	EuryAW (D)	FW		Interspecific	Bobori et al. (2001)
Clupeiformes	Engraulidae	*Anchovia surinamensis*	EurySW	EuryFW	Miocene	Interspecific	Lovejoy et al. (2006)
Clupeiformes	Engraulidae	*Jurengraulis juruensis*	EurySW	FW	Miocene	Interspecific and intergeneric	Lovejoy et al. (2006)
Cyprinodontiformes	Fundulidae	*Fundulus* spp.	EurySW	FW		Interspecific	Whitehead (2010)
Gasterosteiformes	Gasterosteidae	*Gasterosteus aculeatus*	SW, euryAW (D)	FW	Pleistocene to present	Intraspecific	Klepaker (1993); McKinnon and Rundle (2002); Bell et al. (2004); Gelmond et al. (2009)

Order	Family	Genera/Species	Salinity	Salinity	Period	Divergence	Reference
Gobiiformes	Gobiidae	*Economidichthys, Knipowitschia, Orsinogobius, Padogobius, Proterorhinus*	SW, EurySW	FW	Miocene	Interspecific and intergeneric	Economidis and Miller (1990); Miller (1990)
Myliobatiformes	Pomatotry-gonidae	*Pomatotrygon, Paratrygon, Plesiotrygon*	SW	FW	Miocene	Interspecific and intergeneric	Lovejoy et al. (2006)
Perciformes	Sciaenidae	*Plagioscion, Pachypops, Pachyurus, Petilipinnis*	EurySW	EuryFW, FW	Miocene	Interspecific and intergeneric	Lovejoy et al. (2006)
Salmoniformes	Galaxiidae	*Galaxias auratus, G. tanycephalus, G. truttaceus*	EuryAW (D)	FW	Pleistocene to recent	Interspecific and intraspecific	Ovenden and White (1990); Ovenden et al. (1993)
Salmoniformes	Galaxiidae	*Galaxias vulgaris* complex	EuryAW (D)	FW	Pliocene	Interspecific	Waters and Wallis (2001a, b)
Salmoniformes	Salmonidae	*Oncorhynchus, Salmo, Salvelinus* spp.	EuryAW (D)	FW		Intraspecific	Hendry et al. (2003a)
Siluriformes	Ariidae	*Notarius, Catharops, Potamarius, Arius, Cephalocassis, Hemiarius, Neoarius, Potamosilurus, Cinetodus,*	EurySW	FW		Interspecific and intergeneric	Betancur-R (2010)

(Continued)

Table 10.5 (Continued)

Order	Family	Taxon	Ancestral halohabitat	Derived halohabitat	Timing of transition	Taxonomic level of diversification	References
		"Sciades", *Brustarius*, *Pachyula*, *Doiichthys*, *Nedystoma*, *Nempteryx*, *Cochlefelis*					
Tetraodontiformes	Tetraodontidae	*Auriglobus*, *Carinotetraodon*, *Colomesus*, *Tetraodon*	EurySW, EuryAW	FW	Miocene to recent	Interspecific and intergeneric	Yamanoue et al. (2011)

A selection of studies are listed that documented diversification within a taxon in halohabitat or salinity tolerance. For each taxon, the order and family, the ancestral and derived halohabitat (SW: saltwater; BW: brackish water; EurySW: saltwater and brackish water; EuryAW: saltwater, brackish water, and freshwater; D: diadromy; EuryFW: brackish water and freshwater; FW: freshwater), the time at which the transitions occurred, and the taxonomic level (diversification within species: intraspecific; transition giving rise to new species or genera: interspecific, intergeneric, etc.) are presented.

Two euryhaline-migratory species of *Galaxias* (*G. truttaceus* and *G. brevipinnis*), a southern hemisphere genus of salmoniform, have undergone repeated transitions to FW, giving rise to species complexes in Tasmania and New Zealand's South Island (Table 10.5). *Galaxias auratus* and *G. tanycephalus* inhabit lake clusters in separate drainage basins of Tasmania, and are each extremely similar to *G. truttaceus*, with which they form a well-defined clade (Ovenden et al., 1993). *Galaxias truttaceus* itself has several landlocked populations (Ovenden and White, 1990). Resolution of relationships among the three nominal species and reconstruction of the isolating events have been hampered by bottleneck- or founder effect-induced reductions in genetic diversity of the landlocked species and genetic variability of the migratory progenitor (Ovenden et al., 1993), but it appears that all landlocking events occurred in the past 100,000 years. Phylogenetic resolution and paleoreconstruction has been more successful for the more diverse *G. vulgaris* New Zealand complex of landlocked species, which arose from the diadromous *G. brevipinnis*. A well-resolved phylogeny for the group indicates that nine stenohaline–FW species arose from three separate losses of migration; this conclusion required the assumption that migration was the basal condition for the group, which is supported on other lines of evidence (Waters and Wallis, 2001a). Time since divergence estimates and geological evidence indicates that a 2–4-million-year-old (Pliocene) uplift of the South Island's mountain range was the process that isolated previously migratory populations from the sea (Waters and Wallis, 2001b).

The FW habitat is plesiomorphic for taxa in larger FW groups, such as the catfishes. Two catfish families, Ariidae and Plotosidae, consist largely of euryhaline–SW species. Phylogenetic analysis securely places this as the derived halohabitat within the Siluriformes, and was independently derived for each family or superfamily (Sullivan et al., 2006). Transition to FW occurred 10–15 times within the Ariidae, yielding 16 partially or fully FW genera (Table 10.5) that are located in every region where marine ariids are found (Betancur-R, 2010). In this group, the proclivity to evolve FW habitat occupation appears to reflect a tendency to stenohalinity that was not lost in the SW ancestors.

The Anguilliformes provide an example of a large group in which euryhaline taxa evolved from stenohaline–SW ancestors. A recent phylogenetic analysis of the Anguilliformes strongly supports an SW origin of this group. Catadromy (hence developmental amphihalinity) evolved once in the order, and is a synapomorphy for the family Anguillidae and its single genus *Anguilla* of 16 species, all of which are catadromous (Inoue et al., 2010).

Another family-wide analysis, for the pufferfishes Tetraodontidae, finds that the derived FW lineages, occurring repeatedly on different continents, are well dispersed across the phylogeny (Yamanoue et al., 2011). Habitat optimized on the phylogenetic tree indicates that the coastal SW habitat is ancestral for the family (Table 10.5). Stenohaline FW puffers occur on South America, Southeast Asia, and Central Africa; divergence time estimates suggest that the transition to FW occurred first in Asia (Eocene, up to 78 mya) and most recently in South America (Miocene or more recent). These transitions have given rise to 29 species in four genera.

Taxon-wide data on salinity tolerance are much harder to come by than taxon-wide data on halohabitat, and thus the study by Whitehead (2010) on the frequency, distribution, and timing of transitions in tolerance euryhalinity within the killifish genus *Fundulus* is unique and valuable (Table 10.5). For the most part (23 species), this study was able to use salinity tolerance data that had been collected using the gradual experimental design; data on halohabitat were used for two additional species, providing character data for about 75% of the species in the genus. Upper salinity tolerance data (all species were tolerant of FW) resolved into three groups, consisting of relatively stenohaline (limit 20–26 ppt), intermediate (60–75 ppt), and tolerant (80–115 ppt). Mapping of physiological characters on the phylogeny indicated that the tolerant physiology is basal and that there have been five independent transitions to less tolerant states. The effort to reconstruct ancestral physiology was challenged by high transition rates among character states, so that the state of deep nodes could not be attributed with confidence. This problem is likely to arise frequently in such analyses, in groups that have undergone rapid diversification (i.e. an adaptive radiation) associated with changes in salinity tolerance.

Some areas such as the Amazon have been hotspots for transitions to FW, which can often be explained by large-scale events such as marine incursions that acted on multiple euryhaline groups simultaneously. The Amazon basin is richly endowed with FW derivatives of SW fishes, offering at least 39 genera in 17 largely SW families within 14 orders. In an effort to clarify the timing and mechanism of origins of these groups, Lovejoy et al. (2006) tested predictions arising from the hypothesis that Miocene marine incursions, which established a large system of brackish lakes, promoted transition. Their analysis, combining phylogeny, geology, the fossil record, and biogeography, supported the Miocene incursion model for multiple groups including potamotrygonids, engraulids, belonids, hemirhamphids, and sciaenids (Table 10.5). A genus of puffers also occurs in the Amazon Basin and an independent analysis of time of divergence for the FW species from its sister taxon is consistent with the Miocene marine incursion model (Yamanoue et al., 2011).

The Mediterranean Basin is another region with a large number of SW- or euryhaline-to-FW transitions that are attributable to geological history.

In comparison to the rest of Europe, the Mediterranean Basin has a relatively large number of fish species and a high degree of endemism; endemism is especially high in the eastern portion of the region as represented by Greek collection sites (Ferreira et al., 2007). Including introduced and diadromous species, 135–162 fish species inhabit Greece's FW (Bobori and Economidis, 2006; Oikonomou et al., 2007). At least 13 of these species represent relatively recent transitions from SW to FW habitats (Table 10.5). FW species that are clearly derived from euryhaline relatives include two species of landlocked shad (*Alosa macedonica* and *A. vistonica*) (Bobori et al., 2001), a blenny (*Salaria fluvatilis*) (Zander, 1974), and nine species in five genera of goby (Economidis and Miller, 1990; Miller, 1990). Differentiation of the euryhaline silverside *Atherina boyeri* in Hellenic lakes has already been noted. *Pungitius hellenicus* is a critically endangered species of stickleback that is endemic to a small region of FW springs and associated wetlands (Keivany et al., 1997) and is the only member of its genus to be stenohaline. Many of these transitions can be attributed to the dynamic history of salinity transitions in the region. The most detailed reconstruction of diversification upon transition to FW in the Mediterranean basin has been outlined for gobies (Economidis and Miller, 1990; Miller, 1990). In this reconstruction, separation between the ancestor of *Economidichthys+Knipowitschia* and *Pomatoschistus* occurred during the middle-Miocene closure of the brackish Sarmatic Sea, an event that represented the onset of Ponto-Caspian endemism. Separation between *Economidichthys* and *Knipowitschia* occurred during the late-Miocene Messinian salinity crisis.

Postglacial changes in the distribution of surface FW and the elevation of landmasses are primarily responsible for the landlocking of euryhaline species at high latitudes and some cases of lower latitude landlocking. Most of the existing stenohaline FW populations of three-spined stickleback were isolated from ancestral SW habitat as a result of glacial retreat and isostatic rebound, wherein landmasses rose in elevation when relieved of masses of ice (Bell and Foster, 1994). Changes in sea level during and after the Pleistocene created lagoons and promoted diversification of silversides in southern Brazil (Beheregaray et al., 2002).

A high incidence of FW derivations in some regions may be attributable to ecological, in addition to or instead of, physical–geographic factors. The Usumacinta River of Mexico and Guatemala harbors multiple independent incidences of FW derivation. High calcium concentration in the water of this karstic region may essentially lower the physiological hurdle that must be surmounted for colonization from BW (Lovejoy and Collette, 2001). In general, the extent of diversification that occurs in FW following colonization by SW forms will be dependent on factors such as the diversity of habitats, the intensity of competition from already established FW species (Miller, 1966; Betancur-R et al., 2012), and the availability of refuge from predation (e.g. widespread albeit seasonal areas of hypoxic water in the Amazon) (Anjos et al., 2008).

While cases of euryhaline differentiation associated with transitions to another halohabitat have been emphasized here, diversification within a euryhaline halohabitat has also been documented. A species complex of gynogenetic unisexual silversides arose from repeated hybridizations between female *Menidia peninsulae* and males of a congener, probably *M. beryllina* (Echelle et al., 1989). It is likely that the unisexual complex arose early in the divergence between the two euryhaline parental species, because gynogenetic lines will arise when parental species differ in regulation of meiosis but do not differ to the extent that hybrid offspring would have markedly lower fertility or viability. Such diversifying contact between populations early in the speciation process is quite consistent with arguments summarized above regarding why estuarine environments could serve as incubators of evolutionary novelty.

3.4. Adaptation upon Transitions in Halohabitat

Intraspecific divergence in morphology, behavior, physiology, and life history occurs between euryhaline forms and their counterparts in FW and SW. Adaptive morphological and behavioral changes are associated with changes in predator regimen and prey field (McKinnon and Rundle, 2002; Bell et al., 2004; Palkovacs and Post, 2009) and reproductive substrates (Beheregaray and Levy, 2000). Morphometric analysis has revealed body shape differences between euryhaline and stenohaline forms (Klepaker, 1993; Gelmond et al., 2009; Fluker et al., 2011). Change in water chemistry (particularly lower availability of calcium) can also affect body form via direct effects (phenotypic plasticity) and heritable effects on ion uptake and deposition; although the role of water chemistry in selecting heritable differences in calcium regulation is implied by studies demonstrating growth differences between armor gene alleles in stickleback (Barrett et al., 2008), to the authors' knowledge this has not been further tested. Differences between euryhaline and FW forms have been found in salinity tolerance (Dunson and Travis, 1991; Foote et al., 1992; Plaut, 1998; Purcell et al., 2008; McCairns and Bernatchez, 2010), expression patterns of loci associated with osmoregulation (Nilsen et al., 2007; McCairns and Bernatchez, 2010; Whitehead et al., 2011), and gene sequence in osmoregulation loci or regions closely associated with such loci (implicating positive selection for change in coding regions) (Hohenlohe et al., 2010; DeFaveri et al., 2011; Czesny et al., 2012). At least in three-spined stickleback, life history trait changes upon landlocking include reduction in clutch mass (g), clutch size (number of eggs), and reproductive allocation (proportion of body mass devoted to reproduction) (Baker et al., 2008). A shift in reproductive timing has been observed: lacustrine populations of *Galaxias truttaceus* shifted from autumn spawning to spring spawning (Ovenden and White, 1990), ostensibly in response to strong overwinter mortality selection on early life stages. Within the salmonids there is an apparent evolutionary progression to acquiring salinity tolerance earlier in development (McCormick, 2013, Chapter 5, this volume).

Where they are in contact, euryhaline and FW populations or sister species may be reproductively isolated, providing a necessary condition for speciation. Morphological changes associated with the transition facilitate prezygotic isolation, particularly in species with intersexual selection. An FW (*Lucania goodei*) and a euryhaline (*L. parva*) species of killifish co-occur in some locations in Florida, USA. Genetic differences between the species are small, and prezygotic (behavioral) isolation maintains the species boundary; no loss of viability in hybrids has been found despite demonstrable differentiation between the species in salinity tolerance (Fuller et al., 2007). Conversely, in a contact zone between euryhaline and FW forms of stickleback where hybrids are common, prezygotic isolation appears to be weak but genetic evidence suggests that there is a robust postzygotic barrier (Honma and Tamura, 1984; Jones et al., 2006).

4. CONVERGENCE AND EURYHALINITY

Euryhalinity has arisen multiple times within the ray-finned fishes. The fossil record indicates that there were multiple independent transitions to FW halohabitat within the Actinopterygii, each of which required prior capability of functioning in intermediate salinity levels. The phylogenetic distribution of halohabitat types among extant fishes indicates that euryhalinity was pervasive, if not common, among basal ray-finned fishes. Hence, it seems that euryhalinity was an ancestral condition or was readily derived. The phylogeny suggests that subsequent lineages were less euryhaline, however. Only a few orders branching from intermediate reaches of the actinopterygian tree are thoroughly euryhaline. Percomorphs present some increased affiliation with BW and some orders within the

Percomorphacea are quite estuarine. Therefore, judging from the macro-evolutionary pattern of halohabitat use, euryhalinity happened multiple times: euryhaline ostariophysans arose from stenohaline FW ancestors and euryhaline percomorphs arose from stenohaline SW ancestors. The dataset on physiological tolerance also supports the multiple-origin model for euryhalinity. Most ostariophysans have demonstrably narrow values for tolerance breadth, but the capacity for dealing with BW and salt water appears in some derived families (although to the authors' knowledge no euryhaline ostariophysans, such as the marine catfishes, have been subjected to salinity tolerance testing, it is virtually certain that this would demonstrate that they are derived outliers from their order's distribution). The picture is not so clear for the percomorphs, because little is known about the tolerance limits of the stenohaline SW haloniche, which is the inferred ancestral condition. With existing data, it is not possible to determine whether the phylogenetic pattern of halohabitat use for the percomorphs and their precursors reflected a stenohaline SW physiology, in which case the physiological capacity to handle BW and FW was derived independently of more basal actinopterygians, or alternatively whether the physiological capacity for euryhalinity was maintained in spite of the stenohaline habitat use.

Comparative studies on the mechanisms of euryhalinity among multiple taxa and at multiple levels of biological integration are needed. We have a good grasp of how changing salinity levels are physiologically accommodated for model species of most major fish taxa (Edwards and Marshall, 2013, Chapter 1, this volume), yet even within this limited representation there is evident variability in response mechanisms (Zydlewski and Wilkie, 2013, Chapter 6; Marshall, 2013, Chapter 8, this volume). Differences in the genetic and physiological mechanisms of euryhalinity should reflect phylogenetic legacies and will shed light on alternative evolutionary pathways to broad halotolerance.

5. CONCLUSIONS AND PERSPECTIVES

Comparatively little attention has been directed at evolutionary changes and consequences associated with the salt–fresh habitat transition in vertebrates, relative to the aquatic–terrestrial transition. The movement of fishes (and/or their predecessors) into FW, which required the capability of dealing with a broad range of salinity levels, had substantial macroevolutionary repercussions. Implications for colonization of land and the origin of Tetrapoda aside, the colonization of FW habitats initiated a quantum leap in diversification. Despite the vanishingly small amount of FW habitat relative to SW habitat, extant FW fish species diversity is comparable to SW fish diversity (Horn, 1972) and within-species genetic differentiation is greater in FW fish species than in SW fish species (Ward et al., 1994). This disparity is attributable to the greater restrictions of gene flow among locations in FW habitats, the greater spatial heterogeneity of habitat, and the lower productivity of FW, which reduces sustainable population size and increases the potency of genetic drift.

Euryhalinity has accordingly been nominated as a key innovation (Lee and Bell, 1999), meaning a trait whose evolution enables exploitation of a new adaptive zone, triggering cladogenesis (Galis, 2001). Does the evolution of halotolerance consistently promote diversification or adaptive radiation into new halohabitats? This question can be addressed by mapping physiological capability on phylogenies as in Whitehead (2010). Given high variability among families, a comparative analysis that spanned several closely related families would be valuable. Does diversification go both ways? The evolutionary history of ariid catfish (Betancur-R, 2010) is unique, at least to date, in documenting bidirectional diversification.

A peculiar feature of euryhalinity meriting further study, in the context of the thesis that it has played a significant role in the diversification of vertebrates, is its apparent rarity. If it is indeed a potent generator of biological diversity, it is also transitional: it ushers in a round of cladogenesis seemingly resulting in stenohaline taxa. The rarity of euryhaline species may reflect substantial fitness costs of plasticity (and costs of migration, in the case of diadromous fishes) that are exceeded by benefits under special circumstances, so that traits promoting euryhalinity are rapidly lost if they are not under strong selection. Thorough study of the circumstances in which the benefits of broad salinity tolerance exceed the costs will require analysis of biotic interactions such as competition, because the outcome of interactions in one set of abiotic conditions may be reversed under another set (Dunson and Travis, 1991). Another factor contributing to the rarity of euryhaline species is the rarity and mobility of estuarine habitat, owing to its restriction to a narrow and dynamic coastal zone and changing sea levels. Any particular estuary is geologically young (McLusky, 1989). Habitat rarity and mobility are both features that could limit its inhabitants to a short evolutionary lifespan.

Euryhalinity is a graded feature that shows variability in its upper and lower limits among teleosts (Fig. 10.4). Based on efforts to summarize it, the salinity tolerance literature does not support a simple expectation that the transition from stenohalinity to euryhalinity (or the reverse) is quantized, requiring only the addition or deactivation of a single switch that activates ion absorption or secretion, water uptake or elimination. How is physiological capability tuned to environmental demands – does halotolerance breadth reliably indicate the range of salinity to which a population is exposed? Which genetic and physiological components of the response to changing salinity are most decisive in limiting capability?

Judging from a broad phylogenetic view of halohabitat, euryhalinity was a lost trait for a considerable period of actinopterygian evolution, and then was rediscovered. Does this reflect physiological capacity? In particular, are most SW fishes stenohaline? How do pathways promoting broad tolerance differ among major groups that independently underwent transition among halohabitats? In other words, how do the genetic and physiological bases for evolutionary euryhalinity vary among broad taxa? It is to be hoped that more analyses using the phylogenetically rigorous comparative approach will incorporate measures of salinity tolerance to determine whether broad tolerance of species inhabiting FW or SW plays a role in the evolution of euryhalinity. In other words, can euryhaline species evolve as easily from stenohaline species with narrow halotolerance as from those with broad halotolerance?

Our present limited view of FW colonization events in the fossil record is bound to improve. In early vertebrate evolution it seems that the boundary between SW and FW was easily breached. We know little about the business of the early euryhaline fishes. Were they migrants? What habitats did they frequent? Did occupation of FW precede or coincide with the Devonian rise of terrestrial plants? Our present limited view of FW colonization events in the fossil record is bound to improve.

ACKNOWLEDGMENTS

This chapter was completed while ETS was supported by a fellowship from the Fulbright Foundation – Greece. The assistance of Charis Apostolidis, Andrew Bush, Arne Christensen, the staff at FishBase, Paul Lewis, John G. Lundberg, William Marshall, Ricardo Betancur-R., and Konstantine Stergiou is gratefully acknowledged. Any use of trade, product, or firm names is for descriptive purposes only and does not imply endorsement by the U.S. Government.

Table 10.S1

Data on halotolerance by order, family, and species.

Order	Family	Species	Limits estimated				Reference
			Larva		Juv&Ad		
			Dir	Grad	Dir	Grad	
Acipenseriformes	Acipenseridae	Acipenser naccarii	2		4		Cataldi et al. (1999)
		Huso huso			1		Farabi et al. (2007)
Albuliformes	Albulidae	Albula sp.	2		6		Pfeiler (1981)
Atheriniformes	Atherinidae	Chirostoma promelas	2	2			Martinez-Palacios et al. (2008)
		Craterocephalus stercusmuscarum				2	Williams and Williams (1991)
		Leuresthes sardine	4		2		Reynolds and Thomson (1974)
		Leuresthes tenuis	2		2		Reynolds et al. (1976)
		Menidia beryllina			2		Hubbs et al. (1971)
		Odontesthes bonariensis			2		Tsuzuki et al. (2000)
		Odontesthes hatcheri			2		Tsuzuki et al. (2000)
	Melanotaeniidae	Melanotaenia splendida			2	2	Williams and Williams (1991)
Characiformes	Characidae	Astyanax bimaculatus				1	Chung (1999)
Clupeiformes	Clupeidae	Clupea harengus	1				Yin and Blaxter (1987)
Cypriniformes	Cobitidae	Cobitis taenia		2			Bohlen (1999)
	Cyprinidae	Aristichthys nobilis			2		Garcia et al. (1999)
		Barbus callensis			2		Kraiem and Pattee (1988)
		Carassius auratus			2	2	Jasim (1988)
					1		Schofield and Nico (2009)
					2	2	Threader and Houston (1983)
		Catla catla			2		Ghosh et al. (1973)
					2		Chervinski (1977b)
		Ctenopharyngodon idella				4	Kilambi and Zdinak (1980)
					2		Maceina and Shireman (1979)

(*Continued*)

Table 10.S1 (Continued)

Order	Family	Species	Limits estimated				Reference
			Larva		Juv&Ad		
			Dir	Grad	Dir	Grad	
		Cyprinus carpio			2		Abo Hegab and Hanke (1982)
					2		Geddes (1979)
		Danio rerio			2		Dou et al. (2006)
		Hybognathus placitus			2		Ostrand and Wilde (2001)
		Hypophthalmichthys molitrix			2		Chervinski (1977b)
		Labeo rohita			2		Ghosh et al. (1973)
						2	Pillai et al. (2003)
		Notropis buccula			2		Ostrand and Wilde (2001)
		Notropis oxyrhynchus			2		Ostrand and Wilde (2001)
		Pogonichthys macrolepidotus				2	Young and Cech (1996)
		Ptychocheilus lucius			2		Nelson and Flickinger (1992)
		Puntius conchonius			2		Nazneen and Begum (1981)
		Puntius sophore			2		Nazneen and Begum (1981)
		Rutilus rutilus			2		Schofield et al. (2006)
Cyprinodontiformes	Aplocheilidae	Aplocheilus panchax			2		Nazneen and Begum (1981)
	Cyprinodontidae	Adinia xenica				2	Nordlie (1987)
		Cyprinodon dearborni			1	1	Chung (1982)
		Cyprinodon rubrofluviatilis			2		Ostrand and Wilde (2001)
		Cyprinodon variegatus				1	Jordan et al. (1993)
		Floridichthys carpio				2	Nordlie and Haney (1993)
		Jordanella floridae				2	Nordlie and Haney (1993)
						2	Nordlie and Haney (1993)
	Fundulidae	Fundulus catenatus				2	Griffith (1974)
		Fundulus chrysotus			2	2	Crego and Peterson (1997)
						2	Griffith (1974)

	Species			Reference
	Fundulus cingulatus		2	Griffith (1974)
	Fundulus confluentus		2	Griffith (1974)
	Fundulus diaphanus		2	Griffith (1974)
	Fundulus grandis		2	Crego and Peterson (1997)
			2	Perschbacher et al. (1990)
	Fundulus heteroclitus		4[a]	Griffith (1974)
	Fundulus jenkinsi		2	Griffith (1974)
	Fundulus kansae	2		Stanley and Fleming (1977)
	Fundulus luciae		2	Griffith (1974)
	Fundulus majalis		2	Griffith (1974)
	Fundulus notatus		2	Griffith (1974)
	Fundulus notti	2		Crego and Peterson (1997)
	Fundulus olivaceus		2	Griffith (1974)
	Fundulus pulvereus		2	Griffith (1974)
	Fundulus rathbuni		2	Griffith (1974)
	Fundulus sciadicus		2	Griffith (1974)
	Fundulus seminolis	2		DiMaggio et al. (2009)
			2	Griffith (1974)
	Fundulus similis		2	Crego and Peterson (1997)
	Fundulus stellifer		2	Griffith (1974)
	Fundulus waccamensis		2	Griffith (1974)
	Fundulus zebrinus		2	Griffith (1974)
	Lucania goodei		2	Ostrand and Wilde (2001)
	Lucania parva		2	Dunson and Travis (1991)
				Dunson and Travis (1991)
Poeciliidae	*Gambusia affinis*		2	Chervinski (1983)
	Poecilia latipinna		2	Nazneen and Begum (1981)
			2	Nordlie and Walsh (1989)
	Poecilia reticulata		6	Shikano and Fujio (1998)
	Xiphophorus helleri		2	Dou et al. (2006)

(Continued)

Table 10.S1 (Continued)

Order	Family	Species	Larva Dir	Larva Grad	Juv&Ad Dir	Juv&Ad Grad	Reference
Esociformes	Esocidae	Esox lucius			2		Jacobsen et al. (2007)
						2	Jørgensen et al. (2010)
Gadiformes	Gadidae	Gadus morhua			1		Provencher et al. (1993)
			1				Yin and Blaxter (1987)
Gasterosteiformes	Gasterosteidae	Gasterosteus aculeatus			1		Campeau et al. (1984)
		Gasterosteus wheatlandi			1		Campeau et al. (1984)
	Syngnathidae	Hippocampus kuda			1		Hilomen-Garcia et al. (2001)
Mugiliformes	Mugilidae	Chelon labrosus			1		Chervinski (1977a)
						1	Hotos and Vlahos (1998)
		Liza aurata			2		Chervinski (1975)
		Liza haematocheila	2			1	Bulli and Kulikova (2006)
		Liza saliens			2	1	Chervinski (1977a)
		Mugil cephalus			1	1	Hotos and Vlahos (1998)
Osmeriformes	Osmeridae	Hypomesus nipponensis			2		Swanson et al. (2000)
		Hypomesus transpacificus			2		Swanson et al. (2000)
	Retropinnidae	Retropinna semoni			2		Williams and Williams (1991)
Perciformes	Ambassidae	Chanda commersonii			2		Rajasekharan Nair and Balakrishnan Nair (1984)
		Chanda thomassi			2		Rajasekharan Nair and Balakrishnan Nair (1984)
	Anarhichadidae	Anarhichas lupus			1		Le Francois et al. (2003)
	Blenniidae	Parablennius sanguinolentus				1	Plaut (1999)
		Salaria fluviatilis				1	Plaut (1998)

Family	Species			Reference
	Salaria pavo		1	Plaut (1998)
Centropomidae	*Centropomus parallelus*	1		Tsuzuki et al. (2007)
Cichlidae	*Hemichromis letourneuxi*	2	2	Langston et al. (2010)
	Oreochromis aureus		2	Lutz et al. (2010)
		2		Watanabe et al. (1985)
	Oreochromis mossambicus	2	2	Lutz et al. (2010)
		2	2	Lemarie et al. (2004)
	Oreochromis niloticus			Li and Li (1999)
			2	Li et al. (2008)
			2	Lutz et al. (2010)
			2	Watanabe et al. (1985)
	Sarotherodon melanotheron	2	2	Lemarie et al. (2004)
			2	Li et al. (2008)
Eleotridae	*Dormitator maculatus*	2	2	Nordlie et al. (1992)
	Hypseleotris klunzingeri	2	2	Williams and Williams (1991)
Gobiidae	*Boleophthalmus boddaerti*	2		Ip et al. (1991)
	Gobiosoma robustum	2		Schöfer (1979)
	Luciogobius pallidus	2		Hirashima and Takahashi (2008)
	Microgobius gulosus	2		Schöfer (1979)
	Rhinogobius sp1	2		Hirashima and Tachihara (2000)
	Rhinogobius sp2	2		Hirashima and Tachihara (2000)
Lutjanidae	*Lutjanus argentimaculatus*	1		Estudillo et al. (2000)
Moronidae	*Dicentrarchus labrax*	1		Dalla Via et al. (1998)
				Marino et al. (1994)
		2		Varsamos et al. (2001)
Osphronemidae	*Trichogaster trichopterus*	2		Dou et al. (2006)
Percichthyidae	*Maccullochella peelii peelii*	2		Mellor and Fotedar (2005)
Percidae	*Perca fluviatilis*	2		Bein and Ribi (1994)
Rachycentridae	*Rachycentron canadum*	2		Faulk and Holt (2006)
Sciaenidae	*Cynoscion nebulosus*	3		Banks et al. (1991)
Serranidae	*Centropristis striata*	1		Young et al. (2006)
Siganidae	*Siganus rivulatus*			Saoud et al. (2007)

(*Continued*)

Table 10.S1 (Continued)

Order	Family	Species	Larva Dir	Larva Grad	Juv&Ad Dir	Juv&Ad Grad	Reference
	Sparidae	Acanthopagrus butcheri				1	Partridge and Jenkins (2002)
Pleuronectiformes	Teraponidae	Bidyanus bidyanus			2		Guo et al. (1995)
	Paralichthyidae	Paralichthys californicus					Madon (2002)
		Paralichthys dentatus					Malloy and Targett (1991)
		Paralichthys lethostigma		1			Cai et al. (2007)
			1				Daniels et al. (1996)
							Smith et al. (1999)
		Paralichthys olivaceus			2		Wang et al. (2000)
		Paralichthys orbignyanus	2			1	Sampaio et al. (2007)
	Pleuronectidae	Microstomus achne			1		Wada et al. (2007)
		Platichthys bicoloratus			1		Wada et al. (2007)
		Platichthys flesus			1		Arnold-Reed and Balment (1991)
			1				Yin and Blaxter (1987)
		Platichthys stellatus			1		Takeda and Tanaka (2007)
					1		Wada et al. (2007)
		Pseudopleuronectes yokohamae			1		Wada et al. (2007)
		Verasper variegatus			1		Wada et al. (2007)
Salmoniformes	Scophthalmidae	Scophthalmus maximus			2		Mu and Song (2005)
	Salmonidae	Coregonus nasus			3		de March (1989)
		Oncorhynchus tshawytscha			1		Taylor (1990)
		Salvelinus alpinus			1		Dempson (1993)
					1		Staurnes et al. (1992)
Scorpaeniformes	Cottidae	Cottus asper				1	Henriksson et al. (2008)
		Leptocottus armatus				1	Henriksson et al. (2008)

Order	Family	Species	Dir	Grad	Reference
Siluriformes	Callichthyidae	*Callichthys callichthys*	2		Mol (1994)
		Hoplosternum littorale	2		Mol (1994)
		Megalechus thoracata	2		Mol (1994)
	Clariidae	*Clarias gariepinus*	2		Britz and Hecht (1989)
				2	Odo and Inyang (2001)
		Clarias lazera		2	Chervinski (1984)
				2	Clay (1977)
		Heterobranchus longifilis		2	Fashina-Bombata and Busari (2003)
	Ictaluridae	*Ictalurus catus*		2	Kendall and Schwartz (1968)
		Ictalurus furcatus		2	Allen and Avault (1971)
		Ictalurus punctatus		2	Allen and Avault (1971)
		Pylodictis olivaris		4	Bringolf et al. (2005)
Synbranchiformes	Synbranchidae	*Monopterus albus*		2	Schofield (2003)
Tetraodontiformes	Tetraodontidae	*Sphoeroides greeleyi*		2	Prodocimo and Freire (2001)
		Sphoeroides testudineus		2	Prodocimo and Freire (2001)

For one or more reference on each species, the table provides the number of tolerance limits determined by ontogenetic stage of the subjects and experimental approach (direct in left-hand column and gradual in right-hand column for each stage, e.g. the top row indicates that two direct limits were determined for larvae and four direct limits were determined for juveniles + adults, in *Acipenser naccarii*).

Juv&Ad: juvenile and adolescent; Dir: Direct approach; Grad: Gradual approach.

[a]Two limits reported in this paper for *Fundulus swampinus*, which is a synonym of *Fundulus heteroclitus*.

Table 10.S2
Halotolerance groups defined by cluster analysis, direct-method experiments.

Group	Species	Lower	Upper	Breadth
Euryhaline	*Albula sp.*	2.9	52	49
	Albula sp.	3.3	59	56
	Albula sp.	5.2	63	58
	Cyprinodon rubrofluviatilis	0	46	46
	Fundulus kansae	0.4	44	44
	Fundulus zebrinus	0	43	43
	Gobiosoma robustum	0	55	55
	Leuresthes sardina	5	58	53
	Microgobius gulosus	2	60	58
Stenohaline	*Acipenser naccarii*	0	15	15
	Acipenser naccarii	0	22	22
	Ambassis ambassis	0.45	31	31
	Ameiurus catus	0	14	14
	Aplocheilus panchax	0	10	10
	Barbus callensis	0.5	16	15
	Bidyanus bidyanus	0	17	17
	Boleophthalmus boddarti	1.7	31	29
	Carassius auratus	0	12	12
	Carassius auratus	0	16	16
	Catla catla	0	12	12
	Clarias gariepinus	0.042	13	12
	Clarias gariepinus	0.14	11	11
	Coregonus nasus	0	16	16
	Ctenopharyngodon idella	0.5	10	9.5
	Ctenopharyngodon idella	0	15	15
	Cyprinus carpio	0	17	17
	Cyprinus carpio	0	15	15
	Danio rerio	0	12	12
	Esox lucius	0	12	12
	Fundulus chrysotus	0	26	26
	Fundulus notti	0	17	17
	Fundulus seminolis	0	28	28
	Gambusia affinis	0.4	22	21
	Hemichromis letourneuxi	0	25	25
	Heterobranchus longifilis	0	7	7
	Hybognathus placitus	0	16	16
	Hypophthalmichthys molitrix	0.5	8.8	8.3
	Hypophthalmichthys nobilis	0	7.6	7.6
	Ictalurus furcatus	0	14	14
	Ictalurus punctatus	0	14	14
	Labeo rohita	0	11	11
	Leuresthes tenuis	8.6	38	29
	Lucania goodei	0	25	25
	Melanotaenia splendida	0.3	21	21

(*Continued*)

Table 10.S2 (Continued)

Group	Species	Lower	Upper	Breadth
	Menidia beryllina	0.8	36	35
	Monopterus albus	0.2	17	17
	Notropis buccula	0	18	18
	Notropis oxyrhynchus	0	15	15
	Odontesthes bonariensis	0	25	25
	Odontesthes hatcheri	0	25	25
	Oreochromis aureus	0	20	20
	Oreochromis niloticus	0	20	20
	Oreochromis niloticus	0	14	14
	Oreochromis niloticus	0	20	20
	Parambassis thomassi	0	23	23
	Poecilia latipinna	0	10	10
	Poecilia reticulata	0	34	34
	Poecilia reticulata	0	23	23
	Poecilia reticulata	0	27	27
	Ptychocheilus lucius	0	13	13
	Puntius conchonius	0	8.4	8.4
	Puntius sophore	0	8.4	8.4
	Pylodictis olivaris	0	15	15
	Pylodictis olivaris	0	10	10
	Rutilus rutilus	0	14	14
	Sarotherodon melanotheron	0	34	34
	Scophthalmus maximus	16	38	22
	Trichopodus trichopterus	0	17	17
	Xiphophorus helleri	0	20	20

For two named halotolerance groups identified by centroid cluster analysis, the table provides species, the lower and upper LC_{50} halotolerance limits, and halotolerance breadth.

Table 10.S3

Halotolerance groups defined by cluster analysis, gradual method experiments.

Group	Species	Lower	Upper	Breadth
Euryhaline FW	*Dormitator maculatus*	0	75	75
	Fundulus chrysotus	0	65	65
	Fundulus diaphanus	0	70	70
	Fundulus grandis	0	80	80
	Fundulus jenkinsi	0	74	74
	Fundulus seminolis	0	60	60
	Fundulus waccamensis	0	55	55
	Gambusia affinis	0.4	59	58
	Hemichromis letourneuxi	0	55	55
	Jordanella floridae	0	80	80
	Lucania parva	0	80	80

(*Continued*)

Table 10.S3 (Continued)

Group	Species	Lower	Upper	Breadth
	Poecilia latipinna	0	80	80
	Retropinna semoni	0.3	59	58
Euryhaline	*Adinia xenica*	0	100	100
	Cyprinodon variegatus	0	125	125
	Floridichthys carpio	0	90	90
	Fundulus confluentus	0	99	99
	Fundulus heteroclitus	0	114	114
	Fundulus kansae	0.4	99	99
	Fundulus luciae	0	101	101
	Fundulus majalis	0	99	99
	Fundulus pulvereus	0	101	101
	Fundulus zebrinus	0	89	89
	Sarotherodon melanotheron	0	123	123
Stenohaline	*Carassius auratus*	0	14	14
	Carassius auratus	0	12	12
	Clarias gariepinus	0.14	11	11
	Clarias gariepinus	0.12	23	22
	Craterocephalus stercusmuscarum	0.3	44	43
	Ctenopharyngodon idella	0	16	16
	Ctenopharyngodon idella	0	14	14
	Esox lucius	0	14	14
	Fundulus catenatus	0	24	24
	Fundulus chrysotus	0	20	20
	Fundulus cingulatus	0	23	23
	Fundulus heteroclitus	0	27	27
	Fundulus notatus	0	20	20
	Fundulus notti	0	28	28
	Fundulus olivaceus	0	24	24
	Fundulus rathbuni	0	26	26
	Fundulus sciadicus	0	24	24
	Fundulus seminolis	0	23	23
	Fundulus stellifer	0	21	21
	Hypomesus nipponensis	0	27	27
	Hypomesus transpacificus	0	19	19
	Hypseleotris klunzingeri	0.3	38	38
	Labeo rohita	0	9	9
	Melanotaenia splendida	0.3	30	30
	Monopterus albus	0.3	14	14
	Oreochromis aureus	0.4	38	38
	Oreochromis mossambicus	0.4	47	46
	Oreochromis niloticus	0	46	46
	Oreochromis niloticus	0.4	26	26
	Pogonichthys macrolepidotus	0	19	19
	Pylodictis olivaris	0	16	16

For three named groups identified by centroid cluster analysis, the table provides species, the lower and upper LC_{50} halotolerance limits, and tolerance breadth.

REFERENCES

Abo Hegab, S. and Hanke, W. (1982). Electrolyte changes and volume regulatory processes in the carp (*Cyprinus carpio*) during osmotic-stress. *Comp. Biochem. Physiol.* 71A, 157–164.

Allen, K. O. and Avault, J. W., Jr. (1971). Notes on the relative salinity tolerance of channel and blue catfish. *Prog. Fish.-Cult.* 33, 135–137.

Anjos, M. B., De Oliveira, R. R. and Zuanon, J. (2008). Hypoxic environments as refuge against predatory fish in the Amazonian floodplains. *Braz. J. Biol.* 68, 45–50.

Arnold-Reed, D. E. and Balment, R. J. (1991). Salinity tolerance and its seasonal-variation in the flounder. *Platichthys flesus. Comp. Biochem. Physiol. A* 99, 145–149.

Baker, J. A., Heins, D. C., Foster, S. A. and King, R. W. (2008). An overview of life-history variation in female threespine stickleback. *Behaviour* 145, 579–602.

Ballantyne, J. S. and Fraser, D. I. (2013). Euryhaline elasmobranchs. In *Fish Physiology,* Vol. *32, Euryhaline Fishes* (eds. S. D. McCormick, A. P. Farrell and C. J. Brauner), pp. 125–198. New York: Elsevier.

Ballantyne, J. S., Moyes, C. D. and Moon, T. W. (1987). Compatible and counteracting solutes and the evolution of ion and osmoregulation in fishes. *Can. J. Zool.* 65, 1883–1888.

Bamber, R. N. and Henderson, P. A. (1988). Pre-adaptive plasticity in atherinids and the estuarine seat of teleost evolution. *J. Fish Biol.* 33, 17–23.

Banks, M. A., Holt, G. and Wakeman, J. (1991). Age-linked changes in salinity tolerance of larval spotted seatrout (*Cynoscion nebulosus*, Cuvier). *J. Fish. Biol.* 39, 505–514.

Barbour, C. D. (1973). A biogeographical history of *Chirostoma* (Pisces: Atherinidae): a species flock from the Mexican plateau. *Copeia* 1973, 533–556.

Barrett, R. D. H., Rogers, S. M. and Schluter, D. (2008). Natural selection on a major armor gene in threespine stickleback. *Science* 322, 255–257.

Beheregaray, L. B. and Levy, J. A. (2000). Population genetics of the silverside *Odontesthes argentinensis* (Teleostei, Atherinopsidae): evidence for speciation in an estuary of southern Brazil. *Copeia* 2000, 441–447.

Beheregaray, L. B. and Sunnucks, P. (2001). Fine-scale genetic structure, estuarine colonization and incipient speciation in the marine silverside fish *Odontesthes argentinensis. Mol. Ecol.* 10, 2849–2866.

Beheregaray, L. B., Sunnucks, P. and Briscoe, D. A. (2002). A rapid fish radiation associated with the last sea-level changes in southern Brazil: the silverside *Odontesthes perugiae* complex. *Proc. R. Soc. Lond. B Biol. Sci.* 269, 65–73.

Bein, R. and Ribi, G. (1994). Effects of larval density and salinity on the development of perch larvae (*Perca fluviatilis* L.). *Aquat. Sci.* 56, 97–105.

Bell, M. A. and Foster, S. A. (1994). Introduction to the evolutionary biology of the threespine stickleback. In *The Evolutionary Biology of the Threespine Stickleback* (eds. M. A. Bell and S. A. Foster), pp. 1–27. Oxford: Oxford University Press.

Bell, M. A., Aguirre, W. E. and Buck, N. J. (2004). Twelve years of contemporary armor evolution in a threespine stickleback population. *Evolution* 58, 814–824.

Betancur-R, R. (2010). Molecular phylogenetics supports multiple evolutionary transitions from marine to freshwater habitats in ariid catfishes. *Mol. Phylogenet. Evol.* 55, 249–258.

Betancur-R, R., Orti, G., Stein, A. M., Marceniuk, A. P. and Pyron, R. A. (2012). Apparent signal of competition limiting diversification after ecological transitions from marine to freshwater habitats. *Ecol. Lett.* 15, 822–830.

Bilton, D. T., Paula, J. and Bishop, J. D. D. (2002). Dispersal, genetic differentiation and speciation in estuarine organisms. *Estuar. Coast. Shelf Sci.* 55, 937–952.

Blackburn, J. and Clarke, W. C. (1987). Revised procedure for the 24 hour seawater challenge test to measure seawater adaptability of juvenile salmonids. *Can. Tech. Rep. Fish. Aquat. Sci.* No. 1515, 1–35.

Bloom, D. D., Piller, K. R., Lyons, J., Mercado-Silva, N. and Medina-Nava, M. (2009). Systematics and biogeography of the silverside tribe *Menidiini* (Teleostomi: Atherinopsidae) based on the mitochondrial ND2 gene. *Copeia* 2009, 408–417.

Bobori, D. C. and Economidis, P. S. (2006). Freshwater fishes of Greece: their biodiversity, fisheries and habitats. *Aquat. Ecosyst. Health Manage.* 9, 407–418.

Bobori, D. C., Koutrakis, E. T. and Economidis, P. S. (2001). Shad species in Greek waters – an historical overview and present status. *Bull. Franc. Pêche Piscicult.* 362–363, 1101–1108.

Bohlen, J. (1999). Influence of salinity on early development in the spined loach. *J. Fish Biol.* 55, 189–198.

Brauner, C. J., Gonzales, R. J. and Wilson, J. M. (2013). Extreme environments: hypersaline, alkaline, and ion-poor waters. In *Fish Physiology*, Vol. 32, *Euryhaline Fishes* (eds. S. D. McCormick, A. P. Farrell and C. J. Brauner), pp. 435–476. New York: Elsevier.

Brett, J. R. (1956). Some principles in the thermal requirements of fishes. *Q. Rev. Biol.* 31, 75–87.

Bringolf, R. B., Kwak, T. J., Cope, W. and Larimore, M. S. (2005). Salinity tolerance of flathead catfish: implications for dispersal of introduced populations. *Trans. Am. Fish. Soc.* 134, 927–936.

Britz, P. J. and Hecht, T. (1989). Effects of salinity on growth and survival of African sharptooth catfish (*Clarias gariepinus*) larvae. *J. Appl. Ichthyol.* 5, 194–202.

Bulger, A. J., Hayden, B. P., Monaco, M. E., Nelson, D. M. and McCormick-Ray, M. G. (1993). Biologically-based estuarine salinity zones derived from a multivariate analysis. *Estuaries* 16, 311–322.

Bulli, L. I. and Kulikova, N. I. (2006). Adaptive capacity of larvae of the haarder *Liza haematocheila* (Mugilidae, Mugiliformes) under decreasing salinity of the environment. *J. Ichthyol.* 46, 534–544.

Cai, W., Liu, X., Ma, X., Zhan, W. and Xu, Y. (2007). Tolerance of southern flounder to low salinity with fresh water acclimation. *Mar. Fish. Res.* 28, 31–37.

Campeau, S., Guderley, H. and FitzGerald, G. (1984). Salinity tolerances and preferences of fry of two species of sympatric sticklebacks: possible mechanisms of habitat segregation. *Can. J. Zool.* 62, 1048–1051.

Cataldi, E., Barzaghi, C., Di Marco, P., Boglione, C., Dini, L., McKenzie, D., Bronzi, P. and Cataudella, S. (1999). Some aspects of osmotic and ionic regulation in Adriatic sturgeon *Acipenser naccarii*. 1: Ontogenesis of salinity tolerance. *J. Appl. Ichthyol.* 15, 57–60.

Chervinski, J. (1975). Experimental acclimation of *Liza aurata* (Risso) to freshwater. *Bamidgeh* 27, 49–53.

Chervinski, J. (1977a). Adaptability of *Chelon labrosus* (Risso) and *Liza saliens* (Risso) (Pisces, Mugilidae) to fresh water. *Aquaculture* 11, 75–79.

Chervinski, J. (1977b). Note on the adaptability of silver carp – *Hypophthalmichthys molitrix* (Val.) – and grass carp – *Ctenopharyngodon idella* (Val.) – to various saline concentrations. *Aquaculture* 11, 179–182.

Chervinski, J. (1983). Salinity tolerance of the mosquito fish, *Gambusia affinis* (Baird Girard). *J. Fish. Biol.* 22, 9–11.

Chervinski, J. (1984). Salinity tolerance of young catfish, *Clarias lazera* (Burchell). *J. Fish. Biol.* 25, 147–149.

Chung, K. (1982). Salinity tolerance of tropical salt-marsh fish of Los Patos Lagoon, Venezuela. *Bull. Jpn. Soc. Sci. Fish.* 48, 873.

Chung, K. (1999). Physiological responses of tropical fishes to salinity changes. In *Special Adaptations of Tropical Fish* (eds. J. Nelson and D. MacKinley), pp. 77–84. Bethesda, MD: American Fisheries Society.

Clay, D. (1977). Preliminary observations on salinity tolerance of *Clarias lazera* from Israel. *Bamidgeh* 29, 102–109.

Crego, G. and Peterson (1997). Salinity tolerance of four ecologically distinct species of *Fundulus* (Pisces: Fundulidae) from the northern Gulf of Mexico. *Gulf Mex. Sci.* 15, 45–49.

Czesny, S., Epifanio, J. and Michalak, P. (2012). Genetic divergence between freshwater and marine morphs of alewife (*Alosa pseudoharengus*): a "next-generation" sequencing analysis. *PLOS ONE* 7, e31803.

Dalla Via, J., Villani, P., Gasteiger, E. and Niederstatter, H. (1998). Oxygen consumption in sea bass fingerling *Dicentrarchus labrax* exposed to acute salinity and temperature changes: metabolic basis for maximum stocking density estimations. *Aquaculture* 169, 303–313.

Daniels, H. V., Berlinsky, D. L., Hodson, R. G. and Sullivan, C. V. (1996). Effects of stocking density, salinity, and light intensity on growth and survival of Southern flounder *Paralichthys lethostigma* larvae. *J. World Aquacult, Soc.* 27, 153–159.

Darwin, C. R. (1876). *On the Origin of Species by Means of Natural Selection, or the Preservation of Favoured Races in the Struggle for Life.* London: John Murray.

DeFaveri, J., Shikano, T., Shimada, Y., Goto, A. and Merila, J. (2011). Global analysis of genes involved in freshwater adaptation in threespine sticklebacks (*Gasterosteus aculeatus*). *Evolution* 65, 1800–1807.

de March, B. G. E. (1989). Salinity tolerance of larval and juvenile broad whitefish (Coregonus nasus). *Can. J. Zool.* 67, 2392–2397.

Dempson, J. (1993). Salinity tolerance of freshwater acclimated, small-sized Arctic charr, *Salvelinus alpinus* from northern Labrador. *J. Fish. Biol.* 43, 451–462.

DiMaggio, M. A., Ohs, C. L. and Petty, B. D. (2009). Salinity tolerance of the Seminole killifish, *Fundulus seminolis*, a candidate species for marine baitfish aquaculture. *Aquaculture* 293, 74–80.

Ditrich, H. (2007). The origin of vertebrates: a hypothesis based on kidney development. *Zool. J. Linn. Soc.* **150**, 435–441.

Dodson, J. J. (1997). Fish migration: an evolutionary perspective. In *Behavioral Ecology of Teleost Fishes* (ed. J.-G. J. Godin), pp. 10–36. Oxford: Oxford University Press.

Dou, H., Huang, J., Wang, X., Fan, W. and Liu, L. (2006). Salinity tolerance and salt water acclimation of gourami *Trichogaster trichopterus*. *J. Fish. Sci. China* 13, 775–780.

Doudoroff, P. (1945). The resistance and acclimatization of marine fishes to temperature changes. II. Experiments with *Fundulus* and *Atherinops*. *Biological Bulletin* 88, 194–206.

Dunson, W. A. and Travis, J. (1991). The role of abiotic factors in community organization. *Am. Nat.* **138**, 1067–1091.

Echelle, A. A., Dowling, T. E., Moritz, C. C. and Brown, W. M. (1989). Mitochondrial-DNA diversity and the origin of the *Menidia clarkhubbsi* complex of unisexual fishes (Atherinidae). *Evolution* 43, 984–993.

Economidis, P. S. and Miller, P. J. (1990). Systematics of freshwater gobies from Greece (Teleostei: Gobiidae). *J. Zool.* 221, 125–170.

Edwards, S. L. and Marshall, W. S. (2013). Principles and patterns of osmoregulation and euryhalinity in fishes. In *Fish Physiology*, Vol. 32, *Euryhaline Fishes* (eds. S. D. McCormick, A. P. Farrell and C. J. Brauner), pp. 1–44. New York: Elsevier.

Elliott, M., Whitfield, A. K., Potter, I. C., Blaber, S. J. M., Cyrus, D. P., Nordlie, F. G. and Harrison, T. D. (2007). The guild approach to categorizing estuarine fish assemblages: a global review. *Fish Fish.* 8, 241–268.

Estudillo, C. B., Duray, M. N., Marasigan, E. T. and Emata, A. C. (2000). Salinity tolerance of larvae of the mangrove red snapper (*Lutjanus argentimaculatus*) during ontogeny. *Aquaculture* 190, 155–167.

Evans, D. H. (1984). The role of gill permeability and transport mechanisms in euryhalinity. In *Fish Physiology*, Vol. 10A, *The Physiology of Developing Fish: Eggs and Larvae* (eds. W. S. Hoar and D. J. Randall), pp. 239–283. San Diego: Academic Press.

Evans, D. H., Piermarini, P. M. and Choe, K. P. (2005). The multifunctional fish gill: dominant site of gas exchange, osmoregulation, acid–base regulation, and excretion of nitrogenous waste. *Physiol. Rev.* 85, 97–177.

Farabi, S. M. V., Hajimoradloo, A. and Bahmani, M. (2007). Study on salinity tolerance and some physiological indicator sofion-osmoregulatory system in juvenile beluga, Huso huso (Linnaeus,1758) in the south Caspian Sea: effect of age and size. Iran. J. Fish. Sci. 6, 15–32.

Fashina-Bombata, H. A. and Busari, A. N. (2003). Influence of salinity on the developmental stages of African catfish *Heterobranchus longifilis* (Valenciennes, 1840). *Aquaculture* 224, 213–222.

Faulk, C. K. and Holt, G. J. (2006). Responses of cobia *Rachycentron canadum* larvae to abrupt or gradual changes in salinity. *Aquaculture* 254, 275–283.

Ferreira, T., Oliveira, J., Caiola, N., De Sostoa, A., Casals, F., Cortes, R., Economou, A., Zogaris, S., Garcia-Jalon, D., Ilheu, M., Martinez-Capel, F., Pont, D., Rogers, C. and Prenda, J. (2007). Ecological traits of fish assemblages from Mediterranean Europe and their responses to human disturbance. *Fish. Manage. Ecol.* 14, 473–481.

Fluker, B. L., Pezold, F. and Minton, R. L. (2011). Molecular and morphological divergence in the inland silverside (*Menidia beryllina*) along a freshwater–estuarine interface. *Environ. Biol. Fish.* 91, 311–325.

Fontaine, M. (1975). Physiological mechanisms in the migration of marine and amphihaline fish. *Adv. Mar. Biol.* 13, 241–355.

Foote, C. J., Wood, C. C., Clarke, W. C. and Blackburn, J. (1992). Circannual cycle of seawater adaptability in *Oncorhynchus nerka*: genetic differences between sympatric sockeye salmon and kokanee. *Can. J. Fish. Aquat. Sci.* 49, 99–109.

Francisco, S. M., Cabral, H., Vieira, M. N. and Almada, V. C. (2006). Contrasts in genetic structure and historical demography of marine and riverine populations of *Atherina* at similar geographical scales. *Estuar. Coast. Shelf Sci.* 69, 655–661.

Friedman, M. (2010). Explosive morphological diversification of spiny-finned teleost fishes in the aftermath of the end-Cretaceous extinction. *Proc. R. Soc. B* 277, 1675–1683.

Friedman, M. and Blom, H. (2006). A new actinopterygian from the Famennian of East Greenland and the interrelationships of Devonian ray-finned fishes. *J. Paleontol.* 80, 1186–1204.

Fuller, R. C., McGhee, K. E. and Schrader, M. (2007). Speciation in killifish and the role of salt tolerance. *J. Evol. Biol.* 20, 1962–1975.

Galis, F. (2001). Key innovations and radiations. In *The Character Concept in Evolutionary Biology* (ed. P. W. Günter), pp. 581–605. San Diego: Academic Press.

Garcia, L. M. B., Garcia, C. M. H., Pineda, A. F. S., Gammad, E., Canta, J., Simon, S. P. D., Hilomen-Garcia, G., Gonzal, A. and Santiago, C. (1999). Survival and growth of bighead carp fry exposed to low salinities. *Aquacult. Int.* 7, 241–250.

Geddes, M. (1979). Salinity tolerance and osmotic behaviour of European carp (*Cyprinus carpio* L.) from the River Murray, Australia. *Trans. R. Soc. S. Aust.* 103, 185–189.

Gelmond, O., von Hippel, F. A. and Christy (2009). Rapid ecological speciation in three-spined stickleback *Gasterosteus aculeatus* from Middleton Island, Alaska: the roles of selection and geographic isolation. *J. Fish Biol.* 75, 2037–2051.

Ghosh, A. N., Ghosh, S. R. and Sarkar, N. N. (1973). On the salinity tolerance of fry and fingerlings of Indian major carps. *J. Inland Fish. Soc. India* 5, 215–217.

Green, R. H. (1965). Estimation of tolerance over an indefinite time period. *Ecology* 46, 887.

Griffith, R. W. (1974). Environment and salinity tolerance in the genus *Fundulus*. *Copeia* 1974, 319–331.

Griffith, R. W. (1987). Fresh-water or marine origin of the vertebrates. *Comp. Biochem. Physiol. A* 87, 523–531.

Gunter, G. (1967). Vertebrates in hypersaline waters. *Contrib. Mar. Sci.* 12, 230–241.

Guo, R., Mather, P. and Capra, M. (1995). Salinity tolerance and osmoregulation in the silver perch, *Bidyanus bidyanus* Mitchell (Teraponidae), an endemic Australian freshwater teleost. *Mar. Freshw. Res.* 46, 947–952.

Halstead, L. B. (1985). The vertebrate invasion of fresh water. *Philos. Trans. R. Soc. Lond. B Biol. Sci.* 309, 243–258.

Hamilton, M. A., Russo, R. C. and Thurston, R. V. (1977). Trimmed Spearman–Karber method for estimating median lethal concentrations in toxicity bioassays. *Environ. Sci. Technol.* 11, 714–719.

Heimberg, A. M., Cowper-Sallari, R., Semon, M., Donoghue, P. C. J. and Peterson, K. J. (2010). MicroRNAs reveal the interrelationships of hagfish, lampreys, and gnathostomes and the nature of the ancestral vertebrate. *Proc. Natl Acad. Sci. U. S. A.* 107, 19379–19383.

Hendry, A. P., Bohlin, T., Jonsson, B. and Berg, O. K. (2003a). To sea or not to sea? Anadromy versus non-anadromy in salmonids. In *Evolution Illuminated: Salmon and Their Relatives* (eds. A. P. Hendry and S. C. Stearns), pp. 92–125. Oxford: Oxford University Press.

Hendry, A. P., Castric, V., Kinnison, M. T. and Quinn, T. P. (2003b). The evolution of philopatry and dispersal: homing versus straying in salmonids. In *Evolution Illuminated: Salmon and Their Relatives* (eds. A. P. Hendry and S. C. Stearns), pp. 52–91. Oxford: Oxford University Press.

Henriksson, P., Mandic, M. and Richards, J. G. (2008). The osmorespiratory compromise in sculpins: impaired gas exchange is associated with freshwater tolerance. *Physiol. Biochem. Zool.* 81, 310–319.

Heras, S. and Roldan, M. I. (2011). Phylogenetic inference in *Odontesthes* and *Atherina* (Teleostei: Atheriniformes) with insights into ecological adaptation. *Compt. Rend. Biol.* 334, 273–281.

Hirashima, K. and Takahashi, H. (2008). Early life history of aquarium-held blind well goby *Luciogobius pallidus*, collected from Wakayama Prefecture, Japan. *Jpn. J. Ichthyol.* 55, 121–125.

Hilomen-Garcia, G. V., Reyes, R. D. and Garcia, C. M. H. (2001). Tolerance and growth of juvenile seahorse *Hippocampus kuda* exposed to various salinities. World Aquaculture Society, Louisiana State University, 294.

Hirashima, K. and Tachihara, K. (2000). Embryonic development and morphological changes in larvae and juveniles of two land-locked gobies, *Rhinogobius spp.* (Gobiidae), on Okinawa Island. *Jpn. J. Ichthyol.* 47, 29–41.

Hohenlohe, P. A., Bassham, S., Etter, P. D., Stiffler, N., Johnson, E. A. and Cresko, W. A. (2010). Population genomics of parallel adaptation in threespine stickleback using sequenced RAD tags. *PLOS Genet.* 6, e1000862.

Honma, Y. and Tamura, E. (1984). Anatomical and behavioral differences among threespine sticklebacks: the marine form, the landlocked form and their hybrids. *Acta Zool.* 65, 79–87.

Horn, M. H. (1972). The amount of space available for marine and freshwater fishes. *Fish. Bull. Natl. Ocean. Atmos. Adm.* 70, 1295–1297.

Hotos, G. and Vlahos, N. (1998). Salinity tolerance of *Mugil cephalus* and *Chelon labrosus* (Pisces: Mugilidae) fry in experimental conditions. *Aquaculture* 167, 329–338.

Hubbs, C., Sharp, H. and Schneider, J. (1971). Developmental rates of *Menidia audens* with notes on salt tolerance. *Trans. Am. Fish. Soc.* 100, 603–610.

Hutchinson, G. E. (1960). On evolutionary euryhalinity. *Am. J. Sci.* 258A, 98–103.

Ip, Y. K., Lee, C. G. L., Low, W. P. and Lam, T. J. (1991). Osmoregulation in the mudskipper, *Boleophthalmus boddaerti*. 1. Responses of branchial cation activated and anion stimulated adenosine triphosphatases to changes in salinity. *Fish Physiol. Biochem.* 9, 63–68.

Inoue, J. G., Miya, M., Miller, M. J., Sado, T., Hanel, R., Hatooka, K., Aoyama, J., Minegishi, Y., Nishida, M. and Tsukamoto, K. (2010). Deep-ocean origin of the freshwater eels. *Biol. Lett.* 6, 363–366.

Janvier, P., Halstead, L. B. and Westoll, T. S. (1985). Environmental framework of the diversification of the Osteostraci during the Silurian and Devonian [and discussion]. *Philos. Trans. R. Soc. Lond. B Biol. Sci.* 309, 259–272.

Jacobsen, L., Skov, C., Koed, A. and Berg, S. (2007). Short-term salinity tolerance of northern pike, *Esox lucius*, fry, related to temperature and size. *Fish. Manage. Ecol.* 14, 303–308.

Jasim, B. (1988). Tolerance and adaptation of goldfish *Carassius auratus* (L.) to salinity. *J. Biol. Sci. Res.* 19, 149–154.

Jones, F. C., Brown, C., Pemberton, J. M. and Braithwaite, V. A. (2006). Reproductive isolation in a threespine stickleback hybrid zone. *J. Evol. Biol.* 19, 1531–1544.

Jordan, F., Haney, D. C. and Nordlie, F. G. (1993). Plasma osmotic regulation and routine metabolism in the Eustis pupfish, *Cyprinodon variegatus hubbsi* (Teleostei: Cyprinodontidae). *Copeia* 1993, 784–789.

Jørgensen, A. T., Hansen, B. W., Vismann, B., Jacobsen, L., Skov, C., Berg, S. and Bekkevold, D. (2010). High salinity tolerance in eggs and fry of a brackish *Esox lucius* population. *Fish. Manage. Ecol.* 17, 554–560.

Kefford, B. J., Papas, P. J., Metzeling, L. and Nugegoda, D. (2004). Do laboratory salinity tolerances of freshwater animals correspond with their field salinity? *Environ. Pollut.* 129, 355–362.

Keivany, Y., Nelson, J. S. and Economidis, P. S. (1997). Validity of *Pungitius hellenicus* Stephanidis, 1971 (Teleostei, Gasterosteidae), a stickleback fish from Greece. *Copeia* 1997, 558–564.

Kendall, A. W., Jr. and Schwartz, F. J. (1968). Lethal temperature and salinity tolerances of the white catfish, *Ictalurus catus*, from the Patuxent River, Maryland. *Chesapeake Sci.* 9, 103–108.

Kilambi, R. and Zdinak, A. (1980). The effects of acclimation on the salinity tolerance of grass carp, *Ctenopharyngodon idella* (Cuv. and Val.). *J. Fish Biol.* 16, 171–175.

Kinnison, M. T. and Hendry, A. P. (2003). From macro- to micro-evolution: tempo and mode in salmonid evolution. In *Evolution Illuminated: Salmon and Their Relatives* (eds. A. P. Hendry and S. C. Stearns), pp. 208–231. Oxford: Oxford University Press.

Klepaker, T. (1993). Morphological changes in a marine population of threespined stickleback, *Gasterosteus aculeatus*, recently isolated in fresh water. *Can. J. Zool.* 71, 1251–1258.

Klossa-Kilia, E., Papasotiropoulos, V., Tryfonopoulos, G., Alahiotis, S. and Kilias, G. (2007). Phylogenetic relationships of *Atherina hepsetus* and *Atherina boyeri* (Pisces: Atherinidae) populations from Greece, based on mtDNA sequences. *Biol. J. Linn. Soc.* 92, 151–161.

Kosswig, C. (1967). Tethys and its relation to the peri-Mediterranean faunas of freshwater fishes. In *Aspects of Tethan Biogeography* (eds. C. G. Adams and D. V. Ager), pp. 313–321. Systematics Association Publication 7.

Kraiem, M. and Pattee, E. (1988). Salinity tolerance of the barbel, *Barbus callensis* Valenciennes, 1842 (Pisces, Cyprinidae) and its ecological significance. *Hydrobiologia* 166, 263–267.

Langston, J. N., Schofield, P. J., Hill, J. E. and Loftus, W. F. (2010). Salinity tolerance of the African jewelfish *Hemichromis letourneuxi*, a non-native cichlid in south Florida (USA). *Copeia* 2010, 475–480.

Le Francois, N. R., Lamarre, S. and Blier, P. U. (2003). Evaluation of the adaptability of the Atlantic wolffish (*Anarhichas lupus*) to low and intermediate salinities. *19th (2002) Annual Meeting Aquaculture Association*, Charlottetown, Canada, 15–17.

Lee, C. E. and Bell, M. A. (1999). Causes and consequences of recent freshwater invasions by saltwater animals. *Trends Ecol. Evol.* 14, 284–288.

Lemarie, G., Baroiller, J. F., Clota, F., Lazard, J. and Dosdat, A. (2004). A simple test to estimate the salinity resistance of fish with specific application to *O. niloticus* and *S. melanotheron*. *Aquaculture* 240, 575–587.

Letunic, I. and Bork, P. (2011). Interactive Tree of Life v2: online annotation and display of phylogenetic trees made easy. *Nucleic Acids Res.* 39, W475–W478.

Li, J. and Li, S. (1999). Study on salinity tolerance of GIFT strain of Nile tilapia. *J. Zhejiang Ocean U. (Nat. Sci.)* 18, 107–111.

Li, S. F., Yan, B., Cai, W. Q., Li, T. Y., Jia, J. H. and Zhang, Y. H. (2008). Heterosis and related genetic analysis by SSR for the salt tolerance of reciprocal hybrids between Nile tilapia (*Oreochromis niloticus*) and blackchin tilapia (*Sarotherodon melanotheron*). *J. Fish. Sci. China* 15, 189–197.

Lotan, R. (1971). Osmotic adjustment in the euryhaline teleost *Aphanius dispar* (Cyprinodontidae). *Z. Vergleich. Physiol.* 75, 383–387.

Lovejoy, N. R. and Collette, B. B. (2001). Phylogenetic relationships of New World needlefishes (Teleostei: Belonidae) and the biogeography of transitions between marine and freshwater habitats. *Copeia* 2001, 324–338.

Lovejoy, N. R., Albert, J. S. and Crampton, W. G. R. (2006). Miocene marine incursions and marine/freshwater transitions: evidence from neotropical fishes. *J. S. Am. Earth Sci.* 21, 5–13.

Lutz, C. G., Armas-Rosales, A. M. and Saxton, A. M. (2010). Genetic effects influencing salinity tolerance in six varieties of tilapia (*Oreochromis*) and their reciprocal crosses. *Aquacult. Res.* 41, e770–e780.

Maples, C. G. and Archer, A. W. (1989). The potential of Paleozoic nonmarine trace fossils for paleoecological interpretations. *Palaeogeog. Palaeoclimat. Palaeoecol.* 73, 185–195.

Marshall, W. (2013). Osmoregulation in estuarine and intertidal fishes. In *Fish Physiology*, Vol. 32, *Euryhaline Fishes* (eds. S. D. McCormick, A. P. Farrell and C. J. Brauner), pp. 395–434. New York: Elsevier.

McCairns, R. J. S. and Bernatchez, L. (2010). Adaptive divergence between freshwater and marine sticklebacks: insights into the role of phenotypic plasticity from an integrated analysis of candidate gene expression. *Evolution* 64, 1029–1047.

McCormick, S. D. (2013). Smolt physiology and endocrinology. In *Fish Physiology*, Vol. 32, *Euryhaline Fishes* (eds. S. D. McCormick, A. P. Farrell and C. J. Brauner), pp. 199–251. New York: Elsevier.

McDowall, R. M. (1988). *Diadromy in Fishes*. London: Croom Helm.

McDowall, R. M. (2001). Diadromy, diversity and divergence: implications for speciation processes in fishes. *Fish Fish.* 2, 278–285.

McKinnon, J. S. and Rundle, H. D. (2002). Speciation in nature: the threespine stickleback model systems. *Trends Ecol. Evol.* 17, 480–488.

McLusky, D. S. (1989). *The Estuarine Ecosystem*. New York: Chapman and Hall.

Mellor, P. and Fotedar, R. (2005). Physiological responses of Murray cod (*Maccullochella peelii peelii*) (Mitchell 1839) larvae and juveniles when cultured in inland saline water. *Indian J. Fish.* 52, 249–261.

Miller, P. J. (1990). The endurance of endemism: the Mediterranean freshwater gobies and their prospects for survival. *J. Fish Biol.* 37 (Suppl. A), 145–156.

Miller, R. R. (1966). Geographical distribution of central American freshwater fishes. *Copeia* 1966, 773–802.

Myers, G. S. (1938). Fresh-water fishes and West Indian zoogeography. *Annu. Rep. Bd. Regents Smithson. Inst.* 92, 339–364.

Nazneen, S. and Begum, F. (1981). Salinity tolerance in some freshwater fishes. *Biologia (Pakistan)* 27, 33–38.

Nelson, S. and Flickinger, S. A. (1992). Salinity tolerance of Colorado squawfish, *Ptychocheilus lucius* (Pisces: Cyprinidae). *Hydrobiologia* 246, 165–168.

Nordlie, F. (1987). Salinity tolerance and osmotic regulation in the diamond killifish *Adinia xenica*. *Environ. Biol. Fish.* 20, 229–232.

Nordlie, F. and Haney, D. (1993). Euryhaline adaptations in the fat sleeper, *Dormitator maculatus*. *J. Fish. Biol.* 43, 433–439.

Nordlie, F., Haney, D. and Walsh, S. (1992). Comparisons of salinity tolerances and osmotic regulatory capabilities in populations of sailfin molly (*Poecilia latipinna*) from brackish and fresh waters. *Copeia* 1992, 741–746.

Nelson, J. S. (2006). *Fishes of the World*. New York: John Wiley & Sons.

Nilsen, T. O., Ebbesson, L. O. E., Madsen, S. S., McCormick, S. D., Andersson, E., Bjoernsson, B. T., Prunet, P. and Stefansson, S. O. (2007). Differential expression of gill Na$^+$,K$^+$-ATPase α- and β-subunits, Na$^+$,K$^+$,2Cl$^-$ cotransporter and CFTR anion channel in juvenile anadromous and landlocked Atlantic salmon *Salmo salar*. *J. Exp. Biol.* 210, 2885–2896.

Nordlie, F. G. (1985). Osmotic regulation in the sheepshead minnow *Cyprinodon variegatus* Lacepede. *J. Fish Biol.* 26, 161–170.

Nordlie, F. G. (2009). Environmental influences on regulation of blood plasma/serum components in teleost fishes: a review. *Rev. Fish Biol. Fish.* 19, 481–564.

Nordlie, F. G. and Walsh, S. J. (1989). Adaptive radiation in osmotic regulatory patterns among 3 species of cyprinodontids (Teleostei, Atherinomorpha). *Physiol. Zool.* 62, 1203–1218.

Odo, G. E. and Inyang, N. M. (2001). Growth, feed utilization and survival of African catfish *Clarias gariepinus* (Burchill, 1822) fingerlings reared in tanks at different salinity levels. *J. Aquat. Sci.* 16, 124–126.

Oikonomou, A. N., Giakoumi, S., Vardakas, L., Barbieri-Tseliki, R., Stoumpoudi, M. and Zogaris, S. (2007). The freshwater ichthyofauna of Greece – an update based on a hydrographic basin survey. *Medit. Mar. Sci.* 8, 91–166.

Ostrand, K. G. and Wilde, G. R. (2001). Temperature, dissolved oxygen, and salinity tolerances of five prairie stream fishes and their role in explaining fish assemblage patterns. *Trans. Am. Fish. Soc.* 130, 742–749.

Ovenden, J. R. and White, R. W. G. (1990). Mitochondrial and allozyme genetics of incipient speciation in a landlocked population of *Galaxias truttaceus* (Pisces, Galaxiidae). *Genetics* 124, 701–716.

Ovenden, J. R., White, R. W. G. and Adams, M. (1993). Mitochondrial and allozyme genetics of two Tasmanian galaxiids (*Galaxias auratus* and *G. tanycephalus*, Pisces: Galaxiidae) with restricted lacustrine distributions. *Heredity* 70, 223–230.

Partridge, G. J. and Jenkins, G. I. (2002). The effect of salinity on growth and survival of juvenile black bream (*Acanthopagrus butcheri*). *Aquaculture* 210, 219–230.

Palkovacs, E. P. and Post, D. M. (2009). Experimental evidence that evolutionary divergence in predator foraging traits drives ecological divergence in prey communities. *Ecology* 90, 300–305.

Palkovacs, E. P., Dion, K. B., Post, D. M. and Caccone, A. (2008). Independent evolutionary origins of landlocked alewife populations and rapid parallel evolution of phenotypic traits. *Mol. Ecol.* 17, 582–597.

Perschbacher, P., Aldrich, D. and Strawn, K. (1990). Survival and growth of the early stages of gulf killifish in various salinities. *Prog. Fish.-Cult.* 52, 109–111.

Peterson, M. S. (1988). Comparative physiological ecology of centrarchids in hyposaline environments. *Can. J. Fish. Aquat. Sci.* 45, 827–833.

Pfeiler, E. (1981). Salinity tolerance of leptocephalous larvae and juveniles of the bonefish (Albulidae: *Albula*) from the Gulf of California. *J. Exp. Mar. Biol. Ecol.* 52, 37–45.

Pillai, D., Jose, S., Mohan, M. V. and Joseph, A. (2003). Effect of salinity on growth and survival of rohu, *Labeo rohita* (Ham.) under laboratory and field conditions. *Fish. Technol. (India)* 40, 91–94.

Plaut, I. (1998). Comparison of salinity tolerance and osmoregulation in two closely related species of blennies from different habitats. *Fish Physiol. Biochem.* 19, 181–188.

Plaut, I. (1999). Effects of salinity on survival, osmoregulation, and oxygen consumption in the intertidal blenny, *Parablennius sanguinolentus*. *Copeia* 1999, 775–779.

Post, D. M., Palkovacs, E. P., Schielke, E. G. and Dodson, S. I. (2008). Intraspecific variation in a predator affects community structure and cascading trophic interactions. *Ecology* 89, 2019–2032.

Potter, I. C., Ivantsoff, W., Cameron, R. and Minnard, J. (1986). Life cycles and distribution of atherinids in the marine and estuarine waters of southern Australia. *Hydrobiologia* 139, 23–40.

Prodocimo, V. and Freire, C. A. (2001). Ionic regulation in aglomerular tropical estuarine pufferfishes submitted to sea water dilution. *J. Exp. Mar. Biol. Ecol.* 262, 243–253.

Provencher, L., Munro, J. and Dutil, J. D. (1993). Osmotic performance and survival of Atlantic cod (*Gadus morhua*) at low salinities. *Aquaculture* 116, 219–231.

Purcell, K. M., Hitch, A. T., Klerks, P. L. and Leberg, P. L. (2008). Adaptation as a potential response to sea-level rise: a genetic basis for salinity tolerance in populations of a coastal marsh fish. *Evol. Applic.* 1, 155–160.

Rajasekharan Nair, J. and Balakrishnan Nair, N. (1984). Salinity–temperature interaction in the distribution of two tropical glassy perchlets of the genus *Chanda* Ham (= *Ambassis* Cuv. & Val.). *Comp. Physiol. Ecol.* 9, 245–249.

Reynolds, W. W. and Thomson, D. A. (1974). Temperature and salinity tolerances of young Gulf of California grunion, *Leuresthes sardina* (Atheriniformes: Atherinidae). *J. Mar. Res.* 32, 37–45.

Reynolds, W. W., Thomson, D. A. and Casterlin, M. E. (1976). Temperature and salinity tolerances of larval Californian grunion, *Leuresthes tenuis* (Ayres): a comparison with Gulf grunion, *L. sardina* (Jenkins and Evermann). *J. Exp. Mar. Biol. Ecol.* 24, 73–82.

Rosen, D. E. (1974). Phylogeny and zoogeography of salmoniform fishes and relationships of *Lepidogalaxias salamandroides*. *Bull. Am. Mus. Nat. Hist.* 153, 265–326.

Sampaio, L. A., Freitas, L. S., Okamoto, M. H., Louzada, L. R., Rodrigues, R. V. and Robaldo, R. B. (2007). Effects of salinity on Brazilian flounder *Paralichthys orbignyanus* from fertilization to juvenile settlement. *Aquaculture* 262, 340–346.

Saoud, I. P., Kreydiyyeh, S., Chalfoun, A. and Fakih, M. (2007). Influence of salinity on survival, growth, plasma osmolality and gill Na^+K^+-ATPase activity in the rabbitfish *Siganus rivulatus*. *J. Exp. Mar. Biol. Ecol.* 348, 183–190.

Schöfer, W. (1979). Investigations on the capability of roach (*Rutilus rutilus* L.) to reproduce in brackish water. *Arch. Hydrobiol.* 86, 371–395.

Schofield, P. J. (2003). Salinity tolerance of two gobies (*Microgobius gulosus, Gobiosoma robustum*) from Florida Bay (USA). *Gulf Mex. Sci.* 21, 86–91.

Schofield, P. J. and Nico, L. G. (2009). Salinity tolerance of non-native Asian swamp eels (Teleostei: Synbranchidae) in Florida, USA: comparison of three populations and implications for dispersal. *Environ. Biol. Fish.* 85, 51–59.

Schofield, P. J., Brown, M. E. and Fuller, P. L. (2006). Salinity tolerance of goldfish *Carassius auratus* L., a non-native fish in the United States. *Fla. Sci.* 69, 258–268.

Shikano, T. and Fujio, Y. (1998). Maternal effect on salinity tolerance in newborn guppy *Poecilia reticulata*. *Fish. Sci.* 64, 52–56.

Schultze, H.-P. and Cloutier, R. (1996). Comparison of the Escuminac Formation ichthyofauna with other late Givetian/early Frasnian ichthyofaunas. In *Devonian Fishes and Plants of Miguasha, Quebec, Canada* (eds. H.-P. Schultze and R. Cloutier), pp. 348–368. München: Dr. Friedrich Pfeil.

Scott, D. M., Wilson, R. W. and Brown, J. A. (2007). The osmoregulatory ability of the invasive species sunbleak *Leucaspius delineatus* and topmouth gudgeon *Pseudorasbora parva* at elevated salinities, and their likely dispersal via brackish waters. *J. Fish Biol.* 70, 1606–1614.

Shrimpton, J. M. (2013). Seawater to freshwater transitions in diadromous fishes. In *Fish Physiology*, Vol. 32, *Euryhaline Fishes* (eds. S. D. McCormick, A. P. Farrell and C. J. Brauner), pp. 327–393. New York: Elsevier.

Smith, H. W. (1932). Water regulation and its evolution in the fishes. *Q. Rev. Biol.* 7, 1–26.

Smith, T. I. J., Denson, M., Heyward, L. D., Jenkins, W. and Carter, L. (1999). Salinity effects on early life stages of southern flounder *Paralichthys lethostigma*. *J. World Aquacult. Soc.* 30, 236–244.

Sparks, J. S. and Smith, W. L. (2005). Freshwater fishes, dispersal ability, and nonevidence: "Gondwana life rafts" to the rescue. *Syst. Biol.* 54, 158–165.

Stanley, J. G. and Fleming, W. R. (1977). Failure of seawater-acclimation to alter osmotic toxicity in *Fundulus kansae*. *Comp. Biochem. Physiol. A* 58, 53–56.

Staurnes, M., Sigholt, T., Lysfjord, G. and Gulseth, O. (1992). Difference in the seawater tolerance of anadromous and landlocked populations of Arctic char (*Salvelinus alpinus*). *Can. J. Fish. Aquat. Sci.* 49, 443–447.

Sulikowski, J. A. and Maginniss, L. A. (2001). Effects of environmental dilution on body fluid regulation in the yellow stingray, *Urolophus jamaicensis*. *Comp. Biochem. Physiol. A Mol. Integr. Physiol.* 128, 223–232.

Sullivan, J. P., Lundberg, J. G. and Hardman, M. (2006). A phylogenetic analysis of the major groups of catfishes (Teleostei: Siluriformes) using rag1 and rag2 nuclear gene sequences. *Mol. Phylogenet. Evol.* 41, 636–662.

Swanson, C., Reid, T., Young, P. S. and Cech, J., Jr (2000). Comparative environmental tolerances of threatened delta smelt (*Hypomesus transpacificus*) and introduced wakasagi (*H. nipponensis*) in an altered California estuary. *Oecologia* 123, 384–390.

Takeda, Y. and Tanaka, M. (2007). Freshwater adaptation during larval, juvenile and immature periods of starry flounder *Platichthys stellatus*, stone flounder *Kareius bicoloratus* and their reciprocal hybrids. *J. Fish Biol.* 70, 1470–1483.

Taylor, E. (1990). Variability in agonistic behaviour and salinity tolerance between and within two populations of juvenile Chinook salmon, *Oncorhynchus tshawytscha*, with contrasting life histories. *Can. J. Fish. Aquat. Sci.* 47, 2172–2180.

Threader, R. and Houston, A. H. (1983). Use of NaCl as a reference toxicant for goldfish, *Carassius auratus*. *Can. J. Fish. Aquat. Sci.* 40, 89–92.

Tsuzuki, M. Y., Cerqueira, V. R., Teles, A. and Doneda, S. (2007). Salinity tolerance of laboratory reared juveniles of the fat snook *Centropomus parallelus*. *Braz. J. Oceanogr.* 55, 1–5.

Tsuzuki, M. Y., Aikawa, H., Struessmann, C. A. and Takashima, F. (2000). Physiological responses to salinity increases in the freshwater silversides *Odontesthes bonariensis* and *O. hatcheri* (Pisces, Atherinidae). *Rev. Bras. Oceanogr.* 48, 81–85.

Varsamos, S., Connes, R., Diaz, J. P., Barnabe, G. and Charmantier, G. (2001). Ontogeny of osmoregulation in the European sea bass *Dicentrarchus labrax* L. *Mar. Biol.* 138, 909–915.

Varsamos, S., Nebel, C. and Charmantier, G. (2005). Ontogeny of osmoregulation in postembryonic fish: a review. *Comp. Biochem. Physiol. A Mol. Integr. Physiol.* 141, 401–429.

Vega, G. C. and Wiens, J. J. (2012). Why are there so few fish in the sea? *Proc. R. Soc. B Biol. Sci.* 279, 2323–2329.

Vize, P. D. (2004). A Homeric view of kidney evolution: a reprint of H.W. Smith's classic essay with a new introduction. *Anat. Rec.* 277A, 344–354.

Wada, T., Aritaki, M., Yamashita, Y. and Tanaka, M. (2007). Comparison of low-salinity adaptability and morphological development during the early life history of five pleuronectid flatfishes, and implications for migration and recruitment to their nurseries. *J. Sea Res.* 58, 241–254.

Wang, H., Xu, Y. and Zhang, P. (2000). Salinity tolerance of embryos and yolk-sac larvae of *Paralichthys olivaceus. J. Fish. Sci. China* 7, 21–23.

Ward, R. D., Woodwark, M. and Skibinski, D. O. F. (1994). A comparison of genetic diversity levels in marine, freshwater, and anadromous fishes. *J. Fish Biol.* 44, 213–232.

Watanabe, W. O., Kuo, C.-M. and Huang, M.-C. (1985). The ontogeny of salinity tolerance in the tilapias *Oreochromis aureus, O. niloticus,* and an *O. mossambicus* × *O. niloticus* hybrid, spawned and reared in freshwater. *Aquaculture* 47, 353–367.

Waters, J. M. and Wallis, G. P. (2001a). Cladogenesis and loss of the marine life-history phase in freshwater galaxiid fishes (Osmeriformes: Galaxiidae). *Evolution* 55, 587–597.

Waters, J. M. and Wallis, G. P. (2001b). Mitochondrial DNA phylogenetics of the *Galaxias vulgaris* complex from South Island, New Zealand: rapid radiation of a species flock. *J. Fish Biol.* 58, 1166–1180.

Whitehead, A. (2010). The evolutionary radiation of diverse osmotolerant physiologies in killifish (*Fundulus* sp.). *Evolution* 64, 2070–2085.

Whitehead, A., Roach, J. L., Zhang, S. J. and Galvez, F. (2011). Genomic mechanisms of evolved physiological plasticity in killifish distributed along an environmental salinity gradient. *Proc. Natl Acad. Sci. U. S. A.* 108, 6193–6198.

Wiley, E. O. and Johnson, G. D. (2010). A teleost classification based on monophyletic groups. In *Origin and Phylogenetic Interrelationships of Teleosts* (eds. J. S. Nelson, H.-P. Schultze and M. V. H. Wilson), pp. 123–182. München, Germany: Verlag Dr. Friedrich Pfeil.

Williams, M. D. and Williams, W. (1991). Salinity tolerances of four species of fish from the Murray-Darling River system. *Hydrobiologia* 210, 145–160.

Yamanoue, Y., Miya, M., Doi, H., Mabuchi, K., Sakai, H. and Nishida, M. (2011). Multiple invasions into freshwater by pufferfishes (Teleostei: Tetraodontidae): a mitogenomic perspective. *PLOS ONE* 6 (2), e17410.

Yin, M. and Blaxter, J. H. S. (1987). Temperature, salinity tolerance, and buoyancy during early development and starvation of Clyde and North Sea herring, cod, and flounder larvae. *J. Exp. Mar. Biol. Ecol.* 107, 279–290.

Young, P. S. and Cech, J. J. (1996). Environmental tolerances and requirements of splittail. *Trans. Am. Fish. Soc.* 125, 664–678.

Young, S. P., Smith, T. I. J. and Tomasso, J. R. (2006). Survival and water balance of black sea bass held in a range of salinities and calcium-enhanced environments after abrupt salinity change. *Aquaculture* 258, 646–649.

Zander, C. D. (1974). Evolution of Blennioidei in the Mediterranean Sea. *Rapp. P.-V. Reun. Comm. Int. Explor. Sci. Mer. Mediterr. Monaco* 22 (7), 57.

Zydlewski, J. and Wilkie, M. P. (2013). Freshwater to seawater transitions in migratory fishes. In *Fish Physiology,* Vol. 32, *Euryhaline Fishes* (eds. S. D. McCormick, A. P. Farrell and C. J. Brauner), pp. 253–326. New York: Elsevier.

INDEX

OTHER VOLUMES IN THE FISH PHYSIOLOGY SERIES

Printed and bound by CPI Group (UK) Ltd, Croydon, CR0 4YY

08/05/2025

01864956-0001